DIGITAL ELECTRONICS

A Simplified Approach

DIGITAL ELECTRONICS

A Simplified Approach

Robert D. Thompson

Upper Saddle River, New Jersey
Columbus, Ohio

Library of Congress Cataloging in Publication Data

Thompson, Robert D.
  Digital electronics : a simplified approach / Robert D. Thompson.—1st ed.
    p. cm.
  ISBN 0-13-505694-2
  1. Digital electronics.   I. Title.

TK7868.D5 T495 2001
621.381—dc21

                                        00-029822

Vice President and Publisher: Dave Garza
Editor in Chief: Stephen Helba
Acquisitions Editor: Scott J. Sambucci
Production Editor: Rex Davidson
Design Coordinator: Karrie Converse-Jones
Cover Designer: Dean Barrett
Cover art: Dean Barrett
Production Coordination: York Production Services
Production Manager: Pat Tonneman
Marketing Manager: Ben Leonard

This book was set in Times Roman by York Graphic Services, Inc. It was printed and bound by Courier Westford, Inc. The cover was printed by Phoenix Color Corp.

10 9 8 7 6 5 4 3 2 1
ISBN: 0-13-505694-2

Dedication

To my wife, Barbara F. Thompson, whose patience and love made another book possible.

PREFACE

Digital Electronics: A Simplified Approach was written for the reader who is interested in understanding digital logic operation. It presents the prerequisite material for advanced microprocessor/microcomputer courses. A good foundation in semiconductors and circuits is helpful, but not mandatory, for learning most of the material in this textbook.

Information in this book is presented using a **developmental approach** to learning. It places learning by rote in antiquity where it belongs. Instead, the reader is taught how to **derive** the knowledge necessary to understand operational and design concepts.

Each chapter starts with a brief introduction, a list of important terms, and chapter objectives. The chapter objectives are broad by design and tell the reader what the ultimate outcome of achieving the section objectives will produce. The introduction and list of important terms let the reader know generally where he or she is going and what to look for along the way.

Each section within a chapter starts with objectives that support the overall chapter objectives. These section objectives let the reader know exactly what should be learned in the section. Each section ends with review questions that ensure the section objectives have been met.

Internal summaries are incorporated throughout the book. These summaries may be placed after each section, or they may include two or more sections if the material so dictates. Internal summary questions follow each internal summary.

There is a chapter summary at the end of each chapter. In addition, end-of-chapter questions cover the most important points in the chapter. Most chapters contain a **practical applications section** and a **troubleshooting section.** These sections are used to in corporate real-world applications and problems.

Critical thinking questions and problems at the end of each chapter are marked **CT.** These questions/problems generally require the logical application of previously learned material or the application of several different principles or pieces of information and are sometimes more difficult than standard questions in terms of deriving an answer.

The information set off in colored blocks and titled **In-Depth Look** in the text is optional but often beneficial. The information delves into basic operation of a device or circuit. If operation of that device or circuit is already understood, the reader can then skip forward to the end of the in-depth material.

I wish to thank the following reviewers of the manuscript: Dale Blackburn, ECPI College of Technology; Steve Coe; Gene Dunlap, ITT Technical Institute; Doug Fuller, Humber College of Applied Arts & Technology; Joe Gryniuk, Lake Washington Technical College; Anthony Hearn, Community College of Philadelphia; David Longobardi; William Mack, Harrisburg Area Community College; Robert Martin, Northern VA Community College; Byron Paul; Dr. Robert A. Powell; Lee Rosenthal, Fairleigh Dickinson University; and Larry J. Wheeler, PSE&G Nuclear.

Contents

1 INTRODUCTION

GENERAL HISTORY

The vacuum tube was invented shortly after the turn of the twentieth century. Early computers used thousands of these large, power-hungry devices to perform their computations.

The ENIAC (Electronic Numerical Integrator and Calculator) computer was unveiled in 1946. It weighed in at 30 tons and filled a two-car garage. Reliable operation of the ENIAC usually spanned about seven minutes before one of its 18,000 vacuum tubes failed. The ENIAC was programmed by the wiring connected to various circuits in the machine. Reprogramming required rewiring thousands of wires to accomplish a new task. Needless to say, reprogramming was a very time-consuming process. In spite of these disadvantages, the ENIAC could perform about 5000 additions or subtractions per second when it was operating properly.

The transistor was invented at Bell Telephone Laboratories in 1948. This historic event occurred without much fanfare or media attention. Little did anyone realize what an impact the invention would have in the future.

The innocent years of the 1950s were mostly a time of leisure in America. They brought Eisenhower to the presidency, yet they brought the Korean conflict. Television came to millions of American homes. McDonald's became a household word, and Holiday Inn became the motel for many travelers. The 1950s brought a revolution in music—Elvis style. Names such as Milton Berle, Marlon Brando, and Marilyn Monroe were catapulted into history.

The early fifties also brought the 5-ton UNIVAC I (Universal Automatic Computer). The UNIVAC I was sold to the U.S. Census Bureau. These early computers, based on vacuum-tube technology, are classified as **first-generation computers.** They are characterized by their huge size and the excessive heat generated by the tubes. Most modern pocket calculators contain as much computing power as these early computers.

Second-generation computers (1959–1964) utilized the more reliable diode and transistor. Those computers were much smaller, faster, cheaper, and more reliable than their vacuum tube precursors.

Third-generation computers (1965–1970) utilized integrated circuit (IC) technology. The process of manufacturing hundreds or even thousands of transistors on a sliver of silicon culminates in an integrated circuit. Naturally, faster processing, more memory capability, and smaller machines evolved with the advent of IC technology.

The microprocessor is the identifying feature of the **now generation of computers.** Microprocessors have increased in capability and speed continuously over the past three decades. They are known as "computers-on-a-chip." Technological advances in manufacturing techniques have allowed more and more to be integrated into less and less space. For example, Intel's 4004 microprocessor contains 2300 transistors; the 486DX2 microprocessor contains 1.2 million transistors; the Intel Pentium processor contains 3.1 million; and the powerful Intel Pentium II contains 7.5 million transistors.

Microprocessors fade into history faster than popular new songs. Intel's 4004, Apple's 6502, Motorola's 6800, and Zilog's Z-80 microprocessors have all come and gone. They certainly contributed their share to the history of personal computers (PCs) and digital electronics.

From early vacuum tubes to transistors (1950s), from transistors to integrated circuits (1960s), and from integrated circuits to microprocessors (1970s), the computer has evolved into the small, miraculous machine it is today.

Since the advent of Intel's first microprocessor in 1971, the manufacture of digital electronic equipment has exploded beyond most people's wildest dreams. That first microprocessor, the Intel 4004, was a 4-bit chip that could execute 60,000 instructions per second. Intel's 8086 microprocessor appeared on the scene in 1978. Their 486DX microprocessor, introduced in 1991, is a 32-bit chip capable of executing 41 million instructions per second.

Today's high-performance digital systems, such as the PC, will one day be museum artifacts. That day is probably not too far off for many of today's PCs. Yesterday's digital prototype circuits are constantly being improved and placed into production. Faster microprocessors and memory devices loom just over the horizon. The PC evolution and revolution are well-documented history. However, great strides are being made constantly in many other areas of digital technology.

For example, a quick walk around a music store will verify compact discs (CDs) have replaced records in most instances. These CDs are recorded using digital techniques, and the audio quality is presently unsurpassed. Digital audio tapes (DATs), digital compact cassettes (DCCs), and mini discs (MDs) have all recently redefined the industry's capabilities. Computer-controlled robots have drastically changed manufacturing facilities. Onboard computers control major engine functions in automobiles. In addition, navigational aids are installed in many new automobiles. The medical field utilizes state-of-the-art digital equipment in hospitals and doctor's offices. Security systems are installed in most new homes. All of this has come about because of digital technology.

Our lives are touched many times daily by digital technology—by alarm clocks, pagers, mobile phones, microwave ovens, music, television, video games, PCs, and the Internet, to name just a few. The list is almost endless and is continuously getting longer.

The past few years have seen remarkable improvements in microcircuit technology. The ICs of the 1990s are far more complex than the circuit boards of the 1960s and 1970s. Future demands on industry will dictate higher-density, faster circuits, and they are not demands that will be easily met. These demands come from the public's fascination with portable electronic devices as well as their desire to do more and more with their home computers.

The task that lies directly ahead of us in the electronics field is to master the theory and application of digital electronics. The remainder of this book is dedicated to that task. In some cases, learning to design a digital circuit allows for a better understanding of its operation. Design of the circuit will be explained in these cases.

ANALOG/DIGITAL SIGNALS

A first step in the direction of mastering the subject requires comparison of the term "digital" with the term "analog". **Analog signals** vary continuously within some range of values. The signal in Fig. 1–1 is an ac sine wave that contains an infinite number of voltage levels. **Digital signals,** such as the signal in Fig. 1–2, are made up of discrete levels. This signal contains two voltage levels, which are referred to in digital technology as **logic low (Logic 0)** and **logic high (Logic 1).** It is easy to see in Fig. 1–2 that the logic levels can be used to represent the binary levels of 0 and 1. It is these low (0) and high (1) levels that a digital circuit understands. That is why *binary is often called the machine-level language of computers and other digital circuits.*

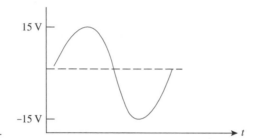

FIGURE 1-1   AC sine wave.

Figure 1–3 shows two methods that can be used to produce Logic 0 and Logic 1 levels. In Fig. 1–3(a), $V_{out} = 5$ V with the switch open and 0 V with it closed. The resistor pulls V_{out} up to 5 V when the switch is open in this arrangement. In Fig. 1–3(b), $V_{out} = 5$ V with the switch closed and 0 V with it open. The resistor pulls V_{out} down to 0 V when the switch is open in this arrangement. The convention utilizing a Logic 1 as a **high** and a Logic 0 as a **low** is called **positive logic** and is the type of logic used in this textbook. Figure 1–3 is used for illustrative purposes. Normally, the low and high digital signals are generated in digital circuits.

FIGURE 1-2   Digital signal.

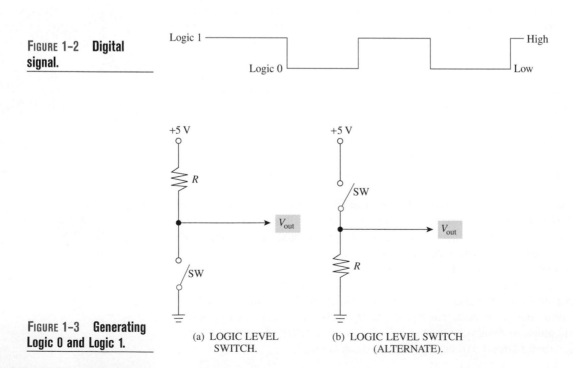

FIGURE 1-3   Generating Logic 0 and Logic 1.

(a) LOGIC LEVEL SWITCH.

(b) LOGIC LEVEL SWITCH (ALTERNATE).

DIGITAL IC CLASSIFICATIONS

Many digital circuits presently in use are available in integrated circuits (ICs), which are often called chips. These ICs are classified in accordance with the number of logic gates they contain:

1. **Small-Scale Integration (SSI)** includes all digital ICs with fewer than 12 gates. Many of the logic gates presented in the text fall in this classification.
2. **Medium-Scale Integration (MSI)** classifies ICs that contain 12 to 99 gates. Chapter 10 includes some of the more popular MSI ICs.
3. **Large-Scale Integration (LSI)** includes ICs that contain 99 to 2999 gates. Many of the programmable logic devices (PLDs) discussed in the text in Chapter 13 fall in this classification.
4. **Very Large-Scale Integration (VLSI)** classifies ICs that contain 3000 to 99,999 gates.
5. **Ultra Large-Scale Integration (ULSI)** includes all digital ICs that contain more than 100,000 gates.

CHAPTER-BY-CHAPTER LOOK AT THIS TEXTBOOK

Chapter 2 of this text provides an in-depth look at the **binary** (base-2), **octal** (base-8), and **hexadecimal** (base-16) **numbering systems.** Since we normally think in decimal (ten digits), a short review of the decimal numbering system appears in the first section of this chapter. This review provides a foundation for some of the basic concepts presented in the sections covering binary, octal, and hexadecimal numbers. As previously explained, digital systems operate with Logic 0 and Logic 1 levels. These levels are used to represent certain conditions or to represent data or instructions in a system. Since digital systems operate from this binary machine-level language, our binary study includes sequential counting as well as conversion from binary numbers to decimal numbers and vice versa. Octal and hexadecimal numbering systems are actually just shorthand for the bulky binary numbers encountered in digital systems. Therefore, conversion from the often clumsy binary numbers to the more convenient octal or hexadecimal numbers is presented in this chapter. Chapter 2 also contains several special codes that are used in digital technology.

Chapter 3 presents **logic gates—AND, OR, NAND, NOR,** and **NOT** gates. These basic gates are the foundation for all digital circuits. They are used to **implement** decision-making combinational logic circuits and sequential logic circuits. **Combinational logic circuits** have no memory capability; therefore, their output depends entirely upon their present inputs. **Sequential logic circuits** have memory, so their output depends not only upon their present inputs but also upon the previous sequence of inputs. To implement a circuit is to build it, to construct the actual hardware product. The logic gates can be used to solve logical problems by making decisions based on their current inputs. For example, if Condition *A* is true *and* Condition *B* is true, the AND gate will output a high, indicating the inputs are *all* true. An OR logic gate would output a high level (true output) if Condition *A* were true *or* Condition *B* were true. The chapter includes static (steady-state inputs) analysis and dynamic (changing inputs) analysis of logic circuits, in addition to troubleshooting.

Boolean algebra, known simply as the mathematics of logic circuits, is presented in **Chapter 4.** Boolean algebra is used to describe logical functions with mathematical symbols. Using these symbols allows the design of digital circuits to be simplified mathematically. Simplification results in fewer gates and fewer ICs used in circuit construction. This results in several more benefits—less real estate on a circuit board is used, less power is consumed, and circuit cost is lowered. In addition, using Boolean algebra equations for logic circuits allows the technician to analyze circuit operation and to troubleshoot defective circuits, because the Boolean equations "talk" in a "Here's how I work" format. Chapter 4 starts with the mathematical symbols used to represent logic functions, progresses to how to interpret and simplify Boolean expressions and use truth tables, and concludes with final logic circuit design.

By the time you reach Chapter 5, you should be very familiar with basic logic gate operation. **Chapter 5** presents various logic gates connected together to perform specific functions. The chapter starts with exclusive-OR and exclusive-NOR gate operation. It proceeds to an error detection system using a parity generator and a parity checker such as those used to test memory in your home computer when you turn it on. Actual detection, control, decoder, and selection circuits are analyzed to round out this chapter.

Latch and **flip-flop** circuits are covered in detail in **Chapter 6.** The material in this chapter is as important to a solid foundation in digital theory as are the logic gates of Chapter 3. The latches and flip-flops are often used to store or transfer data. In reality, a latch circuit or flip-flop is nothing more than a combinational logic circuit with a memory capability. Flip-flops allow the design of **sequential logic circuits** such as up- and down-counters. The most popular types of latch and flip-flop ICs are discussed in detail in this chapter, which concludes with the practical applications and troubleshooting of flip-flops.

Counting operations and frequency division are accomplished by the **counters** covered in **Chapter 7.** Two basic types of counters–asynchronous and synchronous–are presented from both an operational and design point of view. A combination of these two types, the hybrid counter, is also introduced and analyzed. Counters are designed using the flip-flops that are explained in Chapter 6. An example of frequency division circuits used to obtain the **time-of-day (TOD) clock tick** in a computer is presented in the applications section of this chapter. In addition, a counter that will count from 000 to 999 is presented.

Chapter 8 introduces **registers.** A register is a group of latches or flip-flops (Chapter 6) used to store, transfer, or shift data. The different classifications of registers are presented, and several circuits show how registers can be used as code detectors or multiply/divide circuits.

Chapter 9 has two basic purposes—presentation of binary arithmetic and binary arithmetic circuits. A general review of decimal arithmetic is touched upon in this chapter as an introduction to **binary arithmetic.** Once the mechanics of basic binary arithmetic are covered, the operation of **binary arithmetic circuits** is analyzed. Half-adders, full-adders, and adder/subtractor circuits are all presented. Binary-coded decimal (BCD) arithmetic is presented, and a BCD adder circuit is analyzed. Finally, an arithmetic logic unit (ALU) IC is discussed and examples of both its arithmetic and logical operations are analyzed.

As mentioned in the IC classifications section of this chapter, medium-scale-integration (MSI) ICs contain 12 to 99 logic gates. Many different MSI ICs are used in digital electronics. Several important MSI circuits are presented and analyzed in **Chapter 10**— among them **decoders, encoders, multiplexers, demultiplexers,** and **magnitude comparators.** The practical applications and troubleshooting sections include MSI ICs used in computer address decoding circuits, display decoders, logic function generators, and parallel-to-serial and serial-to-parallel data conversion circuits. A few other types of MSI circuits are presented in the textbook in the chapters that detail their specific operation. For example, MSI counters are discussed in Chapter 7 and MSI registers in Chapter 8.

Digital ICs are available in several different technologies (families). The most popular, transistor-to-transistor logic (TTL) and complementary metal-oxide semiconductor (CMOS) are presented in **Chapter 11.** Analysis of interface of the two technologies (TTL circuits to CMOS circuits and CMOS circuits to TTL circuits) reveals they often require special circuitry to condition the TTL logic signal for compatibility with the CMOS circuit or the CMOS signal for compatibility with the TTL circuit. Interface requirements are discussed in this chapter also. Special gates and circuits with 3-state outputs are analyzed in detail. Data converters are also presented. Data converters are used to convert data formats from digital to analog and analog to digital. It is this conversion that provides the connecting element between the analog and digital worlds.

Chapter 12 provides introductory concepts for memory devices. Read-only memory (ROM) and random-access memory (RAM) are discussed in detail. The chapter presents the basics of these types of memory, how the memory circuits work, and how they are organized and addressed.

Programmable logic devices (PLDs) are introduced via symbology in **Chapter 13.** "Programmable logic device" is a generic term that refers to ICs with uncommitted logic arrays. Two categories of PLDs, programmable logic array (PLA), and Programmable Array Logic (PAL^R), are presented in detail. Finally, field programmable gate array (FPGA) devices are analyzed. These devices provide logic gate capacities in a register-rich environment.

This book also contains several appendices. The technology of logic families covers transistor-to-transistor logic (TTL) and complementary MOS (CMOS) devices; this information is in **Appendix A.** The information in Appendix A is supplemental in nature and may be omitted without detriment to your education. Selected manufacturer's data sheets appear in **Appendix B,** and an overview of the IEEE logic symbols used in the text appears in **Appendix C.**

2 NUMBERING SYSTEMS AND CODES

Topics Covered in this Chapter

Introduction

A study of numbering systems is relatively important at this point because digital circuits only understand "on" or "off" conditions. These conditions can be represented by voltage levels, but they must be readily convertible to numbers. Since only two conditions must be met in a digital circuit (off and on), the binary numbering system is used. This system uses only the numbers 0 and 1.

Binary is characterized or based on the number 2. In the binary (base-2) numbering system, the two logic levels used to represent lows and highs are Logic 0 and Logic 1 respectively. The levels are represented in digital circuits by **bits.** The word "bit" was derived from the words *b*inary dig*it*. Since binary is used in all digital systems, a method must be available to convert a decimal number to a binary number. In addition, the reverse procedure is necessary to take the outputs of digital circuits in binary form and make them readable in decimal form. This chapter presents methods to convert binary numbers to decimal numbers and vice versa.

American Standard Code for Information Interchange (ASCII)

Binary

Binary-Coded Decimal (BCD)

Binary-Coded Octal (BCO)

Binary-Coded Hexadecimal (BCH)

Bit

Byte

Decrement

8-4-2-1 Code

Excess-3 (XS3) Code

Gray Code

Hexadecimal

Increment

Least Significant Bit (LSB)/Digit (LSD)

Modulus (MOD)

Most Significant Bit (MSB)/Digit (MSD)

Nibble

Octal

Positional-Weighted Numbering System

Propagation Delay Time

Radix

Radix Division

Radix Multiplication

Sum of Weights

Chapter Objectives

1. Count sequentially in binary, octal, and hexadecimal.

2. Given a number in a specified base, convert it to decimal, binary, octal, or hexadecimal.

3. Given binary-coded decimal numbers, convert them to decimal numbers and vice versa.

4. Given gray code numbers, convert them to binary numbers and vice versa.

5. State the purpose for BCD, gray, and excess-3 codes, and the ASCII.

It will soon become evident that working with binary numbers of any magnitude can be a laborious task. Large binary numbers are awkward to manipulate. A digital system can handle them without a problem, but we can't. Thus, two additional numbering systems are often used in addition to the binary system when working with digital circuits. **Octal** is a **base-8** numbering system that utilizes only the digits 0 through 7. **Hexadecimal** is a **base-16** numbering system that uses the digits 0 through 9 and the letters A through F. These two numbering systems provide a shorthand for handling the bulkier binary numbers.

The **radix** of a numbering system is its base, and it is written as a **subscript** to a number. Since several different numbering systems are used in this chapter, the radix (base) of the system will always be identified for binary, octal, and hexadecimal numbers. For example, if the number 3 is an octal number, it will be written as $3_{(8)}$, and if it is a hexadecimal number, $3_{(16)}$. The radix of a numbering system identifies which numbering system is being used. In addition, it identifies the total quantity of different symbols used in the system. If no subscript is present in this text, the number is a decimal number unless specifically stated otherwise.

The need for octal and hexadecimal numbers in addition to decimal and binary numbers will become evident as this chapter progresses. To keep our perspective, we will constantly be comparing binary, octal, and hexadecimal numbers to their equivalent decimal numbers.

When studying digital systems it is common to use decimal, binary, and octal or hexadecimal numbers. Data entered on the keyboard of a computer is automatically converted to binary data inside the system. If the number 9 key is depressed on the keyboard, the number is immediately converted to a binary code that represents the number 9 and sent to the system board in the computer. That number will be stored in a memory location within the computer. The particular memory location is assigned an address that may contain 16, 20, or even more binary digits (bits). To simplify reading such a large binary address in technical data, it is normally converted to hexadecimal (hex).

The primary function of this chapter is to familiarize you with the binary, octal, and hexadecimal numbering systems. Conversion between one numbering system and another will be covered in detail. The most important objective in this chapter is to teach you to convert a number in a particular base to the other three bases. For example, given a binary number, you should be able to convert it to its equivalent decimal, octal, and hexadecimal numbers.

In addition, some binary codes used in digital systems will be presented. Some codes have been developed to make circuits user friendly and to improve reliability and operation. Others have been developed for standardization and for detecting errors. These codes (binary-coded decimal, or BCD, gray code, excess-3 code, and the American Standard Code for Information Interchange, ASCII) are presented in detail in the latter part of the chapter.

SECTION 2-1: DECIMAL NUMBERING SYSTEM (BRIEF REVIEW)

The decimal numbering system is a **positional-weighted numbering system.** This means that each digit position has a specific weight (value). For example, the digit 5 represents different values depending on its location with respect to the decimal point. The numbers 0.5, 5, and 500 all contain a 5, but each 5 has a different place value. As you already know, the decimal system uses ten different basic symbols: 0, 1, 2, 3, 4, 5, 6, 7, 8, and 9. Each of these symbols is called a **digit.** For instance, if the value of the number 60,328.4 is to be determined by positional weight, it is done as follows (refer to Table 2–1 for positional weights):

$$
\begin{array}{cccccccc}
10^4 & & 10^3 & & 10^2 & & 10^1 & & 10^0 & \cdot & 10^{-1} \\
6 & & 0 & & 3 & & 2 & & 8 & \cdot & 4
\end{array}
$$
$$(6 \times 10^4) + (0 \times 10^3) + (3 \times 10^2) + (2 \times 10^1) + (8 \times 10^0) + (4 \times 10^{-1})$$

									TABLE 2-1
Powers of 10	10^5	10^4	10^3	10^2	10^1	10^0	10^{-1}	10^{-2}	**Decimal Numbering System Positional Weights**
Positional Weights	100,000	10,000	1000	100	10	1	0.1	0.01	

The place value of a digit is determined by its position in the number with respect to the decimal point. The place value multiplies by the radix of the numbering system each sequential move to the left of the decimal point. It divides by the radix each sequential move to the right of the decimal point.

In this example, the weights of each position are summed together to determine the decimal number. This procedure is referred to as the **sum-of-weights** method.

$$
\begin{aligned}
6 \times 10,000 \;\; (10^4) &= 60,000.0 \\
+\,0 \times 1,000 \;\; (10^3) &= +\quad 0.0 \\
+\,3 \times 100 \;\; (10^2) &= +\quad 300.0 \\
+\,2 \times 10 \;\; (10^1) &= +\quad 20.0 \\
+\,8 \times 1 \;\; (10^0) &= +\quad 8.0 \\
+\,4 \times 0.1 \;\; (10^{-1}) &= +\quad 0.4 \\
\hline
&\quad\; 60,328.4
\end{aligned}
$$

A digit in a column is worth that digit times the weight of the column. Note in the example that the 0 has no value, but it serves as a placeholder for the positional weight of 1000 (10^3). Although it is unnecessary for this process to be accomplished using decimal numbers because we are so used to using them, it can be necessary when using other numbering systems. The reason for this necessity is that we do not think in terms of base-2, base-8, or base-16 systems.

In the number $60,328.4_{(10)}$, the **most significant digit (MSD)** is the 6. The MSD is the digit that carries the most weight (value) in the number. It is normally the digit to the extreme left in a decimal number.

The **least significant digit (LSD)** in the number is the 4. The LSD is the digit that carries the least weight in a number. This will always be the digit in the units column (10^0 = units) of a whole number. In a whole number (integer), the decimal point is assumed to be to the right of the number if it is not shown. The decimal point is used to separate the integral portion of a number from the fractional portion.

As we progress, keep in mind that all positional-weighted numbering systems are similar—they have a base, point, least and most significant digits/bits, and place values.

SECTION 2–2: BINARY NUMBERING SYSTEM

OBJECTIVES

1. Count sequentially in binary.
2. Given decimal numbers, convert them to binary numbers.
3. Given binary numbers, convert them to decimal numbers.

The binary numbering system, like the decimal system, is a positional-weighted system. Since the binary system only uses 0 and 1, it is a base-2 system. The weights of each place are determined in the same manner as they are in the decimal system. The positional weights are shown in Table 2–2. Like the decimal system, the binary weight table starts

TABLE 2-2 Binary Numbering System Positional Weights	Powers of 2	2^8	2^7	2^6	2^5	2^4	2^3	2^2	2^1	2^0	2^{-1}	2^{-2}
	Positional Weights*	256	128	64	32	16	8	4	2	1	0.5	0.25

*In decimal.

with the base (2) to the zero power on the integer side of the number. Any number raised to the zero power is one. Every number in the binary numbering system represents an appropriate factor times a power of 2.

To count in binary, compare the numbers shown in Table 2–3. If the positional weights as shown in Table 2–2 are checked and added using the sum-of-weights procedure, the binary counts will make sense. Check the count of $0011_{(2)}$:

$$
\begin{array}{cccc}
2^3 & 2^2 & 2^1 & 2^0 \\
0 & 0 & 1 & 1_{(2)} \\
 & & (1 \times 2^1) & + (1 \times 2^0) = \\
 & & (1 \times 2) & + (1 \times 1) = \\
 & & 2 & + \quad 1 \quad = 3
\end{array}
$$

Note: The two 0s to the left of $11_{(2)}$ have no value and can be ignored, although there are times when they must be present. The next **increment** (up-count) from $0011_{(2)}$ is $0100_{(2)}$. The next **decrement** (down-count) from $0011_{(2)}$ is $0010_{(2)}$.

As illustrated in Table 2–3, it takes 4 bits to count to 15. To determine the number of bits required to count to any number, use the following formula:

TABLE 2-3 Decimal and Binary Counts	Decimal	Binary
	0	0000
	1	$0001 = 2^0$
	2	$0010 = 2^1$
	3	0011
	4	$0100 = 2^2$
	5	0101
	6	0110
	7	0111
	8	$1000 = 2^3$
	9	1001
	10	1010
	11	1011
	12	1100
	13	1101
	14	1110
	15	1111
	16	$10000 = 2^4$
	32	$100000 = 2^5$
	64	$1000000 = 2^6$
	128	$10000000 = 2^7$
	256	$100000000 = 2^8$
	512	$1000000000 = 2^9$
	1024	$10000000000 = 2^{10}$

Maximum count $= 2^n - 1$

where n = number of bits required.

$$\begin{aligned}\text{Maximum count} &= 2^n - 1 \\ &= 2^4 - 1 \\ &= 16 - 1 \\ &= 15\end{aligned}$$

This indicates mathematically that it takes 4 bits to count to 15. Although it is a guess at this point to plug some unknown power of 2 into the calculator, it won't take long to learn a couple of reference points. Mainly, $2^{10} = 1024$ and $2^{20} = 1,048,576$. With these numbers as a reference, guessing becomes easy if you realize every power increment doubles the answer.

If you don't desire to guess, the number of bits required can be calculated. For example, a binary count up to $15_{(10)} (1111_{(2)})$ is

```
0000
0001
0010
0011
0100
0101
0110
0111
1000
1001
1010
1011
1100
1101
1110
1111
```

This count is exactly as it would be produced from a binary counter that had the ability to count up to $1111_{(2)}$. Although 0000 is not a count, it is the starting state for the counter. Therefore, the number of states produced by the counter is 16, because its count starts at the 0000 state, while its maximum count is 15. The maximum number of states produced by a counter is defined as its modulus (MOD). This is calculated as 2^n, where n is the number of bits used. The maximum count $(2^n - 1)$ is one less because the $0000_{(2)}$ state is not a count.

If n represents the number of bits required, it can be calculated as follows:

$$n = \log \text{ desired MOD} \div \log 2$$

where n = number of bits required.

The problem posed was to determine the number of bits required to count to $15_{(10)}$ in binary. The MOD of the circuit is $16_{(10)}$ including the zero state.

$$\begin{aligned}n &= \log 16 \div \log 2 \\ &= 4\end{aligned}$$

How many bits are required to count to 1023?

$$\begin{aligned}n &= \log 1024 \div \log 2 \\ &= 10\end{aligned}$$

Note: The desired MOD is the maximum count *plus 1*.

$$\begin{aligned}
\text{Maximum count} &= 2^n - 1 \\
&= 2^{10} - 1 \\
&= 1024 - 1 \\
&= 1023
\end{aligned}$$

Thus, it takes 10 bits to count to 1023.

How many bits are required to count to 1999?

$$\begin{aligned}
n &= \log 2000 \div \log 2 \\
&= 10.965 \text{ (round up to next whole number)}
\end{aligned}$$

$$\begin{aligned}
\text{Maximum count} &= 2^n - 1 \\
&= 2^{11} - 1 \\
&= 2048 - 1 \\
&= 2047
\end{aligned}$$

As illustrated in the preceding problem, it takes 11 bits to count to 2047. This is greater than the desired maximum count of 1999, but as previously proven, 10 bits will only allow a count up to 1023, so 11 bits will have to be used. By using a counter that produces 11 bits, a count from 0 ($00000000000_{(2)}$) to 2047 ($11111111111_{(2)}$) could be obtained. The result of $2^n - 1$ must be equal to or greater than the desired maximum count to produce that count in binary. If the result is greater than the desired count, the counter can be forced to quit counting early. This "short-counting" is referred to as **truncating** a counter, and will be discussed in Chapter 7.

Binary-to-Decimal Conversion

If it were necessary to determine the decimal equivalent of $1101_{(2)}$, one method that could be employed is the **sum-of-weights** method previously discussed. To use this method, *add the place value weights of each bit position that contains a 1.*

$$\begin{array}{cccc}
2^3 & 2^2 & 2^1 & 2^0 \\
1 & 1 & 0 & 1_{(2)} \\
(1 \times 2^3) & + \ (1 \times 2^2) & + & (1 \times 2^0)
\end{array}$$

Note that the 2^1 bit position was skipped in the addition because that particular position contains a 0. Refer to Table 2–2 to determine the positional weights of 2^3, 2^2 and 2^0.

$$\begin{array}{ccccccc}
(1 \times 8) & + & (1 \times 4) & + & (1 \times 1) & = & \\
8 & + & 4 & + & 1 & = & 13_{(10)}
\end{array}$$

In this example, the positional values containing a 1 can simply be added together. The sum-of-weights method is reliable, although it gets cumbersome with relatively large binary numbers.

In the number $1101_{(2)}$, we can assume the binary point and the least significant bit (LSB) are on the right, and the most significant bit (MSB) is on the left. This assumption can be made throughout the text unless otherwise indicated. However, it is not unusual in some digital circuit diagrams for the LSB to be on the left. When this occurs in the text, the LSB and/or MSB will always be annotated.

Another method that can be used to convert binary numbers to decimal numbers is called the **radix multiplication** method. Once learned, this method can be applied to octal- and hexadecimal-to-decimal conversions also. Radix multiplication is the process used to convert a binary, octal, or hex whole number to a decimal number. This is accomplished

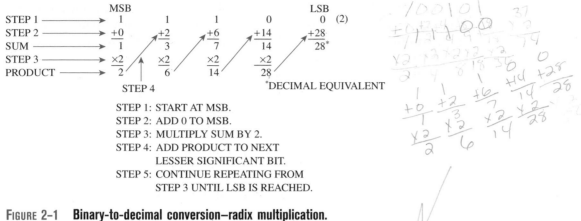

STEP 1: START AT MSB.
STEP 2: ADD 0 TO MSB.
STEP 3: MULTIPLY SUM BY 2.
STEP 4: ADD PRODUCT TO NEXT
          LESSER SIGNIFICANT BIT.
STEP 5: CONTINUE REPEATING FROM
          STEP 3 UNTIL LSB IS REACHED.

FIGURE 2–1   **Binary-to-decimal conversion–radix multiplication.**

by repeated multiplication using the radix of the numbering system from which we are converting. As an example, let's convert $11100_{(2)}$ to decimal as shown in Fig. 2–1.

STEP 1:   Start at the MSB of the binary number to be converted.

STEP 2:   Add 0 to the MSB.

STEP 3:   Multiply the sum by 2.

STEP 4:   Add the product to the next lesser significant bit.

STEP 5:   Continue repeating this process from step 3 until the LSB is reached. *Note:* Do not multiply by 2 after the product is added to the LSB. This last addition produces the decimal equivalent of the binary number.

This method of binary-to-decimal conversion is called radix multiplication because of the repeated multiplication of the sums by the radix of the binary numbering system. The answer shown in the example can be proven by using the sum-of-weights method.

$$
\begin{array}{ccccc}
2^4 & 2^3 & 2^2 & 2^1 & 2^0 \\
1 & 1 & 1 & 0 & 0_{(2)} \\
(1 \times 2^4) & + \ (1 \times 2^3) & + \ (1 \times 2^2) & = & \\
(1 \times 16) & + \ (1 \times 8) & + \ (1 \times 4) & = & \\
16 & + \quad 8 & + \quad 4 & = 28_{(10)} &
\end{array}
$$

Let's solve another problem to ensure a thorough understanding of the mechanics of radix multiplication.

PROBLEM

$1\ 0\ 0\ 1\ 0\ 1_{(2)} = $ _____ $_{(10)}$

$$
\begin{array}{cccccc}
1 & 0 & 0 & 1 & 0 & 1 \\
+0 & +2 & +4 & +8 & +18 & +36 \\
\hline
1 & 2 & 4 & 9 & 18 & 37 \\
\times 2 & \times 2 & \times 2 & \times 2 & \times 2 & \\
\hline
2 & 4 & 8 & 18 & 36 &
\end{array}
$$
    Decimal equivalent

The decimal equivalent of $100101_{(2)}$ is 37.

The radix multiplication procedure illustrated on the previous page is used on the integer portion of a binary number. If that number contains a fraction, either the **sum-of-weights** or **radix division** method will have to be used to convert the binary fraction to a decimal fraction.

The sum-of-weights method is presented first:

$$
\begin{aligned}
0.11_{(2)} \quad &= \quad \underline{\hspace{3cm}}_{(10)} \\
(1 \times 2^{-1}) \ + \ (1 \times 2^{-2}) \ &= \\
(1 \times 0.5) \ + \ (1 \times 0.25) \ &= \\
0.5 \quad + \quad 0.25 \quad &= 0.75 \\
0.11_{(2)} \quad &= \quad 0.75_{(10)}
\end{aligned}
$$

In lieu of this procedure, **radix division** could be used. This procedure consists of repeated division by 2.

0 . 1 1
$1 \div 2$
$1 . 5 \div 2$

0.75

Step 1: Bring down the LSB.
Step 2: Divide by 2.
Step 3: Bring down the quotient and the next bit. Divide by 2.
Step 4: Bring down the quotient.
Decimal equivalent

$0.11_{(2)} = \mathbf{0.75_{(10)}}$

Convert $0.10\dot{1}_{(2)}$ to its decimal equivalent.

0 . 1 0 1
$1 \div 2$
$0 . 5 \div 2$
$1 . 25 \div 2$

0.625

$0.101_{(2)} = \mathbf{0.625_{(10)}}$

Decimal-to-Binary Conversion

It is just as important to be able to convert decimal numbers to binary numbers as it is to convert binary numbers to decimal numbers. The radix division method will be used to convert whole numbers from decimal to binary. Radix division is the process used to convert an integer decimal number to another base number (binary, octal, or hex). This is done by repeated division by the radix of the numbering system to which we are converting. When using this method to convert decimal to binary, repeated division by 2 is accomplished. As an example, convert $14_{(10)}$ to its binary equivalent as shown in Fig. 2–2. Start with radix division in mind and proceed as follows:

STEP 1: Divide the decimal number by 2.

STEP 2: Place the whole number part of the quotient to the left of and adjacent to the decimal number you divided.

STEP 3: Put the remainder above the decimal number you divided.

STEP 4: Divide the whole number obtained in step 2 by 2. Repeat steps 2, 3, and 4 until the quotient is 1. When the quotient equals 1, bring the 1 up as indicated by the arrow in the example. This last 1 will always be the MSB of the binary answer. The **remainders** obtained during each division are the binary equivalent to the decimal number. *Note*: If you are dividing with a calculator, the remainder will always be 0 or 0.5, depending upon whether the decimal number is even or odd. A remainder of 0.5 equals an actual remainder of 1 in the binary answer because 0.5 of the base-2 numbering system is 1.

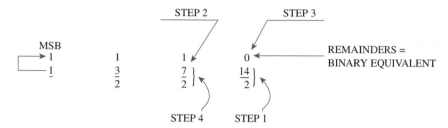

STEP 1: DIVIDE THE DECIMAL NUMBER BY 2.
STEP 2: PLACE WHOLE NUMBER PART OF QUOTIENT
TO LEFT OF NUMBER YOU DIVIDED.
STEP 3: PUT REMAINDER ABOVE DECIMAL NUMBER
YOU DIVIDED.
STEP 4: DIVIDE WHOLE NUMBER FROM STEP 2 BY 2.
REPEAT STEPS 2, 3, AND 4 UNTIL
QUOTIENT IS 1. BRING 1 UP.

FigurE 2–2   **Decimal-to-binary conversion–radix division.**

$$14_{(10)} = \mathbf{1110}_{(2)}$$

Convert $92_{(10)}$ to binary using the steps outlined.

$$
\begin{array}{cccccccc}
\rightarrow 1 & 0 & 1 & 1 & 1 & 0 & 0^* & \text{Binary equivalent} \\
1 & \dfrac{2}{2} & \dfrac{5}{2} & \dfrac{11}{2} & \dfrac{23}{2} & \dfrac{46}{2} & \dfrac{92}{2} &
\end{array}
$$

Note: When 2 is divided into 2 in the last division, the whole number part of the quotient is 1, which is placed to the left of the number you divided and the remainder is 0. Don't forget, the last quotient must be brought up as shown.

$$92_{(10)} = \mathbf{1011100}_{(2)}$$

Since radix division was used on the integer, radix multiplication must be used to convert the fractional side of a decimal number to its binary equivalent.

Convert $0.875_{(10)}$ to its binary equivalent.

```
        0.875
         ×2
MSB   1.750
         ×2
      1.500
         ×2
LSB   1.000
```

The procedure consists of repeated multiplication of the fraction by 2. The number 1 in the product of the first multiplication by 2 is the MSB of the binary equivalent number. The next multiplication (0.750×2) produces another 1, and the last multiplication (0.500×2) produces a 1, which is the LSB. The 1s to the left of the decimal point form the binary equivalent as shown. Thus,

$0.875_{(10)} = \mathbf{0.111_{(2)}}.$

Convert $0.375_{(10)}$ to its binary equivalent.

$$
\begin{array}{l}
\quad 0\,.\,3\,7\,5 \\
\quad \underline{\times\,2} \\
\quad 0\,.\,7\,5\,0 \\
\quad \underline{\times\,2} \\
\quad 1\,.\,5\,0\,0 \\
\quad \underline{\times\,2} \\
LSB \quad 1\,.\,0\,0\,0 \\
\quad 0.375_{(10)} = \mathbf{0.011_{(2)}}
\end{array}
$$

The LSB is the right-most bit in the answer $(0.011_{(2)})$. It will always be the lower bit in the multiplication problem.

Using a Calculator

Why bother learning the intricate procedures of converting from one number base to another when a calculator will solve the problem?

Let's assume you have a calculator on which you can designate a base other than base-10. This tool can save you a lot of time when converting numbers. Its use for number conversion is recommended once the basics are learned, but there are times when it can't help.

First, try converting a number that contains a fraction from one base to another. The nondecimal number bases of most calculators operate for integers only. You probably just found this out.

Now try converting $4096_{(10)}$ to binary on the calculator. Unless you have a pretty expensive calculator, you probably got an "error" message. If your calculator can display only nine characters, then you will be limited in another area. $111111111_{(2)} = 511_{(10)}$. This is the highest binary-to-decimal number conversion that can be accomplished with nine bits. This is a severe limitation considering today's technology.

If you still think it might not be necessary to learn these number conversion procedures, consider this scenario: You went to take an entry-level test for an electronic technician's position and you either didn't have a calculator or you were not allowed to use it during the test. Enough said?

Section 2–2: Review Questions

Answers are given at the end of the chapter.

A. The binary numbering system is a positional-weighted numbering system.
(1) True
(2) False

B. To *decrement* a count by one means to _____ by one.
    (1) count up
    (2) count down

C. What is the next binary count up from $111_{(2)}$?
D. What procedure can be used to convert a binary number such as 1011 to its decimal equivalent?
E. What is the maximum decimal equivalent count that can be obtained with 8 bits?
F. Define the term "modulus."

Convert the following binary numbers to their decimal equivalents:

G. $101_{(2)} =$ _____ $_{(10)}$
H. $1111_{(2)} =$ _____ $_{(10)}$
I. $1101_{(2)} =$ _____ $_{(10)}$
J. $1010100_{(2)} =$ _____ $_{(10)}$
K. $100001_{(2)} =$ _____ $_{(10)}$
L. $11101110_{(2)} =$ _____ $_{(10)}$
M. $11.11_{(2)} =$ _____ $_{(10)}$

Convert the following decimal numbers to their binary equivalents:

N. $4_{(10)} =$ _____ $_{(2)}$
O. $10_{(10)} =$ _____ $_{(2)}$
P. $15_{(10)} -$ _____ $_{(2)}$
Q. $16_{(10)} =$ _____ $_{(2)}$
R. $20_{(10)} =$ _____ $_{(2)}$
S. $140_{(10)} =$ _____ $_{(2)}$
T. $375_{(10)} =$ _____ $_{(2)}$
U. $23.5625_{(10)} =$ _____ $_{(2)}$
V. $12.375_{(10)} =$ _____ $_{(2)}$

SECTION 2-3: OCTAL NUMBERING SYSTEM

OBJECTIVES

1. Count sequentially in octal.
2. Given octal numbers, convert them to decimal numbers.
3. Given decimal numbers, convert them to octal numbers.
4. Given octal numbers, convert them to binary numbers and vice versa.

Note: This section may be omitted without detriment to your studies in the rest of this textbook.

"Octal" means "based on eight." This positional-weighted numbering system uses the numbers 0, 1, 2, 3, 4, 5, 6, and 7. The weights (place values in decimal) are shown in Table 2–4.

Powers of 8	8^5	8^4	8^3	8^2	8^1	8^0	8^{-1}	8^{-2}	**TABLE 2-4**
Positional Weights	32,768	4096	512	64	8	1	0.125	0.015625	**Octal Numbering System Positional Weights**

TABLE 2-5	Decimal	Octal	Decimal	Octal
Decimal and Octal Counts	0	0	13	15
	1	1	14	16
	2	2	15	17
	3	3	16	20
	4	4	17	21
	5	5	18	22
	6	6	19	23
	7	7	20	24
	8	10	21	25
	9	11	22	26
	10	12	23	27
	11	13	24	30
	12	14	25	31

When we are counting in decimal and the highest digit (9) is reached, we automatically move to the next higher place value—9 to 10. The number $10_{(10)}$ actually indicates 1 ten (10^1) and 0 units (10^0). When we are counting in octal and the highest digit (7) is reached, we must do the same thing—move to the next higher place value. Therefore, an octal count would appear as 4, 5, 6, 7, 10, 11, and so on. The $10_{(8)}$ number indicates 1 eight (8^1) and 0 units (8^0). The count of $11_{(8)}$ indicates $(1 \times 8^1) + (1 \times 8^0)$, which equals 9 in decimal. Compare the decimal and octal counts shown in Table 2–5.

Since octal numbers are sometimes used in digital systems' technical data, conversion between the octal numbers and the more familiar decimal equivalents is mandatory.

Octal-to-Decimal Conversion

The sum-of-weights method may again be used to convert an octal number to its decimal equivalent. If the octal number $26_{(8)}$ is converted to decimal, Table 2–5 indicates the decimal equivalent is $22_{(10)}$. Let's ensure the table is correct.

$$
\begin{array}{cc}
8^1 & 8^0 \\
2 & 6_{(8)} \\
\end{array}
$$
$$(2 \times 8^1) + (6 \times 8^0) =$$
$$(2 \times 8) + (6 \times 1) =$$
$$16 \quad + \quad 6 \quad = 22_{(10)}$$

The radix multiplication method may also be used for whole-number conversion from octal to decimal in lieu of the sum-of-weights method. The procedure is the same as that used in binary-to-decimal conversion. The only difference is the radix is now 8. Thus, repeated multiplication by 8 must be accomplished. Let's use the same octal number as was used in the sum-of-weights exercise.

Convert $26_{(8)}$ to its decimal equivalent.

Follow the steps listed for the binary-to-decimal conversion in Fig. 2–1, *except* multiply the sum by 8 in step 3.

$$
\begin{array}{cc}
2 & 6_{(8)} \\
+0 & +16 \\
\hline
2 & 22 \\
\times 8 & \\
\hline
16 & \\
\end{array}
$$
Decimal equivalent

$$26_{(8)} = \mathbf{22_{(10)}}$$

Now convert $377_{(8)}$ to its decimal equivalent.

$$
\begin{array}{ccc}
3 & 7 & 7_{(8)} \\
+0 & +24 & +248 \\
\hline
3 \nearrow & 31 \nearrow & 255 \qquad \text{Decimal equivalent} \\
\times 8 \diagup & \times 8 \diagup & \\
\hline
24 \diagup & 248 \diagup &
\end{array}
$$

$377_{(8)} = \mathbf{255_{(10)}}$

For the fraction side of an octal number, the radix division by 8 procedure must be employed.

Convert $0.625_{(8)}$ to its decimal equivalent.

$$
\begin{array}{l}
0 \;.\; 6 \; 2 \; 5 \\
\qquad\qquad 5 \div 8 \\
\quad 2 \;.\; 6 \; 2 \; 5 \div 8 \\
6 \qquad .\; 3 \; 2 \; 8 \; 1 \; 2 \; 5 \div 8 \\
\qquad\qquad 0 \;.\; 7 \; 9 \; 1
\end{array}
$$

$0.625_{(8)} = \mathbf{0.791_{(10)}}$

Decimal-to-Octal Conversion

The radix division method of converting a whole decimal number to its octal equivalent will be illustrated. The procedure listed in Fig. 2–2 is comparable for this conversion with one exception—divide by 8.

Convert $100_{(10)}$ to its octal equivalent.

$$
\begin{array}{ccc}
\rightarrow 1 & 4 & 4 \qquad \text{Octal equivalent} \\
1 & 12 & 100_{(10)} \\
\cline{1-3}
& 8 & 8
\end{array}
$$

$100_{(10)} = \mathbf{144_{(8)}}$

Note: Using a calculator to perform the $100 \div 8$ results in 12.5. The 0.5 remainder equals 4 because it is actually 0.5 of the base of the numbering system to which we are converting. Regardless of the remainder, it must always be multiplied times the base of the numbering system to which we are converting to put it in its proper form. Once in proper form, the remainders make up the octal number.

Convert $79_{(10)}$ to its octal equivalent.

$$
\begin{array}{ccc}
\rightarrow 1 & 1 & 7 \qquad \text{Octal equivalent} \\
1 & 9 & 79 \\
\cline{1-3}
& 8 & 8
\end{array}
$$

$79_{(10)} = \mathbf{117_{(8)}}$

Note: The remainder on a calculator from the $79 \div 8$ problem is 0.875. Again, this remainder times 8 produces the actual octal remainder for the LSD portion of the octal answer.

To convert on the fractional side, the procedure is reversed from the whole number side—use radix multiplication to convert the decimal number to its octal equivalent.

$$0.625_{(10)} = \underline{\hspace{2cm}}_{(8)}$$

$$\begin{array}{r} 0.625 \\ \times\ 8 \\ \hline 5.000 \end{array}$$

$$0.625_{(10)} = \mathbf{0.5}_{(8)}$$

Octal-to-Binary Conversion

One of the major advantages of using the octal numbering system is the relative ease with which it can be converted directly to binary.

Since 7 is the highest digit in the octal numbering system, all of the octal digits can be coded into binary by using groups of three bits as shown in Table 2–6. The formula for maximum count as previously presented is $2^n - 1$, and $2^3 - 1 = 7$.

To convert any octal number to its binary equivalent, code each octal digit as a group of three bits as illustrated:

$$144_{(8)} = \underline{\hspace{2cm}}_{(2)}$$

$$\begin{array}{ccc} 1 & 4 & 4_{(8)} \\ \hline 001 & 100 & 100_{(2)} \end{array}$$

The resultant groups of bits form the binary answer. That $144_{(8)} = \mathbf{1100100}_{(2)}$ can be proven by converting each of these numbers to decimal using radix multiplication.

$$\begin{array}{ccc} 1 & 4 & 4_{(8)} \\ +0 & +8 & +96 \\ \hline 1 & 12 & 100 \\ \times 8 & \times 8 & \\ \hline 8 & 96 & \end{array} \quad \text{Decimal equivalent}$$

$$\begin{array}{ccccccc} 1 & 1 & 0 & 0 & 1 & 0 & 0_{(2)} \\ +0 & +2 & +6 & +12 & +24 & +50 & +100 \\ \hline 1 & 3 & 6 & 12 & 25 & 50 & 100 \\ \times 2 & \times 2 & \times 2 & \times 2 & \times 2 & \times 2 & \\ \hline 2 & 6 & 12 & 24 & 50 & 100 & \end{array} \quad \text{Decimal equivalent}$$

TABLE 2-6	Octal	Binary
Binary-Coded Octal (BCO)	0	000
	1	001
	2	010
	3	011
	4	100
	5	101
	6	110
	7	111

$$144_{(8)} \quad = \mathbf{100_{(10)}}$$
$$1100100_{(2)} = \mathbf{100_{(10)}}$$

The simple procedure of writing each octal digit in a group of three bits produces what is referred to as **binary-coded octal (BCO).** As the name implies, BCO is an octal number that has been coded in binary. It is this shortcut that makes using octal numbers in digital systems popular.

Convert $4075_{(8)}$ to its binary equivalent.

$$\underbrace{4}_{100} \quad \underbrace{0}_{000} \quad \underbrace{7}_{111} \quad \underbrace{5_{(8)}}_{101_{(2)}}$$

$$4075_{(8)} = \mathbf{100000111101_{(2)}}$$

Binary-to-Octal Conversion

If the process used to obtain the BCO number is reversed, an octal number can be obtained directly from a binary number.

Convert $10101110_{(2)}$ to its octal equivalent.

$$\underline{010} \quad \underline{101} \quad \underline{110}_{(2)}$$
$$2 \qquad 5 \qquad 6_{(8)}$$

Starting at the LSB, group the binary bits in groups of three. Next, annotate the octal digit represented by each group of three bits. The annotated number represents the octal equivalent of the binary number.

Convert $1111.11_{(2)}$ to its octal equivalent.

$$\underline{001} \quad \underline{111} \quad . \quad \underline{110}$$
$$1 \qquad 7 \quad . \quad 6$$

$$1111.11_{(2)} = \mathbf{17.6_{(8)}}$$

In the conversion illustrated, the fractional side of the binary number was coded into an octal digit. Note that a 0 was added to form a 3-bit group to the right of the octal point. Always set up groups of three bits starting at the octal point and moving away. If the 0 had not been added to the fractional group of bits, the answer might have appeared to be a 3.

Section 2–3: Review Questions

Answers are given at the end of the chapter.

A. Octal means based on _____ .
   (1) 8
   (2) 10
   (3) 16

B. The next sequential count up from $17_{(8)}$ is _____ $_{(8)}$.

Convert the following octal numbers to their decimal equivalents:

C. $17_{(8)} =$ _____ $_{(10)}$
D. $26_{(8)} =$ _____ $_{(10)}$
E. $100_{(8)} =$ _____ $_{(10)}$
F. $567_{(8)} =$ _____ $_{(10)}$
G. $1251_{(8)} =$ _____ $_{(10)}$

Convert the following decimal numbers to their octal equivalents:

H. $10_{(10)} =$ _____ $_{(8)}$
I. $26_{(10)} =$ _____ $_{(8)}$
J. $100_{(10)} =$ _____ $_{(8)}$
K. $331_{(10)} =$ _____ $_{(8)}$
L. $101_{(10)} =$ _____ $_{(8)}$

Convert the following octal numbers to their binary equivalents:

M. $10_{(8)} =$ _____ $_{(2)}$
N. $32_{(8)} =$ _____ $_{(2)}$
O. $147_{(8)} =$ _____ $_{(2)}$

Convert the following binary numbers to their octal equivalents:

P. $101110111_{(2)} =$ _____ $_{(8)}$
Q. $1101010000_{(2)} =$ _____ $_{(8)}$
R. $1011.1_{(2)} =$ _____ $_{(8)}$

SECTION 2-4: HEXADECIMAL NUMBERING SYSTEM

OBJECTIVES

1. Count sequentially in hexadecimal.
2. Given hexadecimal numbers, convert them to decimal, binary, and octal numbers.
3. Given numbers expressed in a base other than hexadecimal, convert them to hexadecimal numbers.

With the advent of personal computers (PCs) in the late 1970s and their constant rise in popularity with consumers, faster clock speeds and larger word sizes and addresses were a natural progression.

Imagine what it would be like to read technical literature explaining the operation of a PC that constantly referred to memory addresses such as $11111110000010101101_{(2)}$. This kind of data would be mind-boggling. These large binary numbers, even though commonplace in today's technology, require being put into a usable and understandable format. Hexadecimal is the numbering system that overcomes the cumbersome binary numbers and solves this problem.

The hexadecimal (hex) numbering system (base 16) requires sixteen characters. In addition to the numbers 0 through 9, the letters A through F are used to complete the 16-character requirement. In hexadecimal, the A, B, C, D, E, and F should not be thought of as letters of the alphabet. They should be thought of as numbers, where $A_{(16)} = 10_{(10)}$, $B_{(16)} = 11_{(10)}$, and so on, as illustrated in Table 2–7.

Decimal	Hexadecimal	Decimal	Hexadecimal	TABLE 2-7
0	0	17	11	**Decimal and Hexadecimal Counts**
1	1	18	12	
2	2	19	13	
3	3	20	14	
4	4	21	15	
5	5	22	16	
6	6	23	17	
7	7	24	18	
8	8	25	19	
9	9	26	1A	
10	A	27	1B	
11	B	28	1C	
12	C	29	1D	
13	D	30	1E	
14	E	31	1F	
15	F	32	20	
16	10	33	21	

When counting in hex, the same basic rule followed in the other numbering systems applies. That is, when the highest number, which is F in hex, is reached, we must move to the next higher place value. Study the counts of decimal versus hex in Table 2–7. Check the positional weights shown in Table 2–8 to assure the count sequence is correct and that you understand it.

Hexadecimal-to-Decimal Conversion

Like its binary and octal counterparts, a method must be provided to convert hex numbers to their decimal equivalents. The system could be the sum-of-weights system, but we will skip demonstrating that method for the more popular radix multiplication method for the integer portion of the hex number.

Since by now you should be adept using this procedure, a couple of examples will suffice. The procedure is identical to the binary-to-decimal and octal-to-decimal conversions except the repeated multiplication is by 16.

Convert the hex number 1F to its decimal equivalent.

$$1F_{(16)} = \underline{\hspace{2cm}}_{(10)}$$

$$
\begin{array}{cc}
1 & F \\
+0 & +16 \\
\hline
1 & 31 \quad \text{Decimal equivalent} \\
\times 16 & \\
\hline
16 & \\
\end{array}
$$

$$1F_{(16)} = \mathbf{31_{(10)}}$$

Powers of 16	16^5	16^4	16^3	16^2	16^1	16^0	16^{-1}	TABLE 2-8
Positional Weights	1,048,576	65,536	4096	256	16	1	0.0625	**Hexadecimal Numbering System Positional Weights**

To become thoroughly proficient at converting hex to decimal, you must commit to memory the decimal values of A through F. In the preceding problem, $F = 15_{(10)}$.

Convert $1CE_{(16)}$ to its decimal equivalent.

$$
\begin{array}{ccc}
1 & C & E \\
+0 & +16 & +448 \\
\hline
1 & 28 & 462 \qquad \text{Decimal equivalent} \\
\times 16 & \times 16 & \\
\hline
16 & 448 &
\end{array}
$$

$1CE_{(16)} = \mathbf{462_{(10)}}$

For the fractional side of a hex number, convert $0.B4_{(16)}$ to its decimal equivalent by using the radix division method of conversion.

$$
\begin{array}{c}
0\,.B\quad 4 \\
\big\downarrow \\
4 \div 16 \\
1\;\;1.2\;\;5 \div 16 \\
0.7\,0\,3
\end{array}
$$

$0.B4_{(16)} = \mathbf{0.703_{(10)}}$

Decimal-to-Hexadecimal Conversion

The procedure used for decimal-to-hex conversion is radix division for the whole numbers.

Convert the decimal number 692 to its hex equivalent.

$$
\begin{array}{ccc}
2 & B & 4 \qquad \text{Hex equivalent} \\
2 & 43 & 692 \\
\hline
& 16 & 16
\end{array}
$$

$692_{(10)} = \mathbf{2B4_{(16)}}$

In this example, the problem $692 \div 16$ produced 43.25. The 0.25 remainder equals 4 when it is multiplied times the base of the numbering system to which we are converting. The $43 \div 16$ problem yielded 2.6875. The 0.6875 remainder equals 11 when converted to hex. The key here is to remember that $11 = B$.

On the other side of the hex point, convert $0.75_{(10)}$ to its hex equivalent.

$$
\begin{array}{c}
0.75_{(10)} \\
\times 16 \\
\hline
12.00 = C_{(16)}
\end{array}
$$

$0.75_{(10)} = \mathbf{0.C_{(16)}}$

Confusion sometimes overwhelms one when all of this number conversion information is presented in rapid succession. Don't feel alone if you have this feeling. Some of the problems are when to divide and when to multiply and by what to divide or multiply. Do you multiply by the radix of the numbering system *from* which you are converting or *to* which you are converting? Or do you divide?

The following general guidelines should help clarify these problems.

1. *Radix multiplication and radix division can only be used to convert to or from the decimal numbering system.* There are no shortcuts, other than calculator conversion, that can be used.
2. *Never multiply or divide by ten.* Since conversion is always to a decimal number or from a decimal number, one of the numbers involved with the conversion will always be base 10. Therefore, always use the base of the other number to multiply or divide by.
3. Remember, *when converting "from" decimal, divide the integer side of the number.* Therefore, *when converting "to" decimal, multiply the integer portion of the number.*
4. Always use the opposite radix procedure on the fractional side of a number.

Hexadecimal-to-Binary Conversion

F is the highest character in the hexadecimal numbering system. Since $F = 15_{(10)}$, how many bits would be required to code a hex number into binary? Maximum count $= 2^n - 1 = 2^4 - 1 = 15$. Thus, four bits are required. All of the hex numbers can be coded into binary by using groups of four bits as illustrated in Table 2–9.

The decimal number $692 = 2B4_{(16)}$. Convert the hex number 2B4 to its binary equivalent.

$$2B4_{(16)} = \underline{\hspace{3cm}}_{(2)}$$

$$\underbrace{2}\quad \underbrace{B}\quad \underbrace{4}_{(16)}$$
$$0010\quad 1011\quad 0100_{(2)}$$
$$2B4_{(16)} = \mathbf{1010110100}_{(2)}$$

Hexadecimal	Binary	TABLE 2-9
0	0000	**Binary-Coded**
1	0001	**Hexadecimal**
2	0010	**(BCH)**
3	0011	
4	0100	
5	0101	
6	0110	
7	0111	
8	1000	
9	1001	
A	1010	
B	1011	
C	1100	
D	1101	
E	1110	
F	1111	

The groups of four bits shown yield the binary equivalent to the hex number. This is nothing more than a hex number that is coded in binary. To accomplish this procedure, code each hex digit as a group of four bits. This shorthand method of conversion is called **binary-coded hexadecimal (BCH)** and provides a direct conversion method to go from hex to binary.

Binary-to-Hexadecimal Conversion

The inverse of the BCH procedure allows conversion of a binary number to its hex equivalent.

Convert $11101011000001_{(2)}$ to its hex equivalent.

$$\underline{0011} \quad \underline{1010} \quad \underline{1100} \quad \underline{0001}_{(2)}$$
$$3 \qquad A \qquad C \qquad 1_{(16)}$$

The conversion starts at the LSB of the binary number, and the binary numbers have been separated into groups of four bits. Note that two 0s were added to the left side of the binary number to make the last group of four bits. Once this has been completed, write the hex number for each group of four bits. This yields the hex equivalent number.

More Shortcuts with BCO and BCH

Often the shortcuts employing BCO and BCH can be applied as real time-savers. If the octal number $1257_{(8)}$ requires conversion to hex, different methods can be employed to obtain the desired results. First, radix multiplication will take the octal number to its decimal equivalent.

$$
\begin{array}{cccc}
1 & 2 & 5 & 7_{(8)} \\
+0 & +8 & +80 & +680 \\
\hline
1 & 10 & 85 & 687 \\
\times 8 & \times 8 & \times 8 \\
\hline
8 & 80 & 680
\end{array}
\qquad \text{Decimal equivalent}
$$

Next, the decimal equivalent derived above can be converted to its hex equivalent by radix division.

$$
\begin{array}{ccc}
2 & A & F \\
2 & 42 & 687_{(10)} \\
& \overline{16} & \overline{16}
\end{array}
\qquad \text{Hex equivalent}
$$

It would have been much simpler to *code the octal number in binary with groups of three bits, and then decode that binary number into groups of four bits to arrive at the hex equivalent number.*

$$
\begin{array}{cccc}
1 & 2 & 5 & 7_{(8)} \\
\underline{001} & \underline{010} & \underline{101} & \underline{111} \\
2 & & A & F_{(16)}
\end{array}
$$

This procedure only takes a minute. Coding octal and hex to binary and decoding binary to octal or hex will come in extremely handy when working with memory addressing in computers. In addition, data in computers is normally given in hex numbers.

Hexadecimal is more widely used in computers than octal because PCs typically operate with data and addresses in multiples of four. This makes it very convenient to group four bits and give that group a hex number. A group of four bits is called a **nibble.** When eight bits are used, they are referred to as a **byte** of data.

As a final problem, convert $70A.B_{(16)}$ to its octal equivalent.

$$\begin{array}{ccccc} 7 & 0 & A & . & B_{(16)} \\ \underline{0111} & \underline{0000} & \underline{1010} & . & \underline{1011}\ \underline{0000} \\ 3 & 4 & 1 & 2 & .\quad 5 \quad 4_{(8)} \end{array}$$

$70A.B_{(16)} = \mathbf{3412.54_{(8)}}$

This type of direct conversion saves a lot of time. In this procedure, *the hex number was coded in binary groups of four bits and then decoded into groups of three bits to arrive at the octal number*. The group of four low bits to the right of the binary point have no weight and were added so groups of three bits could be obtained when decoding to the octal number.

This section started with a bulky $11111110000010101101_{(2)}$ binary address. It is easy to see that an $FE0AD_{(16)}$ hex address is much more manageable.

Section 2–4: Review Questions

Answers are given at the end of the chapter.

A.   What is the next sequential count *up* from $1FF_{(16)}$?
B.   What is the next sequential count *down* from $1FF_{(16)}$?

Convert the following hexadecimal numbers to their decimal equivalents.

C. $CA_{(16)} = $ _____ (10)
D. $102_{(16)} = $ _____ (10)
E. $BB6_{(16)} = $ _____ (10)
F. $12A5_{(16)} = $ _____ (10)

Convert the following decimal numbers to their hex equivalents.

G. $273_{(10)} = $ _____ (16)
H. $462_{(10)} = $ _____ (16)
I. $681_{(10)} = $ _____ (16)
J. $1000.75_{(10)} = $ _____ (16)

Convert the following hex numbers to their binary equivalents.

K. $1A_{(16)} = $ _____ (2)
L. $2B3_{(16)} = $ _____ (2)
M. $3FF_{(16)} = $ _____ (2)
N. $33B2_{(16)} = $ _____ (2)

Convert the following binary numbers to their hex equivalents.

O. $110100001_{(2)} = $ _____ (16)
P. $1111010111101_{(2)} = $ _____ (16)

Convert the following octal numbers to their hex equivalents.

Q. $21_{(8)} = $ —————$_{(16)}$
R. $604_{(8)} = $ —————$_{(16)}$

SECTIONS 2–1 THROUGH 2–4: INTERNAL SUMMARY

The **binary** numbering system is a base-2 system, **octal** is a base-8 system, and **hexadecimal** is a base-16 system. Octal and hexadecimal numbers provide a simple method of determining what a binary number actually is without having to keep track of a multitude of binary digits (bits).

The **sum-of-weights** method may be used to determine the decimal equivalent value of any positional-weighted number. To determine the decimal value of a number using this method, multiply each digit in the number times its place value and add the products. The resultant sum will equal the decimal value of the number.

Radix conversion procedures are depicted in Fig. 2–3.

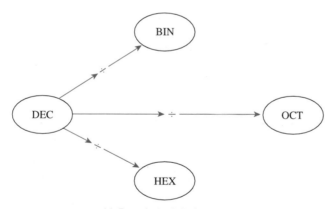

(a) From decimal divide integer.

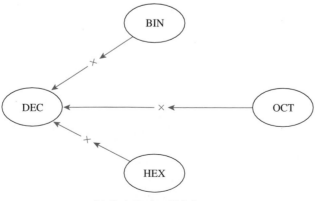

(b) To decimal multiply integer.

FIGURE 2–3   **Radix conversion procedures.**

Radix division can be used to convert a decimal whole number to its binary, octal, or hexadecimal equivalent. This procedure is accomplished by repeated radix division. Use radix multiplication on the opposite side of the decimal point.

Radix multiplication can be used to convert binary, octal, and hex whole numbers to their decimal equivalents. This procedure is accomplished by repeated radix multiplication. Use the opposite radix procedure, radix division in this instance, on the fractional portion of the number.

The radix multiplication and radix division procedures can only take a number to or from decimal. For example, neither of these procedures can be directly used to convert a binary number to octal. The binary number could be converted to decimal using radix multiplication, and then the decimal number could be converted to octal using radix division, but conversion in this manner is time-consuming.

Octal-to-binary conversion can be done directly by coding each octal digit as a **group of three bits (BCO)**. Reversing this process allows for a simplified binary-to-octal conversion.

Hex-to-binary and binary-to-hex conversions are done as described above by using **groups of four bits (BCH).**

Number conversion from octal to hex or hex to octal is easy to accomplish by using BCO or BCH to encode the number in binary and then decode it in the proper bit group size.

Table 2-10 provides a comparison of decimal, binary, octal, and hexadecimal counts.

Decimal	Binary	Octal	Hexadecimal	
				TABLE 2-10
0	00000	0	0	**Decimal, Binary,**
1	00001	1	1	**Octal, and Hex**
2	00010	2	2	**Counts**
3	00011	3	3	
4	00100	4	4	
5	00101	5	5	
6	00110	6	6	
7	00111	7	7	
8	01000	10	8	
9	01001	11	9	
10	01010	12	A	
11	01011	13	B	
12	01100	14	C	
13	01101	15	D	
14	01110	16	E	
15	01111	17	F	
16	10000	20	10	
17	10001	21	11	
18	10010	22	12	
19	10011	23	13	
20	10100	24	14	
21	10101	25	15	
22	10110	26	16	
23	10111	27	17	
24	11000	30	18	
25	11001	31	19	
26	11010	32	1A	

SECTIONS 2-1 THROUGH 2-4: INTERNAL SUMMARY QUESTIONS

Answers are given at the end of the chapter.

1. The octal numbering system is a base-_____ system.
   a. 2
   b. 8
   c. 10
   d. 16

2. The next sequential count up from $10000_{(2)}$ is
   a. $11111_{(2)}$
   b. $11110_{(2)}$
   c. $10001_{(2)}$
   d. $1111_{(2)}$

3. The next sequential count up from $17_{(8)}$ is
   a. $16_{(8)}$
   b. $18_{(8)}$
   c. $20_{(8)}$
   d. $30_{(8)}$

4. The hexadecimal numbering system is a base-_____ system.
   a. 2
   b. 8
   c. 10
   d. 16

5. The next sequential count down (decrement) from $1A_{(16)}$ is
   a. $18_{(16)}$
   b. $19_{(16)}$
   c. $25_{(16)}$
   d. $26_{(16)}$

6. When converting a whole number from octal to decimal, the radix _____ procedure may be used.
   a. division
   b. addition
   c. subtraction
   d. multiplication

7. The hexadecimal numbering system is a positional-weighted system.
   a. True
   b. False

8. Octal numbers may be coded for direct conversion to binary by using groups of _____ bits.
   a. 2
   b. 3
   c. 4
   d. 6

9. The hex number E represents the decimal number _____.
   a. 9
   b. 10
   c. 14
   d. 15

10. A byte of data consists of _____ bits.
    a. 1
    b. 4
    c. 8
    d. 16

Convert the following numbers to the base indicated:

11. $21_{(10)} =$ _____ $_{(2)}$
12. $46_{(10)} =$ _____ $_{(2)}$
13. $250_{(10)} =$ _____ $_{(2)}$
14. $105_{(10)} =$ _____ $_{(2)}$
15. $10111_{(2)} =$ _____ $_{(10)}$
16. $101011_{(2)} =$ _____ $_{(10)}$
17. $11111_{(2)} =$ _____ $_{(10)}$
18. $111110101_{(2)} =$ _____ $_{(10)}$
19. $125_{(8)} =$ _____ $_{(10)}$
20. $733_{(8)} =$ _____ $_{(10)}$
21. $FFD_{(16)} =$ _____ $_{(2)}$
22. $3B6_{(16)} =$ _____ $_{(10)}$
23. $255_{(10)} =$ _____ $_{(16)}$
24. $950_{(10)} =$ _____ $_{(8)}$
25. $1AB_{(16)} =$ _____ $_{(8)}$
26. $0.A8_{(16)} =$ _____ $_{(10)}$
27. $0.5625_{(10)} =$ _____ $_{(8)}$
28. $10110.1_{(2)} =$ _____ $_{(8)}$

SECTION 2–5: BINARY-CODED DECIMAL

OBJECTIVES

1. State the advantages and disadvantages of using BCD numbers.
2. Given decimal numbers, code them to BCD numbers and vice versa.

Many different codes have been developed for use in digital systems to enhance their capabilities and improve and standardize, to a degree, their operations. A **code** is a group of bits used to represent a digit, letter, or symbol. In most instances, the code is not positional weighted. One of these codes is **binary-coded decimal (BCD).**

Direct conversions from octal to binary and binary to octal were presented in Section 2–3. Hex-to-binary and binary-to-hex conversions were illustrated in Section 2–4. These conversion methods using BCO and BCH were straightforward and easy to understand, primarily because all octal digits (0–7) can be represented with groups of three bits. Also, all hex characters (0–F) can be represented with groups of four bits.

The decimal digits 0 through 9 can be coded with bits in the same manner as BCH. Like hex numbers, the decimal numbers require groups of four bits because $8_{(10)}$ and $9_{(10)}$ must be coded as 1000 and 1001. This leaves six 4-bit groups that can occur that are not valid BCD numbers. Binary-coded decimal numbers, including the invalid BCD numbers, are shown in Table 2–11. The subject of these invalid BCD numbers will be discussed shortly.

When a hex number was coded in binary (BCH), the resultant coded number was the binary equivalent to the hex number. When a decimal number is coded in binary (BCD), the resultant number is *not* binary.

TABLE 2-11	Decimal	Binary-Coded Decimal (BCD)
Binary-Coded Decimal (BCD)	0	0000
	1	0001
	2	0010
	3	0011
	4	0100
	5	0101
	6	0110
	7	0111
	8	1000
	9	1001
	—	1010*
	—	1011*
	—	1100*
	—	1101*
	—	1110*
	—	1111*

*Invalid BCD numbers—Not used.

First, check out the following example to see why we can't code decimal to binary and achieve the same results obtained in BCH. In the following example, $36_{(10)}$ has been coded using 4-bit groups of bits, but the resultant answer is not the binary equivalent of 36:

$36_{(10)} = 0011\ 0110$

$36_{(10)}$ *does not equal* $0011\ 0110_{(2)}$. This binary number ($110110_{(2)}$) equals $54_{(10)}$.

Decimal-to-BCD Conversion

If *each decimal digit is coded into a group of four bits as shown, we can label the resultant binary-coded decimal:*

$36_{(10)} = 0011\ 0110_{(BCD)}$

It is very important to annotate the BCD subscript to differentiate this number from a binary number. In BCH this was unnecessary because the result was binary. Now code $92_{(10)}$ to BCD.

$$\underbrace{9}_{1001} \quad \underbrace{2_{(10)}}_{0010_{(BCD)}}$$

BCD-to-Decimal Conversion

The reverse of this procedure will take a BCD number back to its decimal equivalent. Start at the LSB and group the bits into groups of four as shown. Next, write the decimal digit that is represented by each group of four bits.

$$\underbrace{1001}_{9} \quad \underbrace{0010}_{2_{(10)}}{}_{(BCD)}$$

It is apparent that a BCD number is easy to read. Compare reading the number $1001110_{(2)}$ to its comparable $0111\ 1000_{(BCD)}$.

$$1001110_{(2)} = 78_{(10)} = 0111\ 1000_{(BCD)}$$

It is a breeze to read the BCD number as $78_{(10)}$. However, the binary number would have to be converted to decimal with a calculator, by radix multiplication, or with a sum-of-weights procedure to determine its value.

Since each digit is encoded with a group of four bits, BCD is sometimes referred to as the **8-4-2-1 code**. This relates directly to the weight of the bits within each group of bits. However, as previously proven, BCD is not a positional-weighted numbering system.

Though BCD numbers are easy to recognize and code and decode, they have certain disadvantages. As mentioned earlier, some 4-bit groups of numbers are not used in BCD. These groups form invalid BCD numbers. To illustrate, $10_{(10)} = 0001\ 0000_{(BCD)}$. Therefore, 1010 cannot be a valid BCD number.

The decimal number 16 is represented here in binary and BCD. One of the prime disadvantages of BCD is that it takes more bits to represent a decimal number in BCD than it does in binary.

$$16_{(10)} = 10000_{(2)} = 0001\ 0110_{(BCD)}$$

Since it is mandatory to code each decimal digit as a group of "four" bits, it is easy to see that it takes more bits to code the decimal number 16 to BCD than it does to convert it to binary.

The disadvantage of more bits for coding purposes brings about other disadvantages. If BCD is used in a digital system that contains memory, more memory space is required. Processing more bits also results in more power consumption and reduced efficiency.

BCD is no different from pure binary to a digital circuit. There is no way to identify the BCD subscript in a system. Therefore, if you are adding binary numbers the digital system's adder sums them as indicated in the rules for binary addition below. But, what if these numbers are BCD numbers? The system operates the same. Thus, the results of adding BCD numbers in a system would be devastating without additional circuitry to correct some inherent mistakes that occur.

The rules for binary addition are quite simple:

$$
\begin{array}{cccc}
0 & 0 & 1 & 1 \\
+0 & +1 & +0 & +1 \\
\hline
0 & 1 & 1 & 10* \\
\end{array}
$$

You can see by applying the rules of binary addition that

$$
\begin{array}{rcl}
0101_{(2)} & = & 5_{(10)} \\
+0101_{(2)} & = & +5_{(10)} \\
\hline
1010_{(2)} & = & 10_{(10)} \\
\end{array}
$$

On the other hand, what would happen if these were BCD numbers? The adder circuit doesn't know and performs its functions in the same manner. A 1010 sum is an invalid BCD number. Thus, when adding BCD numbers an invalid sum must be detected and a correction factor inserted to make the sum valid. The point for now is that BCD

---

*As a tool in binary addition, we usually say $1 + 1 = 0$ with a **carry** of 1.

arithmetic circuits require additional circuitry to detect an invalid sum and then more circuitry to correct that sum to a usable BCD number. These arithmetic circuits will be presented in Chapter 9.

Section 2–5: Review Questions

Answers are given at the end of the chapter.

A. State the advantages of BCD.
B. State the disadvantages of BCD.
C. Code $29_{(10)}$ in BCD.
D. Code $1246_{(10)}$ in BCD.
E. Decode $0111\ 1000_{(BCD)}$ to decimal.
F. Decode $0010\ 0011\ 0111_{(BCD)}$ to decimal.

SECTION 2–6: GRAY CODE

OBJECTIVES

1. State the purpose for gray code.
2. Given binary numbers, convert them to gray code and vice versa.

Another code often used in digital systems to improve reliability is the **gray code.** This code is sometimes called the **error-minimizing code** or the **reflected code.**

The code was designed so that *successive counts result in only one bit changing states.* This is shown in Table 2–12 in the gray code chart. The term **reflected code** was derived from the fact that the numbers from $8_{(10)}$ down are an exact reflection of the numbers from $7_{(10)}$ up if the first bit is disregarded. Note the last three bits in counts 7 and 8 = X100, 6 and 9 = X101, 5 and 10 = X111, and so on.

The gray code is *not* a weighted code; thus it is not an arithmetic code. The binary counts shown in Table 2–12 were added for comparison purposes only. If you select any gray code number and move to its next increment count or decrement count, you will see

TABLE 2-12 Gray Code	Decimal	Binary	Gray Code
	0	0000	0000
	1	0001	0001
	2	0010	0011
	3	0011	0010
	4	0100	0110
	5	0101	0111
	6	0110	0101
	7	0111	0100
	8	1000	1100
	9	1001	1101
	10	1010	1111
	11	1011	1110
	12	1100	1010
	13	1101	1011
	14	1110	1001
	15	1111	1000

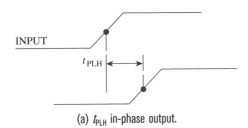

(a) t_{PLH} in-phase output.

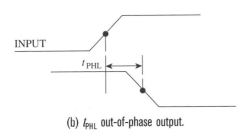

(b) t_{PHL} out-of-phase output.

FIGURE 2–4   **Propagation delay time.**

that only one bit changes states. This is not true in the binary count. In some binary counts, an increment of one count results in all bits changing states. Compare a binary increment from $0111_{(2)}$ to $1000_{(2)}$. Often bits changing states present a problem since they do not change from high to low at the same rate as they change from low to high.

To expand on this subject, think back to basic transistor theory. It takes x amount of time to drive a transistor into saturation. When transistors are driven into saturation, as they are in many digital circuits, they are driven past the saturation point. Therefore, it takes longer to cut them off when they are in saturation than it does to drive them to saturation from cutoff. This is evidenced by comparing high-to-low propagation delay times to low-to-high propagation delay times in a data book. These delay times are illustrated in Fig. 2–4. **Propagation delay time** is the time between a specified point on the input waveform and a specified point on the output waveform when the output is changing from one level (low or high) to the other level. In layman's terms, this is nothing more than the amount of time that elapses between when the input requests the circuit to do something and when the circuit gets it done.

Take a 7400 quad 2-input NAND gate and check its propagation delay times. The **high-to-low propagation delay time t_{PHL}** is typically 7 ns from input to output. On the other hand, the **low-to-high propagation delay time t_{PLH}** is typically 11 ns. The preceding times are for a standard TTL NAND gate. These figures prove that it takes longer to cut off a transistor (low-to-high transition) when it is saturated than it does to drive it into saturation from cutoff (high-to-low transition).

What does this disparity in propagation delay times mean in terms of circuit operation? Consider the count sequence of an up-counter as shown.

BINARY COUNT	ACTUAL COUNT SEQUENCE		GRAY CODE
0011	0011		0010
	0000	Short-duration miscount	
0100	0100		0110
	0000	Short-duration miscount	
0101	0101		0111

The short-duration miscount results from the faster t_{PHL} of the high-to-low count transition. To put it simply, the 1s get to 0 before the 0s get to 1. Although this problem can be masked out in decoding circuits, it is minimized when the gray code is used because

FIGURE 2–5 **Gray code–to–binary conversion.**

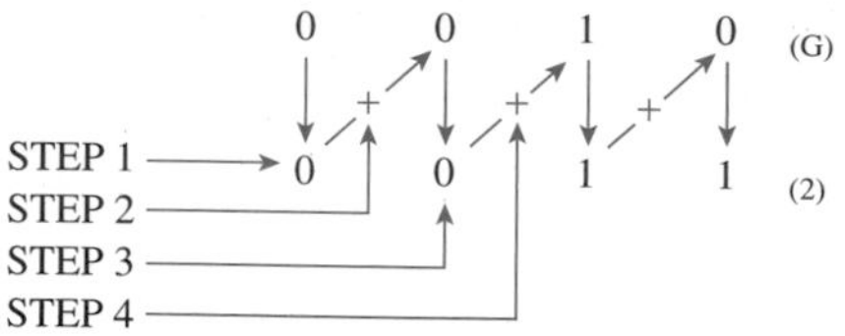

STEP 1: BRING DOWN MSB.
STEP 2: ADD MSB DIAGONALLY TO NEXT
          LESSER SIGNIFICANT BIT.
STEP 3: PLACE SUM BELOW BIT
          ADDED TO. NOTE: IF A CARRY
          IS GENERATED, DISCARD IT.
STEP 4: ADD SUM DIAGONALLY TO
          NEXT LESSER SIGNIFICANT BIT.
          REPEAT STEPS 3 AND 4 UNTIL
          ALL GRAY CODE BITS HAVE BEEN
          ADDED.

only one bit changes states as the circuit counts. For the same reason, the gray code is also routinely used when converting mechanical data such as angular shaft displacement, rpm, and direction of rotation to electrical data.

Gray-to-Binary Conversion

To convert a gray code number to its equivalent binary number, follow the steps illustrated in Fig. 2–5. *Note:* Step 1 in the figure states "Bring down MSB." Since the gray code is a nonweighted code, using the terminology "MSB" is not entirely appropriate. However, it is used here to indicate the position of the bit to be brought down.

Convert $0010_{(G)}$ to its binary equivalent.

$0010_{(G)} = \mathbf{0011}_{(2)}$

No carries were generated in the example problem in Fig. 2–5. If a carry is generated, as shown in the following example, *the carry must be discarded.*

Convert $0111_{(G)}$ to its binary equivalent.

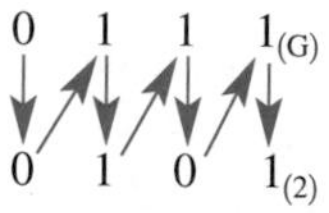

$1 + 1 = 0$ with carry of 1 that is discarded.

$0111_{(G)} = \mathbf{0101}_{(2)}$

Binary-to-Gray Conversion

To convert a binary number to its gray code equivalent, follow the steps listed in Fig. 2–6.

Convert $0100_{(2)}$ to gray code.
If a carry is generated, it is again discarded.

$0100_{(2)} = \mathbf{0110}_{(G)}$

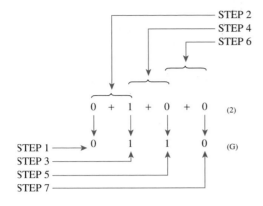

STEP 1: BRING DOWN MSB.
STEP 2: ADD MSB TO ADJACENT BIT.
STEP 3: PLACE SUM BELOW ADJACENT
BIT. NOTE: IF A CARRY IS
GENERATED, DISREGARD IT.
STEP 4: ADD SECOND PAIR OF BITS.
STEP 5: PLACE SUM BELOW LAST
BIT ADDED.
STEP 6: ADD THIRD PAIR OF BITS.
STEP 7: PLACE SUM BELOW LAST
BIT ADDED.

FIGURE 2-6   **Binary-to-gray code conversion.**

Convert $1111_{(2)}$ to gray code.

$$1 \longrightarrow 1 \longrightarrow 1 \longrightarrow 1_{(2)}$$
$$1 \quad\quad 0 \quad\quad 0 \quad\quad 0_{(G)}$$

$$1111_{(2)} = \mathbf{1000}_{(G)}$$

All of the additions in this example are $1 + 1 = 0$ with a carry of 1 that is discarded.

A simple way of remembering whether to add adjacent or diagonal numbers is to learn this memory aid, "Binary-to-gray, add across the way."

Although the example gray code numbers in this section contain four bits, the code can have any number of bits.

Section 2-6: Review Questions

Answers are given at the end of the chapter.

A. What is the primary purpose for the gray code?
B. The gray code is a positional-weighted code.
   (1) True
   (2) False
C. Convert $0101_{(2)}$ to gray code.
D. Convert $11011001_{(2)}$ to gray code.
E. Convert $1101_{(G)}$ to its binary equivalent.
F. Convert $10110110_{(G)}$ to its binary equivalent.

SECTION 2-7: OTHER SPECIAL CODES

OBJECTIVES

1. Given decimal numbers, code them to XS3 code and vice versa.
2. State the meaning and use of ASCII.

Excess-3 Code (XS3)

The **excess-3 code** was designed for use in arithmetic circuits because it is a **self-complementing code.** The code was developed by adding 3 to each decimal digit and then coding the sum in a group of four bits, exactly as BCD numbers were formed.

$$
\begin{array}{cc}
5 & 8 \\
+3 & +3 \\
\hline
8 = 1000_{(XS3)} & 11 = 1011_{(XS3)}
\end{array}
$$

The XS3 code is shown in Table 2–13. The 4-bit groups 0000, 0001, 0010, 1101, 1110, and 1111 are not used in the XS3 code.

To code multiple-digit decimal numbers to XS3, add 3 to each digit and then code the sum to a 4-bit binary number.

Code $68_{(10)}$ to XS3.

$$
\begin{array}{cc}
6 & 8 \\
+3 & +3 \\
\hline
9 & 11
\end{array}
$$

$$1001 \quad 1011_{(XS3)}$$

Convert $509_{(10)}$ to XS3.

$$
\begin{array}{ccc}
5 & 0 & 9 \\
+3 & +3 & +3 \\
\hline
8 & 3 & 12
\end{array}
$$

$$1000 \quad 0011 \quad 1100_{(XS3)}$$

The reverse procedure can be used to convert an XS3 number to decimal. Write the decimal digit for each XS3 4-bit group and then subtract 3.

$$0110 \quad 1001 \quad 0111_{(XS3)}$$

$$
\begin{array}{ccc}
6 & 9 & 7 \\
-3 & -3 & -3 \\
\hline
3 & 6 & 4_{(10)}
\end{array}
$$

$$0110 \quad 1001 \quad 0111_{(XS3)} = \mathbf{364}_{(10)}$$

Although the XS3 code is a nonweighted code, its self-complementing property makes it useful in arithmetic circuits. Its use will be presented in Chapter 9.

TABLE 2-13	Decimal	Excess-3 Code
Excess-3 (XS3) Code	0	0011
	1	0100
	2	0101
	3	0110
	4	0111
	5	1000
	6	1001
	7	1010
	8	1011
	9	1100

American Standard Code for Information Interchange

The PC brought to light, among many other things, a dire need for standardization in the digital world. For one thing, a standard code was needed to represent data from a keyboard. This standardization was mandated to allow interchange of system hardware. Before this standardization, each manufacturer developed codes to represent keyboard data. This left systems with almost total incompatibility to be interfaced with other systems, and hardware interchange was difficult if not impossible to accomplish.

Since the **ASCII** (pronounced "as-key") encodes the letters in the alphabet as well as numbers, it is an **alphanumeric code.** It is a 7-bit code. Seven bits allow representation of 128 (2^7) different characters and commands. The ASCII is used to transfer information in many computers. The 128 codes allow representation of all uppercase and lowercase letters, the decimal numbers 0–9, punctuation marks, and other special symbols. Additionally, the ASCII contains some command codes for formatting text. All of these characters and commands, along with their code number and binary representation are shown in Table 2–14. Table 2–15 defines the characters used in ASCII.

A point of interest is to compare the upper- and lowercase letters "A" and "a" in Table 2–14. The uppercase "A" is 0100 0001 (hex 41) while the lowercase "a" is 0110 0001 (hex 61). The two 7-bit codes differ by only one bit.

One more point of interest should be made about Table 2–14. The binary number for each code contains 8 bits. The MSB of every binary code is a zero since the ASCII is a 7-bit code. However, this eighth bit can be used, and it is in most modern computers. The code is called the **extended ASCII** when the eighth bit is used. The extended ASCII allows for 128 additional graphics characters that can be used in a system.

Section 2–7: Review Questions

Answers are given at the end of the chapter.

A. Code $49_{(10)}$ to XS3.
B. Code $188_{(10)}$ to XS3.
C. Convert $1100\ 0110_{(XS3)}$ to its decimal equivalent.
D. What do the letters ASCII stand for?

Code	Character	Binary	Hex	TABLE 2-14
0	NUL	0000 0000	00	**American Standard Code for Information Interchange**
1	SOH	0000 0001	01	
2	STX	0000 0010	02	
3	ETX	0000 0011	03	
4	EOT	0000 0100	04	
5	ENQ	0000 0101	05	
6	ACK	0000 0110	06	
7	BEL	0000 0111	07	
8	BS	0000 1000	08	
9	HT	0000 1001	09	
10	LF	0000 1010	0A	
11	VT	0000 1011	0B	
12	FF	0000 1100	0C	
13	CR	0000 1101	0D	
14	SO	0000 1110	0E	
15	SI	0000 1111	0F	
16	DLE	0001 0000	10	

	Code	Character	Binary	Hex
TABLE 2-14 (cont'd) **American Standard Code for Information Interchange**	17	DC1	0001 0001	11
	18	DC2	0001 0010	12
	19	DC3	0001 0011	13
	20	DC4	0001 0100	14
	21	NAK	0001 0101	15
	22	SYN	0001 0110	16
	23	ETB	0001 0111	17
	24	CAN	0001 1000	18
	25	EM	0001 1001	19
	26	SUB	0001 1010	1A
	27	ESC	0001 1011	1B
	28	FS	0001 1100	1C
	29	GS	0001 1101	1D
	30	RS	0001 1110	1E
	31	US	0001 1111	1F
	32	SP	0010 0000	20
	33	!	0010 0001	21
	34	"	0010 0010	22
	35	#	0010 0011	23
	36	$	0010 0100	24
	37	%	0010 0101	25
	38	&	0010 0110	26
	39	'	0010 0111	27
	40	(	0010 1000	28
	41	)	0010 1001	29
	42	*	0010 1010	2A
	43	+	0010 1011	2B
	44	,	0010 1100	2C
	45	−	0010 1101	2D
	46	.	0010 1110	2E
	47	/	0010 1111	2F
	48	0	0011 0000	30
	49	1	0011 0001	31
	50	2	0011 0010	32
	51	3	0011 0011	33
	52	4	0011 0100	34
	53	5	0011 0101	35
	54	6	0011 0110	36
	55	7	0011 0111	37
	56	8	0011 1000	38
	57	9	0011 1001	39
	58	:	0011 1010	3A
	59	;	0011 1011	3B
	60	<	0011 1100	3C
	61	=	0011 1101	3D
	62	>	0011 1110	3E
	63	?	0011 1111	3F
	64	@	0100 0000	40
	65	A	0100 0001	41
	66	B	0100 0010	42
	67	C	0100 0011	43
	68	D	0100 0100	44
	69	E	0100 0101	45
	70	F	0100 0110	46
	71	G	0100 0111	47
	72	H	0100 1000	48

Code	Character	Binary	Hex
73	I	0100 1001	49
74	J	0100 1010	4A
75	K	0100 1011	4B
76	L	0100 1100	4C
77	M	0100 1101	4D
78	N	0100 1110	4E
79	O	0100 1111	4F
80	P	0101 0000	50
81	Q	0101 0001	51
82	R	0101 0010	52
83	S	0101 0011	53
84	T	0101 0100	54
85	U	0101 0101	55
86	V	0101 0110	56
87	W	0101 0111	57
88	X	0101 1000	58
89	Y	0101 1001	59
90	Z	0101 1010	5A
91	[	0101 1011	5B
92	\	0101 1100	5C
93	]	0101 1101	5D
94	^	0101 1110	5E
95	—	0101 1111	5F
96	`	0110 0000	60
97	a	0110 0001	61
98	b	0110 0010	62
99	c	0110 0011	63
100	d	0110 0100	64
101	e	0110 0101	65
102	f	0110 0110	66
103	g	0110 0111	67
104	h	0110 1000	68
105	i	0110 1001	69
106	j	0110 1010	6A
107	k	0110 1011	6B
108	l	0110 1100	6C
109	m	0110 1101	6D
110	n	0110 1110	6E
111	o	0110 1111	6F
112	p	0111 0000	70
113	q	0111 0001	71
114	r	0111 0010	72
115	s	0111 0011	73
116	t	0111 0100	74
117	u	0111 0101	75
118	v	0111 0110	76
119	w	0111 0111	77
120	x	0111 1000	78
121	y	0111 1001	79
122	z	0111 1010	7A
123	{	0111 1011	7B
124			0111 1100	7C
125	}	0111 1101	7D
126	~	0111 1110	7E
127	DEL	0111 1111	7F

TABLE 2-14 *(cont'd)*

American Standard Code for Information Interchange

	Character	Meaning	Character	Meaning
TABLE 2-15 **ASCII Character Meaning**	NUL	Null	DC1	Direct Control 1
	SOH	Start of Heading	DC2	Direct Control 2
	STX	Start Text	DC3	Direct Control 3
	ETX	End Text	DC4	Direct Control 4
	EOT	End of Transmission	NAK	Negative Acknowledge
	ENQ	Enquiry	SYN	Synchronous Idle
	ACK	Acknowledge	ETB	End Transmission Block
	BEL	Beep	CAN	Cancel
	BS	Backspace	EM	End of Medium
	HT	Tab	SUB	Substitute
	LF	Line Feed	ESC	Escape
	VT	Cursor Home	FS	Cursor Right
	FF	Form Feed	GS	Cursor Left
	CR	Carriage Return	RS	Cursor Up
	SO	Shift Out	US	Cursor Down
	SI	Shift In	SP	Space
	DLE	Data Link Escape	DEL	Delete

SECTIONS 2-5 THROUGH 2-7: INTERNAL SUMMARY

Many codes have been developed over the years for use in digital systems. Most codes provide some advantages in a system, but they usually are accompanied by drawbacks. Some of the most important codes and their features were presented in this chapter.

Binary-coded decimal (BCD) is used because it is easy to encode a decimal number to BCD and decode the BCD number. It is also easy to read a BCD number.

To encode a decimal number to BCD, each decimal digit is coded as a group of four bits and the result is subscripted BCD. To decode a BCD number, write the decimal digit for each group of four bits. The BCD code is sometimes called the **8-4-2-1 code** because these numbers represent the weights of the bits within each 4-bit group.

There are some disadvantages related to BCD. (1) It takes more bits to represent a decimal number greater than nine in BCD than it does in binary. (2) Utilizing more bits requires more memory space, results in more power consumption, and reduces efficiency. (3) Invalid sums often appear in BCD arithmetic circuits, and these sums must be detected as invalid and then corrected. Both of these actions require additional circuitry.

The **gray code** is called the **error-minimizing code.** Only one bit changes states on each successive count in the gray code. Often when counting in binary, three or four bits change states on a count increment or decrement. For instance, consider the following binary counts: 0011 to 0100, 0111 to 1000, and 1011 to 1100. Since the bits do not change from low to high in the same time required to change from high to low, bit racing results in some false counts. This problem doesn't happen when using the gray code.

Gray code can be changed to binary by (1) bringing down the MSB and (2) adding diagonally up and to the right. Binary can be changed to gray code by (1) bringing down the MSB, and (2) adding adjacent bits. In both of these conversion methods, any carries that are generated during the addition are discarded.

The **excess-3 (XS3) code** is a self-complementing code used in some arithmetic circuits. Decimal numbers are coded to XS3 by adding three to each digit and then coding each digit to a group of four bits as was done in BCH. To decode an XS3 number, write the decimal digit for the 4-bit group and subtract three from that decimal digit.

The **American Standard Code for Information Interchange (ASCII)** provides 128 codes that are used to represent keyboard data as well as some command codes. This code is a 7-bit code. The **extended ASCII** utilizes 8 bits and allows for a total of 256 (2^8) codes. In most computers, the additional codes are used to represent graphics characters.

Sections 2-5 through 2-7: Internal Summary Questions

Answers are given at the end of the chapter.

1. How many bits are required to code a decimal digit to BCD?
   a. 3   c. 8
   b. 4   d. 9

2. How many invalid BCD numbers are there?
   a. 3   c. 6
   b. 4   d. 9

3. When a decimal number such as 311 is coded in binary-coded decimal (BCD), the resultant number equals the binary equivalent to the decimal number.
   a. True
   b. False

4. Code $49_{(10)}$ in BCD.
   a. 100 1001   c. 0100 1001
   b. 111 1100   d. 0111 1100

5. Decode $0011\ 1001\ 1001_{(BCD)}$ to decimal.
   a. 66   c. 612
   b. 399   d. 921

6. Another name for BCD is the 8-4-2-1 code.
   a. True
   b. False

7. $1 + 1 = 1$ in binary addition.
   a. True
   b. False

8. Convert $1010_{(2)}$ to gray code.
   a. A   c. 1100
   b. 10   d. 1111

9. Convert $11101110_{(2)}$ to gray code.
   a. 10011001   c. 1414
   b. 10110100   d. EE

10. Convert $1000_{(gray\ code)}$ to its binary equivalent.
    a. 0111   c. 1100
    b. 1011   d. 1111

11. Convert $10111011_{(gray\ code)}$ to its binary equivalent.
    a. 11100110   c. 11010010
    b. 11101111   d. 11001100

12. The gray code is a weighted code that is often used in arithmetic circuits.
    a. True
    b. False

13. The XS3 code is also called the error-minimizing code.
    a. True
    b. False

14. In which code do successive counts result in only one bit changing states?
    a. XS3   c. Gray
    b. BCD   d. ASCII

15. Code $68_{(10)}$ to XS3.
    a. 0011 0101   c. 0100 0100
    b. 0110 1000   d. 1001 1011

16. Which code is often used to represent the numbers, letters, and punctuation marks on a computer keyboard?
    a. XS3
    b. BCD
    c. Gray
    d. ASCII

17. The standard ASCII contains _____ bits.
    a. 7
    b. 8
    c. 10
    d. 16

18. How many characters and commands can the 8-bit extended ASCII represent?
    a. 127
    b. 128
    c. 255
    d. 256

END OF CHAPTER QUESTIONS/PROBLEMS

Answers are given in the Instructor's Manual.

Section 2-2

1. Write the decimal count sequence 0 through $9_{(10)}$ in binary. (*Note:* Use 4-bit groups of binary digits for each number.)

2. Convert the following binary numbers to their decimal equivalents:
    a. $110_{(2)}$
    b. $1110_{(2)}$
    c. $1111_{(2)}$
    d. $101011_{(2)}$
    e. $101100_{(2)}$
    f. $11110000_{(2)}$
    g. $11011011_{(2)}$
    h. $11101.11_{(2)}$

3. Convert the following decimal numbers to their binary equivalents:
    a. $10_{(10)}$
    b. $20_{(10)}$
    c. $100_{(10)}$
    d. $210_{(10)}$
    e. $260_{(10)}$
    f. $1000_{(10)}$
    g. $2048_{(10)}$
    h. $32.625_{(10)}$

*CT 4. Your task is to design a counter circuit that will count sequentially up to and include the count of $3000_{(10)}$. If you assume one circuit can produce one bit, how many circuits will be required in your design to produce the desired binary count?

CT 5. If the circuit in Question 4 were *not* truncated, the maximum count would be _____. The MOD number of that circuit would be _____ if it were *not* truncated.

Section 2-3

6. Write the decimal count sequence 0 through $10_{(10)}$ in octal:

7. Convert the following octal numbers to their decimal equivalents:
    a. $15_{(8)}$
    b. $27_{(8)}$
    c. $55_{(8)}$
    d. $134_{(8)}$
    e. $171_{(8)}$
    f. $752_{(8)}$
    g. $1750_{(8)}$
    h. $144.6_{(8)}$

8. Convert the following decimal numbers to their octal equivalents:
    a. $21_{(10)}$
    b. $27_{(10)}$
    c. $48_{(10)}$
    d. $95_{(10)}$
    e. $130_{(10)}$
    f. $250_{(10)}$
    g. $1012_{(10)}$
    h. $100.625_{(10)}$

9. Convert the following octal numbers to their binary equivalents:
    a. $15_{(8)}$
    b. $55_{(8)}$
    c. $175_{(8)}$
    d. $402_{(8)}$
    e. $1231_{(8)}$

10. Convert the following binary numbers to their octal equivalents:
    a. $11111_{(2)}$
    b. $101110_{(2)}$
    c. $1111100_{(2)}$
    d. $10000100_{(2)}$
    e. $1111110101_{(2)}$

Section 2-4

11. Write the decimal count sequence 0 through $20_{(10)}$ in hexadecimal.

12. What is the next hexadecimal count up from $39_{(16)}$?

---

*CT: Critical thinking questions/problems at the end of each chapter are marked CT. They generally require a logical application of previously learned material or the application of several different principles or pieces of information and are sometimes more difficult than standard questions in terms of deriving an answer.

13. What is the next sequential hexadecimal count down from $40_{(16)}$?

14. Convert the following hexadecimal numbers to their decimal equivalents:
    a. $D_{(16)}$ e. $FF_{(16)}$
    b. $11_{(16)}$ f. $1AC_{(16)}$
    c. $1B_{(16)}$ g. $1EE_{(16)}$
    d. $20_{(16)}$ h. $FFFF0_{(16)}$

15. Convert the following decimal numbers to their hexadecimal equivalents:
    a. $10_{(10)}$ e. $426_{(10)}$
    b. $24_{(10)}$ f. $5000_{(10)}$
    c. $100_{(10)}$ g. $10,000_{(10)}$
    d. $200_{(10)}$ h. $65,535_{(10)}$

16. Hexadecimal numbers may be coded directly to binary numbers by coding each hex number in a group of _____ bits.

17. Convert the following hexadecimal numbers to their binary equivalents:
    a. $E_{(16)}$
    b. $1C_{(16)}$
    c. $2AA_{(16)}$
    d. $3B2_{(16)}$
    e. $FF0_{(16)}$

18. Convert the following binary numbers to their hexadecimal equivalents:
    a. $101011_{(2)}$
    b. $1111001100_{(2)}$
    c. $10010100000_{(2)}$
    d. $10111110001_{(2)}$
    e. $1111111100001110_{(2)}$

CT 19. A certain memory chip in a digital system contains 8K memory locations. (Each 1K of memory equals 1024 memory locations.) The memory chip's *lowest* (starting) address is $100\ 0000\ 0000\ 0000_{(2)}$. Determine the memory chip's *highest* address in hexadecimal. (*Hint:* Since the starting address is included in the 8K memory locations, the highest address is 8191 memory locations above the lowest address.)

Section 2-5

20. Code the following decimal numbers in BCD:
    a. $8_{(10)}$ d. $100_{(10)}$
    b. $17_{(10)}$ e. $1111_{(10)}$
    c. $99_{(10)}$ f. $5050_{(10)}$

21. Decode the following BCD numbers to decimal:
    a. $0001\ 0101_{(BCD)}$

    b. $0010\ 1001_{(BCD)}$
    c. $1001\ 0000_{(BCD)}$
    d. $0010\ 0010\ 0010_{(BCD)}$
    e. $0001\ 0000\ 0000\ 0000_{(BCD)}$
    f. $0011\ 0000\ 1001\ 0001_{(BCD)}$

22. List two advantages of using BCD instead of binary.

23. List the main disadvantage of using BCD in lieu of binary.

24. What is the main disadvantage of using BCD in arithmetic circuits in digital systems?

Section 2-6

25. What is another name for the gray code?

26. The gray code is a weighted code.
    a. True
    b. False

27. Convert the following gray code numbers to their binary equivalents:
    a. $0011_{(G)}$
    b. $0100_{(G)}$
    c. $1100_{(G)}$
    d. $1101_{(G)}$
    e. $1111_{(G)}$
    f. $01101010_{(G)}$

28. Convert the following binary numbers to gray code:
    a. $1001_{(2)}$ d. $1111_{(2)}$
    b. $1010_{(2)}$ e. $10000_{(2)}$
    c. $1011_{(2)}$ f. $10111_{(2)}$

Section 2-7

29. Convert the following decimal numbers to XS3 code:
    a. $6_{(10)}$
    b. $12_{(10)}$
    c. $66_{(10)}$
    d. $719_{(10)}$

30. Convert the following XS3 codes to decimal:
    a. $0100\ 0011_{(XS3)}$
    b. $0110\ 0110_{(XS3)}$
    c. $0111\ 1000_{(XS3)}$
    d. $1000\ 0101\ 1000_{(XS3)}$

CT 31. The decimal numbers 7 and 8 are stored in ASCII in adjacent memory locations in a computer. Look up the ASCII code for each number and convert it to gray code and to XS3 code.

ANSWERS TO REVIEW QUESTIONS

SECTION 2-2

A. True
B. Count down
C. $1000_{(2)}$
D. Radix multiplication or sum of weights
E. $255_{(10)}$
F. The maximum number of states produced by a counter
G. $5_{(10)}$
H. $15_{(10)}$
I. $13_{(10)}$
J. $84_{(10)}$
K. $33_{(10)}$
L. $238_{(10)}$
M. $3.75_{(10)}$
N. $100_{(2)}$
O. $1010_{(2)}$
P. $1111_{(2)}$
Q. $10000_{(2)}$
R. $10100_{(2)}$
S. $10001100_{(2)}$
T. $101110111_{(2)}$
U. $10111.1001_{(2)}$
V. $1100.011_{(2)}$

SECTION 2-3

A. 8
B. $20_{(8)}$
C. $15_{(10)}$
D. $22_{(10)}$
E. $64_{(10)}$
F. $375_{(10)}$
G. $681_{(10)}$
H. $12_{(8)}$
I. $32_{(8)}$
J. $144_{(8)}$
K. $513_{(8)}$
L. $145_{(8)}$
M. $1000_{(2)}$
N. $11010_{(2)}$
O. $1100111_{(2)}$
P. $567_{(8)}$
Q. $1520_{(8)}$
R. $13.4_{(8)}$

SECTION 2-4

A. $200_{(16)}$
B. $1FE_{(16)}$
C. $202_{(10)}$
D. $258_{(10)}$
E. $2998_{(10)}$
F. $4773_{(10)}$
G. $111_{(16)}$
H. $1CE_{(16)}$
I. $2A9_{(16)}$
J. $3E8.C_{(16)}$
K. $11010_{(2)}$
L. $1010110011_{(2)}$
M. $1111111111_{(2)}$
N. $11001110110010_{(2)}$
O. $1A1_{(16)}$
P. $1EBD_{(16)}$
Q. $11_{(16)}$
R. $184_{(16)}$

SECTION 2-5

A. Easy to recognize
   Easy to code
   Easy to decode
B. Requires more bits and memory space
   Consumes more power
   Less efficient
   Requires additional circuitry in arithmetic operations
C. $0010\ 1001_{(BCD)}$
D. $0001\ 0010\ 0100\ 0110_{(BCD)}$
E. $78_{(10)}$
F. $237_{(10)}$

SECTION 2-6

A. Minimize errors
B. False
C. $0111_{(G)}$
D. $10110101_{(G)}$
E. $1001_{(2)}$
F. $11011011_{(2)}$

SECTION 2-7

A. $0111\ 1100_{(XS3)}$
B $0100\ 1011\ 1011_{(XS3)}$
C. 93
D. American Standard Code for Information Interchange

ANSWERS TO INTERNAL SUMMARY QUESTIONS

SECTIONS 2-1 THROUGH 2-4

1. b
2. c
3. c
4. d
5. b
6. d
7. a
8. b
9. c
10. c
11. $10101_{(2)}$
12. $101110_{(2)}$
13. $11111010_{(2)}$
14. $1101001_{(2)}$
15. $23_{(10)}$
16. $43_{(10)}$
17. $31_{(10)}$
18. $501_{(10)}$
19. $85_{(10)}$
20. $475_{(10)}$
21. $111111111101_{(2)}$
22. $950_{(10)}$
23. $FF_{(16)}$
24. $1666_{(8)}$
25. $653_{(8)}$
26. $0.65625_{(10)}$
27. $0.44_{(8)}$
28. $26.4_{(8)}$

SECTIONS 2-5 THROUGH 2-7

1. b
2. c
3. b
4. c
5. b
6. a
7. b
8. d
9. a
10. d
11. c
12. b
13. b
14. c
15. d
16. d
17. a
18. d

3 LOGIC GATES

Introduction

Although the theory of operation of computer circuits and other types of digital equipment may seem complicated, their operation is relatively simple and very logical. This simplicity and logicality come primarily from the logic gates used in digital equipment.

Logic gates are the basic building blocks for all digital technology. They form the decision-making circuits employed in digital systems. These gates solve logical problems by making decisions based on a set of input conditions. A thorough knowledge of logic gate operation will make your journey through this textbook smooth and easy.

Active-High Signal

Active-Low Signal

AND Gate

Asserted

Combinational Logic Circuits

Complementary Metal-Oxide Semiconductor (CMOS)

"Don't Care" Inputs

Dynamic Operation

Enable/Enabler

Floating Inputs

Inhibit/Inhibitor

Inverter

Mnemonic

NAND Gate

Negation

Negation Indicator

NOR Gate

NOT Gate

OR Gate

Overbar

Pull-Up Resistor

Sequential Logic Circuits

Short Logic

Short Logic

Static Operation

Transistor-to-Transistor Logic (TTL)

Truth Table

Unused Inputs

Chapter Objectives

1. Given a logic gate symbol, identify the symbol and determine the gate's output under stated input conditions.

2. Given a logic gate symbol, draw and complete a truth table.

3. Given a logic gate symbol, identify the enabler and inhibitor of the gate.

4. Given a simple circuit containing logic gates, troubleshoot the circuit to isolate the malfunction to an IC based on specified symptoms.

The basic logic gates we will encounter in this chapter are often connected together to form the combinational logic circuits presented in depth in Chapter 5. Understanding gate logic is prerequisite to understanding the operation of flip-flops, decoders, encoders, binary adders, multiplexers, and a multitude of other circuits found in most digital systems.

There are only three logical functions that you must be concerned with to fully understand logic circuits. These functions are the **AND, OR,** and **NOT.** A good understanding of these three logic functions will allow you to develop a thorough knowledge of other types of logic circuits. Once the operation of logic gates is grasped, understanding the operation of more advanced combinational logic circuits will be an easy task.

Combinational logic circuits are those circuits that make a decision based entirely on the circuit's inputs. Combinational logic circuits do not contain a memory capability. The combinational logic circuits we will study later are merely combinations of the logic gates presented in this chapter and the exclusive gates presented in Chapter 5. Keep in mind that most of the digital circuits in the remainder of this textbook are based on operation of logic gates.

The **sequential logic circuits** in Chapters 6, 7, and 8 employ flip-flops made of logic gates that are manufactured on integrated circuits (ICs). Sequential logic circuits' outputs depend on a memory capability (last circuit condition) as well as the circuit's present inputs. The memory stores the last circuit condition because the sequential logic circuit's next operation depends on the previously stored data as well as the current inputs to the circuit. For example, a sequential logic circuit can store a binary 1 or 0 from a current input change. More important, it can store (retain) a binary 1 or 0 from an earlier input condition.

Section 3-1: AND Gate

Objectives

1. Identify the logic symbols used for an AND gate.
2. Given an AND gate logic symbol, determine the gate's output under stated input conditions.

An AND gate is used in digital circuits when an output is desired only when *all* of the input variables meet a certain condition. In other words, if Condition *A* is true *and* Condition *B* is true, the gate's output will indicate both inputs are true. If either Condition *A* or Condition *B* or both conditions are false, the gate's output will so indicate.

A simple analogy is to think about a two-door car. An AND gate could be used to indicate both doors of the car are closed. If the driver's door is closed, Condition *A* is true. If the passenger's door is closed, Condition *B* is true. If both of these conditions are true, the output of an AND gate can be used to prevent the open door indicator from illuminating. If a door is open, the AND gate's output would be at a level to turn the open door indicator on.

Short logic is a technique that can be used to make logic circuit analysis easy. Short logic is a simple set of precepts designating the relationship between a logic gate's inputs and output. This shortcut to understanding logic gate operation is designed so that once the short logic of the AND gate and the OR gate has been learned, the logic for any of the other logic gates can be developed.

The logic symbols of an AND gate are shown in Fig. 3–1. The standard symbol shown in Fig. 3–1(a) will normally be used in this text. The ANSI/IEEE symbol shown in Fig. 3–1(b) is a logic symbol that is often used and may some day replace the standard symbol. In each logic symbol, the input lines are *A* and *B*, and the output line is *Y*. These input/output letters are arbitrary and could be named any of a multitude of letters or groups of letters. Inside of the rectangular ANSI/IEEE symbol is an "and" sign (&). This sign is called a general qualifying symbol, and this one denotes the AND gate or AND function. General qualifying symbols are normally placed near the top of ANSI/IEEE symbols.

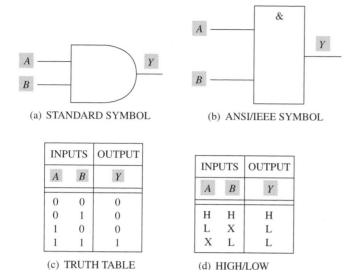

(a) STANDARD SYMBOL

(b) ANSI/IEEE SYMBOL

INPUTS		OUTPUT
A	*B*	*Y*
0	0	0
0	1	0
1	0	0
1	1	1

(c) TRUTH TABLE

INPUTS		OUTPUT
A	*B*	*Y*
H	H	H
L	X	L
X	L	L

(d) HIGH/LOW
TRUTH TABLE

FIGURE 3–1   Two-input AND gate.

The table shown in Fig. 3–1(c) is known as a **truth table** or **function table.** Truth tables are often used in digital electronics to show the operation of logic circuits. A truth table lists every input combination that can exist for a logic gate/circuit and the resulting output for each of those input combinations. As indicated in the truth table, a two-input gate has four possible input combinations. *The number of input combinations for any gate, regardless of the number of inputs, can be calculated using* 2^n. The *n* equals the number of inputs applied to the gate. By raising 2 to the power of the number of inputs, you can always derive the total number of input combinations for a logic gate/circuit. Instead of using 0s and 1s on a truth table, many manufacturers use Ls (low = 0) and Hs (high = 1)—it makes no difference.

Note in the truth table in Fig. 3–1(c) that the gate outputs high only when both inputs *A and B* are high. A Logic 0 into the *A* or *B* input will produce a 0 output. A simpler method of determining this AND gate action is to use **short logic.** The short logic for the AND gate is "any 0 in equals 0 out." In layman's terms, the short logic tells us how the gate functions.

The two-input AND gate can be thought of as two switches in series. To apply power or a signal through the two switches, they would both have to be closed (on). Logic gates are often referred to as switching circuits because the original logic circuits used relays to implement logic functions.

Since the short logic of the AND gate is "any 0 in = 0 out," ponder this point while analyzing the truth table in Fig. 3–1(c). If input *A* is low, output *Y* will be low. Under this input condition, the level of the *B* input is immaterial. With input *A* low, we would refer to the *B* input as a "don't care" condition because the output of the gate will be low whether *B* is low or high.

This same point is true if input *B* is low; the output of the AND gate will be low and input *A* is a don't care input. An *X* is used to indicate the don't care condition. Don't care conditions are shown in the truth table in Fig. 3–1(d). The high/low truth table shown in Fig. 3–1(d) is equivalent to the truth table shown in Fig. 3–1(c).

The circuit shown in Fig. 3–2 will suffice to prove how an AND function can be easily implemented. If inputs *A* and *B* are low, both D_1 and D_2 conduct and current flows through R_D to the +5-V power supply. Therefore, the output at the bottom of R_D is low.

This same condition exists if one input is low and the other is high. For example, if input *A* is low and input *B* is high, D_1 conducts, current flows through R_D, and the output is still low.

When inputs *A* and *B* are both high, D_1 and D_2 are reverse biased and pass no current. Since no current flows through R_D, it drops no voltage, and the output is +5 V (Logic 1). This circuit produces the AND function and proves the short logic "any 0 in = 0 out" is true.

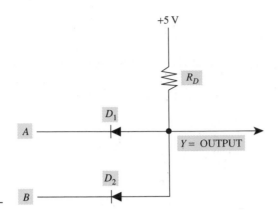

FIGURE 3-2   **Resistor–diode logic circuit.**

An excellent learning aid is derived by annotating the short logic for a gate on the logic symbol as shown in Fig. 3–3(a). This is unnecessary once the short logic of a gate is firmly in your mind—for now, it is best to use the aid. The gate shown is a three-input gate. However, the short logic remains the same as for the two-input gate. In fact, the short logic of an AND gate is "any 0 in = 0 out" for all AND gates regardless of the number of inputs. Eight-input gates are not uncommon in digital logic circuits. The first bit value of the short logic always identifies the *input level* and is read "any 0 in." The second bit value always identifies the *output level* and is read "equals 0 out."

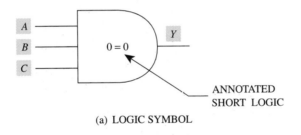

(a) LOGIC SYMBOL

FIGURE 3-3   **Three-input AND gate. *Note the sequential binary count (000–111) used in setting up the truth tables input conditions.**

*INPUTS			OUTPUT
A	B	C	Y
0	0	0	0
0	0	1	0
0	1	0	0
0	1	1	0
1	0	0	0
1	0	1	0
1	1	0	0
1	1	1	1

(b) TRUTH TABLE

=

INPUTS			OUTPUT
A	B	C	Y
H	H	H	H
L	X	X	L
X	L	X	L
X	X	L	L

(c) HIGH/LOW TRUTH TABLE

Also note in the truth table in Fig. 3–3(b) that the short logic input level of 0 for the AND gate represents seven of the eight possible input conditions of a three-input gate. In other words, seven of the eight possible input combinations consist of at least one Logic 0 input. Therefore, as stated in the short logic, the output of the gate is Logic 0 for these seven input conditions. The gate outputs Logic 1 only when *A and B and C* are high. A high/low truth table is shown in Fig. 3–3(c) for your convenience and familiarization.

Section 3-1: Review Questions

Answers are given at the end of the chapter.

A. Draw the standard logic symbol for a two-input AND gate.
B. State the short logic of an AND gate.
C. What is the output of a two-input AND gate if $A = 1$ and $B = 1$?
D. What is the output of a three-input AND gate if $A = 1$, $B = 1$, and $C = 0$?
E. Draw a truth table for a four input AND gate with A, B, C, and D inputs and Y output.

SECTION 3-2: OR GATE

OBJECTIVES

1. Identify the logic symbols used for an OR gate.
2. Given an OR gate logic symbol, determine the gate's output under stated input conditions.

An OR gate is used in digital circuits when an output is desired when *any* of its input variables meets a certain condition. Basically, if Condition A is true *or* Condition B is true, *or both* Conditions A and B are true, the gate's output will indicate at least one of the input conditions is true. If neither of the input conditions is true, the gate's output will so indicate.

The analogy of an open door on an automobile may be used again. In this case, an OR gate will be used to indicate an open door. If the driver's door is open, Condition A is true. If the passenger's door is open, Condition B is true. If both doors are open, Conditions A and B are true. In case of an open door, the output of an OR gate could be used to indicate the condition by turning on the open door indicator.

The logic symbols of an OR gate are shown in Fig. 3–4. Figure 3–4(a) is the standard logic symbol, while (b) is the ANSI/IEEE symbol. The general qualifying symbol (≥ 1) in Fig. 3–4(b) denotes the OR gate or OR function. This particular qualifying symbol is used to indicate at least one active input is needed to activate the output. *Note:* Active-high and active-low signals will be discussed in Sections 3–3, 3–5, and 3–6.

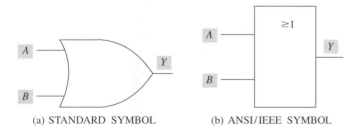

(a) STANDARD SYMBOL   (b) ANSI/IEEE SYMBOL

FIGURE 3-4   Two-input OR gate.

INPUTS		OUTPUT
A	B	Y
0	0	0
0	1	1
1	0	1
1	1	1

=

INPUTS		OUTPUT
A	B	Y
H	X	H
X	H	H
L	L	L

(c) TRUTH TABLE   (d) HIGH/LOW TRUTH TABLE

The truth table in Fig. 3–4(c) indicates that the gate will output high when input *A* or input *B* or both inputs *A* and *B* are high. The gate will output low only when *A* and *B* are low. The two-input OR gate can be thought of as two switches in parallel. To apply power or a signal past the switches, either one or both would have to be closed (on).

Again, a simpler method of stating the axiom for an OR gate is short logic. The short logic for the OR gate is "any 1 in = 1 out." Note in the high/low truth table in Fig. 3–4(d) that when input *A* is high, input *B* is a don't care and vice versa. Think of this statement in terms of the gate's short logic.

A three-input OR gate and its truth tables are shown in Fig. 3–5. The truth table again proves that the short logic for this gate is "any 1 in = 1 out." Keep in mind that *the short logic for any type of gate is always true regardless of the number of inputs to that gate.*

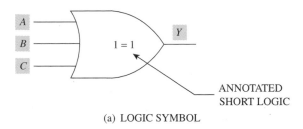

(a) LOGIC SYMBOL

INPUTS			OUTPUT
A	*B*	*C*	*Y*
0	0	0	0
0	0	1	1
0	1	0	1
0	1	1	1
1	0	0	1
1	0	1	1
1	1	0	1
1	1	1	1

(b) TRUTH TABLE

=

INPUTS			OUTPUT
A	*B*	*C*	*Y*
L	L	L	L
H	X	X	H
X	H	X	H
X	X	H	H

(c) HIGH/LOW TRUTH TABLE

FIGURE 3-5   Three-input OR gate.

If you study the combination of two-input OR gates shown in Fig. 3–6, you will see that the circuit provides the same result as the three-input OR gate shown in Fig. 3–5. The short logic proves if *A* or *B* is high, *Y* will be high, thus *Z* will be high. Also, regardless of inputs *A* or *B*, if *C* is high, *Z* will be high. Analysis of the truth tables in Fig. 3–6(b) will prove that all eight input combinations for the three input circuit are present. A truth table for the logic circuit shown in Fig. 3–6(a) is exactly the same as the truth table shown in Fig. 3–5(b).

Section 3-2: Review Questions

Answers are given at the end of the chapter.

A. Draw the standard and ANSI/IEEE logic symbols for a two-input OR gate.
B. State the short logic of an OR gate.
C. What is the output of a two-input OR gate if *A* = 0 and *B* = 0?
D. What is the output of a three-input OR gate if *A* = 1, *B* = 1, and *C* = 0?
E. Draw a truth table for a two-input OR gate with *A* and *B* inputs and *Y* output.

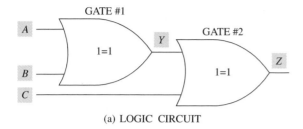

(a) LOGIC CIRCUIT

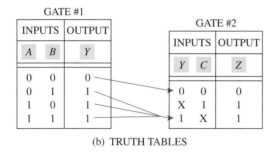

(b) TRUTH TABLES

FIGURE 3-6
Combination of OR gates.

SECTION 3-3: NOT GATE (INVERTER)

OBJECTIVES

1. Identify the logic symbols used to represent an inverter (NOT gate).
2. Given an inverter logic symbol, determine the output under stated input conditions.

The simplest type of digital circuit is an **inverter.** The inverter is often called a **NOT gate.** It is from this gate that we obtain the logical NOT function. The NOT function is merely the inversion of the input signal to its complementary form at the output. In simple terms, it changes a Logic 0 to a Logic 1 and vice versa. This process is often referred to as **negation.** Although this gate is not a decision-making gate, it is still considered a logic gate.

The symbols for a NOT gate are shown in Fig. 3–7. The bubble shown on each of the standard symbols in Fig. 3–7(a) represents inversion and is often referred to as a **negation indicator.** Note in the figure that the bubble can be placed at the input or output of the NOT gate symbol. The significance of where the bubble is placed will be discussed in Chapter 4. In either location, the bubble indicates inversion.

The ANSI/IEEE symbol shown in Fig. 3–7(b) shows one input (A) and one output ($\overline{A}$). The diagonal line drawn at the $\overline{A}$ output is a qualifying symbol that indicates inversion or an active-low output.

The truth table shown in Fig. 3–7(c) indicates that if $A = 0$ (input), then $\overline{A} = 1$ (output), and if $A = 1$, then $\overline{A} = 0$. Since the input to the inverter is labeled A, the output logically must be $\overline{A}$. The overbar placed over the output variable indicates that variable has been complemented (NOTed). $\overline{A}$ is normally read as NOT A. A prime symbol ($'$) is sometimes used in lieu of the overbar. In this case, the signal could be called prime A, NOT A, or bar A—NOT A is preferable.

The input to this circuit was arbitrarily labeled A. It could just as easily have been designated another letter or group of letters. For example, MEMR indicates a Memory Read signal in many computers. If the input to the inverter shown in Fig. 3–7(a) or (b) were labeled MEMR, the output would be labeled $\overline{\text{MEMR}}$.

The letters used to identify signals are sometimes referred to as **mnemonics.** Mnemonics are letters used to name or designate a line or input/output signal in a system. If the letters help identify the usage of a system line they are called mnemonics. The MEMR name above is a good example of a mnemonic.

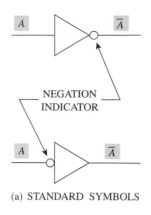

NEGATION
INDICATOR

(a) STANDARD SYMBOLS

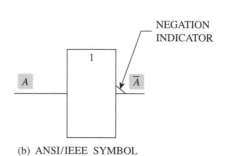

NEGATION
INDICATOR

(b) ANSI/IEEE SYMBOL

INPUT	OUTPUT
A	$\overline{A}$
0	1
1	0

(c) TRUTH TABLE

Figure 3-7   **Inverter.**

Since we only deal with two logic levels in digital electronics, all valid signals must be Logic 0 (low) or Logic 1 (high). Many circuits in a digital system require a particular signal level to clock a device or turn it off or on. Sometimes a low level is required at a certain point in the system to turn off a device. Other points in that same system may require that this signal be inverted to a high level to turn on a gate or some other circuit/device.

A prime example of this is the RESET input to a microprocessor. A microprocessor is the central processing unit (CPU) in a digital computer. It decodes instructions and controls the system. The RESET signal to some microprocessors is referred to as an **active-high** signal. The lack of an overbar over the word RESET indicates that it is an active-high signal. The name implies (and actually means) that this signal must be brought high to reset the microprocessor. The RESET signal is low most of the time during normal microprocessor operation. However, when it is brought high the microprocessor resets. This reset condition causes the microprocessor to immediately terminate its present activity and start over. The reset condition, caused by the RESET signal going high (active), always occurs when computer power is momentarily interrupted.

There are areas in the computer where this same signal must be low to perform a function. When an **active-low** RESET signal is required, the RESET signal is sent through an inverter as shown in Fig. 3–8. The output of the inverter $\overline{(\text{RESET})}$ indicates the signal is now an active-low signal. From the output of the NOT gate, the signal can be routed to all circuits requiring an active-low input to initiate the resetting action.

This example has dealt with the active-high RESET signal and the active-low $\overline{\text{RESET}}$ signal. As shown in Table 3–1, when a signal is at its active level, it is said to be **asserted.**

Figure 3-8   **Reset inverter.**

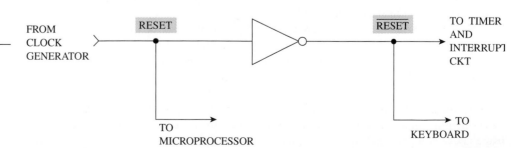

FROM
CLOCK
GENERATOR

RESET

RESET

TO TIMER
AND
INTERRUPT
CKT

TO
MICROPROCESSOR

TO
KEYBOARD

Signal	Active Level	Signal Level		Condition
RESET	H	1	=	Asserted level
RESET	H	0	=	Unasserted level
$\overline{\text{RESET}}$	L	0	=	Asserted level
$\overline{\text{RESET}}$	L	1	=	Unasserted level

TABLE 3-1

Asserted Signal Levels

Note: All signals *without* an overbar are active high. They are asserted when they are at Logic 1. All signals *with* an overbar are active low. They are asserted when they are at Logic 0.

There are times when a signal may accomplish one action when it is high and accomplish another action when it is low. An example of this might be a signal line labeled with the mnemonic RD/$\overline{\text{WR}}$ (READ/NOT WRITE). This signal would allow a computer to read memory when the line is high and write to memory when it is low. In this case, the signal has two active levels.

Section 3–3: Review Questions

Answers are given at the end of the chapter.

A. Draw the standard logic symbol for an inverter.
B. Complete a truth table for an inverter.
C. What does an overbar indicate when placed over a letter?
D. Define "asserted."
E. What is the active level of a signal identified as $\overline{\text{MEMR}}$?

SECTIONS 3-1 THROUGH 3-3: INTERNAL SUMMARY

Logic gates are used as decision-making circuits in digital electronics.

There are only three basic logic functions in digital: **AND, OR,** and **NOT.**

Short logic is a simple set of precepts designating the relationship between a logic gate's inputs and output.

AND gates are used to identify when *all* input conditions to the gate have been met. The short logic of an AND gate is "any 0 in = 0 out."

A truth table (function table) indicates operation of a logic gate/circuit. The table lists every input combination that can exist for a logic gate/circuit and the resulting output for each of the input combinations.

Raising 2 to the power of the number of inputs will always provide the total number of input conditions that can exist for a gate/circuit.

A don't care input to a logic gate indicates it does not matter whether the input is high or low. A don't care input does not affect operation of a gate. Don't care inputs are designated X on a truth table.

OR gates are used to indicate when *any* of the input conditions to the gate has been met. The short logic of an OR gate is "any 1 in = 1 out."

The short logic for a gate is true regardless of the number of inputs to the gate. The AND gate short logic, "any 0 in = 0 out," is true for an eight-input gate just as it is for a two- or three-input gate.

An **inverter,** often called a **NOT** gate, provides inversion (complementation) of a signal. The process of inversion is sometimes referred to as negation. A bubble on a standard gate symbol is used to identify inversion. The bubble is called a negation indicator.

An overbar ($\overline{}$) over a letter or group of letters indicates that signal has been NOTed. $\overline{A}$ is normally read as "NOT A" or "Bar A."

A signal that is **asserted** is at its **active level.** This normally implies that the signal will "do something" when it is asserted. An active-high signal is one that is at Logic 1 when activated. An active-low signal is one that is at Logic 0 when activated. An example of this is the RESET signal. If the signal is labeled RESET, the signal will reset a circuit when it is brought high. If the signal is labeled $\overline{\text{RESET}}$, the overbar indicates the signal is active-low. This indicates the signal will reset a circuit when it is brought low.

SECTIONS 3-1 THROUGH 3-3: INTERNAL SUMMARY QUESTIONS

Answers are given at the end of the chapter.

1. The logic symbol shown in Fig. 3–9 represents a 2-input _____ gate.

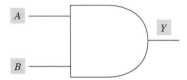

FIGURE 3–9

a. OR
b. AND
c. NOT

2. The short logic of an OR gate is
   a. Any 0 in = 0 out.
   b. Any 0 in = 1 out.
   c. Any 1 in = 1 out.
   d. Any 1 in = 0 out.

3. The ANSI/IEEE symbol shown in Fig. 3–10 represents a/an _____ gate.

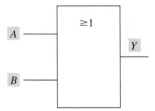

FIGURE 3–10

a. OR
b. AND
c. NOT

4. An AND gate outputs a Logic 1 only when all of its inputs are high.
   a. True
   b. False

5. An OR gate outputs a Logic 1 only when all of its inputs are high.
   a. True
   b. False

6. A five-input logic gate has how many different input combinations?
   a. 4
   b. 8
   c. 16
   d. 32

7. The AND function can be thought of as two switches in parallel.
   a. True
   b. False

8. If input *A* to a two-input AND gate is low, input *B* is a don't care input.
   a. True
   b. False

9. Don't care inputs are indicated with an X on a truth table.
   a. True
   b. False

10. The short logic of an AND gate is
    a. Any 1 in = 0 out.
    b. Any 1 in = 1 out.
    c. Any 0 in = 1 out.
    d. Any 0 in = 0 out.

11. An inverter is often called a NOT gate.
    a. True
    b. False

12. The negation indicator (bubble) on a NOT gate standard symbol can be placed at the gate's input or output.
    a. True
    b. False

13. The logic symbol shown in Fig. 3–11 represents a/an ___ ___ gate.

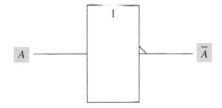

FIGURE 3–11

    a. OR
    b. AND
    c. NOT

14. The acronym MEMR indicates an
    a. Active-low signal
    b. Active-high signal

15. When $\overline{\text{RESET}}$ is at a Logic 0 level it is:
    a. Asserted
    b. Unasserted

SECTION 3-4: AND, OR, AND NOT COMBINATIONAL CIRCUITS

OBJECTIVES

1. State the rule for changing short logic to $\overline{\text{short logic}}$.
2. Given a logic diagram of a combinational logic circuit containing AND, OR, and NOT gates, determine the circuit's output under given input conditions.

From the short logic of an AND gate, we can develop another axiom that will be true for this gate. This concept can be referred to as $\overline{\text{short logic}}$ (note the overbar). The short logic

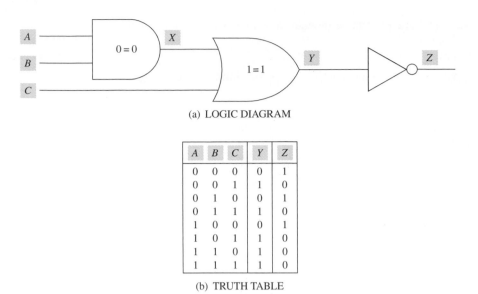

(a) LOGIC DIAGRAM

A	B	C	Y	Z
0	0	0	0	1
0	0	1	1	0
0	1	0	0	1
0	1	1	1	0
1	0	0	0	1
1	0	1	1	0
1	1	0	1	0
1	1	1	1	0

(b) TRUTH TABLE

FIGURE 3–12
Combinational logic circuit.

of an AND gate is "any 0 in = 0 out." The $\overline{\text{short logic}}$ developmental rule requires changing the word "any" to "all" and NOTing the input and output logic levels. Thus, the $\overline{\text{short logic}}$ concept for the AND gate leaves us with "all $\overline{0}$s in = $\overline{0}$ out."

Another fact to keep in mind when dealing with binary numbers is that $\overline{0} = 1$. This indicates that if a signal is NOT 0, then it must be 1. Likewise, $\overline{1} = 0$.

Since $\overline{0} = 1$, the $\overline{\text{short logic}}$ for the AND gate becomes "all 1s in = 1 out." This logic is true for all AND gates. Furthermore, the $\overline{\text{short logic}}$ concept can be applied to all types of logic gates.

If you take an OR gate's $\overline{\text{short logic}}$, you should be able to convert it quickly to $\overline{\text{short logic}}$. The short logic is "any 1 in = 1 out." Conversion leads to "all $\overline{1}$s in = $\overline{1}$ out," which is the same as saying "all 0s in = 0 out." The truth table for an OR gate proves this axiom is true.

The combinational logic circuit in Fig. 3–12 can be analyzed and the Y and Z outputs shown on the truth table determined using short logic. Later we will look at another simple method of analysis, but for now we will use the short logic concept. The short logic has been annotated on the AND and OR gates shown in Fig. 3–12(a) for ease of analysis.

1. Condition $A = 0, B = 0, C = 0$.
   The output of the AND gate (X) is 0. *Note:* This X represents the gate's output and not a don't care condition. Since $X = 0$ and $C = 0$ into the OR gate, the short logic (all 0s in = 0 out) indicates the output of the OR gate is 0. This becomes a 1 out of the inverter (NOT gate) as shown on the truth table in Fig. 3–12(b).
2. Condition $A = 0, B = 0, C = 1$.
   The X output of the AND gate is 0. However, the OR gate's output (Y) is 1 because $C = 1$. Therefore, the X input to the OR gate is actually a don't care in this situation. The Logic 1 output at Y is inverted to become a 0 at Z.
3. Condition $A = 0, B = 1, C = 0$.
   The X output of the AND gate is 0. Since both inputs to the OR gate are 0, its output (Y) is 0. Therefore, the Z output must be 1.

Take a few minutes and analyze the remaining five input combinations for the circuit. If you have no trouble with this analysis and agree with the Y and Z outputs shown in Fig. 3–12(b), it is time to press on. If you have trouble, a quick review of Sections 3–1, 3–2, and 3–3 is in order.

Section 3–4: Review Questions

Answers are given at the end of the chapter.

A. What is the basic rule for changing the short logic of a gate to $\overline{\text{short logic}}$?
B. $\overline{1} = 0$
    (1) True
    (2) False

C. The $\overline{\text{short logic}}$ concept can be applied only to AND gates.
    (1) True
    (2) False

D. What is the $\overline{\text{short logic}}$ of an AND gate?
E. Complete a truth table showing the Y and Z outputs for all input combinations of the circuit shown in Fig. 3–13.

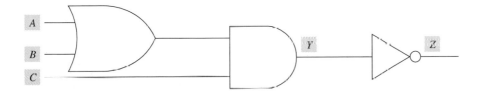

FIGURE 3–13   **Logic circuit diagram.**

SECTION 3–5: NAND GATE

OBJECTIVES

1. Identify the logic symbols for a NAND gate.
2. Given the logic symbol of a NAND gate, complete a truth table for the gate.

There are times in digital circuits where the AND function is required, but an active low logic level is desired at the output. Remember, the *AND* gate's output goes high when all of its inputs are high. Therefore, the AND gate's output is referred to as an active-high output. It would be a simple matter to change the active level of the AND gate output by placing an inverter at the gate's output. This is shown in Fig. 3–14. In this circuit, the output (Y) would be low only when inputs A and B are high. All other input combinations would produce a high output from the NOT gate. A low output from this circuit would be an active output level; a high output would be inactive.

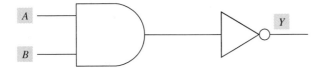

FIGURE 3–14   **AND gate with inverter.**

To solve this problem with a simpler solution than using an extra gate, the NAND gate was designed. The NAND (*Not/AND*) gate symbols and truth table are shown in Fig. 3–15. Note the bubble (negation indicator) at the gate's output on the standard logic symbol shown in Fig. 3–15(a). The bubble indicates that inversion of the ANDed signal is a function of, and internal to, the NAND gate. This bubble replaces the external NOT gate in Fig. 3–14 and makes the output of the NAND gate active low. In the ANSI/IEEE symbol in Fig. 3–15(b), the diagonal line at the Y output is the qualifying symbol that indicates an active-low output.

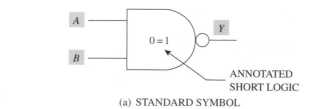

(a) STANDARD SYMBOL

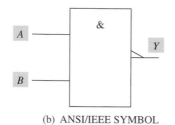

(b) ANSI/IEEE SYMBOL

INPUTS		OUTPUT
A	B	Y
0	0	1
0	1	1
1	0	1
1	1	0

(c) TRUTH TABLE

FIGURE 3–15   Two-input NAND gate.

The only difference between the AND gate and the NAND gate is inversion of the ANDed signal in the NAND gate. Therefore, the short logic for this gate should be easy to derive. The short logic of the AND gate can be used to start the derivation. Since the output signal is inverted after it has been ANDed, the short logic becomes "any 0 in = $\overline{0}$ out." Note that the only difference between the AND gate short logic and the NAND gate short logic is that the output short logic level of the NAND gate has been NOTed. It is easy to remember this derivation if you just look at the bubble or diagonal line on the gate's symbols. "Any 0 in = $\overline{0}$ out" equates to "any 0 in = 1 out." The truth table for the NAND gate shown in Fig. 3–15(c) substantiates this short logic.

This logic can be changed to short logic in exactly the same manner as we changed it for the AND gate—change the "any" to "all" and NOT both the input and output logic levels. Thus, the logic becomes "all $\overline{0}$s in = $\overline{1}$ out." This inverted logic indicates "all 1s in = 0 out."

At this point let's take a look at the AND and NAND gate truth tables in Fig. 3–16. The output levels are exactly opposite for the two gates. Naturally, this is due to the inversion at the NAND gate's output. The output of the AND gate goes active high only when inputs A and B are high. The output of the NAND gate goes active low only when this same condition is met. The other three input combinations to both of these gates leave their outputs at the inactive (unasserted) level.

INPUTS		AND OUTPUT	NAND OUTPUT
A	B	Y	Z
0	0	0	1
0	1	0	1
1	0	0	1
1	1	1	0

FIGURE 3–16   AND/NAND truth table.

AN ANSI/IEEE symbol and truth tables for a four-input NAND gate are shown in Fig. 3–17. Note how the high/low truth table with don't care inputs shown in Fig. 3–17(c) significantly simplifies the truth table shown in Fig. 3–17(b).

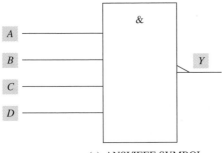

(a) ANSI/IEEE SYMBOL

INPUTS				OUTPUT
A	B	C	D	Y
0	0	0	0	1
0	0	0	1	1
0	0	1	0	1
0	0	1	1	1
0	1	0	0	1
0	1	0	1	1
0	1	1	0	1
0	1	1	1	1
1	0	0	0	1
1	0	0	1	1
1	0	1	0	1
1	0	1	1	1
1	1	0	0	1
1	1	0	1	1
1	1	1	0	1
1	1	1	1	0

(b) TRUTH TABLE

=

INPUTS				OUTPUT
A	B	C	D	Y
H	H	H	H	L
L	X	X	X	H
X	L	X	X	H
X	X	L	X	H
X	X	X	L	H

(c) HIGH/LOW TRUTH TABLE

FIGURE 3-17   Four-input NAND gate.

Section 3-5: Review Questions

Answers are given at the end of the chapter.

A. Draw the standard logic symbol for a two-input <u>NAND</u> gate.
B. State the short logic of a NAND gate. State the <u>short</u> logic of the NAND gate.
C. What is the output of a NAND gate if $A = 0$ and $B = 1$?
D. What is the output of a NAND gate if $A = 1$ and $B = 1$?
E. Draw a truth table for a three-input NAND gate with A, B, and C inputs and Y output.

SECTION 3-6: NOR GATE

OBJECTIVES

1. Identify the logic symbols for a NOR gate.
2. Given the logic symbol of a NOR gate, determine the gate's output under stated input conditions.

Some digital circuits require the OR function to have an active-low output. Remember, the OR gate's output goes active high when any of its inputs are high. Again, an inverter

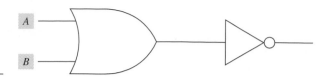

FIGURE 3–18   OR gate with inverter.

as shown in Fig. 3–18 could be used to accomplish this active-level inversion. However, it is much simpler to solve this problem with a NOR (*Not/OR*) gate, where the inversion is accomplished internally.

The NOR gate symbols and truth table are shown in Fig. 3–19. The addition of the bubble (negation indicator) at the gate's output in Fig. 3–19(a) indicates inversion. The bubble indicates that inversion of the ORed signal is a function of, and internal to, the NOR gate. The general qualifying symbol (≥ 1) in Fig. 3–19(b) indicates the OR function, and the diagonal line at the gate's output (*Y*) indicates inversion to the NOR function (active-low output).

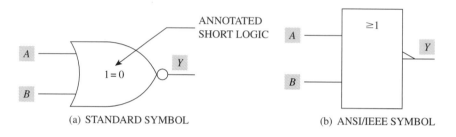

(a) STANDARD SYMBOL   (b) ANSI/IEEE SYMBOL

INPUTS		OUTPUT
A	*B*	*Y*
0	0	1
0	1	0
1	0	0
1	1	0

(c) TRUTH TABLE

FIGURE 3–19   Two-input NOR gate.

The only difference between the OR gate and the NOR gate is inversion of the ORed signal in the NOR gate. The short logic of the NOR gate can be derived using the OR gate's short logic. The OR gate logic, "any 1 in = 1 out," can be changed to "any 1 in = $\overline{1}$ out" because of the NOR gate's inversion. Note, the output logic level in the short logic has been NOTed. Thus, the logic becomes "any 1 in = 0 out" for the NOR gate. The truth table for the NOR gate shown in Fig. 3–19(c) substantiates this logic. Again, it is easy to remember the short logic derivation from the OR gate short logic if you just look at the bubble or diagonal line on the NOR gate's symbol.

This short logic can be changed to short logic following the previously discussed rule. Therefore, the short logic is "all 0s in = 1 out."

The truth table for both the OR and NOR gates is shown in Fig. 3–20 for your comparison. The output levels are opposite due to the NOR gate's inversion. The output of

FIGURE 3–20   OR/NOR truth table.

INPUTS		OR OUTPUT	NOR OUTPUT
A	*B*	*Y*	*Z*
0	0	0	1
0	1	1	0
1	0	1	0
1	1	1	0

the *OR gate* goes *active high* when any of the input conditions are true. The output of the *NOR gate* goes *active low* when any of the input conditions are true.

Section 3–6: Review Questions

Answers are given at the end of the chapter.

A. Draw the standard and ANSI/IEEE logic symbols for a two-input NOR gate.
B. State the short logic of a NOR gate. State the short logic of the NOR gate.
C. What is the output of a two-input NOR gate if $A = 1$ and $B = 1$?
D. What is the output of a two-input NOR gate if $A = 0$ and $B = 0$?
E. What is the output of a four-input NOR gate if $A = 0$, $B = 0$, $C = 0$, and $D = 1$?

SECTIONS 3–4 THROUGH 3–6: INTERNAL SUMMARY

The developmental rule for deriving the short logic of a gate states, *Change the short logic word "any" to "all" and NOT the input and output short logic levels.*

The short logic of gates should be annotated on each gate in a logic diagram to aid circuit analysis until the logic is learned.

$0 = 1$ and $\overline{1} = 0$.

The NAND gate's short logic is "any 0 in – 1 out." This can be derived by using the AND gate's short logic and NOTing the AND gate's *output* logic level. This NOTing is required due to the NAND gate's inversion.

The NAND gate's short logic is "all 1s in = 0 out."

The output logic level of a NAND gate is always the opposite of the output logic level of an AND gate for a given input condition.

The NOR gate's short logic is "any 1 in = 0 out." The short logic is "all 0s in = 1 out."

The output of a NOR gate is always the opposite of an OR gate for a given input condition.

Figure 3–21 presents the short logic and short logic of each of the logic gates we have studied except the NOT gate. The NOT gate operation is simple enough without short logic.

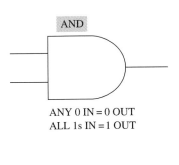

ANY 0 IN = 0 OUT
ALL 1s IN = 1 OUT

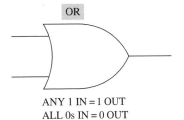

ANY 1 IN = 1 OUT
ALL 0s IN = 0 OUT

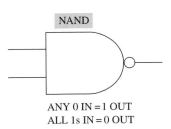

ANY 0 IN = 1 OUT
ALL 1s IN = 0 OUT

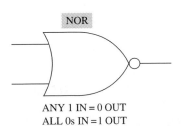

ANY 1 IN = 0 OUT
ALL 0s IN = 1 OUT

FIGURE 3–21   Short logic/short logic review.

SECTION 3-4 THROUGH 3-6: INTERNAL SUMMARY QUESTIONS

Answers are given at the end of the chapter.

1. The $\overline{\text{short logic}}$ of an AND gate is
   a. Any 0 in = 0 out.
   b. Any 1 in = 1 out.
   c. All 0s in = 0 out.
   d. All 1s in = 1 out.

2. The $\overline{\text{short logic}}$ of an OR gate is
   a. Any 0 in = 0 out.
   b. Any 1 in = 1 out.
   c. All 0s in = 0 out.
   d. All 1s in = 1 out.

3. The standard logic symbol shown in Fig. 3–22 represents a/an _____ gate.

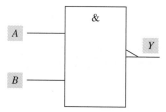

FIGURE 3-22

   a. OR
   b. AND
   c. NOR
   d. NAND

4. The ANSI/IEEE symbol shown in Fig. 3–23 represents a/an _____ gate.

FIGURE 3-23

   a. OR
   b. AND
   c. NOR
   d. NAND

5. The bubble on a NAND gate standard logic symbol indicates inversion.
   a. True
   b. False

6. The short logic of a NOR gate is
   a. Any 0 in = 0 out.
   b. Any 1 in = 1 out.
   c. Any 0 in = 1 out.
   d. Any 1 in = 0 out.

7. The short logic of a NAND gate is
   a. Any 0 in = 0 out.
   b. Any 1 in = 1 out.

c. Any 0 in = 1 out.

d. Any 1 in = 0 out.

8. What is the output of a three-input NAND gate if $A = 0$, $B = 0$, and $C = 0$?

a. Logic 0

b. Logic 1

9. What is the output of a three-input NAND gate if $A = 1$, $B = 1$, and $C = 1$?

a. Logic 0

b. Logic 1

10. What is the output of a three-input NOR gate if $A = 0$, $B = 0$, and $C = 0$?

a. Logic 0

b. Logic 1

SECTION 3-7: DYNAMIC OPERATION OF LOGIC GATES

OBJECTIVES

1. Define static and dynamic operation.
2. Given a logic gate symbol and its input timing waveforms, draw the gate's output waveform.
3. Define enabler and inhibitor.
4. Identify the logic level that will enable or inhibit an AND, OR, NAND, and NOR gate.

Operation of the gates we have presented thus far has been explained with the inputs in a steady-state condition. In other words, the gate inputs were set at certain logic levels and remained there during circuit analysis. This is normally referred to as **static operation.**

In reality, most often the inputs to gates are constantly changing. This is called **dynamic operation.** Dynamic analysis using timing diagrams is not complex if you use short logic or short logic.

Another simple method of learning how logic gates work is presented here as supplemental information to ensure your understanding is complete before dynamic operation is presented.

If inputs and outputs of gates are discussed in terms of highs and lows and related to logic functions, some remarkably simple analogies to the short logic and short logic concepts can be made. Only two things must be kept in mind to perform this analysis:

1. There are only two logic functions to consider—AND and OR.
2. If there is no bubble at an input or output, think of that input or output as *high*. If there is a bubble at an input or output, consider that input or output to be *low*.

The AND gate shown in Fig. 3–24(a) can be analyzed using the two steps above as follows:

If input A is high (no bubble) *and* input B is high (no bubble), the output will be high (no bubble).

The NAND gate shown in Fig. 3–24(b) reveals the following:

If input A is high *and* input B is high, the output will be low (bubble at output).

This analysis was made by referring to the inputs as high because there was no bubble on the inputs of either gate. The output of the AND gate was determined to be high

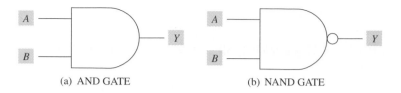

FIGURE 3-24   **Gate operation–review.**

(a) AND GATE   (b) NAND GATE

(c) OR GATE   (d) NOR GATE

if the inputs were both high simply because there was no bubble on the gate's output. The NAND gate's output was determined to be low because there was a bubble on its output.

This procedure can be applied to the OR gate in Fig. 3–24(c) thus:

If input A is high *or* input B is high, the output will be high.

Likewise, the NOR gate shown in Fig. 3–24(d) reveals the following:

If input A is high *or* input B is high, the output will be low.

The preceding analogies relate directly to the AND gate's $\overline{\text{short logic}}$ "all" precept and the OR gate's short logic "any" precept.

AND Gate

Figure 3–25 shows an AND gate along with its input and output waveforms. Our only task here is to prove the output waveform is correct. If it is, we can assume the gate is operating properly.

FIGURE 3-25   **AND gate dynamic operation.**

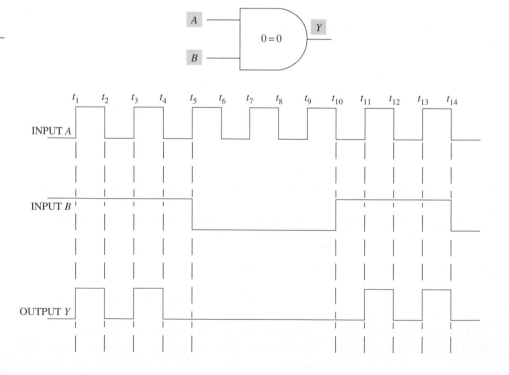

We will begin our analysis at t_1.

t_1-t_2: Both inputs A and B are high; therefore, the output is high. Remember, the logic is "all 1s in = 1 out."

t_2-t_3: Input A is low and the logic, "any 0 in = 0 out," proves the output should be low.

t_3-t_4: Both inputs are high; output should be high.

t_4-t_5: Input A is low; output should be low.

t_5-t_{10}: Input B is low, and short logic indicates the output should be low for the entire time period. In actuality, with input B low during this period, input A is a don't care input.

$t_{10}-t_{11}$: Input A is low; output should be low.

$t_{11}-t_{12}$: Both inputs are high; output should be high.

$t_{12}-t_{13}$: Input A is low; output should be low.

$t_{13}-t_{14}$: Both inputs are high; output should be high.

Once you get used to this type of dynamic analysis, you will automatically look for the short logic condition. For an AND gate, "all 1s in = 1 out" is a relatively easy condition to spot. You can see at a glance in Fig. 3–25 that a high output occurs only between t_1 and t_2, t_3 and t_4, t_{11} and t_{12}, and t_{13} and t_{14}. The analysis of these waveforms can be accomplished for any type of logic gate using the gate's logic.

Two more important points can be derived from Fig. 3–25: First, note the A input is passed to the output of the AND gate with *no phase inversion* when it is allowed through the gate. This is evident if you look at the output between t_1 and t_5 and between t_{10} and t_{14} in Fig. 3–25. Second, the A input is allowed to pass through the gate when the B input is high and prevented from passing through the gate when the B input is low.

In this circuit, we could call the B input a **control input** that determines whether or not the A input signal is allowed to pass through the gate. If the control input at B is high, the signal at the A input passes through to the gate's output with no phase inversion. When we allow this signal to pass through the gate, the gate is **enabled.** When a gate is enabled, it is activated and allowed to pass a signal.

When the control input at B is low, the output is low. Thus, the A input signal does not get to the gate's output. In this case, when the B input is low, the AND gate is **inhibited.** When a gate is inhibited, it is deactivated and its other input(s) cannot pass through to the output. In fact, when the AND gate is inhibited with a Logic 0 on one input, the other input becomes a don't care input.

The logic level that activates a gate and allows it to pass a signal is called the **enabling signal (enabler).** The logic level that deactivates a gate and prevents it from passing a signal is called the **inhibiting signal (inhibitor).**

It is necessary for you to become familiar with which logic levels inhibit and enable certain gates. It is simple to learn these levels because the information is readily available in the short logic of a gate. The input short logic level ("any 0 in" for an AND gate) always indicates the inhibitor for that gate. The inhibitor for the AND gate is Logic 0. The output short logic level ("equals 0 out" for an AND gate) always indicates the output logic level of the gate when it is inhibited.

This basic knowledge can be applied to all of the logic gates we have studied. *The input short logic level always tells us the inhibitor. The output short logic level always tells us the gate's output level when it is inhibited.*

Furthermore, since we are working with binary circuits, only two logic levels exist. Therefore, if a Logic 0 is the inhibitor of the AND gate, then a Logic 1 must be the enabler. Think about it—if it takes a Logic 0 to deactivate a gate, it must take a Logic 1 to activate it. One thing to keep in mind, the Logic 0 is an inhibitor for an AND gate but for some other type of gate it will function as an enabler. That is why you must rely on the gate's short logic to guide you until experience makes this information second nature.

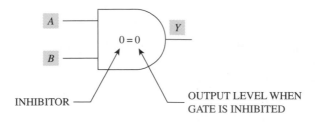

FIGURE 3-26   **AND gate inhibitor/enabler.**

$$\text{ENABLER} = \overline{\text{INHIBITOR}} = \overline{0} = 1$$

If you annotate the short logic of a gate on its logic symbol, as shown in Fig. 3–26, you will instantly know how the gate operates, its inhibitor, its output logic level when it is inhibited, and, by default, its enabler. (*Note:* The enabler is always the complement of the inhibitor.)

OR Gate

Figure 3–27 shows an OR gate and its input and output waveforms. Analysis of this three-input circuit follows:

$t_1 - t_2$: Input A is high and the short logic, "any 1 in= 1 out," proves the output should be high.

$t_2 - t_3$: All three inputs are low; therefore, the output should be low. The $\overline{\text{short logic}}$, "all 0s in = 0 out," proves this output.

$t_3 - t_4$: Input A is high; output should be high.

$t_4 - t_5$: All three inputs are low; therefore, the output should be low.

$t_5 - t_6$: Input A is high; output should be high.

$t_6 - t_7$: All three inputs are low; output should be low.

$t_7 - t_{10}$: Input B is high; thus, the output should be high for the entire period.

$t_{10} - t_{11}$: All three inputs are low; output should be low.

$t_{11} - t_{12}$: Inputs A and C are high; output should be high.

Notice in Fig. 3–27 that it is easy to identify the $\overline{\text{short logic}}$ condition of "all 0s in = 0 out" to determine the gate's output. This condition is true only between t_2 and t_3, t_4 and t_5, t_6 and t_7, and t_{10} and t_{11}. During these intervals, the gate's output is low. At all other times there is at least one high input to produce a high output.

The A input in Fig. 3–27 passes through the gate with no phase inversion when inputs B and C allow it to pass. This is readily apparent if you compare the gate's output to the A input between t_1 and t_7.

In this particular circuit, we could call the B and C inputs the control inputs. These control inputs determine whether the A input signal is allowed to pass through the gate.

It should be apparent from this discussion that if inputs B and C are low, the signal at the A input passes through the gate with no phase inversion. Therefore, Logic 0 inputs at B and C *enable* the gate.

When the control input at B or C is high, the gate is *inhibited* and the A input signal is blocked. During the time interval between t_7 and t_{10}, input B is high, the gate is inhibited, and inputs A and C are don't care inputs. This is also true after t_{11} when input C goes high and inputs A and B become don't care inputs.

It is very important that you remember the points we made about enablers and inhibitors during the AND gate dynamic operation discussion. The points were these: (1) A gate's input short logic level always identifies that gate's inhibitor logic level, and

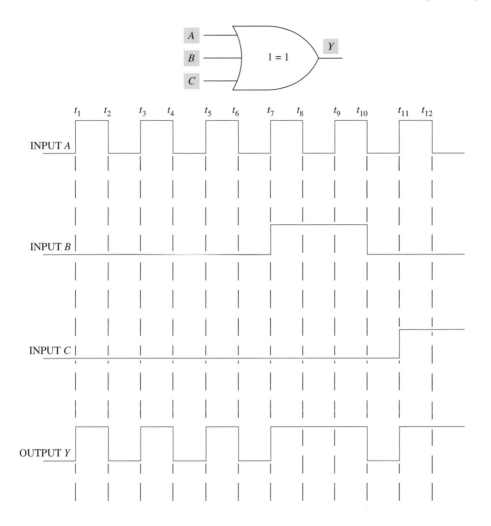

Figure 3-27 **OR gate dynamic operation.**

(2) A gate's output short logic level always identifies that gate's output logic level when it is inhibited.

The short logic of an OR gate tells us that a Logic 1 will inhibit the gate and the gate's output will be high when it is inhibited. The enabler logic level for the gate is Logic 0.

The OR gate's inhibitor (Logic 1) is the opposite level of an AND gate's inhibitor. The OR gate's output level is Logic 1 when it is inhibited (deactivated). The AND gate's output level is Logic 0 when it is inhibited. These input inhibitors and output inhibited levels can get confusing when the NAND and NOR gates are added to the discussion. It is a simple matter to rely on the short logic to identify these levels for you.

It is worth mentioning at this point that we have not talked about gates being "on" or "off" during this discussion of enablers and inhibitors. Many people tend to relate Logic 0 to "off" and Logic 1 to "on" conditions in digital circuits. A problem will be encountered later if this line of thinking is allowed. The problem arrives when internal logic gate analysis is accomplished. During the internal analysis, we will see that a gate's output transistor is "off" when it produces a Logic 1 output and "on" when it produces a Logic 0 output. These are just the opposite levels of those many people think of regarding on and off conditions. This is why confusion can occur if this line of thinking is allowed. It is always best to think of an enabled gate as one that is activated. An inhibited gate is one that is deactivated.

Figure 3–28 is provided with short logic annotated as a pictorial review of the OR gate's operation, inhibitor, output level when inhibited, and enabler.

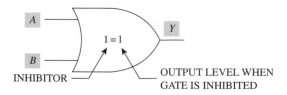

FIGURE 3-28 OR gate inhibitor/enabler.

$$\text{ENABLER} = \overline{\text{INHIBITOR}} = \overline{1} = 0$$

NAND Gate

Figure 3–29 presents a NAND gate along with its A and B inputs and Y output waveforms for dynamic analysis.

t_1–t_2: Inputs A and B are high; therefore, $\overline{\text{short logic}}$, "all 1s in = 0 out" for the NAND gate, indicates the output should be low.

t_2–t_3: Input A is low; therefore, "any 0 in = 1 out" indicates the output should be high.

t_3–t_4: Both inputs are high; output should be low.

t_4–t_5: Input A is low; output should be high.

t_5–t_6: Both inputs are high; output should be low.

t_6–t_7: Input A is low; output should be high.

t_7–t_8: Both inputs are high; output should be low.

t_8–t_{12}: Control input B is low; therefore, output should be high for the entire period.

Notice in Fig. 3–29 that input A passes to the gate's output when the control input at B is high. Also, when the gate is activated the output at Y is the same as the input at A except the output has been inverted. This should be logical because the gate has an inverter at its output.

Since the A input signal is allowed to pass through the gate when the control input is high, a Logic 1 must enable the gate. When input B is low, the gate is deactivated, because a Logic 0 is an inhibitor to the gate. Unlike the AND gate, whose inhibited output logic level is 0, the NAND gate's inhibited output logic level is 1.

FIGURE 3-29 NAND gate dynamic operation.

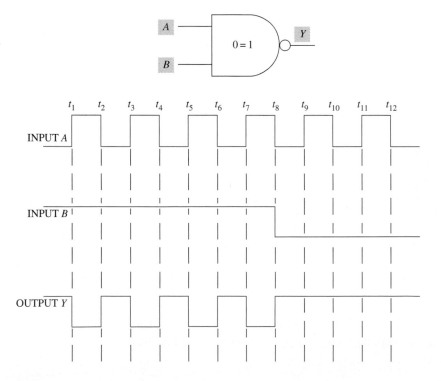

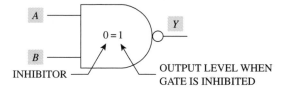

OUTPUT LEVEL WHEN
GATE IS INHIBITED

INHIBITOR

**FIGURE 3–30   NAND
gate inhibitor/enabler.**

$$\text{ENABLER} = \overline{\text{INHIBITOR}} = \overline{0} = 1$$

Figure 3–30 is provided with short logic annotated as a pictorial review of the NAND gate's operation, inhibitor, output level when inhibited, and enabler. Always keep in mind that a gate's short logic tells you its inhibitor and its output inhibited logic level.

NOR Gate

Figure 3–31 presents a NOR gate with its input and output waveforms. Every concept we have applied in this section can be used to analyze dynamic operation of this gate. Apply the concepts we have discussed and ensure you agree with the output waveform shown in Fig. 3–31.

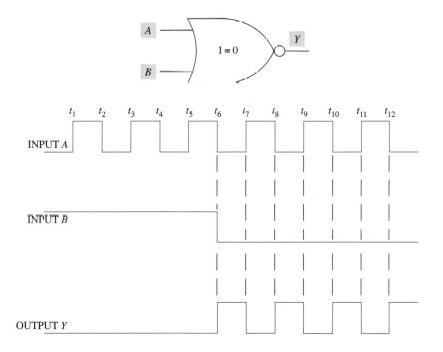

**FIGURE 3–31   NOR gate
dynamic operation.**

Several interesting points can be derived about the NOR gate once your analysis is complete. The output at Y is the inverted A input when the gate is enabled. Since the gate is deactivated when B is high, a Logic 1 on the B input inhibits the gate. Since the inverted signal is present at the output only when the B input is low, a Logic 0 on that input enables the gate. Figure 3–32 is provided with short logic annotated as a pictorial review of gate operation.

**FIGURE 3–32   NOR gate
inhibitor/enabler.**

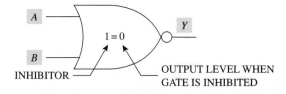

$$\text{ENABLER} = \overline{\text{INHIBITOR}} = \overline{1} = 0$$

Section 3–7: Review Questions

Answers are given at the end of the chapter.

A. Define dynamic logic gate operation.
B. Draw the output waveform of the logic gate shown in Fig. 3–33.

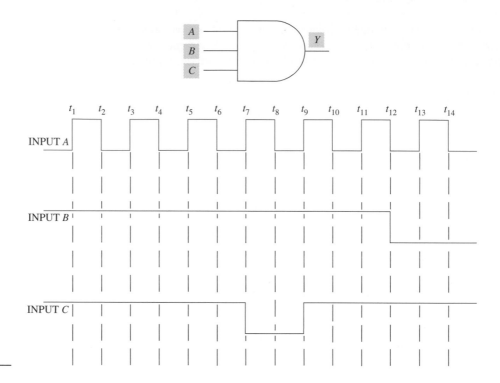

FIGURE 3-33

C. Draw the output waveform of the logic gate shown in Fig. 3–34.

FIGURE 3-34

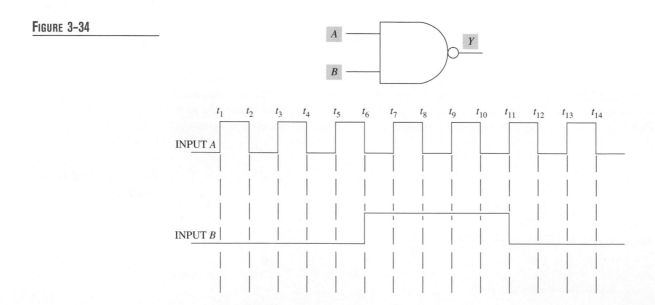

D. Draw the output waveform of the logic gate shown in Fig. 3–35.

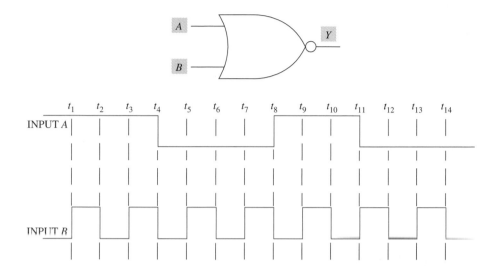

FIGURE 3-35

E. Identify the inhibitor logic level and enabler logic level for each of the gates shown in Fig. 3–36.

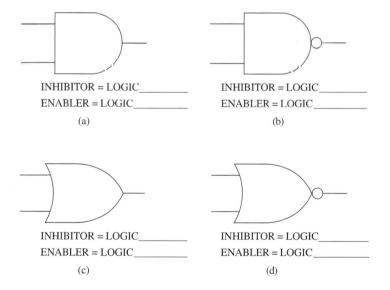

FIGURE 3-36

SECTION 3-7: INTERNAL SUMMARY

Static operation of a logic gate or circuit occurs when the gate's/circuit's inputs are held at constant (steady-state) levels.

Dynamic operation of a logic gate or circuit occurs when the gate's/circuit's inputs are changing.

It is best to use $\overline{\text{short logic}}$ to determine a gate's output level during analysis of dynamic inputs.

The AND gate and OR gate are **noninverting** gates when they are activated.

The NAND gate and NOR gate are **inverting** gates when they are activated.

When a logic gate is enabled, it is activated and allowed to pass a signal to its output. The logic level that activates a gate is called an **enabler** to that gate.

When a logic gate is inhibited, it is deactivated and its other input(s) cannot pass to the gate's output. The logic level that deactivates a gate is called an **inhibitor** to that gate.

Inhibitors and enablers for logic gates can be determined as follows: (1) The input short logic level always identifies the inhibitor. (2) The output short logic level always identifies the gate's output level when it is inhibited. (3) The enabler for a gate is always the inhibitor.

When a gate is inhibited by a certain logic level on one of its inputs, the other inputs are don't care inputs.

Section 3-7: Internal Summary Questions

Answers are given at the end of the chapter.

1. What type of gate operation involves steady-state inputs?
   a. Static
   b. Dynamic

2. The AND gate is an inverting gate.
   a. True
   b. False

3. The NOR gate is an inverting gate.
   a. True
   b. False

4. When a gate is inhibited, it is activated and allowed to pass a signal to its output.
   a. True
   b. False

5. When a two-input AND gate is inhibited by a Logic 0 on one input, the other input is a don't care input.
   a. True
   b. False

6. The output waveform shown in Fig. 3–37 is correct.
   a. True
   b. False

7. The output waveform shown in Fig. 3–38 is correct.
   a. True
   b. False

8. A Logic 0 is an _____ to an AND gate.
   a. enabler
   b. inhibitor

9. A Logic 0 is an _____ to an OR gate.
   a. enabler
   b. inhibitor

10. What is the output level of a two-input NOR gate if $A = 0$ and $B = 1$?
    a. Logic 0
    b. Logic 1

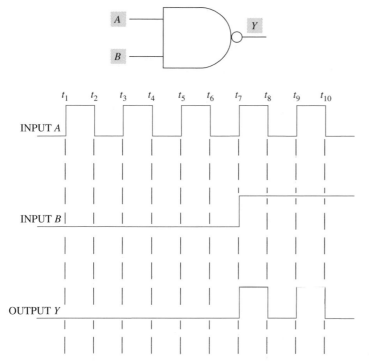

Figure 3–37

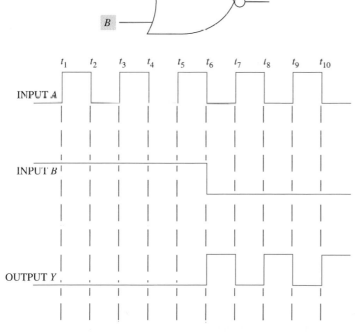

Figure 3–38

Section 3–8: Digital Logic Families

Objectives

1. State the V_{CC}, Logic 0, and Logic 1 voltages for the TTL family.
2. List the primary advantages of the TTL and CMOS families.
3. Define a floating input and state its effect on a gate.

4. State two methods of connecting unused inputs to a gate that will ensure proper operation.

The gates we have been discussing in this chapter are available in two main categories: **bipolar ICs** and **metal-oxide semiconductor (MOS) ICs.**

Although these two main categories are presented in great detail in Appendix A, a few points are presented here so that you will understand enough to begin laboratory experiments using the bipolar and MOS ICs.

The bipolar and MOS ICs are available in several different packages.

The most prevalent packaging technique employs the plastic or ceramic **dual-in-line package (DIP).** The package is hermetically sealed and intended for insertion in circuit boards. Figure 3–39(a) shows a 14-pin DIP digital IC and (b) shows the package outline with pin numbers. The pins are numbered counterclockwise (CCW) as viewed from the top. Pin 1 is located adjacent to and CCW from the notch shown in the figure. In the absence of a notch, a small colored dot, usually white, will be placed directly adjacent to pin 1.

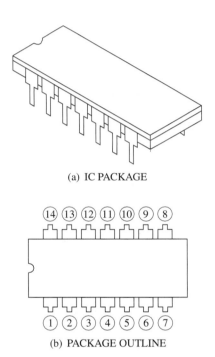

(a) IC PACKAGE

(b) PACKAGE OUTLINE

FIGURE 3–39   14-Pin DIP digital IC.

Several other methods of packaging are currently available. One method utilizes plastic **small outline** packages as shown in Fig. 3–40. A ceramic **flat** package is shown in Fig. 3–41. Finally, a ceramic **chip carrier** package intended for surface mounting on solder lands is shown in Fig. 3–42.

The AND, OR, NOT, NAND, and NOR gates are normally packaged several on an integrated circuit. This type of packaging is classified as small-scale integration (SSI) as long as the number of gates on the IC is less than 12.

An example of an SN7408 quadruple 2-input AND gate is shown in Fig. 3–43. The ANSI/IEEE symbol as used in many commercial data books is shown in Fig. 3–43(a). Figure 3–43(b) shows the logic diagram of the four AND gates contained in this IC. Often the IC is depicted in parts catalogs and replacement guides as shown in Fig. 3–43(c). Note in Fig. 3–43(d) that there are two input pins and one output pin for each of the four AND gates contained in the IC. Use of the pins (1*A*, 1*B*, 1*Y*, etc.) on the IC can be determined by comparing the package outline drawing with the pin configuration diagram or the logic diagram. Many data books show only the package outline drawing and the logic diagram, but pin utilization is easy to discern by comparing one diagram to the other. Also note that ground and V_{CC} are supplied to all four AND gates via pins 7 and 14.

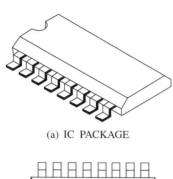

(a) IC PACKAGE

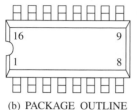

(b) PACKAGE OUTLINE

FIGURE 3–40   16-Pin "small outline" package.

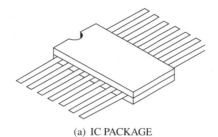

(a) IC PACKAGE

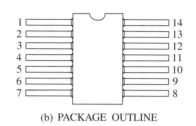

(b) PACKAGE OUTLINE

FIGURE 3–41   14-Pin ceramic flat package.

(a) IC PACKAGE

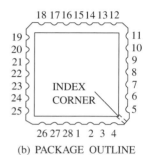

(b) PACKAGE OUTLINE

FIGURE 3–42   28-Terminal ceramic leadless chip carrier package.

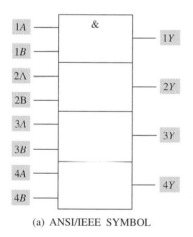

(a) ANSI/IEEE SYMBOL

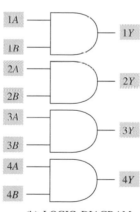

(b) LOGIC DIAGRAM

FIGURE 3–43   SN7408 Quadruple 2-input AND gate.

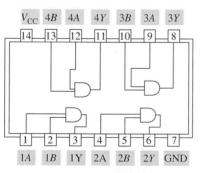

(c) PIN CONFIGURATION DIAGRAM

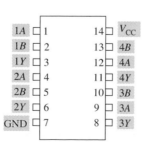

(d) PACKAGE OUTLINE

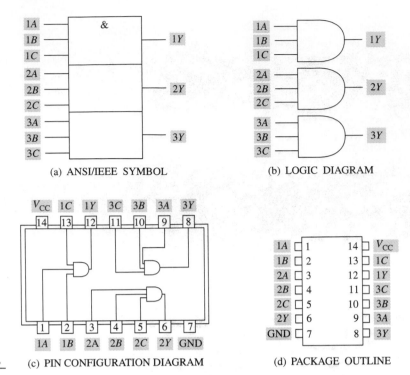

FIGURE 3–44   **SN7411**
Triple 3-input AND gate.

A triple 3-input AND gate (SN7411) is shown in Fig. 3–44, and a dual 4-input AND gate (SN7421) is shown in Fig. 3–45. Notice on the SN7421 that pins 3 and 11 are not used.

An example of an SN7400 quadruple 2-input NAND gate is shown in Fig. 3–46. Again, the ANSI/IEEE symbol, logic diagram, pin configuration diagram, and package outline are shown in the figure.

FIGURE 3–45   **SN7421**
Dual 4-input AND gate.

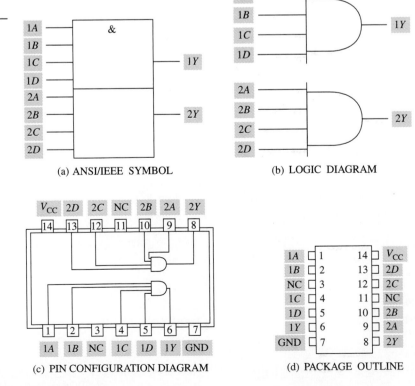

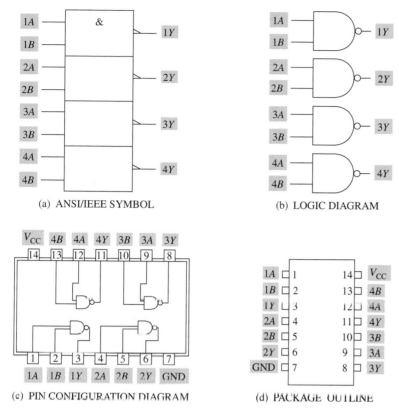

(a) ANSI/IEEE SYMBOL

(b) LOGIC DIAGRAM

(c) PIN CONFIGURATION DIAGRAM

(d) PACKAGE OUTLINE

FIGURE 3-46   SN7400 Quadruple 2-input NAND gate.

Another popular gate IC is the SN7432 OR gate. This IC is shown in Fig. 3-47. Comparison of the package outlines of many of the logic gate ICs reveals that the pin configurations are identical. However, don't fall into a trap because of this. Pins 7 and 14 are not always used for ground and V_{CC}.

A look at the SN7402 NOR gate IC in Fig. 3-48 reveals that the pin configuration of this IC is also different from the aforementioned ICs. Although pins 7 and 14 are used for ground and V_{CC}, note that pin 1 of this IC is an output pin whereas pin 1 of the other ICs we have looked at was an input pin. Every technician and every repair facility must have a data book readily available so IC pinouts can be checked. Make sure you take the time to check the pinouts of various ICs because they can easily be ruined if connected in an improper manner.

The SN7404 Hex Inverter IC is shown in Fig. 3-49. Note with this IC that the pin numbers have drastically changed from the AND, NAND, OR, and NOR gate ICs.

Transistor-To-Transistor Logic Family

The transistor-to-transistor logic (TTL) family uses bipolar transistors to implement logic functions. The TTL family was introduced as a standard product line by Texas Instruments in 1964. It was designated semiconductor network (SN) Series 54 and intended primarily for the military market. Shortly thereafter, Series 74 was commercially available to industry.

The most popular form of digital logic is TTL. This logic family offers low cost, relatively high speed, and a good output drive capability. Drive capabilities are presented in Chapter 11.

There are currently several subfamilies of the TTL family. These subfamilies evolved as a result of design modifications made to offer improvements in switching speed and/or

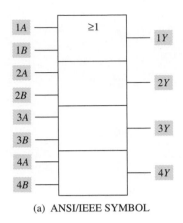
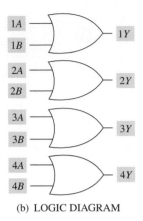

(a) ANSI/IEEE SYMBOL   (b) LOGIC DIAGRAM

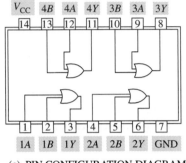

FIGURE 3-47   SN7432 Quadruple 2-input OR gate.

(c) PIN CONFIGURATION DIAGRAM   (d) PACKAGE OUTLINE

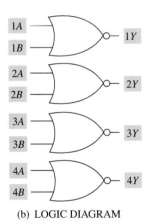

(a) ANSI/IEEE SYMBOL   (b) LOGIC DIAGRAM

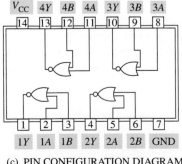
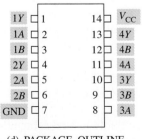

FIGURE 3-48   SN7402 Quadruple 2-input NOR gate.

(c) PIN CONFIGURATION DIAGRAM   (d) PACKAGE OUTLINE

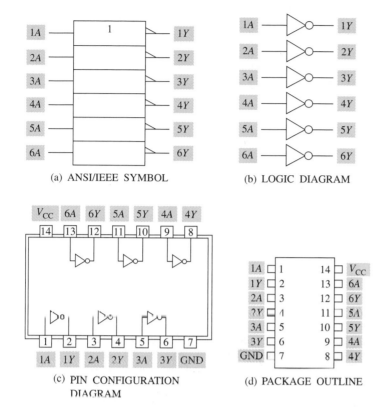

(a) ANSI/IEEE SYMBOL

(b) LOGIC DIAGRAM

(c) PIN CONFIGURATION DIAGRAM

(d) PACKAGE OUTLINE

FIGURE 3-49   SN7404 Hex inverter.

power consumption over the original Series 54/74 families. These subfamilies are identified by letters in the IC's part number. The identifying letters are shown in Table 3–2 for the commercially available Series 74 ICs.

An in-depth look at each of the subfamilies is presented in Appendix A. In addition, a brief historical look will be taken at the SN74HXX (high-speed) and SN74LXX (low-power) subfamilies.

At this point we need to know a few typical characteristics about TTL ICs. Further details will be presented as the need arises.

The nominal V_{CC} supply voltage level for all TTL ICs is $+5$ V. Logic 0 is any voltage from 0 V to 0.8 V. A look at a data sheet for a TTL IC will show that the manufacturer usually specifies a *maximum* output voltage for the low logic level as 0.4 V. Logic 1 is any voltage greater than 2 V. Again, a look at a data sheet will prove that the manufacturer specifies a *minimum* output voltage for the high logic level as 2.4 V. The Logic 0 and Logic 1 voltage levels are depicted in Fig. 3–50. Although the acceptable Logic 0 range is 0 V to 0.8 V, the typical usable range is 0 V to 0.4 V. Likewise, the typical usable range for Logic 1 is 2.4 V to 5 V. The reason for this extra 0.4-V leeway is so that

TTL Subfamilies	Part Number Example
Standard TTL circuits	SN74XX*
Schottky circuits	
Advanced Schottky	SM74ASXX
Low-power Schottky	SN74LSXX
Advanced low-power	
Schottky	SN74ALSXX
Fast	MC74FXX

TABLE 3-2

TTL Subfamilies

*The XX is used in lieu of the actual numbers that identify the type of logic gate on the IC. For example, SN7408 is the part number for a standard TTL quadruple 2-input AND gate.

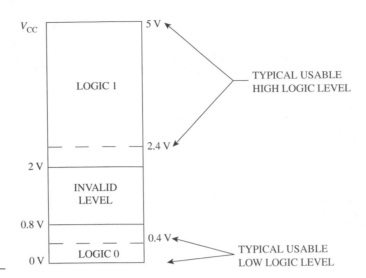

FIGURE 3-50 **TTL logic levels.**

manufacturers do not operate right on the logic level limits. A voltage between 0.8 V and 2 V is invalid. This invalid gap between the maximum low level and the minimum high level gives the device some noise immunity characteristics. The gap prevents noise that might be riding the low-level signal from reaching into the high-level logic range and causing false operation. Similarly, noise riding the high-level signal must reach across the entire gap into the low-level logic range to cause problems. Noise immunity is discussed in Chapter 11.

Complementary Metal-Oxide Semiconductor Logic Family

The complementary metal-oxide semiconductor (CMOS) logic family uses p- and n-channel enhancement-mode transistors to implement logic functions. The logic functions we have discussed in this chapter can be implemented using TTL or MOS technology. MOS ICs were introduced in the mid-1960s. By 1970, many advantages of the MOSFET as an integrated circuit had been discovered.

Research continued at a fast pace throughout the 1970s for several reasons. The ever-present desire to shrink circuit size was paramount. Very high circuit density was accomplished on MOS ICs during the era. This was especially beneficial in MOS large-scale-integration (LSI) memory configurations of the 1970s. The LSI ICs contain up to 9999 gates per chip. This high density advantage is still one of the prime advantages of MOS technology over bipolar technology.

Another important advantage the MOS family has over its bipolar counterpart is low power consumption. This should make sense because the MOSFET is a voltage-controlled device whereas the bipolar family is current controlled.

Several subfamilies of CMOS are currently available. Like the TTL subfamilies, the CMOS variations evolved as improvements were made to the original family to advance product design. These subfamilies are listed in Table 3–3.

TABLE 3-3	CMOS Subfamilies	Part Number Example
CMOS Subfamilies	4000 Series	CD40XX
	Silicon-gate technology	CD74HCXX
	Silicon-gate technology	CD74HCTXX

We need to consider a few characteristics about CMOS ICs at this point. CMOS ICs can be powered with a voltage range of 3 V to 18 V; however, many of these ICs are powered by +5 V to make them compatible with TTL voltages.

The logic-level voltage ranges for CMOS ICs cannot be specified as the TTL voltage levels have been due to the power supply range. Thus, the logic levels are given in percentages of the power supply voltage. All voltages below 30% of V_{DD} are considered to be a Logic 0. All voltages above 70% of V_{DD} are considered to be Logic 1.

Floating Inputs

When an input pin on a logic gate is left unconnected, that input is said to be **floating.** A floating input will cause a TTL gate to act as if the floating input pin were at a Logic 1 level.

In the previous lesson we learned that a Logic 1 will inhibit an OR gate and a NOR gate. Therefore, you should realize that a floating input pin on a TTL OR gate or NOR gate will deactivate the gate.

Figure 3–51(a) shows an OR gate with one of its input pins floating. Since that floating input acts as a Logic 1, the gate is inhibited, and short logic indicates the output would be at a constant high level. This information will be expanded in the troubleshooting discussion in Section 3–9.

A Logic 1 level will enable an AND gate and a NAND gate. Therefore, the effect of a floating input on these gates is entirely different than on the OR and NOR gates. As shown in Fig. 3–51(b), it would be possible to let input A of an AND gate float and the signal at input B would pass to the output. With these input conditions, it would appear the gate is operating properly, but what would happen when a Logic 0 was applied at the A input and never reached the input pin on the gate? The gate would stay enabled, and we could not control it with the A input.

There are times in digital circuits when one pin on an AND or NAND gate might be tied permanently high. This condition will be discussed shortly. At any rate, it seems we could just let that pin float instead of tying it high. Surely, you would think that would result in less power consumption for the gate and the end result should be the same. However, this is not the case.

Let's establish a general rule regarding this subject. *Never* let a logic gate's input float for any reason. The problems caused by a floating input are numerous and sometimes disastrous. One problem relates to a gate's noise immunity. Floating inputs are highly susceptible to random circuit noise. If the noise is picked up on a floating input, erratic gate operation usually occurs. A floating input to a CMOS gate can cause improper operation, but typically increases power consumption—sometimes to the point of burning up the IC. Furthermore, floating inputs to flip-flops (see Chapter 6) sometimes produce improper outputs.

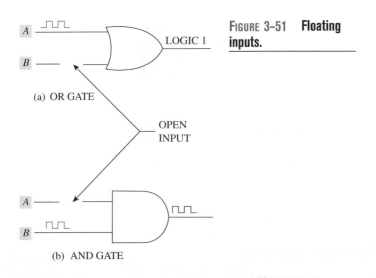

FIGURE 3–51   Floating inputs.

(a) OR GATE

OPEN INPUT

(b) AND GATE

Unused Inputs

As previously mentioned, there are times where the input to an AND or NAND gate might be permanently tied high. Let's take a look at this necessity.

Many times in digital circuit design a 3-input gate is used. As shown in Fig. 3–52, the 3-input AND gate IC (SN74LS15) contains three AND gates. Let's assume that the top AND gate requires the use of all three of its inputs ($1A$, $1B$, and $1C$). However, the other two AND gates require only two inputs. This condition is not unusual in many digital circuits. The problem is what to do with that third input we do not need. We have already established the rule, "Never let the input float."

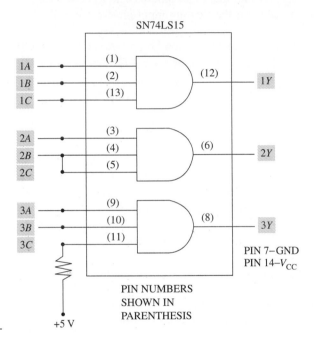

FIGURE 3–52 Triple 3-input AND gate logic diagram.

One method of solving this problem would be to connect the two required inputs to $2A$ and $2B$, and jumper inputs $2C$ and $2B$ together. This is shown on the middle gate in Fig. 3–52.

Another satisfactory method of solving this problem is shown on the lower gate in Fig. 3–52. The $3A$ and $3B$ inputs are connected to their logical inputs. Input $3C$ is tied to +5 V through a resistor called a **pull-up resistor.** The high level at $3C$ enables the gate and allows the $3A$ and $3B$ inputs to control its decision-making process.

Both of these methods of connecting unused inputs will also work for a NAND gate. Furthermore, they will work for the OR and NOR gates with one exception—a Logic 0 is required to enable the OR and NOR gates. Therefore, instead of enabling the gate as shown in the lower gate in Fig. 3–52, the $3C$ input would have to be tied to ground instead of V_{CC} if this were an OR or NOR gate.

Section 3–8: Review Questions

Answers are given at the end of the chapter.

A. What is the nominal V_{CC} supply voltage level for TTL circuits?
B. What are the typical usable Logic 0 voltage limits for TTL?
C. What are the typical usable Logic 1 voltage limits for TTL?
D. State one advantage of the CMOS family over the TTL family.
E. Define a floating input.
F. Should a floating input enable a NAND gate?

G. Is it a good maintenance practice to let an input float to enable a gate?
H. Is it acceptable to tie an unused input pin to an input pin that is being used?
I. Is it acceptable to tie an unused input pin on a NOR gate through a resistor to V_{CC}?
J. Is it acceptable to tie an unused input pin on a NAND gate through a resistor to V_{CC}?

Section 3-9: Troubleshooting Logic Gates

Objectives

1. List the purpose for
    a. A Logic probe.
    b. Logic clips.
    c. A Logic pulser.

2. Given a gate logic symbol or logic diagram with input levels and improper output level specified, determine the possible fault by theoretically troubleshooting.

The advent of integrated circuits in digital systems has done much to improve their reliability. However, problems still occur, and they must be found and repaired. Fortunately, most of these faults (often called "bugs") can be isolated using relatively inexpensive test equipment.

Unfortunately, troubleshooting is not a simple 1-2-3 process that always works and can be quickly learned, but the analytical skills required to debug a circuit can be mastered with a little thought. Logical deduction can be used provided the technician understands how the circuit "should" work under normal operating conditions. Before delving into digital circuit problems, let's take a look at some of the simpler digital troubleshooting equipment available to a technician.

Logic Probe

A logic probe is a technician's handiest troubleshooting aid when isolating faults in digital circuits. There are numerous manufacturers of logic probes, but their basic design and function are similar. A logic probe is shown in Fig. 3–53.

Typically, the probe has a red LED, a green LED, and an amber LED. The red LED indicates a Logic 1 (high) when illuminated. The green LED indicates a Logic 0 (low) when illuminated, and the amber LED indicates a pulsing condition when illuminated. Most often, the amber LED will be flickering (pulsing on and off) when the probe is connected to a pulsing signal line.

The logic probe can be used in digital systems to determine whether a line or input/output pin on an IC is low, high, pulsing, or dead. The probe's power supply leads, not shown in Fig. 3–53, must be connected to the power supply of the circuit under test for the probe to be functional. The probe's black lead must be connected to the test cir-

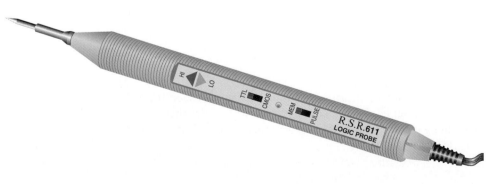

Figure 3–53   Logic probe.

TABLE 3-4	Low (green)	High (red)	Pulse (Amber)	Condition
Logic Probe Indications	0	0	0	No power/open ckt
	0	0	1	Pulsing signal
	0	1	0	Loci 1 (high)
	0	1	1	*Logic 1 pulsing
	1	0	0	Logic 0 (low)
	1	0	1	*Logic 0 pulsing
	1	1	0	No way
	1	1	1	*High/low pulsing

0 = Light off; 1 = light on.
*Any of these indications can be caused by a 50% duty cycle signal.

cuit's ground. The red lead must be connected to the test circuit's $+5$-V power. Since there are usually three LEDs on a logic probe, we can set up a function table as shown in Table 3–4.

The first condition (0-0-0) indicates the pin/line you are checking is dead or has an invalid logic level on it. It is not uncommon to pull one of the probe's power supply leads from the power source when troubleshooting. Therefore, when you encounter this condition (0-0-0), make sure the probe's input power clips are connected to the system's power supply (ground and V_{CC}). Otherwise, you might think a circuit line is dead when in actuality your probe is inoperative.

The other probe conditions are explained in Table 3–4. Condition 1-1-0 cannot exist because a circuit cannot have a Logic 0 and Logic 1 at a given point unless the line/pin is pulsing. In a pulsing condition, the amber LED would also be illuminated.

Extreme care must be taken by a technician when placing the logic probe's tip on a trace or IC pin. Touching two traces or pins at the same time with a probe's tip can short signal levels together that can destroy a chip/circuit. This sometimes happens when the technician allows the probe's tip to slip off a pin.

IC Logic Clips

An IC logic clip is used as an extender to provide easy access to an IC. One type of logic clip is shown in Fig. 3–54. The posts at the top of the extender allow easy signal monitoring or a convenient area to connect test leads. Extender logic clips are available for 14-pin, 16-pin, and 40-pin ICs. This type of clip is often referred to as an IC **glomper clip.**

Logic Pulser

A logic pulser is used to inject logic pulses into a digital circuit/IC. The logic pulser is helpful in isolating circuit problems when used with a logic probe. The pulser may be used to inject a pulse into a circuit, and the logic probe may be used at the same time to detect the pulser's output pulse.

Fault Isolation

In this section we will learn some of the basic troubleshooting techniques required to isolate a fault to an IC. First, let's consider some troubles that can occur in a logic gate. The main thing to keep in mind when troubleshooting is, *What should the gate/circuit do when it is operating properly?*

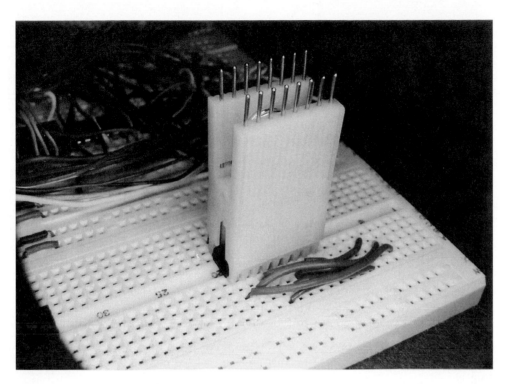

FIGURE 3–54   **IC logic clip. (Photograph by Lynn Post)**

The AND gate shown in Fig. 3–55(a) should produce a low output when either of its inputs is low. It should produce a high output when both of its inputs are high as shown in Fig. 3–55(b).

Now let's analyze the gates shown in Fig. 3–56. We will assume for this exercise that ground (pin 7) and V_{CC} (pin 14) have been checked and the IC is properly powered. Figure 3–56(a) shows an AND gate with two lows in and a high output. We know the output should be low. What could cause this problem?

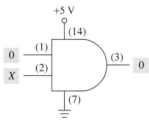

(a) LOW OUTPUT

FIGURE 3–55   **AND gates—proper operation.**

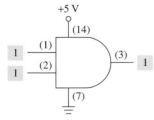

(b) HIGH OUTPUT

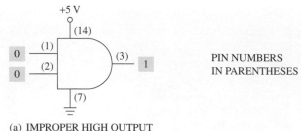

(a) IMPROPER HIGH OUTPUT

PIN NUMBERS
IN PARENTHESES

Figure 3–56
Troubleshooting AND gates.

(b) IMPROPER LOW OUTPUT

(c) IMPROPER HIGH OUTPUT

First, there can be an internal gate problem. If this is the case, we need not be overly concerned about what the actual internal problem is—we merely replace the defective gate.

For our analysis, we will dig a little deeper so that you know what other types of problems might produce this symptom. You will soon see that the problem might not be a defective gate.

If the output at pin 3 of the gate shown in Fig. 3–56(a) is internally shorted to V_{CC}, we would have this symptom. Where would we look if we replaced the AND gate with a new one and the problem remained? There is always the possibility that the output line could be stuck high if that line (trace) is shorted to a high level somewhere in the circuit. This is a plausible cause and is sometimes caused by sloppy soldering.

Another key point to keep in mind during troubleshooting this problem is the gate's output line. Where does it connect? Typically it goes to a **load.** If this AND gate's output is connected to the input of another logic gate, that load gate should be suspect. The input of the load gate could be internally shorted to V_{CC}. This would cause the output line of the driving gate to stick high. Disconnecting the input to the load gate will determine whether the faulty high level is caused by the load gate or something else. If the load gate is disconnected and the high level returns low, the problem is the load gate. If the problem does not go away when the load gate is disconnected, the problem is somewhere else.

Therefore, several possible faults could give us the symptom shown in Fig. 3–56(a). Just being aware that the gate shown doesn't necessarily have to be the problem is enough for now. We will discuss troubleshooting combinational logic circuits in Chapter 5. At that point, we will learn how to determine the actual problem if replacing the gate doesn't repair the system.

Figure 3–56(b) shows a comparable problem. The two highs in should produce a high output, yet the output is stuck low. The output at pin 3 could be internally shorted to ground. If this is the case, replacing the gate will fix the problem. Sometimes the gates are soldered into a circuit and replacing them is not a quick fix—it takes time to desolder a 14-pin IC. If the gate is not the problem, it could be the output trace shorted to ground. Another problem, as previously discussed, could be an input pin on a load gate shorted to ground.

Figure 3–56(c) shows an AND gate with a low and high input and a high output. With a low at pin 1, the input at pin 2 should be a don't care, and the output should be low. The possible faults discussed for Fig. 3–56(a) are all plausible for this circuit. However, if the gate is a TTL AND gate, there is one more problem that could produce this symptom. If the input at pin 1 is open internally, the input is **floating.** Remember that a floating input to a TTL gate causes the gate to act as if the input were at a Logic 1 level. If the input at pin 1 were floating, a high at pin 2 would produce a high output.

One more logical deduction can be made about this problem. If there is a floating input at pin 1 in Fig. 3–56(c), it is an internal gate open. If the open were external, on the trace input to the gate, or if the pin were not making contact with the input line, as shown in Fig. 3–57, there would be no Logic 0 reading at pin 1.

Let's take a detailed look at this situation. If we place the tip of the logic probe on the trace leading to pin 1 of the IC shown in Fig. 3–57, the probe will indicate a Logic 0, but if we place the probe tip directly on pin 1, no LEDs will illuminate. This indicates the problem. Furthermore, it is evidence that the logic probe's tip must be placed directly on the IC's pin instead of on the circuit board's trace to check logic levels. Although pin 1 of the IC shown in Fig. 3–57 is shown bent out, the pin on an improperly seated IC could be bent under the IC and not readily visible to the technician.

This simple type of analysis can be taken a little deeper if you look at the logic circuit shown in Fig. 3–58(a). This circuit could be implemented with one SN7408 quad 2-input AND gate IC. To troubleshoot the circuit, a technician must first determine how it should work if there were no fault. Short logic indicates the Y output can only be high when A, B, C, and D are high. This is depicted in the truth table shown in Fig. 3–58(b).

The very first steps in troubleshooting this IC are to check for power and ground. The four AND gates inside of the IC are independent of each other as shown in Fig. 3–59. One of the AND gates could be faulty while the other three might be operating normally, but all four gates are powered by pin 7 (ground) and pin 14 (V_{CC}). None of the gates will operate if the IC is not properly powered.

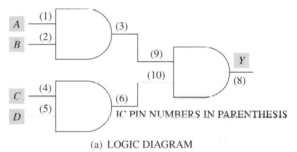

(a) LOGIC DIAGRAM

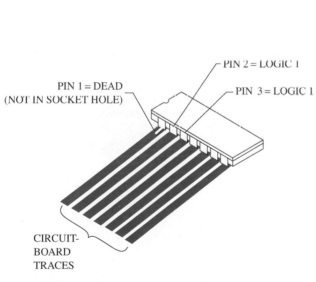

PIN 1 = DEAD
(NOT IN SOCKET HOLE)

PIN 2 = LOGIC 1

PIN 3 = LOGIC 1

CIRCUIT-
BOARD
TRACES

FIGURE 3–57
Troubleshooting ICs.

INPUTS				OUTPUT
A	B	C	D	Y
0	0	0	0	0
0	0	0	1	0
0	0	1	0	0
0	0	1	1	0
0	1	0	0	0
0	1	0	1	0
0	1	1	0	0
0	1	1	1	0
1	0	0	0	0
1	0	0	1	0
1	0	1	0	0
1	0	1	1	0
1	1	0	0	0
1	1	0	1	0
1	1	1	0	0
1	1	1	1	1

(b) TRUTH TABLE

FIGURE 3–58   Combinational logic circuit.

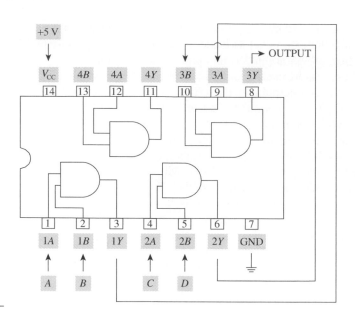

FIGURE 3-59   **SN7408 Quadruple 2-input AND gate (TTL).**

It only takes a few seconds to check these two pins. They are conveniently located at the corners of the IC. Again, it is necessary to observe one precaution while performing this check. Be sure to place the tip of your logic probe at an angle directly on the pin of the IC as shown in Fig. 3–60. Remember, placing the logic probe tip on a circuit trace that leads to an IC pin instead of the actual pin will not prove the proper level is actually applied to the IC.

The green LED on the logic probe should illuminate when the tip is placed on pin 7. The red LED should illuminate when the tip is placed on pin 14. If these two checks are satisfactory, the IC is powered. There is a possibility that the V_{CC} level is not exactly correct or has too much ripple, but for now let's assume it is correct by using the probe.

FIGURE 3-60   **Logic probe check. (Photograph by Lynn Post)**

To troubleshoot the AND gate, we only need to know "any 0 in equals 0 out" and "all 1s in equals 1 out." If the four inputs in Fig. 3–58(a) are all set high, check the output at pin 3 to ensure it is high. If it is, check the output at pin 6. If it is high, ensure the lines (traces) from pin 3 to pin 9 and pin 6 to pin 10 are good by checking the inputs to the output gate at pins 9 and 10. If both of these inputs are high, the output at pin 8 should be high. In most cases, if the inputs at A, B, C, and D are high and the output at Y is not high, the IC is probably bad or is not properly powered. On the other hand, the problem could be the circuit's output trace or the load.

Section 3–9: Review Questions

Answers are given at the end of the chapter.

A. What is the purpose of a logic probe?
B. What logic level is represented by an illuminated red LED on a logic probe?
C. What logic level is represented when the logic probe amber light is on?
D. What is the purpose of a logic pulser?
E. List the possible fault(s) that could cause the erroneous output of the TTL gate shown in Fig. 3–61.

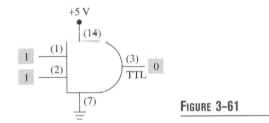

FIGURE 3–61

F. List the possible fault(s) that could cause the erroneous output of the TTL gate shown in Fig. 3–62.

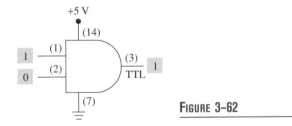

FIGURE 3–62

G. List the possible fault(s) of the output of the TTL gate shown in Fig. 3–63.

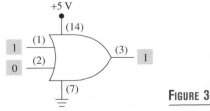

FIGURE 3–63

The two main categories of logic gates are bipolar and MOS.

The most prevalent IC-packaging technique employs DIP.

The transistor-to-transistor logic (TTL) family uses bipolar transistors to implement logic functions.

The TTL logic family offers low cost, relatively high speed, and a good output drive capability.

TTL subfamilies include the advanced Schottky (AS), low-power Schottky (LS), and advanced low-power Schottky (ALS).

The normal voltage levels associated with the TTL family are $V_{CC} = +5$ V; Logic $0 = 0$–0.8 V; Logic $1 = 2$–5 V. Logic levels between 0.8 V and 2 V are invalid. The typical usable logic levels are Logic $0 = 0$–0.4 V; Logic $1 = 2.4$–5 V.

The CMOS logic family uses enhancement-mode MOSFETs to implement logic functions.

The prime advantages of the MOS family are high density and low power consumption.

A **floating input** will cause a TTL gate to function as if the floating input pin were at a Logic 1 level. Never leave a logic gate's input floating (disconnected). *Note:* Unused output pins should not be connected if the output is not required in the circuit. Leave these pins disconnected.

Unused inputs to a logic gate in a circuit may be connected to another input pin or connected to an enabler logic level.

A logic probe is used in digital circuits to determine whether an input/output pin or line is low, high, pulsing, or dead.

A logic pulser can be used to inject logic pulses into a digital circuit/IC.

Fault isolation starts with knowing what the gate or circuit should do when it is operating properly.

SECTIONS 3-8 AND 3-9: INTERNAL SUMMARY QUESTIONS

Answers are given at the end of the chapter.

1. What is the typical Logic 0 voltage limits for TTL gates?
   a. 0 V–0.2 V
   c. 0 V–2 V
   b. 0 V–0.4 V
   d. 0 V–5 V

2. What is one of the main advantages of the TTL family over the MOS family?
   a. High speed
   b. High density
   c. Low power consumption

3. What is one of the main advantages of the MOS family over the TTL family?
   a. High speed
   b. High density
   c. High power consumption

4. A floating input
   a. Will enable a logic gate.
   b. Will inhibit a logic gate.
   c. Acts like a low input to a TTL gate.
   d. Acts like a high input to a TTL gate.

5. An unused input pin on an OR gate can be tied high through a pull-up resistor to enable the gate.
   a. True
   b. False

6. An unused input pin on a NAND gate can be tied high through a pull-up resistor to enable the gate.
   a. True
   b. False

7. Which test equipment is best suited to inject logic pulses into a circuit?
   a. Logic probe
   b. Logic pulser
   c. Extender clip
   d. Test monitor clip

8. Which test equipment is best suited to check for logic levels, pulsing signals, and dead lines in a digital circuit?
   a. Logic probe
   b. Logic pulser
   c. Extender clip
   d. Test monitor clip

9. One possible fault for the circuit shown in Fig. 3–64 is (note the annotated logic levels, which are actual circuit readings)

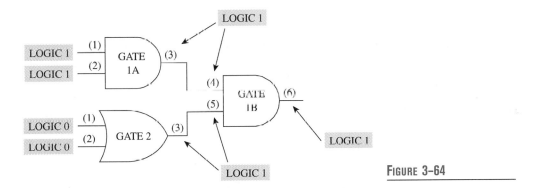

FIGURE 3–64

   a. Gate 1A pin 2 is internally shorted to ground.
   b. Gate 1A pin 2 is internally shorted to V_{CC}.
   c. Gate 1A pin 1 is internally open.
   d. Gate 2 pin 2 is internally open.

10. A possible fault for the circuit shown in Fig. 3–64 is (note the annotated logic levels)
   a. Gate 1B pin 4 is shorted to ground.
   b. Gate 2 pin 3 is shorted to ground.
   c. Gate 2 pin 3 is shorted to V_{CC}.
   d. Gate 1A pin 1 is floating.

SECTION 3–10: PRACTICAL APPLICATION OF LOGIC GATES

OBJECTIVE

Given a logic diagram of the control circuitry for ROM addressing, analyze circuit operation in order to troubleshoot.

A practical circuit using NAND and NOR gates is shown in Fig. 3–65. The objective of this circuit analysis is to explain the operation of the logic gates. The circuit is typical,

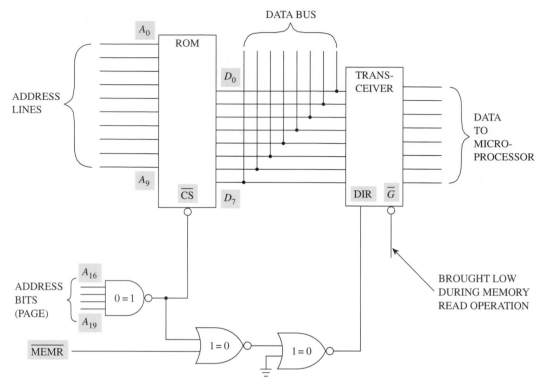

FIGURE 3–65   **Control circuitry for ROM addressing.**

although somewhat simplified, of the control circuitry involved with selecting and addressing the read-only memory (ROM) IC in a computer. Further details of memory ICs and memory addressing will be presented in Chapter 12.

The ROM IC contains data that are necessary for a computer to become operational once power is applied. Other types of memory data are stored in other memory ICs.

The Chip Select ($\overline{\text{CS}}$) input on the ROM IC in Fig. 3–65 must be asserted (brought low) and an address applied to the A_0–A_9 address pins of the chip before data can be read from it. The bidirectional *transceiver* must be enabled (low at $\overline{G}$), and the direction (DIR) pin on the transceiver must be low for data to move through the transceiver from the left side to the right side.

There are four address lines (A_{16}–A_{19}) connected to the NAND gate. The address bits connected to the gate must all be high to provide a low $\overline{\text{CS}}$ input to the ROM IC. This 1-1-1-1 address on the NAND gate's input pins is a hexadecimal F address. In this particular circuit, these four bits are called a **page address.** An F page address on the NAND gate will select this ROM chip and ignore all other memory chips in the system.

With four address lines used for page select, the capability exists to select 0-0-0-0 (hex 0) through 1-1-1-1 (hex F). This represents 16 different page addresses that can be allotted to memory ICs in this computer. Do not relate a page of memory to a page such as the one you are looking at now. A page of computer memory often contains 64 KB (kilobytes) or more of data.

In the system shown in Fig. 3–65, the ROM data is all contained in one IC located at page F. Any other page address (hex 0–E applied to address lines A_{16}–A_{19}) contains at least one Logic 0 among the four address bits. Therefore, the ROM IC can be selected only with page address F.

Once the $\overline{\text{CS}}$ input is asserted, the remaining address lines (A_0–A_9) will select a particular memory location within the ROM IC. It is from that internal address location that we can read eight data bits. These data bits (D_0–D_7) are applied to the transceiver's left side. Assuming $\overline{G}$ is asserted on the transceiver, the DIR pin must be low to allow the data to pass through the transceiver to the microprocessor.

The DIR pin is set properly when an F address is applied to the NAND gate and the Memory Read ($\overline{\text{MEMR}}$) line is asserted. With an F address, the NAND gate's output is low. This low is applied to the top input of the first NOR gate. In addition, if the system is doing a memory read instead of a memory write, the $\overline{\text{MEMR}}$ line will be asserted (low). Since the system cannot write to the ROM chip, the $\overline{\text{MEMR}}$ signal must be low for proper operation.

Two Logic 0s into the first NOR gate will produce a Logic 1 output. This output, when applied to the second NOR gate, will produce a Logic 0 from that gate. This will allow data to pass through the transceiver to the microprocessor.

The annotated short logic and the use of short logic should allow you to analyze the circuit as described above on your own.

Section 3-10: Review Questions

Answers are given at the end of the chapter.

A. What is the purpose of the ground connected to the second NOR gate in Fig. 3–65?
B. What would happen if the $\overline{\text{MEMR}}$ line to the first NOR gate in Fig. 3–65 were open and the gate was a TTL gate?

CHAPTER 3: SUMMARY

Logic gate operation is the foundation for all digital technology.

There are three logic functions in digital: AND, OR, and NOT.

AND gates are used to identify when "all" input conditions to the gate are true. The AND gate's short logic is "any 0 in = 0 out." The AND gate is a noninverting gate whose output is active high.

OR gates are used to identify when "any" of the gate's inputs are true. The OR gate's short logic is "any 1 in = 1 out." The OR gate is a noninverting gate whose output is active high.

Short logic can be derived by taking a gate's short logic, changing the word "any" to "all," and NOTing the input and output short logic levels.

The short logic and short logic for any gate is true regardless of the number of inputs to that gate.

NOT gates are used to invert a logic signal. An overbar over a letter or group of letters in a digital circuit diagram indicates the signal has been NOTed.

An **asserted** signal is one that is at its active level.

NAND gates are merely AND gates whose output is inverted within the gate. The NAND gate's short logic is "any 0 in = 1 out." The NAND gate is an inverting gate whose output is active low.

NOR gates are merely OR gates whose output is internally inverted. The NOR gate's short logic is "any 1 in = 0 out." The NOR gate is an inverting gate whose output is active low.

Logic symbols for all of the gates that are presented in this chapter are shown in Fig. 3–66.

Operation of the logic gates may be thought of in terms of bubbles and lack of bubbles for highs and lows and related to the OR and AND functions. The presence of a bubble on the inputs or output of a gate should be considered low. The lack of a bubble should be considered high. The logic for each gate can be read in this manner with little thought, yet the results are always accurate.

Dynamic operation of logic gates occurs when their inputs are changing. Static operation occurs when a gate's inputs are held constant.

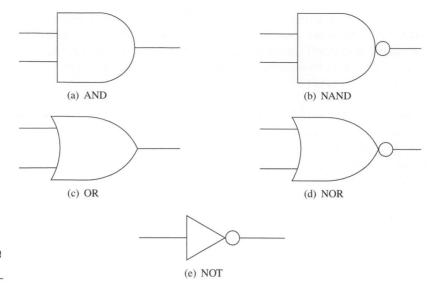

FIGURE 3–66   Logic gate symbols.

(a) AND

(b) NAND

(c) OR

(d) NOR

(e) NOT

A logic gate is **enabled** when it is activated (allowed to pass a signal). It is **inhibited** when it is deactivated (not allowed to pass a signal).

The input short logic level always identifies a logic gate's inhibitor. The enabler is always the inhibitor.

The main categories of logic gates are bipolar and MOS. The bipolar family is almost exclusively TTL. The MOS family uses MOSFETs.

The TTL logic family provides relatively high speed, low cost, and a good output drive capability.

The MOS family provides high packaging density and low power consumption.

A logic gate's inputs, regardless of logic family, should never be left floating. A floating input to a TTL gate acts like a Logic 1 input. A floating input to a MOS gate might result in the chip overheating.

The most important point in troubleshooting logic circuits is knowing how the circuit should work under normal operating conditions. A logic probe can be used to check for logic levels in digital circuits.

CHAPTER 3: END OF CHAPTER QUESTIONS/PROBLEMS

Answers are given in the Instructor's Manual.

SECTION 3-1

1. Draw the standard logic symbol for a 2-input AND gate.

2. Complete the output levels for each input combination shown on the truth table for a 2-input AND gate.

Inputs		Output
A	B	Y
0	0	
0	1	
1	0	
1	1	

3. Draw the ANSI/IEEE symbol for a 3-input AND gate.

4. What will the output logic level of a 3-input AND gate be if $A = 1$, $B = 1$, and $C = 0$?

5. What will the output logic level of a 3-input AND gate be if $A = 1, B = 1$, and $C = 1$?

6. How many input combinations are possible for a 4-input logic gate?

CT 7. What do the Xs on the truth table shown indicate? Complete the output level (Y) for each input combination if this table is for a 2-input AND gate.

Inputs		Output
A	B	Y
H	H	
L	X	
X	L	

Section 3-2

8. Draw the standard logic symbol for a 2-input OR gate.

9. What will the output logic level of a 2-input OR gate be if $A = 1$ and $B = 1$?

10. What will the output logic level of a 2-input OR gate be if $A = 0$ and $B = 0$?

CT 11. Complete the output level for each input combination shown on the truth table for a 3-input OR gate.

Inputs			Output
A	B	C	Y
L	L	L	
H	X	X	
X	H	X	
X	X	H	

12. The circuit shown in Fig. 3–67 functions identically to a 3-input OR gate.
    a. True
    b. False

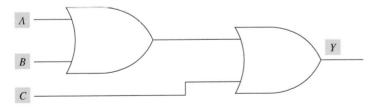

Figure 3-67

Section 3-3

13. Draw the standard logic symbol for a NOT gate.

14. Draw the ANSI/IEEE symbol for a NOT gate.

15. Define "assert" in terms of a digital signal.

16. When a signal labeled READ is at a Logic 1 level, it is
    a. asserted.
    b. unasserted.

Section 3-4

17. What is the output logic level of the circuit shown in Fig. 3–68 when $A = 1, B = 0$, and $C = 0$?

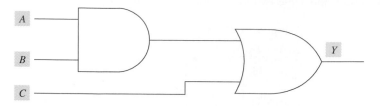

Figure 3-68

18. What is the output logic level of the circuit shown in Fig. 3–68 when $A = 1, B = 1$, and $C = 0$?

19. What is the output logic level of the circuit shown in Fig. 3–68 when $A = 1, B = 0$, and $C = 1$?

20. What is the output logic level of the circuit shown in Fig. 3–69 when $A = 1, B = 0$, and $C = 1$?

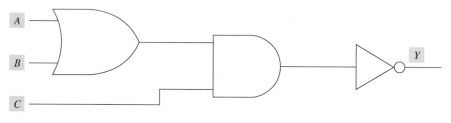

Figure 3-69

21. What is the output logic level of the circuit shown in Fig. 3–69 when $A = 0, B = 0$, and $C = 1$?

22. What is the output logic level of the circuit shown in Fig. 3–69 when $A = 1, B = 0$, and $C = 0$?

23. What is the output logic level of the circuit shown in Fig. 3–69 when $A = 1, B = 1$, and $C = 1$?

CT 24. What is the only input combination to the circuit shown in Fig. 3–70 that will produce a HIGH output?

A = _____   B = _____   C = _____

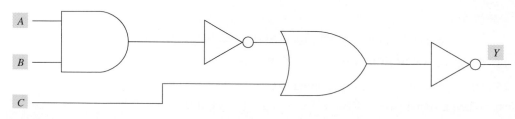

Figure 3-70

Section 3-5

25. Draw the standard logic symbol for a 2-input NAND gate.
26. What is the output logic level of a 2-input NAND gate when $A = 0$ and $B = 0$?
27. What is the output logic level of a 2-input NAND gate when $A = 0$ and $B = 1$?
28. What is the output logic level of a 3-input NAND gate when $A = 1$, $B = 1$, and $C = 1$?
29. Draw a complete truth table for a 3-input NAND gate.

Section 3-6

30. Draw the ANSI/IEEE symbol for a 2-input NOR gate.
31. Draw a truth table for a 2-input NOR gate.
32. What is the output logic level of a 3-input NOR gate when $A = 1$, $B = 1$, and $C = 1$?
33. What is the output logic level of a 3-input NOR gate when $A = 1$, $B = 0$, and $C = 0$?
34. What is the output logic level of a 3-input NOR gate when $A = 0$, $B = 0$, and $C = 0$?

Section 3-7

35. Draw the output waveform of the AND gate shown in Fig. 3-71.

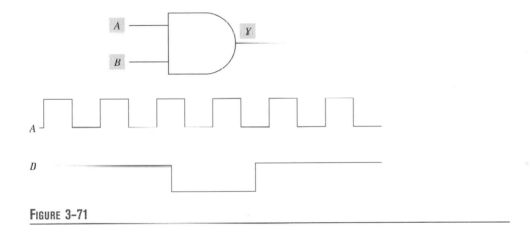

Figure 3-71

36. Draw the output waveform of the OR gate shown in Fig. 3-72.

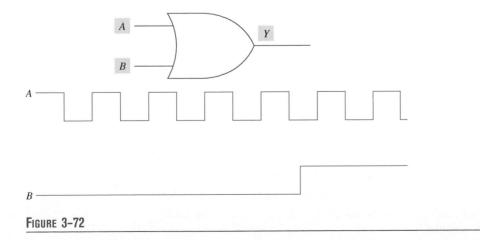

Figure 3-72

37. Draw the output waveform of the NAND gate shown in Fig. 3–73.

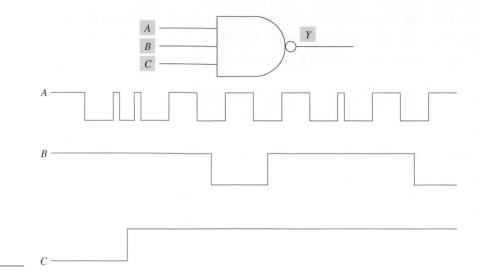

FIGURE 3-73

38. Draw the output waveform of the NOR gate shown in Fig. 3–74.

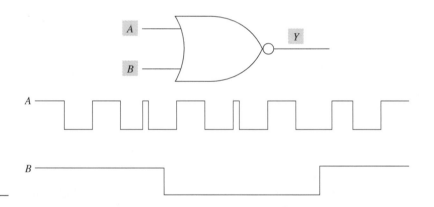

FIGURE 3-74

39. What logic level inhibits an AND gate?
40. What logic level inhibits an OR gate?
41. What logic level enables a NAND gate?
42. What logic level enables a NOR gate?
43. What is the output level of an OR gate when it is inhibited?
44. What is the output level of a NOR gate when it is inhibited?

SECTION 3-8

45. What type of transistor is used in TTL ICs?
46. What is the normal V_{CC} voltage level for TTL ICs?
47. What are the Logic 0 voltage limits for TTL ICs?
48. What are the Logic 1 voltage limits for TTL ICs?
49. What type of transistor is used in CMOS ICs?
50. An open input to a TTL IC acts as a logic _____ level.
51. A floating input to a NOR gate will _____ the gate.
    a. enable
    b. inhibit

52. State two acceptable methods of connecting a 3-input logic gate for use as a 2-input gate.

Section 3-9

53. State the purpose of a logic probe.

54. State the logic probe light indications (light on or off) you would expect to see when checking a line for a normally high signal that is pulsing low.
    Green light _____
    Red light _____
    Amber light _____

CT 55. Which gate in Fig. 3–75 has an open input pin? (*Note:* Assume the logic gates are all TTL and theoretically compare the *normal* desired operation to the *actual* defective circuit operation.)

FIGURE 3-75

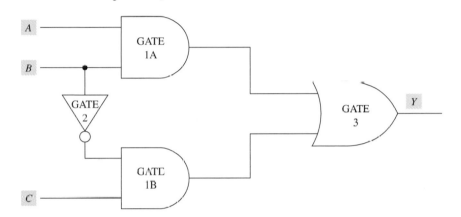

NORMAL			
A	B	C	Y
0	0	0	0
0	0	1	1
0	1	0	0
0	1	1	0
1	0	0	0
1	0	1	1
1	1	0	1
1	1	1	1

ACTUAL			
A	B	C	Y
0	0	0	0
0	0	1	0
0	1	0	0
0	1	1	0
1	0	0	0
1	0	1	0
1	1	0	1
1	1	1	1

INPUTS				OUTPUT
A	B	C	D	Y
0	0	0	0	0
0	0	0	1	0
0	0	1	0	0
0	0	1	1	0
0	1	0	0	0
0	1	0	1	0
0	1	1	0	0
0	1	1	1	0
1	0	0	0	0
1	0	0	1	0
1	0	1	0	0
1	0	1	1	0
1	1	0	0	0
1	1	0	1	0
1	1	1	0	0
1	1	1	1	1

ANSWERS TO R

Section 3-1

A. See Fig. 3–1(a).
B. "Any 0 in = 0 out."
C. Logic 1
D. Logic 0
E. See Fig. 3–76.

Section 3-2

A. See Fig. 3–4(a) and (b).
B. "Any 1 in = 1 out."

FIGURE 3-76

C.   Logic 0
D.   Logic 1
E.   See Fig. 3–4 (c) or (d).

SECTION 3-3

A.   See Fig. 3–7(a).
B.   See Fig. 3–7(c).
C.   The overbar indicates the variable has been complemented (NOTed).
D.   Signal at its active level
E.   Active low

A	B	C	Y	Z
0	0	0	0	1
0	0	1	0	1
0	1	0	0	1
0	1	1	1	0
1	0	0	0	1
1	0	1	1	0
1	1	0	0	1
1	1	1	1	0

FIGURE 3-77

SECTION 3-4

A.   Change the word "any" to "all" and NOT the input and output logic levels.
B.   True
C.   False
D.   "All 1s in = 1 out."
E.   See Fig. 3–77.

SECTION 3-5

A.   See Fig. 3–15(a).
B.   Short logic is "any 0 in = 1 out." $\overline{\text{Short logic is "all 1s in = 0 out."}}$
C.   Logic 1
D.   Logic 0
E.   See Fig. 3–78.

INPUTS			OUTPUT
A	B	C	Y
0	0	0	1
0	0	1	1
0	1	0	1
0	1	1	1
1	0	0	1
1	0	1	1
1	1	0	1
1	1	1	0

OR

INPUTS			OUTPUT
A	B	C	Y
H	H	H	L
L	X	X	H
X	L	X	H
X	X	L	H

FIGURE 3-78

SECTION 3-6

A.   See Fig. 3–19(a) and (b).
B. $\underline{\text{Short logic is "any 1 in = 0 out."}}$
     Short logic is "all 0s in = 1 out."
C.   Logic 0
D.   Logic 1
E.   Logic 0

SECTION 3-7

A.   Operation of a logic gate when its inputs are changing
B.   See Fig. 3–79.
C.   See Fig. 3–80.
D.   See Fig. 3–81.
E.   Fig. 3–36(a)
          Inhibitor = Logic 0
          Enabler = Logic 1

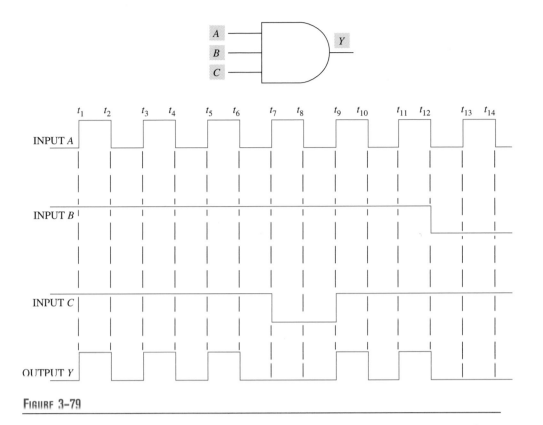

FIGURE 3-79

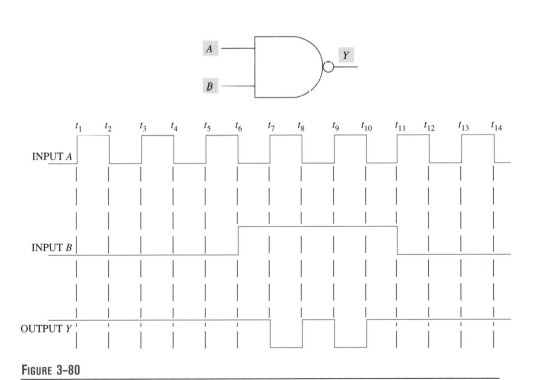

FIGURE 3-80

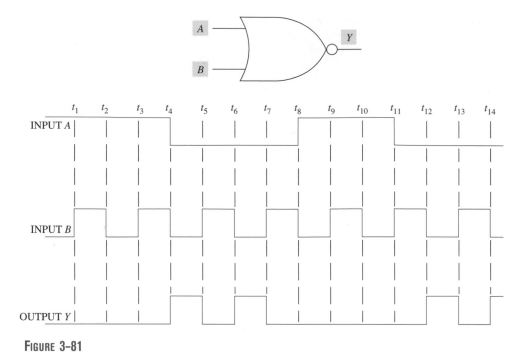

FIGURE 3–81

Fig. 3–36(b)
  Inhibitor = Logic 0
  Enabler = Logic 1
Fig. 3–36(c)
  Inhibitor = Logic 1
  Enabler = Logic 0
Fig. 3–36(d)
  Inhibitor = Logic 1
  Enabler = Logic 0

Section 3-8

A. +5 V
B. 0–0.4 V
C. 2.4–5 V
D. Higher packaging density/lower power consumption
E. An input that is open
F. Yes
G. No
H. Yes
I. No—high level would inhibit the gate.
J. Yes

Section 3-9

A. Troubleshooting—determine whether a line or input/output pin on an IC is low, high, pulsing, or dead.
B. Logic 1
C. Pulsing/changing level
D. Inject logic pulses into a digital circuit/IC.
E. (1) Output pin internally shorted to ground
    (2) Line (trace) from gate output to load shorted to ground
    (3) Load gate input pin shorted to ground

F. (1) Output pin internally shorted to V_{CC}
   (2) Line (trace) from gate output to load shorted to V_{CC}
   (3) Pin 2 floating (open) internally

G. No fault—proper operation

SECTION 3-10

A. Enable the gate.
B. If the $\overline{\text{MEMR}}$ to the first NOR gate were open, that floating input would cause the output of that gate to always be low. Since the inputs to the second NOR gate would always be low, its output would always be high. With Logic 1 applied to the transceiver DIR pin, data can only move from the right to left side. Therefore, we could never read ROM. This would prevent the computer from operating.

ANSWERS TO INTERNAL SUMMARY QUESTIONS

SECTIONS 3-1 THROUGH 3-3

1. b	9. a
2. c	10. d
3. a	11. a
4. a	12. a
5. b	13. c
6. d	14. b
7. b	15. a
8. a	

SECTIONS 3-4 THROUGH 3-6

1. d	6. d
2. c	7. c
3. c	8. b
4. d	9. a
5. a	10. b

SECTION 3-7

1. a	6. b
2. b	7. a
3. a	8. b
4. b	9. a
5. a	10. a

SECTIONS 3-8 AND 3-9

1. b	6. a
2. a	7. b
3. b	8. a
4. d	9. d
5. b	10. c

4 BOOLEAN ALGEBRA AND LOGIC CIRCUITS

Topics Covered in this Chapter

Introduction

In Chapter 2 the binary, octal, and hexadecimal numbering systems were presented. The binary numbering system must be emphasized because digital systems operate with Logic 0 and Logic 1 levels only. The mathematics of logic, called **Boolean algebra,** is presented in this chapter. Like the binary numbering system, Boolean algebra was designed for dealing with two-level functions. It lends itself perfectly to digital circuit analysis. Boolean algebra will provide a simple analytical tool to show how digital circuits function.

IMPORTANT TERMS

Alternate Logic Gate Symbols

Associative Property of Addition

Associative Property of Multiplication

Boolean Algebra

Commutative Property of Addition

Commutative Property of Multiplication

Complement Property

DeMorgan's Theorems

Distributive Property of Multiplication over Addition

Don't Care Inputs

Double Negation Property

Enable Property

Inhibit Property

Karnaugh Map

K-Map Looping

Logical Product

Logical Sum

Product of Sums

Propagation Delay

Redundant Property

Signs of Grouping

Sum of Products

Variable

Vinculum

Chapter Objectives

1. Given a truth table, extract the Boolean expression and simplify it. Implement the simplified expression with the fewest ICs.

2. Identify the alternate logic gate symbol for each logic gate.

3. Simplify Boolean expressions.

4. Given a logic diagram, implement an equivalent logic circuit using only NAND gates or NOR gates.

Boolean algebra is a form of logic that describes logical processes with mathematical symbols. Operation of the logic gates presented in Chapter 3 are described in terms of logic principles in this chapter.

The main benefits of a good understanding of Boolean algebra to a technician are in analysis of digital circuit operation and troubleshooting. Another benefit is in circuit design, where circuits must be implemented with a minimum number of gates. Utilizing a minimum number of gates results in fewer ICs, smaller circuit size, less power consumption, and lower cost.

In 1847 philosopher George Boole presented a paper on symbolic logic. In the 1850s he presented the mathematical theories we know today as Boolean algebra. In 1938 C. E.

Shannon of Bell Telephone Company expanded Boole's theories and derived the Boolean logic (algebra) currently used to analyze logic circuits.

This chapter progresses from the symbols used to represent logic functions in Boolean expressions to the final design of a digital circuit. The intermediate steps between the symbols and final design are covered in detail. These steps include interpreting and simplifying Boolean expressions as well as interpreting truth tables.

SECTION 4-1: BOOLEAN SYMBOLS

OBJECTIVES

1. Write the Boolean expression for an AND, OR, NAND, NOR, and NOT gate.
2. Determine whether a Boolean expression is written to represent a high or low output from a logic gate.

The mathematical symbols used in Boolean algebra are shown in Table 4–1. Boolean algebra uses these mathematical symbols to represent the logical functions accomplished by digital circuits.

TABLE 4-1

Boolean Logic Symbols

Symbol	Example	Meaning
A, B, C X, Y, Z		Letters are used in Boolean expressions to represent the inputs/output of logic gates and circuits. These letters represent variables in a digital circuit, and they must always be low (0) or high (1).
· or ×	$A \cdot B$ $A \times B$ AB	The dot or x signifies the logical product. Many times the logical product is indicated with no symbol. The three examples show A ANDed with B. The symbol should be read *as* AND ($AB = A$ AND B the same as $A \cdot B = A$ AND B).
+	$A + B$	The plus sign indicates the logical sum or OR operation. The example shows A ORed with B. The symbol should be read as OR ($A + B = A$ OR B).
=	$X = A + B$	The equal sign indicates two expressions or variables are identical in value.
$-$	$\overline{A}$ $\overline{A + B}$	The overbar indicates the NOT operation (inversion) which is sometimes called complementation. The examples should be read as NOT A and NOT A OR B (or A OR B NOTed). When the bar is drawn over two or more variables it not only indicates complementation, it is used as a sign of grouping. As a sign of grouping, it collects all terms together to form a single expression and is often called a vinculum.
(□) [□] {□}		The signs of grouping indicate all terms contained within the signs are to be treated as a single expression. When an expression has grouping symbols inside grouping symbols, perform the operations inside the inner grouping symbols first.

AND Function

The AND gate shown in Fig. 4–1 indicates the output of the gate is "$A \cdot B$." The dot is often used to represent the **logical product** of a circuit. The logical product is obtained by ANDing two or more inputs (variables). As noted in Table 4–1, the dot is often left out of the expression, and it is written as AB. The dot is implied, and AB indicates the input variables have been ANDed. *The expression X = AB is known as a Boolean expression.* The letters A, B, and X are known as the **variables** in the expression. The variables must always be low (0) or high (1). The logical product produced by an AND gate is the same as the actual product obtained when multiplying A and B as shown:

$$0 \cdot 0 = 0$$
$$0 \cdot 1 = 0$$
$$1 \cdot 0 = 0$$
$$1 \cdot 1 = 1$$

The Boolean expression AB is written to represent the input conditions that will produce a high output. In this case, if A is high *and* B is high, the gate's output will be high. This relates directly back to the short logic concept for an AND gate.

$$X = A \cdot B$$

FIGURE 4–1   **AND gate/Boolean expression.**

This theory can be expanded to include logic variables that have been inverted. For example, the Boolean expression for the output of the circuit shown in Fig. 4–2(a) is $\overline{A}B$. Another way of drawing this circuit is shown in Fig. 4–2(b). Since the bubble indicates inversion, this is an acceptable way of drawing the circuit. The expression $\overline{A}B$ was derived from the fact that A is NOTed before it is ANDed with B. The expression shows this with an overbar over the A. Since the expression is written for a high output, it must be read as NOT A ANDed with B. This indicates, as shown in the truth table in Fig. 4–2(c), the circuit will output high when A is low *and* B is high.

When reading these Boolean expressions for logic gates or circuits, the expression indicates the output is high when the input conditions are met. *The NOT sign over a variable should always be read as a low value and the lack of a NOT sign over a variable as a high value when reading an expression.*

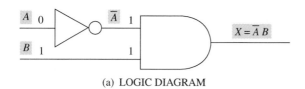

(a) LOGIC DIAGRAM

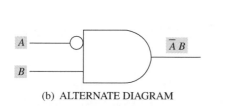

(b) ALTERNATE DIAGRAM

INPUTS		OUTPUT
A	B	X
0	0	0
0	1	1
1	0	0
1	1	0

(c) TRUTH TABLE

FIGURE 4–2   **Inverter-AND gate/Boolean expression.**

OR Function

The OR gate shown in Fig. 4–3 indicates the output of the gate is $A + B$. The plus sign is used to represent the OR function, which is referred to as the **logical sum.** The expression indicates input variable A has been ORed with input variable B.

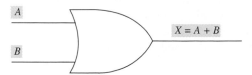

$$X = A + B$$

FIGURE 4–3   **OR gate/Boolean expression.**

The logical sum output of an OR gate with two 1s in is $1 + 1 = 1$. *The logical sum is not equivalent to the actual sum obtained in binary addition.* Remember from the rules of binary addition that $1 + 1 = 0$ with a carry of 1. You should always remember the + sign has two meanings in digital. It will mean binary addition in digital arithmetic circuits. It will mean the logical OR function in Boolean algebra and logic circuits.

Again, the Boolean expression $A + B$ is written to represent the input conditions that will produce a high output. If the expression is read properly, the circuit will output high if A is high *or* B is high.

NAND Function

The NAND gate shown in Fig. 4–4 produces a Boolean expression of $\overline{AB}$. The bar over the entire expression indicates that A and B are ANDed and then NOTed as was discussed in Chapter 3. The bar over the entire expression is often called a **vinculum.** A vinculum used in a Boolean expression is a bar drawn over two or more variables. It is used to indicate that the variables are to be considered as a single variable. For example in the expression $\overline{AB}$, the variables A and B have been NANDed, and the resultant gate output is a single variable.

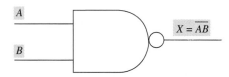

$$X = \overline{AB}$$

FIGURE 4–4   **NAND gate/Boolean expression.**

Since this expression has a NOT sign (vinculum) over AB, it is written to represent the input conditions that will produce a *low* output. For the NAND gate, if A is high *and* B is high, the gate's output will be low. This relates to the short logic of the gate.

Boolean expressions can be expanded for gates with more than two inputs. This is done by adding input letters or mnemonics for each additional input. This is shown in Fig. 4–5 for a 7420 4-input NAND gate.

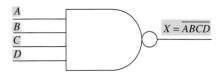

$$X = \overline{ABCD}$$

FIGURE 4–5   **4-input NAND gate/Boolean expression.**

NOR Function

The NOR gate shown in Fig. 4–6 yields a Boolean expression of $\overline{A + B}$. Like the NAND gate Boolean expression, the vinculum over the entire expression indicates A is ORed with

$$X = \overline{A + B}$$

FIGURE 4-6   **NOR gate/Boolean expression.**

B and the result NOTed. The expression represents the input conditions that will produce a low output. In this case if *A* is high *or B* is high, the output will be low.

The only difference between the AND and NAND gate's output expressions is the NOT sign (vinculum) over the expression for the NAND gate. Likewise, the only difference between the OR and NOR gate's output expressions is the NOT sign. In both cases (NAND and NOR), the NOT sign (vinculum) indicates inversion after the logical process but internal to the gate.

The AND (*AB*) and OR (*A* + *B*) Boolean expressions are written for the input conditions that will produce a *high* output. The NAND $(\overline{AB})$ and NOR $(\overline{A + B})$ Boolean expressions are written to indicate the input conditions that will produce a *low* output. In the case of the low output, the vinculum is the symbol that indicates the output will be low if the input conditions are met.

NOT Function

The output of a NOT gate is the complement of the input. If *A* is the input to a NOT gate, $\overline{A}$ (NOT *A*) is the output. Most of the time inverters are used singly to invert a logic level, although at times they may be placed back to back to delay a signal as shown in Fig. 4–7. The signal out of the second inverter is at the same level as the input clock signal, but the signal has been delayed by an amount of time equal to the propagation delay times of the inverters. **Propagation delay** is the amount of time that elapses between when an input signal is applied and when the output signal responds to the input request.

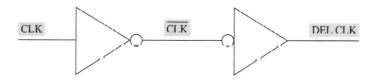

FIGURE 4-7   **Inverters used to delay a signal.**

Inverters can also be placed back to back and used to increase the pulse width of a signal. The circuit shown in Fig. 4–8 is an example. This circuit employs three NOT gates and a NOR gate to increase the duty cycle of a signal from 33% to 50%. The delay shown for each waveform in Fig. 4–8(b) equals the propagation delay of a NOT gate (15 ns in this example). The signals out of the #1 and #3 NOT gates are applied to the NOR gate to produce the 50% duty cycle signal.

Section 4-1: Review Questions

Answers are given at the end of the chapter.

A. Write a Boolean expression for a 2-input AND gate with *A* and *B* inputs.
B. Write a Boolean expression for a 2-input AND gate with *A* and *B* inputs if *B* is NOTed at the input of the gate.
C. Write a Boolean expression for a 2-input OR gate with *A* and *B* inputs.
D. Write a Boolean expression for a 2-input NAND gate with *A* and *B* inputs.
E. Write a Boolean expression for a 2-input NOR gate with *A* and *B* inputs.

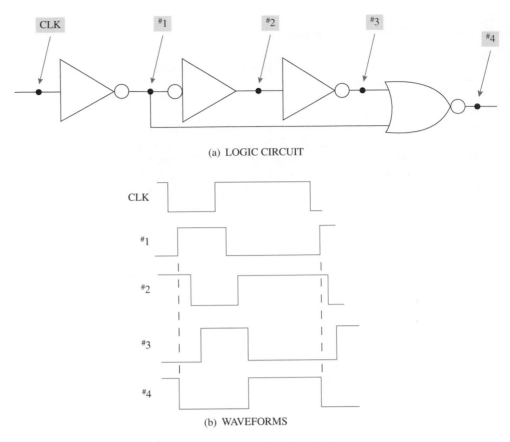

(a) LOGIC CIRCUIT

(b) WAVEFORMS

FIGURE 4-8   **Inverters used to increase duty cycle of a signal.**

F. The expression $A + B$ is written to identify the input conditions that will produce a
    _____ output.
    (1) low
    (2) high

G. The expression $\overline{AB}$ is written to identify the input conditions that will produce a
    _____ output.
    (1) low
    (2) high

H. Define "variable."
I. Define "vinculum."

SECTION 4-2: PROPERTIES OF REAL NUMBERS AND BOOLEAN ALGEBRA

OBJECTIVES

1. Identify the various properties of Boolean algebra.
2. Determine the validity of such additional properties of Boolean Algebra as
    $A + \overline{A}B = A + B$ and $\overline{A} + A\overline{B} = \overline{A} + \overline{B}$.

Many of the rules of algebra you already know apply to Boolean expressions. **Factoring** and **distributing** are among the rules you should be familiar with from previous studies.

Some properties of real numbers in mathematics can also be applied to Boolean algebra. These properties are the **associative properties** of addition and multiplication, the

commutative properties of addition and multiplication; and the **distributive property** of multiplication over addition.

Properties of Real Numbers

Associative Property of Addition

$$(A + B) + C = A + (B + C) = A + B + C$$

This property merely indicates that the order of ORing variables is the same no matter how they are grouped. The term "associative" means independent of the grouping of elements (variables). The first expression includes a sign of grouping (parentheses) that indicates A is ORed with B and the logical sum is then ORed with C. The expression indicates, as shown in Fig. 4–9, that this is equal to ORing B with C and then ORing the logical sum with A. This, incidentally, is the same as ORing all three variables in a 3-input OR gate.

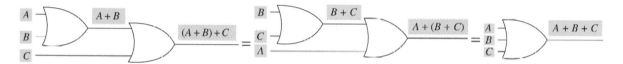

Figure 4-9 Logic circuits showing equality of the associative property of addition.

The **signs of grouping** in the Boolean expressions for the first two circuits shown in Fig. 4–9 were necessary to indicate which operation came first (order of precedence). In the first expression $(A + B) + C$, A and B were ORed first; in the second expression $A + (B + C)$, B and C were ORed first.

Signs of grouping must be used when (1) an OR gate feeds an OR gate, (2) an AND gate feeds an AND gate, and (3) an OR gate feeds an AND gate. No sign of grouping is required when an AND gate feeds an OR gate because the order of operations is the same as basic algebra—logical multiplication precedes logical addition. This is shown in Fig. 4–10.

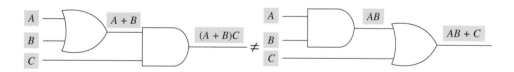

Figure 4-10 Logic circuits showing signs of grouping.

Associative Property of Multiplication

$$(AB)C = A(BC) = ABC$$

This property states the order of ANDing variables is the same regardless of their grouping. The first expression shows A ANDed with B, and the logical product is then ANDed with C. The second expression shows B ANDed with C, and the result then ANDed with A. These two expressions are equal and equivalent to ANDing all three variables in a 3-input AND gate. The equality of the three expressions is shown with gates in Fig. 4–11.

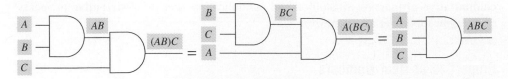

FIGURE 4-11   **Logic circuits showing equality of the associative property of multiplication.**

Commutative Properties of Addition and Multiplication

$$A + B = B + A$$
$$AB = BA$$

The term "commutative" is defined as independent of order. These two properties prove if A is applied to one input of a gate and B is applied to the other, the inputs may be reversed and the output will be identical.

Distributive Property of Multiplication over Addition

$$A(B + C) = AB + AC$$
$$(B + C)A = BA + CA$$

The distributive property proves that ORing variables $(B + C)$ and then ANDing the logical sum with another variable (A) is equal to ANDing the variable (A) with each of the other variables and then ORing their products. The equality of this is shown in Fig. 4–12(a) and (b). The properties of real numbers are listed in Table 4–2.

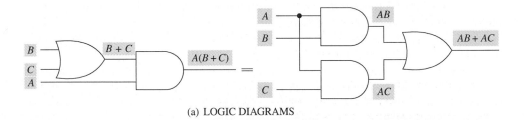

(a) LOGIC DIAGRAMS

FIGURE 4-12   **Logic circuits showing distributive property of multiplication over addition.**

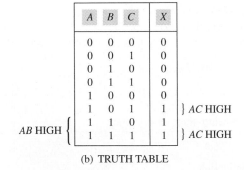

A	B	C	X
0	0	0	0
0	0	1	0
0	1	0	0
0	1	1	0
1	0	0	0
1	0	1	1	} AC HIGH
1	1	0	1
1	1	1	1	} AC HIGH

AB HIGH

(b) TRUTH TABLE

	Property	Name
TABLE 4-2 **Properties of Real Numbers**	$(A + B) + C = A + (B + C) = A + B + C$	Associative Property of Addition
	$(AB)C = A(BC) = ABC$	Associative Property of Multiplication
	$A + B = B + A$	Commutative Property of Addition
	$AB = BA$	Commutative Property of Multiplication
	$A(B + C) = AB + AC$	Distributive Property of Multiplication over Addition

Properties of Boolean Algebra

The properties shown in Fig. 4–13 will be extremely useful in simplifying Boolean expressions. The simplification process reduces the cost of properly implemented circuits. It is well to note at this point that even though A and B are often used to explain the properties of real numbers and Boolean algebra in this text, any letters can be used.

For discussion of the gate properties in Fig. 4–13(a) and (b), the inputs tied to 0 or 1 will be called control inputs.

Inhibit Property

$$A \cdot 0 = 0$$
$$A + 1 = 1$$

The inhibit properties are shown in Fig. 4–13(a). If the control input to the AND gate is low, the output has to be low per the gate's short logic. Remember, a 0 input will always inhibit the AND gate. Therefore, $A \cdot 0 = 0$. With one input to the AND gate low, the other input (A) is actually a don't care input.

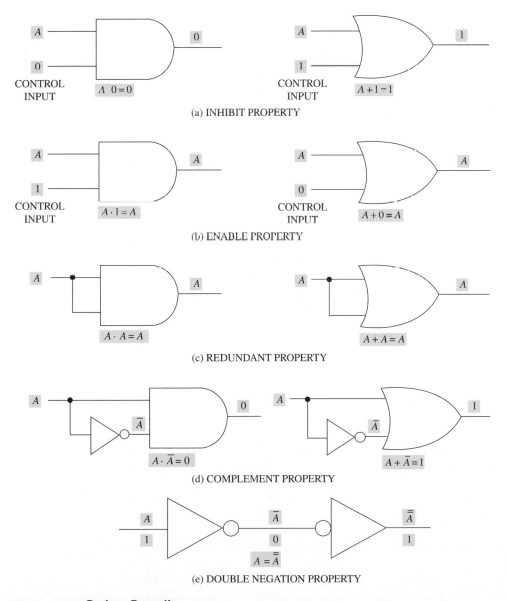

(a) INHIBIT PROPERTY

(b) ENABLE PROPERTY

(c) REDUNDANT PROPERTY

(d) COMPLEMENT PROPERTY

(e) DOUBLE NEGATION PROPERTY

FIGURE 4–13   **Boolean Properties.**

If the control input to the OR gate in Fig. 4–13(a) is high, the gate is inhibited, the output will be high, and A is a don't care input. Thus, the property is $A + 1 = 1$.

The inhibit property provides a general rule we can follow in Boolean simplification. The rule is *anytime a variable in an expression is ANDed or ORed with an inhibitor, the variable can be removed from the expression.*

The inhibit properties can also be applied to multiple-input gates. For example, $A \cdot B \cdot C \cdot 0 = 0$ and $A + B + C + 1 = 1$. This can be directly related to the gate's short logic.

Enable Property

$$A \cdot 1\ = A$$
$$A + 0 = A$$

The enable properties are shown in Fig. 4–13(b). If the control input to the AND gate is high, the gate is enabled and the output will equal A. Thus, $A \cdot 1 = A$.

> Input $A = 1$, control $= 1$, output $= 1 = A$.
> (All 1s in = 1 out.)
>
> Input $A = 0$, control $= 1$, output $= 0 = A$.
> (Any 0 in = 0 out.)

For the OR gate, if the control input is low, the gate is enabled and the output will equal A. Therefore, $A + 0 = A$.

> Input $A = 1$, control $= 0$, output $= 1 = A$.
> (Any 1 in = 1 out.)
>
> Input $A = 0$, control $= 0$, output $= 0 = A$.
> (All 0s in = 0 out.)

The enable property provides us with another general rule we can use when simplifying Boolean expressions. The rule is *anytime a variable in an expression is ANDed or ORed with an enabler, the enabler can be removed from the expression.*

The enable properties can also be applied to multiple-input gates as follows: $\overline{A} \cdot B \cdot C \cdot 1 = \overline{A} \cdot B \cdot C$ and $\overline{A} + \overline{B} + 0 = \overline{A} + \overline{B}$.

Redundant Property

$$A \cdot A\ = A$$
$$A + A = A$$

The redundant property is illustrated in Fig. 4–13(c). If A is high, the AND gate's output is high because both inputs are high. Therefore, the output is equal to A. If A is low, the AND gate's output is low and again equal to A. This proves the property $A \cdot A = A$. The OR gate redundant property $A + A = A$ can be proven in a like manner. If A is high, the output is high and equal to A. If A is low, the output of the OR gate is low and again equal to A.

The redundant property also applies to NOTed values. For example,

$$\overline{A} \cdot \overline{A}\ = \overline{A}$$
$$\overline{A} + \overline{A} = \overline{A}$$

Complement Property

$$A \cdot \overline{A} = 0$$
$$A + \overline{A} = 1$$

The complement property is proven in Fig. 4–13(d). Since the inputs to the AND gate are always complementary, one input must always be low. The low input will inhibit the gate and produce a low output. $A \cdot \overline{A} = 0$. This is true because if $A = 1$, then $\overline{A} = 0$; if $A = 0$, then $\overline{A} = 1$. Similar reasoning can be applied to the OR gate to prove it is always inhibited and $A + \overline{A} = 1$. The premise of this property can be taken directly from the inhibit property.

This property can be applied to make an expression such as $\overline{A}ABC = 0$ or $\overline{A} + A + B + C = 1$. Think of the expressions in terms of the logic gates they represent and they make sense. In the first expression, either $\overline{A}$ or A has to be low. Therefore, the output of this AND gate must be low because "any 0 in = 0 out," and B and C are don't care inputs when input A or $\overline{A}$ is low. The same logical analysis can be performed on the OR gate expression because the gate is inhibited by a Logic 1.

Double Negation Property

$$A = \overline{\overline{A}}$$

The double negation property is illustrated in Fig. 4–13(e). The property proves that if a logic level is double inverted, it is equal to the original input logic level. Thus, double negation signs of *equal length* in a Boolean expression can be canceled. That the bubble is shown at the output of the first NOT gate and at the input of the second one makes no difference—a bubble indicates inversion.

The preceding properties of Boolean algebra are summarized in Table 4–3. The following additional properties of Boolean algebra will be useful in simplifying expressions. Each property will be proven to be true by implementing the circuit and developing a truth table.

1. $A + AB = A$

The logic circuit implementing this expression is shown in Fig. 4–14(a). Circuit analysis under the following input conditions reveals when

$A = 0$ and $B = 0$, the output is 0. The AND gate produces a 0 output and this input to the OR gate, along with the $A = 0$ input, produces a low output.

$A = 0$ and $B = 1$, the circuit output is again 0. The A input to the AND gate inhibits the gate, and it produces a 0 output. This 0 input to the OR gate and the $A = 0$ input cause this gate's output to be 0.

TABLE 4-3	Property	Name
Properties of Boolean Algebra	$A \cdot 0 = 0$	Inhibit
	$A + 1 = 1$	Inhibit
	$A \cdot 1 = A$	Enable
	$A + 0 = A$	Enable
	$A \cdot A = A$	Redundant
	$A + A = A$	Redundant
	$A \cdot \overline{A} = 0$	Complement
	$A + \overline{A} = 1$	Complement
	$A = \overline{\overline{A}}$	Double negation

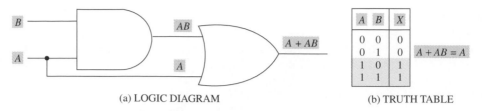

(a) LOGIC DIAGRAM   (b) TRUTH TABLE

FIGURE 4–14   **Implementation of $A + AB$.**

$A = 1$ **and** $B = 0$, the circuit output is high. Anytime the A input to the OR gate is high, the gate is inhibited and it produces a 1 output.

$A = 1$ **and** $B = 1$, the exact same logical analysis of the preceding paragraph can be accomplished. This is true because the gate is inhibited again.

The truth table for this circuit is shown in Fig. 4–14(b). The table reveals $A + AB$ is in fact equal to A. The high output conditions are marked, and they show a high output when A is high regardless of the level of B. This is because when input A is high, the OR gate is inhibited, and it produces a high output. Input AB from the AND gate to the OR gate is a don't care input when A is high.

One last point. This property can be proven mathematically by factoring out the A using the distributive property and then using the inhibit and enable properties as shown:

$A + AB$
$A(1 + B)$ Inhibit property $(B + 1 = 1)$
$A \cdot 1 = A$ Enable property $(A \cdot 1 = A)$

2. $A + \overline{A}B = A + B$

The circuit and truth table for this expression are shown in Fig. 4–15. The input conditions are

$A = 0$ **and** $B = 0$. The $B = 0$ input to the AND gate inhibits the gate, and its output is 0. This 0 and the $A = 0$ input to the OR gate produce a 0 output from the circuit.

$A = 0$ **and** $B = 1$. The $A = 0$ input is inverted prior to being applied to the AND gate. This 1 and the $B = 1$ input to the AND gate produce a high output to the OR gate. Therefore, the OR gate is inhibited, its A input is a don't care, and its output is high.

$A = 1$ **and** $B = 0$ **or** $A = 1$ **and** $B = 1$. Both sets of input conditions produce a high output because $A = 1$ and the OR gate is inhibited.

The truth table in Fig. 4–15(b) is marked to highlight the high outputs because the Boolean expression is written to indicate what conditions will produce a high output. The three high outputs occur when A is high or B is high, exactly the same as an OR gate. Thus, $A + \overline{A}B = A + B$. This can also be proven mathematically by using the $A = A + AB$ property and the redundant, complement, and enable properties.

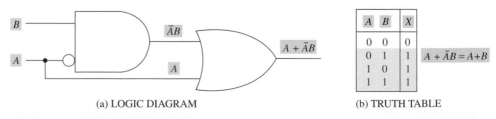

(a) LOGIC DIAGRAM   (b) TRUTH TABLE

FIGURE 4–15   **Implementation of $A + \overline{A}B$.**

3. $\overline{A} + AB = \overline{A} + B$

A similar analysis can be accomplished for this property. The circuit and truth table are shown in Fig. 4–16. Anytime $A = 0$ the OR gate is inhibitied and produces a high output. The OR gate is also inhibited by the AND gate's high output when $A = 1$ and $B = 1$. The truth table verifies the property $\overline{A} + AB = \overline{A} + B$.

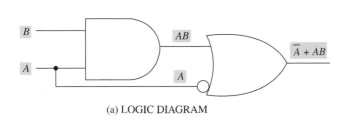

A	B	X	
0	0	1	
0	1	1	
1	0	0	$\overline{A}+AB=\overline{A}+B$
1	1	1	

(a) LOGIC DIAGRAM   (b) TRUTH TABLE

FIGURE 4–16
Implementation of
$\overline{A} + AB$.

4. $A + \overline{A}\,\overline{B} = A + \overline{B}$

The logic circuit used to implement this expression and the truth table are shown in Fig. 4–17.

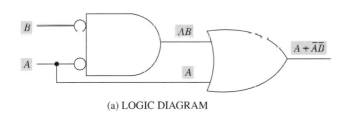

A	B	X	
0	0	1	
0	1	0	$A+\overline{A}\,\overline{B}=A+\overline{B}$
1	0	1	
1	1	1	

(a) LOGIC DIAGRAM   (b) TRUTH TABLE

FIGURE 4–17
Implementation of
$A + \overline{A}\,\overline{B}$.

5. $\overline{A} + A\overline{B} = \overline{A} + \overline{B}$

Refer to Fig. 4–18.

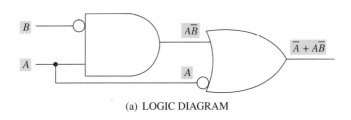

A	B	X	
0	0	1	
0	1	1	
1	0	1	$\overline{A}+A\overline{B}=\overline{A}+\overline{B}$
1	1	0	

(a) LOGIC DIAGRAM   (b) TRUTH TABLE

FIGURE 4–18
Implementation of
$\overline{A} + A\overline{B}$.

6. $\overline{A} + \overline{A}\,\overline{B} = \overline{A}$

Refer to Fig. 4–19.
The equivalency of the preceding three properties can be proven by analyzing the circuits and comparing the results in the truth tables.

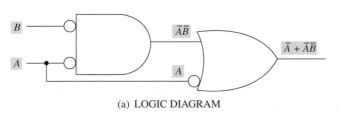

A	B	X	
0	0	1	
0	1	1	
1	0	0	$\overline{A}+\overline{A}\,\overline{B}=\overline{A}$
1	1	0	

(a) LOGIC DIAGRAM   (b) TRUTH TABLE

FIGURE 4–19
Implementation of
$\overline{A} + \overline{A}\,\overline{B}$.

Section 4–2: Review Questions

Answers are given at the end of the chapter.

A. Signs of grouping must be used when an OR gate feeds an OR gate.
   (1) True
   (2) False

B. Signs of grouping must be used when an AND gate feeds an OR gate.
   (1) True
   (2) False

C. Anytime a variable in a Boolean expression is ANDed or ORed with an inhibitor, the variable can be removed from the expression.
   (1) True
   (2) False

D. Anytime a variable in a Boolean expression is ANDed or ORed with an enabler, the enabler can be removed from the expression.
   (1) True
   (2) False

E. $A \cdot \overline{A} = 1$
   (1) True
   (2) False

F. $A + \overline{A} = 1$
   (1) True
   (2) False

G. $A = \overline{\overline{A}}$
   (1) True
   (2) False

H. $A + \overline{A}B = A + B$
   (1) True
   (2) False

I. $\overline{A} + AB = A + B$
   (1) True
   (2) False

SECTION 4–3: DeMorgan's Theorems

OBJECTIVES

1. Use DeMorgan's theorems to simplify Boolean expressions.
2. Use the implications of DeMorgan's theorems to change an inverted-input OR function to a NAND function.

DeMorgan's theorems are the last properties of Boolean algebra for which concern is necessary. Mr. DeMorgan proved "the complement of the OR function is equal to the AND function of the complements." In plain English this means when the OR sum of two variables is complemented $\overline{(A + B)}$, the same results are obtained as if each variable is complemented and then ANDed $(\overline{A}\,\overline{B})$. The implication of the theorem is that a NOR gate's output is equal to an inverted-input AND gate's output. This is shown in Fig. 4–20.

$$\overline{A + B} = \overline{A}\,\overline{B}$$

This theorem can be proven using short logic. If the results of the short logic analysis prove equal, then the functions of the logic gate circuits are equal. The short logic of

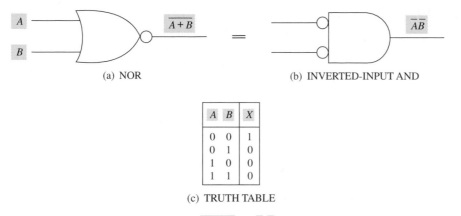

(a) NOR   (b) INVERTED-INPUT AND

A	B	X
0	0	1
0	1	0
1	0	0
1	1	0

(c) TRUTH TABLE

FIGURE 4-20   **DeMorgan's Theorem: $\overline{A + B} = \overline{A}\,\overline{B}$.**

the NOR gate is "any 1 in = 0 out." The inverted-input AND gate's analysis must start with the short logic of an AND gate—"any 0 in = 0 out." Since both input logic levels are inverted at the AND gate's input, the input short logic must be NOTed. This results in "any 1 in = 0 out" and proves the equality described by DeMorgan's theorem.

Another way of analyzing the equality of $\overline{A + B} = \overline{A}\,\overline{B}$ is to read the expressions. The first expression indicates if A is high *or* B is high, the output will be low. These three low outputs are shown on the truth table in Fig. 4–20(c). This leaves only one condition that produces a high output, and that condition is $\overline{A}\,\overline{B}$.

$$\overline{AB} = \overline{A} + \overline{B}$$

DeMorgan also provided the theorem that states, "the complement of the AND function is equal to the OR function of the complements." This theorem indicates the output of a NAND gate $(\overline{AB})$ is equal to the output obtained when each variable is complemented and then ORed $(\overline{A} + \overline{B})$ as shown in Fig. 4–21. This can also be proven with short logic. If the OR gate's short logic is used in the circuit shown in Fig. 4–21(b) and the input logic levels are NOTed, the result is "any 0 in = 1 out." As you know from your study of basic logic gates, this is equal to the NAND gate's short logic.

(a) NAND   (b) INVERTED-INPUT OR

FIGURE 4-21   **DeMorgan's Theorem: $\overline{AB} = \overline{A} + \overline{B}$.**

DeMorgan's theorems provide an invaluable Boolean simplification procedure. *The vinculum in an expression can be broken as long as the logical sign directly under the break is changed* (AND to OR/OR to AND). In DeMorgan's first theorem, $\overline{A + B}$, if the vinculum is broken directly over the OR sign and the sign changed to an AND sign, the altered expression is $\overline{A}\,\overline{B}$. The second DeMorgan theorem, $\overline{AB} = \overline{A} + \overline{B}$, also indicates the vinculum can be broken as long as the sign is changed.

This procedure of breaking the vinculum and changing the logical sign can also be used in an opposite manner. Consider the results when the following steps are applied:

$$\overline{A} \cdot \overline{B}$$

STEP 1:	$\overline{A} + \overline{B}$	Change the sign.
STEP 2:	$\overline{\overline{A}} + \overline{\overline{B}}$	NOT the individual variables (inputs).
STEP 3:	$\overline{\overline{\overline{A}} + \overline{\overline{B}}}$	NOT the entire expression (output).
STEP 4:	$\overline{A + B}$	Cancel all double NOTs.

The original procedure of breaking the vinculum and changing the sign took the expression from a NOR to an inverted-input AND function. The second procedure did just the reverse.

How could the OR expression $\overline{A} + B + \overline{C}$ be changed so that it could be implemented with a NAND gate? Follow the previous steps and refer to Fig. 4–22.

FIGURE 4–22   **Inverse use of DeMorgan's Theorems.**

	$\overline{A} + B + \overline{C}$	OR expression
STEP 1:	$\overline{A} \cdot \overline{B} \cdot \overline{C}$	Change the signs.
STEP 2:	$\overline{\overline{A}} \cdot \overline{B} \cdot \overline{\overline{C}}$	NOT the individual variables.
STEP 3:	$\overline{\overline{\overline{A}} \cdot \overline{B} \cdot \overline{\overline{C}}}$	NOT the entire expression.
STEP 4:	$\overline{A \cdot \overline{B} \cdot C}$	Cancel all double NOTs.

If the resultant expression's $\overline{(A \cdot \overline{B} \cdot C)}$ vinculum is broken, the logical signs changed, and the double NOTs canceled, the original expression will be regained. All of the properties and theorems presented in this chapter are compiled in Table 4–4 for easy reference.

TABLE 4-4	Property/Theorem	Name
Boolean Properties and Theorems	$A \cdot 0 = 0$	Inhibit
	$A + 1 = 1$	Inhibit
	$A \cdot 1 = A$	Enable
	$A + 0 = A$	Enable
	$A \cdot A = A$	Redundant
	$A + A = A$	Redundant
	$A \cdot \overline{A} = 0$	Complement
	$A + \overline{A} = 1$	Complement
	$A = \overline{\overline{A}}$	Double negation
	$A + AB = A$	
	$A + \overline{A}B = A + B$	
	$\overline{A} + AB = \overline{A} + B$	
	$A + \overline{A}\,\overline{B} = A + \overline{B}$	
	$\overline{A} + A\overline{B} = \overline{A} + \overline{B}$	
	$\overline{A} + \overline{A}\,\overline{B} = \overline{A}$	
	$\overline{A + B} = \overline{A}\,\overline{B}$	DeMorgan's Theorem
	$\overline{AB} = \overline{A} + \overline{B}$	DeMorgan's Theorem
	$(A + B) + C = A + (B + C)$	Associative
	$(AB)C = A(BC)$	Associative
	$A + B = B + A$	Commutative
	$AB = BA$	Commutative
	$A(B + C) = AB + AC$	Distributive

Section 4-3: Review Questions

Answers are given at the end of the chapter.

A. $\overline{A + B} = \overline{A}\,\overline{B}$
    (1) True
    (2) False

B. $\overline{AB} = \overline{A} + \overline{B}$
    (1) True
    (2) False

C. Refer to Fig. 4–23. The two gates are equal.
    (1) True
    (2) False

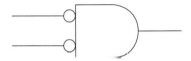   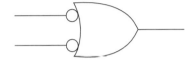

FIGURE 4-23

D. Change the Boolean expression ABC to an expression that can be implemented with a NOR gate and inverters.
E. Change the Boolean expression $X + Y$ to an expression that can be implemented with a NAND gate and inverters.

SECTIONS 4-1 THROUGH 4-3: INTERNAL SUMMARY

The AND gate produces the **logical product** of its inputs. The AND function sign is a dot ($\cdot$) or $\times$. Often the dot or $\times$ is left out of an expression (AB), but it is implied. The Boolean expression for an AND gate is written for a high output.

The OR gate produces the **logical sum.** The OR sign is a plus sign ($A + B$). The Boolean expression for an OR gate is written for a high output.

The NAND gate $(\overline{AB})$ and NOR gate $(\overline{A + B})$ Boolean expressions use **vinculums** over the entire logical product/logical sum. The vinculum identifies the logical function and signifies the output expression is written for a low output.

Remember—A NOT sign over a single variable input to a logic gate indicates the input was NOTed prior to the logical process (e.g., $A + \overline{B}$). A vinculum over the entire expression out of a logic gate indicates inversion after the logical process has been completed (e.g., $\overline{A + B + C}$).

DeMorgan's theorems are

$$\overline{A + B} = \overline{A}\,\overline{B}$$
$$\overline{AB} = \overline{A} + \overline{B}$$

DeMorgan's theorems lead to the following conclusions:

1. A vinculum can be broken over a logical sign provided the sign is changed.
2. The implications of DeMorgan's theorems prove the validity of the following procedure:
    a. Change the sign.
    b. NOT the inputs.
    c. NOT the output.
    d. Cancel all double NOTs.

Review Table 4–4 for the properties of real numbers and Boolean algebra.

SECTIONS 4-1 THROUGH 4-3: INTERNAL SUMMARY QUESTIONS

Answers are given at the end of the chapter.

1. The plus (+) sign in a Boolean expression indicates logical _____.
   a. multiplication
   b. subtraction
   c. addition

2. The Boolean expression for a 3-input AND gate with A, B, and C inputs is
   a. ABC
   b. $\overline{ABC}$
   c. $A + B + C$
   d. $\overline{A + B + C}$

3. The Boolean expression for a 2-input NOR gate with X and Y inputs is
   a. XY
   b. $\overline{XY}$
   c. $X + Y$
   d. $\overline{X + Y}$

4. The expression $A\overline{B}C$ indicates the gate will output high when A is high *and* B is low *and* C is high.
   a. True
   b. False

5. The OR gate produces the logical product.
   a. True
   b. False

6. A vinculum is a bar drawn over two or more variables in a Boolean expression.
   a. True
   b. False

7. The expression $\overline{AB}$ indicates the gate will output high when A is low *and* B is low.
   a. True
   b. False

8. The expression $A + B$ indicates the gate will output high when A is high *or* B is high.
   a. True
   b. False

9. Which property of real numbers indicates $(AB)C = A(BC)$?
   a. Commutative Property of Addition
   b. Associative Property of Addition
   c. Commutative Property of Multiplication
   d. Associative Property of Multiplication

10. Which property of Boolean algebra indicates $A + 0 = A$?
    a. Complement
    b. Redundant
    c. Inhibit
    d. Enable

11. Which property of Boolean algebra states $A \cdot 0 = 0$?
    a. Complement
    b. Redundant
    c. Inhibit
    d. Enable

12. Which property of Boolean algebra indicates $A + 1 = 1$?
    a. Complement
    b. Redundant

   c. Inhibit

   d. Enable

13. Which property of Boolean algebra indicates $A + A = A$?

   a. Complement

   b. Redundant

   c. Inhibit

   d. Enable

14. Which property of Boolean algebra states $A \cdot \overline{A} = 0$?

   a. Complement

   b. Redundant

   c. Inhibit

   d. Enable

15. $\overline{A + B} = \overline{A}\,\overline{B}$

   a. True

   b. False

16. Change the expression $A + B + C + D$ to an expression that can be implemented with a 4-input NAND gate and inverters.

   a. $\overline{\overline{A \cdot B \cdot C \cdot D}}$

   b. $\overline{A \cdot B \cdot C \cdot D}$

   c. $\overline{\overline{A} \cdot \overline{B} \cdot \overline{C} \cdot \overline{D}}$

SECTION 4–4: INTERPRETING BOOLEAN EXPRESSIONS

OBJECTIVES

1. Interpret the meaning of a Boolean expression in terms of inputs/outputs.
2. Given a Boolean expression, determine the input conditions required for a logic gate or circuit to output a specified level.

Before delving into logic circuit design or implementation, it is necessary to interpret a Boolean expression. This interpretation is merely reading the expression to determine what input conditions will produce a certain level output, and we have already done a limited amount of this.

When interpreting a Boolean expression it is important to remember the basics covered in Section 4–1. First, if a Boolean expression is written with a *vinculum over the entire expression*, it is written to indicate the input conditions that will produce a *low output*. Second, if it is written *without a vinculum*, it is written to indicate the input conditions that will produce a *high output*. The expression $\overline{A + C}$ indicates if A is high *or* C is high, the NOR gate will output low. In Section 4–3 DeMorgan's theorems proved the vinculum can be broken over a logical sign if the sign is changed.

For the Boolean expression $\overline{A} + C$, this action results in $\overline{A}\,\overline{C}$ and is read if A is low *and* C is low the circuit will output high. (The individual NOT signs over A and C must be read as low inputs. However, the output is determined to be high since there is no vinculum over the entire expression.) DeMorgan's theorems led us to conclude these two expressions are equal. The truth table in Fig. 4–24(a) can be used as further proof. The first expression, $\overline{(A + C)}$, provides the low outputs on the truth table. The input conditions that produce the low outputs have been highlighted on the truth table. Once all of the lows are plotted for this expression, the remaining conditions must be high. The second expression $(\overline{A}\,\overline{C})$ proves this statement is true.

Although the truth table indicates there are three inputs to the circuit (A, B, and C), the B input is a don't care because it is not applied to this NOR gate.

INPUTS			OUTPUT	
A	B	C	X	
0	0	0	1	A AND C ARE LOW
0	0	1	0	C IS HIGH
0	1	0	1	A AND C ARE LOW
0	1	1	0	C IS HIGH
1	0	0	0	A IS HIGH
1	0	1	0	A/C ARE HIGH
1	1	0	0	A IS HIGH
1	1	1	0	A/C ARE HIGH

(brackets to right of table) $\overline{AC}$ (top group), $\overline{A + C}$ (bottom group)

(a) TRUTH TABLE

(b) NOR GATE SYMBOLS

FIGURE 4-24   Interpreting NOR gate expressions.

It is normally easier to break a vinculum, change the logical sign, and plot high outputs on a truth table. Nevertheless, as shown in the previous example, it is not mandatory. If all high outputs are plotted on a truth table, the remaining input combinations produce low outputs.

One extremely important point needs to be emphasized here. In the expression $\overline{A}\,\overline{C}$, the NOT signs do not mean A and C are low. They indicate *if* A and C are low, the output will be high. The symbol of the inverted-input AND gate in Fig. 4–24(b) should prove this. If A and C are low, they are both inverted to highs at the input of the AND gate; thus, the gate's output will be high.

A little practice with this type of interpretation will enhance your understanding.

The expression $\overline{A} \cdot B \cdot C$ represents the output of a NAND gate with its A input inverted as shown in Fig. 4–25. The expression indicates and should be read, "if A is low *and* B is high *and* C is high, the gate will output low."

$$A + B + \overline{C}$$

A ──○⌐

B ─── (NAND gate with inverted A input) ──○ $\overline{ABC}$

C ───

FIGURE 4-25   Implementation of $\overline{ABC}$.

The expression   represents the output of an OR gate with its C input inverted. The circuit is shown in Fig. 4–26. The expression is read, "if A is high *or* B is high *or* C is low, the gate will output high."

How would you interpret the expression $A \cdot B \cdot \overline{C}(\overline{A} + D)$? We might have a problem reading this one. Part of the expression $(AB\overline{C})$ is written for a high output, and part

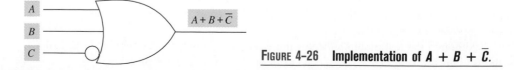

FIGURE 4-26   Implementation of $A + B + \overline{C}$.

of it $(\overline{\overline{A} + D})$ is written for a low output. The expression cannot be read without some simplification. It can be simplified by using the properties and theorems listed in Table 4-4.

$$A \cdot B \cdot \overline{C}(\overline{\overline{A} + D})$$

STEP 1:	$A \cdot B \cdot \overline{C}(\overline{\overline{A}} \cdot \overline{D})$	Break vinculum/change sign.
STEP 2:	$A \cdot B \cdot \overline{C}(A \cdot \overline{D})$	Cancel double NOT.
STEP 3:	$A \cdot A \cdot B \cdot \overline{C} \cdot \overline{D}$	Rearrange.
STEP 4:	$A \cdot B \cdot \overline{C} \cdot \overline{D}$	Redundant property.

This simplified expression is easy to read—if A is high *and* B is high *and* C is low *and* D is low, the circuit's output will be high. As seen in the preceding paragraph, some Boolean expressions are difficult to interpret unless they are simplified. Generally, simplification should take the expression to the point where all vinculums have been broken and their corresponding signs changed. *Upon completion of the simplification process for purposes of interpretation, there may be NOT signs over single-variable terms only. $\overline{A}\,\overline{B}\,C$ is acceptable, but $\overline{ABC}$ is not.* The expression $\overline{A} + \overline{B}$ indicates the gate's output will be low when A is low *or* B is low.

Reading these expressions and breaking vinculums and changing signs should lead to the conclusion that every gate symbol has an **alternate symbol.** The alternate symbol for the NOR gate was developed using DeMorgan's theorem and shown in Fig. 4-24.

Section 4-4: Review Questions

Answers are given at the end of the chapter.

A. The expression $\overline{A}\,\overline{B}\,\overline{C}$ indicates the gate will output low when A is high *and* B is high *and* C is high.
   (1) True
   (2) False

B. The expression $\overline{A + B + C}$ indicates the gate will output low when A is high *or* B is high *or* C is high.
   (1) True
   (2) False

C. The expression $\overline{A \cdot B}$ indicates the gate will output low when A is low *and* B is high.
   (1) True
   (2) False

D. Simplify $\overline{\overline{A + B}}$.

E. Simplify $A \cdot B(\overline{\overline{A} + \overline{B}})$.

F. Simplify $(\overline{A + B + C})D$.

SECTION 4-5: ALTERNATE LOGIC GATE SYMBOLS

OBJECTIVES

1. Identify the alternate logic symbols for the AND, OR, NAND, NOR and NOT gates.
2. Derive an alternate logic gate symbol by following the steps in DeMorgan's theorems.

A Boolean expression of $\overline{\overline{A} \cdot \overline{B}}$ indicates each variable has been NOTed at the AND gate's input and the logical product has been NOTed at its output. The expression states that the gate will output low when A *and* B are low. This is shown in the symbol and truth table in Fig. 4-27.

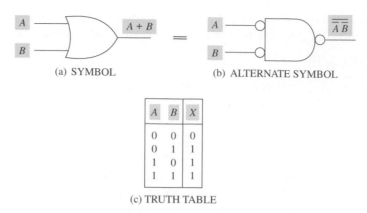

(a) SYMBOL   (b) ALTERNATE SYMBOL

A	B	X
0	0	0
0	1	1
1	0	1
1	1	1

(c) TRUTH TABLE

Figure 4-27   OR function.

Note that the truth table indicates "any 1 in = 1 out." Therefore, the symbol shown in Fig. 4–27(b) is an alternate logic gate symbol for an OR gate. This can be proven with short logic as well as with DeMorgan's theorem.

First, since the gate in Fig. 4–27(b) is an AND gate, start with that gate's short logic. The input levels and the output level are all NOTed. This results in a final logic of "any 1 in = 1 out." Since this is the OR gate short logic, the symbol has to be equivalent to an OR gate.

Second, DeMorgan's theorem may also be used in the expression to prove equality:

$$\overline{\overline{A} \cdot \overline{B}}$$
$$\overline{\overline{A}} + \overline{\overline{B}}$$
$$A + B$$

The standard and alternate logic gate symbols are shown in Fig. 4–28. Take a few minutes and prove the short logic for each symbol is equal to that of its alternate symbol.

(a) AND

(b) OR

(c) NAND

(d) NOR

(e) NOT

Figure 4-28   Standard/alternate logic gate symbols.

Also use DeMorgan's theorem to ensure the Boolean expressions are equivalent. This is not necessary for the NOT gate.

Figure 4–28(a) shows the alternate symbol for the AND gate produces a short logic of "any $\overline{1}$ in $=\overline{1}$ out." This is the short logic of an AND gate. In the Boolean expression $\overline{A} + \overline{B}$, if the vinculum is broken, the logical sign changed, and the double NOTs canceled, the equivalency is again proven.

All of the short logic can be proven by using the basic AND or OR gate's short logic. From this initial logic, NOT the input logic level if bubbles appear at the inputs. Also, NOT the output logic level if a bubble is present at the output.

An alternate logic gate symbol can be derived as shown and explained in Fig. 4–29 by following the steps outlined in DeMorgan's theorems.

1. Change the sign.
2. NOT the input variables.
3. NOT the entire expression.
4. Cancel all double NOTs.

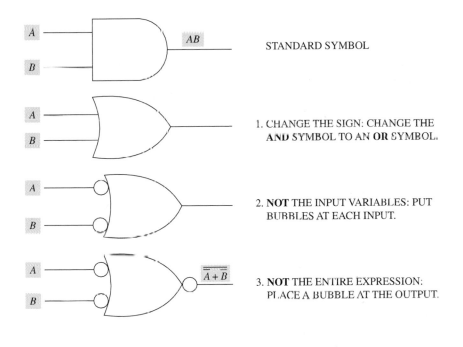

STANDARD SYMBOL

1. CHANGE THE SIGN: CHANGE THE **AND** SYMBOL TO AN **OR** SYMBOL.

2. **NOT** THE INPUT VARIABLES: PUT BUBBLES AT EACH INPUT.

3. **NOT** THE ENTIRE EXPRESSION: PLACE A BUBBLE AT THE OUTPUT.

4. CANCEL ALL DOUBLE **NOTS**. (NOT APPLICABLE FOR THIS SYMBOL.)

FIGURE 4–29 Alternate logic symbol derivation/AND gate.

Some additional gate circuits and their equivalent symbols are shown in Fig. 4–30. *Note*: There is no short logic for a gate with only one of its inputs NOTed. The short logic or short logic of the gate itself can be used once the NOTed input level has been determined. For example, in the upper AND gate shown in Fig. 4–30, the short logic of the AND gate indicates all inputs must be high to get a high output. Therefore, $A = 0$ and $B = 1$ to get a high output.

Instead of using DeMorgan's theorem of breaking the vinculum, let's prove equality of the gate in Fig. 4–31 by using the inverse of his theorem. The procedure is explained in the figure.

Why are alternate logic gate symbols required, since they represent the exact same function? The answer to this question is they are required for determining the output of a gate or circuit with ease. A procedure for reading logic gate outputs was presented in Chapter 3, Section 3–7. In the procedure, only the AND and OR functions were considered. In

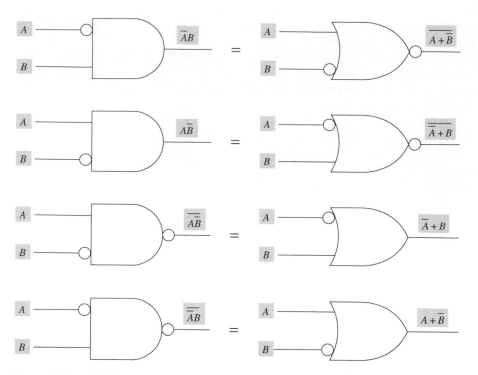

FIGURE 4-30   Additional symbols/alternate symbols.

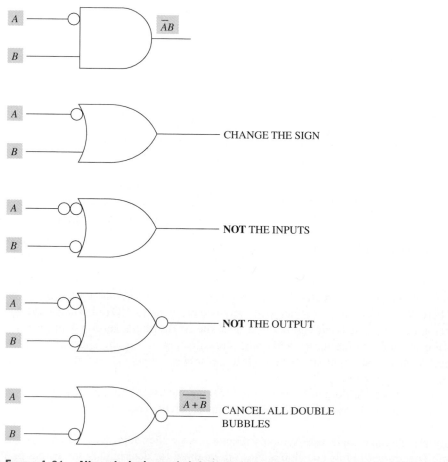

FIGURE 4-31   Alternate logic symbol derivation/procedure.

addition, a bubble on a gate was read as a low and the lack of a bubble was read as a high. This procedure works remarkably well provided the circuits are drawn properly.

Determine the output of the circuit shown in Fig. 4–32(a). If the procedure from Chapter 3 is used, gates #1 and #2 will output low when their inputs are high. Gate #3 will output low when its inputs are high. However, reading gates #1 and #2 provides information about when gate #3's inputs are low, not when they are high. If the circuit is redrawn as shown in Fig. 4–32(b), a much simpler analysis can be performed. The inverted-input OR gate is the alternate logic symbol for a NAND gate.

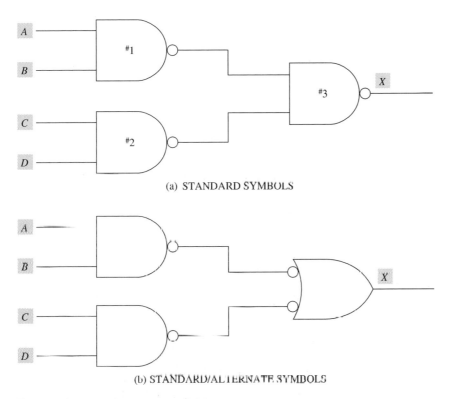

(a) STANDARD SYMBOLS

(b) STANDARD/ALTERNATE SYMBOLS

Figure 4–32   Analyzing logic circuits.

Now read the circuit. Gates #1 and #2 will output low when their inputs are high. Now that we know the circuit input conditions that will produce lows out of gates #1 and #2, we also know *either* of those conditions will produce a high output from the OR gate. Although the symbol shown for the output gate in Fig. 4–32(b) is the alternate symbol for the NAND gate, it was called an OR gate here to aid in determining the *either* condition at its inputs and the high condition that will be produced at its output if either of the inputs is low.

The general rule to follow when drawing logic diagrams is to connect bubble to bubble and no bubble to no bubble. The circuit shown in Fig. 4–33(a) presents the same problem in output interpretation that was encountered in Fig. 4–32(a). The alternate AND gate symbol shown in Fig. 4–33(b) solves that problem. Reading Fig. 4–33(b) from the output toward the input, the circuit will output low when *C* is low *or* when the output of the NOR gate is low. The output of the NOR gate is low when *A or B* is high.

Even though the alternate logic symbol here is for an AND gate, it should be thought of as an inverted-input NOR gate. This allows the logical thinking required to analyze the circuit.

Since all of the input conditions except one $(\overline{A}\,\overline{B}C)$ produce low outputs as shown in the truth table in Fig. 4–33(d), consider redrawing the circuit as shown in Fig. 4–33(c). The NOR gate alternate logic gate symbol has been used in this circuit. The redrawn cir-

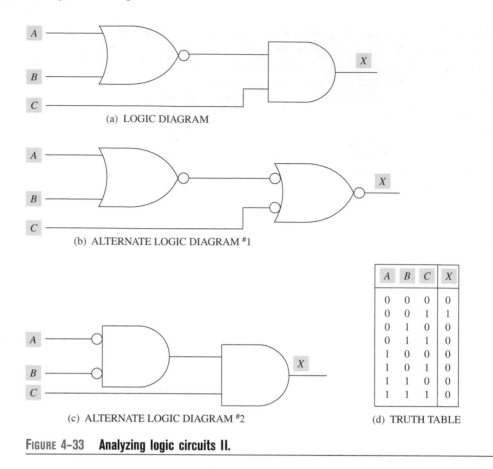

(a) LOGIC DIAGRAM

(b) ALTERNATE LOGIC DIAGRAM #1

(c) ALTERNATE LOGIC DIAGRAM #2

(d) TRUTH TABLE

A	B	C	X
0	0	0	0
0	0	1	1
0	1	0	0
0	1	1	0
1	0	0	0
1	0	1	0
1	1	0	0
1	1	1	0

FIGURE 4–33 Analyzing logic circuits II.

cuit still complies with the bubble-to-bubble/no bubble–to–no bubble rule, and it is very easy to read.

Practice this type of circuit analysis and it won't be long before it becomes second nature. Thereafter, a glance at a circuit will indicate the input conditions that will produce the appropriate output.

The Boolean expression for the circuit shown in Fig. 4–33(b) is $\overline{\overline{A + B} + \overline{C}}$. The expression identifies the low outputs of the circuit. To simplify the expression, cancel the double NOTs over $A + B$ because they are of the same length. This leaves the expression $A + B + \overline{C}$. This expression can be read to indicate the output will be low when A is high *or* B is high *or* C is low. All of these low outputs are indicated on the truth table shown in Fig. 4–33(d). Breaking the remaining vinculum over both + signs produces $\overline{A}\,\overline{B}\,C$. This is the output expression for the high output of the circuit and is easily read from the logic diagram in Fig. 4–33(c). Although either method of determining the output is proper, the Boolean simplification does not need to be accomplished if a technician merely desires to check the circuit's inputs versus output. This can be done easily by interpreting the symbols inputs and outputs using bubbles or lack of bubbles.

Section 4–5: Review Questions

Answers are given at the end of the chapter.

A. Draw the alternate logic gate symbol for each gate shown in Fig. 4–34.
B. Write the unsimplified Boolean expression for each of the alternate logic gate symbols derived in question (A) using inputs A and B.

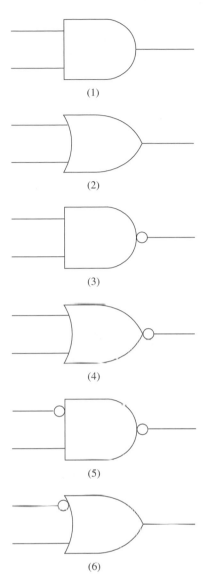

(1)

(2)

(3)

(4)

(5)

(6)

FIGURE 4-34

C. Prove equality of the Boolean expressions written in question (B) to the Boolean expressions of the standard logic symbols.

Refer to the figures indicated and *read* the circuit's inputs that are required to produce low or high outputs. Use only the logic functions and bubbles in this analysis.

D. Fig. 4–35.

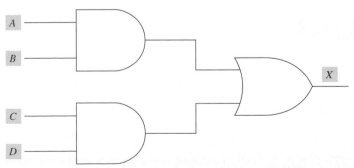

FIGURE 4-35

E. Fig. 4–36.

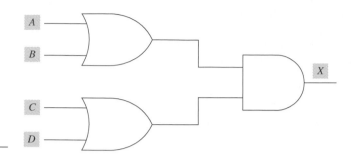

FIGURE 4-36

F. Fig. 4–37.

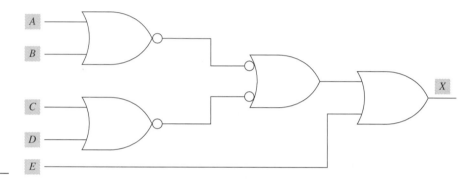

FIGURE 4-37

G. Redraw the circuit shown in Fig. 4–38 using the appropriate alternate logic symbols where necessary.

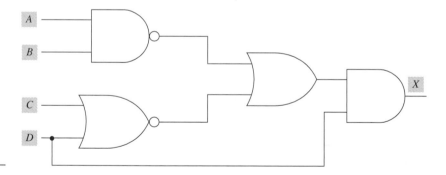

FIGURE 4-38

SECTION 4-6: INTERPRETING TRUTH TABLES

OBJECTIVES

1. Derive a Boolean expression from a truth table.
2. Identify a sum-of-products circuit.
3. Identify a product-of-sums circuit.

Although the preceding section provided additional insight regarding truth tables, more information is needed to fully utilize them in digital design and circuit analysis.

A Boolean expression may be written from a truth table provided the table has been annotated with output conditions. The truth table shown in Fig. 4–39(a) shows a desired high output when A is low *and* B is low and also when A is high *and* B is high.

Two conclusions can be drawn from the preceding statement. First, the $A = 0$ *and* $B = 0$ condition is indicative of an AND gate. Since the Boolean expression derived from this truth table will be written for a high output, it will be $\overline{A}\,\overline{B}$. The use of the NOT signs over A and B reads, "if A is low *and* B is low, the output will be high." Do not write the $A = 0$ and $B = 0$ condition from the truth table as $\overline{AB}$. This expression has an entirely different meaning than $\overline{A}\,\overline{B}$ and is incorrect for this condition. There is another high output on the truth table that must be provided by the circuit. That output will be obtained when A is high *and* B is high. Therefore, the input condition that will produce this high output is $A = 1$ *and* $B = 1$, written as AB.

The second conclusion is drawn from the fact that there are two different conditions in this circuit that require high outputs. These two sets of input conditions must be ORed together. Thus, the Boolean expression for the table is written $\overline{A}\,\overline{B} + AB$. The expression can be read as previously discussed, and it relates directly to the truth table from which it was derived.

The first conclusion proves the horizontal readings taken from a truth table for a high output are for the AND function. The second indicates each "set" of ANDed variables that produces a high output must be ORed with all additional sets where high outputs are desired.

Sum-of-Products Configuration

The circuit used to implement $\overline{A}\,\overline{B} + AB$ is shown in Fig. 4–39(b). This logic circuit is configured in a **sum-of-products (SOP) configuration.** The configuration is called SOP because the OR sum of the products of the AND gate's outputs is produced at the output of the circuit.

When a Boolean expression is derived from a truth table in the manner described, it will always be written in an SOP format if there are two or more high output conditions. Naturally, if there is only one high output, then the expression will be for an AND gate only.

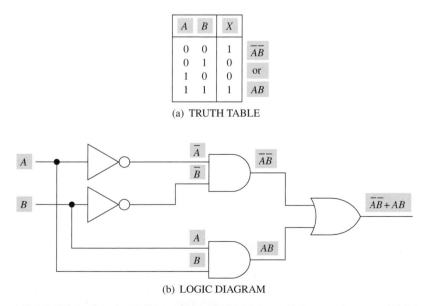

A	B	X	
0	0	1	$\overline{A}\,\overline{B}$
0	1	0	or
1	0	0	
1	1	1	AB

(a) TRUTH TABLE

(b) LOGIC DIAGRAM

FIGURE 4–39   Deriving Boolean expression from truth table.

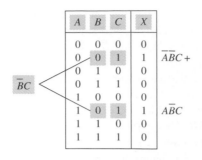

A	B	C	X
0	0	0	0
0	0	1	1	$\overline{A}\,\overline{B}C +$
0	1	0	0
0	1	1	0
1	0	0	0
1	0	1	1	$A\overline{B}C$
1	1	0	0
1	1	1	0

Figure 4-40   Deriving Boolean expression from truth table.

The truth table shown in Fig. 4–40 again requires high outputs from two different sets of input conditions. Note how each set of input conditions for a high output is written in the expression. Where 0s appear under an input letter (A, B, or C), that letter is NOTed. Where 1s are under an input letter, the letter is written in the expression as is. The expression derived from this truth table is $\overline{A}\,\overline{B}C + A\overline{B}C$. The common terms may be factored to simplify this expression.

$\overline{A}\,\overline{B}\,C + A\,\overline{B}C$	SOP expression
$\overline{B}\,C\,(\overline{A} + A)$	Factor out common terms
$\overline{B}\,C\,(1)$	$\overline{A} + A = 1$
$\overline{B}\,C$	$A \cdot 1 = A$

$\overline{B}C$ produces the same high outputs the original expression produced. This is shown on the truth table in Fig. 4–40.

The truth table in Fig. 4–41 yields the following Boolean expression for its high outputs:

$$\overline{A}\,\overline{B}\,\overline{C} + \overline{A}\,\overline{B}C + \overline{A}B\overline{C} + \overline{A}BC + A\overline{B}\,\overline{C} + AB\overline{C} + ABC$$

A	B	C	X
0	0	0	1
0	0	1	1
0	1	0	1
0	1	1	1
1	0	0	1
1	0	1	0
1	1	0	1
1	1	1	1

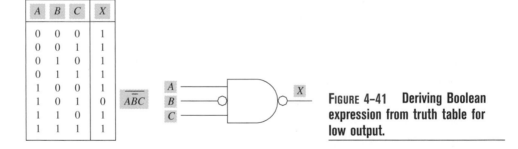

Figure 4-41   Deriving Boolean expression from truth table for low output.

The expression for this truth table could be written for a low output to save a lot of work. The expression, $A \cdot \overline{B} \cdot C$, indicates when the circuit will output low. The remaining outputs will all be high. Actually the expression is $\overline{X} = A\overline{B}C$; however, NOTing each side of the expression produces $\overline{\overline{X}} = \overline{A\overline{B}C}$, which produces $X = \overline{A\overline{B}C}$. If DeMorgan's theorem is applied to the preceding expression, the result is $\overline{A} + B + \overline{C}$. This expression identifies the input conditions that will produce the high outputs.

Another exercise with a truth table should be beneficial. The truth table in Fig. 4–42(a) shows a low output when $X = 1$, $Y = 0$, and $Z = 0$. All other outputs are high. A Boolean expression of $X \cdot \overline{Y} \cdot \overline{Z}$ represents this output condition. The logic circuit is shown in Fig. 4–42(b). When DeMorgan's theorem is applied to the expression, the result is $\overline{X} + Y + Z$. This expression can be verified as accurate by comparing it to the truth table. The table in Fig. 4–42(a) has been annotated with high outputs for the expression $\overline{X} + Y + Z$.

Anytime a Boolean expression is simplified, the simplified expression should be equivalent to the original expression and can be proven on a truth table. The truth table shown in Fig. 4–43 yields an expression of $\overline{A}\,\overline{B}\,\overline{C} + \overline{A}\,\overline{B}C$. The expression can be simplified by factoring out $\overline{A}\,\overline{B}$.

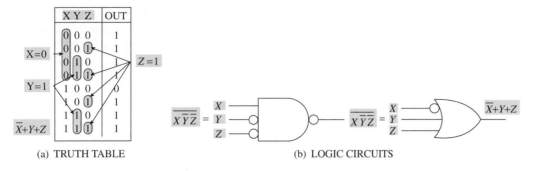

(a) TRUTH TABLE   (b) LOGIC CIRCUITS

FIGURE 4-42   **Implementation of** $\overline{X\,\overline{Y}\,Z} = \overline{X} + Y + \overline{Z}$.

The simplified expression $(\overline{A}\,\overline{B})$ can be proven by going back to the truth table and highlighting or circling the $\overline{A}\,\overline{B}$ conditions that produce high outputs. If the simplified expression produces the same high outputs as the original expression, the expressions are equal. If not, a mistake has been made.

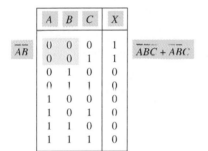

A	B	C	X
0	0	0	1
0	0	1	1
0	1	0	0
0	1	1	0
1	0	0	0
1	0	1	0
1	1	0	0
1	1	1	0

$\overline{A}\overline{B}$ $\overline{A}\overline{B}\overline{C} + \overline{A}\overline{B}C$

FIGURE 4-43   **Truth table used to verify equality of simplified expression.**

Product-of-Sums Configuration

In a sum-of-products (SOP) expression written from a truth table, the set of input variables that produced a high output was ANDed. Then each set of input variables that produced a high output was ORed with other sets that produced high outputs. This always produced an SOP expression because the circuit required to implement the expression is of the AND-OR configuration. An SOP circuit configuration is shown in Fig. 4-44. The output expression for this circuit is $\overline{A}B + AB$.

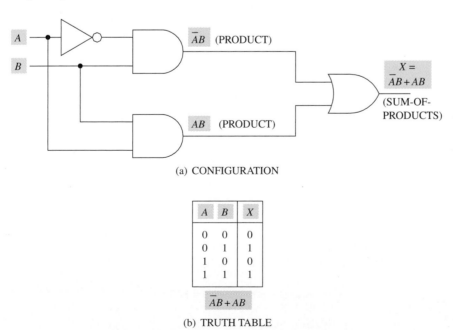

(a) CONFIGURATION

A	B	X
0	0	0
0	1	1
1	0	0
1	1	1

$\overline{A}B + AB$

(b) TRUTH TABLE

FIGURE 4-44   **Sum-of-products circuit.**

Another design technique employs the **product-of-sums (POS) configuration.** As the name implies, this is an OR-AND circuit configuration. A POS circuit configuration is shown in Fig. 4–45. It is called a POS configuration because the AND product of the sums of the OR gate's outputs is produced at the circuit's output. The truth table for this POS circuit is the same as the truth table for a SOP circuit.

A POS expression can be derived from a truth table. However, there are some major differences between extracting a POS expression and an SOP expression from a truth table.

When an SOP expression was extracted from a truth table, the expression was written to represent each high output condition. This is because the OR gate (refer to Fig. 4–44) will output high when any of the sets of input variables produces a high output from the AND gates.

When a POS expression is derived from a truth table, the expression is written to represent every *low output* condition on the truth table. This is because the AND gate (refer to Fig. 4–45) will output low when any of the sets of input variables produce a low output from the OR gates.

In the POS extraction, each variable in a set of input variables that produces a low output is ORed together with other variables in that set. Afterward, each set of ORed variables that produce a low output is ANDed with other sets that produce low outputs. Extraction of the variables that produce low outputs is done in **complementary** form. To clarify this, think about these rules:

1. For each set of input variables on a truth table that produces a *low* output, *write each input variable in complementary form and OR the variables within the set.* For example, the truth table shown in Fig. 4–45(b) shows a low output when $A = 0$ and $B = 0$. Thus, the expression for this low output is written $A + B$. (Each variable is now shown in its complementary form. If it were written in its true form the variables would have been NOTed because A and B in this set of inputs are low.) A low output is also produced when $A = 1$ and $B = 0$. This results in $\overline{A} + B$.
2. *AND each set of ORed input variables together.* The truth table shown in Fig. 4–45(b) produces $(A + B)(\overline{A} + B)$. This final expression is in POS configuration and represents the low outputs on the truth table.

The truth table in Fig. 4–45(b) yields $(A + B)(\overline{A} + B)$ in the POS format and $\overline{\overline{A}\,\overline{B} + A\overline{B}}$ in the SOP format if the vinculum is used to indicate when the circuit's output is low. The second expression is written to show the low output in SOP format. Applying DeMorgan's theorem to the second expression will prove its equality to the first.

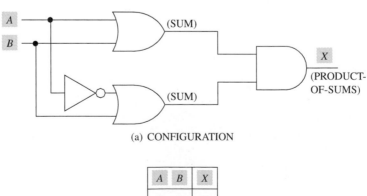

(a) CONFIGURATION

A	B	X
0	0	0
0	1	1
1	0	0
1	1	1

$(A + B)\,(\overline{A} + B)$

(b) TRUTH TABLE

Figure 4–45   Product-of-sums circuit.

A look at the circuit and truth table shown in Fig. 4–45 will explain why the input variables must be extracted in complementary form for a POS expression. If $A = 0$ and $B = 0$, the circuit output will be 0. All 0s into the top OR gate will produce a 0 output, which will inhibit the AND gate and produce a 0 out of the POS circuit. The output expression for a low output from the top OR gate could be written $\overline{\overline{A}\,\overline{B}}$. Using DeMorgan's theorem, this would equate to $A + B$.

The SOP and POS expressions are simplified below to show their equality. The SOP expression is written for a high output.

SOP	**POS**
$\overline{A}B + AB$	$(A + B)(\overline{A} + B)$
$B(\overline{A} + A)$	$A\overline{A} + AB + \overline{A}B + BB$
$B(1)$	$0 + AB + \overline{A}B + B$
B	$AB + \overline{A}B + B$
	$B(A + \overline{A} + 1)$
	$B(1)$
	B

As further proof, let's take a look at the SOP circuit shown in Fig. 4–46. The output of the circuit is $\overline{A}\,\overline{B}C + \overline{A}B\overline{C} + \overline{A}BC + A\overline{B}C + ABC$. Now the POS expression

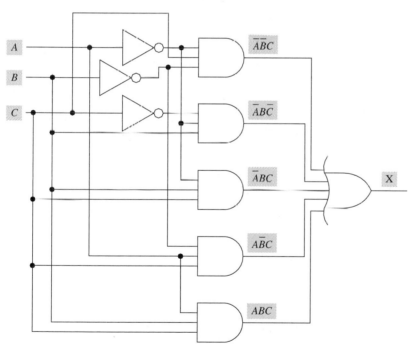

(a) LOGIC DIAGRAM

A	B	C	X
0	0	0	0
0	0	1	1
0	1	0	1
0	1	1	1
1	0	0	0
1	0	1	1
1	1	0	0
1	1	1	1

(b) TRUTH TABLE

FIGURE 4–46   SOP configuration for
$\overline{A}\,\overline{B}C + \overline{A}B\overline{C} + \overline{A}BC + A\overline{B}C + ABC$.

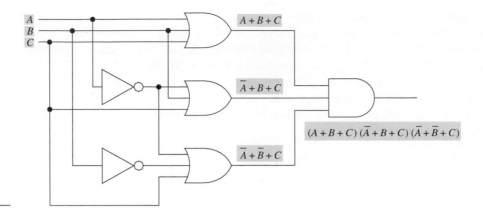

FIGURE 4-47   POS configuration of circuit shown in Fig. 4-46.

will be extracted from the same truth table in Fig. 4–46(b). The expression is $(A + B + C)(\overline{A} + B + C)(\overline{A} + \overline{B} + C)$. The POS circuit is shown in Fig. 4–47. Notice the POS circuit implementation contains fewer gates.

Generally, when a truth table contains only a few high outputs compared to low outputs, an SOP circuit is desired. On the contrary, if a truth table contains only a few low outputs when compared to high outputs, a POS circuit is desired.

The equality of the two expressions and the circuits shown in Fig. 4–46 and Fig. 4–47 is proven below.

SOP

$$\overline{A}\,\overline{B}C + \overline{A}B\overline{C} + \overline{A}BC + A\overline{B}C + ABC$$
$$\overline{A}B(\overline{C} + C) + AC(\overline{B} + B) + \overline{A}\,\overline{B}C$$
$$\overline{A}B(1) + AC(1) + \overline{A}\,\overline{B}C$$
$$\overline{A}B + AC + \overline{A}\,\overline{B}C$$
$$\overline{A}(B + \overline{B}C) + AC$$
$$\overline{A}(B + C) + AC$$
$$\overline{A}B + \overline{A}C + AC$$
$$\overline{A}B + C(\overline{A} + A)$$
$$\overline{A}B + C(1)$$
$$\overline{A}B + C$$

POS

$$(A + B + C)(\overline{A} + B + C)(\overline{A} + \overline{B} + C)$$
$$(A\overline{A} + AB + AC + \overline{A}B + BB + BC + \overline{A}C + BC + CC)(\overline{A} + \overline{B} + C)$$
$$(0 + AB + AC + \overline{A}B + B + BC + \overline{A}C + C)(\overline{A} + \overline{B} + C)$$
$$\overline{A}AB + \overline{A}AC + \overline{A}\,\overline{A}B + \overline{A}B + \overline{A}BC + \overline{A}\,\overline{A}C + \overline{A}C + \overline{B}AB + \overline{B}AC$$
$$\quad + \overline{B}\,\overline{A}B + \overline{B}B + \overline{B}BC + \overline{B}\,\overline{A}C + \overline{B}C + CAB + CAC + C\overline{A}B$$
$$\quad + CB + CBC + C\overline{A}C + CC$$
$$0 + 0 + \overline{A}B + \overline{A}B + \overline{A}BC + \overline{A}C + \overline{A}C + 0 + A\overline{B}C + 0 + 0 + 0$$
$$\quad + \overline{A}\,\overline{B}C + \overline{B}C + ABC + AC + \overline{A}BC + BC + BC + \overline{A}C + C$$
$$\overline{A}B + \overline{A}BC + \overline{A}C + A\overline{B}C + \overline{A}\,\overline{B}C + \overline{B}C + ABC + AC + BC + C$$
$$\overline{A}B(1 + C) + \overline{A}C(1 + \overline{B}) + \overline{B}C(1 + A) + AC(1 + B) + C(1 + B)$$
$$\overline{A}B + \overline{A}C + \overline{B}C + AC + C$$
$$\overline{A}B + C(\overline{A} + \overline{B} + A + 1)$$
$$\overline{A}B + C$$

In addition to SOP and POS circuit configurations, there are sum-of-sums (SOS) and product-of-products (POP) circuit configurations. In an SOS circuit, OR gates feed an OR

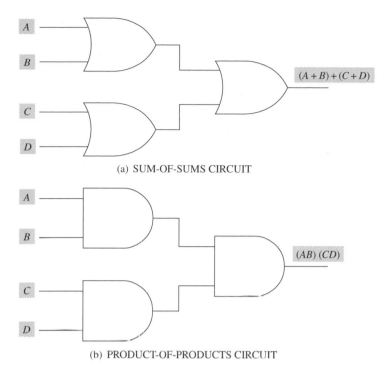

(a) SUM-OF-SUMS CIRCUIT

(b) PRODUCT-OF-PRODUCTS CIRCUIT

FIGURE 4–48   SOS and POP circuit configurations.

gate. In a POP circuit, AND gates feed an AND gate. The SOS and POP circuit configurations are shown in Fig. 4–48.

Section 4–6: Review Questions

Answers are given at the end of the chapter.

Write the SOP Boolean expression for each of the following truth tables and then simplify the expression.

A.

A	B	X
0	0	0
0	1	1
1	0	1
1	1	1

B.

A	B	C	X
0	0	0	0
0	0	1	1
0	1	0	0
0	1	1	1
1	0	0	0
1	0	1	0
1	1	0	1
1	1	1	0

C.

A	B	C	X
0	0	0	1
0	0	1	0
0	1	0	1
0	1	1	0
1	0	0	0
1	0	1	1
1	1	0	0
1	1	1	1

D.

A	B	C	X
0	0	0	0
0	0	1	1
0	1	0	0
0	1	1	1
1	0	0	1
1	0	1	0
1	1	0	1
1	1	1	0

E. Write the Boolean expressions for a *low output* from the truth table shown. Write one expression in the SOP format and another in the POS format.

A	B	C	X
0	0	0	1
0	0	1	1
0	1	0	1
0	1	1	0
1	0	0	0
1	0	1	1
1	1	0	1
1	1	1	1

F. Write a POS expression for the truth table shown.

A	B	X
0	0	1
0	1	1
1	0	0
1	1	0

SECTIONS 4-4 THROUGH 4-6: INTERNAL SUMMARY

When interpreting a Boolean expression, individual NOT signs over variables indicate low logic levels. The lack of NOT signs over variables indicates high logic levels. $\overline{X}\,\overline{Y}Z$ is read "if X is low *and* Y is low *and* Z is high…"

Expressions that contain NOT signs over more than single variables need to be simplified so they can be easily interpreted. In this example, $\overline{X \cdot \overline{Y} \cdot \overline{Z}} = \overline{X} + Y\overline{Z}$, the first expression is simplified by using DeMorgan's theorems so that it can be easily read.

The alternate logic gate symbols are shown in Fig. 4–28. The short logic for a gate's alternate logic symbol is the same as the short logic for that gate.

An alternate logic gate symbol may be derived as follows:

1. Change the sign.
2. NOT the input variables.
3. NOT the entire expression.
4. Cancel all double NOTs.

Boolean expressions written for high outputs from a truth table will be in the SOP configuration. If this set of inputs ($A = 1, B = 0, C = 0$) appears with a high output on a truth table, it is written as $A\overline{B}\,\overline{C}$.

Often the SOP expression written from a truth table requires simplification. This process will be explained in greater detail in the next section. Proof of the equality of an expression to its simplified version may be obtained from the truth table for the expression.

An SOP circuit configuration consists of AND gates feeding an OR gate. A POS circuit configuration consists of OR gates feeding an AND gate.

A POS expression can be extracted from a truth table by writing the expression for low outputs. Each variable in the set of variables must be written in its complementary form and the variables must be ORed. Then each set of ORed variables is ANDed together.

The reason SOP expressions ($A\overline{B}C + AB\overline{C}$) and POS expressions [$(\overline{A} + \overline{B})(A + C)$] are so popular is they are simple to extract from truth tables and implement. This will become more evident before the end of this chapter.

SECTIONS 4–4 THROUGH 4–6: INTERNAL SUMMARY QUESTIONS

Answers are given at the end of the chapter.

1. ABC indicates the gate will output high when A is high *and* B is high *and* C is high.
   a. True
   b. False

2. $\overline{A}B$ indicates the gate will output low when A is low *and* B is high.
   a. True
   b. False

3. $\overline{A}B\overline{C}$ indicates the gate will output high when A is low *and* B is high *and* C is low.
   a. True
   b. False

4. $\overline{AB}$ indicates the gate will output low when A is high *and* B is high.
   a. True
   b. False

5. $\overline{\overline{A}} \cdot \overline{\overline{B}}$ indicates the gate will output low when A is high *and* B is high.
   a. True
   b. False

6. $A + \overline{B}$ indicates the gate will output high when A is high *or* B is low.
   a. True
   b. False

7. Simplify the expression $A \cdot B \cdot C(\overline{\overline{A} + \overline{B}})$ so it can be interpreted for a high output.
   a. $A + B + C$
   b. $\overline{A + B + C}$
   c. ABC
   d. $\overline{ABC}$

8. Match each of the logic symbols shown in Fig. 4–49 to its equivalent alternate logic gate symbol.

9. Match the logic gate symbol shown in Fig. 4–50 with its alternate logic gate symbol.

10. What condition will cause the gate represented by the expression $A + B + C$ to output low?
    a. A *and* B *and* C must be low.
    b. A *and* B *and* C must be high.

11. Identify the Boolean expression for the truth table shown.

    a. $A + B$
    b. $\overline{A + B}$
    c. $\overline{A}B + A\overline{B}$
    d. $\overline{A}B + A\overline{B}$

A	B	X
0	0	0
0	1	1
1	0	1
1	1	0

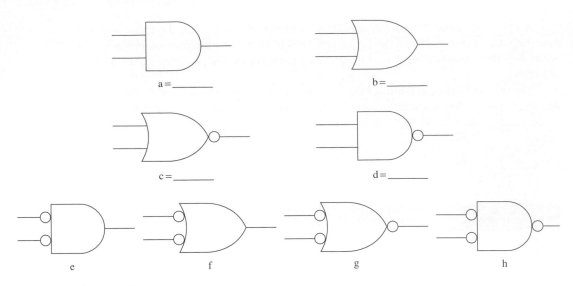

Figure 4-49

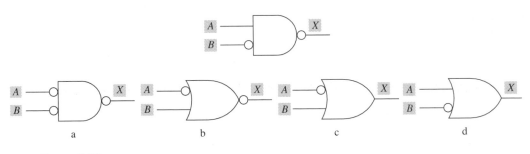

Figure 4-50

12. Identify the Boolean expression for the truth table shown.

    a. $\overline{A}\,\overline{B}\,\overline{C} + \overline{A}B\overline{C} + ABC$

    b. $ABC \cdot A\overline{B}\,\overline{C} \cdot \overline{A}\,\overline{B}\,\overline{C}$

    c. $\overline{A}\,\overline{B}\,\overline{C} + \overline{A}\,\overline{B}C + A\overline{B}\,\overline{C}$

    d. $\overline{A} + \overline{B} + \overline{C}$

A	B	C	X
0	0	0	1
0	0	1	1
0	1	0	0
0	1	1	0
1	0	0	1
1	0	1	0
1	1	0	0
1	1	1	0

13. Simplify the Boolean expression obtained in question 12.

    a. $\overline{A}\,\overline{B} + A\overline{B}\,\overline{C}$

    b. $\overline{B}\,\overline{C} + \overline{A}\,\overline{B}C$

    c. $\overline{A}\,\overline{B} + BC$

    d. $\overline{A}\,\overline{B} + \overline{B}\,\overline{C}$

14. Identify the configuration of the circuit shown in Fig. 4–51.

    a. Sum of sums

    b. Sum of products

    c. Product of sums

    d. Product of products

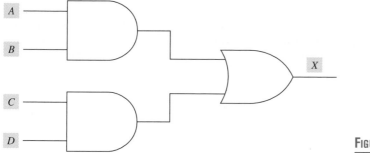

FIGURE 4-51

15. Identify the Boolean expression for the low outputs for the truth table shown.

a. $\overline{A}\,\overline{B}\,\overline{C} + A\overline{B}\,\overline{C}$
b. $(A + B + C)(\overline{A} + B + C)$
c. $(\overline{A}\,\overline{B}\,\overline{C})(A\overline{B}\,\overline{C})$

A	B	C	X
0	0	0	0
0	0	1	1
0	1	0	1
0	1	1	1
1	0	0	0
1	0	1	1
1	1	0	1
1	1	1	1

SECTION 4-7: SIMPLIFYING BOOLEAN EXPRESSIONS WITH BOOLEAN ALGEBRA

OBJECTIVE

Simplify Boolean expressions using the properties of real numbers and Boolean algebra as well as DeMorgan's theorems.

All of the information presented in the preceding sections will now be put together and used to simplify Boolean expressions. A review of Table 4–1 and a copy of Table 4–4 will be helpful in this section. All of the properties of real numbers and the properties of Boolean algebra and DeMorgan's theorems will be employed in the simplification process.

Sometimes when a Boolean expression is factored one way it will yield a different answer than if it is factored another way. Normally the two answers, though different, will yield the same results on a truth table. It is good practice to go back to the truth table to prove the equality of the simplified expression to the original expression, as was done in the last section.

Practice is the best teacher for Boolean simplification once the basics are known. Several problems will be solved and the step-by-step simplification process outlined on the following pages.

PROBLEM 4-1

1. $\overline{A}\,\overline{B}\,\overline{C} + \overline{A}\,\overline{B}C + A\overline{B}\,\overline{C}$ Original expression
2. $\overline{A}\,\overline{B}(\overline{C} + C) + A\overline{B}\,\overline{C}$ Factor out common variable $\overline{A}\,\overline{B}$.
3. $\overline{A}\,\overline{B}\,(1) + A\overline{B}\,\overline{C}$ Complement property
4. $\overline{A}\,\overline{B} + A\overline{B}\,\overline{C}$ Enable property
5. $\overline{B}(\overline{A} + A\overline{C})$ Factor out common variable $\overline{B}$.
6. $\overline{B}(\overline{A} + \overline{C})$ Theorem: $\overline{A} + A\overline{B} = \overline{A} + \overline{B}$.
7. $\overline{A}\,\overline{B} + \overline{B}\,\overline{C}$ Distribute $\overline{B}$.

COMMENTS

STEP 2: $\overline{A}\,\overline{B}$ are common variables in the first two sets of inputs in the expression that can be factored out. Note: $\overline{B}\,\overline{C}$ are also common variables in the first and third sets of inputs. If $\overline{B}\,\overline{C}$ were factored out and the remaining simplification steps followed, the results would be identical.

STEP 3: $\overline{C} + C$ (from step 2) = 1.

STEP 4: $\overline{A}\,\overline{B}1 = \overline{A}\,\overline{B}$

STEP 5: $\overline{B}$ is a common variable in the two sets of inputs in step 4. Thus, $\overline{B}$ may be factored out.

STEP 6: Apply the applicable theorem $(\overline{A} + A\overline{B} = \overline{A} + \overline{B})$. The solution obtained in this step could be considered the proper answer. However, it is best to distribute $\overline{B}$ so the expression can be read without difficulty.

STEP 7: Distributing $\overline{B}$ results in a sum-of-products expression that can be easily read.

The equality of $\overline{A}\,\overline{B} + \overline{B}\,\overline{C}$ to $\overline{A}\,\overline{B}\,\overline{C} + \overline{A}\,\overline{B}C + A\overline{B}\,\overline{C}$ can be proven with a truth table.

PROBLEM 4-2

1. $(\overline{A} + B)(B + \overline{C})$ Original expression
2. $\overline{A}B + \overline{A}\,\overline{C} + BB + B\overline{C}$ Expand the expression.
3. $\overline{A}B + \overline{A}\,\overline{C} + B + B\overline{C}$ Redundant property $(BB = B)$
4. $B(\overline{A} + 1 + \overline{C}) + \overline{A}\,\overline{C}$ Factor out common variable B.
5. $B(1) + \overline{A}\,\overline{C}$ Inhibit property $(A + 1 = 1)$
6. $B + \overline{A}\,\overline{C}$ Enable property $(A \cdot 1 = A)$

PROBLEM 4-3

1. $X + \overline{X}Y + \overline{X}YZ$ Original expression
2. $X + \overline{X}Y(1 + Z)$ Factor out common variable $\overline{X}Y$.
3. $X + \overline{X}Y(1)$ Inhibit property
4. $X + \overline{X}Y$ Enable property
5. $X + Y$ Theorem: $A + \overline{A}B = A + B$

PROBLEM 4-4

$A \cdot B \cdot C + \overline{A \cdot B \cdot C}$ Original expression

1   Complement property (*Note:* The second ABC variable is treated as one NOTed variable.)

PROBLEM 4-5

$\overline{\overline{X} \cdot \overline{Y} \cdot Z} + \overline{X \cdot Y \cdot \overline{Z}}$ Original expression

STEP 1: $\overline{\overline{X}} + \overline{\overline{Y}} + \overline{Z} + \overline{X} + \overline{Y} + \overline{\overline{Z}}$ DeMorgan's theorem (break vinculum/change signs)

STEP 2: $X + \overline{X} + Y + \overline{Y} + Z + \overline{Z}$ Double negation/commutative property

STEP 3: $1 + 1 + 1$ Complement property

STEP 4: 1 Inhibit property

PROBLEM 4-6

	$\overline{A \cdot \overline{B} \cdot C + A + \overline{C}}$	Original expression
STEP 1:	$\overline{A} + \overline{\overline{B}} + \overline{C} + \overline{A} \cdot \overline{\overline{C}}$	DeMorgan's theorem
STEP 2:	$\overline{A} + B + \overline{C} + \overline{A} \cdot C$	Double negation
STEP 3:	$\overline{A}(1 + C) + B + \overline{C}$	Factor out common variable.
STEP 4:	$\overline{A} \cdot (1) + B + \overline{C}$	Inhibit property
STEP 5:	$\overline{A} + B + \overline{C}$	Enable property

PROBLEM 4-7

	$\overline{A} \cdot \overline{B} \cdot \overline{C} + A \cdot \overline{B} \cdot C + \overline{A} \cdot B \cdot \overline{C} + A \cdot \overline{B} \cdot \overline{C}$	Original expression
STEP 1:	$(\overline{A} \cdot \overline{B} \cdot \overline{C} + \overline{A} \cdot \overline{B} \cdot C) + (\overline{A} \cdot \overline{B} \cdot \overline{C} + \overline{A} \cdot B \cdot \overline{C}) + (\overline{A} \cdot \overline{B} \cdot \overline{C} + A \cdot \overline{B} \cdot \overline{C})$	Factor in $\overline{A}\,\overline{B}\,\overline{C}$ because $A + A = A.$
STEP 2:	$\overline{A}\,\overline{B}(\overline{C} + C) + \overline{A}\,\overline{C}(\overline{B} + B) + \overline{B}\,\overline{C}(\overline{A} + A)$	Factor out common variables.
STEP 3:	$\overline{A}\,\overline{B}(1) + \overline{A}\,\overline{C}(1) + \overline{B}\,\overline{C}(1)$	Complement property
STEP 4:	$\overline{A}\,\overline{B} + \overline{A}\,\overline{C} + \overline{B}\,\overline{C}$	Enable property

Note in the original expression that $\overline{A}\,\overline{B}\,\overline{C}$ appeared only once. However, if the expression were left in its original form it could not have been simplified past $\overline{A}\,\overline{B} + \overline{A}B\overline{C} + A\overline{B}\,\overline{C}$ or $\overline{A}\,\overline{B} + \overline{C}\,(\overline{A}B + A\overline{B})$ without using some additional properties of Boolean algebra. Any set of variables in a Boolean expression can be added two (or more) times if necessary to extend simplification because $A + A = A$ and $A + A + A = A$ (redundant property). Factoring in $\overline{A}\,\overline{B}\,\overline{C}$ with two additional terms does not change the expression, yet it allows complete simplification in an easy manner. *Note:* The parentheses added in step 1 are unnecessary—they are added as an aid.

PROBLEM 4-8

	$A \cdot B \cdot \overline{C} + A \cdot \overline{B} \cdot C + \overline{A} \cdot B \cdot C + A \cdot B \cdot C$	Original expression
STEP 1:	$(A \cdot B \cdot C + A \cdot B \cdot \overline{C}) + (A \cdot B \cdot C + A \cdot \overline{B} \cdot C) + (A \cdot B \cdot C + \overline{A} \cdot B \cdot C)$	Factor in ABC.
STEP 2:	$A \cdot B(C + \overline{C}) + A \cdot C(B + \overline{B}) + B \cdot C(A + \overline{A})$	Factor out common variables.
STEP 3:	$A \cdot B(1) + A \cdot C(1) + B \cdot C(1)$	Complement property
STEP 4:	$A \cdot B + A \cdot C + B \cdot C$	Enable property

The same procedure of factoring in one of the original terms two additional times is used here. Refer to Problem 4–7 if necessary.

PROBLEM 4-9

	$(\overline{A} + B + \overline{C})(A + B + \overline{C})$	Original expression
Step 1:	$\overline{A}A + \overline{A}B + \overline{A}\,\overline{C} + BA + BB + B\overline{C} + \overline{C}A + \overline{C}B + \overline{C}\,\overline{C}$	Expand
Step 2:	$0 + \overline{A}B + \overline{A}\,\overline{C} + AB + B + B\overline{C} + A\overline{C} + B\overline{C} + \overline{C}$	Complement/ redundant properties
Step 3:	$\overline{A}B + \overline{A}\,\overline{C} + AB + B + B\overline{C} + A\overline{C} + \overline{C}$	Enable/- redundant properties
Step 4:	$B(\overline{A} + A + 1 + \overline{C}) + \overline{C}(\overline{A} + A + 1)$	Factor out common variables.
Step 5:	$B(1) + \overline{C}(1)$	Inhibit property
Step 6:	$B + \overline{C}$	Enable property

PROBLEM 4-10

	$WX\overline{Y} + \overline{W}X\overline{Y} + X\overline{Y}Z$	Original expression
Step 1:	$X\overline{Y}(W + \overline{W} + Z)$	Factor out common variables.
Step 2:	$X\overline{Y}(1)$	Complement property
Step 3:	$X\overline{Y}$	Enable property

PROBLEM 4-11

	$\overline{\overline{WXYZ}}$	Original expression
Step 1:	$\overline{\overline{W}} + \overline{X} + \overline{Y} + \overline{\overline{Z}}$	DeMorgan's theorem
Step 2:	$W + \overline{X} + \overline{Y} + Z$	Double negation property

PROBLEM 4-12

	$\overline{\overline{\overline{ABC}}}$	Original expression
Step 1:	$\overline{\overline{A}\,\overline{B}} + \overline{C}$	DeMorgan's theorem
Step 2:	$\overline{A}B + \overline{C}$	Double negation property

In Problem 4–12, the double NOTs over AB were canceled in step 2. There is no reason to break vinculums of equal length in lieu of canceling them. Breaking each individual vinculum and changing the sign after each break will result in the same expression as canceling the double NOTs. The following exercise should show the futility of using DeMorgan's theorem instead of just canceling the equal-length vinculums.

$$\overline{A + B + C}$$
$$\overline{\overline{A} \cdot \overline{B} \cdot \overline{C}}$$
$$\overline{\overline{A}} + \overline{\overline{B}} + \overline{\overline{C}}$$
$$A + B + C$$

Section 4–7: Review Questions

Answers are given at the end of the chapter.
Simplify the following Boolean expressions:

A. $ABC + AB\overline{C}$
B. $A + \overline{A}B$
C. $AB + \overline{A}\overline{B}$
D. $(A + B)(B + C)$
E. $\overline{A}BC + A\overline{B}C + A\overline{B}\,\overline{C} + \overline{A}\,\overline{B}\,\overline{C} + ABC$
F. $\overline{A}BC + A\overline{B}C$
G. $\overline{XY} + XYZ$
H. $\overline{ABCD}$

Section 4–8: Simplifying Boolean Expressions with Karnaugh Maps

Objective

Simplify Boolean expressions using Karnaugh maps.

The ability to simplify the Boolean expressions presented in the preceding section depended on a thorough knowledge of the properties of real numbers and Boolean algebra as well as DeMorgan's theorems.

A **Karnaugh map** (K-map) is an excellent tool that may be used to simplify Boolean expressions. The mapping procedure is simple, and knowledge of the Boolean rules and theorems is unnecessary. This is not to say that our study of simplification using Boolean algebra was wasted, as there are times when a K-map cannot be used.

A K-map is nothing more than a graphical representation of a truth table. It is very useful when designing a circuit from a truth table or when an SOP expression for the desired circuit is available.

A K-map for a 2-input circuit and a truth table are shown in Fig. 4–52. A 1 has been placed on the K-map directly from the truth table for each set of input variables that produces a high output. A 0 has been placed on the K-map for each set of input variables

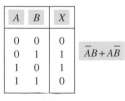

A	B	X
0	0	0
0	1	1
1	0	1
1	1	0

$\overline{A}B + A\overline{B}$

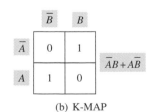

	$\overline{B}$	B
$\overline{A}$	0	1
A	1	0

$\overline{A}B + A\overline{B}$

(a) TRUTH TABLE   (b) K-MAP

Figure 4–52 2-Input K-map.

that produces a low output. Every output condition from the truth table has been plotted on the map. The first condition on the truth table is $A = 0$, $B = 0$, and $X = 0$. This is plotted on the K-map as a zero horizontally adjacent to the $\overline{A}$ row and vertically adjacent to the $\overline{B}$ column. The other input conditions are plotted similarly on the map with their corresponding outputs. The resulting Boolean expression from the truth table, $\overline{A}B + A\overline{B}$, is in the SOP format, and this particular expression cannot be simplified.

Another example of a 2-input circuit truth table is shown in Fig. 4–53(a) for further explanation of the mapping process. The expression $\overline{A}\,\overline{B} + \overline{A}B$ has been plotted on the K-map.

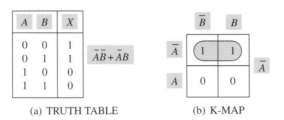

(a) TRUTH TABLE   (b) K-MAP

FIGURE 4-53   2-Input K-map with loop.

Any expression plotted on a K-map may be simplified by looping horizontally and/or vertically adjacent 1s. As shown on the map in Fig. 4–53(b), once the looping has been completed, *all complementary variables can be eliminated from the original expression.* This will result in a simplified, yet equivalent expression. In Fig. 4–53(b), the $\overline{A}$ that is horizontally adjacent to the loop stays in the output expression. Since a B and $\overline{B}$ appear vertically adjacent to the horizontal loop, they may be eliminated. This is the same as algebraically simplifying:

$$\overline{A}\,\overline{B} + \overline{A}B$$
$$\overline{A}(\overline{B} + B)$$
$$\overline{A}(1)$$
$$\overline{A}$$

Looping two 1s together as was done in the preceding problem is called **looping a pair.** Keep in mind that looping can only be accomplished with horizontally and/or vertically adjacent 1s. Once a loop is formed, all variables that are complementary (A and $\overline{A}$ or B and $\overline{B}$) and adjacent to the loop are canceled.

A K-map for a 3-input circuit is shown in Fig. 4–54. The correlation of the squares on the K-map to the truth table is shown on the map. The layout of the map is very important. If it is not properly laid out, it will not produce the correct simplified expression. Note on the map that each adjacent square differs by only one variable. The top left square ($\overline{A}\,\overline{B}\,\overline{C}$) differs from the next square down ($\overline{A}B\overline{C}$) by only the B variable. This is true of all horizontally and vertically adjacent squares on the K-map.

To further expand on this subject, consider the layout of the 4-input map shown in Fig. 4–55. If you pick a square on the map and move one square away either horizontally or vertically, the adjacent square differs from the selected square by only one variable. If $AB\overline{C}D$ (square 10) is selected, the horizontally adjacent squares are $ABCD$ (square 9) and

	$\overline{C}$	C
$\overline{A}\,\overline{B}$	$\overline{A}\,\overline{B}\,\overline{C}$	$\overline{A}\,\overline{B}C$
$\overline{A}B$	$\overline{A}B\overline{C}$	$\overline{A}BC$
AB	$AB\overline{C}$	ABC
$A\overline{B}$	$A\overline{B}\,\overline{C}$	$A\overline{B}C$

FIGURE 4-54   Layout of a 3-input Karnaugh map.

	$\overline{C}\overline{D}$	$\overline{C}D$	CD	$C\overline{D}$
$\overline{A}\overline{B}$	0	1	2	3
$\overline{A}B$	7	6	5	4
AB	8	9	10	11
$A\overline{B}$	15	14	13	12

BINARY				DEC	GRAY			
A	B	C	D		A	B	C	D
0	0	0	0	0	0	0	0	0
0	0	0	1	1	0	0	0	1
0	0	1	0	2	0	0	1	1
0	0	1	1	3	0	0	1	0
0	1	0	0	4	0	1	1	0
0	1	0	1	5	0	1	1	1
0	1	1	0	6	0	1	0	1
0	1	1	1	7	0	1	0	0
1	0	0	0	8	1	1	0	0
1	0	0	1	9	1	1	0	1
1	0	1	0	10	1	1	1	1
1	0	1	1	11	1	1	1	0
1	1	0	0	12	1	0	1	0
1	1	0	1	13	1	0	1	1
1	1	1	0	14	1	0	0	1
1	1	1	1	15	1	0	0	0

FIGURE 4–55 Gray code layout of 4-input K-map.

$ABCD$ (square 11). Each adjacent square contains only one variable that is different from the selected square. The same is true if the vertically adjacent square $\overline{A}BCD$ (square 5) or $A\overline{B}CD$ (square 13) is selected. This is true because the map is laid out in **gray code format** as shown in this figure.

In fact, the lower left square $A\overline{B}\,\overline{C}\,\overline{D}$ is considered adjacent to the upper left square $\overline{A}\,\overline{B}\,\overline{C}\,\overline{D}$ because the two expressions differ by only one variable. Therefore, the bottom row of the map can be rolled up to be adjacent to the top row of the map for looping purposes. It is important when using the K-map to realize these rows are adjacent.

Furthermore, the left column is adjacent to the right column because a set of variables in the left column differs by only one variable from its adjacent set of variables on the opposite side of the map in the right column. Thus, when determining adjacency for looping purposes, the map may also be rolled to make the right and left columns adjacent.

The adjacencies on the K-map are depicted in Fig. 4–56. Figure 4–56(a) proves the four outer-corner blocks of the map are adjacent because any corner differs from its horizontally or vertically adjacent corner by only one variable. The adjacencies are also proven in Fig. 4–56(b) for the top and bottom rows of the map. They are also proven for the left and right columns in Fig. 4–56(c).

Although layout of the K-map is very important, its layout can vary. The only thing to keep in mind is *the gray code format must be used*. Several variations of layout of the map are shown in Fig. 4–57. Two pairs of ones have been looped on this map to form a quad. More on this subject follows shortly. The point here is every map in the figure is laid out differently, but the gray code format has been followed. On every map, the looping has yielded the same answer ($B\overline{C}$). Another point that can be made here is the 0s have not been shown on the map. Although we will show them in the remainder of the maps in this chapter, it is not necessary when looping 1s. The remainder of the maps used for

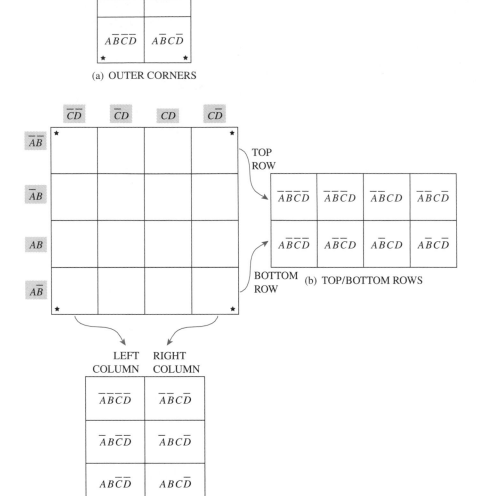

(a) OUTER CORNERS

(b) TOP/BOTTOM ROWS

(c) LEFT/RIGHT COLUMNS

FIGURE 4–56 **K-map adjacencies.**

sum-of-products expressions in this chapter will be laid out as shown originally in Fig. 4–54 and 4–55. Several examples of looping are illustrated in Fig. 4–58 on page 156.

Looping the pair of 1s in Fig. 4–58(a) results in the simplified expression $\overline{A}\,\overline{C}$. Since the loop is horizontally adjacent to $\overline{A}\,\overline{B}$ and $\overline{A}B$, the complementary Bs are eliminated. The loop is vertically adjacent to the $\overline{C}$ column: thus this variable also remains in the output expression.

There are two pairs of 1s on the K-map in Fig. 4–58(b). The upper left pair of 1s yielded $\overline{A}\,\overline{C}$ as explained for the preceding K-map. The right pair of ones is horizontally adjacent to the $\overline{A}B$ and the AB rows. Therefore, the complementary As are eliminated. Similar to the outputs extracted from a truth table, the resultant expressions from each pair of looped 1s are ORed together to produce $\overline{A}\,\overline{C} + BC$. Since all of the 1s on the K-map have been paired, do not pair the horizontally adjacent $\overline{A}B\overline{C}$ and $AB\overline{C}$ 1s because this would put an extraneous set of variables $(\overline{A}B)$ in the output expression, which is unnecessary and merely complicates the output expression.

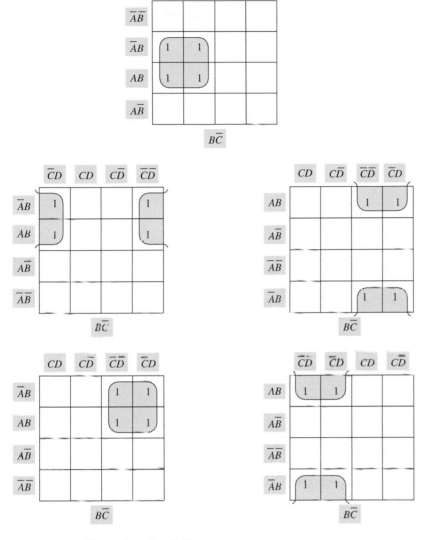

FIGURE 4-57   K-map layout variations.

The K-map shown in Fig. 4–58(c) shows two 1s. Remember, since the expression $\overline{A}\,\overline{B}\,\overline{C}$ only differs by one variable from $A\,\overline{B}\,\overline{C}$, the two 1s are considered to be vertically adjacent.

The map may always be rolled to make the top and bottom adjacent. Since $\overline{A}\,\overline{B}$ appears horizontal to one of the 1s in the pair and $A\overline{B}$ appears horizontal to the other 1, the As are eliminated from the output expression, leaving $\overline{B}\,\overline{C}$.

Three high output conditions are shown on the K-map in Fig. 4–58(d). This K-map illustrates a situation that did not exist on any of the previous maps. First, if the horizontally adjacent 1s are paired, the resultant expression is $\overline{A}B$, because the Cs appear in complementary form and are eliminated. If the remaining 1 (ABC) were not looped into the existing pair as shown, the output expression would be $\overline{A}B + ABC$. As you already know, this expression can be simplified by factoring out the B. If the remaining 1 on the K-map is looped into the existing pair, one variable is eliminated and the output expression is $\overline{A}B + BC$. It is necessary to loop all horizontally and vertically adjacent 1s in order to obtain the simplest output expression. However, as explained for the K-map in Fig. 4–58(b), do not double-loop 1s that have already been looped.

The K-map shown in Fig. 4–58(e) shows one pair of horizontally adjacent 1s, which yield AB after they are paired. The remaining 1 cannot be looped because there

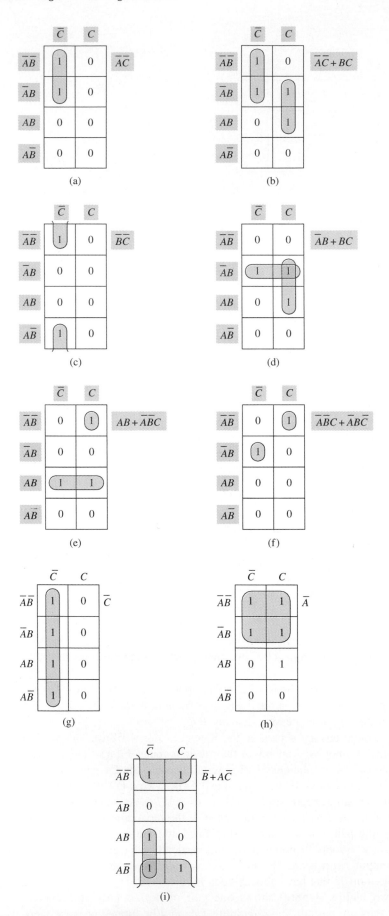

FIGURE 4-58 **Looping Karnaugh maps for 3-input circuits.**

are no horizontally or vertically adjacent 1s. This one yields $\overline{A}\,\overline{B}C$ because it can't be looped, and the output expression is $AB + \overline{A}\,\overline{B}\,C$. This expression cannot be further simplified.

The expression shown for the K-map in Fig. 4–58(f) cannot be simplified. No horizontal or vertical loops can be drawn on the map. Diagonally adjacent 1s cannot be paired.

The K-map in Fig. 4–58(g) presents another option that can be taken when working with these maps. Looping of four horizontally and/or vertically adjacent 1s is acceptable. Looping a group of four adjacent 1s is called looping a *quad*. Looping these four vertically adjacent 1s puts the loop adjacent to two As, two $\overline{A}$s, two Bs, and two $\overline{B}$s. Therefore, all of the As and Bs are eliminated, leaving an output expression of $\overline{C}$.

Looping a *pair* of 1s on the 3-input K-map always eliminated *one* variable from the output expression. Looping a *quad* on a 3-input K-map will always eliminate *two* variables from the output expression.

The K-map in Fig. 4–58(h) contains another quad that can be looped. All of the 1s on the map are either horizontally or vertically adjacent. Since the quad loop is horizontally adjacent to $\overline{A}\,\overline{B}$ and $\overline{A}B$, the Bs are eliminated. Also, the loop is vertically adjacent to $\overline{C}$ and C, so this variable is eliminated. This leaves an output expression of $\overline{A}$.

A quad also exists on the K-map in Fig. 4–58(i). Since the $\overline{A}\,\overline{B}$ row is adjacent to the $\overline{A}B$ row, the complementary As are eliminated by the quad loop. Also, the $\overline{C}$ and C columns are vertically adjacent to the quad loop, so the Cs are eliminated. The output expression for the quad loop is $\overline{B}$. The remaining 1 at $AB\overline{C}$ can be paired into the quad loop to eliminate the B variable as shown. The output expression for the map is $\overline{B} + A\overline{C}$.

A blank K-map for a 4-input circuit is shown in Fig. 4–59. Don't forget that the bottom and top rows are considered to be adjacent. Also, the left and right columns are considered to be adjacent. This rolling principle of adjacency will be shown in some of the following mapping examples.

The K-map in Fig. 4–60(a) shows one pair of 1s that can be looped. The pair is horizontally adjacent to the AB row, so these two variables cannot be eliminated. The pair is vertically adjacent to the $\overline{C}D$ column and the CD column. Since the Cs appear in complementary form, they are eliminated. The resultant expression is ABD.

A quad exists in Fig. 4–60(b). The quad eliminates all of the Cs and Ds and leaves the output expression $\overline{A}B$. Two quads are present in Fig. 4–60(c). The top quad results in $\overline{B}D$, and the bottom quad results in AD. The final simplified expression is $\overline{B}D + AD$. If a decision had been made to loop the three pairs on this K-map, the resultant expression would be $\overline{A}\,\overline{B}D + ABD + A\overline{B}D$. This expression can be further simplified, and this indicates improper looping. *Always loop as large a group of 1s as possible.*

So far we have looped pairs and quads. The K-map in Fig. 4–60(d) provides an opportunity to loop eight 1s **(octet).** The resultant expression is B, and three of the variables have been eliminated.

Looping on 3- and 4-input K-maps must be done in *pairs, quads,* or *octets.* Always loop the largest group(s) of 1s first because this eliminates the most variables. Loop single 1s into horizontally or vertically adjacent loops when possible.

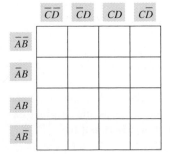

FIGURE 4-59   K-map for 4-input circuits.

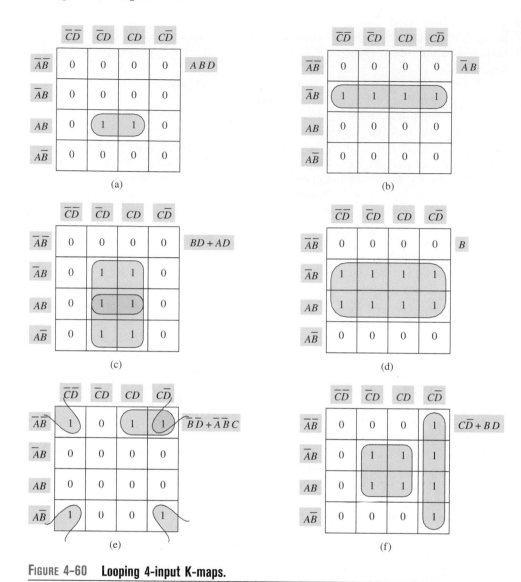

FIGURE 4-60   Looping 4-input K-maps.

In Fig. 4–60(e), the four 1s on the outer corners can be looped to form a quad because the sides and top and bottom of the K-map are considered to be adjacent. Once the quad is looped, the remaining $\overline{A}\,\overline{B}CD = 1$ can be paired into the quad for further simplification. Loop the two quads in Fig. 4–60(f) and ensure you obtain the resultant expression $(C\overline{D} + BD)$.

One final exercise in looping on 4-input K-maps is shown in Fig. 4–61. This exercise includes extracting a Boolean expression from the truth table in Fig. 4–61(a), simplifying the expression by using the K-map in Fig. 4–61(b) or (c), and implementing the circuit as shown in Fig. 4–61(d) or (e).

The Boolean expression as extracted from the truth table has been plotted on both K-maps. The K-map shown in Fig. 4–61(b) has three vertical pairs looped, which yield $\overline{A}\,\overline{CD} + BCD + ACD$ (Expression 1). The K-map in Fig. 4–61(c) has been looped in a different but acceptable manner. This K-map yields $\overline{A}\,\overline{CD} + \overline{A}BD + ACD$ (Expression 2). Note that the middle sets of variables in the two simplified expressions are different.

To prove the two simplified expressions are equivalent, the truth table has been checked for Expressions 1 and 2 to ensure the results are the same as the original expression. Implementation of both expressions is shown in Fig. 4–61(d) and (e).

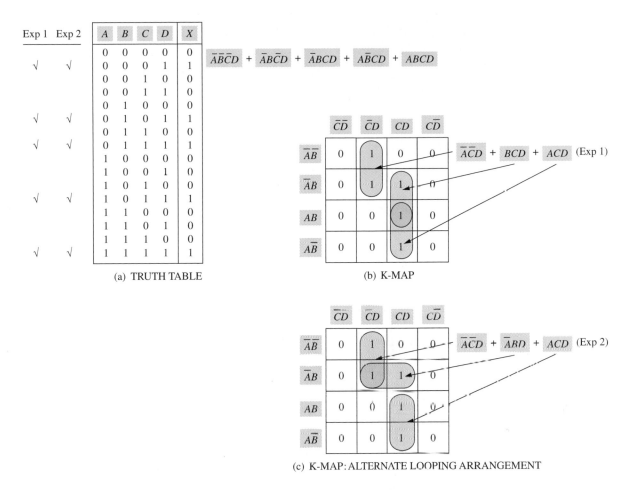

(a) TRUTH TABLE (b) K-MAP

(c) K-MAP: ALTERNATE LOOPING ARRANGEMENT

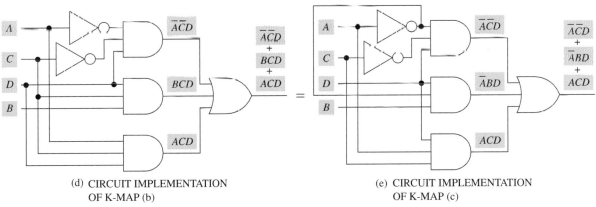

(d) CIRCUIT IMPLEMENTATION
OF K-MAP (b)

(e) CIRCUIT IMPLEMENTATION
OF K-MAP (c)

FIGURE 4–61 Final looping exercise for 4-input K-map.

A 5-input K-map and truth table are shown in Fig. 4–62. Typically, problems involving more than four inputs are solved with Karnaugh-mapping computer programs. Nonetheless, the map is shown for your benefit.

The only exceptions to placing 1s and 0s and looping on a 5-input K-map are

1. All sets of inputs that contain an E are plotted on the upper plane, and all sets of inputs that contain a $\overline{E}$ are plotted on the lower plane.
2. Looping is accomplished between the two planes when possible to eliminate the Es.

A	B	C	D	E	X
0	0	0	0	0	0
0	0	0	0	1	0
0	0	0	1	0	0
0	0	0	1	1	0
0	0	1	0	0	1
0	0	1	0	1	1
0	0	1	1	0	0
0	0	1	1	1	0
0	1	0	0	0	1
0	1	0	0	1	1
0	1	0	1	0	1
0	1	0	1	1	1
0	1	1	0	0	0
0	1	1	0	1	0
0	1	1	1	0	0
0	1	1	1	1	0
1	0	0	0	0	0
1	0	0	0	1	0
1	0	0	1	0	0
1	0	0	1	1	0
1	0	1	0	0	0
1	0	1	0	1	0
1	0	1	1	0	0
1	0	1	1	1	0
1	1	0	0	0	1
1	1	0	0	1	1
1	1	0	1	0	1
1	1	0	1	1	1
1	1	1	0	0	0
1	1	1	0	1	0
1	1	1	1	0	0
1	1	1	1	1	0

(a) TRUTH TABLE

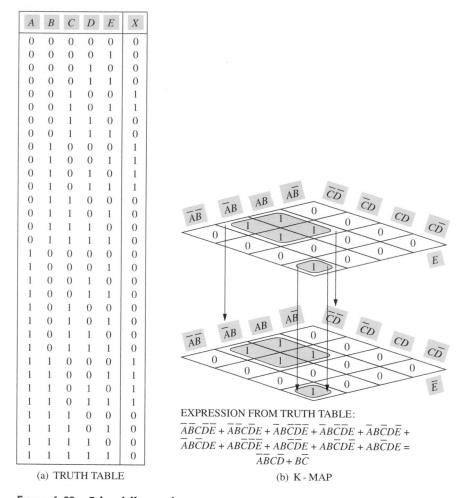

EXPRESSION FROM TRUTH TABLE:

$$\overline{A}\,\overline{B}C\overline{D}\,\overline{E} + \overline{A}\,\overline{B}C\overline{D}E + \overline{A}B\overline{C}\,\overline{D}\,\overline{E} + \overline{A}B\overline{C}\,\overline{D}E + \overline{A}B\overline{C}D\overline{E} +$$
$$\overline{A}B\overline{C}DE + AB\overline{C}\,\overline{D}\,\overline{E} + AB\overline{C}\,\overline{D}E + AB\overline{C}D\overline{E} + AB\overline{C}DE =$$
$$\overline{A}\,\overline{B}C\overline{D} + B\overline{C}$$

(b) K - MAP

FIGURE 4-62 5-Input Karnaugh map.

Don't Care Inputs

There are conditions in digital systems that are immaterial to a certain task. For example, the RAM addresses in a personal computer might be $00000_{(16)}$ to $9FFFF_{(16)}$. The first digit (MSD) of these hex numbers represents the page address of a 64 KB page of memory. Therefore, page addresses of $0000_{(2)}$ (page 0) through $1001_{(2)}$ (page 9) are valid addresses for RAM. These binary numbers represent the first hex digit in each hex address. Page addresses of $1010_{(2)}$ (page A) through $1111_{(2)}$ (page F) would not affect the RAM circuitry. This indicates that the page A through page F addresses would be considered *don't care* inputs to this circuitry.

There are numerous other examples of don't care inputs. If a code or input cannot exist, such as 1100 BCD or 0010 Excess-3 code, certainly that number is a don't care condition for a circuit.

Don't care conditions are indicated on truth tables and K-maps with an X. Since an X signifies a don't care input, it can be considered on a K-map as a low or high. The object is to make the X fit the map for the most convenient looping.

The K-map in Fig. 4–63(a) shows an X at the $\overline{A}\,\overline{B}C$ location. If the X is made a 1, a quad exists, and the resultant expression is $\overline{A}$. If the X were made a 0, the expression would have been $\overline{A}\,\overline{B} + \overline{A}\,\overline{C}$. It is evident which X-level produced the simplest expression.

The K-map in Fig. 4–63(b) contains one X. Since this X is not adjacent to any ones for looping purposes, it is best to make it a 0.

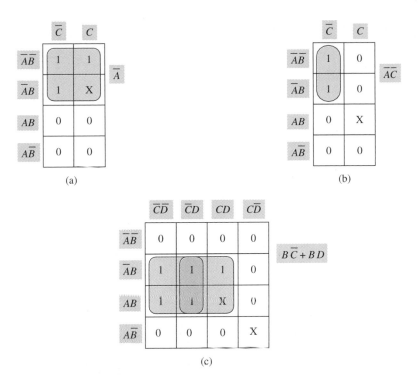

Figure 4-63   Plotting don't care conditions on K-maps.

Figure 4–63(c) contains two don't care conditions. The quad of 1s yields $B\overline{C}$. If the X in the CD column is made a 1, another quad can be looped, which yields BD. The X in the lower right-hand corner is not adjacent to anything to loop, so it should be considered a 0.

Information regarding extraction of a POS expression from a truth table was presented in Section 4–6. Since a K-map is nothing more than a graphical representation of the truth table, there is a procedure that allows extraction of a POS simplified expression from a K-map. Knowledge of this procedure is optional. It will be discussed in Design Problem 4–3 in Section 4–10.

Section 4–8: Review Questions

Answers are given at the end of the chapter.

A. Lay out a 3-input K-map.
B. Lay out a 4-input K-map.
C. Simplify the following SOP expressions using a K-map:

(1) $A\overline{B} + AB$

(2) $A\overline{B}C + ABC$

(3) $\overline{A}\,\overline{B}\,\overline{C} + \overline{A}B\overline{C} + \overline{A}BC + A\overline{B}\,\overline{C} + ABC$

(4) $\overline{A}\,\overline{B}\,\overline{C} + \overline{A}\,BC + \overline{A}BC + A\,\overline{B}\,\overline{C} + AB\overline{C} + \overline{A}B\overline{C}$

(5) $\overline{A}\,\overline{B}\,\overline{C} + \overline{A}\,BC + \overline{A}B\overline{C} + \overline{A}BC + A\overline{B}\,\overline{C} + ABC$

(6) $\overline{A}\,\overline{B}\,\overline{C}\,\overline{D} + \overline{A}\,BCD + AB\overline{C}\,\overline{D} + ABCD$

(7) $\overline{A}B\overline{C}D + AB\overline{C}D + A\overline{B}\,\overline{C}\,\overline{D} + A\overline{B}\,CD$

(8) $\overline{A}\,BC\overline{D} + \overline{A}B\overline{C}\,\overline{D} + \overline{A}BCD + AB\overline{C}D + A\overline{B}\,CD + A\overline{B}CD + A\overline{B}C\overline{D}$

D. Plot the conditions from the truth table on a K-map and simplify.

A	B	C	D	X
0	0	0	0	1
0	0	0	1	0
0	0	1	0	0
0	0	1	1	0
0	1	0	0	0
0	1	0	1	1
0	1	1	0	0
0	1	1	1	1
1	0	0	0	X
1	0	0	1	0
1	0	1	0	0
1	0	1	1	0
1	1	0	0	0
1	1	0	1	1
1	1	1	0	X
1	1	1	1	X

SECTIONS 4-7 AND 4-8: INTERNAL SUMMARY

It is good practice to take a simplified Boolean expression to a truth table to prove equality to the original expression.

Cancel all double NOTs and vinculums of equal length as soon as they appear in the simplification process. It is best to break the longest vinculum first when using DeMorgan's theorems.

Sets of variables in a SOP expression can be added into the expression numerous times for simplification purposes without affecting the expression.

Karnaugh maps (K-maps) provide a graphical representation of a truth table. Two-, three-, and four-input maps are common.

Looping on K-maps is done in **pairs, quads,** and **octets.** All horizontally and vertically adjacent 1s may be looped as long as they form a pair, quad, or octet. Complementary variables that appear adjacent to a loop may be eliminated.

The bottom row of a K-map is considered adjacent to the top row. Likewise, the left side is considered adjacent to the right side for looping purposes.

Always loop the largest group of 1s first.

Don't care conditions appear on a truth table and K-map as X's. An X can be considered a high or low level. The plotted X value should enhance the map for the most convenient looping that will result in the simplest expression.

SECTIONS 4-7 AND 4-8: INTERNAL SUMMARY QUESTIONS

Answers are given at the end of the chapter.

Simplify the following expressions using Boolean algebra:

1. $\overline{AB} + C$

2. $(A + B)(A + C)$

3. $AB\overline{C} + \overline{\overline{A} \, \overline{B}C}$

4. $A\overline{B} \, \overline{C} + AB\overline{C} + \overline{A} \, B\overline{C} + A\overline{B} \, C$

5. $(\overline{X} + Y + \overline{Z})(X + Y + \overline{Z})$

6. $\overline{X}Y\overline{Z} + XY\overline{Z} + \overline{X}YZ + XYZ$

7. $\overline{A\overline{B}C + \overline{A + \overline{C}}}$

8. $(X\overline{\overline{Y}}\overline{Z} + YZ)(X\overline{Y})$

9. $\overline{(A + B)(\overline{A} + C)}$

10. $\overline{\overline{A}} + B + C + (A\overline{B}\,\overline{C}D) + \overline{\overline{A} + \overline{E}}$

Simplify the following expressions using K-maps.

11. $\overline{AB}\overline{C} + AB\overline{C} + ABC$

12. $\overline{A}\,\overline{B}\,\overline{C} + \overline{A}B\overline{C} + A\overline{B}\,\overline{C}$

13. $\overline{A}\,\overline{B}\,\overline{C} + \overline{A}\,\overline{B}C + \overline{A}BC + A\overline{B}\,\overline{C} + A\overline{B}C$

14. $\overline{A}\,\overline{B}\,\overline{C}D + \overline{A}\,\overline{B}CD + \overline{A}\,BC\overline{D}$

15. $\overline{A}\,\overline{B}\,\overline{C}\,\overline{D} + \overline{A}\,\overline{B}\,\overline{C}D + \overline{A}\,\overline{B}\,CD + \overline{A}\,\overline{B}\,C\overline{D} + \overline{A}B\overline{C}\,\overline{D} + \overline{A}\,B\overline{C}\,D + \overline{A}\,BCD + \overline{A}\,BC\overline{D}$

Plot the data from the following truth tables on K-maps and simplify the expressions:

16.

A	B	C	X
0	0	0	1
0	0	1	0
0	1	0	1
0	1	1	0
1	0	0	X
1	0	1	X
1	1	0	1
1	1	1	0

17.

A	B	C	D	X
0	0	0	0	1
0	0	0	1	X
0	0	1	0	0
0	0	1	1	0
0	1	0	0	1
0	1	0	1	1
0	1	1	0	0
0	1	1	1	0
1	0	0	0	0
1	0	0	1	0
1	0	1	0	0
1	0	1	1	0
1	1	0	0	1
1	1	0	1	1
1	1	1	0	0
1	1	1	1	0

18.

A	B	C	D	X
0	0	0	0	1
0	0	0	1	1
0	0	1	0	1
0	0	1	1	0
0	1	0	0	0
0	1	0	1	0
0	1	1	0	0
0	1	1	1	0
1	0	0	0	X
1	0	0	1	0
1	0	1	0	1
1	0	1	1	0
1	1	0	0	0
1	1	0	1	0
1	1	1	0	0
1	1	1	1	0

SECTION 4-9: MULTIPLE FUNCTIONS OF NAND AND NOR GATES

OBJECTIVES

1. Implement logic functions with NAND gates or NOR gates.
2. Implement SOP and POS circuit configurations using only NAND gates or NOR gates.

Implementing a logic circuit with the fewest gates has been emphasized throughout this chapter. That is the main reason Boolean expressions were simplified. Now it is time to shift gears and emphasize implementation of these circuits using the fewest ICs.

If you compare complexity of the schematic diagrams of NAND (7400), NOR (7402), AND (7408), and OR (7432) 2-input gates, you will discover the NAND gate is the simplest. Since NAND gates require fewer components than other logic gates, they are cheaper.

In fact, NAND gates are so popular they come in 8-, 12-, and 13-input versions in addition to the more common 2-, 3-, and 4-input versions.

Both NAND and NOR gates are **multifunctional.** They can be used to implement any logic function. Although NAND implementation is more popular, both types will be discussed.

NAND Implementation of Logic Functions

Implementation of the NOT function using a NAND gate and NOR gate is shown in Fig. 4–64. The information provided in Chapter 3 proved these two gates are inverting gates.

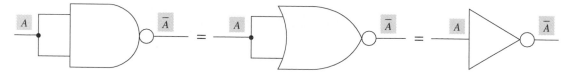

Figure 4-64   NOT function implemented with NAND/NOR gates.

If the inputs to the NAND gate are tied together as shown, both inputs will be high or both inputs will be low at any given time. The NAND logic, "any 0 in = 1 out and all 1s in = 0 out" indicates the inversion capability of this gate.

Likewise, the logic of the NOR gate proves it will invert if its inputs are tied together. Regardless of the number of inputs a NAND or NOR gate has, it functions as a NOT gate as long as *all* of its inputs are tied together.

Figure 4–65 illustrates implementation of the AND function using NAND gates. Since the second NAND gate is performing the NOT function, it cancels the negation at the output of the first NAND gate by double NOTing the output of that gate. The Boolean expression for the circuit is that of the AND function once the double NOTs are canceled.

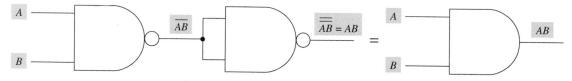

Figure 4-65   AND function implemented with NAND gates.

The logic diagram in Fig. 4–66 shows implementation of the OR function employing only NAND gates. This one is easy to remember if the alternate logic symbol for the

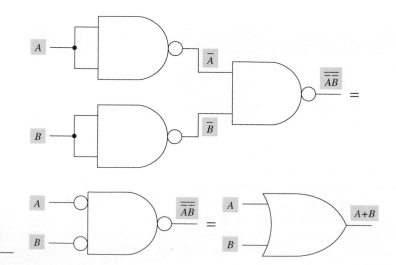

Figure 4-66   OR function implemented with NAND gates.

OR gate is envisioned. The alternate symbol is an inverted-input NAND gate as shown. The Boolean expression from the NAND circuit equates to the OR function.

Figure 4–67 substantiates that the NOR function can be implemented with NAND gates. An inverter has been added to the output of the NAND-implemented OR function of Fig. 4–66 to produce the NOR function.

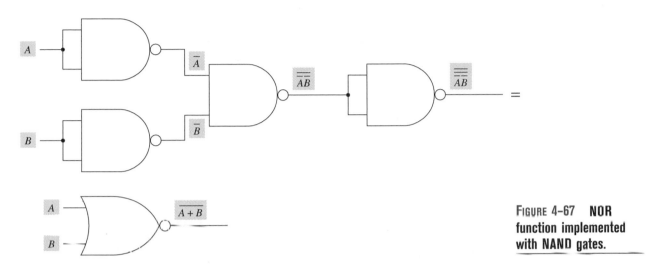

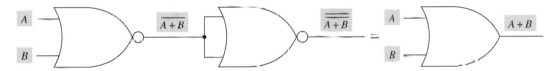

FIGURE 4-67   NOR function implemented with NAND gates.

NOR Implementation of Logic Functions

Implementation of the OR function using NOR gates is depicted in Fig. 4–68. The second NOR gate functions as a NOT gate to double NOT the signal and produce the OR function.

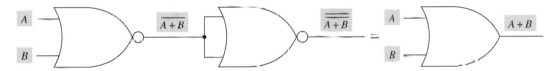

FIGURE 4-68   OR function implemented with NOR gates.

Figure 4–69 illustrates how the AND function can be implemented with NOR gates. If the alternate logic gate symbol for the AND gate is used, implementation should be easy to remember. If DeMorgan's theorem is applied to the NOR-implemented circuit's output expression, the equality to an AND gate is proven.

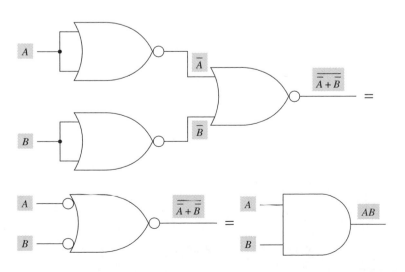

FIGURE 4-69   AND function implemented with NOR gates.

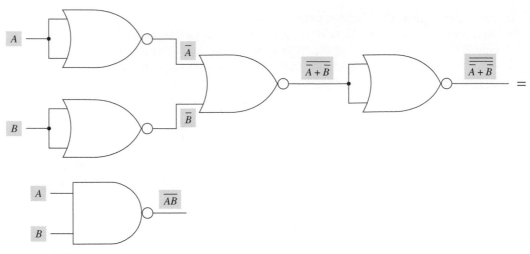

FIGURE 4–70   **NAND function implemented with NOR gates.**

The NAND function can be implemented with NOR gates as shown in Fig. 4–70. Compare implementation of the AND function (Fig. 4–69) to that of the NAND function (Fig. 4–70). A 3-input OR function implemented with NAND gates is shown in Fig. 4–71(a). A 3-input AND function implemented with NOR gates is shown in Fig. 4–71(b).

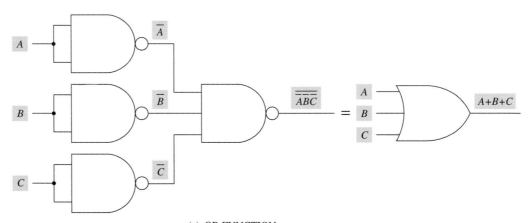

(a) OR FUNCTION
IMPLEMENTED WITH NAND GATES

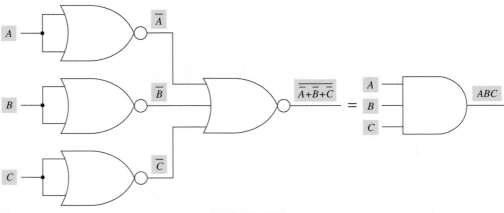

FIGURE 4–71   **3-input logic functions.**

(b) AND FUNCTION
IMPLEMENTED WITH NOR GATES

It might appear from some of the figures that implementation using NAND or NOR gates requires more gates. This may sometimes be the case, but we are concerned with the number of ICs required to implement a circuit more than we are the number of gates. It is time to put all of this information together and see if we can accomplish this circuit minimization.

Sum-of-Product and Product-of-Sum Circuit Importance

The SOP and POS circuit configurations have been related to truth tables and K-maps and have been stressed throughout much of this chapter. This is not to imply that sum-of-sums (SOS) and product-of-products (POP) circuits in addition to mixes of these circuits are not important. Everything has its place.

The emphasis on SOP and POS comes from the ease with which expressions can be extracted from truth tables and simplified, and the simplicity with which these configurations can be implemented with NAND and NOR gates.

By this point in your study you should be very familiar with all of the alternate logic gate symbols. The NAND and NOR gate standard and alternate symbols are shown in Fig. 4–72 for your review.

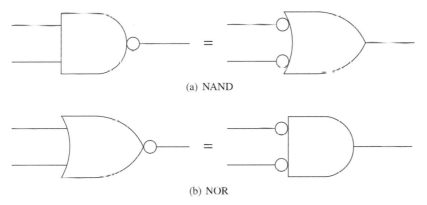

(a) NAND

(b) NOR

FIGURE 4–72   **NAND/NOR logic symbols.**

SOP Circuit Implementation

Let's implement the SOP circuit shown in Fig. 4–73(a) with NAND gates. Each gate function in the SOP circuit is implemented with NAND gates in Fig. 4–73(b) as previously described. However, the NAND-implemented circuit contains two pairs of back-to-back inverters that are unnecessary. Since each pair of back-to-back inverters double NOTs the signal, they can be eliminated as shown in Fig. 4–73(c).

It is this straightforward gate-for-gate implementation that makes constructing SOP circuits with NAND gates so popular. Instead of using two ICs, AND and OR as shown in Fig. 4–73(a), only one IC is required. The Boolean expressions annotated on each drawing indicate the equivalency of implementing with NAND gates.

This implementation is easier to understand if you look at the AND-OR SOP circuit shown in Fig. 4–74. In this figure the alternate OR gate symbol is used. If you envision moving the OR gate input bubbles to the outputs of the AND gates, NAND implementation becomes self-evident.

Take a couple of minutes and implement the circuit shown in Fig. 4–73(a) with NOR gates. The project would require eight NOR gates. This would require two ICs and is not a good idea.

Typically the NAND-implemented SOP circuit is drawn as shown in Fig. 4–75. This is in keeping with the bubble-to-bubble principle used for ease of circuit analysis. In fact, the two bubbles on each input gate's output line effectively cancel. This leaves an AND-OR SOP circuit visually even though it is implemented with NAND gates.

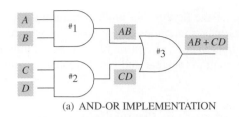

(a) AND-OR IMPLEMENTATION

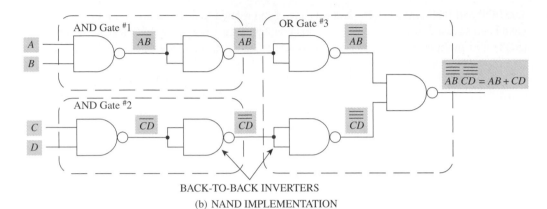

BACK-TO-BACK INVERTERS

(b) NAND IMPLEMENTATION

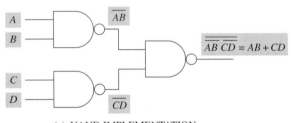

(c) NAND IMPLEMENTATION
WITH REDUNDANT INVERTERS ELIMINATED

FIGURE 4–73   SOP circuit.

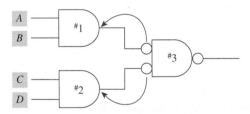

FIGURE 4–74   SOP circuit with AND/alternate OR gate symbols.

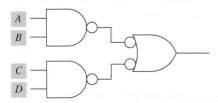

FIGURE 4–75   SOP circuit with NAND/alternate NAND gate symbols.

POS Circuit Implementation

The counterpart of the NAND-implemented SOP circuit is the NOR-implemented POS circuit.

A POS circuit is shown in Fig. 4–76(a). This circuit has been implemented using only NOR gates in Fig. 4–76(b). Again, note that the NOR-implemented circuit contains back-to-back inverters that double NOT the signal. Since these inverters are needless, they are eliminated from the POS circuit in Fig. 4–76(c). The Boolean expressions on each drawing show the equivalency of the three circuits in Fig. 4–76.

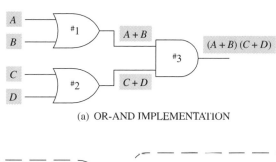

(a) OR-AND IMPLEMENTATION

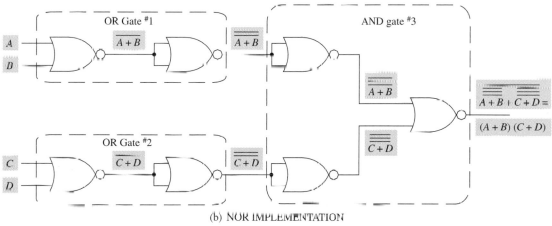

(b) NOR IMPLEMENTATION

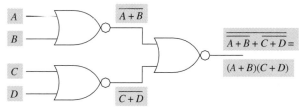

(c) NOR IMPLEMENTATION WITH REDUNDANT INVERTERS ELIMINATED

FIGURE 4–76   POS circuit.

The simplicity of gate-for-gate implementation of the POS circuit configuration with NOR gates creates its popularity.

Figure 4–77 illustrates a concept previously presented in the NAND-implemented SOP circuit. If you move the input bubbles of the alternate AND gate symbol to the outputs of the OR gates, NOR implementation of this circuit becomes obvious.

The NOR-implemented POS circuit will be drawn as shown in Fig. 4–78 for ease of analysis.

Since the SOP and POS expressions can be derived directly from a truth table and simplified with a K-map, these two circuit configurations are extremely popular. Adding

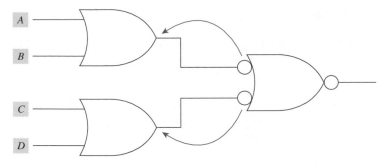

FIGURE 4–77   **POS circuit with OR/alternate AND gate symbols.**

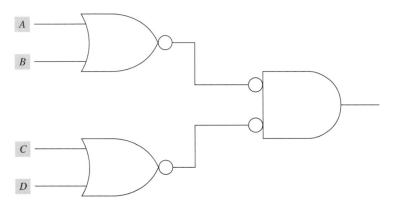

FIGURE 4–78   **POS circuit with NOR/alternate NOR gate symbols.**

to this popularity is the fact that either configuration can be simply implemented with appropriate NAND or NOR gates. This results in fewer ICs and decreased circuit cost.

Section 4–9: Review Questions

Answers are given at the end of the chapter.

A. Implement each of the following logic functions using only NAND gates.
   (1) NOT
   (2) AND
   (3) OR
   (4) NOR

B. Implement each of the following logic functions using only NOR gates.
   (1) NOT
   (2) AND
   (3) OR
   (4) NAND

C. Implement the circuit shown in Fig. 4–79 with all NAND gates.

D. Implement the circuit shown in Fig. 4–80 with all NOR gates.

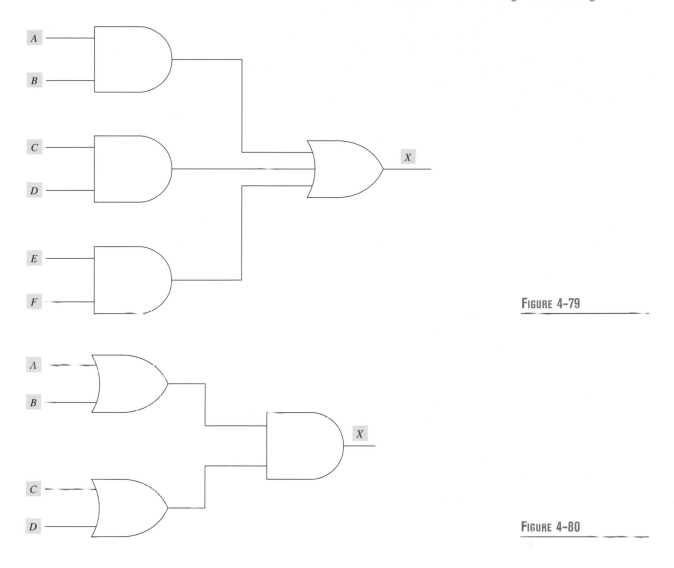

FIGURE 4–79

FIGURE 4–80

OBJECTIVE

Given a Boolean expression or truth table, implement a logic circuit using the fewest ICs.

Our study of Boolean algebra, alternate logic gate symbols, truth tables, Karnaugh maps, and the multiple functions of NAND and NOR gates has prepared us for final logic circuit design. Here we will apply what we have learned to design a logic circuit in its simplest form.

Implementing Logic Circuits From Boolean Expressions

If a Boolean expression is available for a desired logic function, only two steps are required to implement the circuit.

1. Simplify the expression if possible.
2. Implement the circuit with the fewest ICs.

For practice, the expression $\overline{A}B + CD$ will be implemented. Inspection of this expression reveals the two input variable sets ($\overline{A}B$ and CD) are ORed together. Hence, the output gate of the logic circuit will be a 2-input OR gate. Draw a 2-input OR gate and label each input with one of the input variable sets as shown in Fig. 4–81(a).

Next, the input variable sets consist of ANDed terms. This indicates the OR gate is preceded by 2-input AND gates. Once these gates are drawn, label their outputs and connect them to the inputs of the OR gate as shown in Fig. 4–81(b).

The last step is to connect the AND gates to their appropriate inputs. Since the output of one of the AND gates is $\overline{A}B$, the A input to that gate must have an inverter placed in series with it. The other AND gate inputs may be connected directly to their respective inputs as shown in Fig. 4–81(c).

Since this is an SOP circuit, it can be implemented with NAND logic using a gate-for-gate swap. The NAND implemented circuit is shown in Fig. 4–82. This circuit can be implemented with one 7400 quad 2-input IC. The original version of the circuit would have required three ICs (AND, OR, and NOT).

There is always one problem that appears at this point. The problem is whether to simplify a Boolean expression or not. The expression $\overline{A}B + CD$ cannot be simplified. Nonetheless, it was changed to $\overline{\overline{A}\,\overline{B}\,\overline{CD}}$ for NAND implementation in Fig. 4–82 to reduce the number of ICs. Although a gate-for-gate swap was accomplished, the original expression could have been changed using the implications of DeMorgan's theorem:

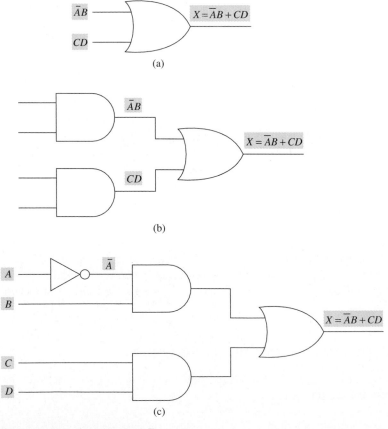

(a)

(b)

(c)

FIGURE 4–81   Implementing $\overline{A}B + CD$.

$\overline{AB} + CD$

Change the sign.

$\overline{AB} \cdot \overline{CD}$

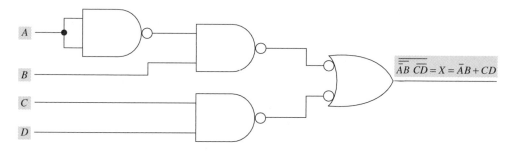

FIGURE 4–82   **Implementing $\overline{AB} + CD$ with NAND gates.**

NOT the individual inputs. Note the $\overline{AB}$ and CD inputs are treated as individual sets of inputs when NOTing.

$\overline{\overline{AB}} \cdot \overline{\overline{CD}}$

NOT the entire expression.

$\overline{\overline{\overline{A} \overline{B}} \cdot \overline{CD}}$

Regardless, the expression $\overline{AB} + CD$ is the only expression that can be written for the circuit in Fig. 4–81 (c). Also, the expression $\overline{\overline{\overline{A}\,\overline{B}} + \overline{\overline{CD}}}$ is the only expression that can be written for the circuit in Fig. 4–82. The point is only one expression will fit a given logic circuit. There may be several equivalent and in some cases simpler expressions, but an expression indicates exactly how a circuit is implemented. Thus, *when writing an expression for a given circuit, do not simplify.*

Implementing Logic Circuits From Truth Tables

If a circuit is to be designed from a given expression or truth table, simplification is requisite to implementation for reasons of economy. Several examples will be provided in the design problems.

DESIGN PROBLEM 4–1

Let's determine the best method to implement a logic circuit from the truth table shown in Fig. 4–83(a). The truth table yields a Boolean expression of $A\overline{B}C + AB\overline{C} + ABC$. If this expression is simplified on the K-map of Fig. 4–83(b), the result is $AB + AC$. This expression is implemented in its true form in Fig. 4–83(c). Implementation in this manner would require only three gates, yet it requires two different ICs. Since it is of the SOP configuration, it can be implemented gate for gate with NAND logic as shown in Fig. 4–83(d). This implementation also requires three gates, but only one IC. The circuit has not been shown with NOR logic because it would take eight gates to implement it.

It would appear the simplest method of implementation would be with the NAND gates shown in Fig. 4–83(d). However, basic Boolean algebra leads to another expression. The expression $AB + AC$ can also be factored to $A(B + C)$. Since $AB + AC = A(B + C)$, the second expression will now be investigated. It is implemented in its true form in Fig. 4–84(a). This expression requires only two gates, which in its true form is a simpler implementation than the original expression. However, it does require an OR and an AND IC. The circuit is shown implemented with NAND logic in Fig. 4–84(b) to show the infeasibility of this type of implementation. Five gates are required here.

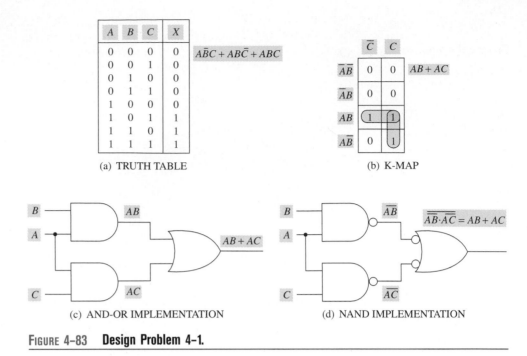

A	B	C	X
0	0	0	0
0	0	1	0
0	1	0	0
0	1	1	0
1	0	0	0
1	0	1	1
1	1	0	1
1	1	1	1

$A\bar{B}C + AB\bar{C} + ABC$

(a) TRUTH TABLE

	$\bar{C}$	C
$\bar{A}\bar{B}$	0	0
$\bar{A}B$	0	0
AB	1	1
$A\bar{B}$	0	1

$AB + AC$

(b) K-MAP

(c) AND-OR IMPLEMENTATION

(d) NAND IMPLEMENTATION

FIGURE 4-83 Design Problem 4-1.

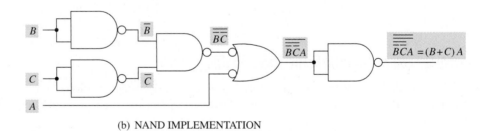

(a) OR-AND IMPLEMENTATION

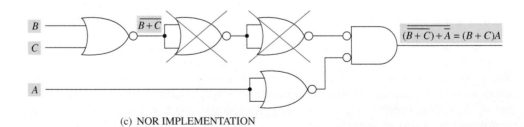

(b) NAND IMPLEMENTATION

(c) NOR IMPLEMENTATION

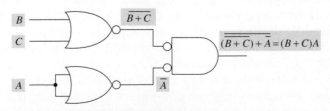

(d) NOR IMPLEMENTATION
WITH REDUNDANT INVERTERS ELIMINATED

FIGURE 4-84 Design Problem 4-1: Alternate solution.

The simplest implementation for this expression uses NOR logic and is shown in Fig. 4–84(c). The circuit has been drawn as previously illustrated in Section 4–9. Although the back-to-back inverters were drawn to reduce the possibility of confusion, they have been crossed out to indicate they would be eliminated in actual circuit construction. The circuit is redrawn in Fig. 4–84(d) without the back-to-back inverters. Like its NAND counterpart for $AB + AC$ implementation, this circuit requires only three gates and one IC. Either implementation, Fig. 4–83(d) or Fig. 4–84(d), is acceptable. All of the logic circuits presented in this discussion have been annotated with Boolean expressions to show their equality.

DESIGN PROBLEM 4-2

Implement the Boolean expression $A\bar{B}C + \bar{A}(B + C)$. The circuit is shown implemented directly from the expression in Fig. 4–85(a). The Boolean expression can be expanded into an SOP expression by distributing the $\bar{A}$. This results in $A\bar{B}C + \bar{A}B + \bar{A}C$. If this expression is taken to a K-map, it can be further reduced to $\bar{A}B + \bar{B}C$ as shown in Fig. 4–85(b).

This presents an opportunity to increase our knowledge of K-mapping procedures. To this point, we have not plotted any two-input expressions on a three-input map. In the expression $A\bar{B}C + \bar{A}B + \bar{A}C$, the first set of input variables presents no problem and is plotted in the lower right-hand block of the K-map in Fig. 4–85(b). The next set of input variables, $\bar{A}B$, implies C may be low or high. Thus, C must be plotted on the K-map as a low $(\bar{A}B\bar{C})$ and a high $(\bar{A}BC)$. The last set of input variables, $\bar{A}C$, must also be plotted on the K-map with B low $(\bar{A}\,\bar{B}C)$ and B high $(\bar{A}BC)$. This last $\bar{A}BC$ term is redundant and has already been plotted. Once looping has been completed, the resultant simplified expression is $\bar{A}B + \bar{B}C$.

Even though the C was plotted both low and high in the input term $\bar{A}B$, the terms cannot be considered don't care inputs. The Cs in this set of inputs are don't cares, but the $\bar{A}B$ term is not a don't care, and it must be plotted as described above. This is true because the Cs are complementary and they cancel, but the $\bar{A}B$ term remains.

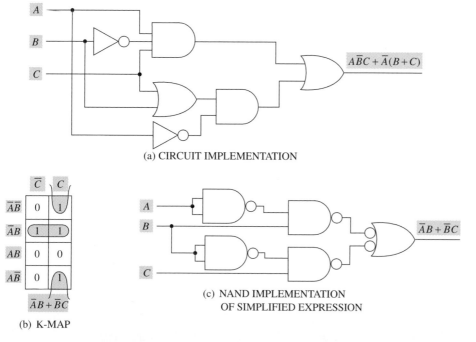

(a) CIRCUIT IMPLEMENTATION

(b) K-MAP

(c) NAND IMPLEMENTATION
OF SIMPLIFIED EXPRESSION

FIGURE 4-85   Design Problem 4-2.

Simplification of this problem algebraically is

$$A\overline{B}C + \overline{A}(B + C)$$
$$A\overline{B}C + \overline{A}B + \overline{A}C$$
$$A\overline{B}C + \overline{A}B\overline{C} + \overline{A}BC + \overline{A}\,\overline{B}C + \overline{A}BC$$

(This step incorporates $\overline{A}B$ factored in with C both low and high and $\overline{A}C$ factored in with B both low and high.)

$$A\overline{B}C + \overline{A}B\overline{C} + \overline{A}BC + \overline{A}\,\overline{B}C$$
$$\overline{A}B(\overline{C} + C) + \overline{B}C(\overline{A} + A)$$
$$\overline{A}B(1) + \overline{B}C(1)$$
$$\overline{A}B + \overline{B}C$$

This final simplified expression is shown implemented with NAND logic in Fig. 4–85(c).

DESIGN PROBLEM 4-3

Suppose a logic circuit had to be designed for the truth table shown in Fig. 4–86. The Boolean expression in SOP form would be $\overline{A}B\overline{C}\,\overline{D} + \overline{A}B\overline{C}D + \overline{A}BC\overline{D} + \overline{A}BCD + A\overline{B}\,\overline{C}D + A\overline{B}C\overline{D} + A\overline{B}CD + AB\overline{C}\,\overline{D} + AB\overline{C}D + ABC\overline{D} + ABCD$. The expression could be written in POS form as $(A + B + C + D)(A + B + C + \overline{D})(A + B + \overline{C} + D)(A + B + \overline{C} + \overline{D})(\overline{A} + B + C + D)$. In either case, a K-map will put the expression in suitable simplified form.

An SOP and a POS K-map have been completed to compare circuit implementations. The SOP K-map in Fig. 4–87(a) produces an expression of $B + AC + AD$. This expression is implemented in Fig. 4–87(b).

Care must be exercised when considering this circuit for NAND logic implementation. A gate-for-gate swap cannot be made because B is not ANDed in the original circuit. Thus, B must be inverted in the NAND logic circuit for proper implementation. This is shown in Fig. 4–87(c).

The POS K-map in Fig. 4–88(a) yields an expression of $(A + B)(B + C + D)$. This expression is implemented in Fig. 4–88(b). This circuit can be implemented in POS configuration with a gate-for-gate swap using NOR gates. The NOR gate circuit is shown in Fig. 4–88(c).

A	B	C	D	X
0	0	0	0	0
0	0	0	1	0
0	0	1	0	0
0	0	1	1	0
0	1	0	0	1
0	1	0	1	1
0	1	1	0	1
0	1	1	1	1
1	0	0	0	0
1	0	0	1	1
1	0	1	0	1
1	0	1	1	1
1	1	0	0	1
1	1	0	1	1
1	1	1	0	1
1	1	1	1	1

FIGURE 4–86   **Design Problem 4-3: Truth table.**

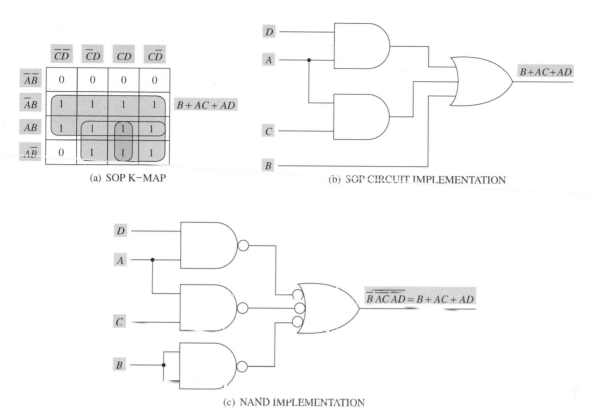

(a) SOP K–MAP

(b) SOP CIRCUIT IMPLEMENTATION

(c) NAND IMPLEMENTATION

FIGURE 4–87   Design Problem 4-3.

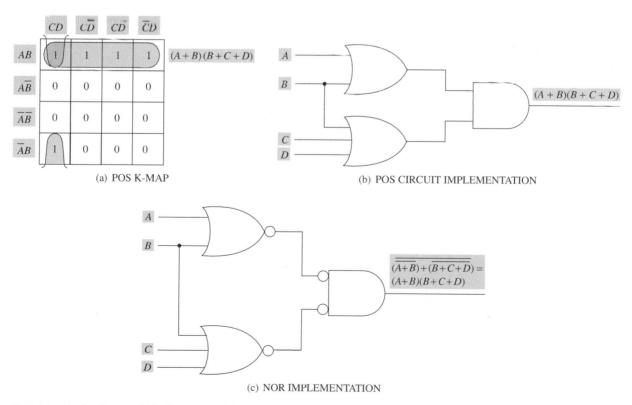

(a) POS K-MAP

(b) POS CIRCUIT IMPLEMENTATION

(c) NOR IMPLEMENTATION

FIGURE 4–88   Design Problem 4-3: Continued.

Although extraction of a POS simplified expression from a K-map is not mandatory at this level, a brief explanation of the process is provided below.

The POS K-map in Fig. 4–88(a) is laid out in negated form in respect to the SOP K-map previously used. For every low on the SOP K-map layout, there is a high on the POS K-map layout. For example, $\overline{A}\,\overline{B}$ on the SOP map layout is AB on the POS map layout; $\overline{A}B$ is $A\overline{B}$, and so on.

Once the POS map is laid out, the 0s on the SOP map are plotted as 1s on the POS map. Looping is the same for both types of maps. The difference is in the resultant expression obtained from the loop. The complemented variables are still eliminated as they were before. However, the variables remaining after the looping process has been completed must be *ORed* when extracted from the POS map. The quad in Fig. 4–88(a) yields $A + B$, and the pair yields $B + C + D$ because the As are eliminated. Each set of variables remaining after looping is ANDed with the remaining set(s). Therefore, the simplified expression is $(A + B)(B + C + D)$.

The original SOP simplified expression $(B + AC + AD)$ could be factored to produce $B + A(C + D)$. This circuit is shown in Fig. 4–89. Its NAND and NOR logic implementations consist of too many gates to be considered as a final design solution. The simplest design implementation is the NOR gate circuit shown in Fig. 4–88(c).

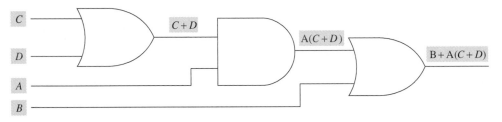

FIGURE 4-89   **Design Problem 4-3: Continued.**

DESIGN PROBLEM 4-4

The next problem is to implement a circuit for the expression $\overline{AB + \overline{BC} \cdot \overline{CD}}$. This expression cannot be taken to a K-map, so it must be simplified using Boolean algebra. The steps of simplification are

$$\overline{AB + \overline{BC} \cdot \overline{CD}}$$
$$\overline{AB} \cdot \overline{\overline{BC}} \cdot \overline{\overline{CD}}$$
$$(\overline{A} + \overline{B})(B + \overline{C})(\overline{C} + D)$$
$$(\overline{AB} + \overline{A}\,\overline{C} + B\overline{B} + \overline{B}\,\overline{C})(\overline{C} + D)$$
$$(\overline{AB} + \overline{A}\,\overline{C} + 0 + \overline{B}\,\overline{C})(\overline{C} + D)$$
$$\overline{AB}\overline{C} + \overline{A}\,\overline{C}\,\overline{C} + \overline{B}\,\overline{C}\,\overline{C} + \overline{A}BD + \overline{A}\,\overline{C}D + \overline{B}\,\overline{C}D$$
$$\overline{AB}\overline{C} + \overline{A}\,\overline{C} + \overline{B}\,\overline{C} + \overline{A}BD + \overline{A}\,\overline{C}D + \overline{B}\,\overline{C}D$$
$$\overline{A}\,\overline{C}(B + 1 + D) + \overline{B}\,\overline{C}(1 + D) + \overline{A}BD$$
$$\overline{A}\,\overline{C}(1) + \overline{B}\,\overline{C}(1) + \overline{A}BD$$
$$\overline{A}\,\overline{C} + \overline{B}\,\overline{C} + \overline{A}BD$$

Since the simplified expression is in SOP form, the circuit can be implemented with NAND gates as shown in Fig. 4-90. Implementation would require two ICs—a quad two–input NAND (7400) and a triple three-input NAND (7410) as depicted in the figure. Note that one section of the 7410 has all three of its inputs tied together. It is functioning as a NOT gate.

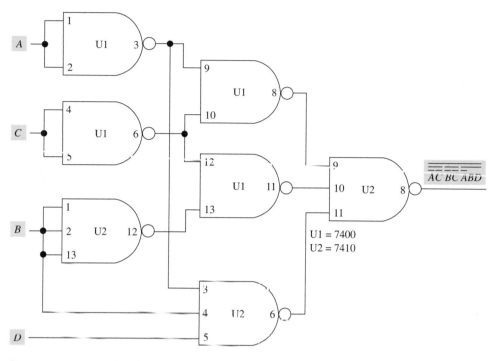

$$U1 = 7400$$
$$U2 = 7410$$

FIGURE 4 00   Design Problem 4-4.

DESIGN PROBLEM 4-5

Design a 4-input circuit that will function as a **majority detector.** The circuit should output high when a majority of the inputs are high.

The first step is to complete a truth table and mark high outputs for every set of input conditions that contains three or four 1s. This is shown in Fig. 4–91(a). Next, plot the Boolean expression from the truth table on a K-map as shown in Fig. 4–91(b) and simplify the expression. This results in $ABD + BCD + ABC + ACD$ because four pairs of 1s can be looped on the map. Implementation is straightforward as shown in Fig. 4–91(c) because this is an SOP expression. The circuit requires one triple 3-input NAND gate (7410) and one dual 4-input NAND gate (7420).

DESIGN PROBLEM 4-6

Let's design a circuit that will output high when an invalid BCD number occurs. Remember from our discussion of binary-coded decimal numbers that 4-bit groups of bits from $1010_{(2)}$ through $1111_{(2)}$ form invalid BCD numbers. The truth table for this circuit is laid out in Fig. 4–92(a). The K-map is shown in Fig. 4–92(b). Note the truth table has been checked and marked to ensure the simplified expression $(AB + AC)$ is equivalent to the circuit's high output requirements. Standard implementation of the circuit appears in Fig. 4–92(c) and (d). The circuit could be implemented with NAND or NOR gates.

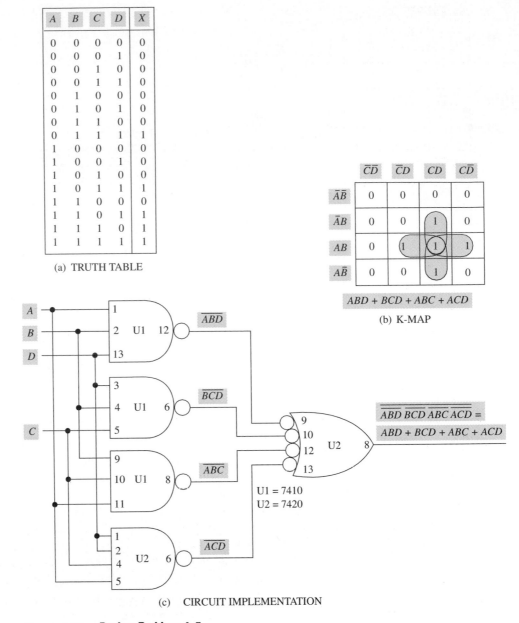

A	B	C	D	X
0	0	0	0	0
0	0	0	1	0
0	0	1	0	0
0	0	1	1	0
0	1	0	0	0
0	1	0	1	0
0	1	1	0	0
0	1	1	1	1
1	0	0	0	0
1	0	0	1	0
1	0	1	0	0
1	0	1	1	1
1	1	0	0	0
1	1	0	1	1
1	1	1	0	1
1	1	1	1	1

(a) TRUTH TABLE

(b) K-MAP

(c) CIRCUIT IMPLEMENTATION

FIGURE 4–91   **Design Problem 4–5.**

SECTIONS 4-9 AND 4-10: INTERNAL SUMMARY

NAND gates or NOR gates can be used to implement all logic functions. Implementation using these gates is summarized in Fig. 4–93 on page 182.

An SOP circuit can be implemented gate for gate using NAND gates. This makes this type of implementation particularly appealing because a Boolean expression written for a high output from a truth table is in SOP form.

A POS circuit can be implemented gate for gate using NOR gates. It is this straight-forward gate swapping that makes this implementation popular. A POS Boolean expression can be extracted directly from a truth table.

When writing a Boolean expression for a logic circuit, do not simplify the expression. For example, the Boolean expression $\overline{AB} + C$ represents the logic circuit shown in

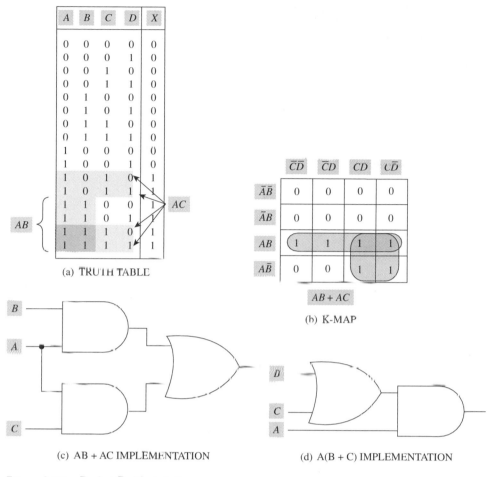

(a) TRUTH TABLE

(b) K-MAP

(c) AB + AC IMPLEMENTATION

(d) A(B + C) IMPLEMENTATION

FIGURE 4-92   Design Problem 4-6.

Fig. 4–94. The expression can be simplified so that it can be easily read as $\overline{A} + B + C$. These two expressions are indentical in accordance with DeMorgan's theorem.

Several design problems were presented in the last section. The general rules followed to implement the simplified circuits were these:

1. Simplify the expression with Boolean algebra or a K-map.
2. Compare the simplified expression to a truth table to ensure equality of the two expressions.
3. Implement the logic circuit with the fewest ICs.

SECTIONS 4-9 AND 4-10: INTERNAL SUMMARY QUESTIONS

Answers are given at the end of the chapter.

1. Identify the AND function in Fig. 4–95 (a, b, c, or d) on page 183.
2. Identify the OR function in Fig. 4–96 (a, b, c, or d) on page 183.
3. Which type of gate can be used to implement an SOP circuit configuration with the fewest ICs?
   a. OR
   b. AND
   c. NOR
   d. NAND

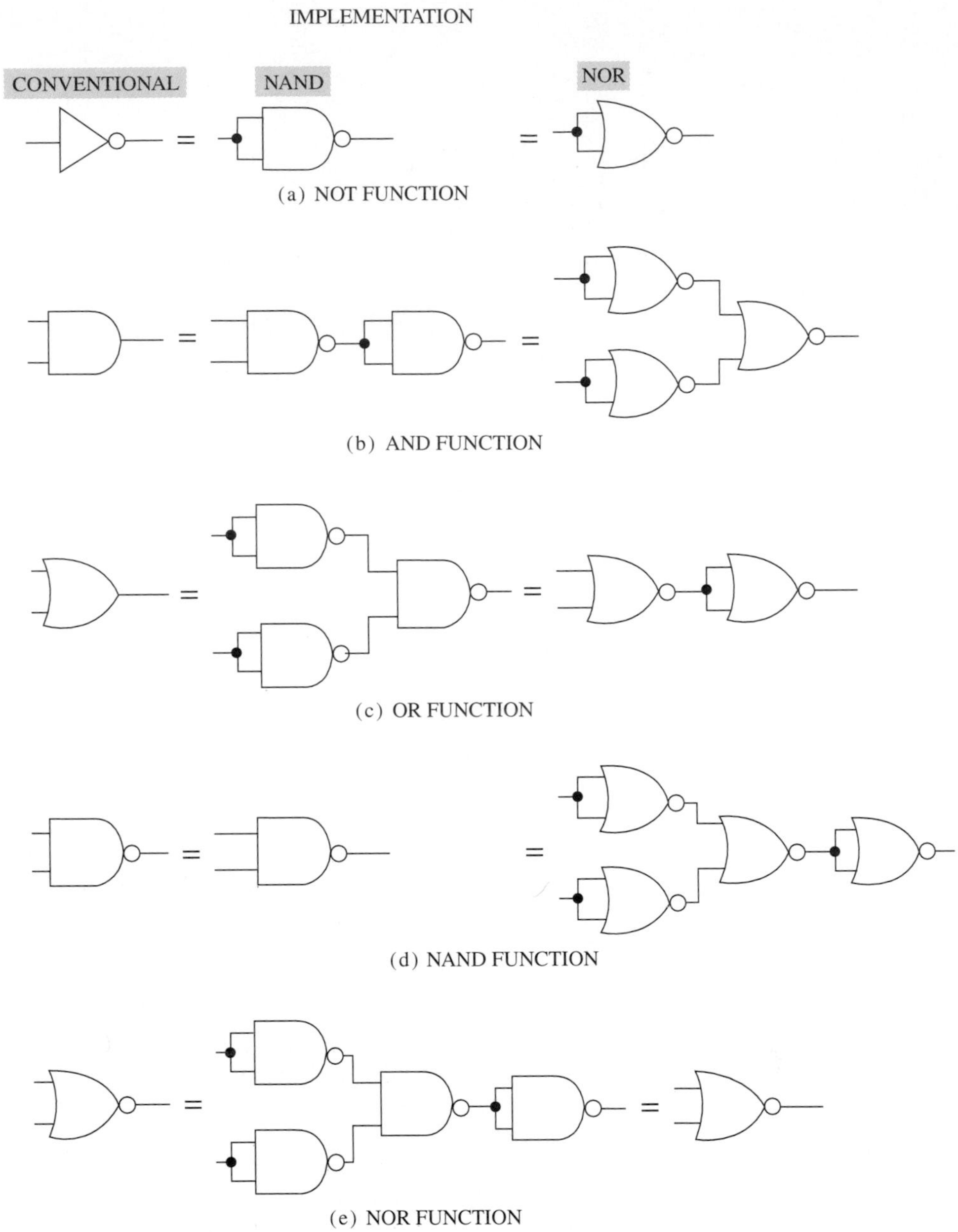

IMPLEMENTATION

CONVENTIONAL   NAND   NOR

(a) NOT FUNCTION

(b) AND FUNCTION

(c) OR FUNCTION

(d) NAND FUNCTION

(e) NOR FUNCTION

FIGURE 4-93   NAND/NOR logic implementation.

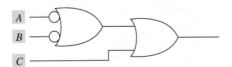

FIGURE 4-94   Implementation of $\overline{AB} + C$.

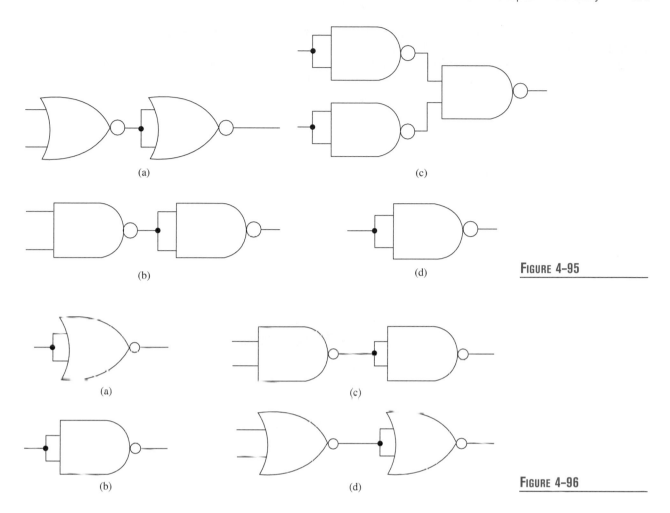

FIGURE 4-95

FIGURE 4-96

4. Which type of gate can be used to implement a POS circuit configuration with the fewest ICs?
   a. OR
   b. AND
   c. NOR
   d. NAND

5. SOP and POS truth tables are laid out identically.
   a. True
   b. False

CHAPTER 4: SUMMARY

Our study in this chapter commenced with Boolean symbols and expressions for AND (e.g., $A \cdot B$), OR $(A + B)$, NAND $(\overline{AB})$, NOR $(\overline{A + B})$, and NOT $(\overline{A})$ gates.

The properties of real numbers and Boolean algebra as well as DeMorgan's theorems are summarized in Table 4–4. DeMorgan's theorems proved $\overline{A + B} = \overline{A}\,\overline{B}$ and $\overline{AB} = \overline{A} + \overline{B}$. These theorems provided a procedure whereby the vinculum in any Boolean expression can be broken provided the logical sign under the break is changed. The implications of these theorems also al-

lowed us to (1) change the sign, (2) NOT the individual variable (inputs), (3) NOT the entire expression (output), and, if applicable, (4) cancel all double NOTs.

A Boolean expression with a vinculum over the entire expression is written to indicate a *low* output. A Boolean expression without a vinculum is written to indicate a *high* output.

There is an alternate logic gate symbol for each standard logic gate. These symbols may be derived from the implications of DeMorgan's theorems by following the four steps listed above.

When drawing logic diagrams, every effort should be made to connect bubble to bubble and no bubble to no bubble using alternate gate symbols when necessary. When extracting an expression from a truth table, high outputs produce an SOP expression and low outputs produce a POS expression.

Karnaugh maps present graphical representations of truth tables. These maps are used to simplify Boolean expressions without getting involved with the mathematics of Boolean algebra. **Looping** on a map is done with horizontally and/or vertically adjacent 1s. Looping is done is **pairs, quads,** and **octets,** and all complementary variables within a loop are eliminated. The top row and bottom row of a K-map are considered to be adjacent for looping purposes. Likewise, the left column and right column are considered to be adjacent. Always loop as large a group of 1s as possible. **Don't care** variables may be plotted on a map as lows or highs and should be plotted to provide the most convenient looping.

NAND gates and NOR gates are multifunctional. Every logic function can be implemented with NAND gates or NOR gates. SOP circuits implemented with NAND gates and POS circuits implemented with NOR gates "usually" represent the simplest implementation.

CHAPTER 4: END OF CHAPTER QUESTIONS/PROBLEMS

Answers are given in the Instructor's Manual.

SECTION 4-1

1. Write the output Boolean expression for a 2-input AND gate with A and B inputs.
2. Write the output Boolean expression for a 3-input OR gate with A, B, and C inputs.
3. The Boolean expression $\overline{ABC}$ represents the output of a 3-input _____ gate.
4. The Boolean expression $\overline{A + B}$ indicates the output of the gate will be low when _____.
5. The Boolean expression $A + B + C$ indicates the output of the gate will be high when _____.
6. Write the output expression of an inverter if its input is XY.

SECTION 4-2

7. Explain what the Associative Property of Addition — $(A + B) + C = A + (B + C) = A + B + C$ — means.
8. The Boolean expression $A \cdot B \cdot 1 = $ _____.
9. Complete the following:
   a. $A(B + C) = $ _____
   b. $A \cdot 0 = $ _____
   c. $X + 1 = $ _____
   d. $A \cdot 1 = $ _____
   e. $X + 0 = $ _____
   f. $X \cdot X = $ _____
   g. $\overline{X} + \overline{X} = $ _____
   h. $A \cdot \overline{A} = $ _____
   i. $A + \overline{A} = $ _____
   j. $A + AB = $ _____
   k. $A + \overline{A}B = $ _____
   l. $\overline{A} + \overline{A} \, \overline{B} = $ _____

SECTION 4-3

10. Complete the following theorems:
    a. $\overline{A + B} = $ _____
    b. $\overline{AB} = $ _____

11. Change the Boolean expression AB to an expression that can be implemented with a NOR gate and NOT gates.
12. Change the Boolean expression $A + B + C$ to an expression that can be implemented with a NAND gate and NOT gates.

SECTION 4-4

13. The expression $\overline{XY}\overline{Z}$ indicates the circuit will output high when _____ .
14. The expression $\overline{\overline{X}}\,\overline{Y}$ indicates the circuit will output _____ when _____ .
15. The expression $\overline{X} + \overline{Y}$ indicates the circuit will output _____ when _____ .
16. Simplify the expression $W\overline{X}Y(\overline{\overline{W} + Z})$.

SECTION 4-5

17. Draw the alternate logic symbol for an *AND* gate.
18. Draw the alternate logic symbol for an *OR* gate.
19. Draw the alternate logic symbol for a *NAND* gate.
20. Draw the alternate logic symbol for a *NOR* gate.
21. Draw the alternate logic symbol for the one shown in Fig. 4–97.

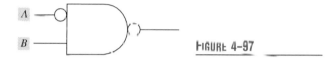

FIGURE 4-97

22. Redraw the circuit shown in Fig. 4–98 using appropriate standard and/or alternate logic symbols.

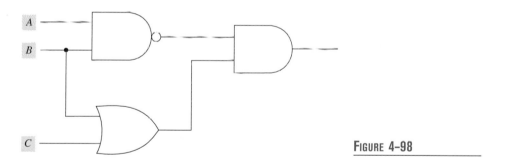

FIGURE 4-98

CT 23. Redraw the circuit shown in Fig. 4–99 using appropriate alternate logic symbols where applicable.

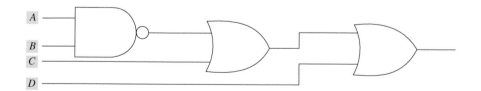

FIGURE 4-99

SECTION 4-6

24. Write the sum-of-products (SOP) expression for each of the following truth tables:

a.

A	B	X
0	0	1
0	1	1
1	0	1
1	1	0

b.

A	B	C	X
0	0	0	0
0	0	1	1
0	1	0	0
0	1	1	0
1	0	0	1
1	0	1	1
1	1	0	0
1	1	1	1

c.

A	B	C	X
0	0	0	1
0	0	1	1
0	1	0	0
0	1	1	0
1	0	0	0
1	0	1	0
1	1	0	1
1	1	1	1

d.

A	B	C	X
0	0	0	0
0	0	1	0
0	1	0	1
0	1	1	1
1	0	0	1
1	0	1	0
1	1	0	0
1	1	1	1

CT 25. Write the product-of-sums (POS) expression for each of the following truth tables:

a.

A	B	C	X
0	0	0	1
0	0	1	1
0	1	0	1
0	1	1	0
1	0	0	0
1	0	1	1
1	1	0	1
1	1	1	1

b.

A	B	C	X
0	0	0	1
0	0	1	1
0	1	0	0
0	1	1	1
1	0	0	1
1	0	1	1
1	1	0	0
1	1	1	0

26. Identify the configuration of the logic circuit shown in Fig. 4–100.

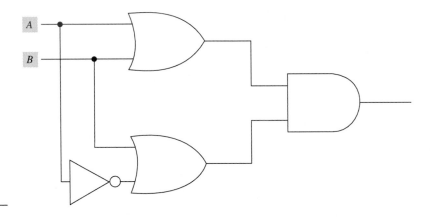

FIGURE 4-100

27. Write the Boolean expression for the logic circuit shown in Fig. 4–100.

SECTION 4-7

28. Simplify the following expressions using Boolean algebra:
    a. $\overline{A}\,\overline{B}\,\overline{C} + \overline{A}B\overline{C}$
    b. $\overline{A}\,\overline{B}\,\overline{C} + \overline{A}\,\overline{B}C + AB\overline{C} + ABC$
    c. $\overline{A}B\overline{C} + A\overline{B}C + \overline{A}BC + A\overline{B}\,\overline{C}$
    d. $(X + Y)(X + Z)$
    e. $XY + \overline{XY}$
    f. $\overline{ABC} + \overline{A} + C$
    g. $(\overline{X} + Y + \overline{Z})(X + Y + \overline{Z})$
    h. $A\,\overline{\overline{B}\,\overline{C}}\,D$
    i. $X\,\overline{Y}\,\overline{\overline{Z}}$

SECTION 4-8

29. Plot the high outputs from the truth table on a K-map and simplify the resultant expression by looping.

a.

A	B	C	X
0	0	0	1
0	0	1	1
0	1	0	1
0	1	1	0
1	0	0	0
1	0	1	0
1	1	0	0
1	1	1	0

b.

A	B	C	X
0	0	0	0
0	0	1	1
0	1	0	0
0	1	1	0
1	0	0	1
1	0	1	1
1	1	0	0
1	1	1	0

c.

A	B	C	X
0	0	0	1
0	0	1	1
0	1	0	0
0	1	1	1
1	0	0	0
1	0	1	X
1	1	0	0
1	1	1	1

d.

A	B	C	X
0	0	0	1
0	0	1	1
0	1	0	0
0	1	1	0
1	0	0	X
1	0	1	X
1	1	0	0
1	1	1	0

30. Plot the high outputs from the truth table on a K-map and simplify the resultant expression by looping.

a.

A	B	C	D	X
0	0	0	0	1
0	0	0	1	1
0	0	1	0	0
0	0	1	1	0
0	1	0	0	0
0	1	0	1	0
0	1	1	0	0
0	1	1	1	0
1	0	0	0	0
1	0	0	1	0
1	0	1	0	0
1	0	1	1	0
1	1	0	0	0
1	1	0	1	0
1	1	1	0	1
1	1	1	1	1

b.

A	B	C	D	X
0	0	0	0	0
0	0	0	1	0
0	0	1	0	0
0	0	1	1	0
0	1	0	0	0
0	1	0	1	0
0	1	1	0	0
0	1	1	1	0
1	0	0	0	0
1	0	0	1	0
1	0	1	0	1
1	0	1	1	1
1	1	0	0	1
1	1	0	1	1
1	1	1	0	1
1	1	1	1	1

c.

A	B	C	D	X
0	0	0	0	1
0	0	0	1	1
0	0	1	0	X
0	0	1	1	0
0	1	0	0	0
0	1	0	1	0
0	1	1	0	1
0	1	1	1	0
1	0	0	0	0
1	0	0	1	0
1	0	1	0	0
1	0	1	1	0
1	1	0	0	0
1	1	0	1	0
1	1	1	0	0
1	1	1	1	0

d.

A	B	C	D	X
0	0	0	0	X
0	0	0	1	0
0	0	1	0	0
0	0	1	1	0
0	1	0	0	0
0	1	0	1	0
0	1	1	0	0
0	1	1	1	0
1	0	0	0	0
1	0	0	1	0
1	0	1	0	1
1	0	1	1	0
1	1	0	0	1
1	1	0	1	1
1	1	1	0	1
1	1	1	1	1

SᴇᴄᴛɪᴏN 4-9

31. Draw the gates shown in Fig. 4–101 using only NAND gates.

    a.   b.   c.

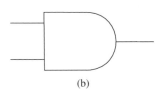

(a)

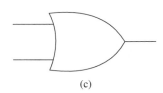

(b)

(c)

Fɪɢᴜʀᴇ 4-101

32. Draw the gates shown in Fig. 4–102 using only NOR gates.

    a.   b.   c.   d.

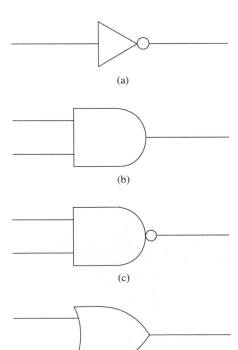

(a)

(b)

(c)

(d)

Fɪɢᴜʀᴇ 4-102

33. Draw $ABC + \overline{CD}$ using NAND logic.
CT 34. Draw the circuit shown in Fig. 4–103 using only NAND gates.
CT 35. Draw the circuit shown in Fig. 4–103 using only NOR gates.

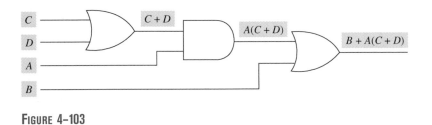

FIGURE 4-103

Section 4-10

CT 36. Plot the low outputs from the truth table on a POS K-map and simplify the resultant expression by looping.

A	B	C	D	X
0	0	0	0	0
0	0	0	1	0
0	0	1	0	0
0	0	1	1	0
0	1	0	0	1
0	1	0	1	1
0	1	1	0	1
0	1	1	1	0
1	0	0	0	1
1	0	0	1	1
1	0	1	0	1
1	0	1	1	1
1	1	0	0	1
1	1	0	1	1
1	1	1	0	1
1	1	1	1	1

37. Write the Boolean expression for the circuit shown in Fig. 4–104 and then simplify the expression.

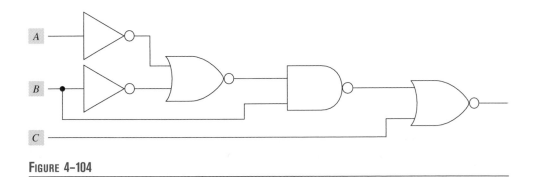

FIGURE 4-104

CT 38. Design a 4-input voter "dissent" circuit whose output will be low when *all* inputs are in 0 states or in 1 states. The circuit's output should be high when any dissenting vote (0) is present unless all votes are 0. Implement the circuit with the fewest ICs. Label the IC number(s) (74XX or 4000-series) on your drawing.
CT 39. Design a 3-input circuit that will output low only on input counts of 1 and 3.

ANSWERS TO REVIEW QUESTIONS

SECTION 4-1

A. $A \times B, A \cdot B, AB$
B. $A\overline{B}$
C. $A + B$
D. $\overline{AB}$
E. $\overline{A + B}$
F. High
G. Low
H. The low and high inputs/outputs to/from a logic gate or circuit
I. A bar drawn over two or more variables in a Boolean expression

SECTION 4-2

A. True   F. True
B. False   G. True
C. True   H. True
D. True   I. False
E. False

SECTION 4-3

A. True   D. $ABC = \overline{\overline{A} + \overline{B} + \overline{C}}$
B. True   E. $X + Y = \overline{\overline{X}\,\overline{Y}}$
C. False

SECTION 4-4

A. False   D. $A + B$
B. True   E. AB
C. True   F. $\overline{A}\,\overline{B}\,\overline{CD}$

SECTION 4-5

A. See Fig. 4–28 and Fig. 4–30.
B. (1) $\overline{\overline{A} + \overline{B}}$
    (2) $\overline{\overline{A}\,\overline{B}}$
    (3) $\overline{A} + \overline{B}$
    (4) $\overline{A}\,\overline{B}$
    (5) $A + \overline{B}$
    (6) $\overline{A\overline{B}}$

C. Use DeMorgan's theorem to break the vinculum and change the sign; or use the inverse of his theorem. This will prove equality if the expressions written in question (B.) are correct.
D. The circuit will output high when *A and B* are high *or* when *C and D* are high: $AB + CD$.
E. The circuit will output high when *A or B* is high *and* *C or D* is high: $AC + AD + BC + BD$.
F. The circuit will output high if *E* is high *or A or B or C or D* is high: $A + B + C + D + E$.
G. See Fig. 4–105.

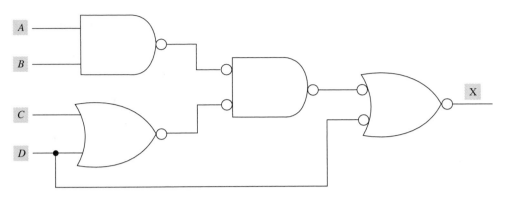

FIGURE 4-105

SECTION 4-6

A. $\overline{A}B + A\overline{B} + AB = A + B$
B. $\overline{A}\,\overline{B}C + \overline{A}BC + AB\overline{C} = \overline{A}C + AB\overline{C}$
C. $\overline{A}\,\overline{B}\,\overline{C} + \overline{A}B\overline{C} + A\overline{B}C + ABC = \overline{A}\,\overline{C} + AC$
D. $\overline{A}\,\overline{B}C + AB\overline{C} + A\overline{B}\,\overline{C} + AB\overline{C} = AC + A\overline{C}$
E. $\overline{ABC + AB\,\overline{C}} = (A + \overline{B} + \overline{C})(A + B + C)$
   Note: The first expression was written as if extracting an SOP expression where the vinculum indicates a low output. The second expression was written using the rules of POS extraction. The expressions are equal.
F. $(\overline{A} + B)(\overline{A} + \overline{B})$

SECTION 4-7

A. AB
B. $A + B$
C. 1
D. $B + AC$
E. $BC + \overline{B}\,\overline{C} + A\overline{B}$ or $BC + \overline{B}\,\overline{C} + AC$
F. 1
G. $\overline{X} + \overline{Y} + \overline{Z}$
H. $AB + \overline{C} + \overline{D}$

SECTION 4-8

A. See Fig. 4–54. *Note:* The variables marked inside of each square on this map are unnecessary for your drawing.
B. See Fig. 4–59.
C. See Fig. 4–106.
D. See Fig. 4–107.

SECTION 4-9

A. See Fig. 4-93.
B. See Fig. 4-93.
C. See Fig. 4–108.
D. See Fig. 4–109.

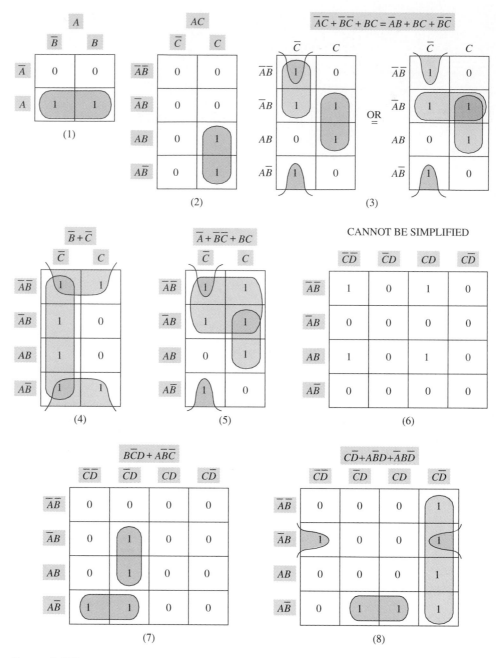

Figure 4-106

Figure 4-107

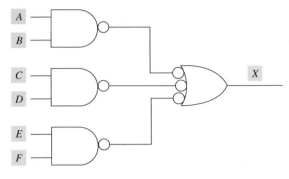

FIGURE 4-108

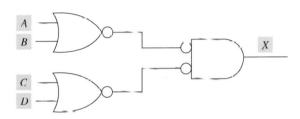

FIGURE 4-109

ANSWERS TO INTERNAL SUMMARY QUESTIONS

SECTIONS 4-1 THROUGH 4-3

1. c	7. b	13. b
2. a	8. a	14. a
3. d	9. d	15. a
4. a	10. d	16. c
5. b	11. c	
6. a	12. c	

SECTIONS 4-4 THROUGH 4-6

1. a	6. a	11. d
2. b	7. c	12. c
3. a	8. g, h, e, f	13. d
4. a	9. c	14. b
5. b	10. a	15. b

SECTIONS 4-7 AND 4-8

1. $\overline{A} + \overline{B} + C$
2. $A + BC$
3. 1
4. $A\overline{B} + B\overline{C}$
5. $Y + \overline{Z}$
6. Y

7. $\overline{A} + B + \overline{C}$
8. $\overline{X} + Y + Z$
9. $\overline{A}\,\overline{B} + A\overline{C}$
10. $A\overline{B}\,\overline{C} + AE$
11. $B\overline{C} + AB$
12. $\overline{A}\,\overline{C} + \overline{B}\,\overline{C}$

13. $\overline{B} + \overline{A}C$
14. $\overline{A}\,\overline{B}D + \overline{A}\,\overline{B}C$
15. $\overline{A}$
16. $\overline{C}$
17. $\overline{A}\,\overline{C} + B\overline{C}$
18. $\overline{B}\,\overline{D} + \overline{A}\,\overline{B}\,\overline{C}$

SECTIONS 4-9 AND 4-10

1. b	4. c
2. d	5. a
3. d	

5 COMBINATIONAL LOGIC CIRCUITS

Introduction

Combinational logic circuits are merely combinations of logic gates that produce outputs dependent upon their input combinations. The circuits designed in Chapter 4 are combinational logic circuits. This chapter is a continuation of Chapter 4 with special emphasis on practical logic circuits.

The output of a combinational logic circuit can be used in many ways. It may be used to turn an alarm on or off, energize a relay, activate a Silicon Controlled Rectifier (SCR), or merely turn on an indicator. It could be used to select or deselect another digital circuit. The combinational logic circuit can be used to check for data errors; add, subtract, or compare binary numbers; detect the presence of a specific binary number or code; convert serial data to parallel data or vice versa; or select or distribute certain binary data.

The **exclusive-OR and exclusive-NOR functions** are the first combinational logic circuits presented in this chapter. The functions have been incorporated in exclusive gates. These gates fulfill requirements not fulfilled by any of the logic gates previously presented.

Boot-up

Demultiplexing

Exclusive-NOR Gate

Exclusive-OR Gate

Hysteresis

Multiplexing

Parity

Parity Bit

Parity Checker

Parity Error

Parity Generator

Parity Scheme

Schmitt-Trigger-Input Circuit

Sigma (Σ)

Terminal Count

Chapter Objectives

1. Given a logic symbol for an exclusive-OR/exclusive-NOR gate, identify the symbol and determine the gate's output when various input combinations are specified.

2. Given a logic diagram of a parity generator/checker and the parity scheme, determine the connections required for proper operation.

3. Identify the symbol for and purpose of Schmitt-trigger-input circuits.

4. Given a logic diagram of a combinational logic circuit and the circuit's inputs, determine its output(s).

5. Troubleshoot combinational logic circuits to a defective logic gate or point in the circuit.

SECTION 5-1: EXCLUSIVE-OR/NOR GATES

OBJECTIVES

1. Identify the logic symbols used to represent exclusive-OR and exclusive-NOR gates.
2. Write a truth table and boolean expression for a specified exclusive gate.
3. Given a logic circuit containing exclusive gates and other logic gates, determine the output(s) when the inputs are given.

Exclusive-OR Gate

The first combinational logic circuit presented in this chapter yields the **exclusive-OR function.** The OR gates that were presented in Chapter 3 are *inclusive* gates. The truth table for the OR gate shows the gate's output is high when A is high *or* B is high *or* when A and B are high. The gate is considered to be inclusive because its high outputs include the A and B high input condition as shown in Fig. 5–1(a). There are many times in digital circuits when the last condition (A and B high inputs/high output) must be excluded. This is accomplished with an **exclusive-OR** gate, often referred to as an **X-OR** gate. There are a multitude of uses for this gate. For example, the rules for binary addition as presented in Chapter 2 are

$$0 + 1 = 1$$
$$0 + 0 = 0$$
$$1 + 0 = 1$$
$$1 + 1 = 0 \text{ with 1 Carry}$$

A	B	Y
0	0	0
0	1	1
1	0	1
1	1	1

(a) OR GATE

A	B	Y
0	0	0
0	1	1
1	0	1
1	1	0

(b) X-OR GATE

 FIGURE 5-1   Truth tables.

As shown in these rules, the outputs of a binary adder require a high output when the inputs are complementary (opposite levels) and a low output when they are equal.

How would you design a logic circuit that would perform the OR function yet exclude the A and B high inputs/high output condition? The first thing that should be done is to complete a truth table showing the circuit requirements. This is shown in Fig. 5–1(b). The truth table yields a Boolean expression of $\overline{A}B + A\overline{B}$. Therefore, the expression required to implement the exclusive-OR function is a sum-of-products expression. The X-OR gate can be implemented as shown in Fig. 5–2(a). The logic symbol for the X-OR gate in Fig. 5–2(b) is similar to the OR gate symbol except for the extra arc at the input. Also shown in the figure is another way to write the $\overline{A}B + A\overline{B}$ expression ($A \oplus B$). The expression is read as "A exclusively ORed with B" and is equal to $\overline{A}B + A\overline{B}$. The circle around the $+$ sign indicates the exclusive function.

As is verifiable in the truth table in Fig. 5–1(b), the short logic for the gate is "any complementary in = 1 out." It can be annotated, if necessary, on the gate as $C = 1$. If one input of an X-OR gate is tied high as shown in Fig. 5–3, the gate functions as an inverter.

Binary addition can be accomplished with an X-OR gate as shown in Fig. 5–4. The sum output is taken from the X-OR gate. The sum is often identified by the Greek letter sigma (Σ), which means "summation." The sum outputs of many adders are depicted Σ_0, Σ_1, and so on in data books. The AND gate was added to the circuit to detect the Carry Out, which occurs when 1 and 1 are added ($1 + 1 = 0$ with 1 Carry). The circuit is called a **half-adder** and can be used to add two binary bits. Its application is limited by the fact that although it can generate a carry out, it cannot add a carry-in bit to its A and B input

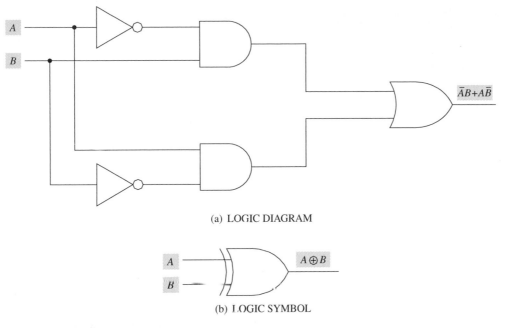

(a) LOGIC DIAGRAM

(b) LOGIC SYMBOL

FIGURE 5-2   **Exclusive-OR gate.**

bits. This is a necessity in binary adders. A **full-adder** circuit with this capability will be presented in Chapter 9. The full adder modifies the half-adder so that a carry-in bit can also be added.

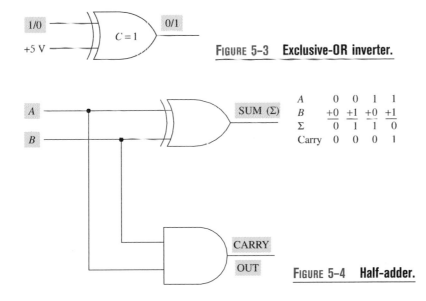

FIGURE 5-3   **Exclusive-OR inverter.**

A	0	0	1	1
B	+0	+1	+0	+1
Σ	0	1	1	0
Carry	0	0	0	1

FIGURE 5-4   **Half-adder.**

Exclusive-NOR Gate

If the output of the X-OR gate is inverted as shown in Fig. 5–5, the output expression for the circuit becomes $\overline{A \oplus B}$. This equation is for an exclusive-NOR (X-NOR) gate and equates to $\overline{A}\,\overline{B} + AB$.

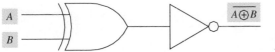

FIGURE 5-5   **X-OR gate and inverter.**

The Boolean expression for the X-NOR gate can be derived from the X-OR expression as follows:

X-OR expression	$\overline{A}B + A\overline{B}$
Expression NOTed (equals the X-NOR function)	$\overline{\overline{A}B + A\overline{B}}$
Break vinculum/change sign	$\overline{\overline{A}B} \cdot \overline{A\overline{B}}$
Break vinculums/change signs	$(\overline{\overline{A}} + \overline{B})(\overline{A} + \overline{\overline{B}})$
Cancel double NOTs	$(A + \overline{B})(\overline{A} + B)$
Distribute	$A\overline{A} + AB + \overline{A}\,\overline{B} + B\overline{B}$
Theorem $X \cdot \overline{X} = 0$	$0 + AB + \overline{A}\,\overline{B} + 0$
Theorem $X + 0 = X$	$AB + \overline{A}\,\overline{B}$

The circuit for the expression derived here can be implemented as shown in Fig. 5–6(a). The truth table in Fig. 5–6(b) shows that the X-NOR gate outputs high when its inputs are equal. The short logic for the X-OR gate can be modified to fit the X-NOR gate by NOTing the output logic level. This can be done because we know the X-NOR function can be obtained from an X-OR gate followed by an inverter. Therefore, "any complementary in = $\overline{1}$ out." Since $\overline{1}$ = 0, the short logic can be annotated on the X-NOR gate as $C = 0$. The logic symbol for the X-NOR gate is shown in Fig. 5–6(c).

The SN74S135 is a quadruple exclusive-OR/NOR gate IC. The logic diagram in Fig. 5–7(a) shows the circuit can function as an X-OR gate when the Control (C) input is low or as an X-NOR gate when C is high. The Boolean expression for the circuit is $\overline{A}\,\overline{B}\,\overline{C} + \overline{A}B\overline{C} + A\,\overline{B}C + ABC$.

The first four input combinations are shown on the function table (Fig. 5–7b) with input C low. The expression $\overline{A}\,\overline{B}\,\overline{C} + \overline{A}B\overline{C}$ represents this circuit and represents the X-OR function. The expression verifies the short logic, "any complementary in = 1 out," when the C input is low.

FIGURE 5–6   Exclusive-NOR gate.

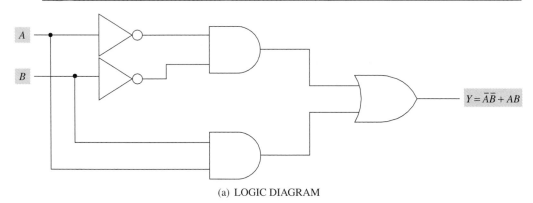

(a) LOGIC DIAGRAM

INPUTS		OUTPUT
A	B	Y
0	0	1
0	1	0
1	0	0
1	1	1

(b) TRUTH TABLE

(c) LOGIC SYMBOL

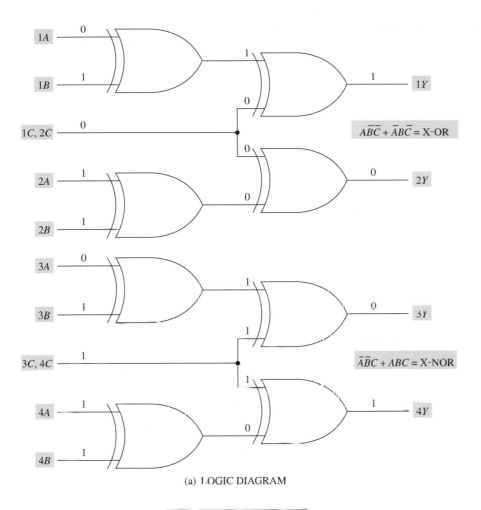

(a) LOGIC DIAGRAM

$A\overline{B}\overline{C} + \overline{A}B\overline{C} = $ X-OR

$\overline{A}\,\overline{B}C + ABC = $ X-NOR

INPUTS			OUTPUT
A	B	C	Y
0	0	0	0
0	1	0	1
1	0	0	1
1	1	0	0
0	0	1	1
0	1	1	0
1	0	1	0
1	1	1	1

X-OR (rows with C=0), X-NOR (rows with C=1)

(b) FUNCTION TABLE

FIGURE 5–7   SN74S135 quadruple exclusive-OR/NOR gate.

The Boolean expression $\overline{A}\,\overline{B}C + ABC$ for the X-NOR function verifies the short logic and shows the output is high when the control input is high and A and B are equal. This can be verified in Fig. 5–7(b).

One of the main uses of an X-NOR gate is a comparator. Its output can be used to indicate when both inputs are equal. The comparison theory of the X-NOR gate will be expanded in Chapter 10 to include outputs from ICs that identify when input A is greater than B or when A is less than B.

Section 5–1: Review Questions

Answers are given at the end of the chapter.

A. Draw the standard logic symbol for an X-OR gate.
B. Write the truth table for an X-OR gate.

C. Write the truth table for an X-NOR gate.

D. Write the Boolean expression for the circuit shown in Fig. 5–8 and simplify to a sum-of-products expression.

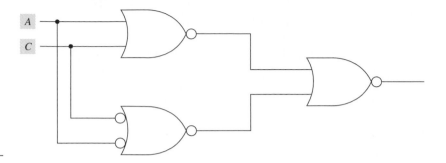

FIGURE 5-8

E. The expression derived in question (D) represents which exclusive function?
   (1) X-OR
   (2) X-NOR

F. Place a NOT gate on one input of an X-OR gate as shown in Fig. 5–9. Algebraically prove this circuit will produce the X-NOR function.

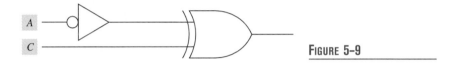

FIGURE 5-9

G. Place a NOT gate on one input of an X-NOR gate as shown in Fig. 5–10. Algebraically prove this circuit will produce the X-OR function.

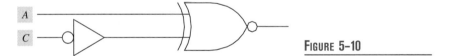

FIGURE 5-10

SECTION 5-2: PARITY

OBJECTIVES

1. Define parity.
2. Given the data information and parity scheme, determine the required level of the parity bit.

Parity Schemes

Parity means the state of being equal. A parity system is used to ensure equality of the received digital data with the transmitted data. A **parity bit** is sometimes used in conjunction with a stream of serial data bits when transmitting information from one computer to another. Transmission of data over long distances is normally accomplished by serially transmitting the data via a **modem** over a telephone line as shown in Fig. 5–11. This is accomplished by converting the binary data to audio. A parity bit is used to establish an **error detection system,** which should ensure accuracy of the data transmitted from one system to another. This is accomplished electronically with a **parity generator** at the transmitting computer and a **parity checker** at the receiving computer.

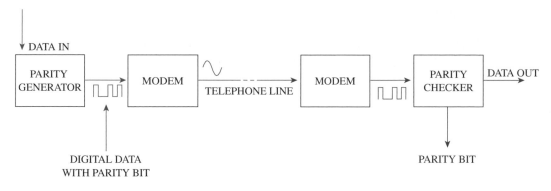

FIGURE 5-11   **Parity system simplified block diagram.**

There are two **parity schemes** that can be used—**odd parity** or **even parity.** Regardless of the scheme selected, the transmitting computer and the receiving computer must be set up for the same scheme.

Odd Parity

If the odd parity scheme is selected, both the parity generator at the transmitting end and the parity checker at the receiving end must be set to generate and detect odd parity. If a 7-bit ASCII bit pattern is to be transmitted, the parity generator will count the number of 1s in the data stream and add a parity bit that will make the total number of 1s transmitted odd. In other words, *a parity generator set up for odd parity guarantees that the transmitting system will always send an odd number of 1s including the parity bit.*

If the ASCII data is $28_{(16)}$, the binary data actually transmitted are 010 1000. The parity generator counts the number of 1s in the data stream and generates a high parity bit (odd parity scheme) to be transmitted with the data as shown:

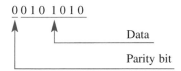

The actual transmitted data is A8 in hexadecimal once the parity bit has been added for transmission. Note the number of 1s in the transmitted data is odd.

If the ASCII data to be transmitted is $2A_{(16)}$, the transmitted binary data are 010 1010. Note there is already an odd number of 1s in the data stream. Therefore, the parity generator produces a low parity bit for transmission with the data.

```
0 0 1 0 1 0 1 0
↑         ↑         Data
|         |_____
|_____ Parity bit
```

When the data are received by the system on the other end of the line, the parity checker counts the total number of 1s including the parity bit. If that number is odd, the parity checker sends the data into the system and disposes of the parity bit as shown in Fig. 5–11. If the number of 1s received is even, the parity checker generates a parity error signal and requests retransmission of the last byte of data because something is wrong.

This error could occur if noise on the telephone line changed a bit level from 0 to 1 or 1 to 0. For example, if a $4A_{(16)}$ 7-bit ASCII data stream and its parity bit were set up for odd parity and transmitted, the data would appear as 0100 1010. If the system were set up for odd parity and noise changed the LSB to a 1 (0100 1011) after transmission, the total number of 1s received would be even. This would cause the parity checker to generate a parity error signal.

Even Parity

A system may be set up for even parity, and this scheme guarantees an even number of 1s, including the parity bit, will always be transmitted. The principle of operation is the same as the odd parity scheme, except the receiving end knows to look for an even number of 1s in each byte of data.

Determine the parity bit required from the parity generator if an even parity scheme is set up and the data are $36_{(16)}$ (011 0110). The parity bit (MSB position) would have to be low. Thus, the transmitted byte would be 0011 0110. Note the total number of 1s is even. If the ASCII code were $37_{(16)}$ (011 0111), the parity bit would be high. The actual transmitted data would be $B7_{(16)}$ (1011 0111) after the parity bit is attached to the data stream.

The parity error detection system has a severe limitation. It cannot detect even numbers of errors. In the preceding example, we transmitted $B7_{(16)}$ in an even parity system. What would happen at the parity checker if noise raised the two low bits high? The received data stream would appear as $FF_{(16)}$ (1111 1111). Notice the total number of 1s is even after the double error. Therefore, the parity checker assumes there is no data error when in fact there is a double error. This limitation is significant because usually the duration of a noise pulse covers many bits and the chance for undetectable double errors or even numbers of errors is significant.

Parity checkers are used in PCs to ensure the Random-Access Memory (RAM) chips are operating properly. RAM is referred to as system memory in a PC. It is the work storage area for data input to the system. RAM ICs and architecture will be presented in detail in Chapter 12. Since data are constantly being stored in and fetched (retrieved) from memory locations in the PC, it is imperative that the memory storage circuits work properly. Parity checking is accomplished during **boot-up** of a computer. Boot-up of the system occurs when power is applied to the computer and it is initialized. During the boot-up RAM test, many bit patterns are written to RAM, parity bits are generated, and data are read back to ensure accuracy. If a data bit is written low into a memory location but the memory circuit is stuck high due to a malfunction, the parity generator/checker within the system will detect the error, generate a parity error signal, and provide a parity error message to the monitor.

Parity Generator

The logic diagram in Fig. 5–12 shows a simple method of implementing a 4-bit parity generator. The X-OR gates are used to add the data input bits $(D_3 \oplus D_2) \oplus (D_1 \oplus D_0)$ as previously explained. The premise of this parity bit generation is simple: (1) carry outputs from this addition are disregarded; (2) when an even number of 1s is added, the sum is 0; and (3) when an odd number of 1s is added, the sum is 1.

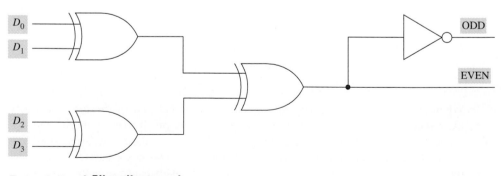

FIGURE 5–12   **4-Bit parity generator.**

Parity Generator/Checker

Figure 5–13(a) shows an SN74180 9-Bit ODD/EVEN Parity Generator/ Checker. The IC can be used for operation in an odd or even parity scheme. The ninth bit (OI or EI) can be used as a parity bit input for parity checking. The function table in Fig. 5–13(b) shows the different functions of the IC. A couple of examples will help clarify the table and circuit operation.

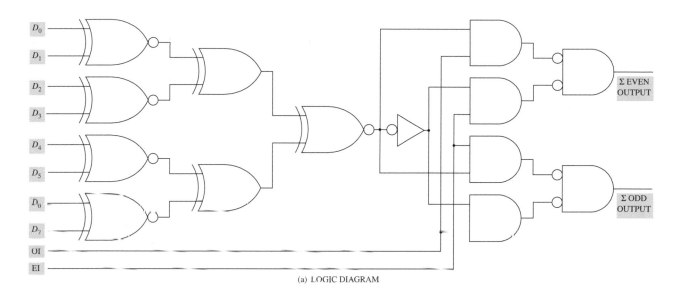

(a) LOGIC DIAGRAM

FUNCTION TABLE				
INPUTS			OUTPUTS	
# OF HIGH INPUTS (D_0–D_7)	EVEN	ODD	Σ EVEN	Σ ODD
0, 2, 4, 6, 8	1	0	1	0
1, 3, 5, 7	1	0	0	1
0, 2, 4, 6, 8	0	1	0	1
1, 3, 5, 7	0	1	1	0
X	1	1	0	0
X	0	0	1	1

(b) FUNCTION TABLE

FIGURE 5–13   SN74180 9-bit odd/even parity generator/checker.

The 74180 logic circuit in Fig. 5–14 is configured in the odd parity scheme as a parity generator. The Odd Input (OI) is connected to +5V (high) and the Even Input (EI) is connected to ground (low). The figure is annotated with data inputs (D_7–D_0) of 00011110. This data input contains an even number of 1s. Tracing the logic levels that are annotated on the figure proves this input combination generates a high Σ ODD output parity bit. Keep in mind that the purpose of an odd parity generator is to ensure an odd number of 1s is transmitted in every group of bits (including the parity bit).

The 74180 circuit in Fig. 5–15 is configured in the odd parity scheme as a parity checker. The difference between the parity generator and the parity checker is at the OI/EI inputs. In the parity checker, the parity bit is applied to the OI pin to detect odd parity. The complement of the parity bit is connected to the EI pin. The received data input bits (without a bit error) and parity bit level from the preceding example are annotated on the figure. The output of the circuit is a high Σ ODD output, which indicates the data received contained an odd number of 1s including the parity bit. Therefore, the received data are correct.

The parity system block diagram shown in Fig. 5–16 depicts the preceding example. The parallel data could be converted to serial data prior to transmission and then reconverted back to parallel data at the receiver. This action is unnecessary for purposes of this explanation.

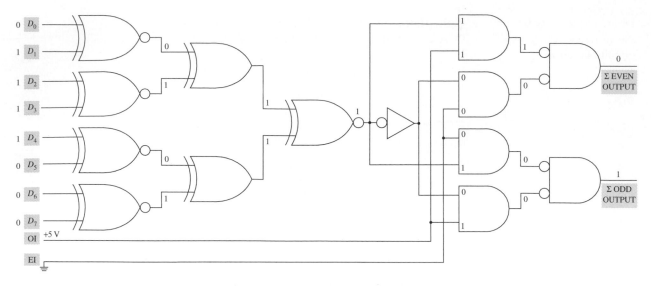

FIGURE 5-14   SN74180–parity generator–odd parity.

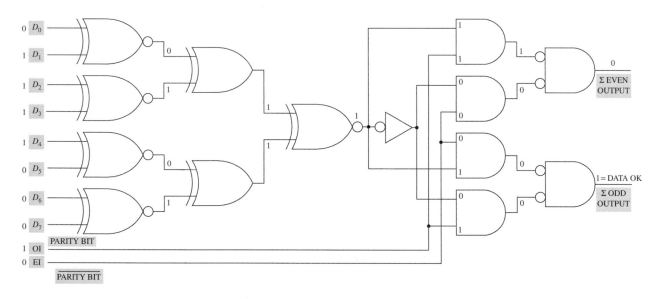

FIGURE 5-15   SN74180–parity checker–odd parity.

Let's repeat the previous data scenario with a **bit error** introduced (Fig. 5–17). The data transmitted is again 00011110. The **received data** D_0 bit has been raised to the 1 logic level. The input data, along with the corresponding odd/even inputs, are annotated on the 74180 parity checker in Fig. 5–17. Tracing the logic levels through the circuit shows the Σ ODD output = 0. This indicates the received data are incorrect.

Section 5-2: Review Questions

Answers are given at the end of the chapter.

A. Designate the proper parity bit level for an odd parity scheme for the following data:
    (1) 0110
    (2) 0111
    (3) 01011111
    (4) 10001000

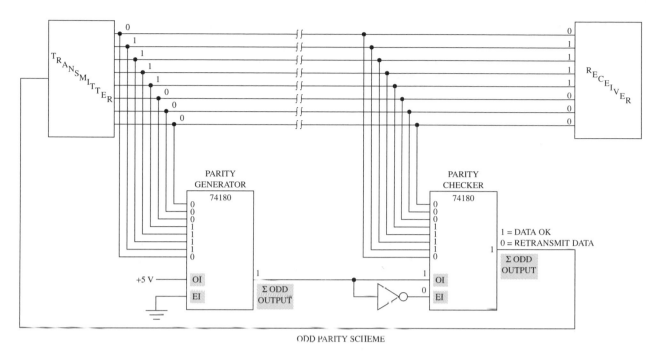

ODD PARITY SCHEME

FIGURE 5-16   **Parity system block diagram.**

B. Designate the proper parity bit level for an even parity scheme for the following data:
    (1) 1010
    (2) 1110
    (3) 10101111
    (4) 11111000

SECTIONS 5-1 AND 5-2: INTERNAL SUMMARY

The **exclusive-OR gate** was designed to produce a high output when the gate's inputs are complementary—$\overline{A}B + A\overline{B} = A \oplus B$. The gate is used in most binary arithmetic and parity circuits.

The **exclusive-NOR gate** was designed to produce a high output when the gate's inputs are equal—$\overline{A}\,\overline{B} + AB = \overline{A \oplus B}$. The gate is often used to compare inputs to determine if they are equal. One example of its use is in a magnitude comparator (Chapter 10).

Parity is a method that is employed to check the validity of data transmitted or written to memory with that of data received or read from memory. Basically, it is an **error detecting scheme.**

Parity requires a generator and a checker. Both must be set up for the same scheme—odd parity or even parity. The established parity scheme guarantees an odd or even number of 1s (high bits), including the parity bit, will always be sent or written to memory. If odd parity is selected, an odd number of 1s must always be sent and received to prevent a parity error signal.

Parity generator/checker circuits are available in ICs, and they may be configured to operate with either parity scheme. The SN74180 IC is an example of a parity generator/checker.

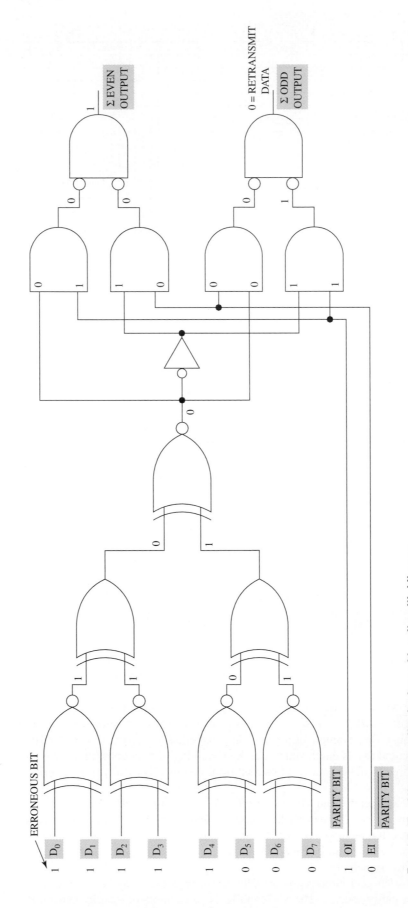

FIGURE 5–17 **SN74180—parity checker—odd parity with bit error.**

Sᴇᴄᴛɪᴏɴꜱ 5–1 ᴀɴᴅ 5–2: Iɴᴛᴇʀɴᴀʟ Sᴜᴍᴍᴀʀʏ Qᴜᴇꜱᴛɪᴏɴꜱ

Answers are given at the end of the chapter.

1. The logic symbol shown in Fig. 5–18 represents a/an _____ gate.
    a. OR   c. X-OR
    b. NOR   d. X-NOR

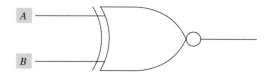

FIGURE 5-18

2. The output Boolean expression of the gate in Fig. 5–18 is
    a. $A + B$ c. $\overline{A \oplus B}$
    b. $A \oplus B$ d. $\overline{A}B + A\overline{B}$

3. The logic symbol shown in Fig. 5–19 represents a/an _____ gate.
    a. OR   c. X-OR
    b. NOR   d. X-NOR

4. The output Boolean expression of the gate in Fig. 5–19 is
    a. $A + B$ c. $\overline{A}B + A\overline{B}$
    b. $A \oplus B$ d. Both b and c

5. What is the output of the gate in Fig. 5–19 when $A = 0$ and $B = 0$?
    a. 0
    b. 1

6. What is the output of the gate in Fig. 5–19 when $A = 0$ and $B = 1$?
    a. 0
    b. 1

7. The gate in Fig. 5–19 can be used to produce the sum (Σ) of bits A and B
    a. True
    b. False

8. The gate in Fig. 5–19 can be used to produce the carry output generated when A and B are added.
    a. True
    b. False

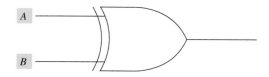

FIGURE 5-19

9. The circuit shown in Fig. 5–20 produces the _____ function.
    a. OR   c. X-OR
    b. NOR   d. X-NOR

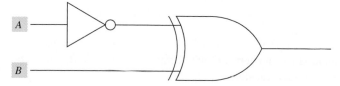

FIGURE 5-20

10. The circuit shown in Fig. 5–21 produces an output equal to
    a. $\overline{A}B + A\overline{B}$ c. $\overline{A}\ \overline{B}$
    b. $\overline{A}\ \overline{B} + AB$ d. AB

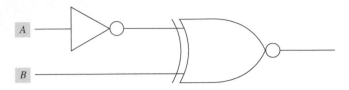

Figure 5–21

11. A parity system is used to ensure equality of the received data with that of the transmitted data.
    a. True
    b. False

12. The odd parity scheme guarantees the number of 0s transmitted will always be odd.
    a. True
    b. False

13. What parity bit must be added to data $78_{(16)}$ for even parity?
    a. 0
    b. 1

14. What parity bit must be added to data $3B_{(16)}$ for odd parity?
    a. 0
    b. 1

15. A parity error detection system cannot detect even numbers of errors such as double or quadruple errors.
    a. True
    b. False

SECTION 5-3: CONTROL CIRCUITS

OBJECTIVES

1. Given a logic diagram of a logic control circuit and the circuit's inputs, determine the circuit's output(s).
2. Given a logic diagram of a logic control circuit, determine the input combinations required to obtain a specifed output.
3. Identify the symbol for and purpose of Schmitt-trigger-input circuits.

Loading data into a digital circuit is depicted in Fig. 5–22(a) and (b). The blocks in the figures represent eight internal circuits, each capable of storing one bit of data. The data can be transferred (loaded) into this circuit **serially** or in **parallel.**

Serial loading in Fig. 5–22(a) requires a clock pulse to load one bit of data. The clock pulse is used to control the loading operation so that timing requirements in the system can be maintained. Once the D_0 bit of data is loaded, another clock pulse is required to load the D_1 bit and shift the D_0 bit one circuit to the right. This same process is repeated until all of the serial data bits (D_0-D_7) have been loaded into the circuit. This action requires a total of eight clock pulses.

Parallel loading data in Fig. 5–22(b) allows all of the data bits (P_0-P_7) to be loaded with one clock pulse. This method of transferring data within a system is often used because it is much faster than serial data transfer.

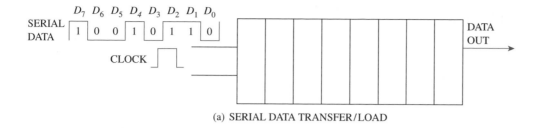

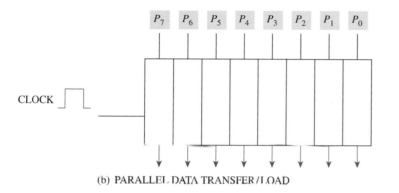

(b) PARALLEL DATA TRANSFER/LOAD

FIGURE 5-22   **Data transfer block diagram.**

There are times when serial data transfer is necessary. One of these times is when data must be transferred over long distances. Serial data transfer requires only one transmission line between the transmitting system and the receiving system. Parallel data transfer would require eight lines to transfer eight data bits (byte of data) simultaneously. This would be cost prohibitive over long distances.

Shift/$\overline{\text{Load}}$ Control Circuit

An example of a combinational logic circuit used to parallel load data into a circuit is shown in Fig. 5–23(a). The Shift/Load input line provides two separate functions. As the mnemonic (Shift/$\overline{\text{Load}}$) implies, when this input line is high, the serial data (D_0) will be *shifted* into circuit X. Note when Shift/$\overline{\text{Load}}$ = 1 both NAND gates are inhibited. This prevents parallel input data (P_0) from being applied to circuit X. The subject of shifting data serially will be expanded in the next control circuit in this section.

When the Shift/$\overline{\text{Load}}$ line is low, the parallel data input (P_0) will be loaded into circuit X. The function table in Fig. 5–23(b) lists the outputs for the usable input combinations to circuit X. For example, when $\overline{\text{Load}}$ = 0 and P_0 = 1, the outputs of the control circuit are $\overline{S}$ = 0 and $\overline{C}$ = 1, which causes circuit X's output to equal 1. Thus, the data at the P_0 input is loaded into the circuit.

Shift Right/Shift Left Control Circuit

The combinational logic circuit shown in Fig. 5–24(a) can be used to shift data to the right from circuit X to circuit Y or to the left from circuit Y to circuit X. Although the overall circuit is somewhat simplified, the use of the shift control circuit will become evident.

The function table in Fig. 5–24(b) shows the required levels of the control inputs (C_0 and C_1) to achieve certain functions:

$C_1 = 0$ **and** $C_0 = 1$: This input combination produces a **shift right operation** in the circuit. Tracing the control input logic levels shows all AND gates except #1 are inhibited. The low outputs from AND gates #2, #3, and #4 act as enablers to the OR gate. The

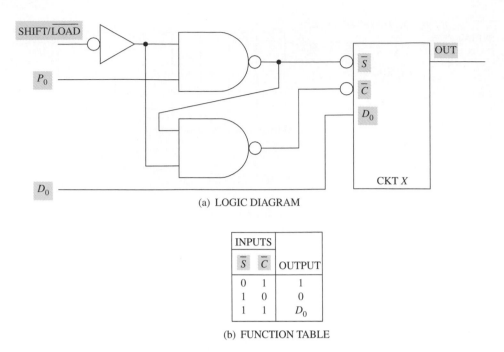

(a) LOGIC DIAGRAM

INPUTS		
$\bar{S}$	$\bar{C}$	OUTPUT
0	1	1
1	0	0
1	1	D_0

(b) FUNCTION TABLE

FIGURE 5-23   **Shift/$\overline{\text{Load}}$ combinational logic circuit.**

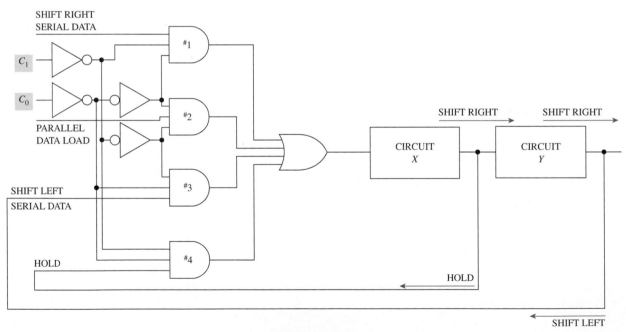

(a) LOGIC DIAGRAM

C_1	C_0	MODE OF OPERATION
0	1	SHIFT RIGHT
1	0	SHIFT LEFT
0	0	HOLD
1	1	PARALLEL LOAD

(b) FUNCTION TABLE

FIGURE 5-24   **Shift Right/Shift Left control circuit.**

#1 AND gate allows the serial data to pass through the control circuitry to circuit X. Another duplicate shift control circuit would allow the data from circuit X to shift right to circuit Y.

$C_1 = 1$ and $C_0 = 0$: This input combination produces a **shift left operation** in the circuit. The control logic causes the data bit in circuit Y to be shifted through AND gate #3 into circuit X. The #3 AND gate is enabled with this input combination, and the remaining AND gates are inhibited.

$C_1 = 0$ and $C_0 = 0$: This input set places the circuit in the **hold mode** of operation. In other words, the data is not shifted in either direction—it is stored. In actuality, the output of circuit X is routed through the enabled #4 AND gate and back to the input of circuit X. You might look at this operation as recirculating the data.

$C_1 = 1$ and $C_0 = 1$: This function allows **parallel loading** of data into the circuit in lieu of serial loading. Note the #2 AND gate is enabled, and the remaining AND gates are inhibited when $C_1 = C_0 = 1$. In this example, one parallel bit would be loaded into circuit X from the parallel data load input. A duplicate control circuit would allow parallel loading of another data bit into circuit Y at the same time.

Reel Direction Control Circuit

The control circuit shown in Fig. 5–25 can be used in a video recorder to control the direction of reel movement. The Reel Play and Reel Reverse signals are generated in the recorder's control circuits. When Reel Play is activated, the direction of reel wind is controlled by the Reel Reverse input because the NAND gate is enabled when Reel Play = 0. If Reel Reverse is high, a low output from the NAND gate turns on the reverse wind circuit. If Reel Reverse is low, a high output from the NAND gate turns on the forward wind circuit.

The NOT gate in the figure contains a symbol not previously used in the text. The symbol inside of the NOT gate represents a **Schmitt-trigger input.**

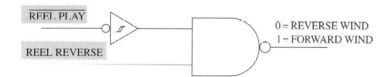

REEL PLAY

REEL REVERSE

0 = REVERSE WIND
1 = FORWARD WIND

FIGURE 5–25   Reel direction control circuit.

Schmitt-Trigger-Input Circuits

Schmitt-trigger-input circuits are used when good, clean digital signals are not available to drive an input. These circuits are capable of accepting slowly changing input signals and converting them into sharply defined, jitter-free output signals. Schmitt-trigger circuits use positive feedback to speed up slow rise and slow fall time signals. The input and output signals for the Schmitt-trigger-input inverter are shown in Fig. 5–26(a). Notice how the rise and fall times of the output signal have been decreased to almost vertical transit times.

Standard TTL circuits will accept 0 V up to +0.8 V as a Logic 0 (low) input and 2 V to 5 V as a Logic 1 (high) input. These levels are shown in Fig. 5–26(b). Keep in mind that unreliable outputs can be produced from an IC when its inputs remain in the **indeterminate logic range** too long. The object of the Schmitt-trigger-input circuit is to get an input signal through the indeterminate logic range as quickly as possible.

The input levels of Schmitt-trigger-input circuits are specified V_{T+} for the positive-going threshold level and V_{T-} for the negative-going threshold level. These values are listed in data books, and they are typically $V_{T-} = +0.6\,\text{V}$ to $+1.1\,\text{V}$ and $V_{T+} = +1.5\,\text{V}$ to $+2\,\text{V}$. The difference between the two threshold levels is called **hysteresis.** Hysteresis is normally about 800 mV.

Several different logic gates and multivibrator circuits are available with Schmitt-trigger-input circuits. Some of the circuits can reliably use signal transition rates as slow as 1 volt-second.

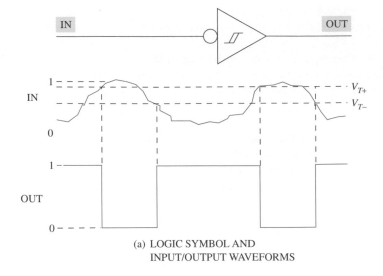

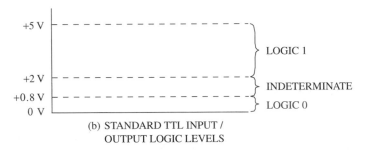

(a) LOGIC SYMBOL AND
INPUT/OUTPUT WAVEFORMS

(b) STANDARD TTL INPUT /
OUTPUT LOGIC LEVELS

FIGURE 5-26   **Schmitt-trigger inverter.**

Section 5-3: Review Questions

Answers are given at the end of the chapter.

A. Draw the symbol for an inverter with a Schmitt-trigger input.
B. What is the purpose of using circuits with Schmitt-trigger inputs?

SECTION 5-4: DETECTION/SELECTION/DISTRIBUTION LOGIC CIRCUITS

OBJECTIVE

Given a logic diagram of a logic control circuit and the circuit's inputs, determine its output(s).

BCD Invalid Sum Detector

Chapter 2 presented a preview of binary-coded decimal (BCD) numbers. A simple arithmetic problem was used to illustrate why invalid BCD sums often appear in BCD arithmetic circuits. Another example is presented here:

$$
\begin{array}{rcl}
0110_{(BCD)} & = & 6_{(10)} \\
+0111_{(BCD)} & = & +7_{(10)} \\
\hline
1101 & = & 13_{(10)}
\end{array}
$$

The problem in this example is that 1101 is not a valid BCD number. The number $13_{(10)}$ in BCD format is $0001\ 0011_{(BCD)}$. Remember, *a BCD number is a group of four bits used to represent one decimal digit*. This indicates that BCD numbers must always be 0 through 9. You know that four bits can be used to represent numbers from 0000 to 1111. This means any sum of 1010 to 1111 generated in a BCD adder is *invalid*. These invalid sums must be detected and then corrected to make them usable. The simple combinational logic circuit shown in Fig. 5–27(a) can be used to solve the detection problem. The correction problem will be solved in Chapter 9.

The Boolean expression for the circuit in Fig. 5–27(a) is $BD + CD$. Keep in mind the expression is written for a high output. It is read "the circuit will produce a high output when B and D are high *or* when C and D are high." Will this expression and logic circuit solve the problem of detecting an invalid sum?

The easiest way to prove the validity of the circuit to detect invalid sums is to compare the expression to a count sequence table. The table is shown in Fig. 5–27(b). The B and D high input combinations are marked on the table with asterisks (*). The C and D high input combinations are marked with pound signs (#). The logic circuit produces high outputs for these conditions. This includes all counts or sums from 1010 to 1111.

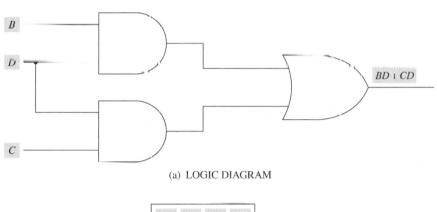

(a) LOGIC DIAGRAM

Q_D	Q_C	Q_B	Q_A	
0	0	0	0	
0	0	0	1	
0	0	1	0	
0	0	1	1	
0	1	0	0	
0	1	0	1	
0	1	1	0	
0	1	1	1	
1	0	0	0	
1	0	0	1	
*	1	0	1	0
*	1	0	1	1
#	1	1	0	0
#	1	1	0	1
*#	1	1	1	0
*#	1	1	1	1

INVALID BCD SUMS

(b) BCD COUNT SEQUENCE AND INVALID SUMS

FIGURE 5-27   BCD invalid sum detector.

Data Selection/Distribution Logic Circuits

A method of implementing a **data selector** logic circuit is shown in Fig. 5–28. The data being brought into the circuit are A_0B_0 and A_1B_1. Naturally, this circuit could be expanded to increase the number of input data bits. The *Select (S)* input is used to select either A_0B_0 or A_1B_1 as the output data.

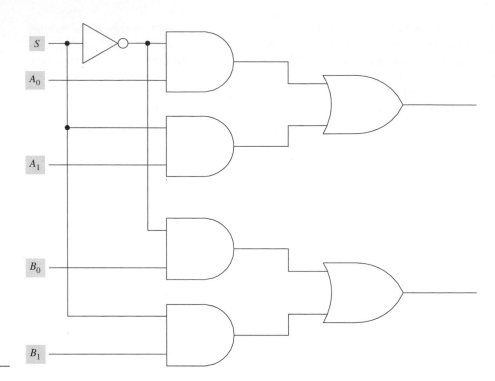

FIGURE 5–28   **Data selector logic circuit.**

The $A_0 B_0$ data are selected and passed through their respective enabled AND gates when the Select input is low. The $A_1 B_1$ AND gates are inhibited when $S = 0$. The $A_1 B_1$ data are selected and passed through their respective enabled AND gates when the Select input is high.

The output expression for the top OR gate in Fig. 5–28 is $\overline{S} \cdot A_0 + S \cdot A_1$. The lower OR gate expression is $\overline{S} \cdot B_0 + S \cdot B_1$. The circuit can be used for time-sharing one output line between two input lines. This is the process of selecting data referred to as **multiplexing** in digital systems.

A **data distributor** combinational logic circuit is shown in Fig. 5–29. Serial data D_0, D_1, D_2, and D_3 are applied to the data input line at a given frequency. The Select inputs (S_1, S_0) are used to control which AND gate is enabled. The enabled AND gate will pass the input data to its output. The Select inputs can be controlled by a binary up-counter. The counter's output must change at the same frequency as the data input changes.

The four serial data input bits will be routed to the four output lines as follows: $D_0 \cdot \overline{S_1}\overline{S_0}$ to the output of AND gate #0; $D_1 \cdot \overline{S_1}S_0$ to the output of AND gate #1; $D_2 \cdot S_1\overline{S_0}$ to the output of AND gate #2; and $D_3 \cdot S_1S_0$ to the output of AND gate #3. Note how the data inputs are routed to a selected AND gate by the binary count of the counter, which is applied to S_1 and S_0. Timing of the data input signal and the select control circuit is critical for proper circuit operation.

The data distributor in Fig. 5–29 can be used to convert serial data to parallel data as shown. The process of data distribution is called **demultiplexing** in digital systems.

Section 5–4: Review Questions

Answers are given at the end of the chapter.

A. Write the Boolean expressions for the two OR gate outputs of the logic circuit in Fig. 5–28.
B. What data are present at the outputs of the logic circuit in Fig. 5–28 when $S = 1, A_0 = B_0 = 1$, and $A_1 = B_1 = 0$?
C. What is meant by the term "multiplexing" in digital circuits?
D. What is meant by the term "demultiplexing"?

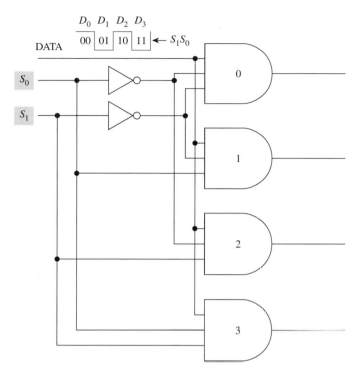

$$D_0 \ D_1 \ D_2 \ D_3$$

FIGURE 5–29 **Data distributor logic circuit.**

OBJECTIVE

Given a logic diagram of a combinational logic circuit and its inputs, determine the circuit's output(s).

Carry Out Logic Circuit

Binary counter circuits are presented in detail in Chapter 7. It is appropriate to mention a few points here so a Carry Out combinational logic circuit can be analyzed in this section.

The maximum count of a 4-bit binary up-counter is $1111_{(2)}$. This count is often called the **terminal count.** The counter will typically start counting in the 0000 state and sequentially count up to 1111 as it is clocked. In some circuits a $\overline{\text{Carry Out}}$ signal is generated on the terminal count. This output can be used to provide a control signal to another 4-bit counter.

Binary counters can be connected in series to increase their maximum count capability. Figure 5–30(a) shows two 4-bit up-counters connected in series. This type of series connection is called **cascading.** A 4-bit binary up-counter can be cascaded with another 4-bit binary up-counter to increase the maximum count from $1111(15_{(10)})$ to $11111111(255_{(10)})$.

The circuit shown in Fig. 5–30(a) shows a clock signal input to produce up-count outputs from each counter. The $\overline{\text{Carry In}}$ $(\overline{C}_{\text{in}})$ signal to each counter must be low for the counter to count. The $\overline{C}_{\text{in}}$ line to the #1 counter is tied low. This up-counter will increment (increase) one count for each clock input pulse. The $\overline{\text{Carry Out}}$ signal from the #1 counter does not go low until the terminal count of 1111 is reached. On a count of 1111, the $\overline{\text{Carry Out}}$ signal is asserted (brought low) and counter #2 will increment one count on the next input clock pulse. This would cause the count to change from $00001111_{(2)}$ to $00010000_{(2)}$ (15 to $16_{(10)}$). The counts of 15, 16, and 17 are annotated at the Q outputs of the counters in Fig. 5–30(a).

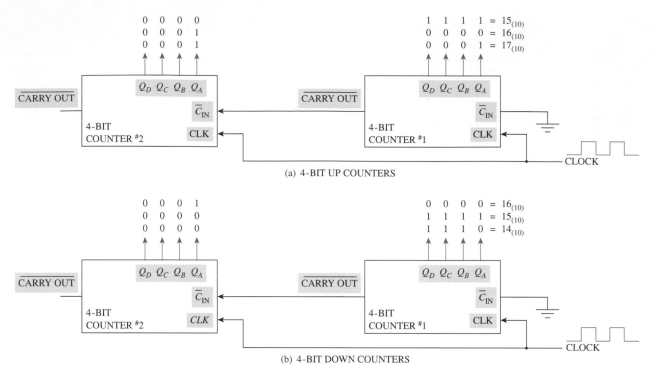

(a) 4-BIT UP COUNTERS

(b) 4-BIT DOWN COUNTERS

FIGURE 5–30 **Cascaded binary counters.**

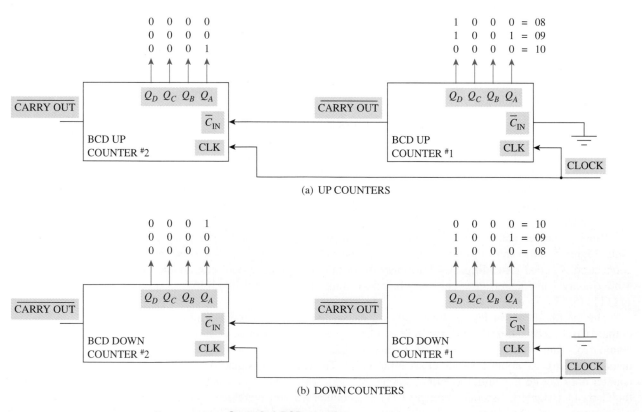

(a) UP COUNTERS

(b) DOWN COUNTERS

FIGURE 5–31 **Cascaded BCD counters.**

The cascaded counters in Fig. 5–30(b) are down-counters. The terminal count of a 4-bit binary down-counter is 0000. This type of counter typically counts down from 1111 to 0000 as it is clocked. Therefore, the Carry Out signal would need to be activated on the count of 0000 to produce a signal to the next 4-bit down-counter. This active signal would cause the #2 counter to decrement (decrease) one count. This effectively is a borrow output from the next counter. For example, the next count down from 00010000 is $00001111_{(2)}$. Thus, activating the $\overline{\text{Carry Out}}$ line on a count of 0000 from the #1 counter causes the counters to decrement from 16 to $15_{(10)}$.

The circuit in the block diagram in Fig. 5–31(a) is a Binary-Coded Decimal (BCD) up-counter. The principles explained for the 4-bit binary up-counter apply here. There is only one difference—the terminal count for the BCD up-counter is 1001. If a BCD up-counter is cascaded with another BCD up-counter, the count capability is increased to $1001\ 1001_{(\text{BCD})}\ (99_{(10)})$.

The $\overline{\text{Carry Out}}$ signal, activated on the BCD count of 1001, will cause the *#2 BCD counter* to increment from $\underline{0000}\ 1001_{(\text{BCD})}$ to $\underline{0001}\ 0000_{(\text{BCD})}$. This produces BCD counts equivalent to 09 and 10 in decimal. These counts are annotated in Fig. 5–31(a).

The terminal down count for the cascaded BCD down-counters in Fig. 5–31(b) is $0000_{(\text{BCD})}$. This terminal BCD count causes the $\overline{\text{Carry Out}}$ signal to be activated and the #2 BCD down-counter to decrement one count as shown in Fig. 5 31(b). The BCD mode of operation is called the **decade** mode of operation in the control circuit discussed next.

The analysis presented here is for a combinational logic circuit that will generate an active-low Carry Out signal on the terminal count. *The signal must be activated at 0000 on a down-count, at 1111 on a 4-bit binary up-count, and at 1001 on a BCD up-count.*

The combinational logic circuit shown in Fig. 5–32(a) is used to generate the active-low $\overline{\text{Carry Out}}$ signal. Some points to keep in mind during this analysis: The counters will not count unless $\overline{C}_{\text{in}} = 0$; the #1, #2, and #3 logic gates are used exclusively during down-count operation; and the #4 through #9 gates are used exclusively for up-count operation. The circuit is redrawn using some alternate gate symbols in Fig. 5–32(b).

4-Bit Binary Up-Count Operation.

Figure 5–32(b) has been annotated with input logic levels for 4-bit binary up-counting in Fig. 5–33. The $\overline{\text{Carry Out}}$ signal in the figure will be activated when the following input conditions are met:

$$\overline{C}_{\text{in}} = 0$$
$$B\text{in}/\overline{\text{Decade}} = 1$$
$$\text{Up}/\overline{\text{Down}} = 1$$
$$Q_3 - Q_0 = 1111 \quad \text{(Terminal count)}$$

The Up/$\overline{\text{Down}}$ = 1 input in Fig. 5–33 enables the NOR gate (#7) after the signal has been inverted by the NOT gate (#8). The Bin/$\overline{\text{Decade}}$ = 1 input enables the AND gate (#6). The terminal count of 1111 produces $\overline{Q}_0 = 0$, $\overline{Q}_1 = 0$, $\overline{Q}_2 = 0$, and $\overline{Q}_3 = 0$. Tracing the annotated logic levels through Fig. 5–33 results in the Carry Out signal being brought low (asserted/activated) when these conditions are applied.

Down-Count Operation.

The following input signals (Fig. 5–34) are required to activate the $\overline{\text{Carry Out}}$ signal during down-count operation:

$$\overline{C}_{\text{in}} = 0$$
$$B\text{in}/\overline{\text{Decade}} = \text{X}$$
$$\text{Up}/\overline{\text{Down}} = 0$$
$$Q_3 - Q_0 = 0000 \quad \text{(Terminal Count)}$$

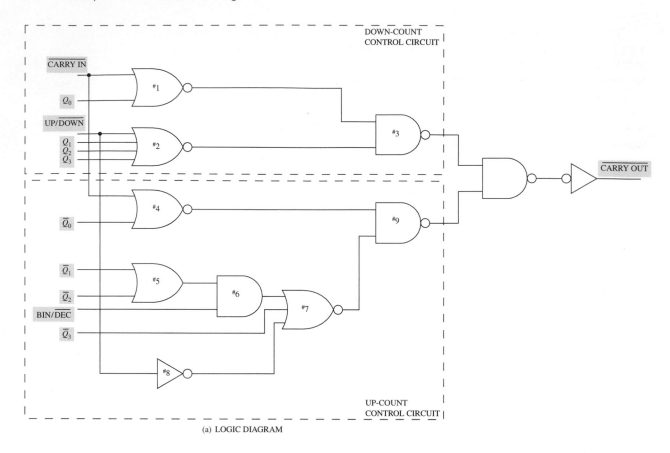

(a) LOGIC DIAGRAM

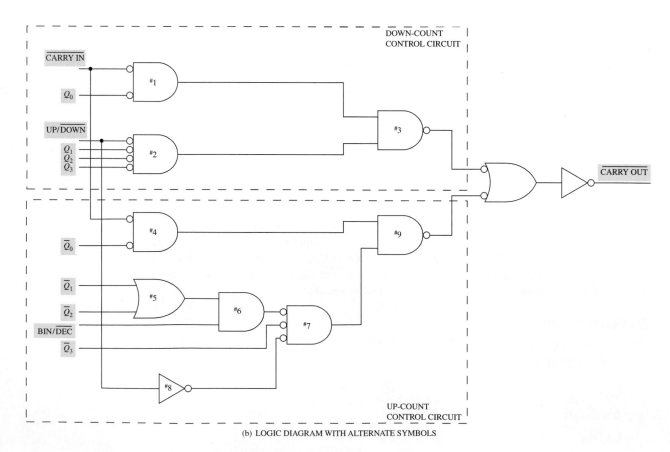

(b) LOGIC DIAGRAM WITH ALTERNATE SYMBOLS

FIGURE 5-32 **Carry out control circuit.**

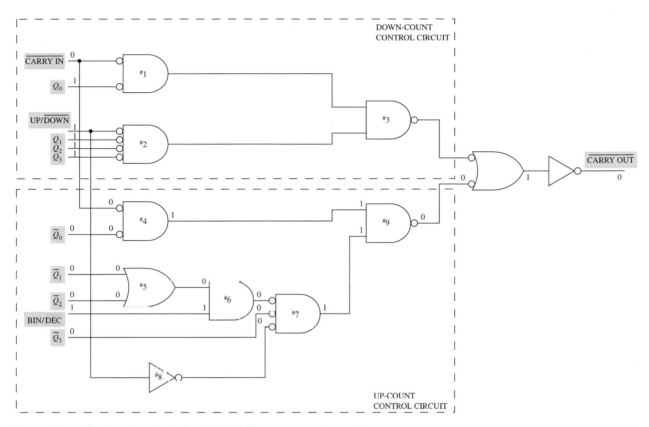

FIGURE 5-33   Carry out control circuit—4-bit binary up-count operation.

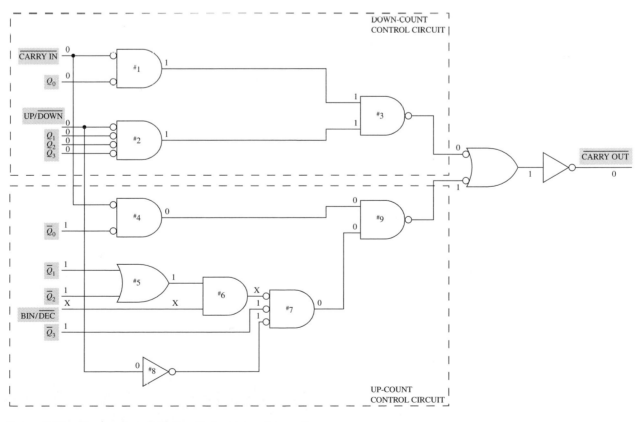

FIGURE 5-34   Carry out control circuit—down-count operation.

Notice the Up/$\overline{\text{Down}}$ = 0 input removes the lower part of the circuit (up-count control circuit gates #4 through #9) by inhibiting NOR gate #7. Gate #7's low output inhibits gate #9.

The logic levels producing $\overline{\text{Carry Out}}$ = 0 are annotated in Fig. 5–34. This down-count configuration requires $Q_3 - Q_0$ = 0000 as the terminal count required to activate $\overline{\text{Carry Out}}$. Note the Bin/Decade input is a don't care input in the down-count configuration because NOR gate (#7) is inhibited by the Up/$\overline{\text{Down}}$ = 0 input. Keep in mind the terminal count is 0000 during both BCD and binary down-counting. This mode of operation will be used for troubleshooting the control circuit in the next section.

BCD Up-Count Operation.

Figure 5–35 has been annotated with the input conditions required to activate $\overline{\text{Carry Out}}$ during BCD up-count operation. Normal operation of a BCD up-counter produces the following count:

Q_3	Q_2	Q_1	Q_0	
0	0	0	0	
0	0	0	1	
0	0	1	0	
0	0	1	1	
0	1	0	0	
0	1	0	1	
0	1	1	0	
0	1	1	1	
1	0	0	0	
1	0	0	1	(Terminal count)
0	0	0	0	

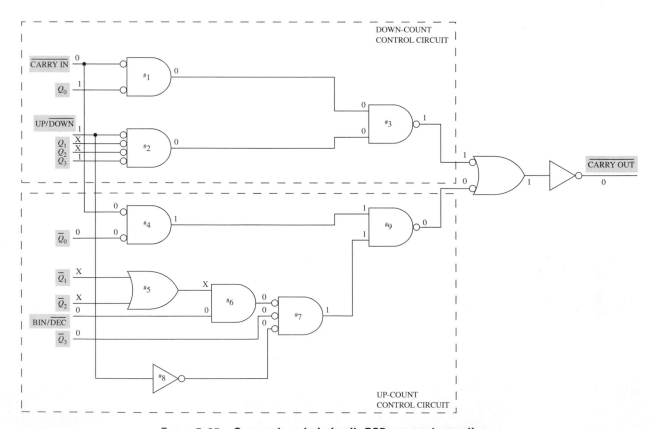

FIGURE 5–35   Carry out control circuit–BCD up-count operation.

The count shows $Q_0 = 1$ and $Q_3 = 1$ during the terminal count. During this count, $\overline{Q_0} = 0$ and $\overline{Q_3} = 0$. These two inputs are used by gates #4 and #7 to determine the terminal count is 1001 instead of 1111 as it was in the binary up-counter.

The input conditions for BCD up-count operation are

$$\overline{C}_{in} \qquad = 0$$

$$Bin/\overline{Decade} = 0$$

$$Up/\overline{Down} \quad = 1$$

$$\overline{Q_0} = \overline{Q_3} \quad = 0 \quad \text{(Terminal count 1001)}$$

The $\overline{Q_1}$ and $\overline{Q_2}$ inputs to the OR gate (#5) were used in binary up-count operation to indicate, in part, when the terminal count of 1111 was reached. These inputs are don't care inputs in this configuration because the AND gate (#6) is inhibited by the Bin/$\overline{Decade}$ = 0 input. This causes the generation of a low Carry Out on the count of 1001.

Rom Address Decoder Logic Circuit

Some of the general concepts of selecting and addressing the Read-Only Memory (ROM) IC in a computer were presented in Section 3–10. The circuit shown in Fig. 5–36 expands those concepts. A review of the Chapter 3 material explaining basic ROM addressing takes only a couple of minutes and will be beneficial during the following discussion.

The circuit in Fig. 5–36 expands the memory capability of the Chapter 3 circuit to six ROM ICs. NAND gates #2 through #7 are used to select the desired ROM IC. Two conditions must be met before these NAND gates are enabled. Address bits A_{16} through A_{19} must be high to produce a low out of their respective page address NAND gate. This F page address allows communication with one of the ROM chips providing the memory read ($\overline{MEMR}$) signal is low. The low from the page address NAND gate produces a ROM Page ($\overline{ROMPG}$) signal, which is inverted to a high and then applied to AND gate #8. The $\overline{MEMR}$ low signal is also inverted and applied to the same AND gate. The output of AND gate #8 is high when A_{16} through A_{19} are high and $\overline{MEMR}$ is low. This high output enables all of the decoding NAND gates. The output of AND gate #8 is low, and all of the decoding gate's outputs will be inactive (high) if the page address is not 1111 or if $\overline{MEMR}$ is not active.

The A_{13}, A_{14}, and A_{15} address input bits in Fig. 5–36 select and activate the proper decoding gate's output. If these three address bits are high, tracing the logic levels shows the #7 NAND gate's output is activated (low). This selects ROM chip #6 because Chip Select ($\overline{CS}$) = 0 on that chip. The remaining address input lines (A_0–A_{12}) select a specific memory location within the selected ROM IC.

Data (D_0–D_7) are placed on the data bus when the proper ROM IC is selected and an address is applied to the IC. The $\overline{G}$ enable and DIR input pins on the *transceiver* must both be low to enable the IC to pass data from the ROM memory location to the microprocessor. A **transceiver** is a transmitter/receiver circuit that allows data movement in one direction or in the opposite direction when the chip is enabled. Direction of the data movement is controlled by control signal DIR. The microprocessor is disconnected from the bus by the transceiver when the $\overline{G}$ input pin is high.

The direction of data movement through the transceiver is controlled by the three NOR gates below the decoding NAND gates (#0 − #7) in the figure. The data flow is from left to right through the transceiver when the DIR input pin is low. This condition exists during a ROM read operation because $\overline{ROMPG}$ = 0 and $\overline{MEMR}$ = 0 into the NOR gate directly under NAND gate #7. The high output from this NOR gate inhibits the next NOR gate and sets the DIR input low. This condition also exists during Input/Output Read ($\overline{IOR}$) operations when $\overline{IOR}$ = 0 and address bits A_8 and A_9 = 0. Address bits A_8 and A_9 are used to enable a support chip decoder for I/O chip select operation in some systems.

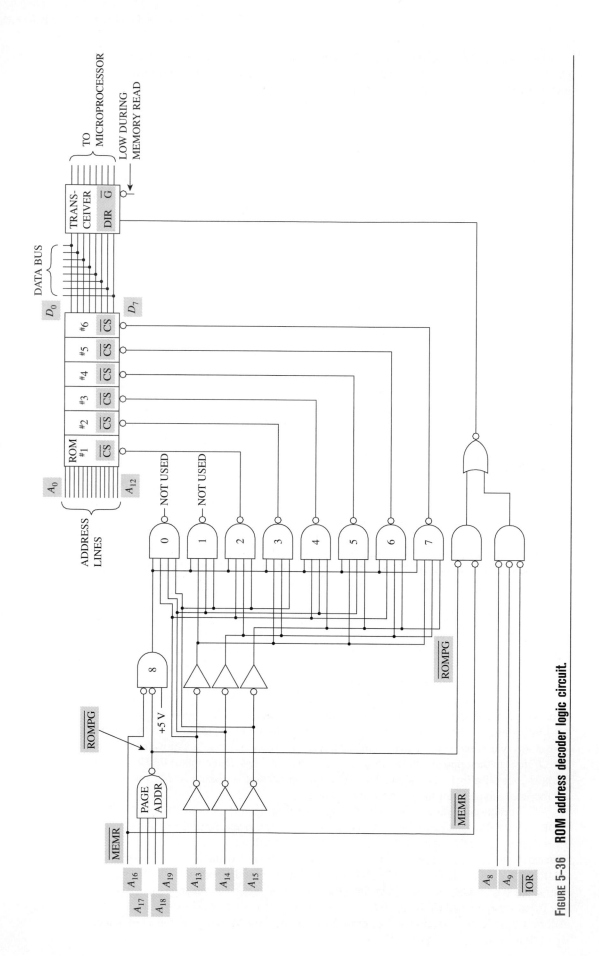

FIGURE 5-36 ROM address decoder logic circuit.

The ROM address decoder logic circuit just presented included Input/Output Read ($\overline{\text{IOR}}$) and Memory Read ($\overline{\text{MEMR}}$) input signals. These signals can be generated by the combinational logic circuit shown in Fig. 5–37.

The signals are generated from three microprocessor output signals—Read ($\overline{\text{RD}}$), Write ($\overline{\text{WR}}$), and Input/Output Memory (IO/$\overline{\text{M}}$). The $\overline{\text{RD}}$ signal is an active-low signal used to indicate the microprocessor is reading an I/O or memory location when the signal is active (low). The $\overline{\text{WR}}$ signal indicates the microprocessor is writing to an I/O or memory location when it is active. The IO/$\overline{\text{M}}$ signal identifies the current microprocessor operation as an I/O access (IO/$\overline{\text{M}}$ = 1) or a memory access (IO/$\overline{\text{M}}$ = 0).

These three microprocessor signals are used in Fig. 5–37 to generate read or write signals specifically for memory chips or I/O devices. Operation of the circuit is simple enough to forgo an explanation.

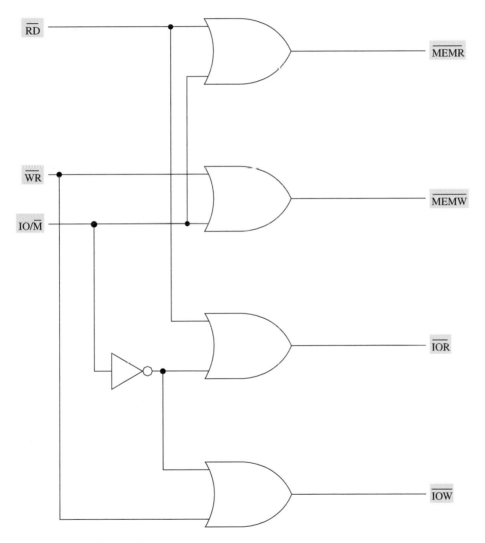

FIGURE 5–37   **Memory and I/O Read/Write signals.**

Section 5-5: Review Questions

Answers are given at the end of the chapter.

A. What mode of operation is selected in Fig. 5–32(b) when $\overline{\text{Carry In}}$ = 0, Up/$\overline{\text{Down}}$ = 1, Bin/$\overline{\text{Dec}}$ = 1?

B. What mode of operation is selected in Fig. 5–32(b) when $\overline{\text{Carry In}}$ = 0, Up/$\overline{\text{Down}}$ = 0, Bin/$\overline{\text{Dec}}$ = 0?

C. What happens in the circuit in Fig. 5–32(b) when $\overline{\text{Carry In}} = 1$?

D. What page address must be applied to the ROM address circuit in Fig. 5–36 to assert (activate) the $\overline{\text{ROMPG}}$ signal?

E. Which ROM chip in Fig. 5–36 is selected when $\overline{\text{MEMR}} = 0$ and

$$A_{19}\, A_{18}\, A_{17}\, A_{16}\, A_{15}\, A_{14}\, A_{13}\, A_{12}$$
$$1 \quad 1 \quad 1 \quad 1 \quad 0 \quad 0 \quad 1 \quad X$$

F. Which ROM chip in Fig. 5–36 is selected when $\overline{\text{MEMR}} = 0$ and

$$A_{19}\, A_{18}\, A_{17}\, A_{16}\, A_{15}\, A_{14}\, A_{13}\, A_{12}$$
$$1 \quad 1 \quad 1 \quad 1 \quad 1 \quad 0 \quad 0 \quad X$$

G. Which ROM chip in Fig. 5–36 is selected when $\overline{\text{MEMR}} = 0$ and

$$A_{19}\, A_{18}\, A_{17}\, A_{16}\, A_{15}\, A_{14}\, A_{13}\, A_{12}$$
$$0 \quad 0 \quad 0 \quad 0 \quad 1 \quad 1 \quad 1 \quad X$$

H. Define a transceiver.

I. What are the output levels at $\overline{\text{MEMR}}$, $\overline{\text{MEMW}}$, $\overline{\text{IOR}}$, and $\overline{\text{IOW}}$ in Fig. 5–37 when the microprocessor input signals are $\overline{\text{RD}} = 0$, $\overline{\text{WR}} = 1$, and $\text{IO}/\overline{\text{M}} = 0$?

SECTION 5-6: TROUBLESHOOTING

OBJECTIVES

1. Given the description of a problem or specific malfunction in a combinational logic circuit, identify the most likely output results.

2. Given the improper output result(s) of a defective combinational logic circuit, determine the most likely malfunction.

Sections 5–3, 5–4, and 5–5 provided detailed operation of numerous combinational logic circuits. These sections provided the knowledge necessary for the troubleshooting presented in this section. A technician must always know how a circuit should work during normal operation before attempting to troubleshoot.

The carry out control circuit presented in the previous section is used in this troubleshooting example. One mode of operation of the carry out control circuit is presented in detail in this section. When troubleshooting the circuit in this mode of operation is thoroughly understood, analysis of the other three modes of operation may be performed and compared to the summary tables provided at the end of this section. You may wish to compare your knowledge and analytical skills with the results in the troubleshooting summary tables provided for each mode of operation without going through the remainder of this discussion. If problems are encountered in your analysis of the down-count mode of operation, which is summarized in Table 5–2, you are encouraged to read the following information.

Table 5–1 shows the mode control inputs that establish proper operation of the circuit for a given count (binary or BCD and up or down). The control circuit is shown in Fig. 5–38. The circuit has been slightly modified when compared to Fig. 5–32(b) so that it can be easily implemented with available ICs.

A counter is connected to the Q_3, Q_2, Q_1, and Q_0 inputs to the control circuit. The counter must have the capability to count up or down in binary or BCD. The $\overline{\text{Carry In}}$ input must be low for the circuit to function in any of its four modes of operation. If $\overline{\text{Carry In}} = 1$, the outputs of the 7402 NOR gates at pins 1 and 4 in Fig. 5–38 will be low. These two signals will inhibit two of the 7400 NAND gates, causing pins 3 and 6 to be high. These signals will produce a low at pin 8 of the NAND gate. This low is inverted at pin 4 of the NOT gate. This condition ($\overline{\text{Carry In}} = 1$) will prevent the $\overline{\text{Carry Out}}$ ($\overline{C}_{\text{out}}$) signal from going low.

Bin/$\overline{\text{Dec}}$	Up/$\overline{\text{Down}}$	Operation Mode with $\overline{C}_{in}$ = 0
X	0	BCD/4-bit binary down-count (Note 1)
0	1	BCD up-count (Note 2)
1	1	4-bit binary up-count (Note 3)

Note 1: $\overline{C}_{out}$ = 0 when Q_3–Q_0 = 0000
Note 2: $\overline{C}_{out}$ = 0 when $\overline{Q}_3$–$\overline{Q}_0$ = 0   (First occurs on count 1001)
Note 3: $\overline{C}_{out}$ = 0 when Q_3–Q_0 = 1111

TABLE 5–1

**Carry Out
Control Circuit
Mode Control**

Down-Count Operation

The Bin/$\overline{\text{Dec}}$ input can be low or high (X, don't care) as long as $\overline{\text{Carry In}}$ = 0 and Up/$\overline{\text{Down}}$ = 0 to establish proper operation for this mode. When a problem is encountered in this circuit, the technician should first try to isolate the problem to the *down-count control circuit* or the *up-count control circuit*. Sometimes this is not a simple matter because the 7402 NOR gate is used in both the up- and down-count sections. At any rate, when a problem exists, check V_{CC} and ground on each IC in the appropriate control circuit. Keep in mind $\overline{C}_{out}$ should go low only on the terminal count of 0000. All of the problems for the control circuit refer to Fig. 5–38.

7402 pin 2 or 3.

If pin 2 or pin 3 of the 7402 NOR gate is open, the $\overline{\text{Carry Out}}$ signal will not go low. The effect of opening pin 2 is the same as making $\overline{\text{Carry In}}$ = 1. However, it is the low applied to the Up/$\overline{\text{Down}}$ input, which is inverted and applied to the 7427 3-input NOR gate, that inhibits that gate and produces a low at pin 12. This signal, which is applied to pin 5 of the NAND gate, inhibits the NAND and produces a high at pin 6 to force pin 8 of the gate low. This results from both pins 9 and 10 of the 7400 gate being high. Therefore, $\overline{C}_{out}$ will never be activated.

7402 pin 1.

An open at pin 1 of the 2-input NOR gate produces a slightly different result. The $\overline{C}_{out}$ signal will be activated (low) on both 0001 and 0000 counts. The incorrect low output occurs on the BCD count of 0001. This happens because the input to the 7400 NAND gate at pin 1 is floating (high). Pin 2 of the gate is high as long as $Q_1 = Q_2 = Q_3 = 0$. This condition exists on both counts of 0001 and 0000.

7402 pin 4, 5, or 6.

An open at any of these pins acts like a high. Since this section of the 7402 is part of the up-count control circuit, normal circuit operation will be observed if any of these pins are open as long as the circuit is configured for down-count operation.

7402 pin 11, 12, or 13.

If either 7402 NOR gate input pin 11 or pin 12 is open, $\overline{C}_{out}$ will not go low because the NOR gate's output will stay low during all counts. Therefore, the NAND gate is inhibited, and its output at pin 3 is held permanently high. This high, along with the normal high from pin 6 of the NAND gate, will keep its output at pin 8 permanently low. Therefore, $\overline{C}_{out}$ cannot go low.

An open at pin 13 of the 7402 gate produces a low at $\overline{C}_{out}$ on every even count because pin 2 of the 7400 NAND gate is floating. Pin 1 of that gate goes high and produces a low at pin 3 on every even count because $Q_0 = 0$ on the even counts.

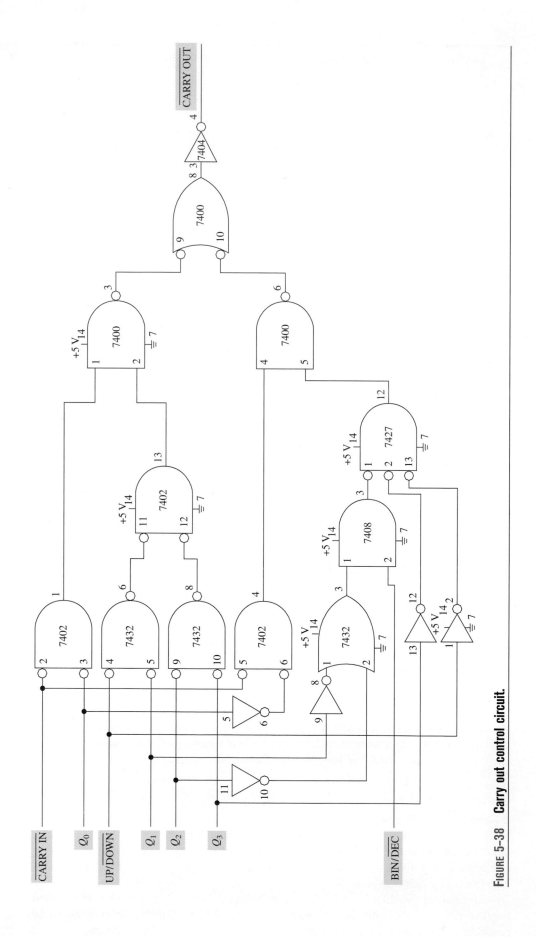

FIGURE 5-38   Carry out control circuit.

7402 pin 7 or 14.

The $\overline{C}_{out}$ signal will usually stay active (low) on all counts with an open ground (pin 7) or open V_{CC} (pin 14) input. Pin 3 of the 7400 NAND gate is always low with V_{CC} or ground removed from the NOR gate. This inhibits the output NAND gate, which produces a high output from that gate. This high output is inverted by the NOT gate, which keeps $\overline{C}_{out}$ low on all counts. A problem with power or ground should be detected early in the troubleshooting process because these levels should be checked early in the procedure.

7432 pin 1, 2, or 3.

A problem at this OR gate will not affect circuit operation because the circuit is configured for down-count operation. The 7427 NOR gate is inhibited by the inverted Up/$\overline{Down}$ = 0 input at pin 13. Therefore, pins 1 and 2 of the 7427 are don't care inputs in this configuration.

7432 pin 4, 5, or 6.

An open at pin 4, 5, or 6 of the OR gate will prevent $\overline{C}_{out}$ from going low on the terminal count. An open at either pin 4 or 5 will inhibit the 7432 OR gate, pin 6 will be high, and pin 11 of the 7402 NOR gate will be stuck high. The same effect is produced when pin 6 of the 7432 is open because pin 11 of the NOR gate is floating. With pin 11 high or floating, a low is produced at pin 13, which inhibits the NAND gate and produces a high at pin 3 of that gate. This high is applied to pin 9 of the NAND gate and a high is applied to pin 10. The high at pin 10 is originated by the Up/$\overline{Down}$ input being low in this configuration. The high inputs at pins 9 and 10 keep the pin 8 output low and cause $\overline{C}_{out}$ to be high on all counts.

7432 pin 8, 9, or 10.

The same line of reasoning used in the previous example may be applied for this section of the OR gate. Again, $\overline{C}_{out}$ will never go active.

7432 pin 7 or 14.

Ground or V_{CC} problems may result in erratic operation. Removal of V_{CC} will produce a high (inactive) $\overline{C}_{out}$ signal on all counts.

7408 pin 1, 2, 3, 7, or 14.

Problems with this AND gate will not affect circuit operation when in the down-count mode.

7400 pin 1, 2, or 3.

An open at pin 1 of this NAND gate produces a low at $\overline{C}_{out}$ on the counts of 0001 and 0000. See the discussion of the 7402 pin 1 open problem for a detailed explanation. An open at pin 2 of this NAND gate produces a low at $\overline{C}_{out}$ on every even count. See the 7402 pin 13 discussion. An open at pin 3 will prevent $\overline{C}_{out}$ from going low on the terminal count. During normal down-count operation, pin 10 of the 7400 NAND gate is high. An open at pin 3 of this gate causes pin 9 to float. The two high inputs cause pin 8 to be held low on all counts, and this keeps $\overline{C}_{out}$ high on all counts.

7400 pin 4, 5, or 6.

During down-count operation pin 5 of the NAND gate is low, the gate is inhibited, and pin 6 is high, which enables the output NAND gate. Therefore, an open at pin 4 or pin 6 results in normal circuit operation, with $\overline{C}_{out}$ going active on the terminal count of 0000.

An open at pin 5 produces an active (low) $\overline{C_{out}}$ on all odd counts in addition to the low produced on the terminal count of 0000. The low produced on the terminal count is a result of normal down-count control circuit operation because there is no malfunction in this circuit. However, an open at pin 5 enables the NAND gate. Pin 6 of the 7402 NOR gate is low on all odd counts, so it produces a high output on these counts. This causes pin 6 of the NAND gate to go low on every odd count. Thus, $\overline{C_{out}}$ goes low on all odd counts and on the terminal count.

7400 pin 8, 9, or 10.

An open at an input or output pin of this NAND gate produces three distinct problems. It can easily be seen that an open at pin 8 of this gate will cause a floating input to the output NOT gate and $\overline{C_{out}}$ will be low on all counts. An open at pin 10 results in normal circuit operation because this pin is normally high when Up/$\overline{Down}$ is low. An open at pin 9 will prevent $\overline{C_{out}}$ from going active on the terminal count. This is because pin 10 is held high during down-count operation and pin 9 is floating which causes the NAND gate output at pin 8 to stay low and $\overline{C_{out}}$ to stay high on all counts.

7400 pin 7 or 14.

Here again the results of an open ground can produce erratic operation. The circuit may even function normally at times with the ground pin open. An open at pin 14 will cause the output of the NAND gate to be dead, which is comparable to a floating input at pin 3 of the NOT gate. This will cause $\overline{C_{out}}$ to be low on all counts.

7427 pin 1, 2, 7, 12, 13, or 14.

An open at any of the input pins of this 3–input NOR gate will result in normal circuit operation because the gate is inhibited during down-count operation. However, an open at the output pin (pin 12) produces an active $\overline{C_{out}}$ on all odd counts and on the terminal count of 0000. See the explanation regarding pin 5 of the 7400 gate for details.

A power/ground (pin 14 or 7) problem produces erratic operation, with $\overline{C_{out}}$ being activated on various counts. These pins should be checked in one of the early troubleshooting steps.

7404 Inverter.

Most of the inverters are located in the up-count control circuit. Problems with the inverters in this section of the circuit typically have no effect on the $\overline{C_{out}}$ signal. The inverters that do affect $\overline{C_{out}}$ during the down-count mode are discussed next.

An open at pin 1 places a low at pin 13 of the 7427, which is normally inhibited during down-count operation. If the Bin/$\overline{Dec}$ input is low and pin 1 is open, $\overline{C_{out}}$ will be activated on several odd counts as shown in Table 5–2. An open at pin 3 of the NOT gate keeps pin 4 low and $\overline{C_{out}}$ active on all counts. An open at pin 4 leaves $\overline{C_{out}}$ dead on all counts. $\overline{C_{out}}$ will also be dead if V_{CC} is removed from the NOT gate.

Table 5–2 provides a summary of the troubles discussed in this section. The power and ground pins of the ICs are not listed in the table because these problems sometimes produce erratic operation and sometimes there is no indication of a problem at all.

A more detailed troubleshooting analysis could take us to the point of shorting input and output pins to high levels and low levels. However, little will be gained from this time-consuming exercise if the preceding analysis is understood.

Table 5–3 provides a summary of troubles and symptoms for the control circuit in Fig. 5–38 when it is set up for BCD up-count operation. The control switches are set as follows for this mode of operation:

$\overline{\text{Carry In}}$ = 0

Bin/$\overline{\text{Dec}}$ = 0

Up/$\overline{\text{Down}}$ = 1

Table 5–4 provides a summary of troubles for 4-bit binary up-count operation. The control switches are set as indicated for this mode of operation:

$\overline{\text{Carry In}}$ = 0

Bin/$\overline{\text{Dec}}$ = 1

Up/$\overline{\text{Down}}$ = 1

IC	Pin # Open	$\overline{C}_{out}$ Symptom—Down-Count
7402	1	Low on 0001 and 0000 counts
	2	High on all counts
	3	High on all counts
	4	Normal operation
	5	Normal operation
	6	Normal operation
	11	High on all counts
	12	High on all counts
	13	Low on all even counts
7432	1	Normal operation
	2	Normal operation
	3	Normal operation
	4	High on all counts
	5	High on all counts
	6	High on all counts
	8	High on all counts
	9	High on all counts
	10	High on all counts
7408	1	Normal operation
	2	Normal operation
	3	Normal operation
7400	1	Low on 0001 and 0000 counts
	2	Low on all even counts
	3	High on all counts
	4	Normal operation
	5	Low on all odd counts and terminal count
	6	Normal operation
	8	Low on all counts
	9	High on all counts
	10	Normal operation
7427	1	Normal operation
	2	Normal operation
	12	Low on all odd counts and terminal count
	13	Normal operation
7404	1	Low on 1111 and terminal count when Bin/$\overline{\text{Dec}}$ = 1 Low on 1111, 1101, 1011, and 1001 counts plus terminal count when Bin/$\overline{\text{Dec}}$ = 0
	2	Normal operation
	3	Low on all counts
	4	Dead
	5–13	Normal operation

TABLE 5-2

Troubleshooting Summary—Carry Out Control Circuit (Down-Count Operation)

TABLE 5-3	IC	Pin # Open	$\overline{C}_{out}$ Symptom—BCD Up-Count
Troubleshooting Summary— Carry Out Control Circuit (BCD Up-Count Operation)	7402	1	Normal operation
		2	Normal operation
		3	Normal operation
		4	Low on 1000 and 1001 counts
		5	High on all counts
		6	High on all counts
		11	Normal operation
		12	Normal operation
		13	Low on all even counts and terminal count
	7432	1–10	Normal operation
	7408	1	Normal operation
		2	High on all counts
		3	High on all counts
	7400	1	Normal operation
		2	Low on all even counts and terminal count
		3	Normal operation
		4	Low on 1000 and 1001 counts
		5	Low on all odd counts
		6	High on all counts
		8	Low on all counts
		9	Normal operation
		10	High on all counts
	7427	1	High on all counts
		2	High on all counts
		12	Low on all odd counts
		13	High on all counts
	7404	1	Normal operation
		2	High on all counts
		3	Low on all counts
		4	Dead
		5	Low on 1000 and 1001 counts
		6	High on all counts
		8	Normal operation
		9	Normal operation
		10	Normal operation
		11	Normal operation
		12	High on all counts
		13	Low on all odd counts

Section 5–6: Review Questions

Answers are given at the end of the chapter.

A. The control circuit in Fig. 5–38 is set up for down-count operation when $\overline{C}_{in} = 0$ and Up/$\overline{\text{Down}}$ = 0. What happens to $\overline{C}_{out}$ when the 7402 (NOR gate) output pin 4 is open?

B. With the circuit in Fig. 5–38 set up for down-count operation, what happens to $\overline{C}_{out}$ when the 7402 input pin 12 is open?

C. What mode of operation is selected in Fig. 5–38 when $\overline{C}_{in} = 0$, Bin/$\overline{\text{Dec}}$ = 0, Up/$\overline{\text{Down}}$ = 1?

D. If the circuit in Fig. 5–38 is set up as stated in question (C), what happens to $\overline{C}_{out}$ when pin 1 of the 7432 OR gate is open?

E. The control circuit in Fig. 5–38 is set up with $\overline{C}_{in} = 0$, Bin/$\overline{\text{Dec}}$ = 1, and Up/$\overline{\text{Down}}$ = 0. What happens to $\overline{C}_{out}$ if input pin 5 to the 7400 NAND gate is open?

IC	Pin # Open	$\overline{C}_{out}$ Symptom—BCD Up-Count
7402	1	Normal operation
	2	Normal operation
	3	Normal operation
	4	Low on 1110 and 1111 counts
	5	High on all counts
	6	High on all counts
	11	Normal operation
	12	Normal operation
	13	Low on all even counts and terminal count
7432	1	High on all counts
	2	High on all counts
	3	High on all counts
	4	Normal operation
	5	Normal operation
	6	Normal operation
	8	Normal operation
	9	Normal operation
	10	Normal operation
7408	1	High on all counts
	2	Normal operation
	3	High on all counts
7400	1	Normal operation
	2	Low on all even counts plus terminal count
	3	Normal operation
	4	Low on 1110 and 1111 counts
	5	Low on all odd counts
	6	High on all counts
	8	Low on all counts
	9	Normal operation
	10	High on all counts
7427	1	High on all counts
	2	High on all counts
	12	Low on all odd counts
	13	High on all counts
7404	1	Normal operation
	2	High on all counts
	3	Low on all counts
	4	Dead
	5	Low on 1110 and 1111 counts
	6	High on all counts
	8	High on all counts
	9	Low on 1101 and 1111 counts
	10	High on all counts
	11	Low on 1011 and 1111 counts
	12	High on all counts
	13	Low on 0111 and 1111 counts

TABLE 5-4

Troubleshooting Summary— Carry Out Control Circuit (4-Bit Binary Up-Count Operation)

SECTIONS 5-3 THROUGH 5-6: INTERNAL SUMMARY

A thorough understanding of logic gate operation and the ability to write and understand Boolean expressions for logic gates make combinational logic circuit analysis a relatively simple process. The material presented in these sections proves the following must be known regarding each type of logic gate:

■ Standard and alternate logic symbols

■ Short logic and short logic

- Truth table
- Boolean expression
- Enabler and inhibitor

Data selection in digital circuits is known as **multiplexing;** data distribution is known as **demultiplexing.** The ICs used to accomplish these functions will be addressed in Chapter 10. For now a general understanding of the logic circuits presented in Section 5–4 will suffice.

Operation of the carry out control circuit described in Section 5–5 is a prime example of putting it all together. It should be evident that troubleshooting this circuit cannot be accomplished until its mode of operation has been set and operation of the circuit is understood.

Sections 5-3 through 5-6: Internal Summary Questions

Answers are given at the end of the chapter.

1. The logic circuit in Fig. 5–39 will output high when
   a. A, B, C, and D are low.
   b. A, B, C, and D are high.
   c. E is high.
   d. Both b and c are correct.

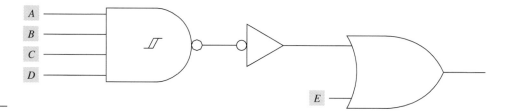

FIGURE 5-39

2. The symbol within the NAND gate in Fig. 5–39 represents a/an _____ NAND gate.
   a. Standard
   b. High-voltage
   c. Open-collector output
   d. Schmitt-trigger input

3. The logic circuit in Fig. 5–40 will output high on
   a. All odd counts
   b. All even counts
   c. All odd counts plus counts of 1100 and 1110
   d. All even counts plus counts of 0001 and 0011

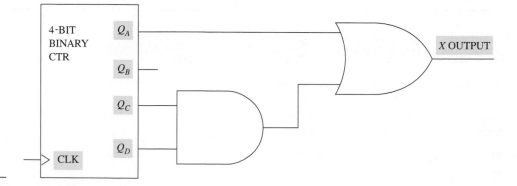

FIGURE 5-40

4. The logic circuit shown in Fig. 5–41 will detect invalid BCD sum numbers 1010 through 1111 if the circuit's outputs that are feeding this logic circuit are Q_D, Q_C, Q_B and Q_A.
   a. True
   b. False

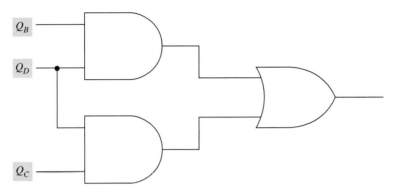

FIGURE 5–41

5. What data are present at the outputs of the logic circuit in Fig. 5–42 when $S = 0$, $A_0 = B_0 = 1$, and $A_1 = B_1 = 0$?
   a. No data
   b. $A_0 = 1$ and $B_0 = 1$.
   c. $A_1 = 0$ and $B_1 = 0$.

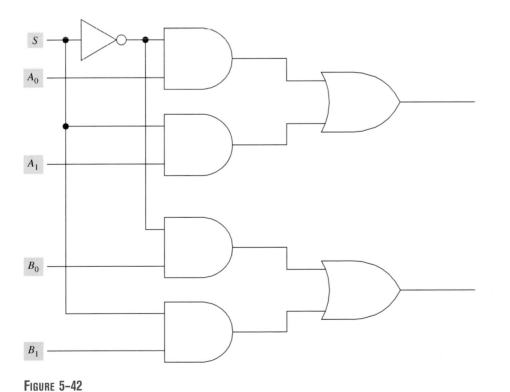

FIGURE 5–42

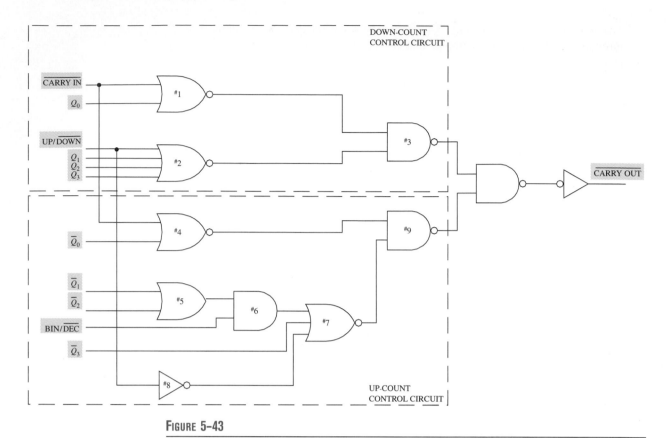

FIGURE 5–43

6. What mode of operation is selected in Fig. 5–43 when $\overline{\text{Carry In}}$ = 0, Up/$\overline{\text{Down}}$ = 0, and Bin/$\overline{\text{Dec}}$ = 0?
   a. Down-count
   b. BCD up-count
   c. 4-bit binary up-count

7. What mode of operation is selected in Fig. 5–43 when $\overline{\text{Carry In}}$ = 0, Up/$\overline{\text{Down}}$ = 1, and Bin/$\overline{\text{Dec}}$ = 1?
   a. Down-count
   b. BCD up-count
   c. 4-bit binary up-count

8. On what count will $\overline{C}_{\text{out}}$ be activated for the mode of operation set up in question 7?
   a. 0000
   b. 1001
   c. 1111
   d. None—$\overline{C}_{\text{out}}$ will stay high

9. What mode of operation is selected in Fig. 5–43 when $\overline{\text{Carry In}}$ = 0, Up/$\overline{\text{Down}}$ = 0, and Bin/$\overline{\text{Dec}}$ = 1?
   a. Down-count
   b. BCD up-count
   c. 4-bit binary up-count

10. On what count will $\overline{C}_{\text{out}}$ be activated for the mode of operation set up in question 9?
    a. 0000
    b. 1001
    c. 1111
    d. None—$\overline{C}_{\text{out}}$ will stay high

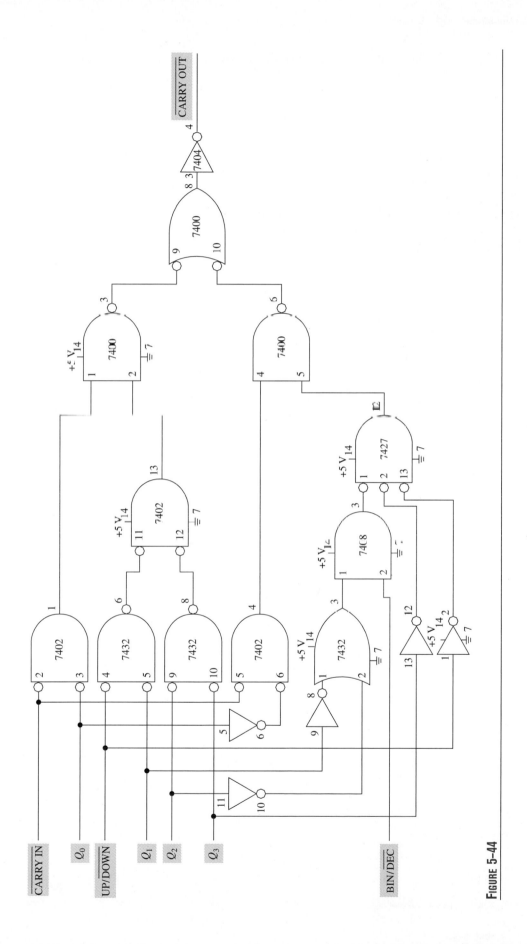

FIGURE 5-44

11. On what count will $\overline{C}_{out}$ be activated if the $\overline{\text{Carry In}}$ input in Fig. 5–43 is high?
    a. 0000
    b. 1001
    c. 1111
    d. None—$\overline{C}_{out}$ will stay high

12. The control circuit in Fig. 5–44 is set up for down-count operation when $\overline{C}_{in} = 0$, $\text{Bin}/\overline{\text{Dec}} = 1$, and $\text{Up}/\overline{\text{Down}} = 0$. On what count will $\overline{C}_{out}$ be activated if pin 3 of the 7400 NAND gate is open?
    a. 0000
    b. None—$\overline{C}_{out}$ will be high on all counts.
    c. $\overline{C}_{out}$ will be active on all even counts.
    d. $\overline{C}_{out}$ will be active on all counts.

13. The control circuit in Fig. 5–44 is set up for down-count operation. On what count will $\overline{C}_{out}$ be activated if pin 2 of the 7400 NAND gate is open?
    a. 0000
    b. $\overline{C}_{out}$ will be active on all counts.
    c. None—$\overline{C}_{out}$ will be high on all counts.
    d. $\overline{C}_{out}$ will be active on all even counts.

14. Which output signal of the circuit in Fig. 5–45 is activated when the microprocessor input signals are $\overline{\text{RD}} = 1$, $\overline{\text{WR}} = 0$, and $\text{IO}/\overline{\text{M}} = 1$?
    a. $\overline{\text{IOR}}$
    b. $\overline{\text{IOW}}$
    c. $\overline{\text{MEMR}}$
    d. $\overline{\text{MEMW}}$

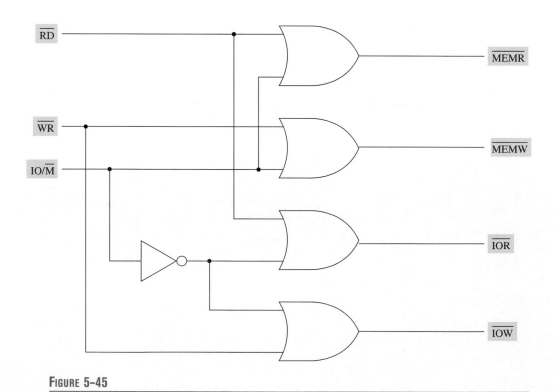

FIGURE 5-45

CHAPTER 5: SUMMARY

Exclusive-OR (X-OR) and **exclusive-NOR (X-NOR)** gates have only two inputs. The X-OR gate outputs a high when its inputs are not equal (complementary in = 1 out). Thus, the X-OR gate's output is always the sum of its inputs if the carry is disregarded. The X-NOR gate outputs a high when its two inputs are equal (complementary in = 0 out).

Parity is an error detection scheme established in a digital system to ensure equality of received data to that of transmitted data. The parity scheme may be **even** or **odd.** If an error is detected, a parity error signal is asserted to so indicate.

Sections 5–3 through 5–5 contain operational analysis of many **combinational logic circuits.** It should be apparent at this point that an understanding of basic logic gate operation makes analysis of these circuits straightforward.

The last section of this chapter covered troubleshooting the combinational logic circuits presented in the earlier sections of the chapter. An understanding of circuit operation simplifies this process tremendously. A logical troubleshooting analysis includes a comparison of what is happening to what should be happening.

CHAPTER 5: END OF CHAPTER QUESTIONS/PROBLEMS

Answers are given in the Instructor's Manual.

SECTION 5-1

1. Draw the standard logic symbol for an X-OR gate.

2. Write the Boolean expression for the output of an X-OR gate whose inputs are A and B.

3. Draw the output of an X-OR gate with the A and B inputs shown in Fig. 5–46.

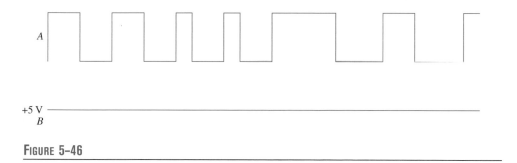

FIGURE 5-46

4. Change the B input of the X-OR gate waveforms in Fig. 5–46 to a logic low (0 volts) and draw the output.

5. Identify the logic symbol shown in Fig. 5–47.

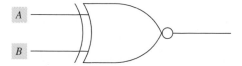

FIGURE 5-47

6. Draw the output of the gate shown in Fig. 5–47 with the *A* and *B* inputs that are shown in Fig. 5–48. (*Note:* Input *B* is at a low logic level.)

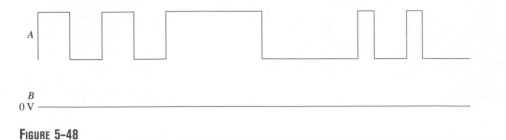

B
0 V

FIGURE 5-48

7. Write the Boolean expression for the output of the gate shown in Fig. 5–47.

CT 8. What logic function is performed by the exclusive gate IC (SN74S135) in Fig. 5–49 when the control input (1C, 2C) is high?

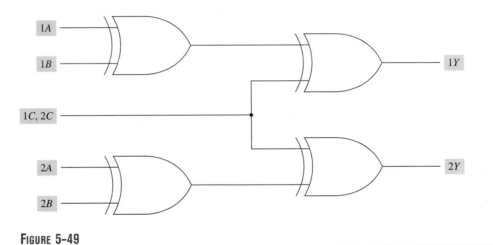

FIGURE 5-49

9. What logic function is performed by the logic circuit shown in Fig. 5–50?

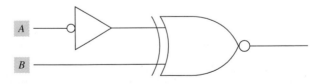

FIGURE 5-50

SECTION 5-2

10. Determine the level of the parity bit from a parity generator set up for even parity when the data are $46_{(16)}$.

11. If the data in question 10 is 7-bit ASCII, what byte of data would actually be transmitted to the receiver?

CT 12. The odd/even parity generator (SN74180) shown in Fig. 5–51 is set up for which parity scheme?

CT 13. What are the Σ EVEN and Σ ODD output levels of the parity generator in Fig. 5–51?

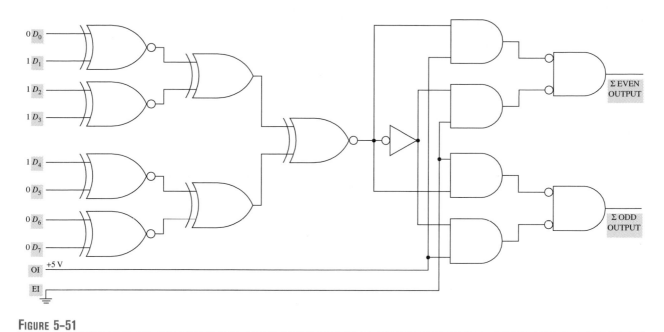

FIGURE 5–51

SECTION 5–3

CT 14. Determine the mode of operation of the logic circuit in Fig. 5–52 when
   a. $C_1 = 0$ and $C_0 = 1$.
   b. $C_1 = 1$ and $C_0 = 0$.
   c. $C_1 = 0$ and $C_0 = 0$.
   d. $C_1 = 1$ and $C_0 = 1$.

15. State the purpose of using circuits/gates with Schmitt-trigger-input circuits.

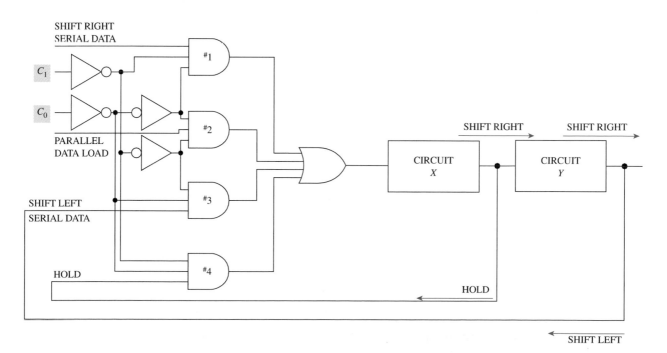

FIGURE 5–52

16. Define the term "multiplexing."

17. What data are selected and passed to the outputs when the Select input of Fig. 5–53 is high?

18. Write the Boolean expression for the X output of Fig. 5–53.

19. Write the Boolean expression for the Y output of Fig. 5–53.

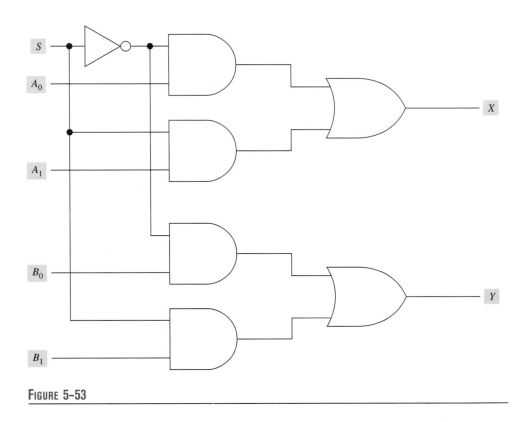

FIGURE 5-53

SECTION 5-5

CT 20. The carry out control circuit in Fig. 5–54 has been annotated with input logic levels. Determine the level of the $\overline{\text{Carry Out}}$ output signal.

CT 21. Which ROM chip (if any) in Fig. 5–55 is selected when $\overline{\text{MEMR}} = 1$ and the 20-bit memory address (A_{19}–A_0) is FFFFF?

CT 22. Which ROM chip (if any) in Fig. 5–55 is selected when $\overline{\text{MEMR}} = 0$ and the 20-bit memory address is FE000?

SECTION 5-6

23. What mode of operation is selected in Fig. 5–56 when $\overline{\text{Carry In}} = 0$, Up/$\overline{\text{Down}} = 0$, and Bin/$\overline{\text{Dec}} = 1$?

CT 24. What $\overline{\text{Carry Out}}$ signal symptom will be observed in Fig. 5–56 (all TTL gates) with the setup as stated in question 23 and
   a. Pin 10 of the 7432 is open?
   b. Pin 1 of the 7408 is open?
   c. Pin 1 of the 7400 is open?
   d. Pin 3 of the 7404 is open?

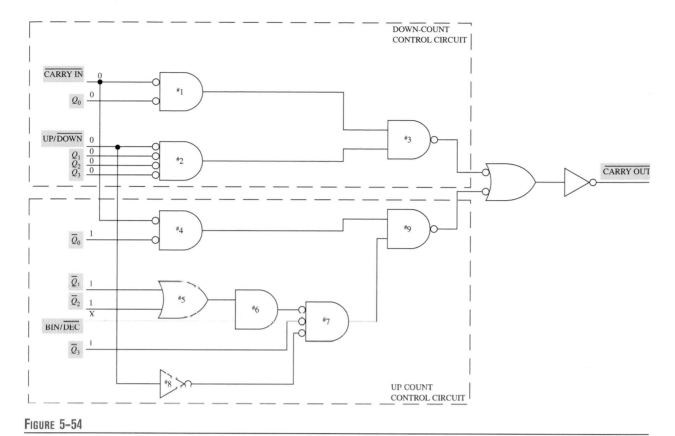

FIGURE 5-54

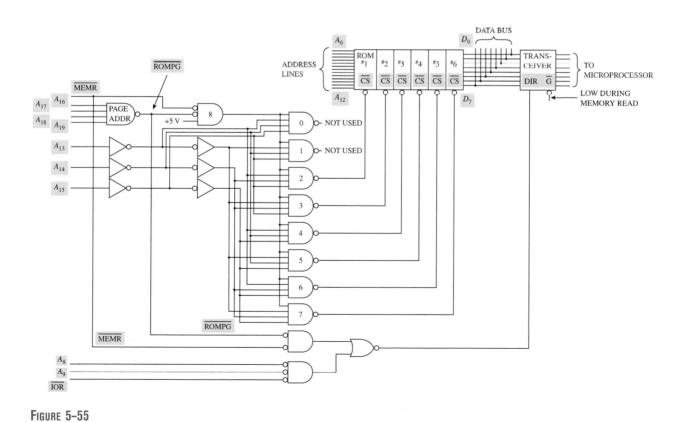

FIGURE 5-55

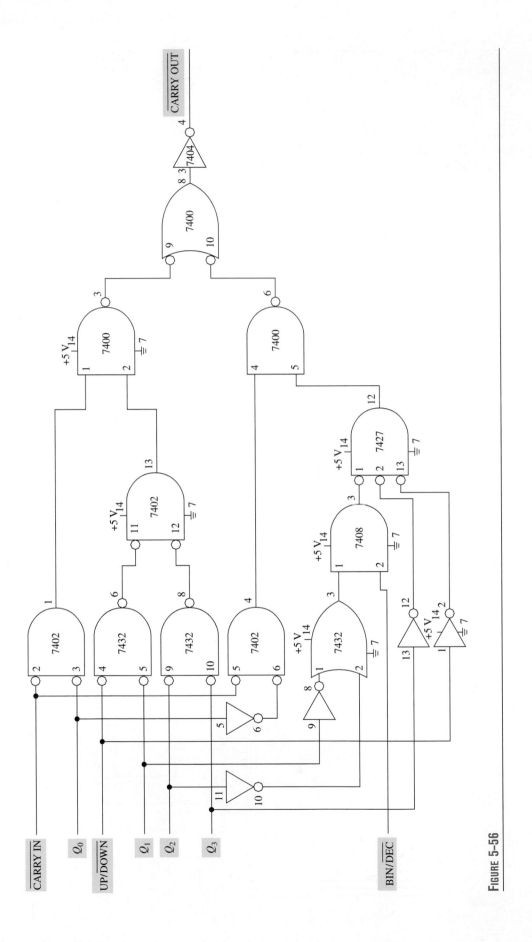

FIGURE 5-56

ANSWERS TO REVIEW QUESTIONS

SECTION 5-1

A. See Fig. 5–2(b).

B.

A	B	Y
0	0	0
0	1	1
1	0	1
1	1	0

C.

A	B	Y
0	0	1
0	1	0
1	0	0
1	1	1

D. $\overline{\overline{A + C} + \overline{\overline{A} + \overline{C}}} =$

$(A + C)(\overline{A} + \overline{C}) =$

$A\overline{A} + A\overline{C} + \overline{A}C + C\overline{C} -$

$0 + A\overline{C} + \overline{A}C + 0 -$

$A\overline{C} + \overline{A}C$

E. X-OR

F. $A\overline{C} + \overline{A}C$ – X-OR gate

$\overline{\overline{A}C + \overline{A}\,\overline{C}}$ = with NOT gate

$AC + \overline{A}\,\overline{C}$ = X-NOR function

G. $AC + \overline{A}\,\overline{C}$ = X-NOR gate

$\overline{A\overline{C} + \overline{A}\,\overline{\overline{C}}}$ = with NOT gate

$A\overline{C} + \overline{A}C$ = X-OR function

SECTION 5-2

A. (1) 1
   (2) 0
   (3) 1
   (4) 1

B. (1) 0
   (2) 1
   (3) 0
   (4) 1

SECTION 5-3

A. See Fig. 5–26(a).

B. Schmitt-trigger-input circuits convert slow-changing input signals to sharply defined, jitter-free square or rectangular wave output signals.

SECTION 5-4

A. $\overline{S}A_0 + SA_1$
   $\overline{S}B_0 + SB_1$

B. $A_1 = 0$ and $B_1 = 0$.

C. Selecting data

D. Distributing data

SECTION 5-5

A. 4-bit binary up-count operation

B. BCD down-count operation

C. Carry Out cannot go active (low) on the terminal count.

D. $F_{(16)} (1111_{(?)})$

E. None; #1 NAND gate's output is active, but it is not used.

F. #3 ROM chip

G. None; all decoding gates are inhibited (disabled).

H. A transmitter/receiver circuit that allows data movement in one direction or the other when it is enabled

I. $\overline{MEMR} = 0$ $\overline{IOR} = 1$
   $\overline{MEMW} = 1$ $\overline{IOW} = 1$

SECTION 5-6

A. Normal operation—$\overline{C}_{out}$ activates on count of 0000.

B. $\overline{C}_{out}$ will not go low on terminal count.

C. BCD up-count

D. Normal operation—$\overline{C}_{out}$ activates on count of 1001.

E. $\overline{C}_{out}$ goes low on all odd counts and the terminal count.

ANSWERS TO INTERNAL SUMMARY QUESTIONS

SECTIONS 5-1 AND 5-2

1. d		9. d	
2. c		10. a	
3. c		11. a	
4. d		12. b	
5. a		13. a	
6. b		14. a	
7. a		15. a	
8. b			

SECTIONS 5-3 THROUGH 5-6

1. d		8. c	
2. d		9. a	
3. c		10. a	
4. a		11. d	
5. b		12. b	
6. a		13. d	
7. c		14. b	

CHAPTER

6 LATCH AND FLIP-FLOP CIRCUITS

Introduction

The combinational logic circuits presented in Chapter 5 did not have the ability to store information. Their output depended on the state of the inputs at any given instant in time. Yet many digital systems require a storage capability. Computers are prime examples of digital systems that need memory.

Data in a digital system is information. This information may be in the form of a program (set of instructions) used by the computer, or it may be the numbers, letters, and symbols on a computer keyboard. In any case, this data needs to be stored in the computer's memory. The data must be in a language that the computer understands. This language, called **machine-level language,** is composed of two levels (states). Since only two levels require representation, binary data are used.

Special digital circuits that have the ability to retain data are used to meet a system's memory requirement. These circuits are **sequential logic circuits.** A sequential logic

Active Clock Transition

Active-High Latch

Active-Low Latch

Asserted

Asynchronous

Binary Counter

Bistable

CLEAR/RESET State

$\overline{\text{CLR}}$

Data

Data Lockout

D-Type Flip-Flop

Dynamic Input Indicator

Edge Detector

Edge-Triggered

Flip-Flop

Gated Latch

Hold Time

INVALID State

J-K Flip-Flop

J-K Master–Slave Flip-Flop

Latch

Machine-Level Language

Modulus (MOD)

NGT

PGT

Postponed Output Indicator

$\overline{\text{PRE}}$

Pulse-Triggered

Register

RETAIN State (Hold) (NC)

SET State

Setup Time

Sequential Logic Circuit

Shift Register

State Indicator

State Table

Steering Gates

Switch Debouncer

Synchronous

Toggle

Transparent Latch

Chapter Objectives

1. Given a latch symbol, identify the symbol and determine its output under stated input conditions.

2. Identify various flip-flop symbols.

3. Given a flip-flop logic symbol and its control and clock input waveforms, determine the flip-flop's output.

4. Given a logic diagram of a circuit containing flip-flops and specific trouble symptoms, troubleshoot the circuit to isolate a malfunction.

circuit's output depends on its previous state (condition) in addition to its current inputs. This is accomplished by using **feedback** from the circuit's outputs back to its inputs.

Latches and **flip-flops** are the types of circuits used in digital systems to store information. The latches and flip-flops discussed in the remainder of this chapter form the foundation of the circuits required to store or transfer data in a digital system. In actuality, a latch circuit or flip-flop is nothing more than a combinational logic circuit designed to have a memory capability. A basic latch forms the core of all flip-flop circuits, and flip-flops are the foundation of memory. Flip-flops are also used extensively in counting operations, frequency division, transferring data, error detection, and microprocessors, as well as many digital control circuits.

Knowledge of flip-flop operation is integral to further advancement in the field of digital study. In fact, it is well known among educators that a thorough knowledge of logic gate and flip-flop operation will make your trek down the digital knowledge road short and downhill all the way.

Latch Introduction

A latch circuit is a bistable device comparable to the transistor bistable multivibrator you studied in circuits. As you remember, *bistable* indicates that the latch has two stable states. These two latch states are called the **SET** state and the **CLEAR** state. Once a latch is put in one of these states it will remain in that state until forced to change states by another input signal.

The transistor multivibrator circuits you previously studied produce square or rectangular waveforms. The problem with these circuits is that the capacitors and resistors that control the output waveform's duration often cause some exponential rise and fall of the waveform. These curves, as well as relatively slow rise and fall times, cause the outputs of transistor multivibrators to be noncompatible with most digital circuits.

There are two basic types of latch circuits: the NAND-gate latch (shown in Fig. 6–1a) and the NOR-gate latch. These latches are formed by using cross-coupled inverting logic

FIGURE 6-1   **Active-low latch.**

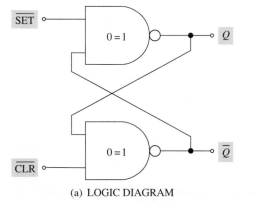

(a) LOGIC DIAGRAM

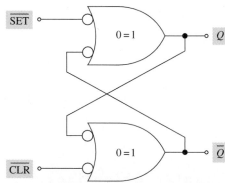

(b) LOGIC DIAGRAM USING
ALTERNATE NAND GATE SYMBOLS

$\overline{SET}$	$\overline{CLR}$	Q	$\overline{Q}$	STATE
0	1			
1	0			
1	1			
0	0			

(c) STATE TABLE

gates. The cross coupling provides the **feedback** necessary for the latch circuit to retain (store) data. A latch constructed with NAND gates is referred to as an **active-low latch,** and a latch constructed with NOR gates is called an **active-high latch.** The active-low and active-high references are derived from the input logic level that must be applied to the latch to put it in a certain state. This will be explained shortly. The NAND-gate latch in Fig. 6–1(a) shows that the latch has two outputs. One of these outputs is labeled Q. The other output, which is the complement of Q, is labeled $\overline{Q}$.

A latch circuit can have only two valid output conditions. One of these conditions is the SET state, where output $Q = 1$ and output $\overline{Q} = 0$. The other condition is the CLEAR state, where $Q = 0$ and $\overline{Q} = 1$. Since the latch is designed to normally have complementary outputs (Q and $\overline{Q}$), it is only necessary to remember that Q is high in the SET state and Q is low in the CLEAR state. $\overline{Q}$ of course, will normally be the opposite level. The $\overline{Q}$ output is a convenience for circuit designers, is often not used in digital circuits, and is sometimes not available as an output on a flip-flop IC. The CLEAR state is also referred to as the RESET state, and in this text the terms are synonymous and are used interchangeably. The latches are sometimes called S-C (SET–CLEAR) latches or S-R (SET–RESET) latches. Since a latch or flip-flop only has a SET state or a CLEAR state, it can store only *one* bit of data. Latch circuits are typically used to store binary information on a temporary basis.

Gated-latch circuits will also be presented in this chapter. Gated latches provide the circuit designer an opportunity to synchronize latch outputs to system requirements.

SECTION 6–1: ACTIVE-LOW LATCH

OBJECTIVES

1. Identify and determine the Q and $\overline{Q}$ outputs of a NAND-gate latch under stated input conditions.
2. Identify the logic symbol for an active-low latch.

The NAND-gate latch shown in Fig. 6–1(a) has two inputs that are labeled $\overline{\text{SET}}$ and $\overline{\text{CLR}}$. The NOT signs over the input labels indicate that the latch is an *active-low input latch.* This means that a Logic 0 must be applied to the input labeled with the name of the state in which we desire to put the latch. This will soon become evident as a state table is developed for the latch.

A **state table** is comparable to the truth table used for logic gates. It is referred to as a state table because it reflects the different states in which the latch can be placed using various input combinations. Since the latch has two inputs, we know that four possible input conditions can exist. This knowledge allows us to design a state table for the latch as shown in Fig. 6–1(c). The state table will be completed as the lesson progresses to show the levels of Q and $\overline{Q}$ for each of the four possible input conditions. In addition, each of these states will be named. It has already been established that if $Q = 1$ and $\overline{Q} = 0$ the circuit is in the SET state. Furthermore, if $Q = 0$ and $\overline{Q} = 1$ the circuit is in the CLEAR (RESET) state. Keep in mind the latch must be put in the SET state to store a binary 1 data bit at the Q output. It must be put in the CLEAR state to store a binary 0 at the Q output.

Since the NAND-gate latch is an active-low latch, the alternate logic gate symbol for the NAND gate will be used in this discussion. The latch using the alternate symbols is shown in Fig. 6–1(b). It is desirable to use the alternate logic gate symbols because it is easy to read the bubbles. The short logic is annotated on the gates.

If $\overline{\text{SET}} = \mathbf{0}$ and $\overline{\text{CLR}} = \mathbf{1},$ as shown in Fig. 6–2(a), the use of short logic allows us to first derive that the output of gate G1 is Logic 1. This is true because $\overline{\text{SET}} = 0$ and "any 0 in = 1 out" of gate G1. Once the Q output has been determined, the feedback from

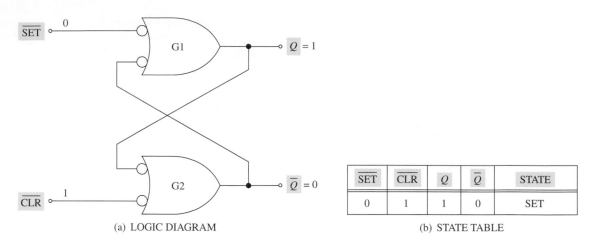

(a) LOGIC DIAGRAM

SET	CLR	Q	$\overline{Q}$	STATE
0	1	1	0	SET

(b) STATE TABLE

FIGURE 6-2   **Active-low latch: SET state.**

this gate can be traced back to the input of gate G2. This Logic 1 feedback input and the $\overline{\text{CLR}} = 1$ input produce a low out of gate G2. This set of inputs places the latch in the SET state because $Q = 1$ and $\overline{Q} = 0$. The input levels ($\overline{\text{SET}}/\overline{\text{CLR}}$), output levels ($Q/\overline{Q}$), and latch state are shown on the state table in Fig. 6–2(b). Note that it took a Logic 0 on the $\overline{\text{SET}}$ input and a Logic 1 on the $\overline{\text{CLR}}$ input to put the latch in the SET state. The latch is known as an active-low latch for this reason. As explained in Chapter 3, when the $\overline{\text{SET}}$ input is brought low, the input is said to be asserted. **Asserting** an input merely means to bring that input to its active level. To put the NAND-gate latch in the SET state, the $\overline{\text{SET}}$ input must be asserted (brought low) while the $\overline{\text{CLR}}$ input remains inactive (high).

If $\overline{\textbf{SET}} = \textbf{1}$ and $\overline{\textbf{CLR}} = \textbf{0}$, as shown in Fig. 6–3(a), the use of short logic will allow us to derive the output of gate G2. The output of this gate is Logic 1. Once the $\overline{Q}$ output has been determined, the feedback from this gate can be traced as previously discussed. This Logic 1 feedback input and the $\overline{\text{SET}} = 1$ input produce a low out of gate G1. This input combination puts the latch in the CLEAR state because $Q = 0$ and $\overline{Q} = 1$. The input/output levels and circuit state are shown on the state table in Fig. 6–3(b). Note that it took a Logic 0 on the $\overline{\text{CLR}}$ input while the $\overline{\text{SET}}$ input was inactive to put the latch in the CLEAR state.

If $\overline{\textbf{SET}} = \textbf{1}$ and $\overline{\textbf{CLR}} = \textbf{1}$, as shown in Fig. 6–4(a), the use of short logic is beneficial only after an output condition (state) is assumed. Let's assume that the circuit is in the SET state as shown. Either state can be assumed. To determine latch operation, trace the $\overline{Q} = 0$ feedback signal to the input of gate G1. This low input causes the Q output of

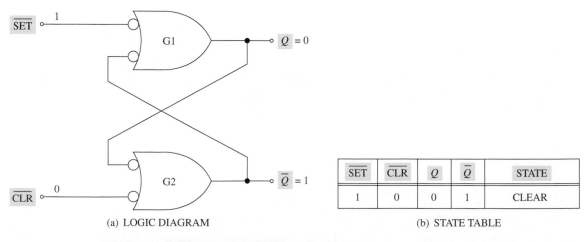

(a) LOGIC DIAGRAM

SET	CLR	Q	$\overline{Q}$	STATE
1	0	0	1	CLEAR

(b) STATE TABLE

FIGURE 6-3   **Active-low latch: CLEAR state.**

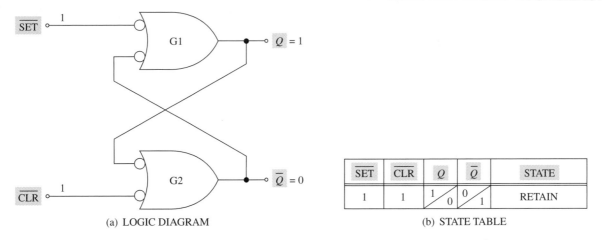

(a) LOGIC DIAGRAM

$\overline{SET}$	$\overline{CLR}$	Q	$\overline{Q}$	STATE
1	1	1 / 0	0 / 1	RETAIN

(b) STATE TABLE

FIGURE 6-4   Active-low latch: RETAIN state.

the gate to stay at its Logic 1 level. Tracing the $Q = 1$ feedback to the input of gate G2 gives us two Logic 1s into that gate, which make its output stay at Logic 0. The two Logic 1s applied to this latch cause it to retain the condition it was in when the two inputs were brought high (inactive). This is called the RETAIN state or *HOLD* mode of operation. The state table in Fig. 6-4(b) shows that if the latch were in the SET state when two Logic 1s were applied, it would retain the SET condition. Likewise, if the latch were in the CLEAR state when two Logic 1s were applied, it would retain the CLEAR state. The Q and $\overline{Q}$ outputs shown on the state table are sometimes labeled NC (No Change) instead of $Q = 1/0$ and $\overline{Q} = 0/1$ for the RETAIN state.

Remember the Logic 1 inputs are inactive inputs to the *active-low latch*. Thus, the latch should remain inactive when two Logic 1s are applied—it should retain the data it had from its previous state. It is this attribute that gives latch circuits their memory capability. In other words, a latch can be put in the SET state or the CLEAR state, and then its inputs can be brought to an inactive level (Logic 1 for the NAND gate latch) to retain the stored data.

It was previously stated that a latch circuit can have only two valid output states (SET or CLEAR). There is no contradiction to this statement by introducing the RETAIN state. This state merely retains the SET or CLEAR condition that the latch was in when its inputs were brought to the inactive level.

If $\overline{SET} = 0$ and $\overline{CLR} = 0$, as shown in Fig. 6-5(a), the use of short logic again quickly helps us develop a state table. The short logic of a NAND gate indicates that 0s

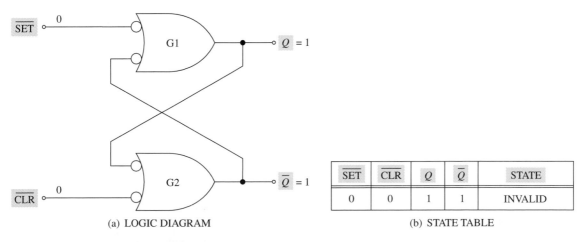

(a) LOGIC DIAGRAM

$\overline{SET}$	$\overline{CLR}$	Q	$\overline{Q}$	STATE
0	0	1	1	INVALID

(b) STATE TABLE

FIGURE 6-5   Active-low latch: INVALID state.

into each logic gate will produce a 1 out of each gate. The outputs of latch circuits are normally complementary (Q and $\overline{Q}$); therefore, equal outputs are considered INVALID. In addition, the state table shows Q to be high when both $\overline{SET}$ and $\overline{CLR}$ are asserted. However, this high output may not persist when the data inputs are both brought high to their inactive level. Therefore, the Q output is *unpredictable* when this situation occurs. This is a condition to be avoided and is thus referred to as an INVALID state. As shown in the state table in Fig. 6–5(b), the input condition of $\overline{SET} = 0$ and $\overline{CLR} = 0$ produces this INVALID output. This INVALID output condition is caused by asserting both inputs simultaneously.

If the inputs to this latch are analyzed, this should make sense. Since the circuit is an active-low latch, the $\overline{SET}$ input can be read as follows: "If this input is asserted (brought low), the circuit will be put in the SET state." When the $\overline{CLR}$ input is asserted, it can be read in like manner. Thus, if both inputs are asserted the circuit is asked to set and clear at the same time. Since the outputs should be complementary and persistent, this output condition must be considered INVALID.

An exercise in analyzing input waveforms will aid your understanding of active-low latch operation. Refer to Fig. 6–6 for the following analysis:

t_0-t_1: $\overline{SET} = 1$ and $\overline{CLR} = 0$. The output will be $Q = 0$ and $\overline{Q} = 1$ (CLEAR state) because the $\overline{CLR}$ input is asserted and the $\overline{SET}$ input is inactive (resting).

t_1-t_2: $\overline{SET} = 0$ and $\overline{CLR} = 1$. This input combination puts the latch in the SET state because $\overline{SET}$ is asserted and the $\overline{CLR}$ input is inactive.

t_2-t_3: The $\overline{CLR}$ input is again asserted, and the latch goes to the CLEAR state.

t_3-t_4: The $\overline{SET}$ input is asserted, and the latch goes to the SET state. Note that the input combinations were complementary up to t_4.

t_4: Both inputs are brought inactive ($\overline{SET}$ goes high and $\overline{CLR}$ remains high). Thus, the latch retains the SET state that it was placed in at t_3.

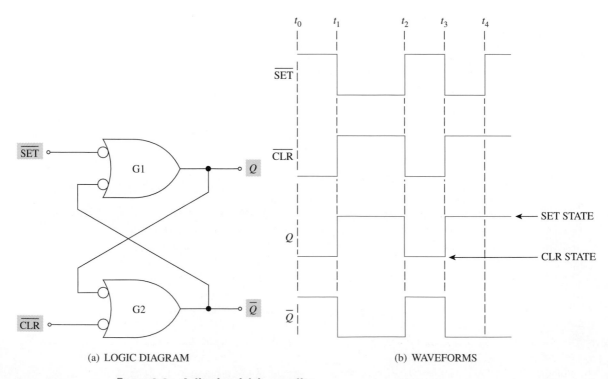

(a) LOGIC DIAGRAM   (b) WAVEFORMS

FIGURE 6-6   **Active low latch operation.**

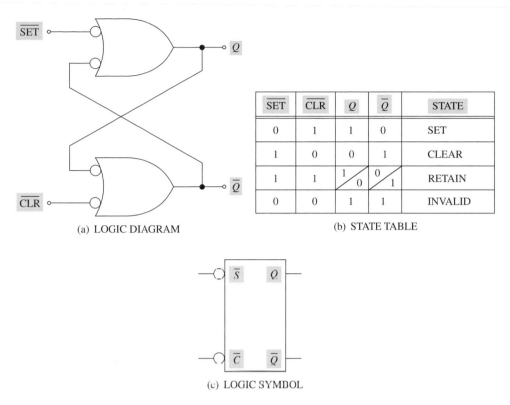

(a) LOGIC DIAGRAM

SET	CLR	Q	Q̄	STATE
0	1	1	0	SET
1	0	0	1	CLEAR
1	1	1/0	0/1	RETAIN
0	0	1	1	INVALID

(b) STATE TABLE

(c) LOGIC SYMBOL

FIGURE 6-7   **Active-low latch.**

It should be evident from the foregoing discussion that it always takes complementary data inputs to put a latch in the SET state or the CLEAR state. A look at the completed state table for the active-low latch shown in Fig. 6-7(b) verifies this fact.

The logic symbol for an active-low latch is shown in Fig. 6-7(c). Although all of the bubbles we have used in the text thus far have indicated inversion, it is extremely important to note on this logic symbol that the bubbles shown on the inputs are **state indicators.** A bubble on an *input* to a latch or flip-flop logic symbol indicates that the circuit is an active-low circuit.

In an active-low latch (Fig. 6-7a), a Logic 0 must be applied to the input labeled with the name of the state in which we desire to put the latch. The other input must be inactive. The bubble merely indicates that the state of the input level required to set or clear the latch is Logic 0. Also note on the logic symbol that the $\bar{S}$ and $\bar{C}$ input letters are NOTed. This also indicates that the input is active low, and the extra marking with the NOT signs might be looked at as insurance.

You should keep in mind that there are many manufacturers of ICs in the world, and these latch markings are not always standardized. Some of the alternate ways of marking active-low latches are shown in Fig. 6-8.

The NOT signs in Fig. 6-8(a) over S and C denote an active-low latch even though the bubbles are not present. Although not usually used, this symbol may be seen in some digital books or schematics. As shown in Fig. 6-8(b), many times the NOT signs are not placed over the input letters, but the bubbles always indicate active-low inputs. Figure 6-8(c) shows the inputs definitely active-low with both NOT signs and bubbles. Note that both outputs are labeled Q in Fig. 6-8(c). The bottom output, which is $\bar{Q}$, has an inverting bubble to indicate Q is NOTed (inverted) to produce $\bar{Q}$. A bubble on this output is an inverter, but the inverter is part of the latch circuit. However, the input bubbles on latch and flip-flop logic symbols are active-low state indicators. The ANSI/IEEE symbol is shown in Fig. 6-8(d). The qualifying symbols at the $\bar{S}$ and $\bar{C}$ inputs indicate these inputs are active-low. They are used in place of the bubbles shown at the inputs of Fig. 6-8(b) and (c). The $\bar{Q}$ outputs are not shown on the ANSI/IEEE symbol.

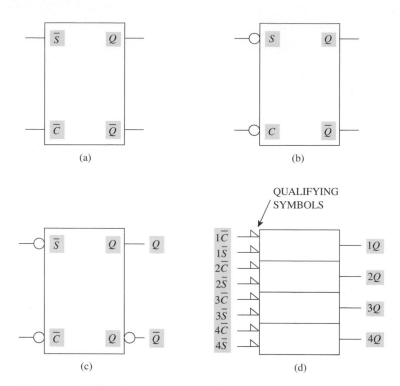

FIGURE 6-8 **Active-low latch logic symbols.**

Another type of state table design can be used in lieu of the one shown in Fig. 6–7(b). The use of the table shown in Fig. 6–9(b) should ensure complete understanding of the active-low latch. Active (A) and Inactive (I) letters are used in place of 0s and 1s for the inputs. All you have to do is look at the logic symbol shown in Fig. 6–9(a) to determine that active (A = 0) is low and inactive (I = 1) is high because this is an active-low input latch. From this starting point, it should be easy to verify the Q and $\overline{Q}$ output levels shown on the state table.

As shown in Fig. 6–9(b), If $\overline{\text{SET}}$ is active and $\overline{\text{CLR}}$ is inactive, the circuit will go to the SET state. If $\overline{\text{CLR}}$ is active and $\overline{\text{SET}}$ is inactive, the circuit will go to the CLEAR state. If neither input is active, the circuit will not change from its previous condition (RETAIN state). If both inputs are activated, the circuit's outputs are INVALID. The SN74279 Quadruple $\overline{\text{S}}$–$\overline{\text{R}}$ latch IC is a good example of commercially available active-low latches.

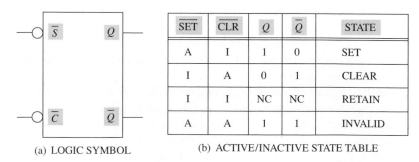

$\overline{\text{SET}}$	$\overline{\text{CLR}}$	Q	$\overline{Q}$	STATE
A	I	1	0	SET
I	A	0	1	CLEAR
I	I	NC	NC	RETAIN
A	A	1	1	INVALID

(a) LOGIC SYMBOL (b) ACTIVE/INACTIVE STATE TABLE

FIGURE 6-9 **Active-low latch.**

Switch Debouncer–Latch Application

Switch contact bounce offers some unique problems in digital circuits. Switches similar to the one shown in Fig. 6–10(a) are often used in experiments to apply a Logic 0 or Logic 1 to the inputs of digital circuits. The problem with this switch, as with all mechanical

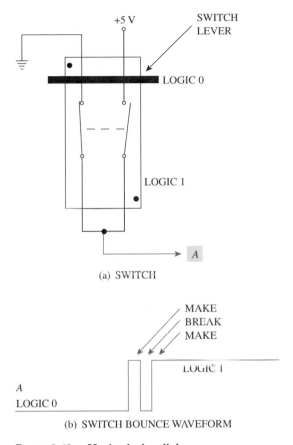

(a) SWITCH

(b) SWITCH BOUNCE WAVEFORM

FIGURE 6-10   **Mechanical switch.**

switches, is contact bounce. For example, assume the switch shown in the figure is currently providing a Logic 0 to input *A* of a digital circuit.

If the switch lever is moved down to produce a Logic 1 output, intermittent contact is usually made before firm contact is established. Basically, the switch arcs in a make–break–make pattern before closure is complete. This contact bounce is shown in Fig. 6–10(b). These sporadic levels can result in false triggering or miscounting in digital circuits. This is especially true if this type of switch is being used to enable or clock a synchronous latch or a flip-flop.

The active-low latch shown in Fig. 6–11(a) can be used to eliminate the effects of the sporadic logic levels introduced into a circuit by switch bounce.

The *Q* output of the circuit shown is low with the switch in the position shown. This condition is the result of $\overline{\text{CLR}}$ being asserted (low) and $\overline{\text{SET}}$ being inactive (high). The $\overline{Q}$ output of the latch is not used and is labeled not connected (NC). The resistors in the circuit are used to pull the voltage level on an input to a Logic 1 when the switch contact to that input is open. If the pull-up resistors were not used in this circuit, the floating input from the open switch contact could result in unreliable latch operation. Thus, the pull-up resistors pull the inputs up to +5 V when the switch contact to that particular input is open. This ensures the latch has a valid Logic 1 input when required.

When the switch lever is moved down, the contact on the left opens and the one on the right closes. Once the contact on the right gets very close, it will probably arc closed. This action will place a low on the $\overline{\text{SET}}$ input and set the latch. The other switch contact is now open, so a Logic 1 is applied to the $\overline{\text{CLR}}$ input from the pull-up resistor. If the arc on the right switch breaks momentarily, the $\overline{\text{SET}}$ input is returned high and the latch is put in the RETAIN state. Therefore, the latch retains the SET state that it was placed in with the first contact arc. Any further arcs will request the latch to go to the SET state, where it currently is, or to retain that state. The *Q* output due to this switch movement is shown in Fig. 6–11(b). All contact bounce effects have been eliminated from the latch output.

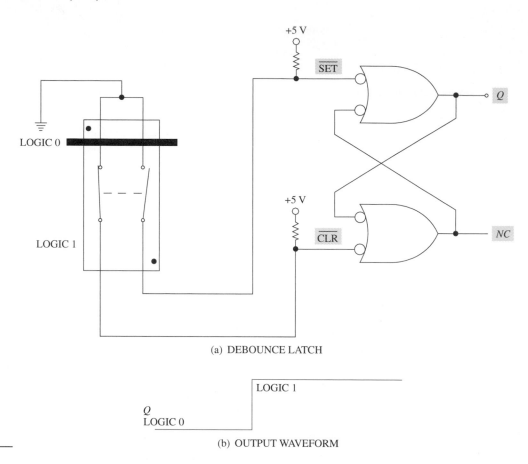

(a) DEBOUNCE LATCH

FIGURE 6-11 **Switch debounce circuit.**

Q
LOGIC 0

LOGIC 1

(b) OUTPUT WAVEFORM

Section 6–1: Review Questions

Answers are given at the end of the chapter.

A. What is the level of the Q output when a latch is in the CLEAR state?
B. The two inputs to an active-low latch must be equal to set the latch.
   (1) True
   (2) False
C. An active-low latch is constructed using what type of logic gates?
D. What levels must be applied to the $\overline{\text{SET}}$ and $\overline{\text{CLEAR}}$ inputs of an active-low latch to put it in the SET state?
E. The bubbles shown on the inputs of latch logic symbols are state indicators.
   (1) True
   (2) False

SECTION 6–2: ACTIVE-HIGH LATCH

OBJECTIVES

1. Identify and determine the Q and $\overline{Q}$ outputs of a NOR-gate latch under stated input conditions.
2. Identify the logic symbol for an active-high latch.

The NOR-gate latch shown in Fig. 6–12 has two inputs that are labeled SET and CLR. Note that there are no NOT signs over the input labels of this latch. This lack of NOT

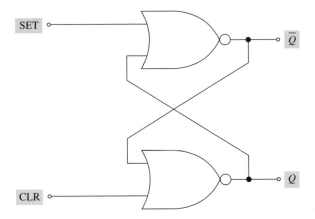

SET

CLR

FIGURE 6-12   **Active-high latch.**

signs indicates that the latch is an *active-high latch*. In an active-high latch, a Logic 1 must be applied to the input labeled with the name of the state in which we desire to put the latch. The other input must be inactive. Remember from the previous section that the inputs must be complementary to put a latch in the SET or CLEAR state.

The method for analyzing this latch is comparable to the one used in the active-low latch analysis. If **SET = 1** and **CLR = 0,** as shown in Fig. 6-13(a), the use of short logic allows us to first derive that the output of gate G1 is low. Once the $\overline{Q}$ output has been determined, the feedback from this gate can be traced to the input of gate G2. This Logic 0 feedback input and the CLR = 0 input produce a high out of gate G2. This combination of inputs places the latch in the SET state because $Q = 1$ and $\overline{Q} = 0$. The input/output levels and latch condition are shown on the state table in Fig. 6-13(b). It took a Logic 1 on the SET input with a Logic 0 on the CLR input to put the latch in the SET state. For this reason the latch is said to be an active-high input latch. In other words, when the SET input is asserted and the CLR input is left inactive, the latch goes to the SET condition.

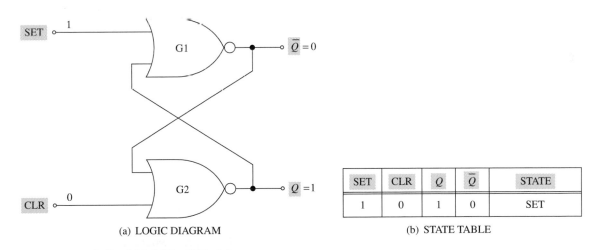

SET	CLR	Q	$\overline{Q}$	STATE
1	0	1	0	SET

(a) LOGIC DIAGRAM   (b) STATE TABLE

FIGURE 6-13   **Active-high latch: SET state.**

If **SET = 0** and **CLR = 1,** as shown in Fig. 6-14(a), the CLR input is asserted and the latch should go to the CLEAR state. A short exercise will prove this statement to be true. The CLR = 1 input to gate G2 forces its output low. The feedback from this gate to gate G1 and the SET input produce a high out of gate G1. Since $Q = 0$, the latch is in the CLEAR state. These conditions are shown on the state table in Fig. 6-14(b). Note that it took a Logic 1 on the CLR input to clear the latch.

If **SET = 0** and **CLR = 0,** as shown in Fig. 6-15(a), the use of short logic cannot be applied to circuit operation until an output condition is assumed. One thing to keep

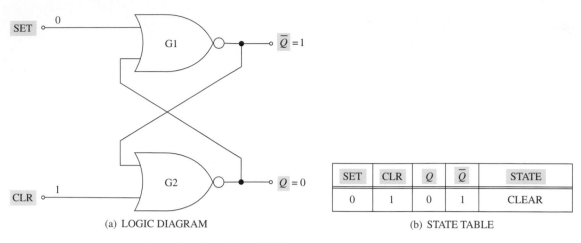

(a) LOGIC DIAGRAM

(b) STATE TABLE

SET	CLR	Q	$\overline{Q}$	STATE
0	1	0	1	CLEAR

FIGURE 6-14   Active-high latch: CLEAR state.

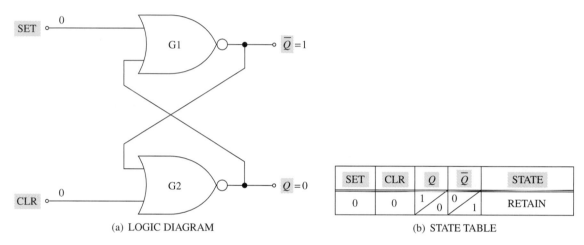

(a) LOGIC DIAGRAM

(b) STATE TABLE

SET	CLR	Q	$\overline{Q}$	STATE
0	0	1 / 0	0 / 1	RETAIN

FIGURE 6-15   Active-high latch: RETAIN state.

in mind at this point is that either output state can be assumed—you cannot be wrong. Let's assume that the circuit is in the CLEAR state as shown. To determine latch operation, we must trace the $\overline{Q} = 1$ feedback signal to the input of gate G2. This will force the output of gate G2 to remain at Logic 0. Tracing the $Q = 0$ feedback to the input of gate G1 and the SET = 0 input produces two Logic 0s into that gate. These inputs make G1's output remain at Logic 1. Thus, the two Logic 0s applied to this latch cause it to retain the state it was in when the two inputs were brought low (inactive). These inputs (SET = 0 and CLR = 0) put the latch in the RETAIN state. The state table in Fig. 6–15(b) shows that if the latch were in the SET state, it would retain the SET condition with two Logic 0s in. The Logic 0s are inactive inputs to this active-high latch. Common sense dictates the latch should remain inactive (RETAIN state) if neither one of its inputs is asserted. In other words, if we don't ask the latch to change states (do something) by asserting (activating) one of its inputs, it will rest (retain the previous condition).

If **SET = 1** and **CLR = 1,** as shown in Fig. 6–16(a), the use of short logic quickly proves that both outputs will be Logic 0. This condition, shown on the state table in Fig. 6–16(b), is an INVALID condition. The state table shows Q is low when both inputs are activated simultaneously. This low output may change to a high output when the SET and CLR inputs are brought inactive. This condition makes the Q output unpredictable and INVALID.

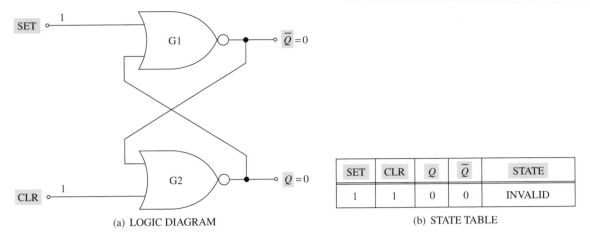

(a) LOGIC DIAGRAM

SET	CLR	Q	$\bar{Q}$	STATE
1	1	0	0	INVALID

(b) STATE TABLE

FIGURE 6-16   Active-high latch: INVALID state.

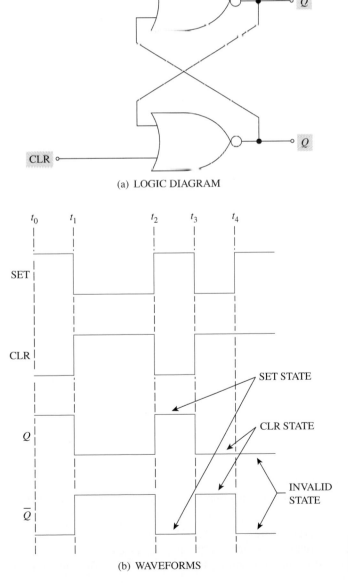

(a) LOGIC DIAGRAM

(b) WAVEFORMS

FIGURE 6-17   Active-high latch operation.

A short exercise analyzing the input waveforms to an active-high latch will enhance your understanding of this circuit. Refer to Fig. 6–17 for the following analysis:

t_0-t_1: SET = 1 and CLR = 0. Therefore, $Q = 1$ and $\overline{Q} = 0$ because the SET input is asserted and the CLR input is inactive. This puts the latch in the SET state.

t_1-t_2: SET = 0 and CLR = 1. This input change puts the latch in the CLEAR state ($Q = 0$) because the CLR input is asserted (high).

t_2-t_3: The SET input is again asserted at t_2, and the latch returns to the SET state.

t_3-t_4: The CLR input is asserted and the latch clears. The inputs have been complementary up to this point.

t_4: Both inputs are asserted and both latch outputs become Logic 0. This represents the INVALID state.

The state table shown in Fig. 6–18(b) summarizes complete operation of the NOR-gate latch shown in Fig. 6–18(a). The logic symbol for this active-high latch is shown in Fig. 6–18(c). Note the absence of NOT signs over the S and C inputs as well as the lack of bubbles on the inputs on the logic symbol. The lack of NOT signs and bubbles indicates an active-high latch.

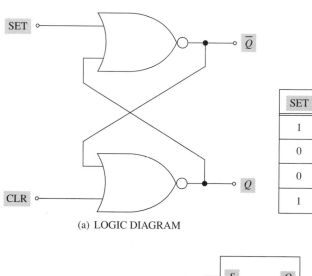

(a) LOGIC DIAGRAM

SET	CLR	Q	$\overline{Q}$	STATE
1	0	1	0	SET
0	1	0	1	CLEAR
0	0	1 / 0	0 / 1	RETAIN
1	1	0	0	INVALID

(b) STATE TABLE

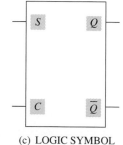

(c) LOGIC SYMBOL

FIGURE 6-18   **Active-high latch.**

Section 6–2: Review Questions

Answers are given at the end of the chapter.

A. What is the level of the $\overline{Q}$ output of an active-high latch when it is in the SET state?

B. The inputs to an active-high latch must be complementary to set or clear the latch.
   (1) True
   (2) False

C. An active-high latch is constructed using what type of logic gates?

D. What levels must be applied to the SET and CLEAR inputs of a NOR-gate latch to put it in the SET state?

E. In what state is an active-high latch when $S = 0$ and $C = 0$?
F. Draw a state table for an active-high latch using Active (A) and Inactive (I) for the SET and CLEAR inputs instead of Logic 0 and Logic 1. (Refer to Fig. 6–9 for an active-low latch state table designed in this manner).

Section 6-3: Gated Latch

Objectives

1. Identify and determine the Q output of gated latches.
2. Identify the logic symbols used for various gated latches.

The active-low and active-high latches presented in the first two sections of this chapter are *asynchronous*—not clock dependent. Asserting one input on these latches caused the output to change states almost instantaneously with the input change. The small difference in time between the input change request and the actual output response of the latch is due to the circuit's inherent **propagation delay.** The importance of this delay will be presented shortly.

A **gated latch** has another input that is used to control when the latch can change states. This gate (G) input is often called an **enable input.** Since the timing of the output level changes is controlled by the enable input, the gated latch is **synchronous.**

Gated S-C Latch

The logic diagram and symbol for a gated S-C latch is shown in Fig. 6–19. The addition of the two **steering gates** at the front of the latch in Fig. 6–19(a) is the method used to allow synchronization. The steering gates steer the signal inputs (SET and CLR) to the

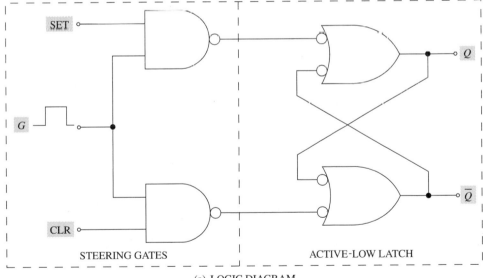

(a) LOGIC DIAGRAM

FIGURE 6-19   Gated S-C latch.

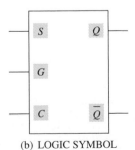

(b) LOGIC SYMBOL

latch or prevent them from reaching it. The steering gates are enabled when the G input is high, and they pass the SET and CLR data inputs to the latch. Since the NAND-gate latch is active low, the SET and CLR inputs to this circuit are active high because they are inverted by the NAND steering gates. This is illustrated in Fig. 6–19(b) by the labels S and C without overbars and by the lack of bubbles on the symbol's inputs.

The steering gates are inhibited when the G input is low. The high outputs from the steering gates when they are inhibited are inactive inputs to the latch. This causes the latch to retain its last condition. In other words, when the enable input goes low, the information that was present on the data inputs at that instant in time (high-to-low transition) is stored at Q.

Analysis of the gated-latch logic diagram in Fig. 6–19(a) will produce the state table shown in Fig. 6–20.

INPUTS			OUTPUTS		
S	C	G	Q	$\overline{Q}$	STATE
1	0	1	1	0	SET
0	1	1	0	1	CLEAR
0	0	1	1/0	0/1	RETAIN
1	1	1	1	1	INVALID
X	X	0	1/0	0/1	RETAIN

FIGURE 6-20   Gated S-C latch state table.

The $S = 1$, $C = 0$ and $S = 0$, $C = 1$ input combinations, when applied with $G = 1$, will produce the SET and CLEAR output conditions respectively. Furthermore, since the circuit is active high at its inputs, $S = 0$ and $C = 0$ represent the inactive input levels that will result in the RETAIN mode of operation. The circuit produces an INVALID output when both inputs are asserted while $G = 1$. Finally, if the enable (G) input is low, the NAND steering gates are inhibited and the latch is in the RETAIN mode the same as when $S = C = 0$.

A timing diagram for the gated S-C latch is shown in Fig. 6–21. Analysis of the circuit follows:

t_1: $G = 0$, so the SET and CLR inputs are don't cares. This condition puts the latch in the RETAIN state. The initial condition of the circuit is the CLEAR state. *Note: This assumption of $Q = 0$ initially will be made throughout the remainder of this text. Any exception to this assumption will be clearly indicated.*

t_2: $G = 1$, so the steering gates are enabled, SET is already active, and CLR is inactive. The latch goes to the SET state.

t_3: $G = 0$, and the latch retains the SET state even though the SET and CLR inputs change.

t_4: $G = 1$, SET is inactive, CLR is active, and the latch clears.

FIGURE 6-21   Gated S-C latch waveform analysis.

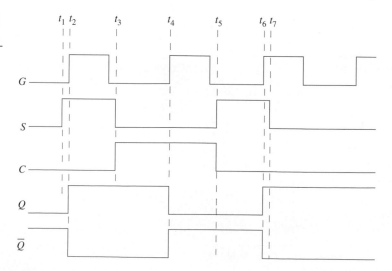

t_5: $G = 0$ = RETAIN state.

t_6: $G = 1$, SET is active, CLR is inactive, and the latch goes to the SET state.

t_7: $G = 1$, $S = 0$, and $C = 0$ = RETAIN state.

Gated-D Latch

The biggest problem encountered with all of the latch circuits that have been analyzed thus far is the INVALID output condition. The predictability of the output of a latch is of utmost importance. Thus, the preceding latch circuits all have this severe defect.

The INVALID output condition is a relatively simple problem to overcome. There are only three output states desired from any latch—SET or CLEAR and RETAIN. The gated S-C latch state table in Fig. 6–20 offers some insight to solution of this problem. The state table shows two different input combinations will produce the RETAIN state. Since the enable (G) input can be used to force the RETAIN state when $G = 0$, the $S = 0$ and $C = 0$ input combination is unnecessary to achieve this condition. If the equal input combination of $S = 0$ and $C = 0$ is unnecessary, and $S = 1$ and $C = 1$ (INVALID state) is undesired, it is a simple task to ensure these input combinations never occur. The resultant circuit used to prevent these input combinations is depicted in Fig. 6–22(a), and its logic symbol is shown in Fig. 6–22(b).

This circuit is known as a **D-type latch** and is normally a gated latch. The **data-** or **delay-type latch** is different from the gated S-C latch in only one respect. To ensure equal data input combinations are never applied to the steering gates, only one data input is applied to the circuit. This input is applied to one steering gate, and it is inverted prior to being applied to the other steering gate.

This configuration allows only complementary inputs to the steering gates. Thus when they are enabled ($G = 1$), the latch will be in the SET or CLEAR state. When the G input is dropped low, the state of the latch at this enable pulse transition is stored (NC— No Change—mode) as shown in Fig. 6–22(c).

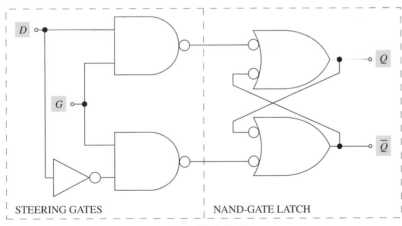

STEERING GATES   NAND-GATE LATCH

(a) LOGIC DIAGRAM

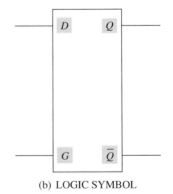

(b) LOGIC SYMBOL

INPUTS		OUTPUTS	
D	G	Q	$\overline{Q}$
1	1	1	0
0	1	0	1
X	0	NC	NC

(c) STATE TABLE

Figure 6-22   D-type latch.

The state table indicates Q will follow the D input when the circuit is enabled. It also shows the circuit is in the No Change (NC) mode when $G = 0$. The NC mode equates directly to the RETAIN mode. This D-type latch is often called a **transparent latch** because Q follows D when the circuit is enabled. A timing diagram depicting operation of a D-type latch is shown in Fig. 6–23.

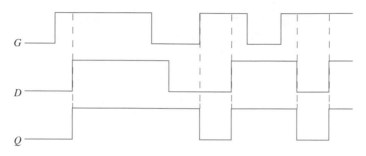

FIGURE 6-23   D-type latch waveforms.

If it is desirable to enable the latch on the *negative* duration of the input enable pulse an inverter can be placed on the G-input line. This is shown in Fig. 6–24(a). If the actual circuit is enabled with a negative pulse it would be drawn as shown in Fig. 6–24(b). The bubble indicates active-low gating. The input pin in this case is labeled ($\overline{G}$).

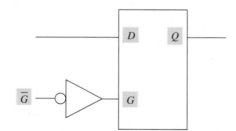

(a) NEGATIVE ENABLED (EXTERNAL NOT GATE)

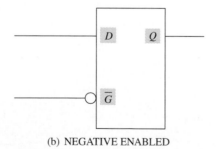

(b) NEGATIVE ENABLED   **FIGURE 6-24   Gated D-type latch.**

4-Bit Bistable Latch

The SN74LS75 and SN74LS77 Four-Bit Bistable Latch ICs are good examples of two other methods used to implement latch circuits. The SN74LS75 latch shown in Fig. 6–25 operates the same as the gated-D latch. That is, Q follows D when the circuit is enabled. It latches (stores) data when the enable input goes low. This is shown on the state table in Fig. 6–25(b). This 16-pin IC provides Q and $\overline{Q}$ outputs. This is shown in the logic diagram in Fig. 6–25(a) and on the ANSI/IEEE symbol for the latch in Fig. 6–25(c). Only one latch of the four available on this IC is shown in the logic diagram. Operation of the latch can be discerned as follows:

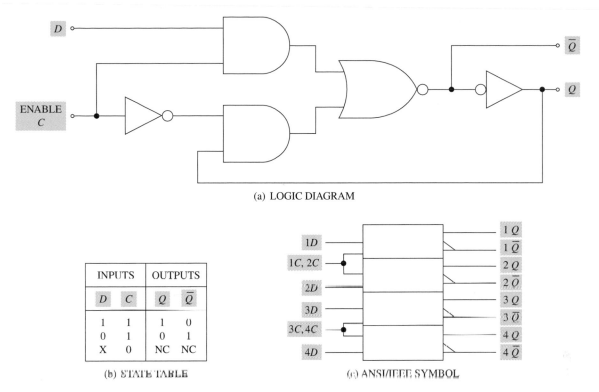

(a) LOGIC DIAGRAM

INPUTS		OUTPUTS	
D	C	Q	$\bar{Q}$
1	1	1	0
0	1	0	1
X	0	NC	NC

(b) STATE TABLE

(c) ANSI/IEEE SYMBOL

FIGURE 6-25   **SN74LS75 four-bit bistable latch.**

When the enable $(C) = 0$, the top AND gate is inhibited, and data is prevented from entering the latch. The C input is inverted and enables the bottom AND gate. This allows the Q output level to control the NOR gate so that the current latch condition is retained.

When the enable $= 1$, the bottom AND gate is inhibited and the top one is enabled. This allows the data input to control the NOR gate, which causes Q to follow D.

The SN74LS77 Four-Bit Bistable Latch provides another slight variation. This IC does not provide a $\bar{Q}$ output. The logic diagram of this latch is shown in Fig. 6–26. The state table is identical to that of the SN74LS75, and the ANSI/IEEE symbol is the same if the $\bar{Q}$ outputs are removed from the SN74LS75 symbol. The small difference in the two latches is the missing $\bar{Q}$ output from the SN74LS77. This allows use of a 14-pin IC package.

The SN74116 Dual 4-Bit Latch with Clear Input is shown in Fig. 6–27. This latch employs yet another variation in latch implementation. Only one latch of the eight available on this IC is shown in Fig. 6–27(a). The state table is shown in Fig. 6–27(b), and the ANSI/IEEE symbol is illustrated in Fig. 6–27(c).

FIGURE 6-26   **SN74LS77 4-bit bistable latch logic diagram.**

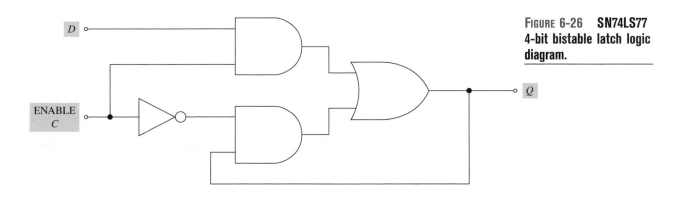

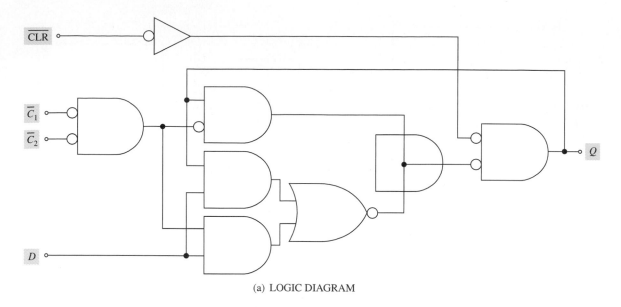

(a) LOGIC DIAGRAM

INPUTS				OUTPUT
$\overline{\text{CLR}}$	ENABLE		DATA	Q
	$\overline{C_1}$	$\overline{C_2}$		
1	0	0	1	1
1	0	0	0	0
1	X	1	X	NC
1	1	X	X	NC
0	X	X	X	0

(b) STATE TABLE

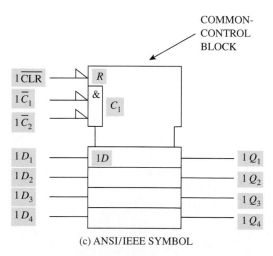

(c) ANSI/IEEE SYMBOL

FIGURE 6-27   SN74116 dual 4-bit latch.

Each 4-bit latch on this IC has an *asynchronous* CLEAR input signal. This allows re-setting of the latch independent of the enable and data inputs. Since the $\overline{\text{CLR}}$ input is tied directly to the output gate, this NOR gate is inhibited when the input is asserted. This will drive Q low regardless of the other input conditions. The state table indicates all other in-puts are don't cares when $\overline{\text{CLR}}$ is activated. This asynchronous input as well as an addi-tional one will be discussed in more detail in the next section.

Each 4-bit latch also has a gated 2-input enable circuit ($\overline{C}_1$ and $\overline{C}_2$). When both of the enable inputs are low and the $\overline{\text{CLR}}$ input is inactive, Q will follow D. Careful analysis of the logic diagram will prove this statement to be true. The latch has been redrawn in Fig. 6–28(a) for the first step of the following discussion.

With $\overline{\text{CLR}} = 1$, $\overline{C}_1 = \overline{C}_2 = 0$, and data = 1, the circuit reveals gate G4 is enabled by the output of gate G1. This allows the high data input to pass through G4 to the input of G5. Since this high input to G5 inhibits the gate, the other input to G5 from G3 is a don't care. The output of G5 goes low when it is inhibited. This drops the bottom input to G6 low. The AND symbol between G5 and G6 represents a **wired-AND** circuit, and for now it may be considered as a normal AND gate. (The inputs are the lines entering the top and bottom of the symbol.) With both inputs to the NOR gate G6 low, its Q output is high. This same procedure can be used to prove Q follows D if D changes to a low level.

If either or both of the enable inputs are taken high, the circuit goes to the RETAIN state. The logic levels for this condition with Q high are shown on Fig. 6–28(b). If $\overline{C}_1$ is taken high, G4 is inhibited and the input data is prevented from reaching the output. Gate G2 is enabled and feedback from Q puts the circuit is the No Change (NC) mode.

It would be good practice at this point to clear Q by asserting the $\overline{\text{CLR}}$ momentarily, then take $\overline{C}_2$ high and prove the circuit retains $Q = 0$. The logic levels for this practice are shown in Fig. 6–28(c).

Let's return briefly to Fig. 6–27(c) and discuss the ANSI/IEEE symbol for this latch. Only one section of the 4-bit latch is shown in Fig. 6 27(a). The distinctive shape of the ANSI/IEEE symbol indicates the portion on the top of the symbol is a **common-control block.** Typically, an input to this block represents an input to each element below the common-control block. In this case the elements are the four latches. Thus, the 1 $\overline{\text{CLR}}$ input to this block indicates each of the 4-bit latches in this section can be asynchronously cleared. The small box within the common-control box contains the AND function symbol. The *qualifying symbols* on the 1 $\overline{C}_1$ and 1 $\overline{C}_2$ input lines indicate active-low inputs are required here and *both* of the inputs must be low to produce the high C_1 internally to enable the four latches.

Section 6–3: Review Questions

Answers are given at the end of the chapter.

A. Gated latches are _____.
   (1) synchronous
   (2) asynchronous

B. Q follows D when the enable input of the latch shown in Fig. 6–22 is _____.
   (1) high
   (2) low

C. The data input to the latch shown in Fig. 6–24(b) is latched (stored) at the Q output when the $\overline{G}$ input is _____.
   (1) high
   (2) low

D. Q should be _____ for the *initial state* when analyzing latches.
   (1) high
   (2) low

E. The D-type latch is often called a transparent latch.
   (1) True
   (2) False

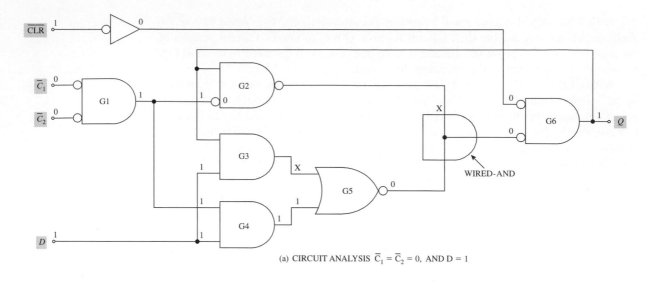

(a) CIRCUIT ANALYSIS $\overline{C}_1 = \overline{C}_2 = 0$, AND D = 1

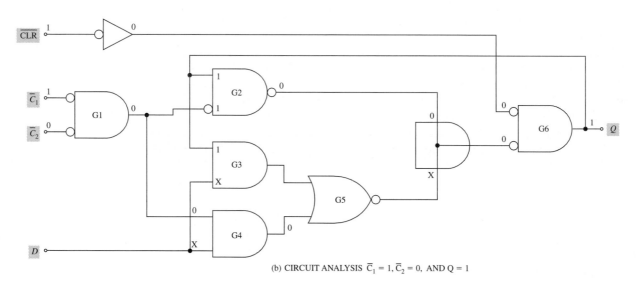

(b) CIRCUIT ANALYSIS $\overline{C}_1 = 1, \overline{C}_2 = 0$, AND Q = 1

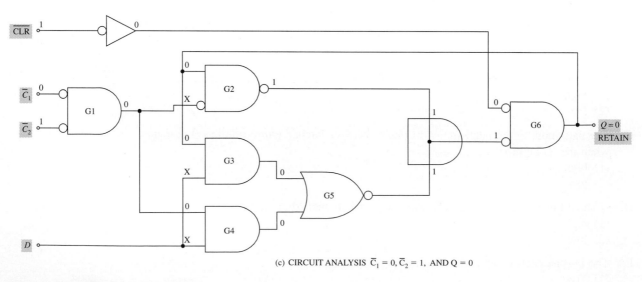

(c) CIRCUIT ANALYSIS $\overline{C}_1 = 0, \overline{C}_2 = 1$, AND Q = 0

Figure 6-28 **SN74116 dual 4-bit latch.**

SECTIONS 6-1 THROUGH 6-3: INTERNAL SUMMARY

Latches and flip-flops are bistable devices that can be used to store or transfer binary data.

Machine-level language is binary.

A NAND-gate latch is an active-low latch.

A NOR-gate latch is an active-high latch.

The NAND- and NOR-gate latches that do not contain steering gate circuits are asynchronous.

There is no reason to try to memorize the input/output states that were explained in the preceding sections. All you need to keep in mind to determine latch operation is this:

1. The inputs to a NAND- or NOR-gate latch must be complementary to *set* or *clear* it.
2. Asserting only one input to an asynchronous latch will put it in the state that the asserted input is labeled.
3. Asserting both inputs will put an asynchronous latch in an INVALID state.
4. Leaving both inputs to an asynchronous latch inactive will put it in the RETAIN state.

An active-low latch requires $\overline{S} = 0$ and $\overline{C} = 1$ inputs to put it in the SET state. In an active-low latch, a Logic 0 must be applied to the input labeled with the name of the state in which we desire to put the latch, and the other input must be inactive.

An active-low latch requires $\overline{S} = 1$ and $\overline{C} = 0$ inputs to put it in the CLEAR state.

An active-low latch will be in the RETAIN state when $\overline{S} = 1$ and $\overline{C} = 1$. Logic 1 is the inactive level for an active-low latch. Naturally, if we leave both inputs inactive, we are not asking the latch to do anything, so it retains its previous condition.

An active-low latch will have an INVALID output if both inputs are asserted (low) simultaneously. This fact is evident if you remember the active-low latch is constructed with NAND gates, and the short logic of a NAND gate is "any 0 in = 1 out." Thus, two 0s into the latch equal two 1s out.

An active-high latch requires $S = 1$ and $C = 0$ inputs to put it in the SET state. Always keep in mind in an active-high latch that a Logic 1 must be applied to the input labeled with the name of the state in which we desire to put the latch, and the other input must be inactive.

An active-high latch requires $S = 0$ and $C = 1$ inputs to put it in the CLEAR state.

An active-high latch will be in the RETAIN state when $S = 0$ and $C = 0$ (both inputs are inactive).

An active-high latch will have an INVALID output if both inputs are asserted (high).

A gated latch is synchronous. Synchronism is attained by the addition of steering gates.

Q follows D when a gated latch is enabled. The latch is in the RETAIN state when it is not enabled.

SECTIONS 6-1 THROUGH 6-3: INTERNAL SUMMARY QUESTIONS

Answers are given at the end of the chapter.

1. Latches are _____ circuits.
   a. astable
   b. bistable
   c. monostable

2. Asserting an input to a circuit means to bring that input to its active level.
   a. True
   b. False

3. To set or clear a latch, the inputs must be complementary.
  a. True
  b. False

4. Activating only one input to a latch will put it in the state that the asserted input is labeled.
  a. True
  b. False

5. Activating both inputs will put a latch in the RETAIN state.
  a. True
  b. False

6. What is the level of Q when an active-low latch is in the SET state?
  a. Low (0)
  b. High (1)

7. What must the input levels of an active-low latch be to put the latch in the CLEAR state?
  a. $\overline{S} = 0, \overline{C} = 0$
  b. $\overline{S} = 1, \overline{C} = 1$
  c. $\overline{S} = 0, \overline{C} = 1$
  d. $\overline{S} = 1, \overline{C} = 0$

8. What is the state of an active-low latch when its inputs are $\overline{S} = 1$ and $\overline{C} = 1$?
  a. SET
  b. CLEAR
  c. RETAIN
  d. INVALID

9. Which logic symbol in Fig. 6–29 represents an active-low latch?
  a.
  b.

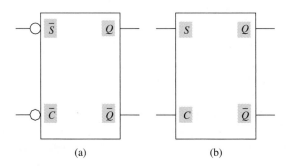

(a)   (b)

FIGURE 6-29

10. What is the level of $\overline{Q}$ when an active-high latch is in the CLEAR state?
  a. Low (0)
  b. High (1)

11. An active-high latch is constructed using cross-coupled _____ gates.
  a. OR
  b. AND
  c. NOR
  d. NAND

12. What must the input levels of a NOR-gate latch be to put it in the SET state?
  a. $S = 0, C = 0$
  b. $S = 1, C = 1$
  c. $S = 0, C = 1$
  d. $S = 1, C = 0$

13. An active-high latch will be in the _____ state when $S = 1$ and $C = 1$.
    a. SET
    b. CLEAR
    c. RETAIN
    d. INVALID

14. What is the output condition of a NOR-gate latch when $S = 0$ and $C = 1$?
    a. SET
    b. CLEAR
    c. RETAIN
    d. INVALID

15. Basic machine-level language utilizes _____ numbers.
    a. hex
    b. octal
    c. binary
    d. decimal

16. Gated latches are asynchronous.
    a. True
    b. False

17. The gated D latch shown in Fig. 6–30 is in the _____ state.
    a. SET
    b. RESET
    c. CLEAR
    d. RETAIN

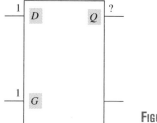

FIGURE 6-30

18. The latch shown in Fig. 6–30 is in the _____ state if the G input is dropped low.
    a. CLEAR
    b. RETAIN
    c. INVALID

Flip-Flop Introduction

Most digital systems utilize intricate, extremely stable timing circuits to synchronize operation. For example, if an operator depresses a key on the keyboard of a computer, this input to the computer is asynchronous because it is not timed with the system's clock. Therefore, a method must be used to bring this asynchronous input into the system synchronously. This synchronization circuitry often employs flip-flops.

Flip-flops, like their latch counterparts, are bistable devices. They can store only one bit of data. The difference between flip-flops and latches is the method used to trigger them. Latch circuits are **pulse triggered.** They are enabled on the positive duration of an enable input pulse or on the negative duration. Flip-flops are **edge-triggered** devices. The input used to trigger a flip-flop to the SET or CLEAR state is called a Clock input. Flip-flops are triggered to change states only on the active transition of the clock pulse from

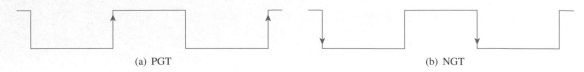

(a) PGT   (b) NGT

FIGURE 6-31   Clock (synchronization) pulses.

low to high or high to low. This synchronous input causes the flip-flop to operate with exact coincidence in time or rate. Take the example of depressing a key on the keyboard of a computer and imagine the havoc this would cause in the computer if that input were not synchronized with everything else that was happening in the system.

The output of a flip-flop can respond to the data inputs only on an **active transition (edge)** of the input clock pulse. If the flip-flop is clocked on the low-to-high clock transition, as shown in Fig. 6–31(a), it is positive-edge triggered. This is usually referred to as the **Positive-Going Transition (PGT)** of the clock pulse. If the flip-flop is clocked on the high-to-low clock transition, as shown in Fig. 6–31(b), the circuit is negative-edge triggered. This is normally called the **Negative-Going Transition (NGT)** of the clock pulse.

The steering gates in a flip-flop will allow the data inputs to reach the latch only during the short-duration **active clock transition.** This will be the NGT or PGT of the clock pulse depending on the type of flip-flop.

SECTION 6-4: D-TYPE FLIP-FLOP

OBJECTIVES

1. Identify and determine the Q and $\overline{Q}$ outputs of D-type flip-flops under stated input conditions.
2. Identify the logic symbols for negative- and positive-edge-triggered D-type flip-flops.

The D-type flip-flop has only one data input pin. This circuit is identical to the D-type latch previously discussed with one exception—the flip-flop contains an **edge-detector** circuit at the Clock input. The D-type flip-flop logic diagram is shown in Fig. 6–32(a).

Positive-Edge-Triggered D-type flip-flop

The edge-detector circuit is shown in block diagram form for now. It will be presented and analyzed shortly. If the circuit in the block represents a **positive-edge detector,** it will produce a short-duration positive pulse on the PGT of the clock pulse. This short-duration pulse is shown below the CLK waveform in Fig.6–32(a) and is labeled PGT pulse. This pulse is used to enable the steering gates. Operation of the circuit is identical to the D-type latch from this point.

When the PGT of the clock pulse enables the steering gates, Q follows the D input. When the short-duration positive clock pulse is not present, the low input from the edge detector inhibits the steering gates. This puts the NAND-gate latch in the RETAIN state. The clock input line can rest low for an extended period to store data in the flip-flop.

Operation of this D-type flip-flop is shown in the state table in Fig. 6–32(b). Notice on the state table that the CLK has an arrow pointing up in two of the three flip-flop conditions. This indicates that Q follows D only on the PGT of the clock pulse.

The logic symbol for a positive-edge-triggered D-type flip-flop is shown in Fig. 6–32(c). The triangle at the CLK input is a **dynamic input indicator.** This indicates the flip-flop is edge triggered. Since there is no bubble adjacent to this dynamic input indicator, this flip-flop is positive-edge triggered. The dynamic input indicator may be thought of as a delta (Δ) sign lying on its side. The term delta implies "change," so the symbol

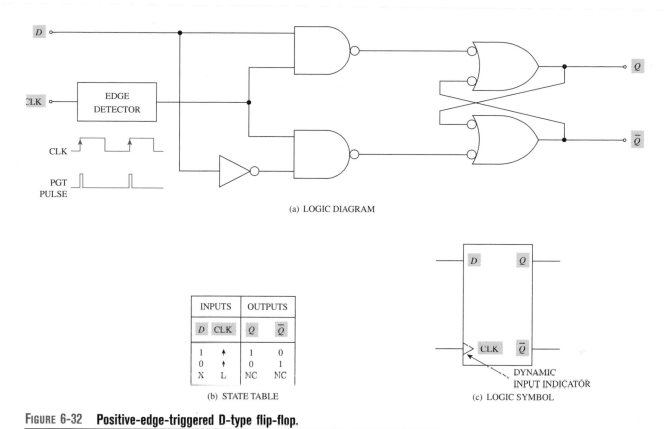

(a) LOGIC DIAGRAM

INPUTS		OUTPUTS	
D	CLK	Q	$\bar{Q}$
1	↑	1	0
0	↑	0	1
X	L	NC	NC

(b) STATE TABLE

(c) LOGIC SYMBOL

FIGURE 6-32   Positive-edge-triggered D-type flip-flop.

should be thought of as one indicating the input must be changing to clock the flip-flop. Since the clock pulse only changes on its low-to-high (up-clock) transition or high-to-low (down-clock) transition, the delta sign must imply edge triggering. In the symbol shown in Fig. 6–32(c), the dynamic input indicator indicates positive-edge triggering.

Edge detection can be accomplished with a simple circuit such as the one shown in Fig. 6–33. This circuit is a positive-edge detector. Edge detection in this circuit is accomplished by ANDing the clock pulse with an inverted, delayed version of itself. The key to this logical operation is that CLK 2 (Fig. 6–33) has been slightly delayed by the inherent propagation delay of the NOT gate as well as inverted. The output of the AND gate is high only when both of its inputs are high. This condition occurs on the PGT of the input clock pulse. Notice the condition does not exist on the NGT of the clock pulse. This narrow PGT pulse is used to enable the steering gates of the circuit in Fig. 6–32(a). The pulse width of the pulse is equal to the propagation delay of the inverter. It is usually 20–25 ns minimum for a TTL IC and is specified as t_w on a data sheet.

A logic symbol and timing diagram for a positive-edge-triggered D-type flip-flop is shown in Fig. 6–34. Note on the timing diagram that Q follows the D input only on the PGT of the clock pulse.

One advantage edge-triggered flip-flops have over level-triggered latches is that the edge-triggered circuits are not as susceptible to noise. The level-triggered latches are susceptible to noise spikes during the entire period of the enable pulse. Therefore, their control inputs (*S/C, D,* or *J/K*) must be held constant for the duration of the enable pulse. On the other hand, the steering gate circuits in edge-triggered flip-flops are enabled only for a very short period of time.

Data Sheet Specifications

This brings up a point that needs to be discussed. The data input to the edge-triggered flip-flops should not be changed during the active clock transition. Unreliable operation will most likely result if this is allowed to happen.

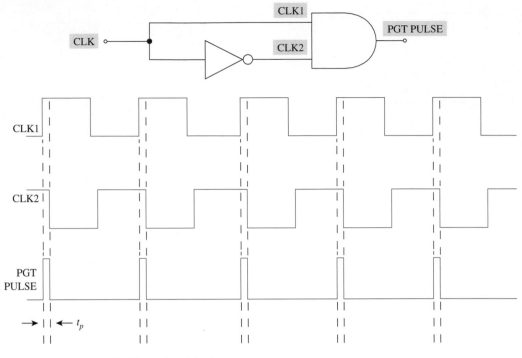

Figure 6-33   Positive-edge detector.

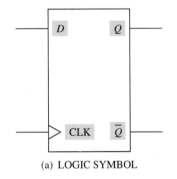

(a) LOGIC SYMBOL

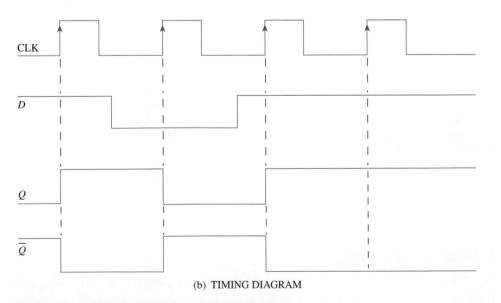

(b) TIMING DIAGRAM

Figure 6-34   Positive-edge-triggered D-type flip-flop.

The data input to the flip-flop in Fig. 6–34 must be present prior to the active clock transition. The amount of time prior to the clock transition is specified on a data sheet as **setup time** (t_{su}). Setup time is the interval of time the data input must be held constant prior to the active clock transition. If this prescribed setup time is met, accurate data recognition at the output is guaranteed. In other words, circuits should be designed so that flip-flops are not requested to change states during a clock transition. Setup time is available in manufacturer's data books.

Hold time (t_h) is the interval of time from the active clock transition to when the input data is no longer required to ensure proper interpretation of the output data. Hold time for most flip-flops is 0; however, it does range up to 5 ns for some flip-flops. Setup time and hold time are illustrated in Fig. 6–35. All of the waveforms are for a positive-edge-triggered flip-flop.

The **maximum clock frequency** (f_{max}) of bistable circuits is another specification of extreme importance to circuit designers. As the name implies, f_{max} is the highest clock rate that can be used to clock a flip-flop and ensure reliable operation.

The inherent propagation delay time of bistable circuits is specified in data books on both the high to low transition and the low to high transitions. The **low-to-high propagation delay time** is referred to as t_{PLH}. The **high-to-low propagation delay time** is referred to as t_{PHL}. Both of these times represent the time delay between specified points on the input and output voltage waveforms. The propagation delay times for the inverting and noninverting functions are shown in Fig. 6–36.

The propagation delay times in a bistable circuit affect the maximum clock frequency of the device. The frequency used to clock a flip-flop cannot be so high that a subsequent clock pulse arrives at the input before the circuit has had sufficient time to respond to the previous one. Some of the more common symbols used on data sheets detailing flip-flop and latch operation are shown in Table 6–1.

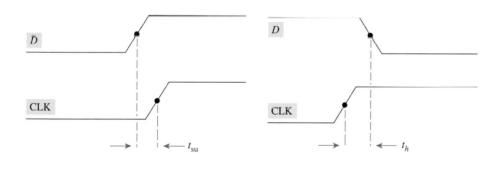

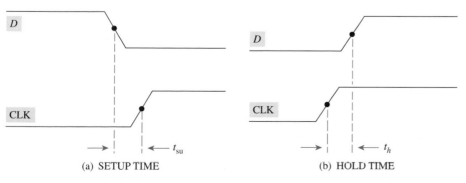

(a) SETUP TIME   (b) HOLD TIME

FIGURE 6–35   **Setup and hold times.**

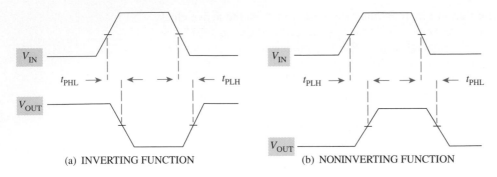

(a) INVERTING FUNCTION   (b) NONINVERTING FUNCTION

FIGURE 6-36   **Propagation delay time.**

TABLE 6-1	Symbol	Meaning
Data Sheet Symbols	H	High level–steady state
	L	Low level–steady state
	↑	Low-to-high transition (PGT)
	↓	High-to-low transition (NGT)
	X	Don't care
	⊓	High-level pulse
	⊔	Low-level pulse
	Toggle	Change states on PGT or NGT

Negative-Edge-Triggered D-Type Flip-Flop

The negative-edge-triggered D-type flip-flop is shown in Fig. 6–37. The symbol for this flip-flop is shown in Fig. 6–37(c). The only difference between this logic symbol and the PGT-triggered flip-flop symbol shown in Fig. 6–34 is the addition of a bubble adjacent to the dynamic input indicator. The bubble indicates negative-edge triggering is employed in this flip-flop.

The logic diagram of the negative-edge-triggered D-type flip-flop is shown in Fig. 6–37(a). The negative-edge detector circuit produces a short-duration positive pulse on the NGT of the CLOCK pulse. The edge detector must produce a positive pulse on the NGT to enable the steering gates in the flip-flop. The CLK input to the edge detector and the NGT pulse it produces are shown in the figure. Notice the internal circuitry (steering gates and latch) of this negative-edge-triggered flip-flop is identical to that of the positive-edge-triggered flip-flop in Fig. 6–32(a).

Since the steering gates are enabled by the short duration positive pulse that occurs on the high-to-low clock transition (NGT), the flip-flop is negative-edge triggered. The Q output follows the D input when the NGT of the clock pulse is applied to the flip-flop. When an NGT is not present, the steering gates are inhibited, and the circuit is in the RETAIN state. It is normal for the clock input to this circuit to rest high for memory operation.

The state table for this flip-flop (Fig. 6–37b) indicates it is negative-edge triggered by the arrows in the CLK column pointing down. The arrow specifies Q follows D *only* on the NGT of the clock pulse.

The logic circuit required to produce the NGT pulse can be implemented as shown in Fig. 6–38. A NOR gate is used in this circuit because a positive pulse is required on the NGT of the clock pulse. Two low inputs are present at the NOR gate for a brief interval when the NGT occurs.

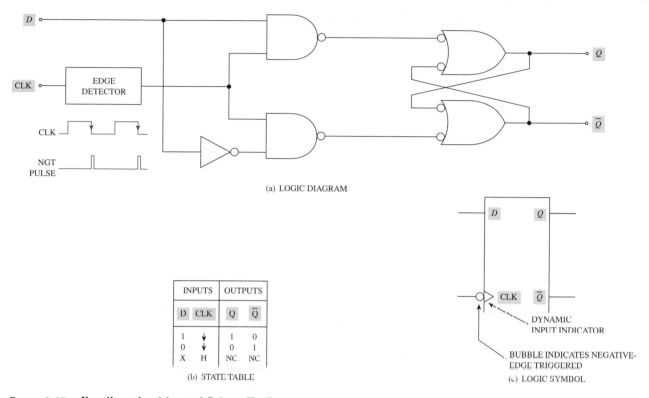

(a) LOGIC DIAGRAM

INPUTS		OUTPUTS	
D	CLK	Q	$\bar{Q}$
1	↓	1	0
0	↓	0	1
X	H	NC	NC

(b) STATE TABLE

DYNAMIC
INPUT INDICATOR

BUBBLE INDICATES NEGATIVE-
EDGE TRIGGERED

(c) LOGIC SYMBOL

Figure 6-37 Negative-edge-triggered D-type flip-flop.

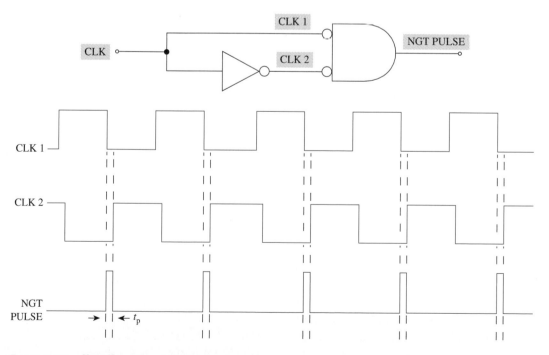

Figure 6-38 Negative-edge detector.

A logic symbol and timing diagram for a negative-edge-triggered D-type flip-flop are shown in Fig. 6–39. Analysis of the Q output waveform reveals Q follows D on the NGT of the clock pulse. The analysis was begun with the flip-flop in the CLEAR state.

Operation of positive-edge and negative-edge triggered D-type flip-flops is summarized by the waveforms in Fig. 6–40. Identical clock and data (D) inputs are applied to each flip-flop. Note the differences of the two Q output waveforms.

These differences come from the fact that one flip-flop is positive-edge triggered while the other is negative-edge triggered. The $\overline{Q}$ outputs of these flip-flops are not shown. This output is not available on some flip-flops. An example of this is the SN74174 Hex D-Type Flip-Flop with CLEAR. This positive-edge triggered flip-flop IC contains six flip-flops with an **asynchronous CLEAR input.** Some flip-flops also have an **asynchronous SET input.**

The D input to a D-type flip-flop is synchronous because it is clocked to Q with the active transition of the clock pulse. The two asynchronous inputs ($\overline{SET}$ and $\overline{CLR}$) referred to in the preceding paragraph are not clock dependent. In fact, these inputs are the overriding inputs that control flip-flop operation when they are activated. The $\overline{CLR}$ input was discussed briefly in the previous section.

Figure 6–41 depicts why these inputs are in charge of circuit operation when either of them is activated. Since they are connected directly to the output NAND gates in the latch, they override the data and clock inputs. The logic diagram shown in Fig. 6–41 is for one-half of an SN7474 Dual D-Type Positive-Edge-Triggered Flip-Flop.

The $\overline{PRE}$ input is labeled $\overline{S}_D$ in some data books. This indicates **direct set.** When this input is activated the flip-flop goes to the SET state. The $\overline{CLR}$ input is sometimes labeled $\overline{C}_D$—direct clear. These asynchronous inputs are provided on many latch and flip-flop ICs since it is impossible to determine the initial state (SET or CLEAR) the circuit will as-

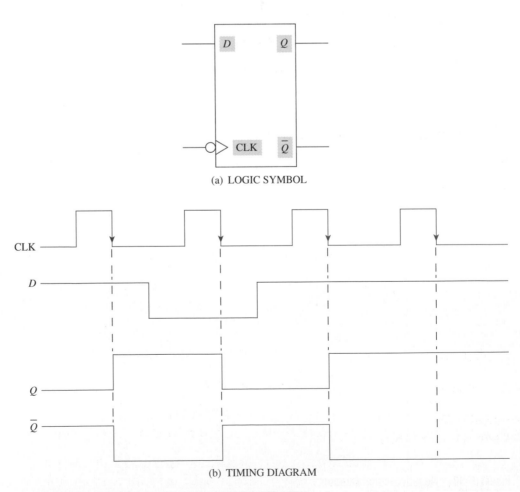

(a) LOGIC SYMBOL

(b) TIMING DIAGRAM

FIGURE 6-39 **Negative-edge-triggered D-type flip-flop.**

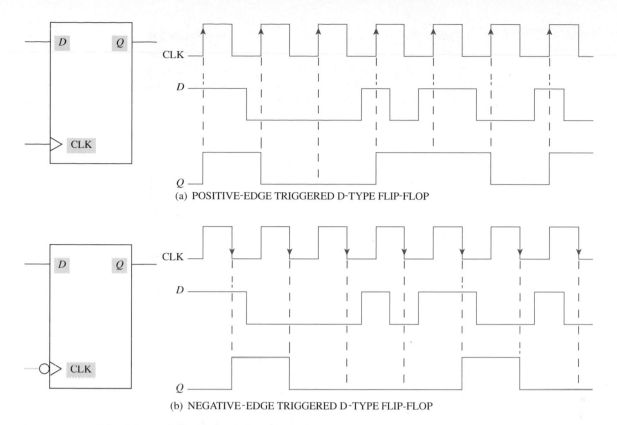

(a) POSITIVE-EDGE TRIGGERED D-TYPE FLIP-FLOP

(b) NEGATIVE-EDGE TRIGGERED D-TYPE FLIP-FLOP

FIGURE 6-40 Edge-triggered flip-flop waveform analysis.

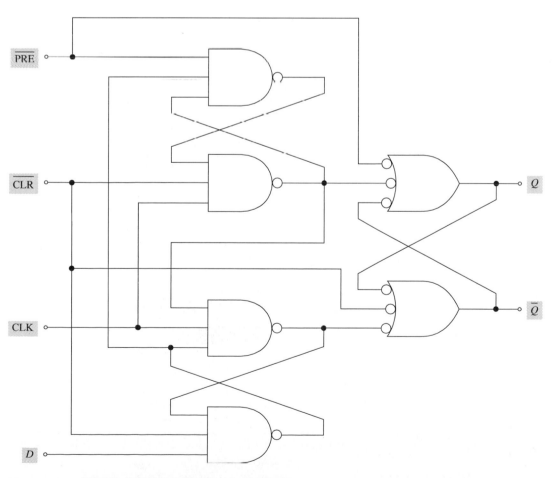

FIGURE 6-41 SN7474 dual D-type flip-flop logic diagram.

sume when power is applied. Sometimes a flip-flop will come up in the CLEAR state; other times it will come up in the SET state. If flip-flops are going to be used to store data, they should all be cleared prior to loading.

The $\overline{\text{PRE}}$ and $\overline{\text{CLR}}$ inputs are both active-low inputs. They should not be activated simultaneously for the same reason both inputs to a NAND-gate latch should not be activated at the same time. Remember, both Q and $\overline{Q}$ go high when this occurs. Also, the condition of the outputs is unpredictable once the $\overline{\text{PRE}}$ and $\overline{\text{CLR}}$ inputs are returned to their inactive level. These two inputs should always be connected to V_{CC} if they are not going to be used to asynchronously control the flip-flop. They should not be left floating. If their use is necessary, they must be connected to a normally high switch or line that can be taken low when direct control of the flip-flop is desired.

Most edge-triggered flip-flops are manufactured with $\overline{\text{PRE}}$ and $\overline{\text{CLR}}$ inputs. If both of these inputs are not available on a flip-flop, usually the $\overline{\text{CLR}}$ input is. There are exceptions, especially in latch circuits.

Figure 6–42 shows the logic symbols and a state table for the SN7474 Dual D-Type Flip-Flop. The ANSI/IEEE symbol is shown in Fig. 6–42(b). The qualifying symbols on the $\overline{\text{PRE}}$ and $\overline{\text{CLR}}$ input lines represent active-low inputs. The state table for this flip-flop has a new look with the addition of the asynchronous inputs.

The first set of input conditions on the state table shows $\overline{\text{PRE}}$ and $\overline{\text{CLR}}$ activated. Both Q and $\overline{Q}$ are high, and a data book will note this condition *unpredictable*. The next condition on the state table shows $\overline{\text{PRE}}$ activated and $\overline{\text{CLR}}$ inactive. The flip-flop goes to the SET state, and the CLK and D inputs are don't cares because $\overline{\text{PRE}}$ is the overriding input. The third condition shows $\overline{\text{CLR}}$ activated. The remainder of the state table has already been discussed.

The waveforms shown in Fig. 6–43 show how the asynchronous inputs can be used to override the synchronous data input and control the flip-flop. Since Q follows D on the

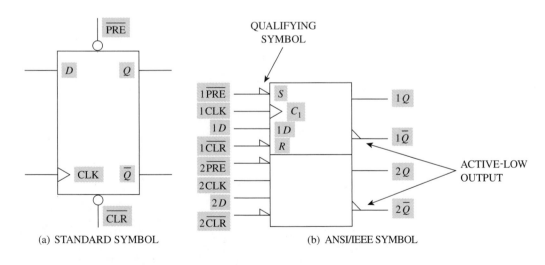

INPUTS				OUTPUTS	
$\overline{\text{PRE}}$	$\overline{\text{CLR}}$	CLK	D	Q	$\overline{Q}$
0	0	X	X	1	1*
0	1	X	X	1	0
1	0	X	X	0	1
1	1	↑	1	1	0
1	1	↑	0	0	1
1	1	0	X	NC	NC

*INVALID

(c) STATE TABLE

FIGURE 6-42 SN7474 dual D-type flip-flop.

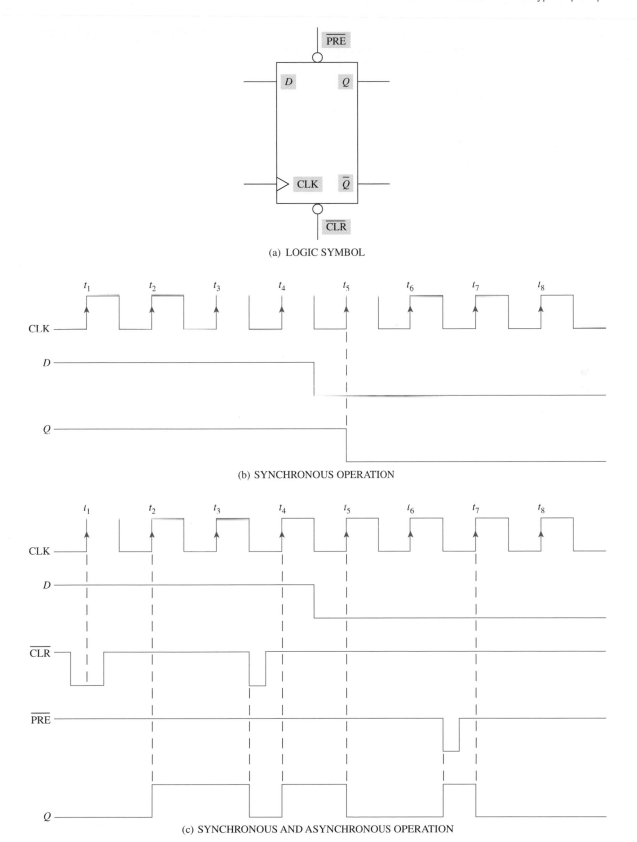

FIGURE 6-43   Dual D-type flip-flop.

PGT of the clock pulse during synchronous operation, the Q output would look like the data input if no asynchronous inputs were activated during operation. This is shown in Fig. 6–43(b) assuming $\overline{PRE} = \overline{CLR} = $ Logic 1. The Q waveform is shifted slightly to the right so that it coincides with the t_5 PGT.

The waveforms in Fig. 6–43(c) show the circuit is asynchronously cleared during t_1 and starts operation in the CLEAR state. Q follows D at t_2, stays high at t_3 since the data input is still high, and is asynchronously cleared again *prior* to t_4. The circuit returns to the SET state at t_4 and clears at t_5 (synchronous operation). It is asynchronously set by activating the $\overline{PRE}$ input prior to t_7 and then returns to the CLEAR state synchronously with the t_7 PGT.

The $\overline{PRE}$ or $\overline{CLR}$ input must be held low for a specific minimum time to ensure asynchronous control of the flip-flop. This time requirement is another data sheet specification that is listed t_w along with the minimum clock duration. This time is typically 25 ns for the TTL family of flip-flops.

Toggle Operation

Figure 6–44(a) shows a D-type flip-flop connected so that the *Q output will change states on every active clock transition*. The switching of a logic level from 1 to 0 or vice versa is referred to as **toggling.** If a flip-flop changes states on every active clock transition, it is operating in the TOGGLE mode of operation.

The waveforms in Fig. 6–44(b) show how this is accomplished. The flip-flop starts out in the CLEAR state. With Q low, $\overline{Q}$ is high, and it is tied to the data input of the flip-flop. Therefore, at t_1, Q goes high. Since Q is high after t_1, $\overline{Q}$ is low and Q will go low

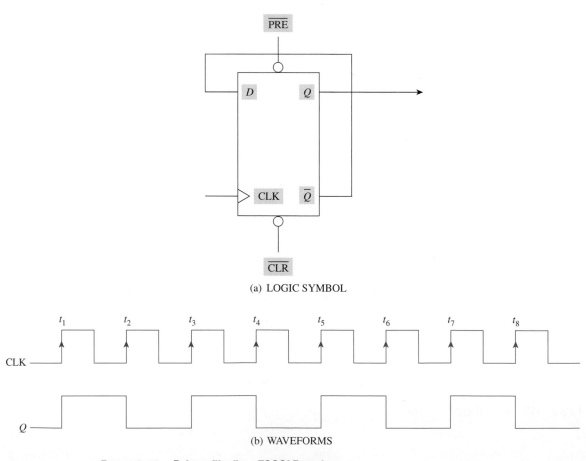

(a) LOGIC SYMBOL

(b) WAVEFORMS

FIGURE 6-44   D-type flip-flop: TOGGLE mode.

on the next PGT at t_2. With the $\overline{Q}$ output controlling the D input, the flip-flop will change states on every PGT of the clock pulse.

The output frequency of this flip-flop is half of the input clock frequency (f_{in}). When a flip-flop is operating in the TOGGLE mode it functions as a **divide-by-2 circuit.**

The logic diagram in Fig. 6–45 shows how flip-flops can be connected to gain higher frequency division capabilities. Flip-flop #1 divides the input clock frequency (f_{in}) of 100 KHz by 2. The Q output of this flip-flop (50 KHz) is used as the CLK input to flip-flop #2. This 50-KHz signal is again divided by 2 to produce 25 KHz. The output signal at Q_2 is the original f_{in} divided by 4. If one more TOGGLE-mode flip-flop were connected to the output at Q_2, its output would be f_{in} divided by 8.

Frequency division is accomplished in this manner in digital circuits. It is this toggle capability that is the basis of binary counting.

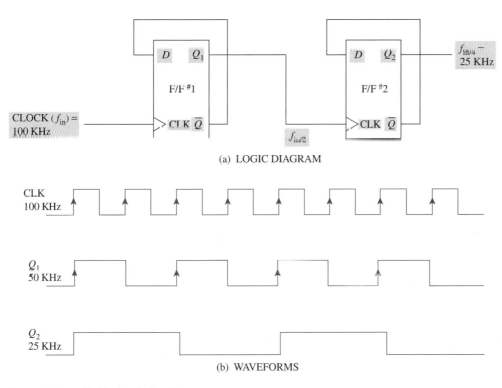

(a) LOGIC DIAGRAM

(b) WAVEFORMS

FIGURE 6-45   **Divide-by-4 circuit.**

Section 6-4: Review Questions

Answers are given at the end of the chapter.

A. Latches are _____ triggered.
   (1) pulse
   (2) edge

B. Flip-flops are _____ triggered.
   (1) pulse
   (2) edge

C. The triangle at the CLK input of a flip-flop indicates
   (1) Active-low input.
   (2) Active-high input.
   (3) Edge triggering.

D. In a negative-edge-triggered D-type flip-flop, *Q* follows *D* on the NGT of the CLK pulse.
   (1) True
   (2) False

E. The data (*D*) input to a D-type flip-flop is an *asynchronous* input.
   (1) True
   (2) False

F. The $\overline{\text{PRE}}$ and $\overline{\text{CLR}}$ inputs to a flip-flop override the data and clock inputs.
   (1) True
   (2) False

G. When the $\overline{\text{PRE}}$ input to a flip-flop is activated, the flip-flop goes to the RESET state.
   (1) True
   (2) False

H. Switching a logic level from 1-to-0 or 0-to-1 is referred to as
   (1) Setting
   (2) Clearing
   (3) Toggling

I. When a flip-flop is operating in the TOGGLE mode, it functions as a divide-by-2 circuit.
   (1) True
   (2) False

SECTION 6-5: J-K FLIP-FLOP

OBJECTIVES

1. Identify J-K flip-flop symbols.
2. Determine the state of a J-K flip-flop when provided the input signals.

The J-K flip-flop is one of the most popular, versatile, and widely used flip-flops in digital circuits. The letters J and K have no significance. The simplified logic diagram of a J-K flip-flop is shown in Fig. 6–46.

The logic diagram looks similar to the diagram of a D-type flip-flop. It contains the edge detector, steering gates, and latch. The asynchronous inputs are not shown in this diagram. It does have two input lines for data inputs (*J* and *K*) in lieu of one. The pri-

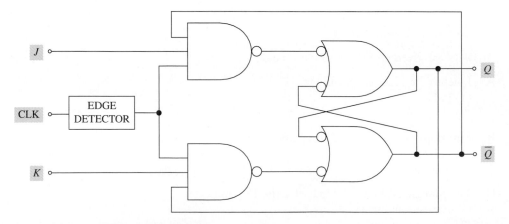

FIGURE 6-46   J-K flip-flop logic diagram.

mary difference between the two flip-flops is that Q and $\overline{Q}$ are tied back to the input steering gates in the J-K. This is done internally on the IC. Without the Q and $\overline{Q}$ feedback, the circuit would be nothing more than an S-C flip-flop. The S-C flip-flop, if one were commercially available, would have an INVALID output if both of its inputs were activated simultaneously. The Q and $\overline{Q}$ feedback in the J-K flip-flop prevent this condition. Instead, the TOGGLE mode of operation is assumed when J and K are both high.

The logic symbols and state table for a negative-edge-triggered J-K flip-flop are shown in Fig. 6-47. It is unnecessary to trace logic levels through the logic diagram because its operation is similar to the flip-flops previously explained. One thing must be kept in mind should you decide to delve into proving the state table by tracing 1s and 0s through the circuit—you must assume an output state before starting your analysis. This is absolutely necessary to determine the initial inputs to the steering gates as well as the latch.

The first three sets of input conditions on the state table in Fig. 6-47(c) relate to the asynchronous $\overline{PRE}$ and $\overline{CLR}$ inputs. Both inputs should not be asserted (brought low) at the same time as is indicated on the state table by the invalid note. The last five conditions on the table show $\overline{PRE}$ and $\overline{CLR}$ inactive.

The flip-flop is in the RETAIN (NC) state when $J = K = 0$ and also when the clock is resting high. The flip-flop would retain its previous state even if the clock were resting low because an NGT is not present.

The primary advantage of the J-K flip-flop is its TOGGLE mode of operation when $J = K = 1$. It functions as the D-type flip-flop did when $\overline{Q}$ was tied back to the D input. In fact, taking $\overline{Q}$ back to the D input of the D-type flip flop also applied $\overline{Q}$ to the inter-

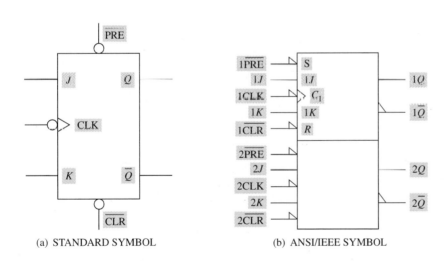

(a) STANDARD SYMBOL   (b) ANSI/IEEE SYMBOL

INPUTS					OUTPUTS	
$\overline{PRE}$	$\overline{CLR}$	CLK	J	K	Q	$\overline{Q}$
0	1	X	X	X	1	0
1	0	X	X	X	0	1
0	0	X	X	X	1*	1*
1	1	↓	0	0	NC	NC
1	1	↓	1	0	1	0
1	1	↓	0	1	0	1
1	1	↓	1	1	TOGGLE	
1	1	1	X	X	NC	NC

*INVALID

(c) STATE TABLE

FIGURE 6-47   SN74LS76 dual J-K flip-flop (negative-edge triggered).

nal inverter and put Q on the other steering gate. Therefore, what was accomplished externally to make the D-type flip-flop toggle is accomplished internally in the J-K flip-flop.

The state table (Fig. 6–47c) shows Q follows J and $\overline{Q}$ follows $\overline{K}$ on the NGT of the clock pulse when the control inputs (J and K) are complementary and $\overline{PRE}$ and $\overline{CLR}$ are inactive.

The SN74LS76 is a Dual J-K Flip-Flop with PRESET and CLEAR. The flip-flop is negative-edge triggered. The symbols shown in Fig. 6–47 (a) and (b) represent this flip-flop. Also, the state table shown in the figure is for the SN74LS76.

The waveforms shown in Fig. 6–48(a) depict operation of the negative-edge-triggered J-K flip-flop. It is beneficial at this point in your studies to relate the waveforms to the state table in Fig. 6–47(c).

The circuit analysis starts with Q low. The flip-flop is synchronously SET by the J and K inputs at t_1. The flip-flop is in the HOLD mode of operation at t_2 and t_3 because $J = K = 0$. $J = 0$ and $K = 1$ prior to the clock NGT at t_4, so the flip-flop clears on this NGT. Note $J = K = 1$ during the t_5, t_6, and t_7 active-clock transitions. Thus, the flip-flop toggles (changes states) on each of these NGTs. $J = 0$ and $K = 1$ at t_8, so the flip-flop clears. It is in the HOLD mode at t_9 and t_{10}.

The waveforms shown in Fig. 6–48(b) show operation of a positive-edge-triggered flip-flop. The state table shown in Fig. 6–47(c) can be used during this circuit analysis. The CLK arrows would point up for a PGT-triggered flip-flop. All other information on the state table is identical. The CLK and control inputs to the PGT-triggered flip-flop were kept the same in Fig. 6–48(b) as they were for the NGT-triggered flip-flop in Fig. 6–48(a). The identical input waveforms allow comparison of operation of the two flip-flops.

An example problem and solution are shown in Fig. 6–49 for a J-K flip-flop. The problem requires determining the Q and $\overline{Q}$ outputs of the J-K flip-flop shown in Fig.

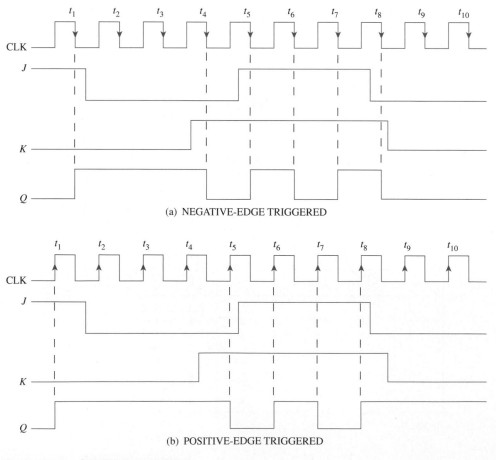

(a) NEGATIVE-EDGE TRIGGERED

(b) POSITIVE-EDGE TRIGGERED

FIGURE 6-48   J-K flip-flop operation.

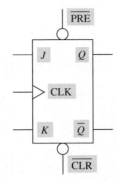

(a) LOGIC SYMBOL

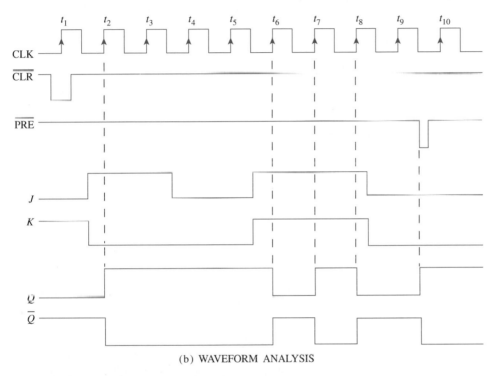

(b) WAVEFORM ANALYSIS

FIGURE 6-49   Positive-edge-triggered J-K flip-flop.

6–49(a). The CLK, *J* and *K* control inputs, asynchronous inputs, and outputs of the flip-flop are shown in Fig. 6–49(b). The circuit is a positive-edge-triggered J-K flip-flop. Here is the solution analysis:

t_1: $\overline{\text{CLR}}$ = 0 = asynchronous clear.

t_2: J = 1 and K = 0 = synchronous set.

t_3: J = 1 and K = 0 = stay set.

t_4: J = 0 and K = 0 = retain SET state.

t_5: J = 0 and K = 0 = retain.

t_6: J = 1 and K = 1 = toggle.

t_7: J = 1 and K = 1 = toggle.

t_8: J = 1 and K = 1 = toggle.

t_9: J = 0 and K = 0 = retain.

t_{9+}: $\overline{\text{PRE}}$ = 0 = asynchronous set.

t_{10}: J = 0 and K = 0 = retain.

Section 6–5: Review Questions

Answers are given at the end of the chapter.

A. Which type of flip-flop has a TOGGLE mode of operation?

What is the state of the flip-flop shown in Fig. 6–49(a) after a PGT when

B. $\overline{PRE} = 0; \overline{CLR} = 1; J = 1; K = 0$?
C. $\overline{PRE} = 1; \overline{CLR} = 1; J = 1; K = 0$?
D. $\overline{PRE} = 1; \overline{CLR} = 1; J = 0; K = 1$?
E. $\overline{PRE} = 1; \overline{CLR} = 0; J = 0; K = 1$?
F. $\overline{PRE} = 1; \overline{CLR} = 1; J = 0; K = 0$?
G. $\overline{PRE} = 1; \overline{CLR} = 1; J = 1; K = 1$?
H. The clock is resting high?
I. $\overline{PRE} = 0$ AND $\overline{CLR} = 0$?
J. The J-K flip-flop shown in Fig. 6–49(a) is
    (1) Negative-edge triggered
    (2) Positive-edge triggered

SECTION 6-6: J-K MASTER-SLAVE FLIP-FLOP

OBJECTIVES

1. Identify symbols used to represent J-K master-slave flip-flops.
2. Given the synchronous and asynchronous inputs to a J-K master-slave flip-flop, determine the outputs.

The last class of flip-flops requiring analysis is the J-K master–slave (MS) flip-flop. Advances in technology have made the use of master–slave flip-flops less and less popular. Input data hold times of zero have placed this type of flip-flop close to obsolescence. However, they are still commercially available and are found in older digital equipment. Thus, their operation is briefly presented in this section.

The J-K MS flip-flop is actually nothing more than two J-K flip-flops connected together as shown in Fig 6–50. One flip-flop is called the **master section** and the other the **slave section.**

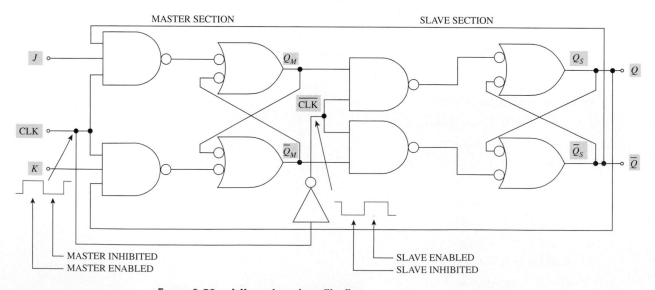

FIGURE 6-50   J-K master–slave flip-flop.

The J-K MS flip-flop operates using a complete clock cycle instead of the edge of a clock pulse. The MS flip-flop is actually pulse triggered. The definition provided in the flip-flop introduction stated latch circuits are pulse triggered and flip-flops are edge triggered. Although this defies, to a degree, our definition of flip-flops presented earlier, bear with the following explanation.

Since the J-K MS flip-flop operates using the complete clock cycle to transfer data, it is classed as a **pulse-triggered device.** The input steering gates are enabled during the entire positive duration of the clock pulse. This, incidently, results in another disadvantage of an MS flip-flop. The J and K inputs must be held at a constant level while the clock is high. Figure 6–50 shows the clock pulse to the slave section is inverted. This indicates the slave section steering gates are inhibited when the master section steering gates are enabled. Figure 6–51 shows the CLK waveform annotated with data transfer times. The figure indicates data is clocked into the master flip-flop when the clock is high (t_1–t_2). Note that when the data in the master section is transferred to the slave (t_2–t_3), the master section steering gates are inhibited.

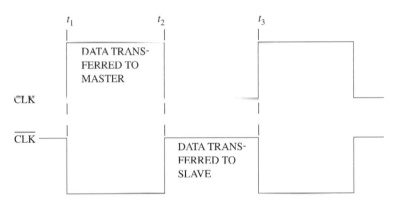

FIGURE 6-51   CLOCK/$\overline{\text{CLOCK}}$ waveforms for J-K master–slave flip-flop.

The SN7476 Dual J-K MS Flip-Flop symbols are shown in Fig 6–52(a) and (b). Remember the data is transferred into the master section on the positive clock pulse and then transferred to the slave on the negative portion of the input clock pulse. The output data actually lags the input data by one-half clock cycle. Therefore, the output is **delayed** or **postponed.** The logic symbol for the SN7476 MS flip-flop has an "inverted backward L" over each of the flip-flop's outputs (Q and $\overline{Q}$). The inverted-backward L is called a **postponed output indicator.** It is used on flip-flop logic symbols to signify "master–slave."

The state table for the SN7476 is shown in Fig. 6–52(c). The state table is comparable to the SN74LS76 negative-edge-triggered flip-flop state table shown in Fig. 6–47(c). The main difference in the two state tables is in the CLK columns. The positive pulses in the SN7476 CLK column indicate the flip-flop is pulse triggered.

Since the master–slave is in reality just two J-K flip-flops, it has a TOGGLE mode of operation when $J = K = 1$.

Operation of the J-K MS flip-flop (Fig. 6–50) is summarized below in relation to the waveforms presented in Fig. 6–53. Both the Q_s and $\overline{Q}_s$ outputs are available from the IC even though $\overline{Q}_s$ is not shown in Fig. 6–53. Q_M is internal and the signal is not available on an output pin.

The flip-flop is asynchronously cleared prior to clock #1.

CLK #1: $J = K = 0$ = retain CLEAR state.

CLK #2: J goes high but master section is inhibited.

CLK #3: Master sets.

CLK #4: Slave sets.

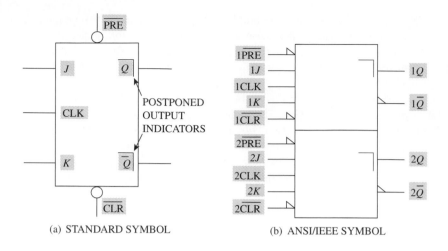

(a) STANDARD SYMBOL   (b) ANSI/IEEE SYMBOL

INPUTS					OUTPUTS	
$\overline{PRE}$	$\overline{CLR}$	CLK	J	K	Q	$\overline{Q}$
0	1	X	X	X	1	0
1	0	X	X	X	0	1
0	0	X	X	X	1*	1*
1	1	⊓	0	0	NC	NC
1	1	⊓	1	0	1	0
1	1	⊓	0	1	0	1
1	1	⊓	1	1	TOGGLE	

*INVALID

(c) STATE TABLE

FIGURE 6-52   SN7476 dual J-K master–slave flip-flop.

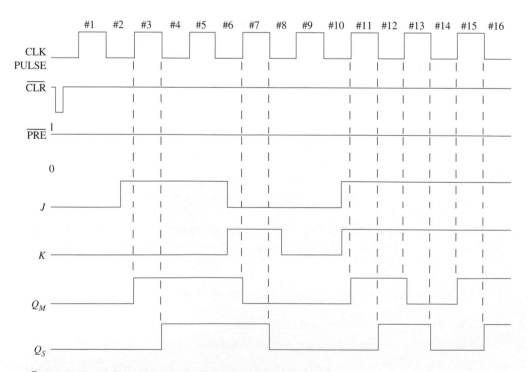

FIGURE 6-53   J-K master–slave flip-flop waveform analysis.

CLK #5: $J = 1$ and $K = 0$, so flip-flop stays SET.

CLK #6: Control inputs change to $J = 0$ and $K = 1$ but master section is inhibited.

CLK #7: Master clears.

CLK #8: Slave clears. Control input K changes so $J = K = 0$ but master section is inhibited.

CLK #9: Master holds CLEAR state.

CLK #10: Slave holds CLEAR state. Control inputs change to $J = K = 1$ but master section is inhibited.

CLK #11: Master toggles.

CLK #12: Slave toggles.

CLK #13: Master toggles.

CLK #14: Slave toggles.

CLK #15: Master toggles.

CLK #16: Slave toggles.

Although the master–slave flip-flop is pulse triggered, data is transferred to the slave section at the negative transition of the input CLK pulse. In essence, this flip-flop functions as a standard negative-edge-triggered flip-flop.

The SN74111 Dual J-K Flip-Flop with Data Lockout is yet another version of the master–slave flip-flop that may be found in some of the older digital equipment. This flip-flop is similar to the master–slave flip-flop discussed in the preceding paragraphs. Its logic symbol is shown in Fig 6–54. The logic symbol shows this master–slave flip-flop is edge triggered. It is this feature that gives the flip-flop its data lockout capability.

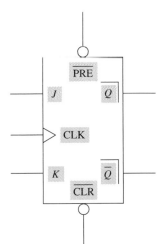

Figure 6-54   SN74111 dual J-K master–slave flip-flop with data lockout.

Since the circuit is edge triggered, the steering gates are enabled for only about 20–25 ns on the PGT of the clock pulse. The J and K inputs may then be changed even though the clock is still high. In a standard master–slave flip-flop, the inputs must be held at a constant level during the period the clock is high.

Section 6-6: Review Questions

Answers are given at the end of the chapter.

A. The *postponed output indicator* shown in Fig. 6 54 identifies the flip-flop as a
_____.

B. The J-K master–slave flip-flop normally operates using a complete clock cycle.
   (1) True
   (2) False
C. What is the state of the flip-flop shown in Fig. 6–52(a) during the *positive clock pulse* when $\overline{PRE} = 1$; $\overline{CLR} = 1$; $J = 1$; $K = 0$. Assume the initial condition is CLEAR prior to the PGT.
D. What is the state of the flip-flop shown in Fig. 6–52(a) during the negative clock pulse immediately following the positive clock pulse of question (C)?
E. The dynamic input and postponed output indicators in Fig. 6–54 indicate this flip-flop is a J-K _____.

SECTIONS 6-4 THROUGH 6-6: INTERNAL SUMMARY

Flip-flops are **bistable devices** that employ **edge triggering.** Their outputs can change state synchronously only on an active transition of the clock pulse.

The flip-flop shown in Fig. 6–55(a) is a **positive-edge-triggered D-type flip-flop.** The one shown in Fig. 6–55(b) is a **negative-edge-triggered D-type flip-flop.** The triangle at the CLK input is a dynamic input indicator that indicates the flip-flop is edge triggered. The lack of a bubble adjacent to the dynamic input indicator in Fig. 6–55(a) indicates the flip-flop is positive-edge triggered. The bubble adjacent to the indicator in Fig. 6–55(b) signifies negative-edge triggering.

The Q outputs of these flip-flops follow the D inputs on the active-clock transition. This is the **PGT** of the clock pulse for the flip-flop in Fig. 6–55(a). It is the **NGT** of the clock pulse for the flip-flop in Fig. 6–55(b). The flip-flops retain the previous state when the clock is resting (low or high).

Setup time and **hold time** must be adhered to for proper flip-flop operation. Setup time is the minimum interval of time the data input(s) must be held constant prior to the active-clock transition. Hold time is the interval of time the data input(s) must be held constant after the active-clock transition.

The $\overline{PRE}$ and $\overline{CLR}$ inputs to flip-flops are **asynchronous**—not clock dependent. These inputs override all other inputs to the flip-flop when they are activated. They are, in a sense, the boss inputs when they are activated. The flip-flop goes to the SET state when $\overline{PRE}$ is activated. It will remain in the SET state as long as the $\overline{PRE}$ input is held low. The flip-flop goes to the CLEAR state when the $\overline{CLR}$ input is activated. These two asynchronous inputs should not be active at the same time. This action would produce an unpre-

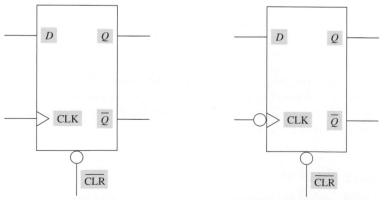

(a) POSITIVE-EDGE TRIGGERED   (b) NEGATIVE-EDGE TRIGGERED

FIGURE 6-55   Edge-triggered D-type flip-flop symbols.

dictable output when the $\overline{\text{PRE}}$ and $\overline{\text{CLR}}$ inputs are returned to their resting level. This condition is undesired and should always be avoided.

"Toggle" means to **change states.** When a flip-flop is operating in the TOGGLE mode of operation, it changes states on every active clock transition. The output frequency of a flip-flop operating in this mode will be half of the input clock frequency. This **divide-by-2** capability provides the basis for all binary counting operations.

The J-K flip-flop is an edge-triggered flip-flop. *Q follows J and $\overline{Q}$ follows K on the active clock transition providing the inputs are complementary.* This indicates the flip-flop is an active-high input circuit. The flip-flop is in the RETAIN mode when $J = K = 0$. It is in the TOGGLE mode when $J = K = 1$.

The J-K master–slave flip-flop consists of two J-K flip-flops cascaded together. The circuit is pulse triggered. The master section will set on the positive clock pulse when $J = 1$ and $K = 0$. The slave section will set when the input clock pulse goes negative. Thus, the output is delayed. This delay is indicated on the master–slave logic symbol by **postponed output indicators** (inverted-backward Ls) over the Q and $\overline{Q}$ output labels.

A master–slave symbol which contains a dynamic input indicator is used to represent an edge-triggered flip-flop. This master-slave flip-flop employs a **data lockout** capability. This capability negates the requirement of having to hold the J and K inputs constant for the entire positive duration of the clock pulse.

SECTIONS 6-4 THROUGH 6-6: INTERNAL SUMMARY QUESTIONS

Answer are given at the end of the Chapter.

1. Identify the symbol shown in Fig. 6–56.
   a. Positive-edge-triggered D-latch
   b. Negative-edge-triggered D-latch
   c. Positive-edge-triggered D-type flip-flop
   d. Negative-edge-triggered D-type flip-flop

2. The Q output of the flip-flop shown in Fig. 6–56 follows the D input on the NGT of the clock pulse.
   a. True
   b. False

3. The flip-flop shown in Fig. 6–56 is in what state when the clock input is low?
   a. SET   c. RETAIN
   b. CLEAR   d. INVALID

4. The triangle at the clock input of the flip-flop shown in Fig. 6–56 is a
   a. Dynamic input indicator
   b. Postponed output indicator

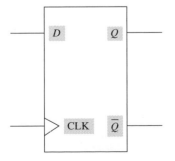

FIGURE 6-56

5. The circuit shown in Fig. 6–57 is a
   a. Positive-edge detector
   b. Negative-edge detector

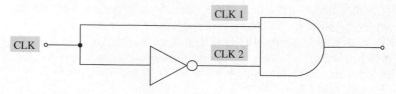

FIGURE 6-57

6. Hold time is the interval of time the data input(s) must be held constant prior to the active-clock transition.
   a. True
   b False

7. The symbol shown in Fig. 6–58 represents a
   a. Positive-pulse-triggered D-type flip-flop
   b. Negative-pulse-triggered D-type flip-flop
   c. Positive-edge-triggered D-type flip-flop
   d. Negative-edge-triggered D-type flip-flop

8. What is the state of the flip-flop shown in Fig. 6–58 after the active-clock transition when

$\overline{PRE} = 1, \overline{CLR} = 0$, and $D = 1$?

   a. SET   c. RETAIN
   b. CLEAR   d. INVALID

9. What is the state of the flip-flop shown in Fig. 6–58 after the active-clock transition when

$\overline{PRE} = 1, \overline{CLR} = 1$, and $D = 1$?

   a. SET   c. RETAIN
   b. CLEAR   d. INVALID

10. Wht is the state of the flip-flop shown in Fig. 6–58 after the active-clock transition when

$\overline{PRE} = 0, \overline{CLR} = 1$, and $D = 0$?

   a. SET   c. RETAIN
   b. CLEAR   d. INVALID

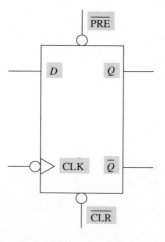

FIGURE 6-58

11. A flip-flop is a divide-by-2 circuit when it is operating in the TOGGLE mode.
    a. True
    b. False

12. The data input to the flip-flop shown in Fig. 6–58 is an asynchronous input.
    a. True
    b. False

13. The symbol shown in Fig. 6–59 represents a
    a. Positive-edge-triggered J-K flip-flop
    b. Negative-edge-triggered J-K flip-flop
    c. Positive-edge-triggered J-K master–slave flip-flop
    d. Negative-edge-triggered J-K master–slave flip-flop

14. What is the state of the flip-flop shown in Fig. 6–59 after the active clock transition when

    $\overline{PRE} = 1, \overline{CLR} = 1, J = 1$, and $K = 0$?

    a. SET   c. RETAIN
    b. CLEAR   d. TOGGLE

15. What is the state of the flip-flop shown in Fig. 6–59 after the active clock transition when

    $\overline{PRE} = 1, \overline{CLR} = 0, J = 1$, and $K = 0$?

    a. SET   c. RETAIN
    b. CLEAR   d. TOGGLE

16. What is the state of the flip-flop shown in Fig. 6–59 after the active clock transition when

    PRE $= 1$, CLR $= 1, J = 1$, and $K = 1$?

    a. SET   c. RETAIN
    b. CLEAR   d. TOGGLE from previous state

17. What is the state of the flip-flop shown in Fig. 6–59 after the active clock transition when

    $\overline{PRE} = 1, \overline{CLR} = 1, J = 0$, and $K = 0$?

    a. SET   c. RETAIN
    b. CLEAR   d. INVALID

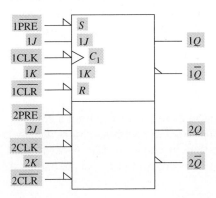

FIGURE 6-59

18. What is the state of the flip-flop shown in Fig. 6–59 after the active clock transition when

$$\overline{PRE} = 1, \overline{CLR} = 1, J = 0, \text{ and } K = 1?$$

a. SET   c. RETAIN
b. CLEAR   d. TOGGLE

19. The symbol shown in Fig. 6–60 represents a

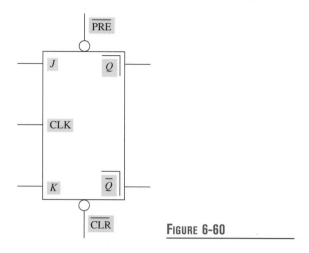

FIGURE 6-60

a. J-K master–slave flip-flop
b. Positive-edge-triggered J-K flip-flop
c. Negative-edge-triggered J-K flip-flop

20. The inverted-backward L over Q and $\overline{Q}$ in Fig. 6–60 is a
a. Dynamic input indicator
b. Postponed output indicator

SECTION 6-7: PRACTICAL APPLICATIONS OF FLIP-FLOPS

OBJECTIVES

1. Identify a basic shift register and counter.
2. Determine the output(s) of a shift register and counter after X clock pulses.

Latches, flip-flops, one-shot multivibrators, and timers are used to form basic **sequential logic circuits.** The introduction to this chapter mentioned several uses for flip-flops. Some of these uses will be presented in this section. A more detailed picture of their uses will be presented in the following chapters.

Shift Register

One of the most common uses of flip-flops is to connect them together to form a **register.** *A register is defined as a group of latches or flip-flops used to transfer or store data.*

A shift register implemented with D-type flip-flops is shown in Fig. 6–61(a). The flip-flops are **cascaded.** Cascading indicates the flip-flops are configured so the output(s) of one flip-flop is the data input(s) to the next.

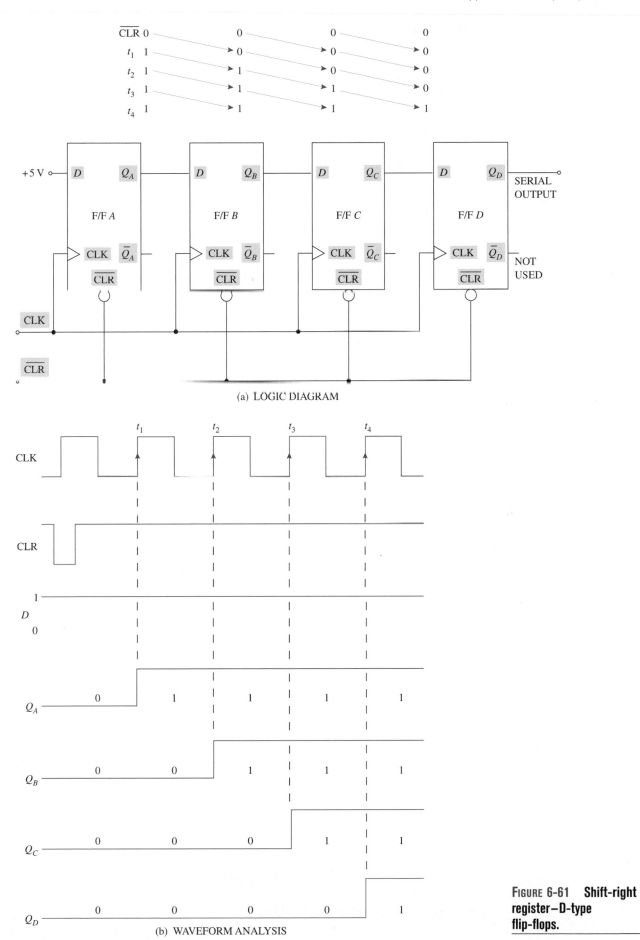

(a) LOGIC DIAGRAM

(b) WAVEFORM ANALYSIS

FIGURE 6-61   Shift-right register—D-type flip-flops.

The external data input to this circuit is connected only to the first flip-flop. It is tied high in this example for explanation purposes. Normally, serial data would be applied, and the data would be changing per the data requirements of the system.

The main point to keep in mind during operation of this register is **setup time.** A flip-flop will respond to the data that is on its input(s) immediately prior to the active-clock transition.

Circuit operation of the shift register shown in Fig. 6–61(a) starts with an asynchronous clear as shown in the waveforms in Fig. 6–61(b). The 0s from the Q outputs are annotated above the flip-flop outputs on Fig. 6–61(a) to indicate this clear condition.

t_1: When the PGT of the clock pulse arrives it is applied to all four flip-flops simultaneously. The high data input applied to flip-flop A will be clocked to its output on this clock transition. The internal propagation delay of this flip-flop might be 20 to 30 ns. Thus, by the time the Q_A output of this flip-flop goes high, the active transition of the clock pulse has come and gone. With this in mind, it is understood that flip-flop B, C, and D will stay in the CLEAR state because they had Logic 0 data inputs during the active clock transition. Note that these and succeeding output conditions are annotated beside the respective clock transition times on the diagram shown in Fig. 6–61(a).

t_2: The PGT at t_2 clocks the high data input again to Q_A. Since the flip-flop was in the SET state, it stays in the SET state. Since Q_A is the data input to flip-flop B, the high on this flip-flop's input will cause it to clock to the SET state at t_2. The data currently in the register after t_2 are 1100.

t_3: Q_A stays SET, Q_B stays SET, and Q_C SETS while Q_D stays in the CLEAR state.

t_4: All flip-flops in the register are now SET.

The 0s and 1s annotated above each flip-flop's output present a picture of circuit operation. Notice how the high data input is shifted into the register one clock pulse at a time. This action is referred to as loading the register. There are four flip-flops in this register, so it takes four clock pulses to load the data. Since the data is shifted in from the Q_A flip-flop to the Q_D flip-flop, this circuit is a shift-right register. Once this register has been loaded, it requires four more clock pulses to transfer the data out of the circuit and down the serial line to some other circuit or device. The waveforms for this register are shown in Fig. 6–61(b). These waveforms are a pictorial of the 0s and 1s annotated on the circuit in Fig. 6–61(a). Figure 6–62 shows how the D-type flip-flop circuit in Fig. 6–61(a) can be implemented with J-K flip-flops. Shift-left registers and left–right shift registers can also be implemented using flip-flops.

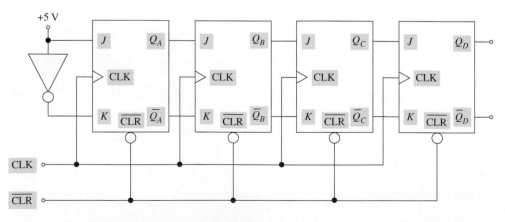

FIGURE 6-62   Shift-right register—J-K flip-flops.

Registers are generally classified by the methods used to **load** data in and **transfer** data out. The register in Fig. 6–62 is classified as a serial-in/serial-out register. Some registers are designed for parallel-in/parallel-out data transfer. This is a subject upon which we will expand in Chapter 8.

Parallel Data Transfer

A parallel-in/parallel-out register is shown in Fig. 6–63. The data input bits ($D_0 - D_5$) are available at the inputs of the D-type flip-flops. If these bits arrived on a **data bus** and the bus needed to be released for other uses, the bits could be loaded into this register with one NGT of the clock pulse and stored. A **data bus** is a group of lines or traces on a circuit board that carry data bits throughout the system.

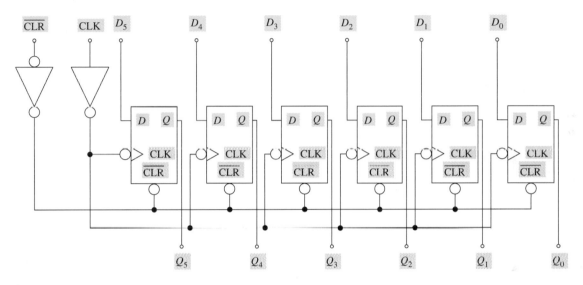

FIGURE 6-63   **Parallel-in/parallel-out register.**

Transferring data in this manner is extremely fast when compared to serially shifting the data through a shift register. If this register were a shift register, it would take six clock pulses to load the data.

The two bubbles on the $\overline{CLR}$ input buffer are **state indicators.** They indicate the $\overline{CLR}$ input to this register is active low and the $\overline{CLR}$ input to each flip-flop is also active low. Also note in the figure that the CLK input is applied through an inverter. This makes the register a PGT-triggered register. Other types of registers such as serial-in/parallel-out and parallel-in/serial-out will be presented in Chapter 8.

Clock-Burst Generator

A **clock-burst generator** is shown in Fig. 6–64. The Q output of the flip-flop is a **train** of clock pulses when the J input is high. The clock pulses are removed from the train when the J input is low. Notice in the figure that the K input is tied to ground. Therefore, when the J input is low, the flip-flop is in the HOLD mode of operation.

The control NAND gate clears the flip-flop every time the Q output and the CLK input are high. This is depicted in Fig. 6–65(a). At t_0 the clock NGT puts the flip-flop in the SET state because J is high and K is low. The clock input goes high halfway between t_0 and t_1. At this instant the CLK input and Q output are high, and the NAND gate asynchronously clears the flip-flop. Notice the pulse width of the positive duration of the Q output is exactly the same as the pulse width of the low clock.

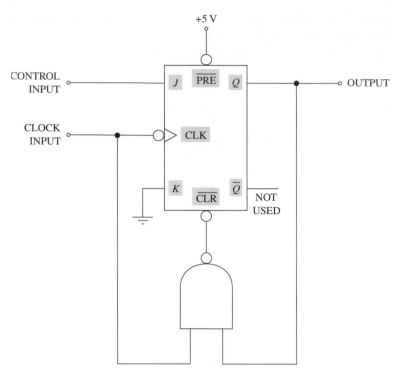

FIGURE 6-64   **Clock-burst generator.**

Since the CLK input in Fig. 6–65(a) is a **50% duty cycle,** this last statement is further expanded by changing the duty cycle to 20% as shown in Fig. 6–65(b). Again, note the positive duration of the output is equal to the low clock pulse width. This is true because the flip-flop is cleared when $CLK = Q = 1$.

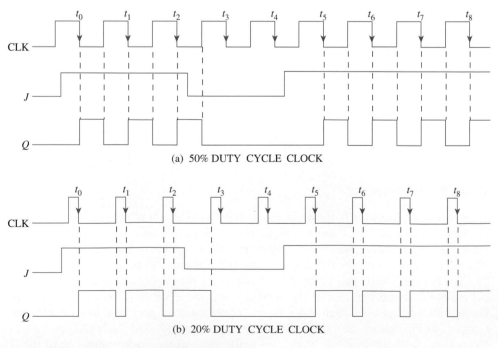

FIGURE 6-65   **Clock-burst generator waveforms.**

Counters

Counters are used in digital circuits to produce numeric counts and/or frequency division.

A basic binary counter is shown in Fig. 6–66(a). The J and K inputs to each flip-flop are tied high. This puts the flip-flops in the TOGGLE mode of operation when they are clocked. The analysis of operation of the D-type and J-K flip-flops in the preceding sections showed they were divide-by-2 circuits when operating in the TOGGLE mode. Every time one of the negative-edge-triggered flip-flops receives an NGT, it toggles (1 to 0 or 0 to 1).

The waveforms for the Q_A, Q_B, and Q_C outputs of this counter are shown in Fig. 6–66(b). The counter is started in the CLEAR state at t_0. Observance of the Q_A output shows that it toggles on every NGT of the input clock pulse. The frequency at this output is $f_{in}/2$.

The Q_A output is tied to the clock input of flip-flop B. The NGTs of this input cause flip-flop B to toggle. Notice in Fig. 6–66(b) that the frequency of the Q_B output is one-half that of the O_A output. Since the original clock frequency has been divided in half two

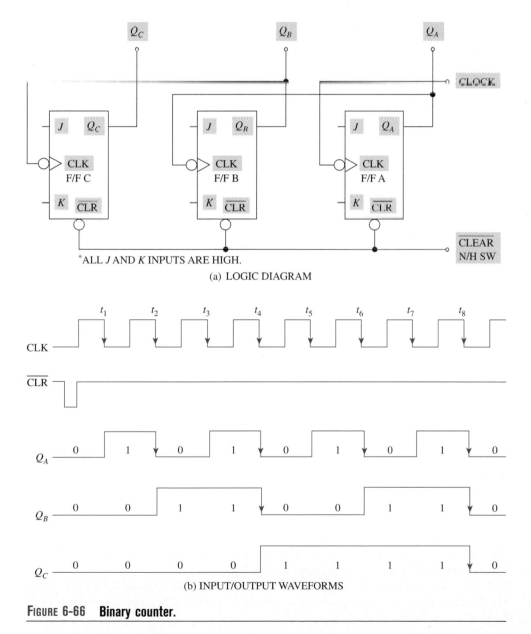

(a) LOGIC DIAGRAM

(b) INPUT/OUTPUT WAVEFORMS

Figure 6-66   Binary counter.

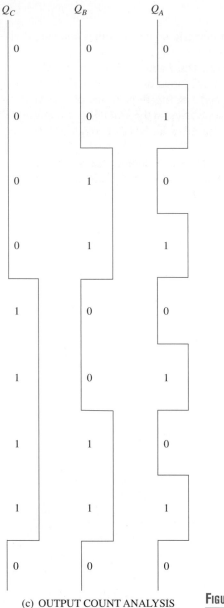

(c) OUTPUT COUNT ANALYSIS

FIGURE 6-66   Continued

times, the Q_B output is $f_{in}/4$. The Q_B output is used to clock flip-flop C. The Q_C output is the Q_B signal divided by 2, so Q_C is $f_{in}/8$.

At time t_8 all three flip-flops are cleared and return to a zero count. The preceding cycle will continue as long as clock pulses are applied to the circuit.

The Q waveforms in Fig. 6–66(b) have been turned 90 degrees in Fig. 6–66(c) so that the binary count output of the counter is readily apparent. Q_A is the LSB and Q_C is the MSB.

The sequential up-count produced by this counter (000, 001, 010, 011, 100, 101, 110, 111) shows that the counter produces eight states including the zero state. The maximum number of states that a counter exhibits is called its modulus, typically abbreviated MOD.

The MOD of a standard counter can be determined by 2^n where n is the number of flip-flops in the counter. The maximum count of a counter can be calculated as MOD-1 because the zero state is not considered a count.

The frequency divide-by capability of a counter at its MSB is equal to its MOD number. The circuit shown in Fig. 6–66(a) is a MOD-8 counter with a maximum count of 7.

The divide-by capability is shown in Fig. 6–66(b). The Q_C output is one complete cycle for eight input clock pulses ($f_{in}/8$).

Counters come in IC packages in many variations. Up/down-counters are available. Typical MOD numbers commercially available are MOD 5, MOD 6, MOD 8, MOD 10, MOD 12, and MOD 16. Any desired MOD can be attained by cascading counters. Counter operation will be detailed in Chapter 7.

Section 6–7: Review Questions

Answers are given at the end of the chapter.

A. Flip-flops are
    (1) Sequential logic circuits
    (2) Combinational logic circuits
B. Define a register.
C. What would the output of the register shown in Fig. 6–62 be if it were cleared, clocked two times, J input to first flip-flop changed to Logic 0, and then clocked two more times?
D. The maximum number of states that a counter can exhibit is called _____ .
E. What is the maximum count of a binary counter that contains four flip-flops?

SECTION 6–8: TROUBLESHOOTING FLIP-FLOPS

OBJECTIVE

State the basic steps required to isloate a malfuncion in a circuit containing flip-flops.

A single flip-flop is shown in Fig. 6–67. The flip-flop is connected in the TOGGLE mode of operation for the first troubleshooting example. If this flip-flop were suspect of causing a problem in a circuit, the very *first step* required in the troubleshooting process is to check V_{CC} and ground. An IC will never work properly if it is not connected to the power supply ground and hot bus. Since the flip-flop is connected in the TOGGLE mode, most of the pins on the IC are tied high. After checking the power supply pins, check the J, K, $\overline{PRE}$, and $\overline{CLR}$ pins to ensure they are all high.

A logic pulser can be used to clock the flip-flop and a logic probe can be used at the same time to check the Q output. Every time the pulser injects a clock signal the logic probe should detect the toggle at the flip-flop's output.

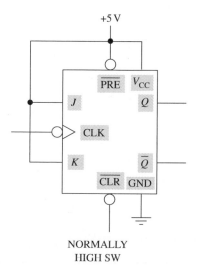

NORMALLY
HIGH SW

FIGURE 6-67   J-K flip-flop–TOGGLE mode.

The shift-right register shown in <u>Fig.</u> 6–68 is a relatively straightforward circuit to troubleshoot. Check V_{CC}, ground, and $\overline{CLR}$ first to ensure they are at the proper level. If the circuit is implemented with discrete flip-flops as shown, clear the flip-flops and then load the register with 1s by clocking the data in one pulse at a time. Check the contents of the register after each clock pulse to ensure the data are clocked in and shifting right. In normal operation with changing input data (dynamic operation), the Q_D serial data output can be monitored with an oscilloscope.

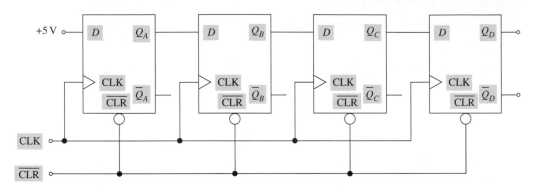

FIGURE 6-68   **Troubleshooting the shift register.**

Several points need to be kept in mind while troubleshooting this circuit. If the CLK input line/trace is broken to any one of the flip-flops, that flip-flop will not change states and this will stop the movement of data in the register. If a line or trace between a Q output and a D input is broken, the flow of data is also stopped. However, with a trace broken between a Q output and a data input, the input to the next flip-flop is floating. If the flip-flop is TTL, that flip-flop will set when it is clocked. If the trace between flip-flops C and D were broken, the first clock pulse after clearing the register would cause the data outputs of the flip-flops to be 1001 $(Q_A–Q_D)$.

The counter shown in Fig. 6–69 is another relatively simple circuit to troubleshoot. The circuit can be checked statically as described at the beginning of this section. An oscilloscope can also be used to troubleshoot this circuit. Apply the clock signal used for the counter to one vertical input of the oscilloscope. Then individually check the output of each flip-flop on the other scope channel in respect to the clock signal.

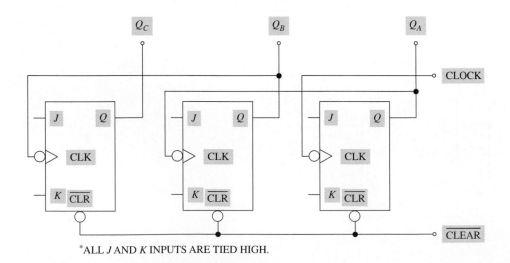

*ALL J AND K INPUTS ARE TIED HIGH.

FIGURE 6-69   **Troubleshooting the binary counter.**

The Q_A output should be $f_{in}/2$, Q_B should be $f_{in}/4$, and Q_c should be $f_{in}/8$. The clock and output waveforms viewed on the scope should appear as they are shown in Fig. 6–66(b).

If the counter does not function properly after preliminary checks, one of the flip-flops might be defective. The preceding checks have ensured V_{CC} and ground are proper, and the control inputs (J and K) and the $\overline{CLR}$ input are all high. The flip-flop is good if a logic pulser will make it toggle.

One more point can be made. The Q outputs are connected to a load. That load may be a decoding gate(s), decoder, or in a laboratory environment it may be LEDs. If the flip-flops check good statically, disconnect the loads. There is a possibility that LEDs will draw enough current to load the flip-flop output to the point the signal is not powerful enough to clock the next flip-flop.

The basic steps required to isolate a malfunction in a register or counter circuit containing flip-flops are these:

1. Check V_{CC} and ground to the circuit.
2. Ensure each flip-flop is powered (V_{CC} and ground).
3. Isolate the malfunction to a flip-flop by checking input data, clocking the flip-flop, and checking its output.
4. If a flip-flop is suspect, check to ensure its asynchronous inputs are inactive.
5. Isolate a flip-flop from its load by disconnecting the load.
6. If a trace on a circuit board is suspect, place a logic pulser on one end of the trace and a logic probe on the other. The probe should respond to the pulser's output if the trace is not broken.
7. Replace the flip-flop if defective or repair the circuit board trace.

Section 6–8: Review Questions

Answers are given at the end of the chapter.

A. The first checks to a flip-flop that is not operating properly should be (select one)
   (1) V_{CC} and ground
   (2) $\overline{PRE}$ and $\overline{CLR}$
   (3) J and K

B. A logic pulser can be used to clock a flip-flop.
   (1) True
   (2) False

C. The frequency output of a flip-flop operating in the TOGGLE mode should be its clock input frequency divided by
   (1) 2
   (2) 4
   (3) 8
   (4) 16

D. After checking V_{CC}, ground, data input(s), $\overline{PRE}$, and $\overline{CLR}$, the load on a flip-flop with a faulty output should be isolated from the flip-flop to ensure it is not causing the malfunction.
   (1) True
   (2) False

CHAPTER 6: SUMMARY

A natural progression can be seen in Fig. 6–70 if each figure is compared to the next figure. Every circuit contains the basic **latch** that is seen in Fig. 6–70(a). The **gated S-C latch** shown in Fig. 6–70(b) adds the steering gates to the first latch so synchronization is achieved. The INVALID state of the gated S-C latch is overcome in Fig. 6–70(c) with the **D-type latch.** The **D-type flip-flop** shown in Fig. 6–70(d) employs edge triggering for precise timing. Finally, the **J-K flip-flop** in Fig. 6–70(e) overcomes the INVALID state completely using feedback that provides the TOGGLE mode of operation. All of these **sequential logic circuits'** out-

puts depend not only on their present inputs, but on the state they were in when those inputs were applied.

A latch or flip-flop is in the SET state when Q is high. It is in the CLEAR state when Q is low. The **active-low latch** in Fig. 6–70(a) is designed using crosscoupled NAND gates. Assuming the inputs are complementary, the latch will set when the $\overline{\text{SET}}$ input is activated. It will clear when the $\overline{\text{CLR}}$ input is activated. The latch retains when neither input is active because the inputs are not asking the latch to do anything. If both inputs are active, the latch has an **unpredictable output** when the inputs are returned to their resting

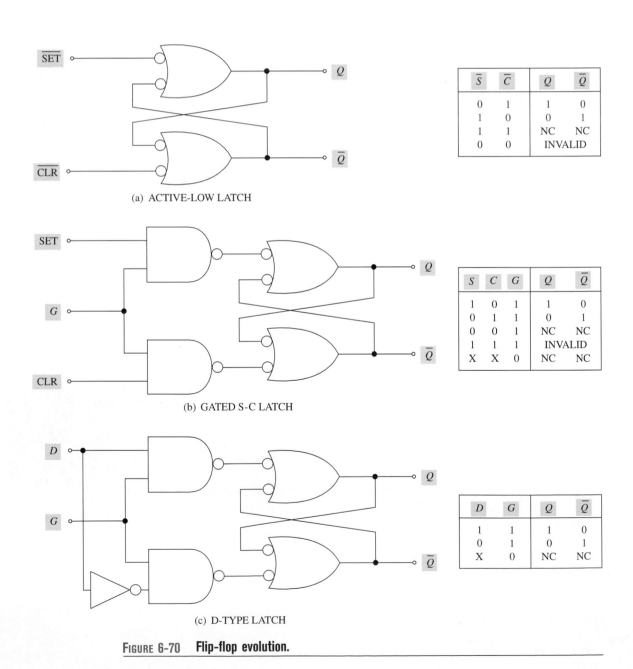

$\overline{S}$	$\overline{C}$	Q	$\overline{Q}$
0	1	1	0
1	0	0	1
1	1	NC	NC
0	0	INVALID	

(a) ACTIVE-LOW LATCH

S	C	G	Q	$\overline{Q}$
1	0	1	1	0
0	1	1	0	1
0	0	1	NC	NC
1	1	1	INVALID	
X	X	0	NC	NC

(b) GATED S-C LATCH

D	G	Q	$\overline{Q}$
1	1	1	0
0	1	0	1
X	0	NC	NC

(c) D-TYPE LATCH

FIGURE 6-70 **Flip-flop evolution.**

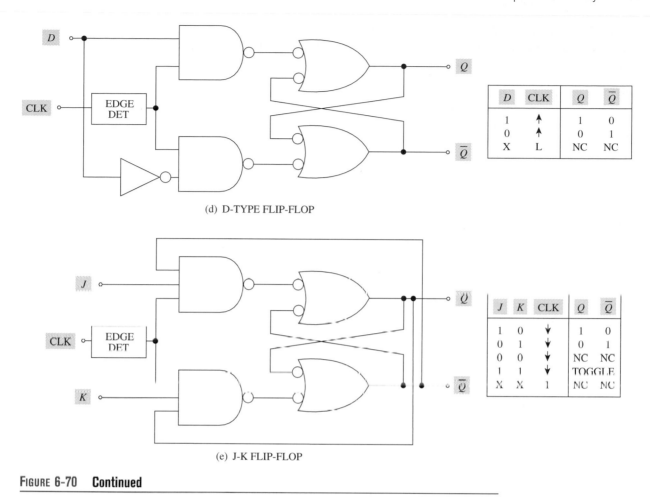

(d) D-TYPE FLIP-FLOP

D	CLK	Q	$\overline{Q}$
1	↑	1	0
0	↑	0	1
X	L	NC	NC

(e) J-K FLIP-FLOP

J	K	CLK	Q	$\overline{Q}$
1	0	↓	1	0
0	1	↓	0	1
0	0	↓	NC	NC
1	1	↓	TOGGLE	
X	X	1	NC	NC

FIGURE 6-70   Continued

state. This state is considered INVALID. These conditions are shown on the state table.

An **active-high latch** can be implemented using two cross-coupled NOR gates. The active-high and active-low latches are **asynchronous** in nature because their outputs respond almost immediately to changes on their inputs.

A **gated latch** is shown in Fig. 6–70(b). A **steering-gate circuit** has been added to the front end of the active-low latch that was depicted in Fig. 6–70(a). The gate (*G*) input in the gated latch is used to enable or inhibit the input steering gates. This action determines when the latch can respond to the SET and CLR inputs. Thus, the gated latch is **synchronous.** The state table summarizes operation of this latch. This type of circuit still has an INVALID output state when it is enabled and both of its inputs are activated.

This undesired INVALID output is overcome with the **D-type latch** shown in Fig. 6–70(c). Activating both inputs to the active-low latch portion of this D-type latch is impossible due to the inverter. The RETAIN (No Change–NC) mode is attained when *G* is low as shown on the state table. *Q follows D when the circuit is enabled* in this latch.

Flip-flops are edge triggered; latches are pulse triggered. The addition of an edge-detector circuit to the input of the D-type latch forms a D-type flip-flop as shown in Fig. 6–70(d). Operation of the D-type flip-flop is similar to operation of the D-type latch. The only difference, as shown on the state table, is that the steering gates in the flip-flop are only enabled for a very short period of time. This is due to the **edge-detector circuit.** The flip-flop can only respond to its inputs during the **active clock transition.** In other words, *Q follows D on the PGT (or NGT) of the clock pulse.*

Another flip-flop variation is the J-K flip-flop shown in Fig. 6–70(e). This circuit has two inputs, but the INVALID output condition doesn't exist because *Q* and $\overline{Q}$ are tied to the input steering gates. This replaces the undesired INVALID output state with the TOGGLE mode of operation. Operation of this flip-flop is summarized on the state table. *The Q output follows J and $\overline{Q}$ follows K on the active clock transition when the J and K inputs are complementary.*

The **J-K master–slave flip-flop** consists of two cascaded J-K flip-flops. The circuit is **pulse triggered.** Data is loaded into the master section when the clock is high. It is transferred to the slave section when the input clock goes low. This flip-flop essentially functions as a negative-edge-triggered flip-flop even though it is pulse triggered.

If the master section of a J-K master–slave flip-flop has an edge-detector circuit, the flip-flop is considered to be one with a **data lockout** capability. This capability allows changing the *J* and *K* inputs while the clock is still high—a feature not allowed in the standard master–slave flip-flop.

Most of the standard and ANSI/IEEE logic symbols for the latches and flip-flops presented in this chapter are shown in Fig. 6–71. It is very important to be able to identify these symbols and relate them to the state tables illustrating their operation.

All flip-flop symbols containing a **dynamic-input indicator** at the clock input are edge triggered. Symbols without this indicator are pulse (level) triggered. The **postponed-output indicator** identifies master–slave flip-flops. The **asynchronous inputs** ($\overline{PRE}$ and $\overline{CLR}$) are overriding inputs when either of them is asserted.

Flip-flops are used in numerous applications in digital circuits. Serially shifting data, controlling clock signals, and binary counting were presented in Section 6–7. Many more applications will be presented in the remainder of this textbook.

Troubleshooting circuits containing flip-flops relies on one main premise—the technician must know what the circuit is supposed to do when it is operating normally. The capability to troubleshoot, isolate a malfunction, and repair a circuit comes with the ability to identify the logic symbols in Fig. 6–71 and complete a state table for each symbol. Proper use of a logic pulser and logic probe to clock and check flip-flop states often leads to isolating a fault.

Many different types of latches and flip-flops are commercially available. No attempt will be made to address operation of other types. However, the knowledge you have gained in this chapter will allow you to pick up any data book and analyze operation of variations of the basic flip-flops. Although there are flip-flops with gated-AND inputs, J-$\overline{K}$ inputs, and other peculiarities, these peculiarities are relatively simple to analyze after the basic flip-flop concepts have been mastered.

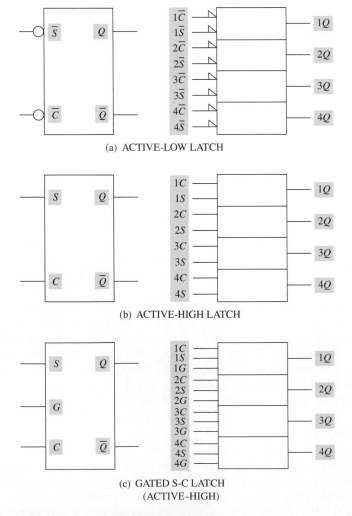

(a) ACTIVE-LOW LATCH

(b) ACTIVE-HIGH LATCH

(c) GATED S-C LATCH
(ACTIVE-HIGH)

FIGURE 6-71 Latch/flip-flop symbols.

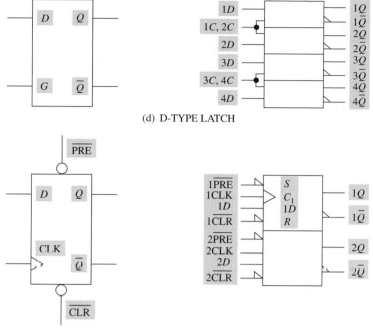

(d) D-TYPE LATCH

(e) POSITIVE-EDGE TRIGGERED
D-TYPE FLIP-FLOP

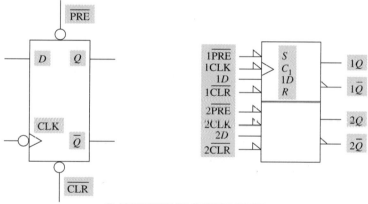

(f) NEGATIVE-EDGE TRIGGERED
D-TYPE FLIP-FLOP

FIGURE 6-71　**Continued**

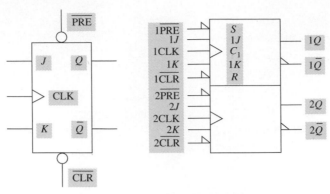

(g) POSITIVE-EDGE TRIGGERED
J-K FLIP-FLOP

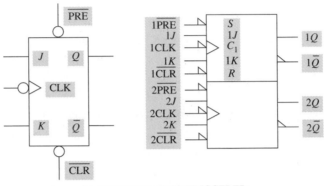

(h) NEGATIVE-EDGE TRIGGERED
J-K FLIP-FLOP

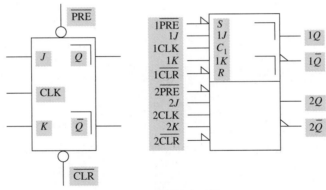

(i) J-K MASTER-SLAVE
FLIP-FLOP

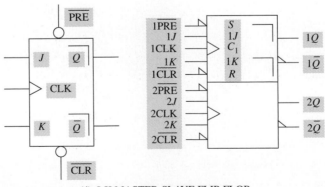

Figure 6-71   **Continued**

(j) J-K MASTER-SLAVE FLIP-FLOP
WITH DATA LOCKOUT

CHAPTER 6: END OF CHAPTER QUESTIONS/PROBLEMS

Answers are given in the Instructor's Manual.

SECTION 6-1

1. The latch shown in Fig. 6–72 is active-_____.
   (a) low
   (b) high

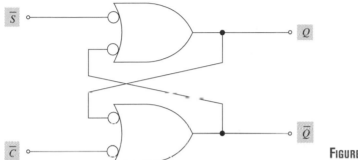

FIGURE 6-72

2. In what state is the latch of Fig. 6–72 when $Q = 1$ and $\overline{Q} = 0$?
3. What logic levels must be applied to the $\overline{S}$ and $\overline{C}$ inputs of the latch in Fig. 6–72 to put it in the CLEAR state?
4. Define "assert."
5. Which input of the latch in Fig. 6–72 must be activated to put the latch in the SET state?
6. What state or mode of operation is the latch of Fig. 6–72 in when $\overline{S} = 1$ and $\overline{C} = 1$?
7. What are the levels of Q and $\overline{Q}$ (Fig. 6–72) when $\overline{S} = 0$ and $\overline{C} = 0$?
8. What is the name of the state of the latch in Fig. 6–72 when $S = 0$ and $\overline{C} = 0$?
9. What is another name for the CLEAR state?
10. How many bits of data can be stored in a latch?
11. What data bit is stored in a latch (Q output) when it is in the CLEAR state?
CT 12. Draw the Q and $\overline{Q}$ outputs of the latch of Fig. 6–72 using the $\overline{S}$ and $\overline{C}$ inputs shown in Fig. 6–73.

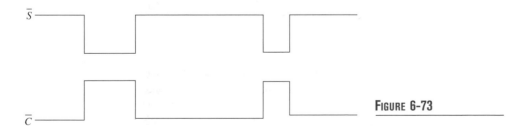

FIGURE 6-73

SECTION 6-2

13. The latch shown in Fig. 6–74 is active-_____.
14. In what state is the latch of Fig. 6–74 if $Q = 0$ and $\overline{Q} = 1$?
15. What logic level(s) must be applied to the S and C inputs of the latch in Fig. 6–74 to put it in the CLEAR state?
16. Which input of the latch in Fig. 6–74 must be activated to put it in the SET state?
17. What state or mode of operation is the latch of Fig. 6–74 in when $S = 0$ and $C = 0$?

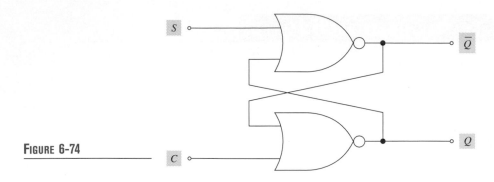

FIGURE 6-74

18. What state or mode of operation is the latch of Fig. 6–74 in when both inputs are activated?

CT 19. Draw the Q and $\overline{Q}$ outputs of the latch of Fig. 6–74 using the S and C inputs shown in Fig. 6–75.

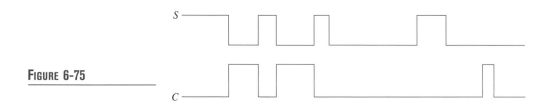

FIGURE 6-75

Section 6-3

20. The gated S-C latch in Fig. 6–76 is an active-_____input circuit.
21. What input levels must be applied to the latch in Fig. 6–76 to put it in the SET state?
    $S = $ _____, $C = $ _____, and $G = $ _____.
22. What input levels must be applied to the latch in Fig. 6–76 to put it in the CLEAR state?
    $S = $ _____, $C = $ _____, and $G = $ _____.
23. Show two different methods to put the latch of Fig. 6–76 in the HOLD mode of operation.
    $S = $ _____, $C = $ _____, and $G = $ _____, or
    $S = $ _____, $C = $ _____, and $G = $ _____.

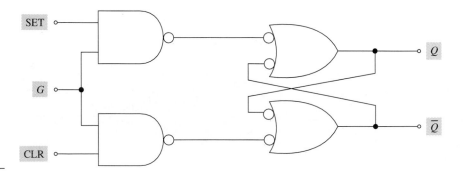

FIGURE 6-76

24. The logic symbol in Fig. 6–77 represents a _____.
25. What is the Q output of the latch in Fig. 6–77 when $D = 1$ and $G = 1$?

CT 26. Draw the Q and $\overline{Q}$ outputs of the latch in Fig. 6–77 using the G and D inputs shown in Fig. 6–78. (Start the waveform analysis with the circuit in the CLEAR state.)

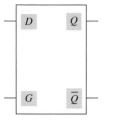

FIGURE 6-77

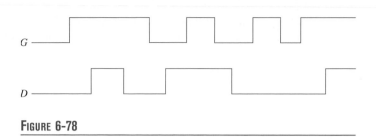

FIGURE 6-78

SECTION 6-4

27. The logic symbol in Fig. 6–79 represents a/an _____.
28. The dynamic input indicator in Fig. 6–79 indicates the flip-flop is _____.
CT 29. Draw the Q output of the flip-flop in Fig. 6–79 using the inputs shown in Fig. 6–80. (Start the waveform analysis with the flip-flop in the RESET state.)

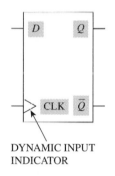

DYNAMIC INPUT
INDICATOR

FIGURE 6-79

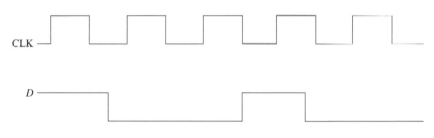

FIGURE 6-80

30. Define setup time as it applies to flip-flops.
31. The logic symbol in Fig. 6–81 represents a/an _____.
32. Draw the Q output of the flip-flop in Fig. 6–81 using the inputs shown in Fig. 6–82. (Start the waveform analysis with the flip-flop in the RESET state and assume $\overline{\text{PRE}}$ and $\overline{\text{CLR}}$ are connected high.)

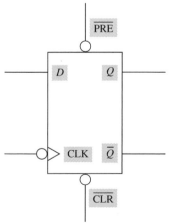

FIGURE 6-81

FIGURE 6-82

CT 33. Draw the Q output of the flip–flop in Fig. 6–81 using the inputs shown in Fig. 6–83.

34. Draw the Q output of the flip-flop in Fig. 6–84.

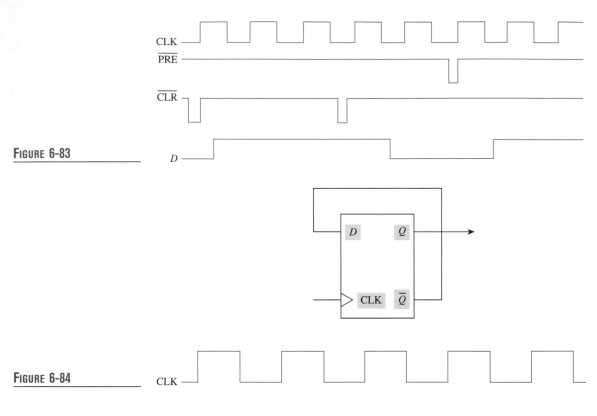

FIGURE 6-83

FIGURE 6-84

SECTION 6-5

35. Complete the Q and $\overline{Q}$ outputs on the table shown in Fig. 6–85.

FIGURE 6-85

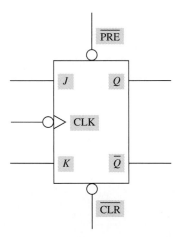

PRE	CLR	CLK	J	K	Q	Q̄
0	1	X	X	X		
1	0	X	X	X		
0	0	X	X	X		
1	1	↓	0	0		
1	1	↓	1	0		
1	1	↓	0	1		
1	1	↓	1	1		
1	1	1	X	X		

CT 36. Draw the *Q* output of the flip-flop in Fig. 6–85 using the inputs shown in Fig. 6-86.

37. Draw the *Q* output of the flip-flop shown in Fig. 6–87 in respect to its input waveforms.

38. Assume the flip-flop in Fig. 6–87 is negative-edge triggered and repeat problem 37.

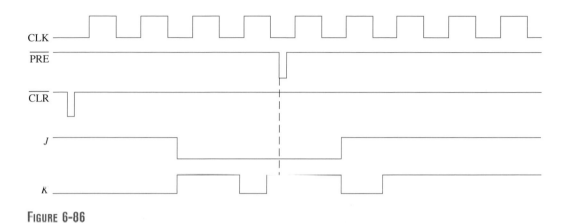

FIGURE 6-86

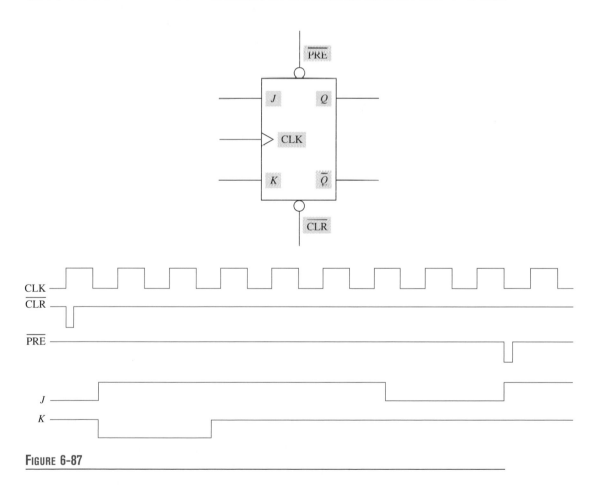

FIGURE 6-87

SECTION 6-6

39. Draw the standard symbol used to represent a positive-pulse-triggered J-K master–slave flip-flop with both of its asynchronous inputs.

CT 40. Draw the *Q* and $\overline{Q}$ outputs of a positive-pulse-triggered J-K master–slave flip-flop using the inputs shown in Fig. 6-88.

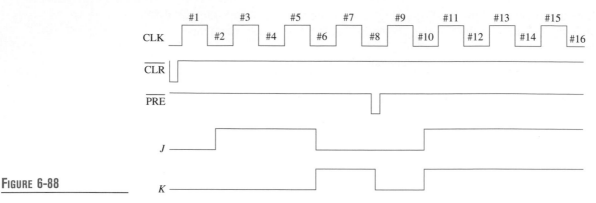

FIGURE 6-88

Sections 6-7 and 6-8

41. Determine the Q_A–Q_F outputs of the circuit shown in Fig. 6–89 after 6 clock pulses. Draw the output waveforms.

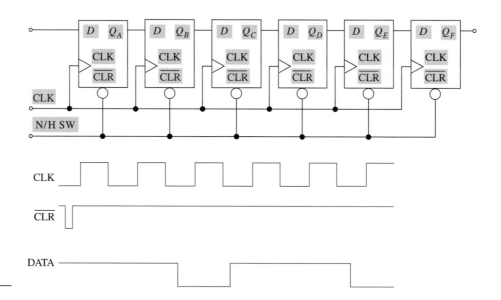

FIGURE 6-89

42. What is the MOD of the circuit shown in Fig. 6–90?
43. What is the maximum count of the circuit in Fig. 6–90?

FIGURE 6-90

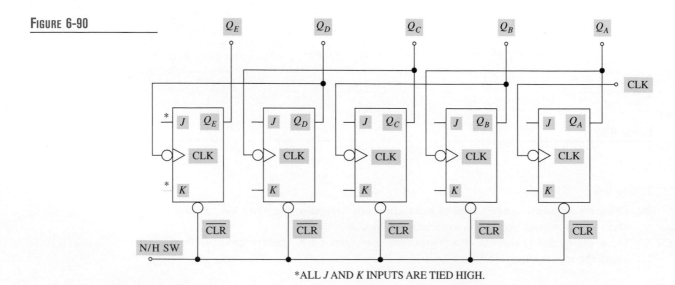

*ALL *J* AND *K* INPUTS ARE TIED HIGH.

CT 44. What would the output of the circuit in Fig. 6–90 be if there were an open between the Q_A output of the first flip-flop and the clock input of the next flip-flop?

45. What would the output of the circuit in Fig. 6–90 be if the $\overline{\text{CLR}}$ input line were shorted to ground?

CT 46. What would the output of the circuit shown in Fig. 6–91 be if the K input pin to flip-flop C were open?

$$\underline{Q_A \; Q_B \; Q_C \; Q_D}$$
$$0 \quad 0 \quad 0 \quad 0$$

t_1
t_2
t_3
t_4
t_5
t_6

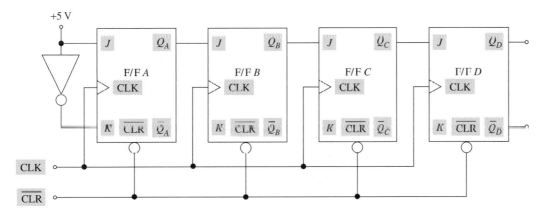

FIGURE 6-91

CT 47. What would the output of the circuit shown in Fig. 6–91 be if the clock input to flip-flop C were open?

$$\underline{Q_A \; Q_B \; Q_C \; Q_D}$$
$$0 \quad 0 \quad 0 \quad 0$$

t_1
t_2
t_3
t_4

48. What is the state of the latch shown in Fig. 6–92 when $A = 1$, $B = 0$, $C = 0$, and CLEAR $= 0$?

49. What is the state of the latch shown in Fig. 6–92 when $A = 0$, $B = 1$, $C = 1$, and CLEAR $= 0$?

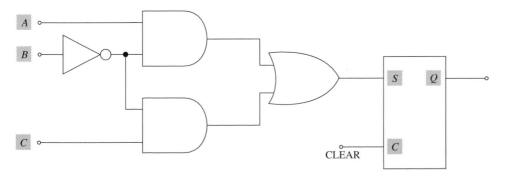

FIGURE 6-92

ANSWERS TO REVIEW QUESTIONS

SECTION 6-1

A. $Q = 0$
B. False
C. NAND
D. $\overline{SET} = 0$
   $\overline{CLR} = 1$
E. True

SECTION 6-2

A. Low (0)
B. True
C. NOR
D. SET = 1
   CLR = 0
E. RETAIN
F.

SET	CLR	Q	$\overline{Q}$	State
A	I	1	0	SET
I	A	0	1	CLEAR
I	I	NC	NC	RETAIN
A	A	0	0	INVALID

SECTION 6-3

A. synchronous
B. high
C. high
D. low
E. True

SECTION 6-4

A. pulse
B. edge
C. Edge triggering
D. True
E. False
F. True
G. False
H. Toggling
I. True

SECTION 6-5

A. J-K flip-flop
B. SET
C. SET
D. CLEAR
E. CLEAR
F. RETAIN previous state
G. TOGGLE
H. RETAIN
I. INVALID
J. Positive-edge triggered

SECTION 6-6

A. master–slave
B. True
C. CLEAR state (Master section is SET but slave section is still CLEAR.)
D. SET
E. MS F/F with data lockout

SECTION 6-7

A. Sequential logic circuits
B. A register is a group of latches or flip-flops used to transfer or store data.
C. 0011 (Q_A–Q_D)
D. modulus or MOD
E. $1111_{(2)} = 15_{(10)}$

SECTION 6-8

A. V_{CC} and ground
B. True
C. 2
D. True

ANSWERS TO INTERNAL SUMMARY QUESTIONS

SECTIONS 6-1 THROUGH 6-3

1 b	10. b
2. a	11. c
3. a	12. d
4. a	13. d
5. b	14. b
6. b	15. c
7. d	16. b
8. c	17. a
9. a	18. b

SECTIONS 6-4 THROUGH 6-6

1. c	11. a
2. b	12. b
3. c	13. b
4. a	14. a
5. a	15. b
6. b	16. d
7. d	17. c
8. b	18. b
9. a	19. a
10. a	20. b

CHAPTER

7 COUNTERS

Topics Covered in this Chapter

Chapter Objectives

1. Define basic counter terminology.

2. Identify a logic diagram or logic symbol of an up- or down-counter and determine whether it is synchronous or asynchronous.

3. Given a logic diagram of a counter, calculate the MOD and frequency divide-by capability at its outputs.

4. Design an asynchronous counter for a specified MOD.

5. Compare synchronous and asynchronous counters in terms of advantages and limitations.

6. Identify and determine the MOD number of a shift-register counter.

7. Given a logic diagram of a counter and problematic symptoms, troubleshoot the circuit to isolate the malfunction.

8. Design a synchronous counter (optional).

Introduction

A **digital counter** is a circuit used to generate binary numbers in a specific count sequence. That sequence is partially controlled by input clock pulses, and it is repetitive as long as these clock pulses are applied. Counters serve two main functions in digital systems—**counting** and **frequency division.**

The flip-flops presented in Chapter 6 form the groundwork for all sequential logic circuits. An example of a basic binary counter was presented in Section 6–7. This counter will be presented again in much more detail and further analyzed in this chapter.

It is the TOGGLE mode of operation that provides the divide-by-2 capability that is integral to all counting operations. When several flip-flops operating in the TOGGLE mode are connected together and clocked, a sequence of counts is produced. This sequence of counts can be used to count events or to divide a frequency to produce a lower frequency. The count sequence may be up or down. It may be used to generate a count that can be used as a truth table sequence to control logic circuits.

Crystal oscillator output frequencies in computers are normally divided to produce usable frequencies in different parts of the computer. For example, some clock control circuits are used to divide the crystal frequency to produce the actual system clock frequency. This frequency is often further divided to produce a frequency low enough to clock other devices. These reduced output frequencies are sometimes required to meet the input frequency requirements of slower operating peripheral ICs.

Counts within a digital computer may also be used to sequentially cycle through memory addresses or to produce time-of-day clock ticks to control the system's calendar. In addition, counters are used in frequency synthesizer applications and in analog-to-digital and digital-to-analog conversion.

Two basic categories of counters will be presented in this chapter: **asynchronous** and **synchronous.** The asynchronous counter is simpler in construction and easier to design than the synchronous counter, but it is limited to slower clock speeds. The asynchronous counter also generates **glitches** (voltage spikes) that might produce erroneous counts. These two disadvantages are overcome to a large degree in synchronous counters. **Hybrid counters** combine clocking schemes of both synchronous and asynchronous counters.

SECTION 7–1: ASYNCHRONOUS COUNTERS

OBJECTIVES

1. Identify an asynchronous (ripple) up- and/or down-counter.
2. Define MOD, asynchronous, increment, and decrement, and calculate the MOD number and maximum count of an asynchronous counter.
3. Determine the count output of an asynchronous up- or down-counter after X clock pulses.

MOD-8 Up-Counter–NGT-Triggered Flip-Flops

The logic diagram of a **MOD-8 asynchronous up-counter** is shown in Fig. 7–1(a). "Modulus (MOD)" was previously defined as "the maximum number of states (conditions) that a counter exhibits during its count sequence." The MOD of this circuit can be determined by raising 2 to the power of the number (n) of flip-flops. In this case, $2^n = 2^3 =$ MOD 8.

The count sequence of this counter is shown in Fig. 7–1(b). The count analysis should commence with the counter in the RESET state. This can be accomplished by connecting a normally high switch to the $\overline{\text{CLR}}$ inputs of the counter and momentarily asserting this input line to force the count outputs to 000. To avoid redundancy, development of the waveforms for this counter will be left to you. If you have any trouble developing the

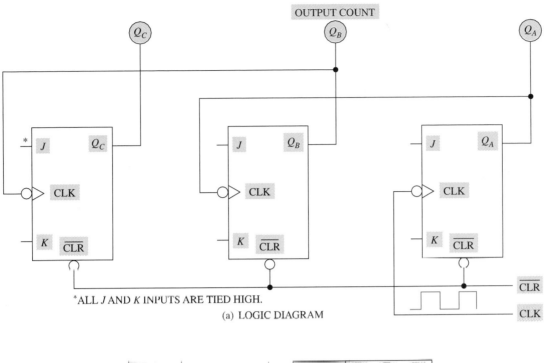

OUTPUT COUNT

*ALL J AND K INPUTS ARE TIED HIGH.

(a) LOGIC DIAGRAM

CLOCK	Q_C	Q_B	Q_A
$\overline{CLR}$	0	0	0
t_1	0	0	1
t_2	0	1	0
t_3	0	1	1
t_4	1	0	0
t_5	1	0	1
t_6	1	1	0
t_7	1	1	1

(b) COUNT SEQUENCE
Q OUTPUTS

CLOCK	$\overline{Q_C}$	$\overline{Q_B}$	$\overline{Q_A}$
$\overline{CLR}$	1	1	1
t_1	1	1	0
t_2	1	0	1
t_3	1	0	0
t_4	0	1	1
t_5	0	1	0
t_6	0	0	1
t_7	0	0	0

(c) COUNT SEQUENCE
$\overline{Q}$ OUTPUTS

FIGURE 7–1   **MOD-8 asynchronous up-counter using NGT-triggered flip-flops.**

waveforms that relate to the count sequence shown in Fig. 7–1(b), please review Section 6–7 before proceeding. Keep in mind during your analysis that the flip-flops in Fig. 7–1(a) toggle only when an NGT of a signal is applied to the CLK input.

The output count is monitored at Q_C, Q_B, and Q_A as indicated in the logic diagram. The Q_A output of this counter is the least significant bit (LSB) and Q_C is the most significant bit (MSB). These outputs are labeled with subscript numbers instead of letters in some data books. Should this be the case, Q_0 is the LSB and the highest subscript number is the MSB. Q_3 would be the MSB in a 4-bit counter, whose outputs would be labeled Q_3, Q_2, Q_1, and Q_0. Count sequences in this text are always monitored from the Q outputs of the circuits. This is done to standardize explanations and problems that are encountered in the text. The $\overline{Q}$ output counts are shown in Fig. 7–1(c). This is done one time only for purposes of this brief explanation. The $\overline{Q}$ outputs show the circuit is counting *down* on these outputs while it is counting *up* on the Q outputs.

The counters in the first part of this chapter are drawn with the LSB on the right and the MSB on the left to aid learning. This allows the count sequences that are presented to be easily interpreted. This is not done in all data books, and there will be times when the MSB is on the right. Some counters in the latter part of this chapter are drawn with the LSB on the left for variety. There is no sense in trying to standardize a text completely when industry has not standardized their drawings, and there is no need to do so.

What is the count of the circuit shown in Fig. 7–1(a) after 3 input clock pulses? This would be a difficult question to answer if you weren't sure where to monitor the output count. It would be impossible to answer if an initial count wasn't known. This initial count for up-counters will be $000_{(2)}$ unless otherwise noted in the text. The answer to the question is $011_{(2)}$, but the answer would be different if the $\overline{Q}$ outputs were producing the count. Another logical reason for monitoring the count at the Q outputs is because they are the only available count output pins on most IC counters.

The term "asynchronous" means "not happening at the same time." The clock input to this counter is applied only to flip-flop A. Note that all three flip-flops in the figure are connected in the TOGGLE mode of operation. Flip-flop B receives its clock input from the output of flip-flop A. The propagation delay of flip-flop A causes the flip-flop B clock input to arrive slightly later than it would if it were applied directly from the clock input to the counter as is the flip-flop A clock. Since the clock input to flip-flop B is delayed and this flip-flop introduces more delay before its output can respond to its input, the delay is increased even further at flip-flop C. This **additive delay** produces a ripple effect in the counter that has resulted in asynchronous counters often being called **ripple counters.** We will expand on this subject when the advantages and disadvantages of asynchronous counters are discussed.

MOD-8 Up-Counter–PGT-Triggered Flip-Flops

Another MOD-8 asynchronous up-counter is implemented in Fig. 7–2. There are two differences between this counter and the preceding one. First, the CLK inputs to flip-flops B and C are connected to the $\overline{Q}$ outputs of flip-flops A and B. In the preceding counter the Q outputs were used to clock the next flip-flop. Second, the flip-flops used in this counter are PGT triggered.

The output count sequence of this counter is identical to that obtained from the counter in Fig. 7–1(a) and is shown in Fig. 7–1(b). The first four clock pulses of the counter shown in Fig. 7–2(a) have been annotated above the flip-flops on the figure to aid the following analysis. The counter is asynchronously cleared first.

t_1: Flip-flop A will toggle when the PGT of the CLK pulse arrives. The Q_A output toggles high, but the $\overline{Q}_A$ output toggles low from 1 to 0 as shown in Fig. 7–2(b). This produces an NGT to the CLK input of flip-flop B. This results in no action by flip-flop B because it is a positive-edge-triggered flip-flop. There will be no change at the output of flip-flop C because it did not receive a PGT as a result of the t_1 input. The count is 001. The output counts of these counters are binary numbers even though the subscript 2 is left off this and following output counts.

t_2: The Q output of flip-flop A toggles again on the PGT of the t_2 clock pulse. This produces a PGT from $\overline{Q}_A$ that causes flip-flop B to toggle. The NGT from $\overline{Q}_B$ causes no change of states at flip-flop C. The count is now 010.

t_3: Again, the Q output of flip-flop A toggles high. Since $\overline{Q}_A$ produces an NGT, no other circuit changes occur. The count is 011.

t_4: The Q output of flip-flop A toggles low, producing a PGT at $\overline{Q}_A$. This causes the Q output of flip-flop B to toggle low, producing a PGT at $\overline{Q}_B$. This causes flip-flop C to toggle high. The count is now 100.

The input of clocks t_5, t_6, and t_7 will cause the counter to sequentially increment to counts 101, 110, and 111. Increment used in this sense means to increase the number or count. This sequential up-count is depicted by the waveforms in Fig. 7–2(b).

The clock pulse arriving at t_8 will toggle flip-flop A and produce PGTs at the other two flip-flop clock inputs that will cause the counter to reset to $000_{(2)}$. A binary counter will reset and repeat its count every 2^n clock pulses. In other words, every eighth input clock pulse will reset this MOD-8 counter to 000, thus preparing it for another count sequence.

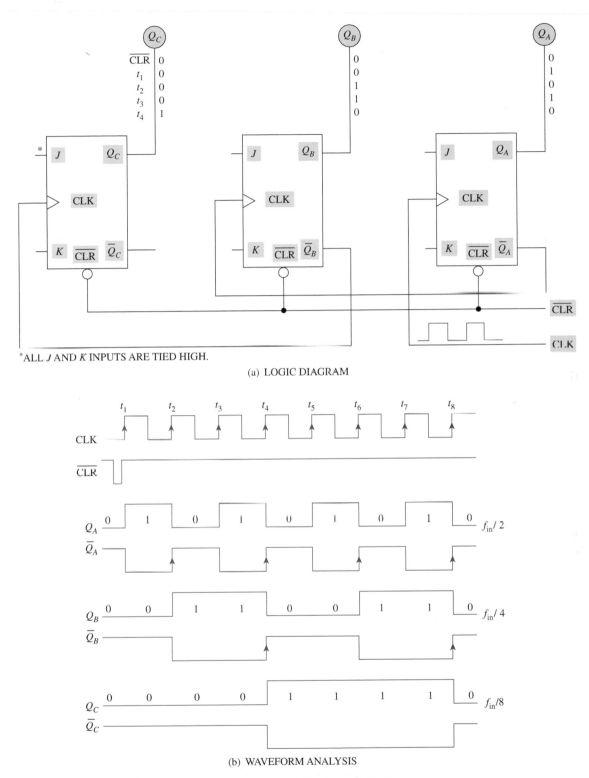

(a) LOGIC DIAGRAM

*ALL *J* AND *K* INPUTS ARE TIED HIGH.

(b) WAVEFORM ANALYSIS

FIGURE 7-2   MOD-8 asynchronous up-counter using PGT-triggered flip-flops.

The MOD number of this circuit is 8 and is verified by the 8 different states it exhibits during its count sequence ($000–111_{(2)}$). The maximum count of $111_{(2)}$ is shown in the waveform count sequence and is calculated as MOD 1. The counts after the active clock transition of each input clock pulse have been annotated on the Q output waveforms in Fig. 7–2(b). The counts are read vertically after each active clock transition with the LSB on the top output waveform (Q_A).

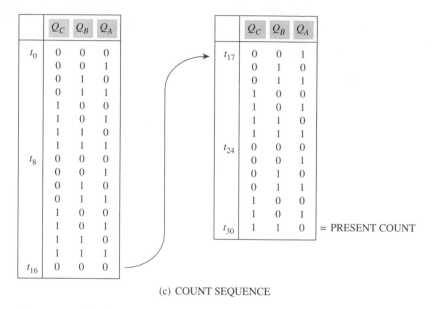

(c) COUNT SEQUENCE

FIGURE 7–2 **Continued**

The waveforms shown in Fig. 7–2(b) also show that the **duty cycle** of the waveforms produced by this counter is 50%. This will be true of outputs when the counter is allowed to count to its full-MOD capability. The figure also shows that the frequency of the Q_A output is the input clock frequency (f_{in}) divided by 2. The Q_B output is its input clock frequency $(\overline{Q_A})$ divided by 2 again. This is equal to $f_{in}/4$. The MSB output at Q_C is again divided by 2, yielding an output frequency of $f_{in}/8$. *The frequency divide-by capability "at the MSB output" of a counter is always equal to its MOD number.* It is for this reason that binary counters are often referred to as **divide-by-*n* counters,** because they produce one complete output cycle for every *n* input clock pulses.

There may be times when the count of a counter needs to be determined after a known number of input clock pulses. What if you had to determine the count of the MOD-8 up-counter shown in Fig. 7–2(a) after 30 clock pulses? One fact is evident if the count sequence in Fig. 7–2(c) is analyzed—every eighth input clock pulse resets the counter to 000. Therefore, the following formula can be used to derive the present count when the total number of input clock pulses is known:

Present count = Total # input CLK/MOD #
30/8 = 3 remainder **6**

The remainder of 6 is the **present count** of the counter after 30 input clock pulses. The integer portion of the answer may be ignored because it represents 3 complete cycles through the count sequence.

If the problem is solved with a calculator, the answer is 3.75. The remainder of 0.75 must be multiplied by the MOD number of the counter because the remainder represents .75 of one complete count sequence (8 clock pulses).

Present count = 30/8 = 3.75
= 0.75 × 8 = **6**

Now determine the count of a MOD-16 counter after 100 input clock pulses.

Present count = 100/16 = 6.25
= 0.25 × 16 = **4**

MOD-8 Up-Counter–D-Type Flip-Flops

A MOD-8 counter using D-type flip-flops is shown in Fig. 7–3 for implementation purposes only. Compare this circuit with the counter shown in Fig. 7–2. Remember, the D-type flip-flop will toggle when its $\overline{Q}$ output is connected to its D input.

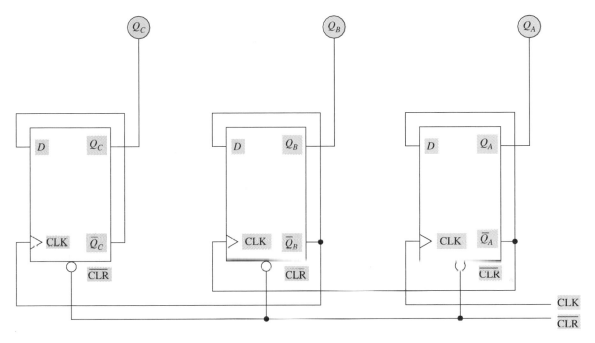

FIGURE 7–3   **MOD-8 asynchronous up-counter using D-type flip-flops.**

MOD-16 Up-Counter

A MOD-16 asynchronous up-counter is shown in Fig. 7–4(a). Its count sequence is shown on the waveforms in Fig. 7–4(b). The MOD ($2^n = 2^4 = 16$) indicates this counter's maximum count is 15 ($1111_{(2)}$). The CLEAR state is assumed at the beginning of the count sequence even though a $\overline{CLR}$ input is not shown in the waveforms. Don't forget, this initial state will be assumed in all subsequent up-counters unless otherwise noted.

MOD-32 Up-Counter

If you were tasked with implementing a MOD-32 asynchronous up-counter, how would you go about it? A MOD-32 up-counter could be implemented by connecting five J-K TOGGLE mode flip-flops in cascade similar to the MOD 16 counter shown in the previous example. Since $2^n = 2^5 = 32$, the maximum count of this counter is 31 ($11111_{(2)}$). The circuit could be implemented with NGT-triggered flip-flops if the Q output were used to clock the next flip-flop. It could be implemented with PGT-triggered flip-flops if the $\overline{Q}$ output were used to clock the next flip-flop.

The divide-by capability of a MOD-32 counter is illustrated in Fig. 7–5 in block diagram form. This capability will double for each additional flip-flop. *The divide-by capability of any output can be determined by raising 2 to the power of the bit position plus 1.* For example, the Q_C output is in bit position 2. Keep in mind that Q_A is in bit position 0 (2^0). Thus, $Q_C = 2^{2+1} = 2^3 = 8$.

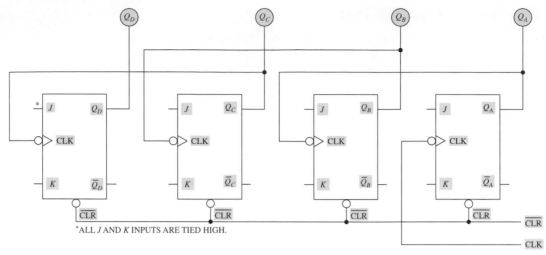

(a) LOGIC DIAGRAM

*ALL J AND K INPUTS ARE TIED HIGH.

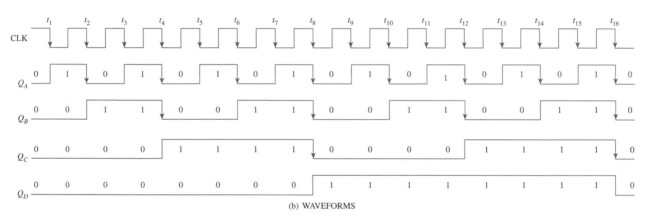

(b) WAVEFORMS

FIGURE 7-4   MOD-16 asynchronous up-counter.

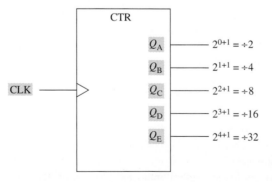

FIGURE 7-5   Counter divide-by capability block diagram.

MOD-8 Down-Counter

The logic diagram of a MOD-8 asynchronous down-counter and its count sequence are depicted in Fig. 7–6. The count sequence commences with 000 (CLEAR state) at the Q outputs.

The NGT of the clock input causes flip-flop A to toggle to the SET state. Since $\overline{Q}_A$ toggles from high to low on this clock input, it produces an NGT to toggle flip-flop B to the SET state. Likewise, $\overline{Q}_B$ toggles from high to low and produces an NGT that toggles flip-flop C to the SET state. The counter sets to the binary count of 111 on this first clock pulse.

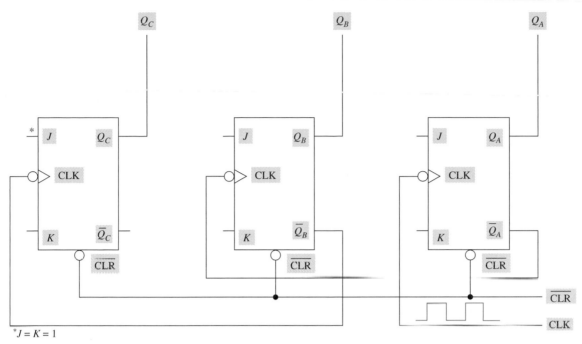

(a) LOGIC DIAGRAM

$^*J = K = 1$

CLOCK	Q_C	Q_B	Q_A
$\overline{\text{CLR}}$	0	0	0
t_1	1	1	1
t_2	1	1	0
t_3	1	0	1
t_4	1	0	0
t_5	0	1	1
t_6	0	1	0
t_7	0	0	1
t_8	0	0	0

(b) COUNT SEQUENCE

Figure 7-6   MOD-8 asynchronous down-counter using NGT-triggered flip-flops.

Analysis of the count sequence reveals the counter **decrements** to the next lower count on each successive clock pulse's negative-going transition. To decrement means to decrease the number or count output of the counter. When the eighth clock pulse arrives, the counter decrements to a terminal count of 000. This prepares the counter for the next count sequence.

This same counter is shown in Fig. 7–7 implemented with PGT-triggered flip-flops. Other than the type of flip-flops used, note the only difference in construction is the use of the Q_A and Q_B outputs to trigger flip-flops B and C respectively.

Two different implementations of a MOD-8 up-counter and two different implementations of a MOD-8 down-counter have been presented in this section. The easiest method that can be used to identify these and other counters without having to rely on memorization is to keep these steps in mind:

1. Determine the MOD number by 2^n.
2. Clear the counter and bring in *one* clock pulse. If the count output goes to 1 (. . . $001_{(2)}$), the counter is an up-counter. If the count goes to maximum (all high outputs— . . . $111_{(2)}$), it is a down-counter.

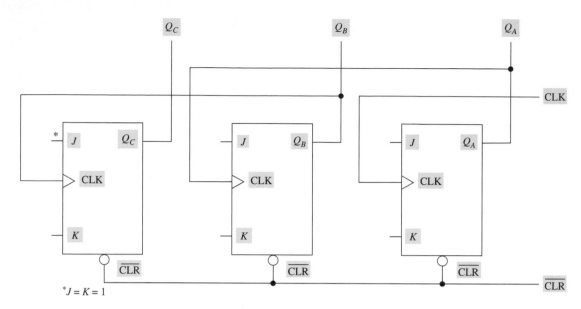

FIGURE 7-7   MOD-8 asynchronous down-counter using PGT-triggered flip-flops.

From the foregoing discussion it should be clear that counters designed for implementation using NGT-triggered flip-flops will not generate the same count sequence if PGT-triggered flip-flops are used. In fact if this implementation error is made, an up-counter becomes a down-counter and vice versa.

How can the number of flip-flops required to implement a certain MOD counter be determined? We learned in Chapter 2 that the number of bits required for a specific MOD can be calculated as log desired MOD ÷ log 2. Since it takes a flip-flop to represent one bit in a counter, the same formula can be used to calculate the number of flip-flops required to implement a certain MOD counter.

$$\text{\# Flip-flops} = \log \text{MOD} \div \log 2$$

Count outputs of counters are often cut short of the full MOD/full-count capability when a lower count or frequency division is desired. This subject will be discussed in the next section.

Section 7-1: Review Questions

Answers are given at the end of the chapter.

A. The modulus of a counter is equal to its maximum count.
    (1) True
    (2) False

B. The MOD of a full-count counter can be determined by raising 2 to the power of the number of flip-flops the counter contains.
    (1) True
    (2) False

C. The MSB of a counter's output is usually identified as Q_0 or Q_A.
    (1) True
    (2) False

D. The term "asynchronous" means "happening at the same time."
    (1) True
    (2) False

E. The propagation delays developed in an asynchronous counter are additive.
   (1) True
   (2) False

F. Asynchronous counters are often referred to as ripple counters.
   (1) True
   (2) False

G. The frequency divide-by capability at the MSB of a counter is always equal to its maximum count.
   (1) True
   (2) False

H. To sequentially increment a counter means to _____ its count by 1.
   (1) increase
   (2) decrease

I. How many flip-flops are required to implement a MOD-64 asynchronous counter?

J. Determine the present count of a MOD-16 up-counter after 155 input clock pulses.

SECTION 7–2: ASYNCHRONOUS TRUNCATED COUNTERS

OBJECTIVES

1. Identify and determine the MOD number and maximum count of asynchronous truncated up-counters.
2. Determine the output frequency of an asynchronous truncated up-counter.
3. Given a specific MOD number, design an asynchronous up-counter.

The MOD-8 and MOD-16 counters in Section 7–1 are **full-count counters.** These counters produce MODs that are equal to an integer power of 2, where the integer is the number of flip-flops. This is expressed in the formula for MOD as 2^n.

It often becomes necessary to have a counter's modulus (frequency divide-by capability) less than an integer power of 2. When a counter is designed to have a reduced MOD number, it is **truncated.** "Truncate" is defined as "to shorten by or as if by cutting off" or "appearing to terminate abruptly." Both definitions fit the description of decreasing the MOD number of a counter. We will define truncate as "short-counting" a counter.

MOD-5 Up-Counter

The logic diagrams in Fig. 7–1 and 7–2 are MOD-8 asynchronous up-counters as previously explained. The logic diagram in Fig. 7–8(a) shows how the MOD-8 up-counter can be modified to change its MOD number.

This circuit has a truncating NAND gate whose output is connected to the $\overline{\text{CLR}}$ input of each flip-flop in lieu of a normally high switch. The inputs to the NAND gate are labeled MR_1 and MR_2 (Master Reset). The remainder of this circuit is identical to the MOD-8 counter in Fig. 7–1.

The Master Reset inputs in Fig. 7–8(a) are connected to the Q_A and Q_C outputs of the flip-flops. The output of the NAND gate will go active (low) when both Q_A and Q_C are high. The count sequence shown in Fig. 7–8(b) indicates this condition ($Q_A = Q_C = 1$) first occurs on a binary count of 101. This count actually occurs but is immediately changed to 000 when the NAND gate's output is activated. The short duration of the 101 count is equal to the propagation delay of the NAND gate plus that of the flip-flops to return to the CLEAR state.

This counter produces 5 states with a maximum count of 4 ($100_{(2)}$). The binary count of 101 is too short in duration to be considered. The waveforms in Fig. 7–8(c) show the

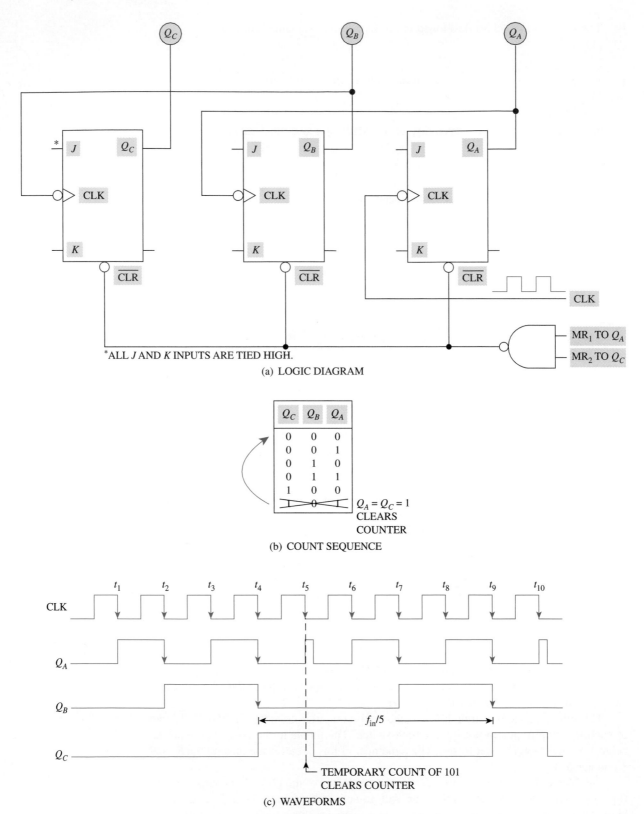

*ALL *J* AND *K* INPUTS ARE TIED HIGH.

(a) LOGIC DIAGRAM

Q_C	Q_B	Q_A
0	0	0
0	0	1
0	1	0
0	1	1
1	0	0
~~1~~	~~0~~	~~0~~

$Q_A = Q_C = 1$
CLEARS
COUNTER

(b) COUNT SEQUENCE

$f_{in}/5$

TEMPORARY COUNT OF 101
CLEARS COUNTER

(c) WAVEFORMS

FIGURE 7–8   **MOD-5 asynchronous up-counter using NGT-triggered flip-flops.**

output at the MSB (Q_C) is the input clock frequency divided by 5. This reinforces the statement that the divide-by capability of a counter at the MSB is equal to its MOD number. Furthermore, the output signal at Q_C is *not* a 50% duty cycle signal because the counter was not allowed to full count to $111_{(2)}$ but was truncated to stop counting short of its normal count.

The count of $101_{(2)}$ from the MOD-5 counter produces a **glitch** at the Q_A output and prolongs the Q_C output by X ns. Although this count cannot be seen on an LED, glitches produced by propagation delays cause problems when the output of an asynchronous counter is decoded. **Decoders** are necessary anytime a counter's output(s) must be converted to an output that is to be used to cause some action in another circuit. A decoder's output will detect a specific count sequence and activate on that count. Decoding circuits will be presented in Section 7–8.

The rule followed to implement the MOD 5 counter is relatively simple. *The rule of truncation for asynchronous up-counters is "tie the bits that are high on the desired MOD number to the truncating circuitry."* In this case, the Q_A and Q_C bits are high on a count of $\underline{101}$.

One additional rule must be followed when designing counters for specific MODs. The number (n) of flip-flops required to implement various MODs must raise 2 to a power that is equal to or greater than the desired MOD ($2^n \geq$ desired MOD).

The MOD-5 counter may also be implemented with PGT-triggered flip-flops as shown in Fig. 7–9. Here the $\overline{Q}$ outputs are used as the asynchronous CLK inputs to flip-flops B and C. The Q_A and Q_C flip-flop outputs are still connected to the MR inputs. Also note that the count output is taken from the Q outputs of the flip-flops. The following design problem is solved to enhance your understanding of an asynchronous truncated up-counter.

MOD-24 Up-Counter

The problem is to design a MOD-24 asynchronous up-counter using NGT-triggered J-K flip-flops. The first step is to determine how many flip-flops are needed. Since four flip-

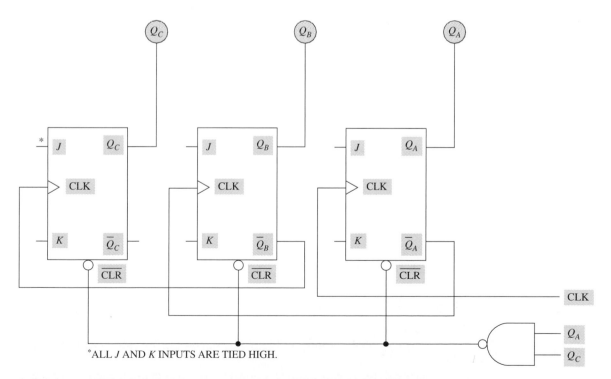

*ALL J AND K INPUTS ARE TIED HIGH.

Figure 7–9   MOD-5 asynchronous up-counter using PGT-triggered flip-flops.

flops will only implement a MOD-16 counter, five flip-flops are required. Instead of guessing what number raised to a certain power will do the job, calculate it as explained in the previous section. Let the letter n represent the number of flip-flops required; n can be calculated as follows:

$$n = \log 24 \div \log 2$$
$$= 1.38 \ldots \div 0.301 \ldots$$
$$= \mathbf{4.585}$$

where 24 is the desired MOD.

The answer indicates more than four flip-flops will be required. Naturally, we will go to the next higher number. Five flip-flops will implement a MOD-32 counter because $2^5 = 32$. Since this MOD is greater than the desired MOD, the counter must be truncated.

The second step is to determine which of the counter outputs must be connected to the truncating circuit. This is accomplished by determining which counter output bits are high on the desired MOD number. The binary number for 24 is 11000. Thus, bits Q_D and Q_E must be connected to the truncating circuit to force the counter to reset after a count of 23 $(10111_{(2)})$.

The resulting circuit is shown in Fig. 7–10. The design problem could have asked for implementation with PGT-triggered J-K flip-flops. This would require that the CLK inputs to the Q_B through Q_E flip-flops be taken from the preceding flip-flop $\overline{Q}$ output. The inputs to the truncating gate would still be connected to Q_D and Q_E.

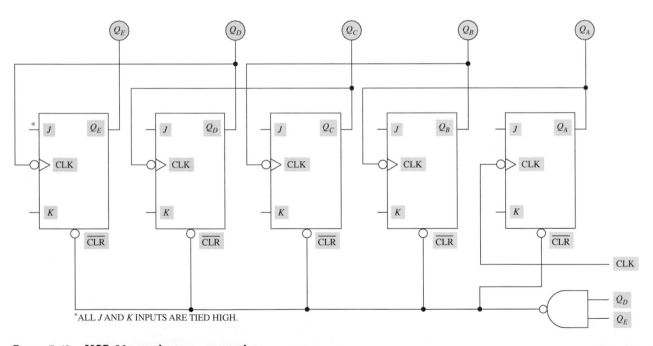

FIGURE 7-10   MOD-24 asynchronous up-counter.

Asynchronous IC Counters

7493/74LS93.

The logic diagram for these commercially available 14-pin IC counters is shown in Fig. 7–11(a). The logic symbol is shown in Fig. 7–11(b), and the function table is shown in Fig. 7–11(c).

This counter can be used as a MOD-8 or MOD-16 up-counter. It can be configured as a *MOD-8 up-counter* if the clock input is applied to the CLK B input pin and the

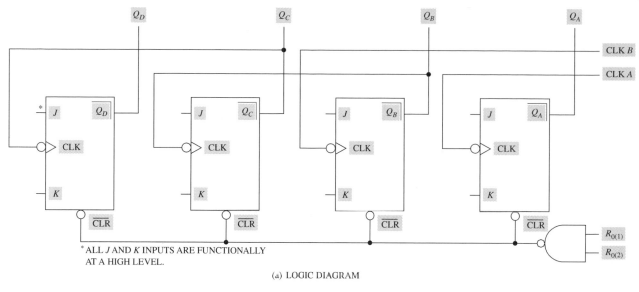

(a) LOGIC DIAGRAM

*ALL J AND K INPUTS ARE FUNCTIONALLY
AT A HIGH LEVEL.

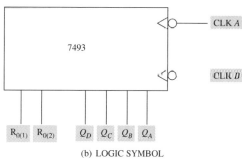

(b) LOGIC SYMBOL

RESET INPUTS		OUTPUTS			
$R_{0(1)}$	$R_{0(2)}$	Q_D	Q_C	Q_B	Q_A
H	H	L	L	L	L
L	X	COUNT			
X	L	COUNT			

NOTES: Q_A IS CONNECTED TO CLK B.
H = HIGH LEVEL (1)
L = LOW LEVEL (0)
X = DON'T CARE

FIGURE 7–11   **7493/74LS93 Asynchronous IC counters.**

RESET inputs ($R_{0(1)}$ and $R_{0(2)}$) are tied to ground. Note in the logic diagram that flip-flop A is not used when the counter is configured in this manner. Since the Q_A output is not used when the counter is clocked at CLK B, the Q_B output becomes the LSB.

The counter can be used as a MOD-16 up-counter if the clock is applied to the CLK A input, the RESET pins are tied to ground, and the Q_A output is externally connected to the CLK B input. Examination of the logic diagram reveals the A flip-flop output is the only flip-flop output in this counter that is not internally connected to the next flip-flop. This gives the counter the versatility to be used as a MOD-8 or MOD-16 up-counter. Figure 7–12 shows the counter in MOD-8 and MOD-16 configurations. The 7493 counter can be truncated to various MOD numbers less than 16. This can be accomplished by following the **rule of truncation** presented earlier in this section.

This counter can be configured as a MOD-6 counter by connecting the bits that are high on the desired MOD number to the truncating circuit ($R_{0(1)}$ and $R_{0(2)}$). First,

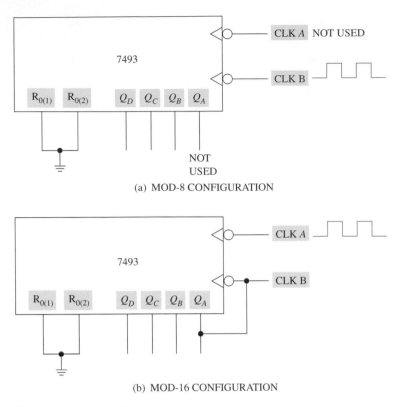

(a) MOD-8 CONFIGURATION

(b) MOD-16 CONFIGURATION

FIGURE 7-12   **7493/74LS93 counter.**

only three of the flip-flops within this IC are required to achieve the desired MOD number. Thus, the counter must be clocked at the CLK B input as shown in Fig. 7–13(a). This removes the A flip-flop from the circuit. Second, because the Q_A output is not used, Q_B is the LSB and Q_D is the MSB. The Q_A output does not need to be connected to anything because it is not used. The binary number 110 indicates Q_D and Q_C are high on the desired MOD. These two bits are tied to the gated-reset input pins. These two RESET inputs are active-high, and they both must be high at the same time to reset the counter. This occurs when the counter reaches the count of $110_{(2)}$, and then it is reset to $000_{(2)}$ almost instantaneously. The 7493 counts to binary 101 and then resets to 000—six states with a maximum count of 101. The frequency at the Q_D output is $f_{in}/6$. The waveforms associated with this circuit are shown in Fig. 7–13(b).

A MOD-9 configuration of the 7493/74LS93 is illustrated in Fig. 7–14. The counter is clocked at CLK A, and external connections have been made between Q_A and CLK B and between Q_A and Q_D and the RESET inputs. The frequency at Q_D is $f_{in}/9$.

One additional problem may be encountered when designing truncated asynchronous IC counters. There are only two RESET input pins. How could a MOD-14 counter be implemented? This MOD number (1110) requires that three of the four output bits be connected to the RESET pins. The simple solution to this problem is to follow the **rule of truncation.** An external AND gate is used in Fig. 7–15 to implement this rule. The count of 14 (1110) will produce highs at Q_D, Q_C, and Q_B, and immediately reset the counter to 0000. The maximum count of 13 (1101) will appear at the counter's output, and then the counter will be reset to 0000 by the truncating circuitry.

7493/74LS93 ANSI/IEEE Symbol.

The ANSI/IEEE symbol for this counter is shown in Fig. 7–16. This counter symbol contains three blocks. The block labeled CTR indicates this is a counter, and this block is called the **common-control block.** This block is separate from the remaining blocks to

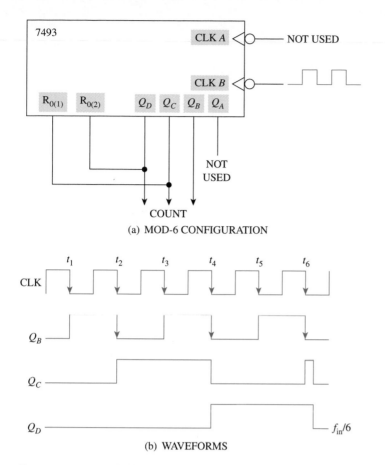

(a) MOD-6 CONFIGURATION

(b) WAVEFORMS

FIGURE 7-13   **7493/74LS93 counter.**

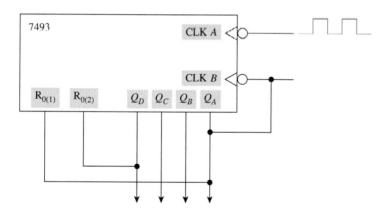

FIGURE 7-14   **7493/74LS93 counter—MOD-9 configuration.**

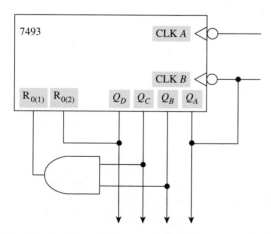

FIGURE 7-15   **7493/74LS93 counter—**
MOD-14 configuration.

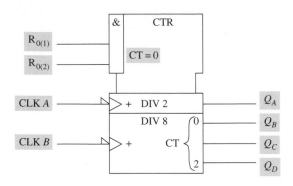

FIGURE 7-16   7493/74LS93 counter—
ANSI/IEEE symbol.

indicate the inputs to this particular block are common to all of the circuits in the counter. The R_0 (RESET) inputs to the counter are active-high and the & symbol indicates both of these inputs must be activated simultaneously to clear the counter. When this is accomplished the count equals 0. This is shown by CT = 0 in the common-control block.

The second block of this ANSI/IEEE symbol is labeled DIV2 and represents flip-flop A in the counter. Since this is a single toggling flip-flop, it is a divide-by-2 circuit. The clock input symbols indicate the clock is active on the negative-going transition. The + notation immediately to the right of the clock inputs indicates the counter will increment one count on each active clock transition.

The third block is labeled DIV8 and represents flip-flops B, C, and D in the counter. The binary grouping symbols 0 and 2 labeled vertically at the Q_B and Q_D outputs represent 2^0 through 2^2 for the DIV 8 section of the counter.

74293/74LS293.

These asynchronous 14-pin IC counters are functionally identical to the 7493/74LS93 counters. The only difference is physical—the arrangement of the IC pins has been changed.

74393/74LS393.

The logic diagram and symbol for these counters are shown in Fig. 7–17(a) and (b). This dual 4-bit binary counter contains eight master–slave flip-flops arranged to implement two individual MOD-16 counters in one 16-pin IC package. Only one of the two counters is shown in the figure. Comparison of the MOD-8/MOD-16 7493 counter logic symbol in Fig. 7–18 with this counter's logic symbol reveals two minor differences. The 74393/74LS393 logic diagram and symbol in Fig. 7–17 show only one clock input for this section of the counter, and all of the flip-flops are asynchronously cascaded internally. Therefore, the MOD-8 capability of the 7493 doesn't exist in the 74393 counter. One other difference is that there is only one active-high clear input.

Cascading Counters

As we have seen, every time a flip-flop toggles in a counter its output frequency is half its input clock frequency. The MOD-16 counter has output frequencies at $Q_A = f_{in}/2$, $Q_B = f_{in}/4$, $Q_C = f_{in}/8$, and $Q_D = f_{in}/16$.

Figure 7–19 shows both sections of the 74393 dual 4-bit binary counter IC. The two sections of the counter have been cascaded by connecting the MSB of the first counter to the clock input of the second section.

The figure shows the MSB output of the second counter is $f_{in}/256$. The input clock frequency to the second counter is $f_{in}/16$. The Q_A output is $f_{in}/32$. This output is internally connected to flip-flop B to produce $f_{in}/64$ at Q_B. Likewise, Q_C is $f_{in}/128$ and Q_D is $f_{in}/256$. *Cascading counters produce a MOD number that is equal to the product of the individual MODs.* For the circuit in Fig. 7–19, MOD-16 × MOD-16 = MOD 256.

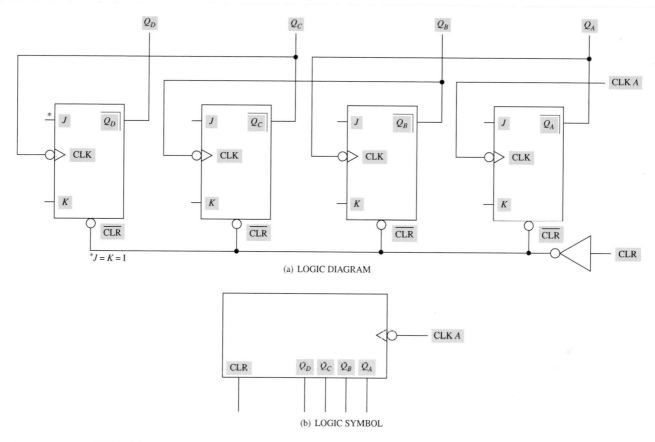

(a) LOGIC DIAGRAM

(b) LOGIC SYMBOL

FIGURE 7-17   **74393/74LS393 counter.**

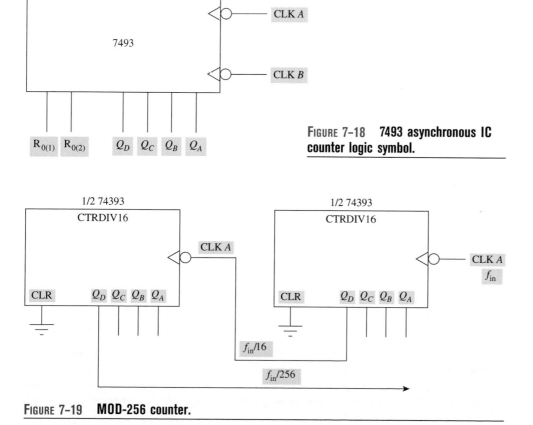

FIGURE 7-18   **7493 asynchronous IC counter logic symbol.**

FIGURE 7-19   **MOD-256 counter.**

A **MOD-60 circuit** is shown in Fig. 7–20(a). The circuit is implemented with a MOD-10 74393 counter and a MOD-6 7493 counter. This circuit can be used as a frequency divide-by-60 circuit because the Q_D output of the MOD-6 counter is $f_{in}/60$.

Figure 7–20(b) shows the count of the MOD-10 counter. Every tenth clock input pulse resets this counter to 0000. Resetting the 74393 at t_{10} produces an NGT at Q_D, which increments the 7493 one count. The 74393 counts to 1001, resets to 0000, and increments the 7493 one count cyclically. The cycle is repeated a total of 6 times, and at t_{60} both counters are cleared to start the process over.

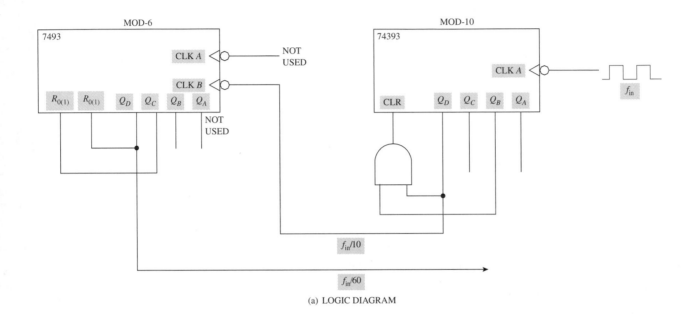

(a) LOGIC DIAGRAM

	Q_D	Q_C	Q_B	Q_A
t_0	0	0	0	0
t_1	0	0	0	1
t_2	0	0	1	0
t_3	0	0	1	1
t_4	0	1	0	0
t_5	0	1	0	1
t_6	0	1	1	0
t_7	0	1	1	1
t_8	1	0	0	0
t_9	1	0	0	1
t_{10}	0	0	0	0

(b) MOD-10 COUNT SEQUENCE

FIGURE 7–20   MOD-60 counter.

Advantages/Disadvantages of Asynchronous Counters

Asynchronous counters are relatively simple to design. They also contain simpler circuitry than synchronous counters, as will be seen shortly. But asynchronous counters have one inherent disadvantage—**additive propagation delays.** Figure 7–21 clearly illustrates this problem. The waveforms shown in this figure are produced by a typical MOD-8 asynchronous counter. The additive propagation delays produce short-duration miscounts that generate problems when decoding the counter's output(s). These miscounts are labeled Temporary States in the figure.

The circuit shown in Fig. 7–22 is a MOD-8 asynchronous up-counter with one decoded output. The AND gate connected to $\overline{Q}_A$, $\overline{Q}_B$, and $\overline{Q}_C$ will decode (detect) the count

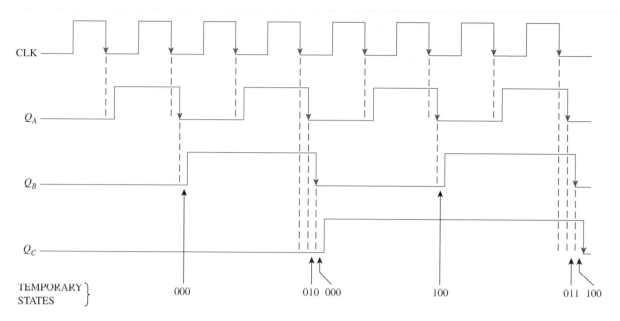

FIGURE 7-21   Asynchronous counter output waveforms illustrating temporary states.

of 000. When $Q_A = Q_B = Q_C = 0$ (count of 000), $\overline{Q}_A$, $\overline{Q}_B$, and $\overline{Q}_C = 1$. Therefore, on the count of 000 the AND gate's output will go high. The gate's output will be low for all other count outputs of this circuit.

The true problem presented by the short-duration miscounts of this counter is that the decoder circuit will detect them and produce an active output. Thus, the decoded output is erroneous. This problem will be solved in Section 7–8.

Another problem that arises due to the additive propagation delays is the clocking frequency of this type of counter is rather limited. The waveforms shown in Fig. 7–21 have been reproduced in Fig. 7–23(a) with time and input clock frequency annotated. The propagation delay of each flip-flop used in this counter example is assumed to be 30 ns.

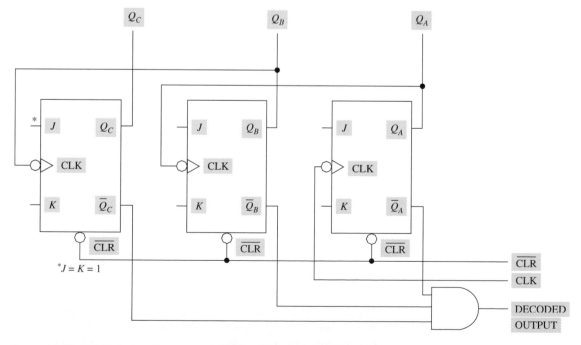

FIGURE 7-22   MOD-8 asynchronous up-counter with 000 count decoder.

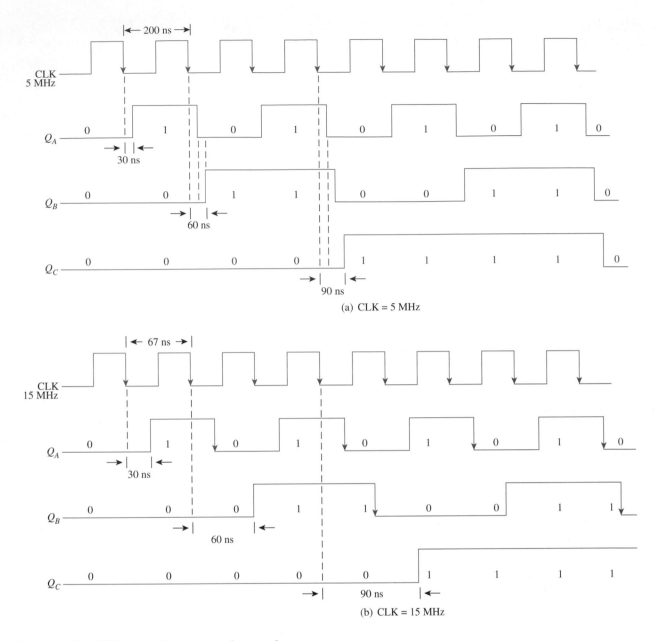

FIGURE 7-23   **MOD-8 asynchronous counter waveforms.**

The set of waveforms in this figure is produced by a 5-MHz input clock frequency. The counts are normal except for the short-duration miscounts, which can be masked out by proper decoding circuits.

The set of waveforms in Fig. 7–23(b) is produced by a 15-MHz input clock frequency. The propagation delay of each flip-flop is again assumed to be 30 ns. Count analysis of this set of waveforms reveals the following counts: 000, 001, 000, 011, 010, 101, 100, 111, and 110 for the first eight input clock pulses. These are not short-duration miscounts; they are full-fledged erroneous counts.

This problem of nonsequential counting stems from the input clock frequency, which in Fig. 7–23(b) is too high. The upper clock frequency of an asynchronous counter must be calculated when designing the counter. Once calculated, the maximum clock frequency cannot be exceeded, as illustrated in Fig. 7–23(b).

The **maximum clock frequency** for an asynchronous counter is calculated as follows:

$$f_{max} = 1/n(t_P)$$

where n = number of flip-flops

t_P = propagation delay of each flip-flop

The MOD-8 counter with a propagation delay of 30 ns per flip-flop in the preceding analysis has a maximum clock frequency of 11.1 MHz.

$$f_{max} = 1/3(30 \text{ ns}) = 1/90 \text{ ns} = \textbf{11.1 MHz}$$

Frequencies below f_{max} will produce some short-duration miscounts, which are normal in asynchronous counters. Frequencies greater than f_{max} will produce erroneous counts.

Section 7-2: Review Questions

Answers are given at the end of the chapter.

A. Define "truncate."
B. Explain the rule of truncation.
C. How many flip-flops are required to implement a MOD-100 asynchronous up-counter?
D. The maximum count of a truncated MOD-7 up-counter is_____.
E. Decoders are required when a counter's outputs must be converted to a signal that will cause some action in a subsequent circuit.
   (1) True
   (2) False

F. Which output bits (Q_D, Q_C, Q_B, and/or Q_A) would need to be connected to the truncating circuitry to implement a MOD-13 counter?
G. What is the MOD number of a circuit implemented with cascaded MOD-10 and MOD-12 counters?
H. Calculate f_{max} for a MOD-16 asynchronous counter implemented with flip-flops that each have a 25-ns propagation delay.

SECTIONS 7-1 AND 7-2: INTERNAL SUMMARY

Asynchronous counters can be identified by the fact that flip-flop clock inputs (except for the LSB flip-flop) come from the preceding flip-flop output. These counters are usually referred to as **ripple counters.**

The MOD number of a counter is its maximum number of states. It can be calculated by raising 2 to the power of the number of flip-flops in the counter providing the counter is not truncated.

The outputs of a counter are taken from the Q outputs of the flip-flops. The Q_A or Q_0 output of a counter is the LSB. The overall frequency divide-by capability at the MSB of a counter is equal to its MOD number.

An up-counter will go to a count of $1_{(10)}$ after it has been cleared and one clock pulse applied. A down-counter will produce all high outputs (maximum count) after it has been cleared and one clock pulse applied.

A counter can be **truncated** to produce MODs less than integer powers of 2. This is accomplished by adding a truncating gate that will clear the counter on the desired MOD number. The **gated-RESET inputs** ($R_{0(1)}$ and $R_{0(2)}$) asynchronously CLEAR all flip-flops when the output of the truncating NAND gate goes low.

The **rule of truncation** for asynchronous counters is "tie the bits that are high on the desired MOD number to the truncating circuitry."

The number of flip-flops required to implement a counter (n) must raise 2 to a power that is equal to or greater than the desired MOD number. If the result is greater than the desired MOD, the counter must be truncated.

Cascading counters will produce an overall MOD number that is equal to the **product** of the individual counter MODs. For example, three cascaded MOD-10 counters will produce a MOD-1000 counter.

Asynchronous counters are simple to design and implement. However, their **additive propagation delays** cause such problems as short-duration miscounts that can be decoded and erroneous counts when clocked with too high an input frequency.

SECTIONS 7-1 AND 7-2: INTERNAL SUMMARY QUESTIONS

Answers are given at the end of the chapter.

1. The MOD number of an up-counter is always equal to its maximum count.
   a. True
   b. False

2. The counter shown in Fig. 7–24 is a/an _____ -counter.
   a. up
   b. down

3. The counter shown in Fig. 7–24 is a MOD-_____ counter.
   a. 2   c. 6
   b. 4   d. 8

4. To decrement a counter means to _____ its count.
   a. increase
   b. decrease

5. The counter shown in Fig. 7–25 is a/an _____ -counter.
   a. up
   b. down

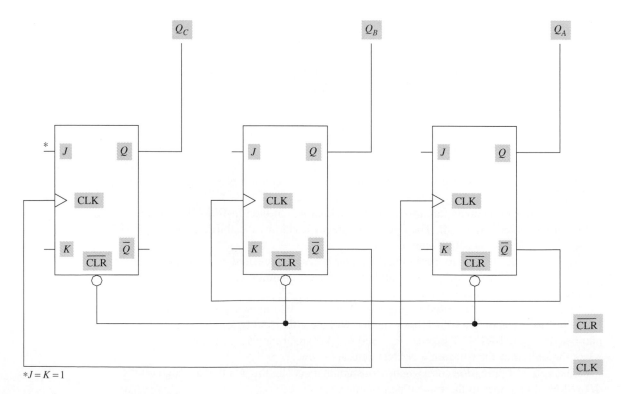

*$J = K = 1$

FIGURE 7-24

6. The MOD number of the counter shown in Fig. 7–25 is
   a. 8   c. 32
   b. 16   d. 64

7. The maximum count of the counter shown in Fig. 7–25 is
   a. 7   c. 16
   b. 15   d. 31

8. The frequency at Q_D of the counter shown in Fig. 7–25 is $f_{in} \div$ _____.
   a. 64   c. 16
   b. 32   d. 8

9. What is the count of the circuit in Fig. 7–25 after 100 clock pulses?
   a. 4   c. 12
   b. 8   d. 16

10. What is the maximum clock frequency that can be used for the counter shown in Fig. 7–25? Assume a 25-ns propagation delay per flip-flop.
    a. 4 MHz   c. 20 MHz
    b. 8 MHz   d. 40 MHz

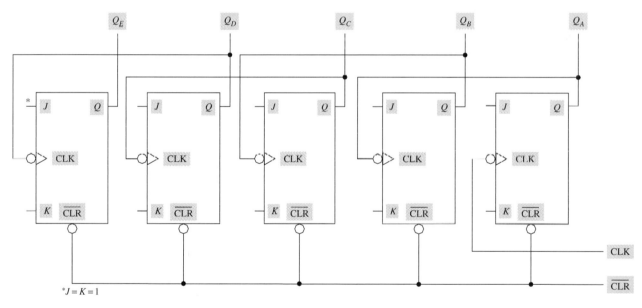

FIGURE 7–25

11. The circuit shown in Fig. 7–26 is a MOD-_____ asynchronous up-counter.
    a. 5   c. 7
    b. 6   d. 8

12. How many flip-flops are required to implement a MOD-65 counter?
    a. 5   c. 7
    b. 6   d. 8

13. What is the MOD number of the 7493 IC shown in Fig. 7–27?
    a. 4   c. 12
    b. 8   d. 16

14. What is the maximum count of the 7493 IC shown in Fig. 7–28?
    a. 7   c. 15
    b. 8   d. 16

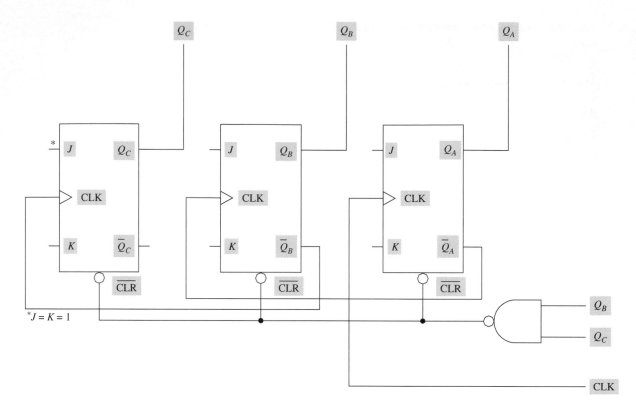

Figure 7–26

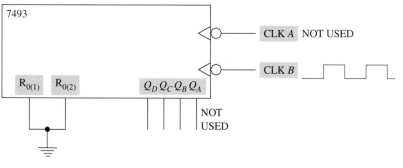

Figure 7–27

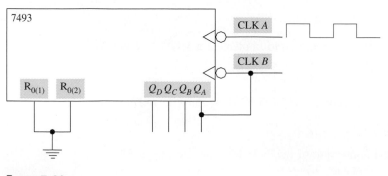

Figure 7–28

15. The ANSI/IEEE symbol shown in Fig. 7–29 contains three main blocks. The top block (labeled CTR) is the
    a. Flip-flop A block
    b. Common-control block
    c. General qualifying block
    d. Dependency notation block

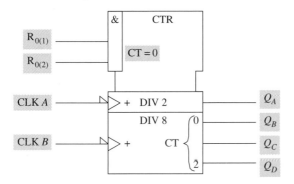

FIGURE 7–29

16. What is the MOD number of the circuit shown in Fig. 7–30?
    a. 12
    b. 36

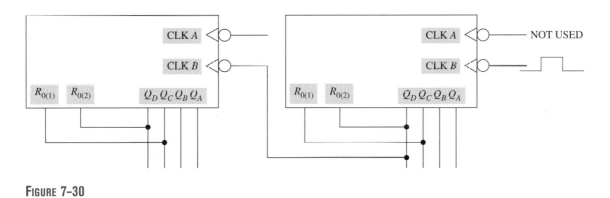

FIGURE 7–30

SECTION 7–3: SYNCHRONOUS COUNTERS

OBJECTIVES

1. Identify and determine the MOD number of synchronous up- and down-counters.
2. State the advantages and disadvantages of synchronous counters when compared to asynchronous counters.

MOD-8 Up-Counter

The term "synchronous" indicates "happening at the same time." A look at the MOD-8 synchronous up-counter in Fig. 7–31(a) reveals all three flip-flops are connected to one clock input line. The short-duration miscounts of asynchronous counters are done away with in synchronous counters because all flip-flops are clocked simultaneously. The clocking allows all flip-flops to change states at the same time. Another factor in synchronous

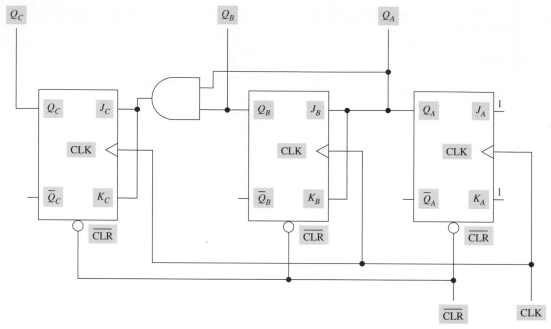

(a) LOGIC DIAGRAM

CLOCK	Q_C	Q_B	Q_A
CLR	0	0	0
t_1	0	0	1
t_2	0	1	0
t_3	0	1	1
t_4	1	0	0
t_5	1	0	1
t_6	1	1	0
t_7	1	1	1
t_8	0	0	0

(b) COUNT SEQUENCE

FIGURE 7–31   **MOD-8 synchronous up-counter.**

counter design is that interchanging NGT-triggered flip-flops for PGT-triggered flip-flops is acceptable and does not affect the counting sequence. However, all of the flip-flops utilized for a given counter must be either NGT or PGT triggered.

The only flip-flop in this type of counter that is hard-wired to toggle on the active transition of every incoming clock pulse is flip-flop A, which produces the LSB. Both J_A and K_A are connected to a Logic 1 level. The LSB in both asynchronous and synchronous counters must toggle on every clock pulse active transition to produce even, odd, even, . . . counts.

The J-K inputs in synchronous counters cannot all be hard-wired to Logic 1 because all of these flip-flops are clocked simultaneously. If they were, the counter in Fig. 7–31(a) would continually toggle from 000 to 111 to 000 on every PGT of the clock pulse.

This leaves the designer with the problem of determining when to allow the proper flip-flop to toggle. This problem will be addressed in detail shortly. For now, let's analyze the circuit in Fig. 7–31(a) and the output required from this up-counter. The count sequence is shown in Fig. 7–31(b).

The count sequence shows the *LSB toggles on every active clock transition*. It also shows that every time the LSB (Q_A) is high (after t_1, t_3, t_5, and t_7) Q_B toggles on the next incoming active clock transition (t_2, t_4, t_6, and t_8). This is accomplished by connecting the J-K inputs to flip-flop B to the Q_A output of flip-flop A.

The next problem to be considered can be clearly seen in the count sequence. The problem is when should flip-flop C toggle. The count sequence shows it must toggle on the t_4 clock pulse input and once again on the t_8 clock input. These two clock inputs are immediately preceded by counts of 011 and 111. In both of these counts, Q_A and Q_B are high. Therefore, an AND gate is connected between the Q_A and Q_B outputs and the J and K inputs of flip-flop C. The AND gate functions as the **TOGGLE control gate.** When a count of 011 or 111 occurs, the output of the AND gate goes high. This places a high on the J and K inputs of flip-flop C and prepares it to toggle on the next active clock transition which occurs at t_4 and t_8. The C flip-flop is the RETAIN mode with $J = K = 0$ when the AND gate's output is low.

Synchronous counters eliminate the additive delays encountered in their asynchronous counterparts. This is because all of the flip-flops in the counter are clocked simultaneously. The maximum clock frequency of the synchronous counter in Fig. 7–31(a) is controlled by the delay of one flip-flop plus the delay of the TOGGLE control AND gate.

If the flip-flops in Fig. 7–31(a) produce delays of 30 ns and the AND gate produces a 17-ns delay, the maximum clocking frequency is approximately 21 MHz.

$$f_{max} = 1/t_P = 1/(30 \text{ ns} + 17 \text{ ns}) = 1/47 \text{ ns} = 21.3 \text{ MHz}$$

where t_P = propagation delay of one flip-flop plus the delay of the AND gate.

The maximum clocking frequency of this synchronous counter is approximately twice that of the MOD-8 asynchronous counter using flip-flops having the same propagation delay times.

A MOD-32 *asynchronous* counter would have a 150-ns propagation delay using these same flip-flops. Its f_{max} would be 6.67 MHz. A *synchronous* MOD-32 counter designed per the previous example would still have an f_{max} of 21.3 MHz. The maximum clocking frequency of asynchronous counters continually decreases as the MOD number is increased. The previous examples show this limiting factor is overcome with synchronous counters because their propagation delays are not additive.

The disadvantage of synchronous counters is that the extra control gate(s) required for proper design increase circuit complexity. This will become increasingly evident as more complex counters are presented.

MOD-16 Up-Counter

A MOD-16 synchronous up-counter is illustrated in Fig. 7–32(a). The TOGGLE control gates are again easy to determine by using the count sequence in Fig. 7–32(b). Flip-flop C must toggle when flip-flop A and flip-flop B outputs are high, and flip-flop D must toggle when the outputs of flip-flops A, B, and C are high. This toggling control is implemented with the two AND gates shown in Fig. 7–32(a). The output waveforms of this counter are shown in Fig. 7–32(c). Notice the flip-flop propagation delays are shown in the drawing (see t_1), but the delays are not additive.

MOD-10 Up-Counter

One of the more popular counters in use today is a MOD-10 up-counter. This counter is called a **BCD** or **decade** counter because it counts 0 through 9. The counter is often used with a decimal readout such as the seven-segment indicator. This MOD-10-up counter logic diagram is shown in Fig. 7–33(a). Its count sequence is shown in Fig. 7–33(b).

The count sequence shows the LSB toggles on the active clock transition of every input clock pulse. It also shows that every time Q_A is high and Q_D is low (after t_1, t_3, t_5, and t_7) Q_B toggles on the next incoming active clock transition (t_2, t_4, t_6, and t_8). The toggling of the B flip-flop is controlled by the AND gate at its J_B input in addition to the Q_A level connected to K_B.

The count sequence also shows that flip-flop C toggles only when $A = B = 1$. This occurs only on the next active-clock transition after the counts of 0011 and 0111. Flip-flop D toggles from low-to-high (Q_D) at t_8. This is because Q_A, Q_B, and Q_C are high after clock

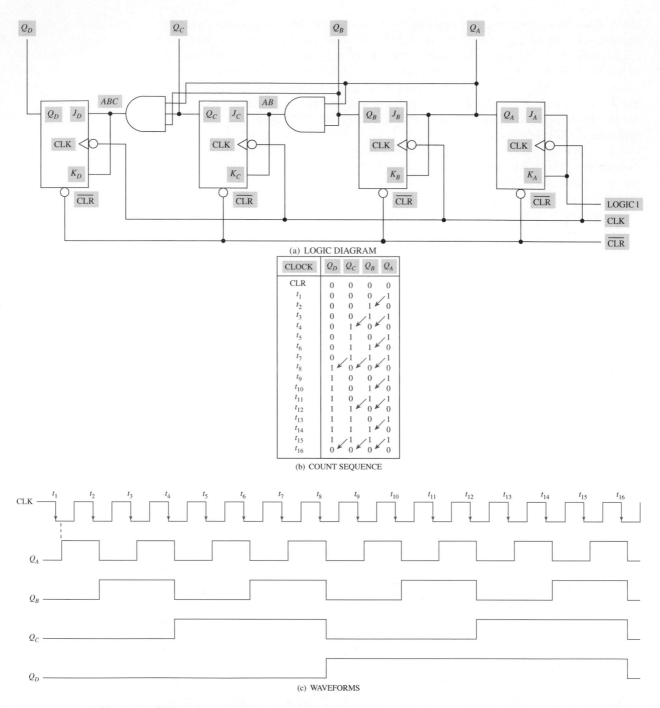

(a) LOGIC DIAGRAM

CLOCK	Q_D	Q_C	Q_B	Q_A
CLR	0	0	0	0
t_1	0	0	0	1
t_2	0	0	1	0
t_3	0	0	1	1
t_4	0	1	0	0
t_5	0	1	0	1
t_6	0	1	1	0
t_7	0	1	1	1
t_8	1	0	0	0
t_9	1	0	0	1
t_{10}	1	0	1	0
t_{11}	1	0	1	1
t_{12}	1	1	0	0
t_{13}	1	1	0	1
t_{14}	1	1	1	0
t_{15}	1	1	1	1
t_{16}	0	0	0	0

(b) COUNT SEQUENCE

(c) WAVEFORMS

FIGURE 7-32   **MOD-16 synchronous up-counter.**

pulse t_7. The only counter change at t_9 is the toggling of the LSB (Q_A) to a high. After t_9 the output of the 3-input AND gate connected to J_D is low because Q_B and Q_C are low. However because K_D is high due to Q_A's previous toggling at t_9, flip-flop D will clear at t_{10}.

The count sequence is derived by controlling the J-K inputs to the flip-flops as follows:

Q_A is derived by $J_A = K_A = 1$, which produces the required toggling action on each active clock transition.

Q_B is derived by $J_B = A\overline{D}$ and $K_B = A$.

Q_C results from $J_C = K_C = AB$, which causes Q_C to toggle after the counts of 00$\underline{1}$ $\underline{1}$ and 01$\underline{1}$ $\underline{1}$.

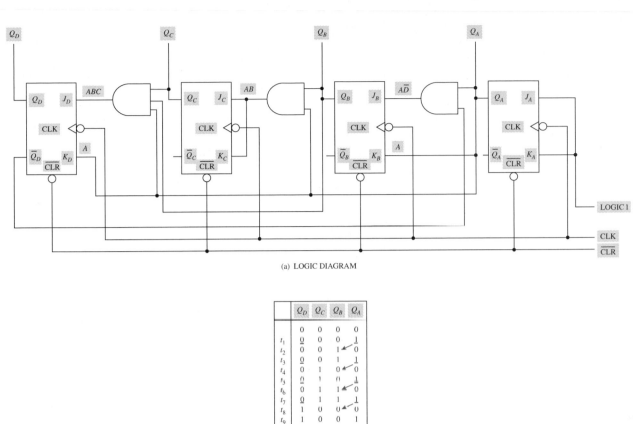

(a) LOGIC DIAGRAM

	Q_D	Q_C	Q_B	Q_A
	0	0	0	0
t_1	0	0	0	1
t_2	0	0	1	0
t_3	0	0	1	1
t_4	0	1	0	0
t_5	0	1	0	1
t_6	0	1	1	0
t_7	0	1	1	1
t_8	1	0	0	0
t_9	1	0	0	1
t_{10}	0	0	0	0

(b) COUNT SEQUENCE

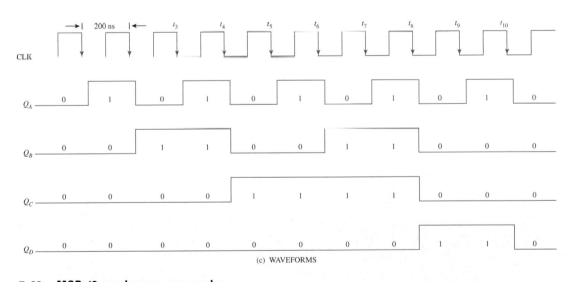

(c) WAVEFORMS

FIGURE 7–33   MOD-10 synchronous up-counter.

Q_D toggles high as a result of $A = B = C = 1$. It returns low after the t_9 active clock transition causes Q_A to toggle high and upon the arrival of the t_{10} clock transition. This clears flip-flop D.

The waveforms for this counter are presented in Fig. 7–33(c). In an asynchronous MOD-10 counter, the short-duration count of 1010 would occur and produce a spike at the Q_B output before the counter cleared. Notice there is no spike generated at t_{10} by this counter—another advantage of synchronous counters.

This section has presented three synchronous up-counters. Design of these types of counters will be presented in Section 7–6.

MOD-8 Down-Counter

A MOD-8 synchronous down-counter logic diagram is shown in Fig. 7–34(a). This down-counter, like its asynchronous counterpart, uses the $\overline{Q}$ outputs of the flip-flops to control the TOGGLE operation.

The LSB flip-flop is connected permanently in the TOGGLE mode of operation. The count sequence depicted in Fig. 7–34(b) shows the output count is taken from the Q out-

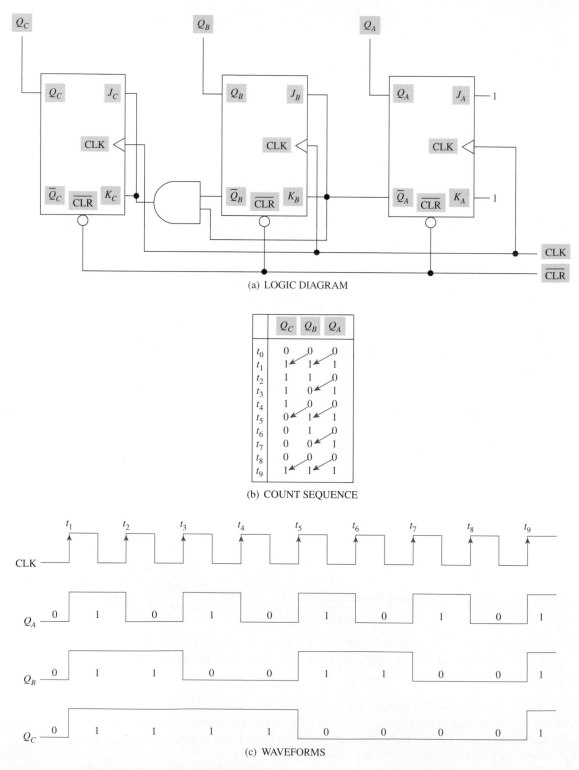

(a) LOGIC DIAGRAM

	Q_C	Q_B	Q_A
t_0	0	0	0
t_1	1	1	1
t_2	1	1	0
t_3	1	0	1
t_4	1	0	0
t_5	0	1	1
t_6	0	1	0
t_7	0	0	1
t_8	0	0	0
t_9	1	1	1

(b) COUNT SEQUENCE

(c) WAVEFORMS

FIGURE 7–34   **MOD-8 synchronous down-counter.**

put of each flip-flop. It also shows that each time the LSB (Q_A) is low (after t_0, t_2, t_4, t_6, and t_8) Q_B toggles on the next active clock transition (t_1, t_3, t_5, t_7, and t_9) because $\overline{Q_A}$ is high during these active clock transitions. Note this action is achieved by connecting J_B and K_B directly to the $\overline{Q_A}$ output.

The count sequence also shows flip-flop C must toggle on clock pulses t_1, t_5, and t_9. These clock pulses are preceded by counts 000, 100, and again by 000. The counts indicate Q_A and Q_B are low during these conditions. Thus, an AND gate is connected to the complement side of flip-flops A and B. Since $\overline{Q_A}$ and $\overline{Q_B}$ are high after t_0, t_4, and t_8, flip-flop C toggles at t_1, t_5, and t_9. Keep in mind that the B and C flip-flops are in the RETAIN mode when $J = K = 0$.

Figure 7–34(c) shows the waveforms for this counter. The output frequencies are $Q_A = f_{in}/2$, $Q_B = f_{in}/4$, and $Q_C = f_{in}/8$. The maximum clocking frequency for this down-counter is calculated in the same manner described for the synchronous up-counter.

The only difference between the synchronous up-counter and the synchronous down-counter is the use of Q or $\overline{Q}$ to control the toggling action. Therefore, it will be a simple matter to add some control gates to allow the counter to be used as an up/down-counter. This modification will be addressed in Section 7-4.

Advantages/Disadvantages of Synchronous Counters

The advantages and disadvantages of synchronous counters compared to asynchronous counters have been discussed as they were encountered in this section. They are summarized here for your convenience.

All flip-flops in a synchronous counter are clocked at the same time, so propagation delays are not additive. This eliminates the short-duration miscounts that are prevalent in asynchronous counters. In addition, the synchronous counter can be clocked at a higher frequency than a comparable MOD number asynchronous counter.

Design of synchronous counters (Section 7–6) is more complex than design of asynchronous counters. Implementation of the synchronous counter requires more gates to control the toggling action. However, this is not considered much of a problem because most of the popular counter versions are available in TTL and CMOS ICs.

Section 7–3: Review Questions

Answers are given at the end of the chapter.

A. Identify the circuit shown in Fig. 7–35.
B. What is the count of the circuit shown in Fig. 7–35 after 3 clock pulses?

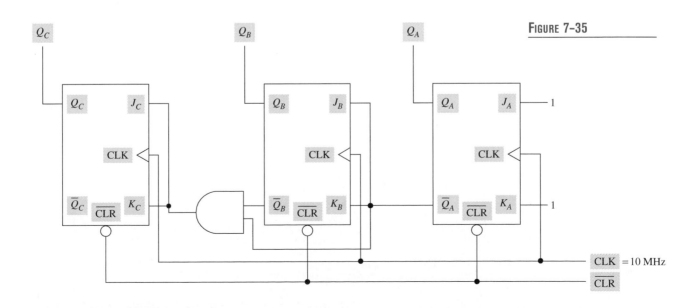

FIGURE 7–35

C. What is the output frequency of the counter shown in Fig. 7–35 at
    (1) Q_A?
    (2) Q_B?
    (3) Q_C?

D. When looking at a logic diagram, what feature characterizes a synchronous counter?
E. List the advantages of synchronous counters over asynchronous counters.

SECTION 7-4: SYNCHRONOUS UP/DOWN-COUNTERS

OBJECTIVE

Identify and determine the MOD number and count sequence (up or down) of synchronous up/down-counters.

Up/down-counters are versatile because they can increment or decrement their counts depending on a control input. Several IC versions of up/down-counters are commercially available. Most of these ICs are also programmable to allow short-counting by modifying the count length with preset inputs. A simple up/down-counter will be presented in this section. IC versions of the programmable up/down-counters will be discussed in the next section.

A quick review will be provided at this point to ensure your understanding of both the up- and down-counters before they are meshed into a single unit. The count sequence (Fig. 7–36) for a MOD-8 synchronous up-counter indicates:

1. The LSB flip-flop must toggle on every active clock transition. This is accomplished by connecting J and K high.
2. The B flip-flop must toggle on the next active clock transition after Q_A has been set high. This is achieved by connecting the B flip-flop J and K inputs to Q_A.
3. The C flip-flop must toggle on the next active clock transition after Q_A and Q_B are set high. This is achieved by connecting Q_A and Q_B to an AND gate and routing its output to the J and K inputs of flip-flop C.

Note: The Q outputs are always used to control the toggling operation in synchronous up-counters.

The count sequence (Fig. 7–36) for a MOD-8 synchronous down-counter indicates:

1. The LSB flip-flop must toggle on every active clock transition. This is accomplished by connecting J and K high.

	UP			DOWN		
	Q_C	Q_B	Q_A	Q_C	Q_B	Q_A
CLR	0	0	0	0	0	0
t_1	0	0	1	1	1	1
t_2	0	1	0	1	1	0
t_3	0	1	1	1	0	1
t_4	1	0	0	1	0	0
t_5	1	0	1	0	1	1
t_6	1	1	0	0	1	0
t_7	1	1	1	0	0	1
t_8	0	0	0	0	0	0

FIGURE 7-36   Up/down-count sequence.

2. The B flip-flop must toggle on the next active clock transition after Q_A is reset low. This is achieved by connecting the B flip-flop J and K inputs to $\overline{Q_A}$.

3. The C flip-flop must toggle on the next active clock transition after both Q_A and Q_B are reset low. This is accomplished by connecting the $\overline{Q_A}$ and $\overline{Q_B}$ outputs to an AND gate and routing its output to the J and K inputs of flip-flop C.

Note: The $\overline{Q}$ outputs are always used to control the toggling operation in synchronous down-counters.

Both the up- and down-count features can be incorporated in a counter with the addition of some logic control gates. The key is to direct the Q or the $\overline{Q}$ outputs to control the flip-flop toggling operation.

A MOD-8 synchronous up/down-counter is shown in Fig. 7–37(a). Operation of this counter as an up-counter and as a down-counter has already been presented. Therefore, only the control circuitry will be discussed in this section. The input control line is labeled with the mnemonic UP/$\overline{\text{DN}}$. This label indicates the counter will count up when the control line is high, and it will count down when the line is low.

When the control line is high (UP), The #1 control gate is enabled and the #2 control gate is inhibited. The low output from the #2 control gate is used to enable the #3 OR gate and inhibit the #5 AND gate. Since both the #2 and #5 gates are inhibited, the $\overline{Q}$ outputs of flip-flops A and B are not used. The enabler applied to the #1 control gate allows Q_A to be passed to the J and K inputs of flip-flop B. It also allows Q_A and Q_B to be routed to the J and K inputs of flip-flop C to control that flip-flop's toggling operation. The use of Q_A and Q_B allows up-count operation.

When the control line is low ($\overline{\text{DN}}$), the #2 control gate is enabled and the #1 control gate is inhibited. This allows the use of $\overline{Q_A}$ and $\overline{Q_B}$ to control the flip-flop's toggling operation and results in *down-count operation.*

If the input control signal level is changed during the counting operation, the counting sequence is reversed. This is illustrated in Fig. 7–37(b). The counter starts in the CLEAR state, increments to a count of 101 when UP/$\overline{\text{DN}}$ = 1, decrements to a count of 000 when UP/$\overline{\text{DN}}$ = 0, and then increments to 110 when the UP/$\overline{\text{DN}}$ control line is returned high as shown by the output waveforms in the figure.

Section 7–4: Review Questions

Answers are given at the end of the chapter.

A. Synchronous down-counters are controlled by using the _____ outputs to control the toggling action.
   (1) $\overline{Q}$
   (2) Q

B. What is the MOD number of the circuit shown in Fig. 7–38?

C. The circuit shown in Fig. 7–38 will count _____ when the UP/$\overline{\text{DN}}$ control input is high.
   (1) up
   (2) down

D. Determine the terminal count (after t_{10}) of the circuit shown in Fig. 7–38. (*Note:* Start the analysis in the CLEAR state.)

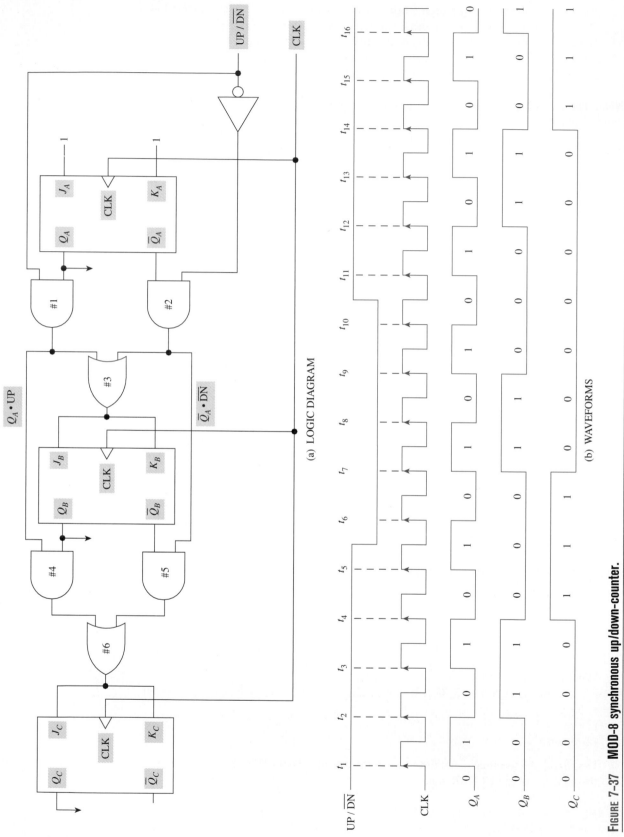

(a) LOGIC DIAGRAM

(b) WAVEFORMS

FIGURE 7-37   MOD-8 synchronous up/down-counter.

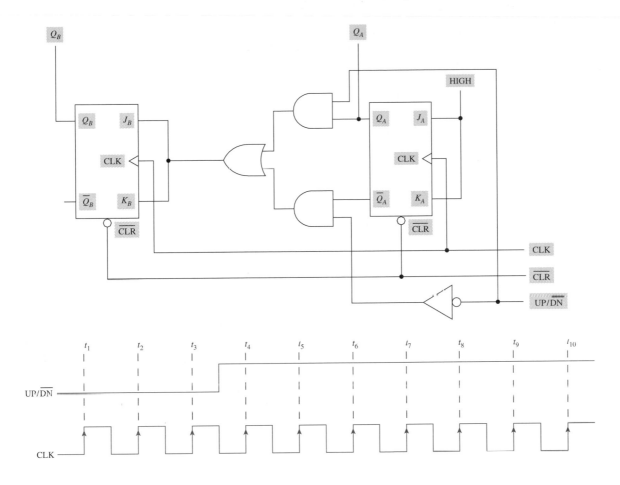

FIGURE 7-38

FIGURE 7-38

SECTION 7-5: SYNCHRONOUS PROGRAMMABLE COUNTERS

OBJECTIVES

1. Given the logic symbol or diagram and function table for a synchronous programmable counter, determine the required inputs to establish count direction and a specified MOD.
2. Given a data book and count requirements, determine the connections and preload data required to cascade two or more counter ICs that will produce the desired counts.

Programmable counters are extremely popular in the digital field as evidenced by the large variety available commercially in ICs. The capability to **preload** a number into a counter to establish its initial count allows flexible programmability in setting up various MOD numbers. After an initial number is preloaded, the normal up- or down-count sequence commences with the preloaded number and continues until the terminal count is reached. For example, if a 3-flip-flop up-counter is preloaded with a $011_{(2)}$, the counter will count 011, 100, 101, 110, and 111. However, if the initial preset number of 011 is not reloaded after the terminal count of 111 is reached, the counter will recycle to 000.

The problem is to repetitively reload the preset number after each terminal count is reached, and it is a relatively simple problem to solve, as will be seen shortly. It has been made even easier to solve thanks to the ingenuity of IC designers. Several IC counters will be presented in this section. First, the basic concept of presetting the initial count into a counter will be presented.

Presettable Counters

Figure 7–39 shows a MOD-8 synchronous programmable up-counter. The addition of the NAND gates to the standard synchronous up-counter allows for asynchronous preloading of a number. The Q outputs of this counter have been designated Q_2, Q_1, and Q_0. This is an acceptable practice and can be noted in most any data book. The output count of the circuit is still taken from the Q side of the flip-flops.

The NAND gates in Fig. 7–39 are enabled when the $\overline{\text{LOAD}}$ input is asserted (brought low). This activates the asynchronous flip-flop inputs ($\overline{\text{PRE}}$ or $\overline{\text{CLR}}$) and overrides the clock. The counter will not count when the $\overline{\text{LOAD}}$ input line is active.

If the binary number 011 ($P_2 = 0$, $P_1 = 1$, and $P_0 = 1$) is set at the preload inputs it will be asynchronously loaded when $\overline{\text{LOAD}}$ is asserted.

When $\overline{\text{LOAD}} = 0$, the 0 input at P_2 will cause a 1 at the $\overline{\text{PRE}}$ input of the Q_2 flip-flop. This is an inactive input to this pin. The inverted 0 out of the NOT gate along with the $\overline{\text{LOAD}}$ enabler causes a 0 out of that NAND gate to be applied to the $\overline{\text{CLR}}$ input of the Q_2 flip-flop. This clears the flip-flop so that $Q_2 = 0$.

With $\overline{\text{LOAD}}$ still asserted, the 1 input at P_1 causes the NAND gate's output to go low at the $\overline{\text{PRE}}$ input of the Q_1 flip-flop. This asserted input sets the flip-flop so that $Q_1 = 1$. At the same time the inverted P_1 bit from the NOT gate causes the output of that NAND gate to stay high, which keeps the $\overline{\text{CLR}}$ input inactive. This same action causes Q_0 to set to 1 when $P_0 = 1$.

The binary number placed on the preload inputs of a programmable counter will always be loaded when the $\overline{\text{LOAD}}$ line is activated. The inverters placed on the preload lines ensure that only one of the flip-flop's asynchronous inputs is activated at any given time.

If the $\overline{\text{LOAD}}$ input is activated per the previous example, the counter goes to a count of 011. Assuming the CLK input is continuously applied, the counter will increment to 111 when $\overline{\text{LOAD}} = 1$. When this terminal count is reached the counter will reset to 000 and recycle. Now comes the problem of repetitively reloading the 011 after every terminal count of 111.

A modified version of the programmable counter just discussed is shown in Fig. 7–40(a). The loading problem has been solved by replacing the NOT gate on the loading line with a NOR gate. The inputs to this gate are connected to the Q_2, Q_1, and Q_0 outputs of the flip-flops. The desired count sequence for this circuit is shown in Fig. 7–40(b). The rule of truncation forms the basis for solving this preload problem. The desired count is 011 to 111, 011 to 111, and so on. If the load gates are not enabled after a count of 111, the counter will not count properly.

The next normal sequential count of this counter after 111 is 000. Thus, when this count occurs the output of the NOR gate goes high, enables the preload gates, and asynchronously reloads the 011 into the counter.

The rule of truncation presented for asynchronous up-counters stated "tie the bits that are high on the desired MOD number to the truncating circuit." This rule cannot be applied per se in programmable counters, but the synthesis can be used. In an asynchronous up-counter, the MOD number was always one higher than the desired maximum count. Thus the MOD number was used for truncating purposes. The same line of thinking applies to these programmable up-counters. The counter in the preceding example was to count to 111 and then reload to 011. Therefore, logical deduction indicates the next count above the desired reload state should be connected to the truncating circuit. In the previous example, that next count was 000, and when it occurred the NOR gate reloaded the 011.

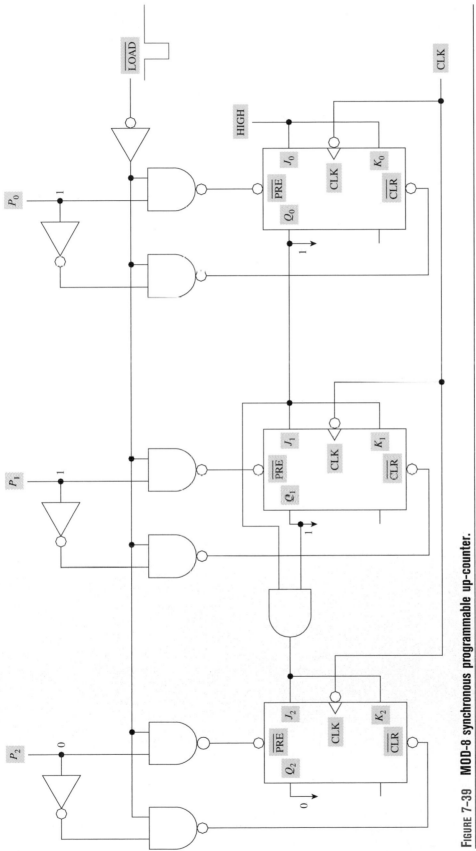

FIGURE 7-39   MOD-8 synchronous programmable up-counter.

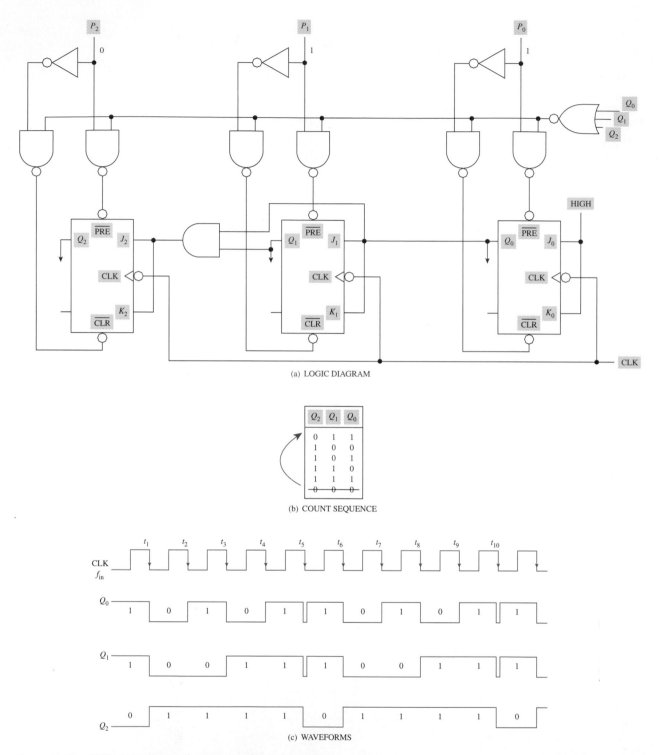

(a) LOGIC DIAGRAM

Q_2	Q_1	Q_0
0	1	1
1	0	0
1	0	1
1	1	0
1	1	1
0	0	0

(b) COUNT SEQUENCE

(c) WAVEFORMS

FIGURE 7–40   **MOD-5 synchronous programmable up-counter.**

The waveforms in Fig. 7–40(c) show the counting action of this circuit. Note the short spikes that occur at Q_0 and Q_1 after t_5 and t_{10}. Since the counter sequence is 011, 100, 101, 110, and 111, it is a MOD-5 counter. This is validated by the Q_2 output waveform in the figure ($f_{in} \div 5$).

Let's assume you had to modify the circuit shown in Fig. 7–40(a) to a MOD-4 counter. What circuit changes would have to be made? The only change necessary to make this a MOD-4 counter is to set the preload inputs to 100 ($P_2 = 1$, $P_1 = 0$, and $P_0 = 0$). This will allow the counter to count 100, 101, 110, and 111 and then recycle to 100 and continue counting. This should clearly show the advantage this counter has over those previously presented.

A MOD-6 synchronous programmable down-counter is shown in Fig. 7-41(a). Operation of the circuit is basically the same as the synchronous programmable up-counter. The count sequence in Fig. 7-41(b) shows the desired count necessary to produce the six states is 101, 100, 011, 010, 001, and 000. Note the next sequential count after 000 is 111. This count must be used to produce a high to enable the preload gates. The NOR gate used in the up-counter won't work is this circuit because its logic is "all 0s in = 1 out." The AND gate was selected because its logic is "all 1s in = 1 out." Thus, when the counter goes to a count of 111, the count of 101 is immediately preloaded by the enabled preload NAND gates.

Synchronous IC Counters

74193/74LS193

This synchronous 4-bit reversible (up/down) binary counter is fully programmable with preset inputs. The MOD may be changed by changing the binary preset number. This counter was designed to be cascaded without the need for external circuitry.

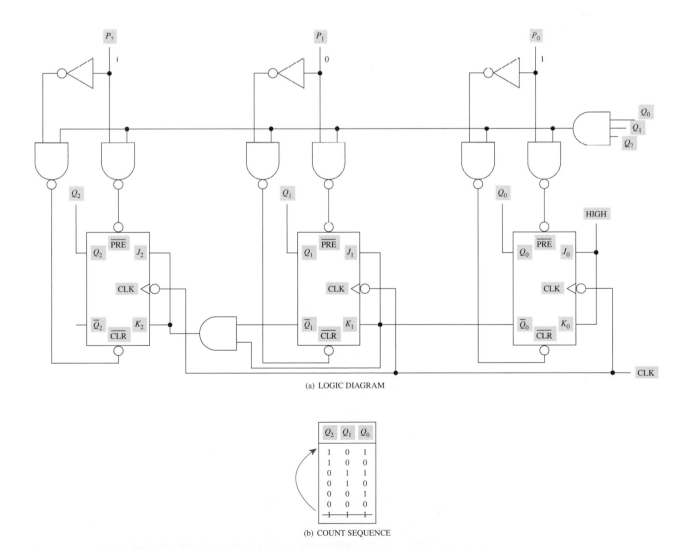

(a) LOGIC DIAGRAM

Q_2	Q_1	Q_0
1	0	1
1	0	0
0	1	1
0	1	0
0	0	1
0	0	0
1	1	1

(b) COUNT SEQUENCE

FIGURE 7-41   **MOD-6 synchronous programmable down-counter.**

Figure 7–42 shows the logic diagram of the counter. The clock input is applied to the Up (pin 5) or Down (pin 4) input to set the count direction. The *unused clock input must be connected high* for proper count operation. If the unused clock input is taken low for any reason the counter will not operate properly. If the unused clock input is taken low on an *even count* (LSB = 0), the counter will stop counting. If it is taken low on an *odd count* (LSB = 1) during up-counting, it will odd-count only.

Although theoretically proving the odd-count process when the clock is taken low provides a good exercise in tracing logic on the diagram, it will be left to you. Suffice it to say the counter will not operate properly if the unused clock input is not tied high. The desired clock signal (up or down) is applied to all four flip-flops simultaneously.

An active-high CLR input (pin 14) resets the counter. It overrides the clock and $\overline{\text{LOAD}}$ inputs. The $\overline{\text{LOAD}}$ input (pin 11) asynchronously loads the desired preset number into the counter to set its MOD when it is activated. This is accomplished by enabling the preload NAND gates as previously explained. The preset numbers are applied at the data input A,

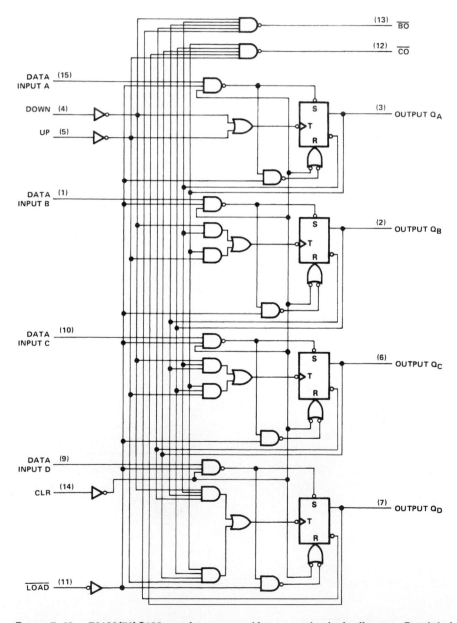

FIGURE 7–42   74193/74LS193 synchronous up-/down-counter logic diagram. *Reprinted by permission of Texas Instruments.*

B, C, and D pins 15, 1, 10, and 9. Data input A is the LSB and data input D is the MSB. The $\overline{\text{LOAD}}$ input overrides the active-clock input.

The counter produces a **Carry Output** ($\overline{\text{CO}}$—pin 12) and a **Borrow Output** ($\overline{\text{BO}}$—pin 13). The active-low Carry Output is generated when an **overflow condition** exists. In other words, every time the terminal count of 1111 is reached during an up-count operation, the Carry Output is activated. Every time the terminal count of 0000 is reached during a down-count operation, the Borrow Output is activated. These two outputs allow simple cascading of the counters. In addition, they provide a built-in preloading control signal, which does away with the requirement for external truncating circuitry.

The count output is taken from Q_A–Q_D (pins 3, 2, 6, and 7). Q_A is the LSB and Q_D is the MSB.

The logic symbol for this counter is shown in Fig. 7–43. This symbol is the one that will be used in digital schematics to represent the counter. The mnemonics on the symbol indicate almost everything a technician needs to know about the counter. However, the technician would have to check the circuit description in a data book to know the unused clock input must be tied high.

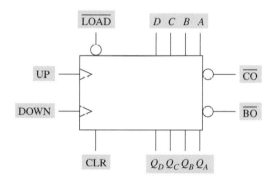

FIGURE 7–43   74193/74LS193 synchronous up-/down-counter logic symbol.

Another point to keep in mind is that different manufacturer's data books often use different pin names from those furnished in this text. Table 7–1 illustrates some of the variations encountered in different data books for this IC. The first column contains the names of the pins used in this text, which were extracted from the *Texas Instruments TTL Logic Data Book*.

Sometimes the pin names furnished in a data book do not quickly identify the use of the pin or whether it is an input or output pin. For example, if you were unaware that the 74193 Up/Down pins were for the clock input, the logic diagram in Fig. 7–42 would solve the problem.

TABLE 7-1	Pin #				Generic Name
74193 Counter—	11	$\overline{\text{LOAD}}$	$\overline{\text{LOAD}}$	$\overline{\text{PL}}$	Parallel load/preset load
Various Pin Names	15	A	A	P_0	Preset number (LSB)
	1	B	B	P_1	Preset number
	10	C	C	P_2	Preset number
	9	D	D	P_3	Preset number (MSB)
	4	Down	Count down	CP_D	Clock—down-count
	5	Up	Count up	CP_U	Clock—up-count
	3	Q_A	Q_A	Q_0	Output count (LSB)
	2	Q_B	Q_B	Q_1	Output count
	6	Q_C	Q_C	Q_2	Output count
	7	Q_D	Q_D	Q_3	Output count (MSB)
	14	CLR	CLEAR	MR	Clear/reset
	13	$\overline{\text{BO}}$	$\overline{\text{Borrow}}$	$\overline{\text{TC}_D}$	Borrow Out/terminal count down
	12	$\overline{\text{CO}}$	$\overline{\text{Carry}}$	$\overline{\text{TC}_U}$	Carry Out/terminal count up

The terminal count circuits for the 74193 are shown in Fig. 7–44. *The $\overline{CO}$ signal is generated when the up-clock is low and the Q_A–Q_D counter outputs are high (1111).* This condition inputs all highs to the NAND gate and causes its output to go low. *The $\overline{BO}$ signal is activated when the down-clock is low and the Q_A–Q_D counter outputs are low.* The 0000 counter output and the low down-clock puts all highs into the NAND gate because the clock pulse is inverted and $\overline{Q}$ outputs are used from the counter. For the following discussion, keep in mind that the input clock pulse (Up or Down) must be on the low portion of its cycle to generate the terminal count.

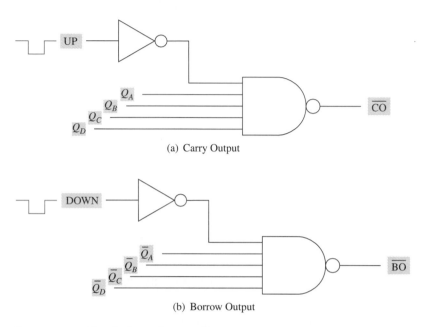

(a) Carry Output

(b) Borrow Output

FIGURE 7–44   Terminal count generation.

A cascaded counter consisting of two MOD-16 74193 ICs is shown in Fig. 7–45(a). The counters are both configured to up-count. The Carry Output of the lower 4-bit counter (Q_A–Q_D) is used to clock the upper 4-bit counter (Q_E–Q_H). The Q outputs of the upper 4-bit counter have been designated Q_E through Q_H for this discussion. The counts of each counter are shown directly below the count output pins. The circuit is a MOD-256 counter that can count from 00000000 to 11111111 ($255_{(10)}$).

The timing waveforms shown in Fig. 7–45(b) show the timing relationship of the $\overline{CO}$ signal to the clock pulse. The PGT of the input up-clock pulse increments the lower 4-bit counter all the way up to 1111 (terminal count for an up-counter). The $\overline{CO}$ signal does not go active immediately upon generation of the terminal count at the PGT of the clock pulse. Instead it goes active half a clock cycle later when the clock pulse goes low. When the $\overline{CO}$ signal is activated, its lagging edge (PGT) increments the upper 4-bit counter to a count of 0001 as shown in the count sequence in Fig. 7–45(a). The $\overline{CO}$'s pulse width is equal to the negative duration of the up-clock input.

A cascaded MOD-256 down-counter implemented with 74193 ICs is shown in Fig. 7–46. The Borrow Output signal of the lower 4-bit counter (Q_A–Q_D) is used to clock the upper 4-bit counter. The counts of each counter are shown below the counter output pins. The down-count shown in the figure has arbitrarily been started at $00100000_{(2)}$. The active-low $\overline{BO}$ signal from the lower counter is generated on the terminal count of 0000 when the clock goes low. This clocks the upper 4-bit down-counter to decrement from 0010 to 0001 and from 0001 to 0000 as shown in the figure.

The MOD number of a programmable counter can normally be determined from the preloaded number and the terminal count. If a $1010_{(2)}$ were preloaded in a typical pro-

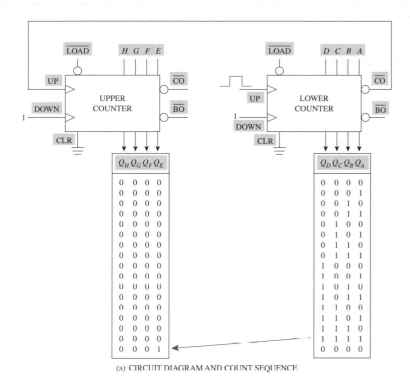

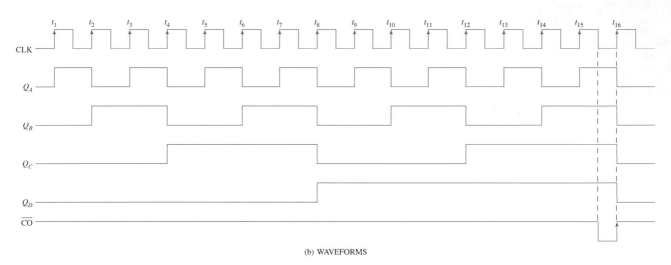

(a) CIRCUIT DIAGRAM AND COUNT SEQUENCE

(b) WAVEFORMS

FIGURE 7-45   74193 ICs cascaded—MOD-256 up-counter.

grammable up-counter, the counter would count 1010, 1011, 1100, 1101, 1110, and 1111, and then return to the preloaded 1010. This would produce a MOD-6 counter.

The 74193 is different when $\overline{CO}$ or $\overline{BO}$ is used to preload the counter. The logic symbol in Fig. 7–47(a) shows the clock pulse is applied to the Up input and the Down input is tied high. This configures the circuit as an up-counter. The Carry Output ($\overline{CO}$) is connected to the $\overline{LOAD}$ input. The $\overline{CO}$ signal goes low during every terminal count of 1111. Therefore, every time the terminal count occurs, $\overline{CO}$ signal is activated, which asserts $\overline{LOAD}$, and the preset 1010 is reloaded into the counter. The difference alluded to earlier stems from the timing of the terminal count and the preload action. Instead of being a MOD-6 counter as mentioned in the previous paragraph, this is a MOD-5 counter.

The timing waveforms for this circuit are shown in Fig. 7–47(b). The waveforms commence with the preload count of 1010 and increment to 1111. When the clock goes low

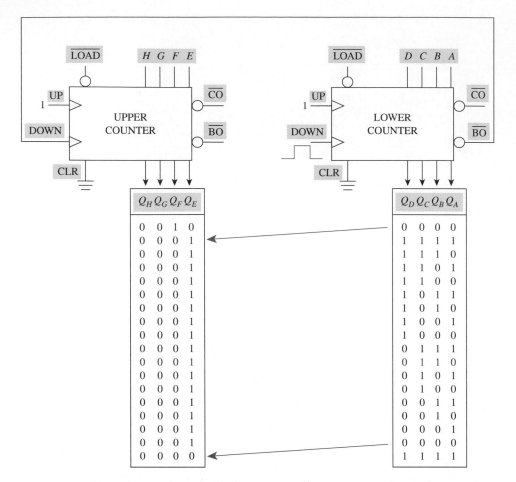

FIGURE 7–46 74193 ICs cascaded—MOD-256 down-counter circuit diagram and count sequence.

after t_5, the $\overline{\text{CO}}$ signal is activated. This loads the 1010 immediately back into the counter. At t_6 the counter increments to 1011.

The terminal count of 1111 and the preload count of 1010 last only one-half clock cycle each. Since these two counts occur during one cycle, the counter is a MOD-5 up-counter. This is shown in the waveform at the Q_C output, which is $f_{\text{in}}/5$. Q_D stays high during the entire count sequence. Even though six different states are shown in the waveforms in Fig. 7–47(b), only five clock pulses put the counter through these six states.

```
1010 – – – – – – – – – – – – – – – – – – – LOAD
1011 – – – – – – – – – – – – – – – – – – – – t₁
1100 – – – – – – – – – – – – – – – – – – – t₂
1101 – – – – – – – – – – – – – – – – – – – t₃
1110 – – – – – – – – – – – – – – – – – – – t₄
1111 – – – – – 1010– – – – – – – – – – – – t₅
```

Thus, the circuit's frequency divide-by capability is 5.

The MOD number of a programmed 74193 up-counter can be calculated by subtracting the preloaded number from the terminal count of 1111.

$$
\begin{array}{rcr}
\text{Terminal count} & = & 1111 \\
\text{Preloaded number} & = & -1010 \\
\hline
\text{MOD} & = & 0101
\end{array}
$$

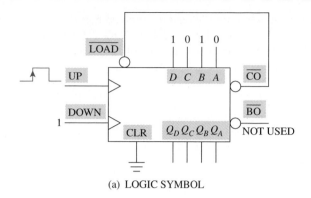

(a) LOGIC SYMBOL

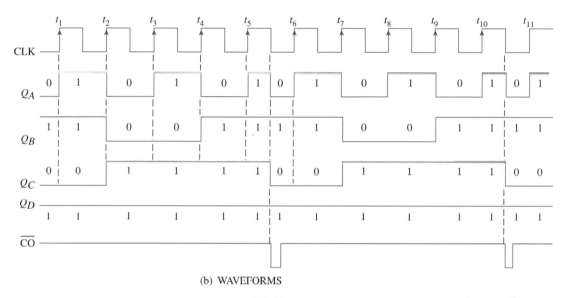

(b) WAVEFORMS

FIGURE 7–47 Programmed 74193 MOD-5 up-counter.

A similar line of thinking can be pursued for the 74193 down-counter. Figure 7–48 shows a MOD-5 down-counter. A binary 0101 is present at the preload inputs, and the $\overline{BO}$ is connected to $\overline{LOAD}$. The counter will count down as shown:

```
0101 – – – – – – – – – – – – – – – – – – – LOAD
0100 – – – – – – – – – – – – – – – – – – – t₁
0011 – – – – – – – – – – – – – – – – – – – t₂
0010 – – – – – – – – – – – – – – – – – – – t₃
0001 – – – – – – – – – – – – – – – – – – – t₄
0000 – – – – – 0101 – – – – – – – – – – – – t₅
```

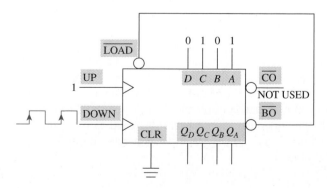

FIGURE 7–48 Programmed 74193 MOD-5 down-counter.

Once again the terminal down-count of 0000 and the preload number of 0101 are present for only one-half clock cycle each. Therefore, this is a MOD-5 down-counter. The MOD number of a programmed 74193 down-counter is equal to the preloaded input binary number.

74193/74LS193 ANSI/IEEE Symbol.   The ANSI/IEEE symbol for the 74193 counter is shown in Fig. 7–49. The fact that it is a MOD-16 counter is shown by CTRDIV 16 in the common-control block. Keep in mind this block indicates its inputs are common to all circuits in the counter. The CT = 0 indicates the count is 0 when the CLR input is high. $\overline{1}$CT = 15 shows the $\overline{CO}$ output is low when the count is 15 *and* the Up input (G_1) is low. The G represents AND dependency. $\overline{2}$CT = 0 shows the $\overline{BO}$ output is low when the count is 0 and the Down input (G_2) is low. The + and − signs in the common-control block indicate up-count or down-count respectively.

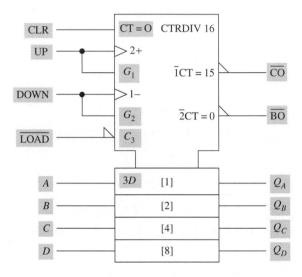

FIGURE 7-49   74193/74LS193 ANSI/IEEE symbol.

The 2 at the + sign shows the Down input must be high (G_2) for the circuit to up-count. Likewise, the "1" at the − sign shows the Up input must be high (G_1) to enable down-counting. This can be related to the previous discussion, when it was stated the unused clock input must be tied high for the circuit to count properly.

This symbol also incorporates **dependency notation.** Dependency notation shows relationships between inputs, outputs, or inputs and outputs in a relatively simple manner. The letter C indicates **control dependency.** The control dependency label C_3 along with the 3D label in flip-flop A indicates all of the flip-flops will be loaded when $\overline{LOAD}$ is brought low. The [1], [2], [4], and [8] indicate the bit position weight of each flip-flop in the counter; Q_A is the LSB and Q_D is the MSB.

Look-Ahead Carry.   Some counters employ a look-ahead carry feature, which allows them to operate at faster clock speeds. Figures 7–50(a) and (b) show two MOD-16 synchronous counters. Both counters are drawn with the LSB (Q_A) on the left. The logic diagram shown in Fig. 7–50(a) has an AND gate connected at its Q_D output to generate a Carry Output when the terminal count of $1111_{(2)}$ is reached. This is similar to the $\overline{CO}$ generated by the previously discussed 74193 IC except this Carry Output is active high.

The logical *AB* output that controls *J* and *K* of flip-flop C can be considered a carry input to that circuit. Also, the logical *ABC* that controls the *J* and *K* inputs of flip-flop D can be called a carry. Although synchronous counters were designed to overcome the cumulative propagation delays of flip-flops, the Carry Output generated in Fig. 7–50(a) hasn't overcome additive delay problems.

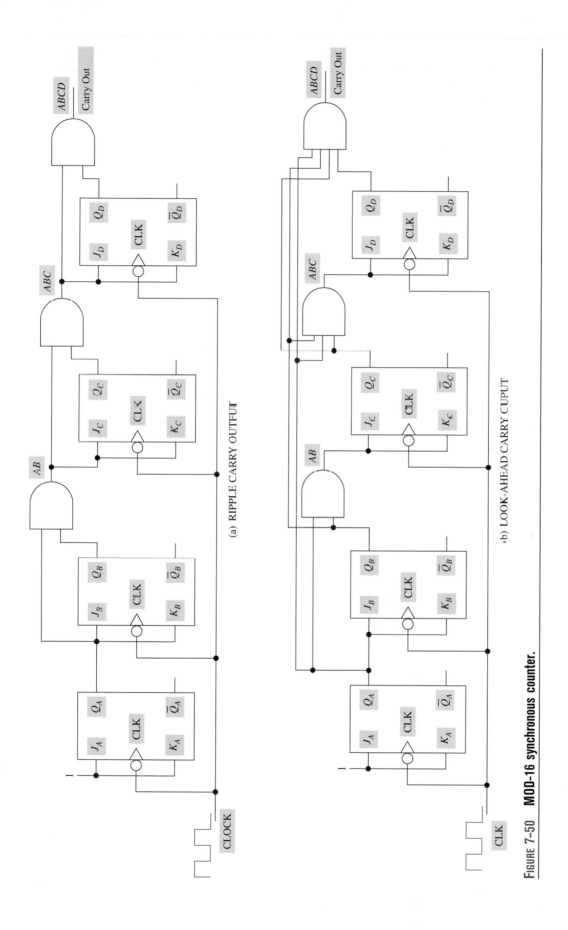

(a) RIPPLE CARRY OUTPUT

(b) LOOK-AHEAD CARRY CUPUT

Figure 7-50 **MOD-16 synchronous counter.**

The Carry Output of this counter is not valid until the outputs of all AND gates are valid. This valid condition does not occur until the sum of all of the AND gate propagation delays has elapsed. This Carry Output must ripple through the counter control gates, much as the count ripples through an asynchronous counter. This limits the upper frequency limit of the counter.

Figure 7–50(b) employs look-ahead carry and overcomes the disadvantage of the previous circuit. The two counters shown are identical with one exception. The AND gate's inputs that generate the look-ahead Carry Output are connected directly to the Q output of each flip-flop in Fig. 7–50(b). This does away with the AND gate additive delays of the ripple Carry Output and allows higher clocking speeds.

74LS163.

This synchronous 4-bit binary counter is presented because of some special features that differentiate it from the previous synchronous counters. The logic symbol and function table for the 74LS163 are presented in Fig. 7–51.

This programmable counter features an internal look-ahead carry circuit for use at high frequencies for fast counting. The look-ahead carry circuit generates a Ripple Carry Output (RCO) that can be used for cascading multiple counters without additional gating.

The function table in Fig. 7–51(b) indicates the $\overline{\text{CLR}}$ input is synchronous (occurs on PGT of the CLK), and it overrides the $\overline{\text{LOAD}}$ and enable inputs (ENT and ENP). This IC has two separate count enable inputs. Both enable inputs (ENP and ENT) must be high to enable counting. **Enable P** is a count-enable input, and **Enable T** is a count-enable input and a Ripple Carry Output control. Presetting a binary number into the counter is also accomplished synchronously on the PGT of the clock pulse when the $\overline{\text{LOAD}}$ input

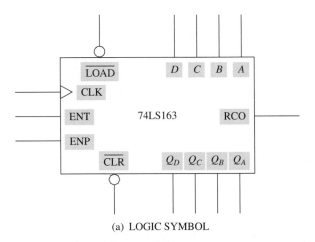

(a) LOGIC SYMBOL

$\overline{\text{CLR}}$	$\overline{\text{LOAD}}$	ENT	ENP	CLK	FUNCTION
0	X	X	X	↑	CLEAR
1	0	X	X	↑	PRELOAD
1	1	0	X	↑	HOLD
1	1	X	0	↑	HOLD
1	1	1	1	↑	COUNT

(b) FUNCTION TABLE

FIGURE 7–51 **74LS163 synchronous 4-bit binary counter.**

is asserted. The data at A, B, C, and D are loaded to the flip-flop outputs on the PGT of the clock after $\overline{\text{LOAD}}$ is taken low provided the $\overline{\text{CLR}}$ input is high.

The counter progresses through its normal up-count on the PGT of the CLK input when the $\overline{\text{CLR}}$ and $\overline{\text{LOAD}}$ inputs are inactive and the enable pulses ENT and ENP are both active (high). If either of the enable inputs is taken low during the count sequence, the counter will stop counting and hold.

Figure 7–52 shows the logic diagram of this counter. The clock input is applied through an inverting buffer (NOT gate) to trigger the counter on the low-to-high clock transition. All flip-flops are clocked simultaneously.

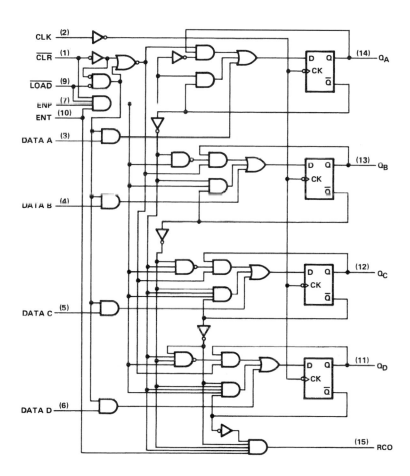

FIGURE 7-52   74LS163 synchronous counter logic diagram. *Reprinted by Permission of Texas Instruments.*

In-Depth Look at Operation:

The LSB flip-flop and its associated count-enable and control gates are redrawn in Fig. 7–53 to illustrate the synchronous action of the $\overline{\text{CLR}}$ input. The numbers in parenthesis on the logic diagram allow tracing logic levels to the D input of flip-flop A. The low at the D input of the flip-flop will be clocked to Q_A on the NGT of the clock input to the flip-flop. Since this clock input has been applied to the flip-flop through a NOT gate, the counter is actually cleared on the PGT of the incoming clock pulse.

The LOAD input can be traced logically through the diagram to show the data A input is loaded to flip-flop A on the PGT of the CLK pulse when $\overline{\text{LOAD}}$ is activated.

Tracing the logic levels through the count-enable and control gates will show why the $\overline{\text{CLR}}$ input must be high (inactive) to load the counter with a preset number. You can also prove the counter goes to the HOLD mode if ENT or ENP is taken low. Last, set the $\overline{\text{CLR}}$, $\overline{\text{LOAD}}$, ENT, and ENP inputs high and trace the logic levels through the diagram to prove flip-flop A will toggle on each clock pulse PGT.

An exercise in tracing logic levels as previously described proves the function table is correct and does much to enhance your understanding of synchronous counter operation.

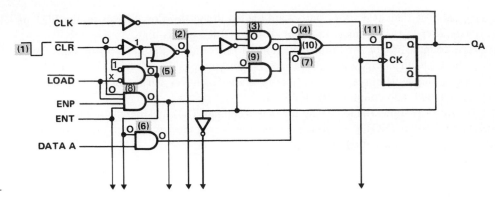

FIGURE 7-53   74LS163 counter–LSB section. Reprinted by Permission of Texas Instruments.

The Ripple Carry Output RCO in Fig. 7–52 goes high on the binary count of 1111. The ENT input is used to enable both counting and the AND gate that generates the RCO pulse. The duration of the RCO pulse is approximately equal to the positive portion of the Q_A output.

Changing MOD Numbers.   The 74LS163 counter can be truncated using conventional truncating circuitry as previously discussed. To truncate this circuit in the conventional manner, the *bits that are high on the desired maximum count must be connected to the NAND gate*. This is shown in the MOD-10 counter in Fig. 7–54(a). The counters that were presented earlier in this chapter have all been cleared asynchronously. Thus, their output count was

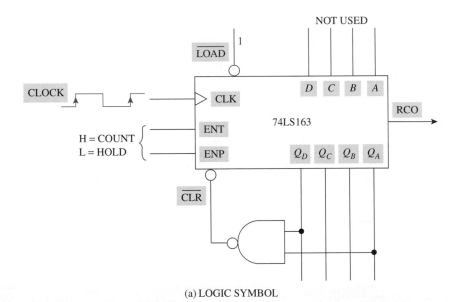

(a) LOGIC SYMBOL

CLK	Q_D	Q_C	Q_B	Q_A
	0	0	0	0
t_1	0	0	0	1
t_2	0	0	1	0
t_3	0	0	1	1
t_4	0	1	0	0
t_5	0	1	0	1
t_6	0	1	1	0
t_7	0	1	1	1
t_8	1	0	0	0
t_9	1	0	0	1
t_{10}	0	0	0	0

FIGURE 7-54   74LS163 MOD-10 counter.

(b) COUNT SEQUENCE

immediately reset to 0000 when the $\overline{\text{CLR}}$ input was activated. However, since the clearing action of this counter is synchronous, the truncating circuit is connected to the bits that are high on the maximum count instead of the bits that are high on the desired MOD number. The count sequence in Fig. 7–54(b) shows that the binary count of 1001 occurs for a full clock cycle even though Q_A and Q_D are tied to the NAND gate. This is because when the count of 1001 occurs, the $\overline{\text{CLR}}$ input is activated, but the actual clearing is done on the next PGT of the clock input. The waveforms illustrating this action are shown in Fig. 7–55.

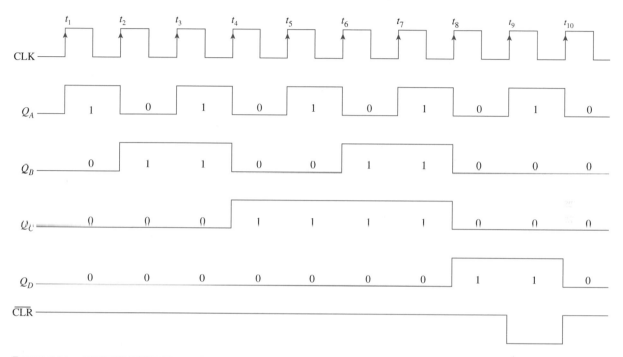

FIGURE 7-55   74LS163 MOD-10 counter waveforms.

Figure 7–56 shows another method of connecting the 74LS163 using preload data to set the counter to MOD-10 operation. A binary 0110 must be preloaded when $\overline{\text{LOAD}}$ is brought active. This is because the preload action, like the clearing action, is synchronous.

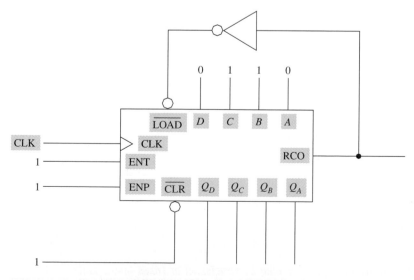

FIGURE 7-56   74LS163 MOD-10 counter using preload.

Modes of Operation.   The logic symbol in Fig. 7–57(a) is set up to show the waveforms in Fig. 7–57(b). The symbol shows the counter with a preset input of $1100_{(2)}$. The waveforms are used to illustrate the different modes of operation of the 74LS163. These waveforms are identical to a 74LS161 binary counter's waveforms with one exception—the $\overline{\text{CLR}}$ input of the 74LS161 4-bit binary counter is asynchronous. The following sequence is depicted in the waveforms for the 74LS163:

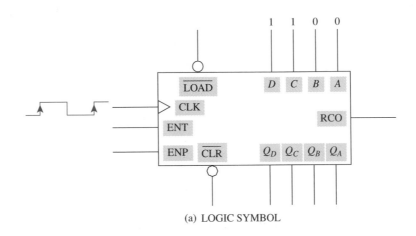

(a) LOGIC SYMBOL

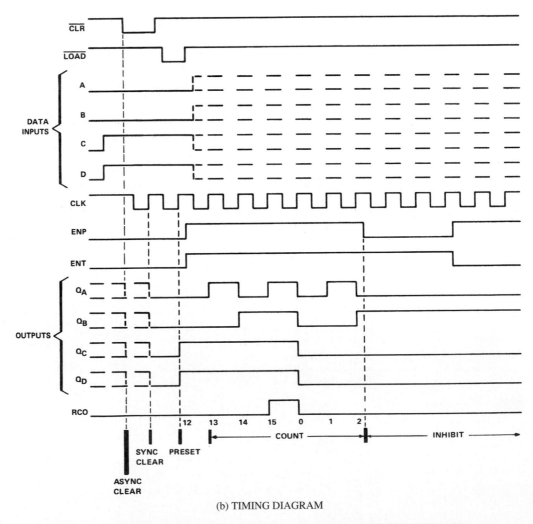

(b) TIMING DIAGRAM

FIGURE 7–57   **74LS163 counter.** *(b) Reprinted by Permission of Texas Instruments.*

1. $\overline{\text{CLR}}$ is activated. The counter output goes to 0000 on the next PGT of the clock pulse.
2. LOAD is activated. The counter is preloaded with a count of 1100 on the next PGT of the clock pulse.
3. ENT and ENP are activated (brought high). The counter counts commencing at 1100 and continues 1101, 1110, 1111, 0000, 0001, and 0010.
4. ENP is brought low shortly after the counter reaches the count of 0010. This places the counter in the HOLD mode by enabling the AND gate connected to the Q_A output and recirculating the Q_A output signal.
5. ENP is activated and ENT is brought low. This keeps the counter in the HOLD mode.

Cascading 74LS163 ICs. Figure 7–58 illustrates cascading two 74LS163 ICs to produce a MOD-256 counter (MOD-16 × MOD-16). This configuration uses the RCO pulse to enable the higher-order counter. The second counter increments one count each time RCO goes high. The 74LS163 data sheet specifies 9 ns (typical) propagation delay from ENT to RCO for this counter family. This delay is additive so the maximum operating frequency (f_{max}) decreases as additional counters are cascaded. The formula for f_{max} in this configuration is $1/(\text{CLK to RCO } t_{\text{PLH}}) + (\text{ENT to RCO } t_{\text{PLH}}) (N - 2) + (\text{ENT } t_{\text{su}})$ when more than two counters are cascaded.

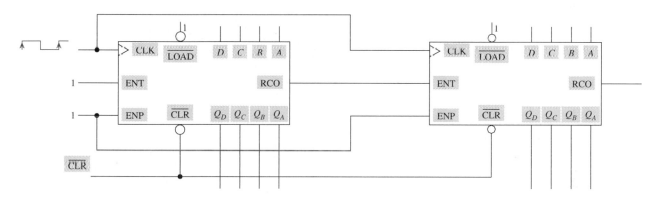

FIGURE 7-58   Cascaded MOD-256–74LS163 ICs–RCO to ENT.

This frequency limitation is overcome in Fig. 7–59 because the propagation delay of ENT to RCO has been eliminated in this configuration. The desired result is obtained because RCO is connected to ENP instead of ENT. Note in the logic diagram for this counter (Fig. 7–52) that ENP is not connected to the RCO pulse AND gate. This leaves the circuit with a maximum operating frequency of $1/(\text{CLK to RCO } t_{\text{PLH}}) +$

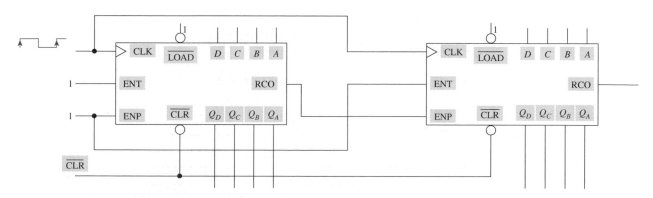

FIGURE 7-59   Cascaded MOD-256–74LS163 ICs–RCO to ENP

(ENP t_{su}). Since the clock inputs are applied simultaneously and ENP is used on the second counter, no additive delays would be encountered if additional 74LS163 ICs were cascaded.

74LS190.

The 74LS190 IC is a synchronous reversible up/down BCD counter. The logic symbol and function table for this counter are shown in Fig. 7–60.

The function table in Fig. 7–60(b) indicates the counter may be asynchronously preset to any BCD digit when $\overline{\text{LOAD}}$ is activated. The circuit will not count unless the Count Enable ($\overline{\text{CTEN}}$) input is low. The Down/Up($\text{D}/\overline{\text{U}}$) input controls the direction of the count. When this input is *high* the counter counts *down* and when it is *low* the counter counts *up*. The $\overline{\text{LOAD}}$ and $\overline{\text{CTEN}}$ inputs must be at the proper level as shown in the function table to allow counting.

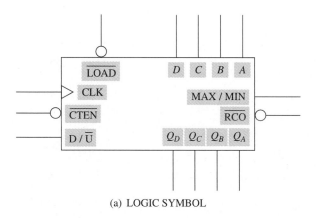

(a) LOGIC SYMBOL

$\overline{\text{LOAD}}$	$\overline{\text{CTEN}}$	$\text{D}/\overline{\text{U}}$	FUNCTION
0	X	X	ASYNCHRONOUS PRELOAD
1	1	X	HOLD
1	0	1	COUNT DOWN
1	0	0	COUNT UP

(b) FUNCTION TABLE

FIGURE 7–60   74LS190 synchronous BCD up/down-counter.

The MOD number of the counter is controlled by presetting the data inputs. The $\overline{\text{RCO}}$ can be used to activate the $\overline{\text{LOAD}}$ input for preloading the counter. Figure 7–61 shows a BCD 0011 set at the preload data inputs. This same configuration was used by connecting the $\overline{\text{CO}}$ to the $\overline{\text{LOAD}}$ input of the 74193 counter. The same loading conditions apply to the 74LS190. The terminal count of 1001 and the preload count of 0011 last only one-half clock cycle each. This makes the counter in Fig. 7–61 a MOD-6 counter:

$$
\begin{array}{llll}
0011 & ------------------ & \overline{\text{LOAD}} \\
0100 & ------------------ & t_1 \\
0101 & ------------------ & t_2 \\
0110 & ------------------ & t_3 \\
0111 & ------------------ & t_4 \\
1000 & ------------------ & t_5 \\
1001 & ----0011--------- & t_6 \\
\end{array}
$$

The 74LS190 counter produces two outputs that can be used for cascading multiple counters. The outputs are the **Maximum/Minimum (MAX/MIN) count output** and the **Ripple Clock Output ($\overline{\text{RCO}}$).**

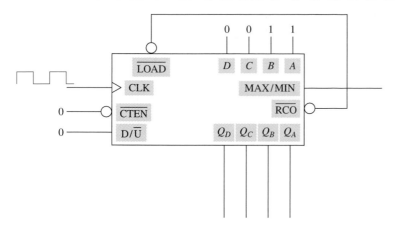

FIGURE 7-61   74LS190 MOD-6 up-counter.

The MAX/MIN count output is an active-high output. The MAX/MIN circuit is shown in Fig. 7-62(a). The output pulse for up-count operation is generated on a binary count of 1001. The output pulse for down-count operation is generated on a binary count of 0000.

The Ripple Clock Output ($\overline{\text{RCO}}$) is an active-low output signal. The logic gate generating this pulse is shown in Fig. 7-62(b). The pulse is generated when a terminal count of 1001 (up-count) or 0000 (down-count) generates a high MAX/MIN count output signal provided $\overline{\text{CTEN}} = 0$ and the clock is on the low portion of its input.

The waveforms for the MAX/MIN and $\overline{\text{RCO}}$ circuits are shown in Fig. 7-63. The up-count operation waveforms are shown in Fig. 7-63(a) and the down-count waveforms are shown in Fig. 7-63(b).

The MAX/MIN output for the up-counter is generated from t_9 to t_{10} in Fig. 7-63(a). The same output is generated for the down-counter from t_{10} to t_{11} in Fig. 7-63(b). Although this may seem like an anomaly, observation of the two sets of waveforms reveals no inconsistency. Every tenth clock pulse will take the counters through one complete

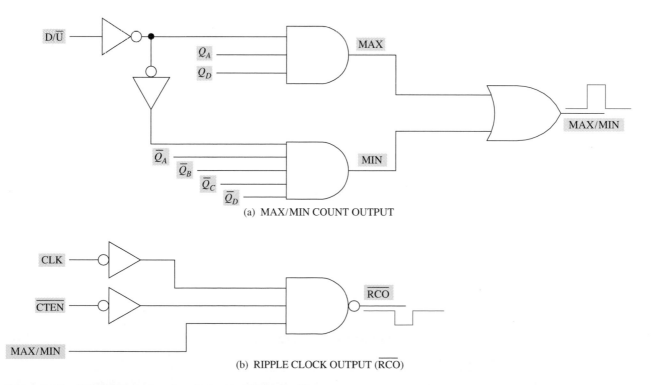

(a) MAX/MIN COUNT OUTPUT

(b) RIPPLE CLOCK OUTPUT ($\overline{\text{RCO}}$)

FIGURE 7-62   74LS190 counter—terminal count logic circuitry.

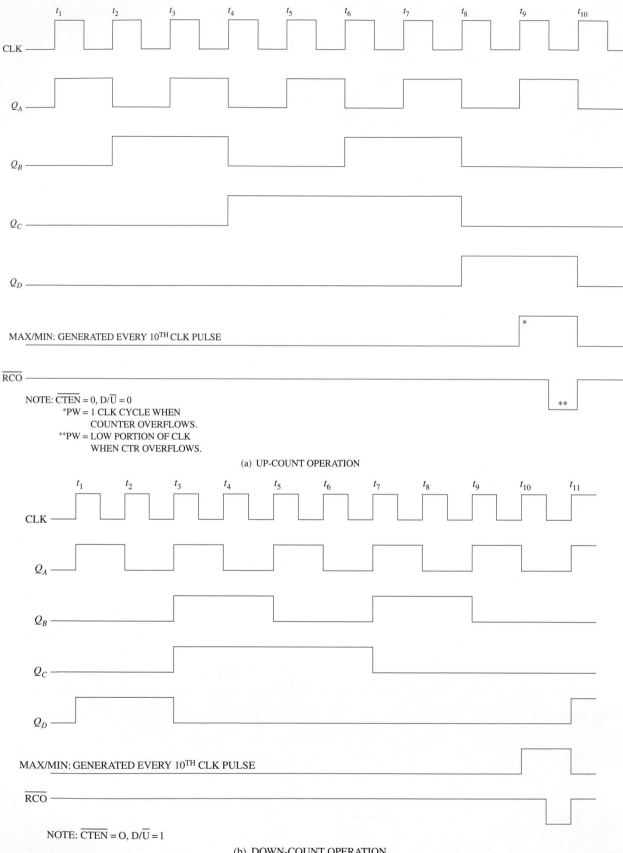

FIGURE 7–63   74LS190 timing diagram.

count cycle. The count sequence for the up-counter commences at 0001 and resets to 0000 at t_{10}. This represents ten states for the counter. The count sequence of the down-counter commences at 1001 and resets to 0000 at t_{10}. Thus, every tenth clock pulse increments the up-counter until MAX/MIN is activated (t_9, t_{19}, t_{29}, etc.). Every tenth clock pulse decrements the down-counter until MAX/MIN is activated (t_{10}, t_{20}, t_{30}, etc.).

Cascading 74LS190 ICs.   Figure 7–64 shows two methods of cascading 74LS190 BCD counters. These configurations produce MOD-100 counters. The configuration shown in Fig. 7–64(a) uses **parallel clocking.** The $\overline{RCO}$ is used to enable the high-order counter by activating $\overline{CTEN}$. Each time the low-order counter reaches a terminal count of 1001, $\overline{RCO}$ is brought low and enables the high-order counter to increment one count.

   The configuration shown in Fig. 7–64(b) uses **serial clocking.** The $\overline{RCO}$ is used to clock the high-order counter. The MAX/MIN count output can be used in lieu of the $\overline{RCO}$. The MAX/MIN output is generated as a look-ahead carry feature, which allows high-speed counting.

   Both of these counters can count to $99_{(10)}$ because they are BCD counters. The counts in Fig. 7–65 show the sequence through the count of $10_{(10)}$. The BCD outputs of these counters can be decoded and displayed on seven-segment indicators like the ones most digital clocks use.

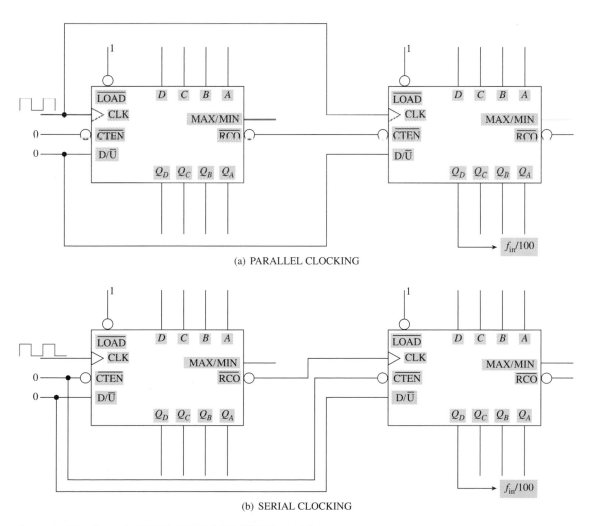

(a) PARALLEL CLOCKING

(b) SERIAL CLOCKING

Figure 7-64   **Cascaded MOD-100 74LS190 BCD up-counter.**

BCD		DEC	INDICATORS
MSD	LSD		
0 0 0 0	0 0 0 0	0 0	
0 0 0 0	0 0 0 1	0 1	
0 0 0 0	0 0 1 0	0 2	
0 0 0 0	0 0 1 1	0 3	
0 0 0 0	0 1 0 0	0 4	
0 0 0 0	0 1 0 1	0 5	
0 0 0 0	0 1 1 0	0 6	
0 0 0 0	0 1 1 1	0 7	
0 0 0 0	1 0 0 0	0 8	
0 0 0 0	1 0 0 1	0 9	
0 0 0 1	0 0 0 0	1 0	

FIGURE 7-65   **Cascaded MOD-100 count sequence.**

4029.

The 4029 IC is a synchronous CMOS binary/decade up/down-counter. The counter contains D-type flip-flops with logic gates to allow proper toggling operation. The logic symbol and function table for the 4029 are shown in Fig. 7–66. The function table for the counter is a bit more complex than the one for the 74LS190, but this is a more versatile counter. The 4029 can be used to count in binary up to 1111 or from 1111 down to 0000. It can also be used to count BCD up or down.

The function table in Fig. 7–66(b) indicates the MOD number of the counter may be set by bringing the Preset Enable (PE) high. Asserting this input asynchronously preloads the binary or BCD number that is available at the preset input pins (P_3–P_0). The PE input overrides all other inputs to the counter when it is activated.

The count direction (up or down) is controlled by the Up/$\overline{\text{Down}}$ input. The counter counts up when this input is high and down when it is low. If the Binary/$\overline{\text{Decade}}$ input is

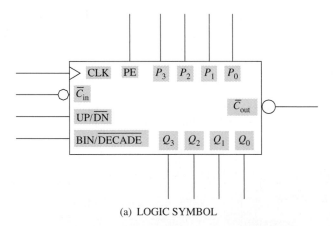

(a) LOGIC SYMBOL

PE	BIN/$\overline{\text{DEC}}$	UP/$\overline{\text{DN}}$	$\overline{C}_{\text{in}}$	CLK	FUNCTION
1	X	X	X	X	PRELOAD
0	1	1	0	↑	COUNT UP–BINARY
0	0	1	0	↑	COUNT UP–DECADE
0	1	0	0	↑	COUNT DOWN–BINARY
0	0	0	0	↑	COUNT DOWN–DECADE
0	X	X	1	X	NO COUNT–HOLD

(b) FUNCTION TABLE

FIGURE 7-66   **4029 binary/decade up/down CMOS counter.**

high and the Up/$\overline{\text{Down}}$ input is high, the counter counts up to binary 1111. If the Binary/$\overline{\text{Decade}}$ input is low and the Up/$\overline{\text{Down}}$ input is high, the counter counts up to 1001. Notice the Carry In ($\overline{C}_{\text{in}}$) input must be low for the counter to count. The counter will stop counting if $\overline{C}_{\text{in}} = 1$ because the toggle enable to the LSB flip-flop prevents toggling in the counter under this condition.

The $\overline{\text{Carry Out}}$ signal goes active low during binary up-count operation when the terminal count of $1111_{(2)}$ is reached. It goes low during decade up-count operation on the terminal count of $1001_{(\text{BCD})}$. The signal is activated at 0000 during down-count operation for both binary and BCD. Generation of the Carry Out ($\overline{C}_{\text{out}}$) signal was discussed in detail in Chapter 5.

The circuit shown in Fig. 7–67 contains two 4029 counters cascaded as a divide-by-25 BCD down-counter. Note the preload inputs are set to $25_{(10)} = 0010\ 0101_{(\text{BCD})}$. The $\overline{C}_{\text{out}}$ of the LSD counter is used as the $\overline{C}_{\text{in}}$ of the MSD counter. When the LSD group of bits reaches 0000, $\overline{C}_{\text{in}}$ of the MSD counter is asserted, and this allows the MSD counter to decrement one count on the next clock pulse PGT. The $\overline{C}_{\text{out}}$ signals from both counters are connected to the truncating NOR gate. When both counters reach their terminal down-counts of 0000, the two lows into the NOR gate produce a high output. This output is tied to both counters' PE inputs. When PE is activated the $25_{(10)}$ is preloaded back into the counters, and the count down process starts over.

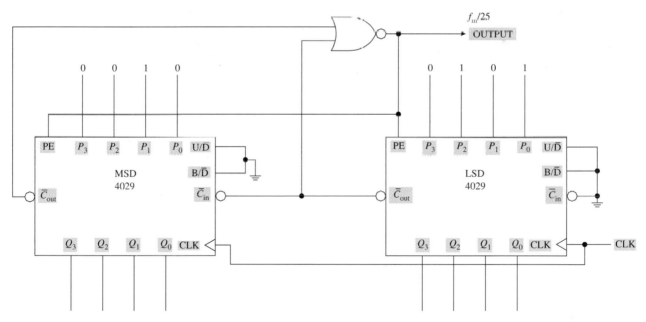

FIGURE 7-67   Cascaded MOD-25 BCD (decade) down-counter using 4029 ICs.

Studying the 4029 timing diagram in Fig. 7–68 will aid your understanding of this counter. The counter starts in the CLEAR state and increments to a count of $1001_{(\text{BCD})}$. The $\overline{C}_{\text{out}}$ signal is activated when this count is reached as the counter is set for decade counting. Prior to the next PGT, the Up/$\overline{\text{Down}}$ input is taken low. This reverses the count direction, and the counter decrements with each PGT down to its terminal count of 0000. This count causes the $\overline{C}_{\text{out}}$ signal to again go low. Prior to the next PGT, $\overline{C}_{\text{in}}$ is taken high, and the counter stops counting. The counter resets to $1001_{(\text{BCD})}$ on the next PGT to prepare for another down-count sequence. However, PE is activated and the counter is asynchronously preloaded with a $0110_{(\text{BCD})}$. The count direction is again reversed to up-count operation when Up/$\overline{\text{Down}}$ is taken high. Analysis of the timing diagram reveals the effects of changing various counter inputs on the output waveforms.

TIMING DIAGRAM

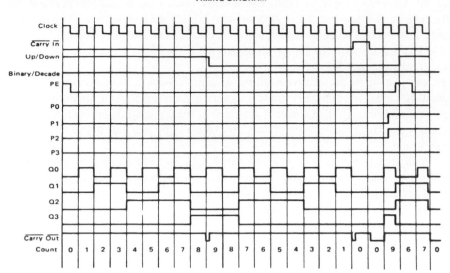

FIGURE 7-68   **4029**
CMOS counter timing
diagram. *Courtesy of*
Motorola.

4040.

The 4040 IC is a CMOS 12-bit binary up-counter. The logic symbol for the IC is shown in Fig. 7–69. The counter contains a wave-shaping circuit for the clock input. This input circuit will take a 120 Vac sine wave, reduce its amplitude, and square it up to make it usable as a clock input to the counter. The 4040 has the capability of counting up to 4095 $(2^{12} - 1)$.

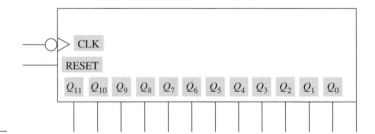

FIGURE 7-69   **4040**
12-bit CMOS binary
counter logic symbol.

The counter is shown in Fig. 7–70 configured as a divide-by-3600 circuit. The input to this circuit is connected directly to a standard 120-Vac, 60-Hz power line. The count of 3600 is $111000010000_{(2)}$. Bits Q_4, Q_9, Q_{10}, and Q_{11} are high on this count, so they are connected to the truncating gate. When this count is reached, the output of the AND gate goes high and asynchronously resets the counter. The divide-by-3600 action takes the 60-Hz input and converts it to one pulse per minute at the output.

A large variety of synchronous programmable counters are commercially available. Counters to fit most every need can be purchased, and they are economical. Data books need to be checked for counter availability and specifications. The next section details synchronous counter design.

Section 7-5: Review Questions

Answers are given at the end of the chapter.

A. The MOD number of a programmable counter can be set by preloading a number into the counter to establish its initial count.
   (1) True
   (2) False

Refer to the appropriate data sheet for the following questions when necessary:

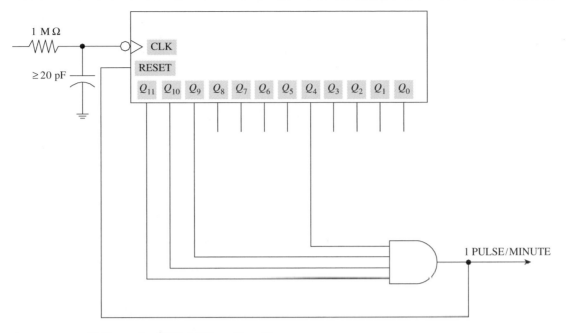

FIGURE 7–70   4040 counter MOD-3600 configuration.

B. The counter shown in Fig. 7–71 will not count when $\overline{\text{LOAD}}$ is asserted.
    (1) True
    (2) False

C. What is the count sequence of the counter shown in Fig. 7–71 if $\overline{\text{LOAD}}$ is activated, then taken high, and the counter is clocked five times?

D. Configure the 74LS193 IC shown in Fig. 7–72 to be a MOD-7 down-counter. Show the input clock pulses.

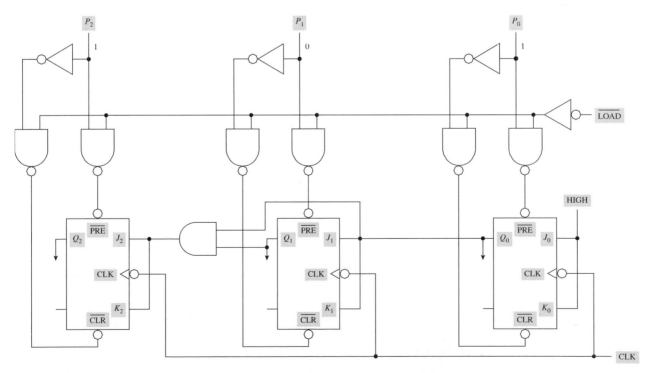

FIGURE 7–71

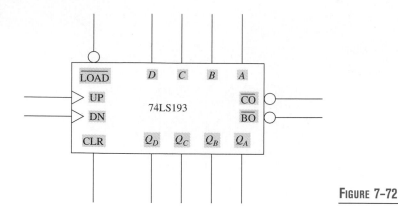

FIGURE 7-72

E. Configure the 74LS193 IC in Fig. 7–72 to be a MOD-7 up-counter. Show the input clock pulses.

F. Show the connections required to cascade 74LS163 ICs to implement a MOD-100 up-counter.

G. The logic symbol for a 74LS190 IC is shown in Fig. 7–73. What logic levels must be applied to the following inputs to make the counter count down in BCD?
   (1) $\overline{\text{LOAD}}$
   (2) $\overline{\text{CTEN}}$
   (3) $\text{D}/\overline{\text{U}}$
   (4) CLK

H. The MAX/MIN output of the 74LS190 is an _____ output. Refer to Fig. 7–73.
   (1) active-low
   (2) active-high

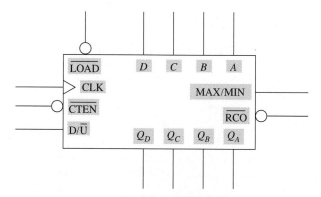

FIGURE 7-73

I. The MAX/MIN output of the 74LS190 is asserted during up-count operation on a count of _____. Refer to Fig. 7–73.
   (1) 0111
   (2) 1001
   (3) 1111

J. The logic symbol for a 4029 CMOS counter is shown in Fig. 7–74. What logic levels must be applied to the following inputs to make the counter count down in binary?
   (1) $\overline{\text{PE}}$
   (2) Bin/$\overline{\text{Decade}}$
   (3) Up/$\overline{\text{Dn}}$
   (4) Clock

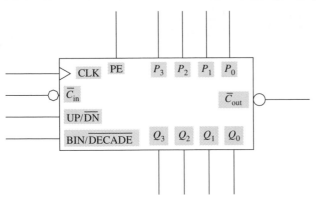

FIGURE 7–74

SECTIONS 7-3 THROUGH 7-5: INTERNAL SUMMARY

Synchronous counters use one common clock line to clock all counter flip-flops simultaneously. Clocking in this manner eliminates the short-duration miscounts that are normally encountered in asynchronous (ripple) counters.

Synchronous counter design and construction are more complex than that of asynchronous counters. The LSB flip-flop is the only flip-flop in the synchronous counter that can be hard-wired to toggle on every active clock transition. This mandates additional gating circuitry to control the toggle action of the remaining flip-flops.

Synchronous counters eliminate the additive delays that are caused by the ripple action of the clock in asynchronous counters. Therefore, their clocking frequencies are higher than comparable asynchronous counters.

A MOD-10 counter is usually referred to as a **BCD** or **decade** counter. The counter produces 0000 through 1001 outputs. These counters are often used to produce outputs that can be decoded and used to drive seven-segment indicators to produce decimal readouts.

Synchronous counters can be designed to **count up or down.** The theory is the same as that of asynchronous counter design in determining count direction. Up/down synchronous counters can be designed using logic gates to control toggling action within the counter by using Q **outputs from the flip-flops for up-counting** and $\overline{Q}$ **outputs for down-counting.**

Many programmable counters allow the user to set count direction and program for binary or decade count operation. The user may also preload data into the counter to control its MOD number. The preload data is input asynchronously in most counters, but synchronous loading is available in some ICs.

Most programmable counters provide outputs that can be used for cascading several ICs without the use of additional gating. Some of these outputs are active-low and some are active-high. Many of the counters employ look-ahead carry, which allows them to operate at higher frequencies.

The main features of synchronous counters have been presented in this chapter. Hybrid counters utilize a combination of synchronous and asynchronous characteristics. These counters will be presented in Section 7–7. Additional features of all types of counters are detailed in manufacturer's data sheets. These data sheets are invaluable references when designing or selecting substitute ICs because there is no way to remember the multitude of details that characterize different counters.

SECTIONS 7-3 THROUGH 7-5: INTERNAL SUMMARY QUESTIONS

Answers are given at the end of the chapter.

Note: The use of manufacturer's data sheets is recommended to assist you in answering these summary questions.

1. A counter is _____ if all of its flip-flops are clocked simultaneously by one common clock input.
   a. synchronous
   b. asynchronous

2. Asynchronous counters can normally be clocked at higher frequencies than comparable synchronous counters of the same MOD.
   a. True
   b. False

3. Synchronous counters employ extra logic gates to control flip-flop toggle operation.
   a. True
   b. False

4. Identify the counter shown in Fig. 7–75.
   a. MOD-10 asynchronous up-counter
   b. MOD-16 asynchronous up-counter
   c. MOD-10 synchronous up-counter
   d. MOD-16 synchronous up-counter

5. Identify the counter shown in Fig. 7–76 (on p. 386).
   a. MOD-8 asynchronous up-counter
   b. MOD-10 asynchronous up-counter
   c. MOD-8 synchronous down-counter
   d. MOD-10 synchronous down-counter

6. What is the MOD number of the counter in Fig. 7–77 (on p. 387)?
   a. MOD-2   c. MOD-5
   b. MOD-3   d. MOD-6

7. The circuit shown in Fig. 7–77 is a/an _____ -counter.
   a. up
   b. down

8. The cascaded counter in Fig. 7–78 (on p. 388) is a MOD- _____ up-counter.
   a. 16   c. 100
   b. 60   d. 256

9. The $\overline{CO}$ of the 74193 counters in Fig. 7–78 is activated on a count of
   a. 0000   c. 1010
   b. 1001   d. 1111

10. The divide-by capability at the MSB of a truncated 74LS193 down-counter is equal to
    a. 10   c. The preset number
    b. 16   d. The terminal count

11. The data at the *A*, *B*, *C*, and *D* inputs of the 74LS163 in Fig. 7–79 (on p. 388) can be loaded on the PGT of the CLK pulse when
    a. $\overline{CLR} = 0$ and $\overline{LOAD} = 0$
    b. $\overline{CLR} = 1$ and $\overline{LOAD} = 0$
    c. $\overline{CLR} = 1$, $\overline{LOAD} = 1$, and ENP = 0
    d. $\overline{CLR} = 1$, $\overline{LOAD} = 1$, ENT = 1, and ENP = 1

12. The 74LS163 in Fig. 7–79 will stop counting when
    a. ENT = 0
    b. ENP = 0
    c. $\overline{CLR} = 0$
    d. All of the above

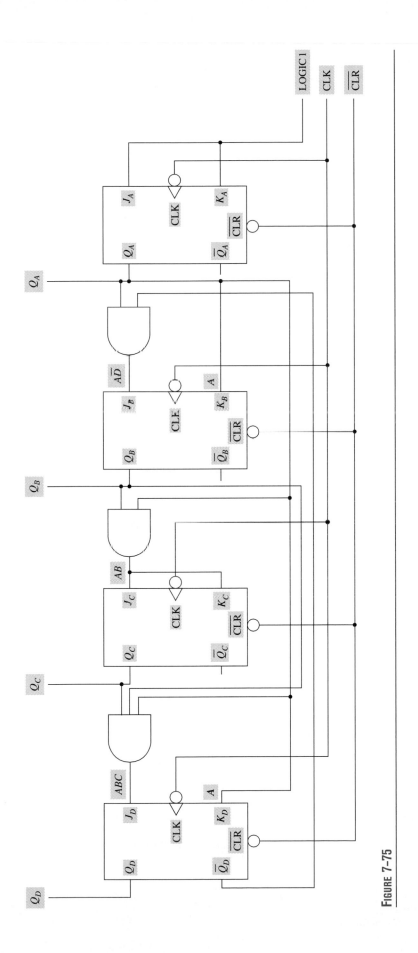

Figure 7-75

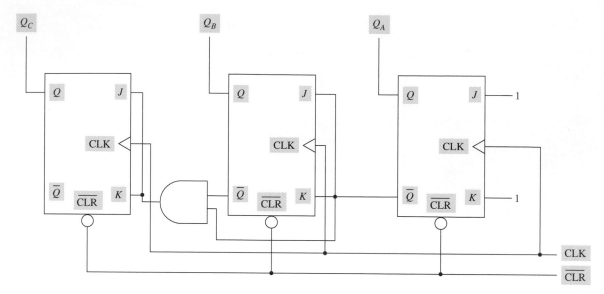

FIGURE 7-76

13. The $\overline{\text{LOAD}}$ input to the 74LS163 in Fig. 7–79 overrides the $\overline{\text{CLR}}$ input.
    a. True
    b. False

14. The circuit shown in Fig. 7–80 is a MOD- _____ counter.
    a. 8   c. 10
    b. 9   d. 12

15. The 74LS190 counter in Fig. 7–81 will count down on the PGT of the clock when
    a. $\overline{\text{LOAD}} = 0$, $\overline{\text{CTEN}} = 0$, and D/$\overline{\text{U}}$ = 0
    b. $\overline{\text{LOAD}} = 1$, $\overline{\text{CTEN}} = 1$, and D/$\overline{\text{U}}$ = 1
    c. $\overline{\text{LOAD}} = 1$, $\overline{\text{CTEN}} = 0$, and D/$\overline{\text{U}}$ = 0
    d. $\overline{\text{LOAD}} = 1$, $\overline{\text{CTEN}} = 0$, and D/$\overline{\text{U}}$ = 1

16. The 74LS190 counter in Fig. 7–81 will stop counting when $\overline{\text{CTEN}}$ is taken high.
    a. True
    b. False

17. The $\overline{\text{RCO}}$ of the 74LS190 in Fig. 7–81 is
    a. Active-low
    b. Active-high

18. What function is the 4029 CMOS counter in Fig. 7–82 performing when PE = 0, Bin/$\overline{\text{Decade}}$ = 1, Up /$\overline{\text{Dn}}$ = 1, and $\overline{C}_{\text{in}}$ = 0?
    a. Count up—decade
    b. Count up—binary
    c. Count down—decade
    d. Count down—binary

19. What function is the 4029 CMOS counter in Fig. 7–82 performing when PE = 0, Bin/$\overline{\text{Decade}}$ = 0, Up /$\overline{\text{Dn}}$ = 0, and $\overline{C}_{\text{in}}$ = 0?
    a. Count up—decade
    b. Count up—binary
    c. Count down—decade
    d. Count down—binary

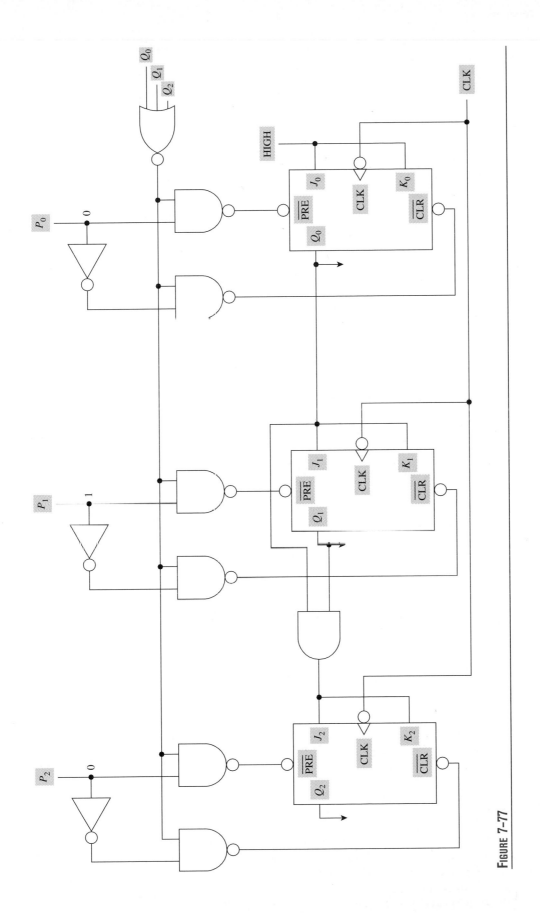

FIGURE 7-77

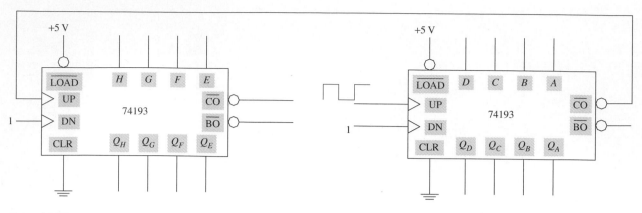

FIGURE 7–78

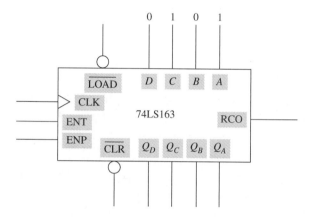

FIGURE 7–79

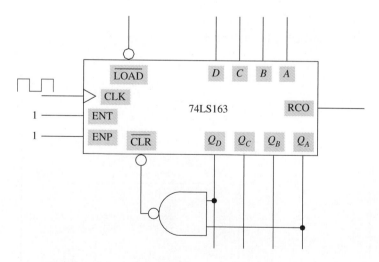

FIGURE 7–80

20. What function is the 4029 CMOS counter in Fig. 7–82 performing when PE = 0, Bin/$\overline{\text{Decade}}$ = 0, Up /$\overline{\text{Dn}}$ = 0, and $\overline{C}_{\text{in}}$ = 1?
    a. No count
    b. Preload data
    c. Count down—binary
    d. Count down—decade

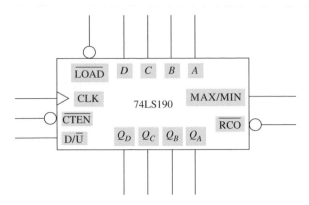

FIGURE 7–81

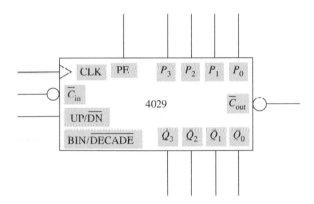

FIGURE 7–82

SECTION 7–6: SYNCHRONOUS COUNTER DESIGN

OBJECTIVE

Given a specfic count sequence or desired MOD number and type of flip-flops to be used, design a synchronous counter.

Note: This section may be omitted in your studies of basic digital theory without detriment to your studies in the rest of this textbook. The information presented in this section is supplemental in nature. It is, however, beneficial to those who will be involved in the design of counters.

MOD-16 Up-Counter Design

It is a relatively easy task to design full-count synchronous counters because the count sequence of the desired count readily shows the toggle control gates required to implement the counter. However, when the desired MOD number is not equal to an integer power of 2, synchronous counter design is a bit more complex. Karnaugh mapping becomes a valuable tool in simplifying these design problems.

Although the design of the MOD-16 synchronous up-counter was relatively simple, its design using K-maps will be accomplished to introduce the design concept.

The initial step in synchronous up-counter design is to complete a **state table.** The present state (count) is referred to as Q^n and is shown in Table 7–2 in the Q^n column for the MOD-16 counter. This table shows the desired count sequence of the counter.

TABLE 7-2	Present State Table	Next State Table
State Table	Q^n	(Q^{n+1})
	0000	0001
	0001	0010
	0010	0011
	0011	0100
	0100	0101
	0101	0110
	0110	0111
	0111	1000
	1000	1001
	1001	1010
	1010	1011
	1011	1100
	1100	1101
	1101	1110
	1110	1111
	1111	0000

The following step is to complete a **next state table.** This next state is called and is also shown in Table 7–2. The next state is the present count sequentially incremented by 1 for this up-counter.

The next design step requires utilization of a **transition table.** A transition table for J-K flip-flops is shown in Table 7–3. This table uses the present state (count) of a flip-flop's Q output (Q^n) to show the levels of J and K required to produce the next flip-flop state (Q^{n+1}). Refer to Fig. 7–83 for the following discussion regarding the transition states shown on the transition table (7–3).

If $Q^n = 0$ and the next state (Q^{n+1}) needs to remain at 0, J *must equal 0*. If K is 0, the flip-flop is in the HOLD mode and will retain $Q = 0$. If K is 1, the flip-flop will still stay in the CLEAR state with $Q = 0$ when it is clocked. Thus, K is a don't care input here. This is shown in Fig. 7–83(a).

If $Q = 0$ and the next state must be 1, the transition table shows J *must equal 1*. If J is 1 and K is 0, the flip-flop goes to the SET state ($Q = 1$) when it is clocked. If $J = 1$ and $K = 1$, the flip-flop will toggle to the SET state when clocked. This indicates K again is a don't care input, and Fig. 7–83(b) shows this transition state.

If $Q = 1$ and the next state must be 0 as shown in Fig. 7–83(c), the transition table indicates K *must be 1*. If this condition is true and J is 0, the flip-flop clears when it is clocked. If $J = 1$ and $K = 1$, the flip-flop will toggle to the CLEAR state when it is clocked. Thus, J is a don't care input.

The last condition on the transition table shows $Q = 1$ and the next state must remain at the 1 level. The logic of the first example applies in this situation also. In either case ($J = 1$ or $J = 0$) the flip-flop will remain in the SET state as shown in Fig. 7–83(d).

TABLE 7-3	Present State	Next State	Required Levels for (Q^{n+1})	
Transition Table for J-K Flip-Flops	Q^n	(Q^{n+1})	J	K
	0	0	0	X
	0	1	1	X
	1	0	X	1
	1	1	X	0

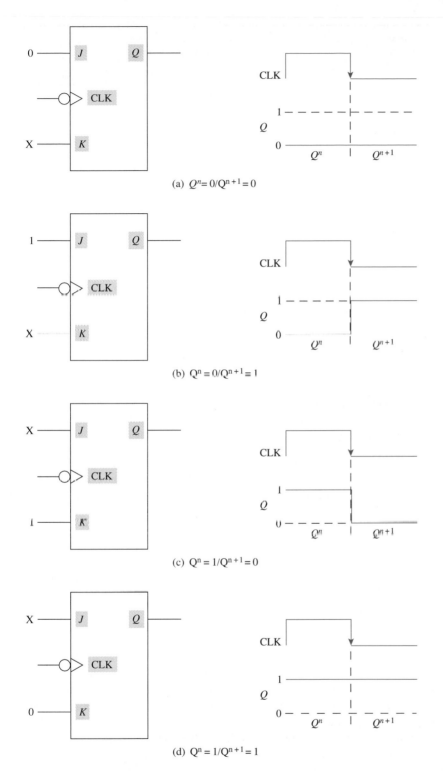

(a) $Q^n = 0 / Q^{n+1} = 0$

(b) $Q^n = 0 / Q^{n+1} = 1$

(c) $Q^n = 1 / Q^{n+1} = 0$

(d) $Q^n = 1 / Q^{n+1} = 1$

FIGURE 7–83   Transition states.

A partial **design table** is shown in Table 7–4. This table incorporates the **present state table** (Q^n) and the **next state table** (Q^{n+1}) with the transition table data entered. Extreme care must be taken when completing the design table. It should be completed one column at a time. Constant use of the transition table (7–3) is required when transferring the transition table data to the design table (7–4).

The partial design table (7–4) has been completed only for a present state of 0000 and the next state of 0001. This is done for the sake of simplicity at this point, and the complete table will be presented in the next table.

TABLE 7-4 **Partial Design Table**	Present State Q^n $Q_D Q_C Q_B Q_A$	Next State Q^{n+1} $Q_D Q_C Q_B Q_A$	Transition Table Data							
			Q_D		Q_C		Q_B		Q_A	
			J	K	J	K	J	K	J	K
	0 0 0 0	0 0 0 1	0	X	0	X	0	X	1	X

Step 4 · Step 3 · Step 2 · Step 1

The J and K data from the transition table (7–3) are transferred to the design table (7–4) in the following manner:

STEP 1: The transition table (7–3) indicates that when $Q_D{}^n = 0$ and $Q_D{}^{n+1} = 0$, $J = 0$, and $K = X$. This data relates to the MSB, so it is placed in the Q_D column in Table 7–4.

STEP 2: When $Q_C{}^n = 0$ and $Q_C{}^{n+1} = 0$, $J = 0$, and $K = X$.

STEP 3: When $Q_B{}^n = 0$ and $Q_B{}^{n+1} = 0$, $J = 0$, and $K = X$.

STEP 4: When $Q_A{}^n = 0$ and $Q_A{}^{n+1} = 1$, $J = 1$, and $K = X$.

This procedure enters the data for each Q output of the present state 0000 (row) across the transition table data portion of the design table. Once this has been accomplished, the remaining data may be entered for each of the remaining present state conditions.

You may be more comfortable completing all of the columnar data one column at a time. In this case, all of the data for Q_D should be completed on every row, then for Q_C, and so on. Either method is acceptable.

A completed design table for a MOD-16 synchronous up-counter is shown in Table 7–5. Let's verify the Q_A column data. When $Q^n = 0000$ and $Q^{n+1} = 0001$, it can be seen that Q_A must change from 0 to 1. The transition table indicates $J = 1$ and $K = X$ to assume the Q^{n+1} condition on an active clock transition. This information has been entered on the design table under the transition table data Q_A columns for J and K. It is important not to transpose the J and K conditions because they will be entered separately on K-maps to complete the circuit design.

The next Q_A count shows $Q^n = 0001$ and $Q^{n+1} = 0010$. The transition table shows $J = X$ and $K = 1$ in order to assume the Q^{n+1} state. This information ($J = X$ and $K = 1$) is entered adjacent to $Q^n = 0001$ in the J and K columns of Q_A. The remainder of this column should be checked to ensure a thorough understanding of this transfer operation before proceeding.

The Q_B column of the design table may be checked after the Q_A column. Then the two remaining columns should be verified. You may prefer to start with the Q_D column data in lieu of the Q_A column as you complete a design table. Once you have developed a procedure you are comfortable with, stick to it.

All of this may seem like a cumbersome procedure, but it goes quickly once you have designed a couple of counters. However, you must be careful because a mistake here will result in an improperly designed counter.

TABLE 7-5 Design Table MOD-16 (J-K Flip-Flops)	Present State Q^n $Q_D Q_C Q_B Q_A$	Next State Q^{n+1} $Q_D Q_C Q_B Q_A$	\multicolumn{8}{c}{Transition Table Data}

Present State $Q_D Q_C Q_B Q_A$	Next State $Q_D Q_C Q_B Q_A$	Q_D J	Q_D K	Q_C J	Q_C K	Q_B J	Q_B K	Q_A J	Q_A K
0000	0001	0	X	0	X	0	X	1	X
0001	0010	0	X	0	X	1	X	X	1
0010	0011	0	X	0	X	X	0	1	X
0011	0100	0	X	1	X	X	1	X	1
0100	0101	0	X	X	0	0	X	1	X
0101	0110	0	X	X	0	1	X	X	1
0110	0111	0	X	X	0	X	0	1	X
0111	1000	1	X	X	1	X	1	X	1
1000	1001	X	0	0	X	0	X	1	X
1001	1010	X	0	0	X	1	X	X	1
1010	1011	X	0	0	X	X	0	1	X
1011	1100	X	0	1	X	X	1	X	1
1100	1101	X	0	X	0	0	X	1	X
1101	1110	X	0	X	0	1	X	X	1
1110	1111	X	0	X	0	X	0	1	X
1111	0000	X	1	X	1	X	1	X	1

The final design step prior to circuit implementation is to transfer the J and K data from the design table (7–5) to K-maps. The maps are used to simplify the J and K data to determine what type of gating circuits are required to implement the synchronous counter.

The K-maps presented in Chapter 4 were laid out as shown in Fig. 7–84. The decimal number placed within each square represents the decimal value of the binary number that square represents using A as the LSB. The adjacencies of the map and the looping process are used to simplify expressions. This can be accomplished as long as the map is designed using the gray code format. Several variations of labeling K-maps were previously presented. The key point to keep in mind when constructing all K-maps is that the gray code format must be adhered to. This format mandates that any horizontally or vertically adjacent squares can differ by only one variable.

	$\overline{C}\overline{D}$	$\overline{C}D$	CD	$C\overline{D}$
$\overline{A}\overline{B}$	0	8	12	4
$\overline{A}B$	2	10	14	6
AB	3	11	15	7
$A\overline{B}$	1	9	13	5

FIGURE 7–84   Standard Karnaugh map.

A variation of the K-map in Fig. 7–84 is presented in Fig. 7–85. The AB terms have been placed adjacent to the columns instead of the rows. The CD terms have been placed adjacent to the rows instead of the columns. Nonetheless, the gray code format has been used.

Another way of drawing this K-map is depicted in Fig. 7–86. The BA terms across the top of the map have been replaced with logic levels, and the B term has been placed before the A term. Likewise, the DC terms on the left side of the map have also been re-

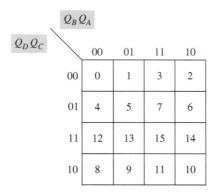

	$\overline{AB}$	$A\overline{B}$	AB	$\overline{A}B$
$\overline{CD}$	0	1	3	2
$C\overline{D}$	4	5	7	6
CD	12	13	15	14
$\overline{C}D$	8	9	11	10

FIGURE 7–85   Karnaugh map variation.

$Q_D Q_C$ \ $Q_B Q_A$

	00	01	11	10
00	0	1	3	2
01	4	5	7	6
11	12	13	15	14
10	8	9	11	10

FIGURE 7–86   Reference Karnaugh map.

placed with logic levels. The decimal value of every block on this map is identical to that of the same block on the K-map in Fig. 7–85. It is easy to see that the decimal values of each block are easier to identify using this map.

These maps are often referred to as **control matrices** or **excitation maps** by many in the digital field. However, they are K-maps and to keep matters simple, they will be called K-maps in this text. The principles of looping previously studied will be used to simplify the data and determine the gating requirements for a synchronous counter.

The map in Fig. 7–86 is called a **reference K-map** because it will be used to determine where to plot the J and K data from the design table. The decimal numbers annotated within each block simplify transferring this data.

The reference map shows how the MOD-16 counter increments through its 16 states. The K-maps for the MOD-16 counter are shown in Fig. 7–87. Notice there are two K-maps for flip-flop A. One map is for the J input (J_A), and one is for the K input (K_A). The cells in the map are used to plot the present state of the counter.

Data for the present state (Q^n) of 0000 in the Q_A column of Table 7–5 indicates $J = 1$. This data is transferred to the K-map #0 block ($\overline{D}\,\overline{C}\,\overline{B}\,\overline{A}$) on the J_A K-map. Likewise, the $K = X$ is entered in the K-map #0 block on the K_A K-map. The reference K-map in Fig. 7–86 should be used to aid this transfer of data from the design table to the K-maps until you are familiar with the numbers of the cells in the K-map.

The data for all 16 J_A and all 16 K_A blocks on the two K-maps for flip-flop A are transferred from the design table to the maps in the manner described above.

Both K-maps for flip-flop A contain all 1s or Xs in their cells. The X conditions are don't cares. Our study of K-map procedures taught us that the X condition could be plotted as a 0 or a 1—whatever fits the map best. The Xs are considered 1s in these two maps. Therefore, all 16 blocks in each map contain 1s. An example containing all 1s was not presented during our study of K-mapping. Using the procedures for simplifying will result in all 16 blocks being looped in a double octet because all 1s are vertically and/or

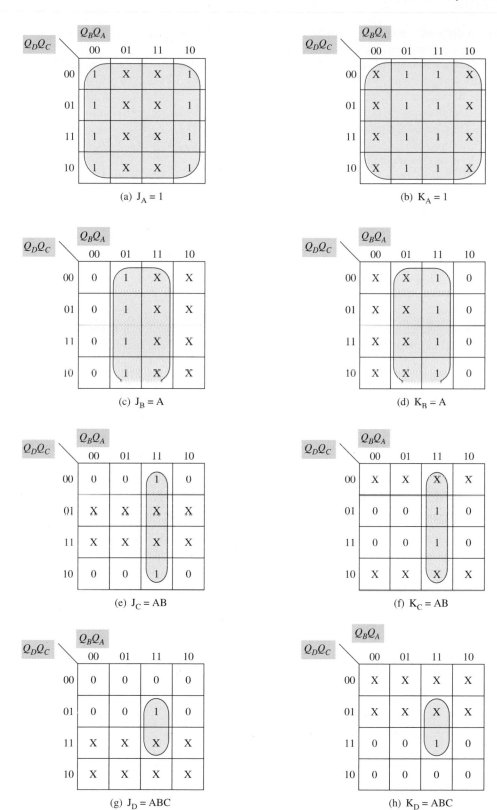

Figure 7–87   MOD-16 up-counter K-maps.

horizontally adjacent. The simplified expression after looping leaves $J_A = 1$ and $K_A = 1$ because all complementary variables are eliminated.

The two K-maps (J_B and K_B) for flip-flop B are completed using the design table data by following the same procedure. The result of looping the 1s and Xs on both maps simplifies the output to $J_B = K_B = A$. The octet loop in Fig. 7–87(c) or (d) for flip-flop B allows canceling all Cs and Ds on the left side and the complementary Bs at the top of the map. This leaves the simplified expression equal to A.

The two K-maps for flip-flop C in Fig. 7–87(e) and (f) produce a quad loop on each. This loop simplifies the output to AB.

The two K-maps for flip-flop D produce a pair on each. This looped pair results in the complementary Ds being canceled and simplifies the output to ABC.

The K-maps in Fig. 7–87 indicate the following counter equations:

$$J_A = 1, \quad K_A = 1$$
$$J_B = A, \quad K_B = A$$
$$J_C = AB, \quad K_C = AB$$
$$J_D = ABC, \quad K_D = ABC$$

Implementation of this MOD-16 synchronous up-counter is shown in Fig. 7–88. Note the counter is connected as indicated by the K-maps of Fig. 7–87 as follows:

J_A and K_A are connected to logic 1.

J_B and K_B are connected directly to the Q_A output of flip-flop A.

J_C and K_C are connected by an AND gate to both Q_A and Q_B.

J_D and K_D are connected by another AND gate to Q_A, Q_B, and Q_C.

MOD-10 Up-Counter Design

The MOD-10 decade counter design is presented in the following paragraphs. The six steps required to design a synchronous counter are presented in Fig. 7–89 as a quick review to enhance your grasp of the steps and to put the design process into one cohesive unit. Although the steps are presented separately in this review, they may be combined as they are in the MOD-10 counter design example.

STEP 1:   *State Table.* This table is used to show the desired count sequence of the counter.

STEP 2:   *Next State Table.* This table is used to indicate the desired next state of the counter after it has been clocked one time.

STEP 3:   *Transition Table.* This table is used to compare the present state (count) of the counter with the desired next state and to indicate the levels of J and K required to produce that next state.

STEP 4:   *Design Table.* This table provides an overall view of the design process by combining steps 1, 2, and 3. This table may be completed as the initial step once implementation practice has made you comfortable with the design process.

STEP 5:   *K-Maps.* The K-maps are used to simplify the levels of J and K that are required at each flip-flop's input to produce the proper counting sequence. The simplified expressions are extracted from the K-maps.

STEP 6:   *Counter Implementation.* The counter is designed using the counter equations simplified on the K-maps.

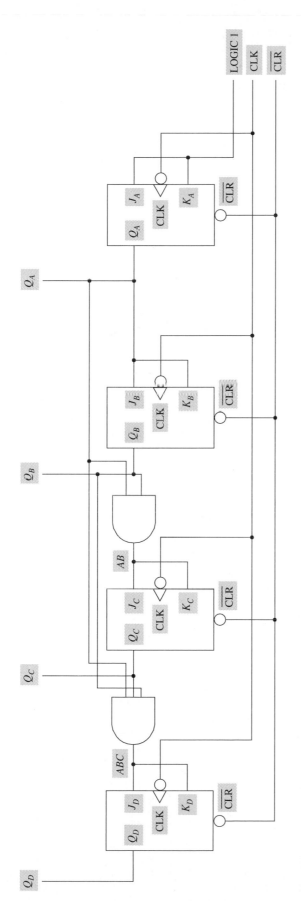

FIGURE 7-88   MOD-16 Synchronous Up-Counter logic diagram.

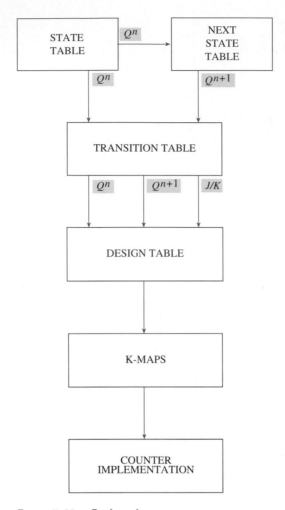

FIGURE 7–89   Design steps.

The state table and the next state table for the MOD-10 up-counter have been completed and incorporated in the **design table** shown in Table 7–6. This completes the first four steps of the design process.

The data on the design table has been mapped in Fig. 7–90. The K-maps contain Xs for all counts above 9. Looping the 1s and Xs produces the following counter equations:

TABLE 7-6	Present State	Next State	Transition Table Data							
Design Table **MOD-10** **(J-K Flip-Flops)**	Q^n	Q^{n+1}	Q_D		Q_C		Q_B		Q_A	
	$Q_D Q_C Q_B Q_A$	$Q_D Q_C Q_B Q_A$	J	K	J	K	J	K	J	K
	0000	0001	0	X	0	X	0	X	1	X
	0001	0010	0	X	0	X	1	X	X	1
	0010	0011	0	X	0	X	X	0	1	X
	0011	0100	0	X	1	X	X	1	X	1
	0100	0101	0	X	X	0	0	X	1	X
	0101	0110	0	X	X	0	1	X	X	1
	0110	0111	0	X	X	0	X	0	1	X
	0111	1000	1	X	X	1	X	1	X	1
	1000	1001	X	0	0	X	0	X	1	X
	1001	0000	X	1	0	X	0	X	X	1

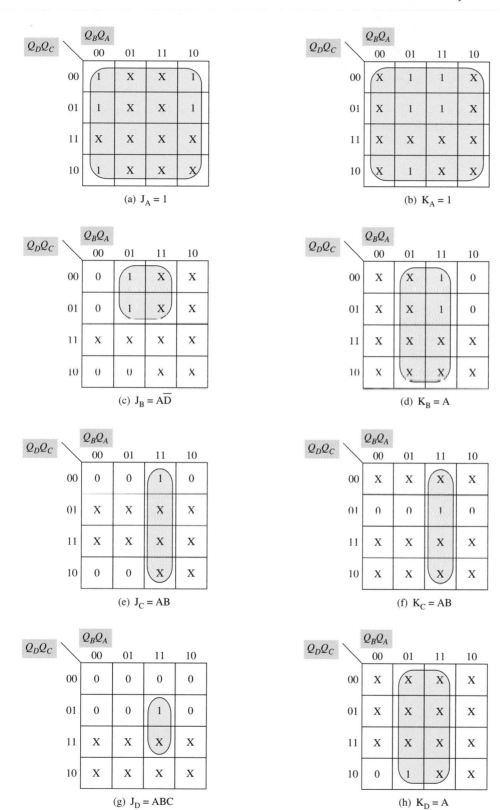

FIGURE 7-90   **MOD-10 up-counter K-maps.**

$$J_A = 1, \quad K_A = 1$$
$$J_B = A\overline{D}, \quad K_B = A$$
$$J_C = AB, \quad K_C = AB$$
$$J_D = ABC, \quad K_D = A$$

Figure 7–91 shows implementation of this MOD-10 synchronous up-counter. Implementation of the counter design equations shows

J_A and K_A are connected to Logic 1.

J_B is connected by an AND gate to Q_A and $\overline{Q}_D$. K_B is connected to Q_A only.

J_C and K_C are connected by an AND gate to the Q_A and Q_B outputs.

J_D is connected by a three-input AND gate to Q_A, Q_B, and Q_C. K_D is connected directly to Q_A.

MOD-6 Up-Counter Design

A useful counter in digital clock circuits is a MOD-6 counter. A MOD-6 counter is often cascaded with a MOD-10 counter to produce a MOD-60 circuit. The usefulness of a MOD-60 circuit in clocks is its ability to take a 60-Hz line signal and convert it to one pulse per second.

The design table for the MOD-6 synchronous up-counter is shown in Table 7–7. The design table has been slightly modified because only three flip-flops are required to implement this counter.

The count of this MOD-6 counter will be 000, 001, 010, 011, 100, 101, 000, and so on. The transition data has been transferred to the K-maps in Fig. 7–92 (on p. 402). The counter equations derived from looping on the maps are

$$J_A = 1, \quad K_A = 1$$
$$J_B = A\overline{C}, \quad K_B = A$$
$$J_C = AB, \quad K_C = A$$

The logic diagram of the circuit is shown in Fig. 7–93 (on p. 403).

Synchronous Counter Design Using D-Type Flip-Flops

Synchronous counters can be designed using D-type flip-flops instead of J-K flip-flops. As previously explained, a D-type flip-flop can be made to toggle when $\overline{Q}$ is connected back to the D input of the flip-flop and the flip-flop is clocked.

The same basic design tools used in J-K flip-flop design may be used when designing with D-type flip-flops. The transition table for implementation with D-type flip-flops has been modified and is presented in Table 7–8. The remaining design tools are the same as those previously used.

Table 7–9 shows the design table for a MOD-6 up-counter. The K-maps in Fig. 7–94 (on p. 404) produce the following counter equations for the D-inputs of the three flip-flops:

$$D_A = \overline{A}$$
$$D_B = \overline{A}B + A\overline{B}\,\overline{C}$$
$$D_C = AB + \overline{A}C$$

The equation for D_A shows $\overline{Q}_A$ must be tied back to the D_A input. This places the LSB flip-flop in the permanent toggle mode. Implementation of the counter is shown in Fig. 7–95. Drawing the sum-of-products circuits as shown at the D_B and D_C inputs of the flip-flops is becoming standard practice in many data books.

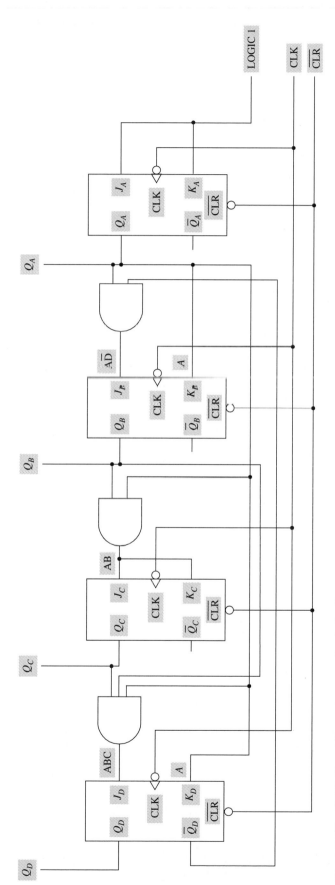

FIGURE 7-91   MOD-10 synchronous up-counter logic diagram.

TABLE 7-7 Design Table MOD-6 (J-K Flip-Flops)	Present State	Next State	Transition Table Data					
	Q^n	Q^{n+1}	Q_C		Q_B		Q_A	
	$Q_C Q_B Q_A$	$Q_C Q_B Q_A$	J	K	J	K	J	K
	000	001	0	X	0	X	1	X
	001	010	0	X	1	X	X	1
	010	011	0	X	X	0	1	X
	011	100	1	X	X	1	X	1
	100	101	X	0	0	X	1	X
	101	000	X	1	0	X	X	1

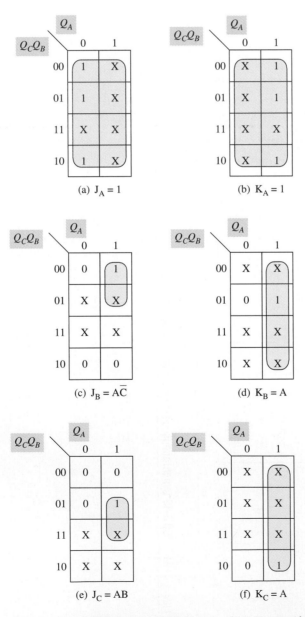

(a) $J_A = 1$ (b) $K_A = 1$

(c) $J_B = A\overline{C}$ (d) $K_B = A$

(e) $J_C = AB$ (f) $K_C = A$

FIGURE 7-92   **MOD-6 up-counter K-maps (J-K flip-flops).**

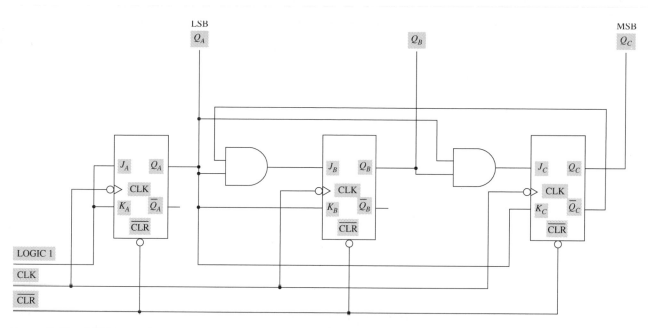

FIGURE 7-93   **MOD-6 synchronous up-counter logic diagram (J-K flip-flops).**

TABLE 7-8	Present State	Next State	Required Levels for Q^{n+1}
Transition Table for D-Type Flip-Flops	Q^n	Q^{n+1}	D
	0	0	0
	0	1	1
	1	1	1
	1	0	0

TABLE 7-9	Present State	Next State	Transition Table Data		
Design Table MOD-6 (D-Type Flip-Flops)	Q^n	Q^{n+1}	Q_C	Q_B	Q_A
	$Q_C Q_B Q_A$	$Q_C Q_B Q_A$	D_C	D_B	D_A
	000	001	0	0	1
	001	010	0	1	0
	010	011	0	1	1
	011	100	1	0	0
	100	101	1	0	1
	101	000	0	0	0

SECTION 7-6: INTERNAL SUMMARY

This section has introduced the concept of sequential circuit design procedures applicable to synchronous counters. The design steps can be used to design other types of counters. For example, down-counters and gray code counters produce a challenge to the novice designer. The key is to follow the steps presented in this section.

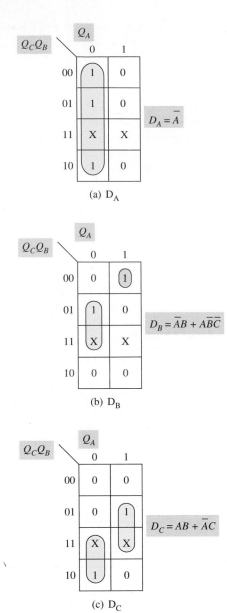

(a) D_A

$D_A = \overline{A}$

(b) D_B

$D_B = \overline{A}B + A\overline{B}\,\overline{C}$

(c) D_C

$D_C = AB + \overline{A}C$

FIGURE 7-94   MOD-6 up-counter K-maps (D-type flip-flops).

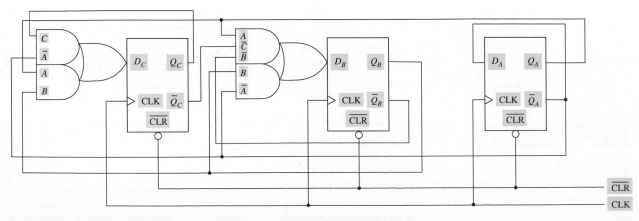

FIGURE 7-95   MOD-6 synchronous up-counter logic diagram (D-type flip-flops).

STEP 1:   Complete the **state table.** This table shows the desired count sequence.

STEP 2:   Complete the **next state table.** This table shows the desired next count of the counter when it is clocked. This next state does not have to be a sequential increment or decrement. Any sequence of counts can be selected.

STEP 3:   Use the **transition table** to indicate what levels of *J* and *K* or *D* are required to produce the desired next state.

STEP 4:   The **design table** may be used to show steps 1, 2, and 3 on one table.

STEP 5:   Simplify the levels of *J* and *K* or *D* on **Karnaugh maps.** The resulting simplified expressions are the counter equations required to implement the counter.

STEP 6:   **Implement** the circuit.

SECTION 7–7: HYBRID COUNTERS

OBJECTIVE

Given a data book and specified hybrid counter, determine the required inputs/connections to derive a symmetrical divide-by-*n* output.

Hybrid counters, as the name implies, incorporate both synchronous and asynchronous clocking methods. These counters can produce **symmetrical divide-by-*n* outputs.** Synchronous and asynchronous counters that are not allowed to count their full-count cycle do not produce symmetrical (50% duty cycle) output signals at their MSB.

The 74LS163 synchronous 4-bit binary counter discussed in Section 7–5 is truncated to MOD-10 in Fig. 7–96(a). The waveforms produced by the counter are shown in Fig. 7–96(b). The Q_D output is $f_{in}/10$, but it is not a symmetrical waveform because the 4-bit counter was not allowed to count its full-count sequence (0000–1111).

Hybrid counters solve this asymmetrical output problem. They incorporate synchronous clocking to some flip-flops and asynchronous clocking to others.

74LS92.

The 74LS92 divide-by-12 or divide-by-6 counter is a prime example of a commercially available hybrid counter. Figure 7–97 shows the logic symbol and diagram of this counter. The A flip-flop is clocked by the CLK *A* input. This flip-flop operates in the TOGGLE mode at all times because *J* and *K* are internally tied high. The Q_A output of this divide-by-2 flip-flop is not connected internally to the remaining flip-flops in the counter. This type of input setup is like the 74LS93 asynchronous counter presented in Section 7–2. This means that Q_A must be externally connected to the CLK *B* input to obtain 4-bit operation. The CLK *B* input is synchronous to flip-flops B and C. The Q_C output of flip-flop C provides an asynchronous clock input to flip-flop D. In this counter it is easy to see that we have synchronous and asynchronous clocking methods being used that cause this IC to be classified as a hybrid counter.

Before delving into 4-bit operation, the circuit will be analyzed in detail for 3-bit operation. The CLK *A* input and Q_A output are not used for 3-bit operation.

MOD-6 Operation.   The CLK *B* input is used for the 3-stage section of the counter. The Q_D output will be a symmetrical divide-by-6 signal. The count sequence of the counter is not important and not sequential. The counter is used for frequency division when a sym-

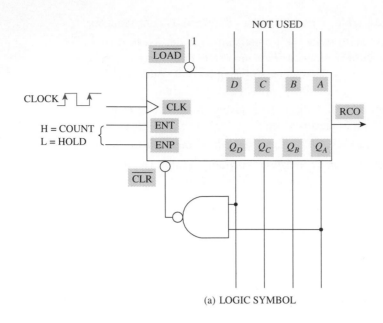

(a) LOGIC SYMBOL

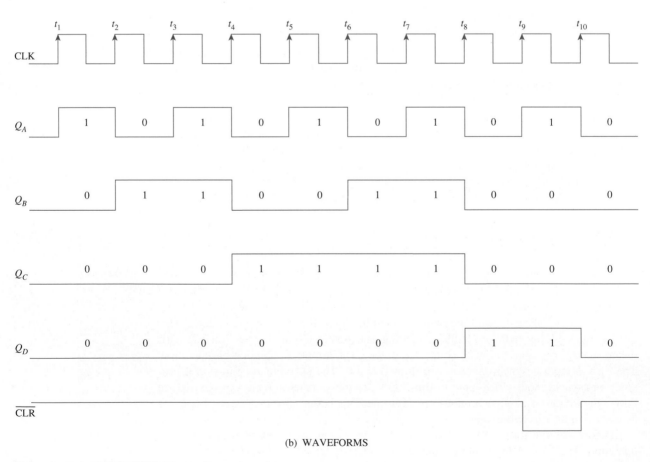

(b) WAVEFORMS

Figure 7-96   74LS163 MOD-10 counter.

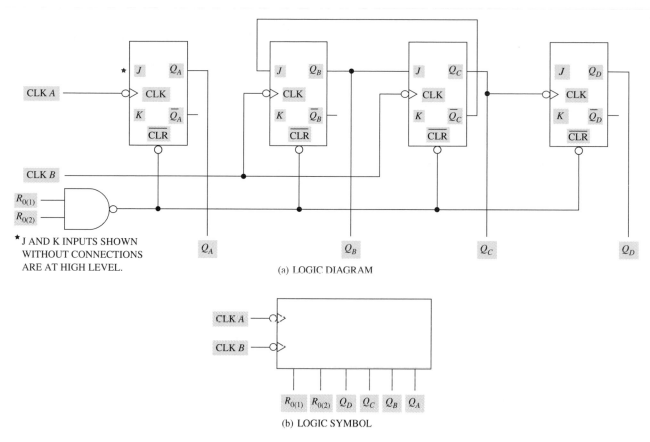

*J AND K INPUTS SHOWN
WITHOUT CONNECTIONS
ARE AT HIGH LEVEL.

(a) LOGIC DIAGRAM

(b) LOGIC SYMBOL

FIGURE 7–97   **74LS92 hybrid counter.**

metrical output is a necessity. The 3-stage section of the counter used for MOD-6 operation has been reproduced in Fig. 7–98(a). The RESET inputs ($R_{0(1)}$ and $R_{0(2)}$) are tied to ground. If either of these inputs is low, the output of the NAND gate will be high, and the circuit will count. Both inputs must be taken high simultaneously to clear the counter.

In-Depth Look at Operation

A count sequence chart is provided in Fig. 7–98(b) to aid this analysis. Refer to the chart during the following discussion.

t_0: Start in the CLEAR state.

t_1: Since the counter is in the CLEAR state prior to t_1, $\overline{Q_C}$ is high and flip-flop B is in the TOGGLE mode of operation when t_1 arrives. Therefore, Q_B goes high at t_1. However, Q_B was low immediately prior to the arrival of t_1. When t_1 arrives the inputs to flip-flop C are $J = 0$ and $K = 1$. Arrival of the NGT of the clock input at t_1 causes no change in Q_C because its data inputs request the CLEAR state and the flip-flop is already cleared. Since there is no change of the Q_C output at t_1, there is no NGT applied to the clock input of the D flip-flop; thus, Q_D dose not change states.

The output is now $Q_B = 1$, $Q_C = 0$, and $Q_D = 0$.
Note: The key to analyzing this counter's operation is to look at the J inputs of flip-flops B and C immediately prior to arrival of the clock NGT. The K inputs of these two flip-flops are tied internally high. Flip-flop D has its J and K inputs tied high, so it will toggle anytime Q_C transitions from high to low (NGT).

t_2: Since $\overline{Q_C}$ is still high after t_1, the B flip-flop is still in the TOGGLE mode of operation when t_2 arrives. This causes Q_B to toggle low at t_2. Keep in mind that Q_B was high immediately prior to the arrival of the NGT of the clock at t_2. This high at the J input of the C flip-flop causes Q_C to toggle high at t_2. Since Q_C toggled from a low to a high level at t_2, the PGT it produces has no affect on flip-flop D because it is a negative-edge-triggered flip-flop. Therefore, Q_D remains low.

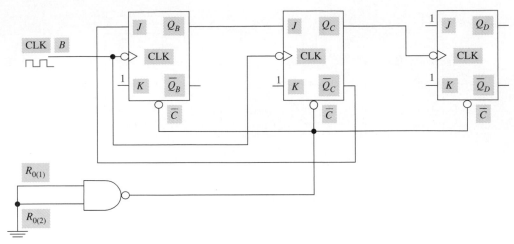

(a) LOGIC DIAGRAM

	Q_B	Q_C	Q_D
CLR	0	0	0
t_1	1	0	0
t_2	0	1	0
t_3	0	0	1
t_4	1	0	1
t_5	0	1	1
t_6	0	0	0

(b) COUNT SEQUENCE

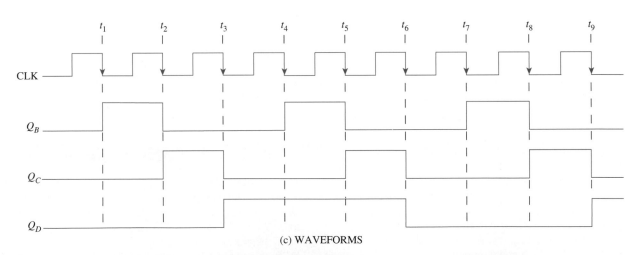

(c) WAVEFORMS

FIGURE 7–98   **74LS92 MOD-6 counter.**

The output is now $Q_B = 0$, $Q_C = 1$, and $Q_D = 0$.

t_3: Since Q_C toggled high at t_2, $\overline{Q}_C$ is now low. This places a logic low on the J input of flip-flop B. This flip-flop was cleared at t_2. Therefore, it will not change states at t_3. The J input to flip-flop C is low immediately prior to t_3. Furthermore, this flip-flop is in the SET state prior to t_3. Thus, the NGT of the clock at t_3 clears flip-flop C. Since the Q_C output goes from high-to-low (NGT) at t_3, this level transition clocks flip-flop D and causes it to toggle.

The output is now $Q_B = 0$, $Q_C = 0$, and $Q_D = 1$.

t_4: Q_C is low and $\overline{Q}_C$ is high immediately prior to arrival of the t_4 clock pulse. This puts flip-flop B in the TOGGLE mode of operation. When the t_4 clock pulse arrives, the output at Q_B will toggle high. Since Q_B was low prior to t_4, the J and K inputs to the C flip-flop require the circuit to clear at t_4. Since it is already cleared, no change will result. Furthermore, since there is no change at the output of Q_C, there is no level transition to clock flip-flop D, and its output stays high.

The output is now $Q_B = 1$, $Q_C = 0$, and $Q_D = 1$.

t_5: Again note that $\overline{Q}_C = 1$ and flip-flop B is in the

TOGGLE mode of operation after t_4. Therefore, arrival of the t_5 NGT will cause Q_B to toggle low. Also, immediately prior to t_5 Q_B was high. This causes Q_C to toggle high at t_5. The Q_C output transition produces a PGT at the input of flip-flop D. This PGT has no affect on flip-flop D.

The output is now $Q_B = 0$, $Q_C = 1$, and $Q_D = 1$.

t_6: $\overline{Q}_C$ is low after t_5. This input requests the B flip-flop to clear at t_6. Since the flip-flop was cleared at t_5, no change occurs at Q_B when the t_6 clock pulse arrives. The J input to the C flip-flop was low prior to the arrival of the clock at t_6. The t_6 clock NGT clears the C flip-flop because J is low and K is high when it arrives. The action of Q_C at t_6, high-to-low transition, clocks flip-flop D and causes Q_D to toggle low.

The output has returned to $Q_B = 0$, $Q_C = 0$, and $Q_D = 0$.

The chart in Fig. 7-98(b) indicates six distinct states including 000. Therefore, this 74LS92 is configured as a MOD-6 counter. The waveforms in Fig. 7-98(c) verify the output at Q_D is the clock input frequency divided-by-6, and this output is symmetrical.

Mod-12 Operation. The 74LS92 is shown configured for divide-by-12 operation in Fig. 7-99(a). The clock input is applied to CLK A, and Q_A has been connected to the CLK B input. This configures the counter for 4-bit operation.

Flip-flop A is internally connected in the TOGGLE mode of operation. This divides the CLK A input frequency by 2 before it is applied to the CLK B input. This divided-by-2 clock input produces outputs from the 3-bit section of the counter that are identical to those obtained during MOD-6 operation *except their output duration is twice as long.* Your basic understanding of time and frequency should make this clear. If the input frequency to the MOD-6 section is divided in half, the time of each output waveform should double. This is evident if you compare the MOD-12 counter output waveforms shown in Fig. 7-99(b) with those of the MOD-6 counter waveforms in Fig. 7-98(c). The Q_D output in Fig. 7-99(b) is a symmetrical output equal to $f_{in}/12$.

74LS90.

The 74LS90 divide-by-10 or divide-by-5 counter is another popular hybrid counter. Figure 7-100(a) shows the logic diagram of the counter. The logic symbol is shown in Fig. 7-100(b). Without delving too deeply into circuit analysis, the highlights of this counter's operation are presented next.

The function table for the 74LS90 is shown in Fig. 7-101 (on p. 412). The table indicates the counter will clear (0000) when both reset-to-zero inputs ($R_{0(1)}$ and $R_{0(2)}$) are asserted as long as one of the R_9 inputs is low. It also shows the counter will set to $1001_{(2)}$ when both of the reset-to-9 inputs ($R_{9(1)}$ and $R_{9(2)}$) are asserted. The R_9 inputs are for use in BCD 9s complement applications (Chapter 9). Since the truncating gates are NAND gates (any 0 in = 1 out), the table shows as long as one input to each gate is low the counter will count.

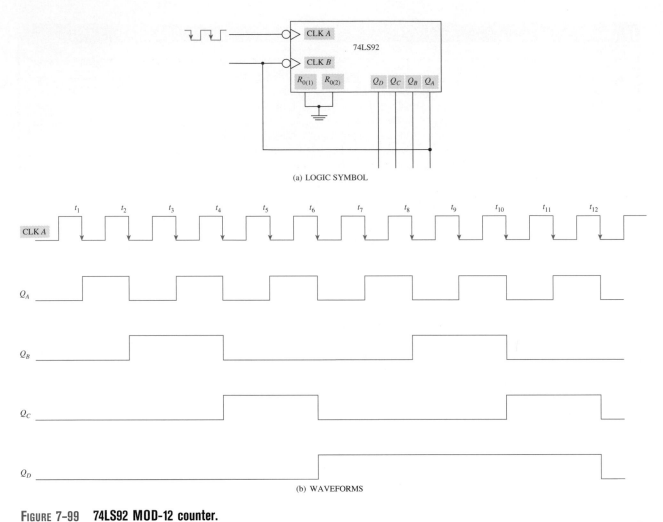

(a) LOGIC SYMBOL

(b) WAVEFORMS

FIGURE 7–99 **74LS92 MOD-12 counter.**

MOD-5 Operation. Figure 7–102(a) shows the configuration for divide-by-5 operation of the 74LS90 IC. The A flip-flop is not used since the clock pulse is applied to the CLK B input. The Q_D output of this configuration is the input frequency divided by 5.

MOD-10 BCD Operation. The counter is shown in Fig. 7–102(b) configured as a BCD counter. The output sequentially up-counts from 0000 to 1001 and then resets to 0000. Although the Q_D output of this counter is $f_{in}/10$, the output is not symmetrical. Typically, this is not important when counting in BCD. If the counter is to be used for frequency division and a symmetrical divide-by-10 output is required, the counter can be reconfigured as shown in Fig. 7–102(c).

MOD-10 Bi-Quinary Operation. The clock input for this configuration is applied to CLK *B*, and the Q_D output is connected to CLK *A*. The counter is now configured as a MOD-5 counter cascaded with a MOD-2 counter. As the name implies, bi-quinary (2-5) operation produces a MOD-10 counter. The count sequence is shown in Fig. 7–103. The count sequence is not sequential or important when the counter is configured for bi-quinary operation. What is important is shown in the count sequence—Q_A is equal to $f_{in}/10$ and symmetrical.

The MOD-5 section of the counter produces the count that is highlighted in Fig. 7–103. Each time the MOD-5 section completes its sequence, an NGT from Q_D is provided to flip-flop A to further divide the counter's output at Q_A by 2.

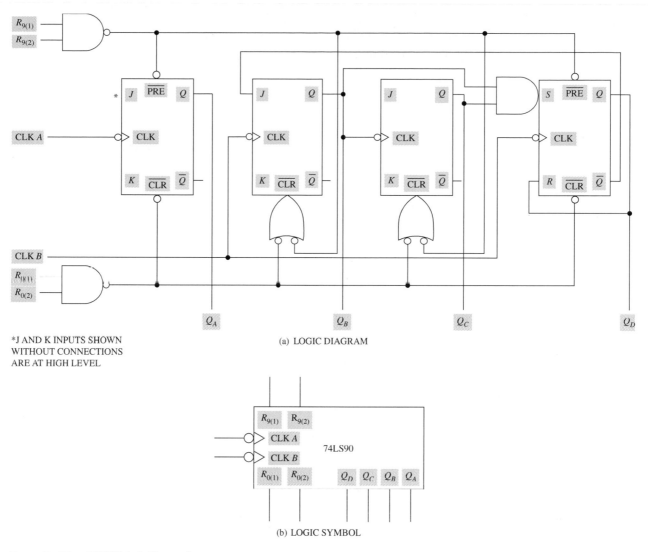

FIGURE 7–100   74LS90 hybrid counter.

74LS290.

This hybrid 14-pin divide-by-10 or divide-by-5 counter is functionally and electrically identical to the 74LS90 counter. Only the arrangement of the pins on the IC has been changed.

Section 7-7: Review Questions

Answers are given at the end of the chapter.

A. What is the primary advantage of a hybrid counter over a synchronous or asynchronous counter?
B. Hybrid counter ICs employ both synchronous and asynchronous internal clocking.
    (1) True
    (2) False
C. Draw the external connections required to make the 74LS92 shown in Fig. 7–104 a divide-by-6 counter. Show the input clock pulses.

RESET INPUTS				OUTPUT			
$R_{0(1)}$	$R_{0(2)}$	$R_{9(1)}$	$R_{9(2)}$	Q_D	Q_C	Q_B	Q_A
1	1	0	X	0	0	0	0
1	1	X	0	0	0	0	0
X	X	1	1	1	0	0	1
X	0	X	0	COUNT			
0	X	0	X	COUNT			
0	X	X	0	COUNT			
X	0	0	X	COUNT			

FIGURE 7–101   74LS90 counter function table.

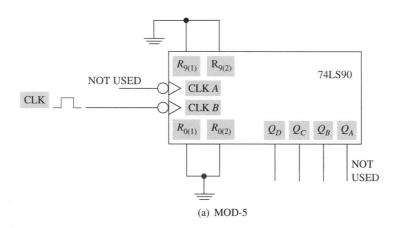

(a) MOD-5

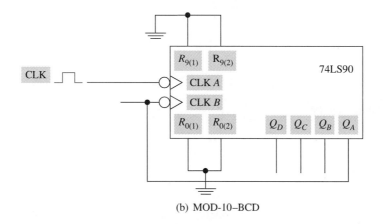

(b) MOD-10–BCD

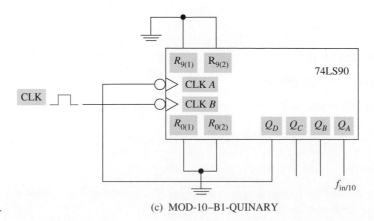

FIGURE 7–102   74LS90 counter.

(c) MOD-10–B1-QUINARY

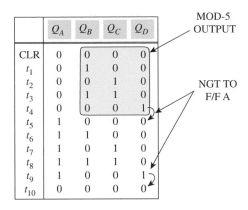

	Q_A	Q_B	Q_C	Q_D
CLR	0	0	0	0
t_1	0	1	0	0
t_2	0	0	1	0
t_3	0	1	1	0
t_4	0	0	0	1
t_5	1	0	0	0
t_6	1	1	0	0
t_7	1	0	1	0
t_8	1	1	1	0
t_9	1	0	0	1
t_{10}	0	0	0	0

MOD-5 OUTPUT

NGT TO F/F A

FIGURE 7–103   74LS90 bi-quinary count sequence.

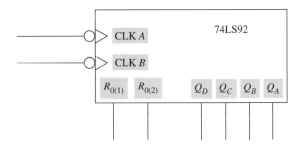

FIGURE 7–104

D. Draw the external connections required to make the 74LS92 shown in Fig. 7–104 a divide-by-12 counter. Show the input clock pulses.

SECTION 7–8: COUNTER DECODING

OBJECTIVES

1. Identify active-high and active-low output decoders.
2. Given the desired decoded outputs (number and level) and a specific IC counter, design the decoding circuit with strobing.

Counter outputs often require **decoding** in order to produce output counts that are usable. **Decoding** is the process of identifying and/or converting a binary number or code. The basics of counter decoding have already been applied in counters that were truncated. The truncating gate was nothing more than a logic gate used to decode a specific count for clearing or presetting the counter.

The output counts of a sequential up/down-counter often must be decoded to enable other digital circuits at a specified time. The output count(s) of a MOD-8 sequential up-counter as shown in Fig. 7–105 cannot be used to initialize an action in a system unless the desired count is decoded. The count sequence of the counter in Fig. 7–105 is 000, 001, 010, 011, 100, 101, 110, 111, 000, and so on.

How could an output count of 110 from the counter be used to activate a circuit that required a high input? A count of 110 produces $Q_C = 1$, $Q_B = 1$, and $\overline{Q}_A = 1$. It is these three counter outputs that must be connected to the decoding gate. The counter in Fig. 7–106 shows how this problem is resolved. The Q_C, Q_B, and $\overline{Q}_A$ outputs of the counter are connected to the inputs of the AND decoding gate. When the count of 110 occurs, all

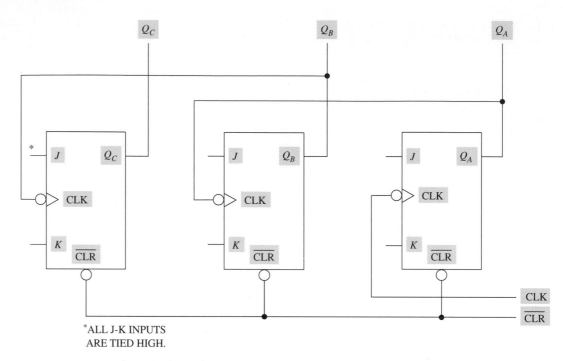

*ALL J-K INPUTS
ARE TIED HIGH.

FIGURE 7-105 MOD-8 asynchronous up-counter.

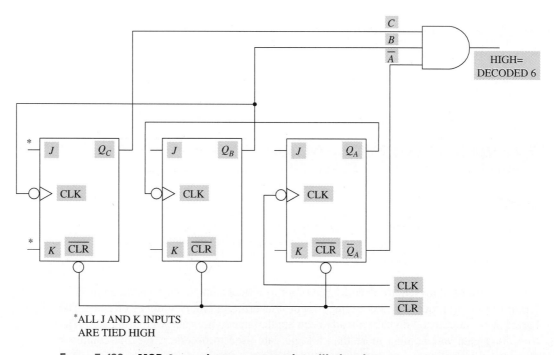

*ALL J AND K INPUTS
ARE TIED HIGH

FIGURE 7-106 MOD-8 asynchronous up-counter with decoder.

of the inputs to the AND gate are high and the output goes high. This type of decoding using an AND gate is called **active-high decoding.**

The rule of truncation cannot be used when decoding a counter. The reason is the rule states "tie the bits that are high . . . to the truncating gate." The problem that would be encountered if the rule were applied to decoding is that only Q_C and Q_B would be connected to the decoding gate. If only these two bits were connected to the decoding gate in the preceding example, the gate would decode both a count of 110 and 111. Thus, the output from each counter flip-flop must be connected to the decoder gate. The guidance here is to tie the bits that are high on the desired decoded count ($\underline{11}0 = Q_C$ and Q_B) to the decoder gate and tie the *complement* of the bits that are low on the desired decoded output count ($11\,\underline{0} = Q_A$) to the decoder as shown in Fig. 7–106.

If the previous problem had asked how an output count of 110 from the counter could be used to activate a circuit that required a *low* input, the answer would have been the same except a NAND gate would have been used instead of the AND gate. The use of NAND gates for decoding produces **active-low output decoding.**

A MOD-4 up-counter is shown with all four of its output states decoded in Fig. 7–107. The output of the $\overline{Q}_B\,\overline{Q}_A$ AND decoder goes high only on a count of 00.

One prevalent problem is encountered when decoding the output of an asynchronous counter. The counter's output typically consists of short-duration miscounts as previously discussed. These counts have little effect if the decoded output is being used to drive LEDs or seven-segment indicators. However, if the decoded output is to be used to clock a circuit or clear or set a flip-flop, the results of decoded miscounts can cause serious problems.

The remedy for this problem is to "mask out" these short-duration miscounts, or glitches. This can be done with a **strobe pulse.** Glitch periods always occur immediately after the active clock transition to the counter. The strobe pulse inhibits the decoding gate(s) during the counter glitch periods and thereby prevents erroneous decoder outputs. The strobe pulse does not prevent the glitches from asynchronous counters from occurring, but it prevents the short-duration miscounts from being decoded. One simple method of generating a strobe pulse is shown in Fig. 7–108(a). A 7493 asynchronous IC up-counter is shown configured as a MOD-8 counter in the figure. The count of 101 is decoded by the active-low output NAND gate.

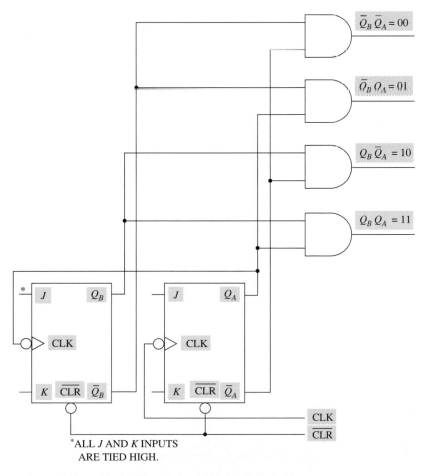

FIGURE 7–107   **MOD-4 up-counter with output decoders.**

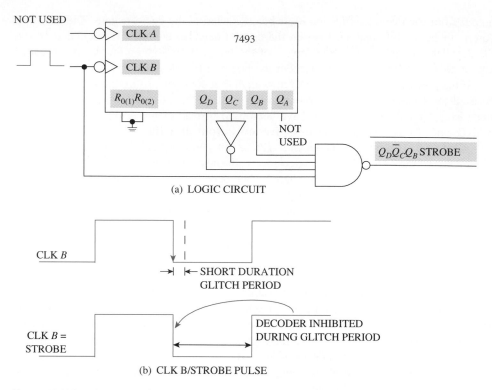

(a) LOGIC CIRCUIT

(b) CLK B/STROBE PULSE

FIGURE 7-108   **7493 asynchronous counter with strobed decoder.**

The desired decoded output occurs when $Q_D = 1$, $Q_C = 0$, and $Q_B = 1$. The Q_A output is not used in this configuration. Since Q_C is low on the desired count, it is inverted by the NOT gate. The Boolean expression at the decoder output indicates a low output when Q_D is high, Q_C is low, Q_B is high, and Strobe is high. The CLK B input is used as the Strobe pulse for the decoder.

The CLK B input and strobe pulse are shown in Fig. 7–108(b). The decoder gate is enabled when the CLK B input is high. The NGT of the CLK B input triggers the counter to increment one count. It is immediately after this active clock transition that glitches can occur. However, during this period the strobe pulse is low and the decoder gate is inhibited.

When numerous count states must be decoded, an IC decoder should be used. Much more information on decoders will be presented when medium-scale integrated circuits are covered in Chapter 10.

Section 7-8: Review Questions

Answers are given at the end of the chapter.

A. Decoding is the process of identifying and/or converting a binary number or code.
 (1) True
 (2) False

B. NAND gates are used when active- _____ decoding is desired.
 (1) low
 (2) high

C.   Strobing decoder gates prevents erroneous decoder outputs.
 (1) True
 (2) False

D. Draw a 3-input AND gate connected to decode a MOD-8 counter's 001 output.
E. Refer to question (D). Is the AND gate's output low or high on the count of 001?
F. Draw a 4-input NAND gate connected to decode a BCD counter's $1001_{(BCD)}$ output.
G. Refer to question (F). Is the NAND gate's output low or high on the count of 1001?

SECTION 7-9: SHIFT-REGISTER COUNTERS

OBJECTIVES

1. Identify and determine the MOD number of Ring and Johnson counters.
2. Given a data book and a specified IC counter, determine the required inputs/connections to derive a specific divide-by-n frequency output.

Shift registers were presented in Section 6–7. The discussion in Section 6–7 was limited to serially shifting data through a shift register. The data shifted through the register was connected only to the first flip-flop. That data was shifted one flip-flop to the right with each input clock pulse. One of the key points emphasized in that section needs to be reemphasized here. A flip-flop will respond to the data that is on its input immediately prior to the active clock transition. This key point relates to a flip flop's **setup time.** A couple of configurations of shift registers make useful counters. These configurations form **Ring counters** and **Johnson counters.**

Ring Counter.

A Ring counter is a shift register whose output is connected back to its own input. This is shown in Fig. 7–109. The bit pattern developed by a Ring counter shifts through the counter stages. This pattern can be used to control a sequence of events such as the events required in computer or video recorder operations.

The Ring counter in Fig. 7–109 is implemented with PGT-triggered D-type flip-flops. Note the output of flip-flop C is connected back to the data input of flip-flop A.

The Ring counter must be *preloaded* with a 1 to function properly. The purpose of the counter is to rotate the 1 through the counter one flip-flop at a time per clock pulse. If the counter were started in the CLEAR state, 0s would be continually rotated through the counter and nothing would be accomplished. The Ring counter in Fig. 7–109 is loaded with a one by momentarily activating the Preload input. This action sets flip-flop A and

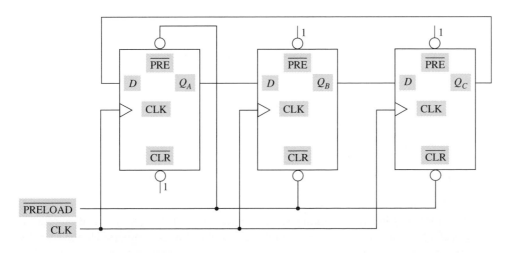

FIGURE 7-109   3-bit Ring counter.

clears flip-flops B and C. The initial count is 100. The data is now serially shifted through the register one flip-flop for each active clock transition. The count sequence is

$$
\begin{array}{ll}
100 - \overline{\text{Preload}} \\
010 - t_1 \\
001 - t_2 \\
100 - t_3
\end{array}
$$

The waveforms in Fig. 7–110 show the frequency at any one of the Q outputs is $f_{\text{in}}/3$. *The MOD number of a Ring counter is equal to the number of flip-flops in the counter.* Each flip-flop in this counter receives the high bit one time for every three clock pulses. The waveforms show that the number of states in the Ring counter is equal to the number of flip-flops. Therefore, the output ($f_{\text{in}}/3$) can be taken from any flip-flop and does not require decoding gates.

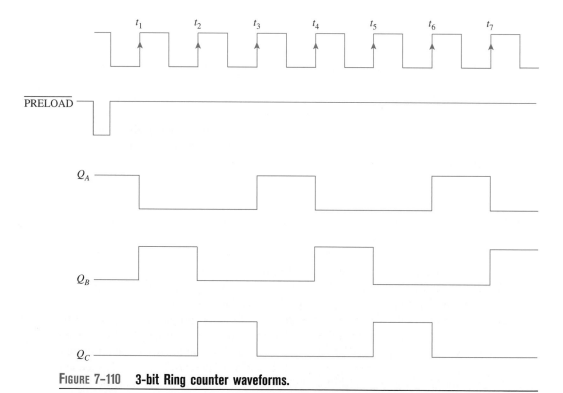

FIGURE 7–110   3-bit Ring counter waveforms.

A MOD-5 Ring counter can be implemented by using five flip-flops. The Ring counter in Fig. 7–111 is implemented with J-K flip-flops. The same preloading setup used in the last Ring counter is used to set one flip-flop and clear the rest.
The count sequence is

10000
01000
00100
00010
00001
10000

Figure 7–112 shows the output waveforms of this Ring counter.
The problem with implementing Ring counters with flip-flops is that they require more flip-flops than standard counters with comparable MOD numbers. However, their use is sometimes advantageous because their outputs do not require decoding gates. This does away with the propagation delays associated with those decoding gates.

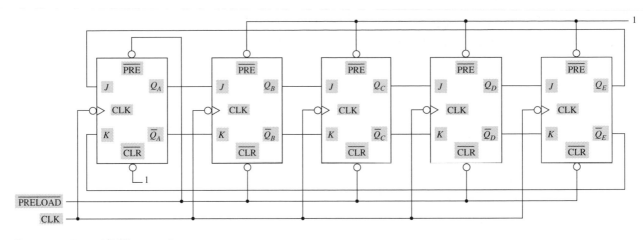

FIGURE 7–111   5-bit Ring counter.

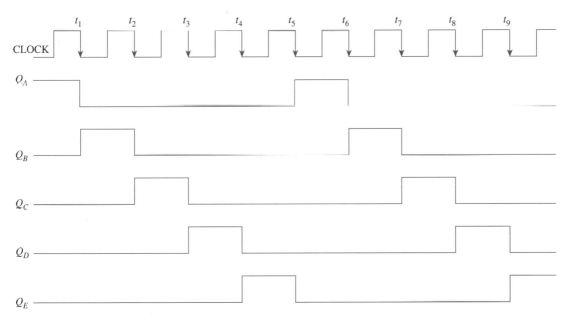

FIGURE 7–112   5-bit Ring counter waveforms.

Johnson Counter.

The 3-bit Ring counter in Fig. 7–109 has been modified to a Johnson counter in Fig. 7–113. The modification consists of connecting $\overline{Q_C}$ back to the data input of flip-flop A in lieu of Q_C.

The counter can be cleared by momentarily asserting the $\overline{\text{CLR}}$ input. Since $\overline{Q_C}$ is connected back to the data input of flip-flop A, that flip-flop will set upon arrival of the first active clock transition. In a sense the counter is self-loading because $\overline{Q}$ produces the bits for the shifting pattern. The bit pattern is:

000	$\overline{\text{CLR}}$
100	t_1
110	t_2
111	t_3
011	t_4
001	t_5
000	t_6

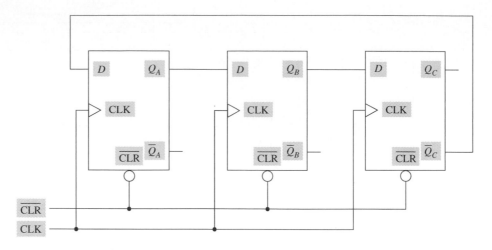

FIGURE 7–113   MOD-6 Johnson counter.

The bit pattern shows the counter fills up with 1s and then fills up with 0s. The pattern shows six distinct states. This indicates *the MOD number of a Johnson counter is equal to two times the number of flip-flops in the counter.*

Figure 7–114(a) shows a MOD-8 Johnson counter with decoding gates to detect any sequence of the counters output. A standard 4-bit binary counter would require a 4-input AND or NAND gate to detect an output condition. The Johnson counter requires only 2-input AND or NAND gate decoders because of its unique bit pattern. The counter is sometimes called a **twisted-ring counter** because of this bit pattern. The pattern shown is evidence of this uniqueness:

Q_A	Q_B	Q_C	Q_D	Decode
0	0	0	0	$\overline{Q_A}\,\overline{Q_D}$
1	0	0	0	$Q_A\overline{Q_B}$
1	1	0	0	$Q_B\overline{Q_C}$
1	1	1	0	$Q_C\overline{Q_D}$
1	1	1	1	$Q_A Q_D$
0	1	1	1	$\overline{Q_A}Q_B$
0	0	1	1	$\overline{Q_B}Q_C$
0	0	0	1	$\overline{Q_C}Q_D$
0	0	0	0	$\overline{Q_A}\,\overline{Q_D}$

The CLEAR condition can be decoded by ANDing or NANDing $\overline{Q_A}$ and $\overline{Q_D}$. Likewise the all-high output condition can be decoded using Q_A and Q_D. All of the remaining states are decoded by ANDing or NANDing the Q output of a flip-flop with the $\overline{Q}$ output of the next flip-flop or by ANDing or NANDing the $\overline{Q}$ output of a flip-flop with the Q output of the next flip-flop. This is true regardless of the number of flip-flops contained in a Johnson counter. The AND gates shown in Fig. 7–114(a) will detect all counter output conditions. The counter and decoder output waveforms are shown in Fig. 7–114(b).

14018.

The 14018 IC is a presettable divide-by-*n* counter that contains five Johnson counter stages. The logic symbol and function table are shown in Fig. 7–115(a) and (b).

The function table shows the complement of the data at a given D input will be clocked to $\overline{Q}$ on the PGT of the clock pulse providing RESET and Preset Enable (PE) are low. It also shows the complement of the JAM input data is asynchronously loaded to $\overline{Q}_n$ when

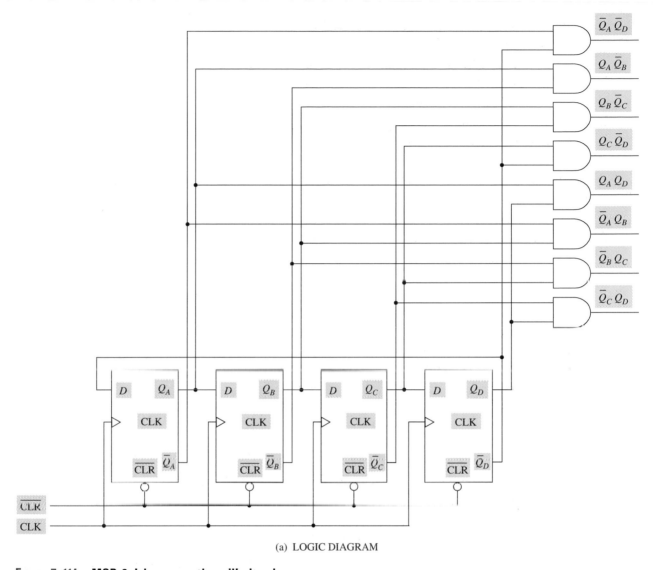

(a) LOGIC DIAGRAM

FIGURE 7–114   **MOD-8 Johnson counter with decoder.**

PE is activated and RESET is inactive. The term "jam" refers to the fact that a condition can be forced into a flip-flop regardless of its data input or the clock. In other words, this is an asynchronous load. The counter will reset $(\overline{Q}_n = 1)$ asynchronously when RESET is activated.

The logic diagram and function select table are shown in Fig. 7–116. The 14018 can be configured to divide by any number from 2 to 10.

14018 Divide-by-10.   The Johnson counter will divide by 10 when $\overline{Q}_5$ is connected to the data input. This is the standard Johnson counter configuration discussed previously for a 5-flip-flop Johnson counter. This configuration is shown in Fig. 7–117.

14018 Divide-by-6.   The counter will divide by 6 when $\overline{Q}_3$ is connected to the data input. This effectively makes the circuit a 3-flip-flop Johnson counter whose MOD number is two times the number of flip-flops.

14018 Divide-by-5.   The function select table indicates $\overline{Q}_3$ must be ANDed with $\overline{Q}_2$ to perform the divide-by-5 function. The circuit in Fig. 7–118 shows the MOD-5 configuration. The circuit uses an external AND gate per instructions in the function select table. The

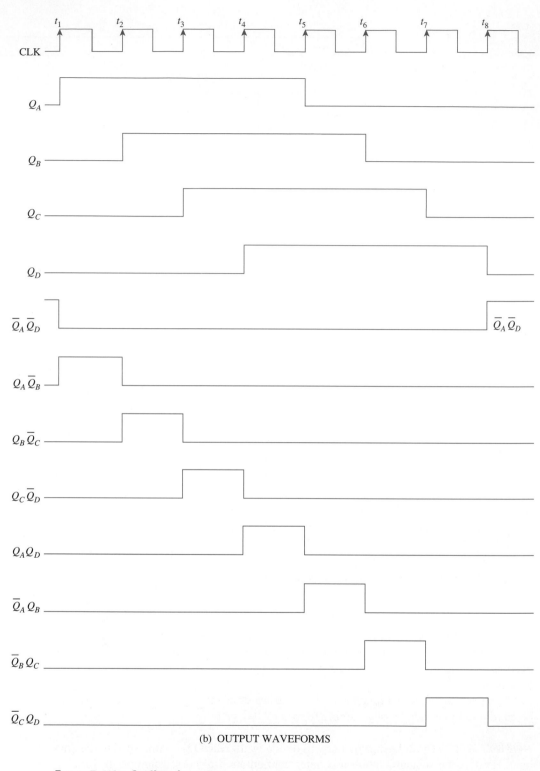

(b) OUTPUT WAVEFORMS

FIGURE 7–114   **Continued**

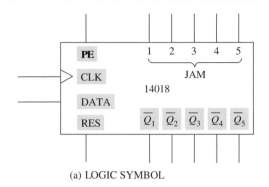

(a) LOGIC SYMBOL

RESET	PE	CLK	JAM IN	$\overline{Q}_n$
0	0	↑	X	$\overline{D}_n^*$
0	1	X	0	1
0	1	X	1	0
1	X	X	X	1

*DATA INPUT FOR N STAGE

(b) FUNCTION TABLE

FIGURE 7-115   14018 presettable divide-by-*n* Johnson counter.

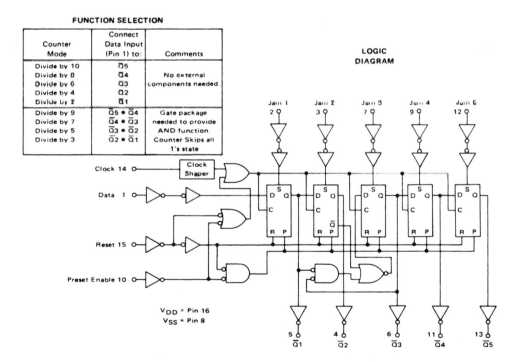

FIGURE 7-116   14018 presettable divide-by-*n* Johnson counter. *Courtesy of Motorola.*

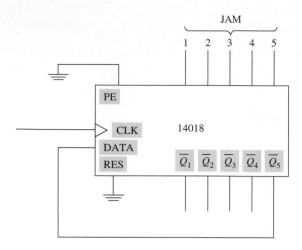

FIGURE 7–117   14018 counter divide-by-10 configuration.

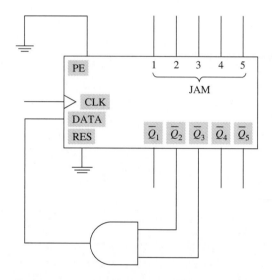

FIGURE 7–118   14018 counter divide-by-5 configuration.

outputs from Q and $\overline{Q}$ are

Q_1	Q_2	Q_3	$\overline{Q}_1$	$\overline{Q}_2$	$\overline{Q}_3$	
0	0	0	1	1	1	AND output places a high on the D input.
1	0	0	0	1	1	
1	1	0	0	0	1	AND output places a low on the D input.
*0	1	1	*1	0	0	
0	0	1	1	1	0	
0	0	0	1	1	1	Recycles

*Note the counter skips the 111 state at the Q outputs, which is the 000 state at the $\overline{Q}$ outputs.

The output of the counter may be taken from $\overline{Q}_1$.

Section 7-9: Review Questions

Answers are given at the end of the chapter.

A. Which 4-flip-flop counter has a modulus of 4?

B. Is it necessary to load a high (1) into one flip-flop of the Ring counter to make it work properly?
C. Is it necessary to load a high (1) into one flip-flop of the Johnson counter to make it work properly?
D. Draw a 3-bit Johnson counter.
E. What is the count of the Johnson counter in question (D) after t_3?
F. What is the modulus of the Johnson counter in question (D)?

SECTIONS 7–7 THROUGH 7–9: INTERNAL SUMMARY

Hybrid counters employ synchronous and asynchronous clocking. The primary advantage of these counters is their ability to produce **symmetrical divide-by-*n*** outputs.

The **74LS92** hybrid counter is a divide-by-12 or divide-by-6 counter. A data book provides the information necessary to determine what connections are required to control the IC's divide-by capability.

The **74LS90** hybrid counter is a divide-by-10 or divide-by-5 counter. This counter can produce two different MOD-10 outputs that depend on the wiring configuration. The counter can produce a normal BCD up-count output (0000–1001). In addition, it can divide by 10 and produce a symmetrical output when configured for **bi-quinary operation.**

Counter outputs require **decoding** when an output is to be used to cause an action in a circuit at a specific time. **Active-high** decoded outputs are produced by **AND gates.** **Active-low** decoded outputs are produced when **NAND gates** are used as the decoders.

The binary number 101 indicates $Q_C = 1$, $Q_B = 0$, and $Q_A = 1$. If these bits were connected to an AND gate, its output Boolean expression would be $Q_C\overline{Q_B}Q_A$. Since Q_B was low on this count, $\overline{Q_B}$ was connected to the AND gate. Note that to decode this output, the bits that are high (Q_C and Q_A) are tied directly to the decoder gate. The complement of the low bits on the desired count must also be connected to the decoder. This is shown in Fig. 7–119. The AND gate will output high only when all of its inputs are high. This condition only occurs on a binary count of 101.

Strobing a decoder circuit inhibits the decoder gates during the counter's glitch period. It is during this period that short-duration miscounts occur. These short-duration miscounts can be masked out from the decoder's output by controlling the decoder with a strobe pulse. The secret is to turn off the decoder during the glitch period, let the counter's output settle, and then turn the decoder back on.

There are two basic types of shift-register counters–Ring counters and Johnson counters:

A **Ring counter** is a shift register whose Q output is connected back to its input. The counter must be preloaded with a 1 to perform frequency division. The MOD number of a Ring counter is equal to the number of flip-flops in the counter.

A **Johnson counter** is a shift register whose $\overline{Q}$ output is connected back to the counter's input. The counter does not require preloading—it can self-start from the CLEAR condition. Once clocking begins, the counter fills up with 1s and then fills up with 0s. The MOD number of a Johnson counter is equal to two times the number of flip-flops in the counter.

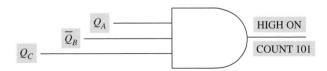

FIGURE 7–119   AND gate decoder for binary 101.

Sections 7-7 through 7-9: Internal Summary Questions

Answers are given at the end of the chapter.

Note: The use of manufacturer's data sheets is recommended to assist you in answering these summary questions.

1. Hybrid counters employ _____ clocking.
   a. synchronous
   b. asynchronous
   c. synchronous and asynchronous

2. The primary advantage of hybrid counters over other types of counters is they can
   a. Be decoded
   b. Be cascaded
   c. Be truncated
   d. Produce symmetrical divide-by-n outputs

3. Hybrid counters are commercially available in IC packages.
   a. True
   b. False

4. The 74LS92 IC in Fig. 7–120 is configured as a divide-by-_____ counter.
   a. 5
   b. 6
   c. 10
   d. 12

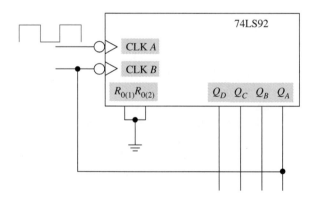

Figure 7–120

5. The 74LS92 IC in Fig. 7–121 is configured as a divide-by-_____ counter.
   a. 5
   b. 6
   c. 10
   d. 12

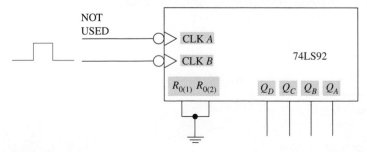

Figure 7–121

6. The 74LS90 counter shown in Fig. 7–122 is configured as a _____ counter.
   a. MOD-5
   b. MOD-16
   c. MOD-10 BCD
   d. MOD-10 bi-quinary

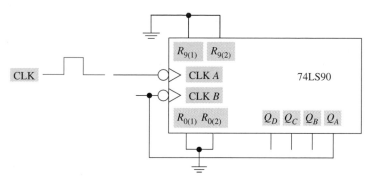

FIGURE 7–122

7. The decoder gate shown in Fig. 7–123 produces an active-_____ output.
   a. low
   b. high

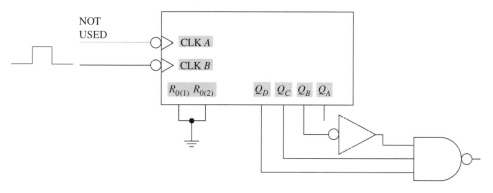

FIGURE 7–123

8. The decoder in Fig. 7–123 will decode the binary count of
   a. 000
   b. 011
   c. 110
   d. 111

9. A specific count from a binary counter must be decoded if an action is required at that specific count.
   a. True
   b. False

10. Strobe pulses are used to clock counters so they do not produce glitches at their outputs.
    a. True
    b. False

11. Identify the circuit shown in Fig. 7–124.
    a. MOD-4 Ring counter
    b. MOD-8 Ring counter
    c. MOD-4 Johnson counter
    d. MOD-8 Johnson counter

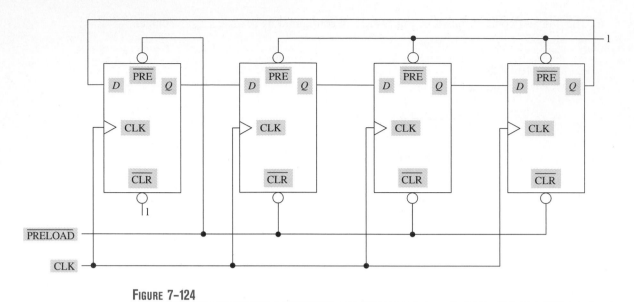

FIGURE 7–124

12. Determine the MOD number of the circuit shown in Fig. 7–125. (*Note:* initially two flip-flops are set and two are cleared.)
    a. 2
    b. 4
    c. 6
    d. 8

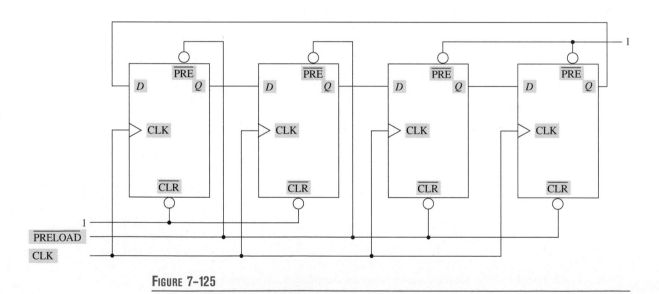

FIGURE 7–125

13. Identify the circuit shown in Fig. 7–126.
    a. Ring counter
    b. Johnson counter
    c. Synchronous counter
    d. Asynchronous counter

14. What is the MOD number of the counter shown in Fig. 7–126?
    a. 3
    b. 6
    c. 8
    d. 16

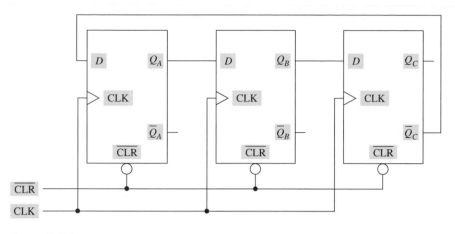

FIGURE 7-126

15. What is the output of the circuit in Fig. 7-126 after four clock pulses? (*Note:* Start in the CLEAR state.)
   a. 110
   b. 111
   c. 011
   d. 001

SECTION 7-10: PRACTICAL APPLICATIONS OF COUNTERS

OBJECTIVE

Implement counters as frequency dividers to design a digital circuit.

Frequency Division

A prime example of frequency division using flip-flops and counters is presented in Fig. 7-127. This system employs frequency division to obtain a **time-of-day (TOD) clock tick** to keep a digital system's clock updated. This time/date information is sometimes used to time-stamp files.

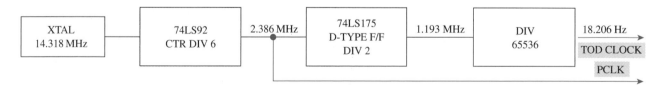

FIGURE 7-127   TOD clock tick circuit block diagram.

The base frequency in this example is derived from a standard 14.31818-MHz crystal. This frequency is divided by 6 to obtain 2.386 MHz. In some systems this frequency is used to clock slower peripheral devices and is called a PCLK (Peripheral Clock) signal. The 2.386 MHz is further divided by 2 to produce 1.193 MHz. This signal is used in the programmable interval timer circuit to obtain the TOD clock tick.

A 16-bit down-counter in the programmable interval timer divides the 1.193-MHz signal by 65,536 to produce 18.206 Hz, which is the **TOD clock tick signal.**

The mathematics of this explanation are

$$14.31818 \text{ MHz}/6 = 2.386 \text{ MHz}$$
$$2.386 \text{ MHz}/2 = 1.193 \text{ MHz}$$
$$1.193 \text{ MHz}/65{,}536 = \textbf{18.206 Hz} \text{ (TOD clock tick)}$$
$$18.206 \text{ Hz}/65{,}536 = 277.8^{-6} \text{ Hz} = 3600 \text{ seconds}$$

The TOD clock circuit uses the 18.206-Hz clock tick to increment a 16-bit up-counter. When the counter reaches $\text{FFFF}_{(16)}$ ($65{,}535_{(10)}$) the hours section of the clock increments by one because 65,536 TOD clock ticks equal one hour (3600 seconds).

Counting

Decimal Counting Unit.

Figure 7–128 shows a decimal counting unit. The circuit counts from 000 to 999 (MOD-1000). It is implemented by cascading three 7490 counter ICs in BCD configuration. The outputs of each counter are decoded in a BCD-to-seven-segment decoder/driver. These decoders will be presented in Chapter 10. Once decoded, the outputs of the decoder/drivers are used to illuminate the proper count on the seven-segment indicators. The count sequence of the units counter (LSD) is:

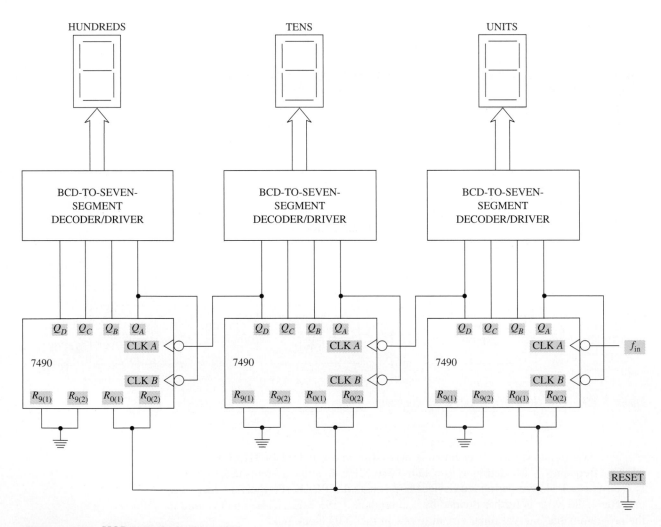

FIGURE 7-128 **MOD-1000 decimal counter.**

Q_D	Q_C	Q_B	Q_A	
0	0	0	0	
0	0	0	1	t_1
0	0	1	0	t_2
0	0	1	1	t_3
0	1	0	0	t_4
0	1	0	1	t_5
0	1	1	0	t_6
0	1	1	1	t_7
1	0	0	0	t_8
1	0	0	1	t_9
*0	0	0	0	t_{10}

The count sequence from t_9 to t_{10} (See * in count sequence) produces an NGT at the Q_D output of the units counter every tenth clock pulse. This NGT increments the 10s counter. After the units and 10s counters have both been incremented to counts of 9(t_{99}), the 100s counter increments one time on the NGT from the Q_D output of the 10s counter. This NGT from the tens counter is produced every hundredth clock pulse. The counters increment to a maximum count of 999 and then reset to 000.

24-Hour Clock

Another application of counting is shown in the 24-hour clock in Fig. 7–129. The clock consists of BCD and truncated counters that control the seconds, minutes, and hours sections of the clock. The counter's outputs are decoded and used to drive the seven-segment indicators as they were in the decimal counting unit.

The **seconds section** of the circuit is clocked by a 1-Hz clock pulse that could be derived from the 60-Hz line frequency available at a wall outlet. This can be accomplished by a divide-by-60 counter. The 60-Hz signal must first be made compatible with TTL or CMOS input requirements. The seconds section must count from 00 to 59. A BCD counter is used to count the LSD (0–9) of the seconds. The BCD counter increments once each second from 0000 to $1001_{(BCD)}$. A MOD-6 counter is used to count the MSD (0–5) of the seconds.

The count produced by the LSD seconds counter is identical to the count shown for the units counter in the decimal counting unit just presented. Again, the count sequence from t_9 to t_{10} produces an NGT at the Q_D output of the LSD seconds counter every 10 seconds. Therefore, the MSD seconds counter increments once every 10 seconds. The MSD counter

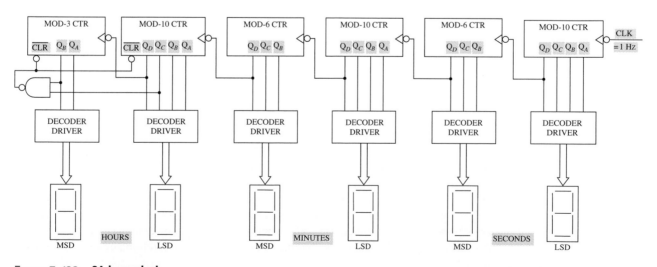

FIGURE 7–129   24-hour clock.

in the seconds section is a MOD-6 counter. This counter will increment once every 10 seconds through the following count sequence:

Q_D	Q_C	Q_B	
0	0	1	t_1
0	1	0	t_2
0	1	1	t_3
1	0	0	t_4
1	0	1	t_5
0	0	0	t_6

This MOD-6 counter increments from 000 to 101 and then resets. Each time the MOD-6 seconds counter resets, an NGT is generated at Q_D. This causes the LSD minutes section of the circuit to increment one count. This action occurs every 60 seconds.

We will forgo a detailed discussion of the minutes section because it is identical to the seconds section. After 59 minutes, both counters in the minutes section reset. When this occurs, an NGT is produced at Q_D of the MOD-6 minutes counter. This increments the LSD hours section by one count.

The hours section of the counter increments once every 60 minutes at the LSD. This continues through 23:59:59. The count of 24 in the hours section is decoded by the NAND gate and resets the hours section to 00.

There are many ways to implement this 24-hour clock. It could easily be modified to a 12-hour clock. The use of IC counters to implement the clock consumes much space on a protoboard. Digital clock circuitry similar to that described is currently available on a single IC chip.

SECTION 7-11: TROUBLESHOOTING COUNTERS

OBJECTIVE

Given a counter symptom for a specified counter, determine the possible faults.

One problem sometimes encountered when troubleshooting counters is **miscounting.** This problem is more prevalent in asynchronous counters because their internal propagation delays are additive. The short-duration miscounts (glitches) produced by asynchronous counters are normal. However, if the maximum operating frequency of a counter is exceeded, miscounts will certainly occur.

Figure 7–130 illustrates the results of exceeding f_{max} for an asynchronous counter. The waveforms are from a MOD-8 up-counter with flip-flop delays of 40 ns each. The total propagation delay in the counter is 120 ns. This delay allows for a maximum operating frequency of 8.3 MHz. The counter in this example is being clocked at 15 MHz. The desired count and actual count in decimal are annotated below the output waveforms. Decreasing the input clock frequency will determine whether the problem of miscounting is related to the incoming clock frequency.

A counter implemented with discrete flip-flops is a simple circuit to troubleshoot. Figure 7–131 shows a MOD-8 asynchronous up-counter. Numerous faulty output conditions can occur in this circuit. Regardless of the fault symptom, it is quick and easy to first check V_{CC} and ground on the flip-flop ICs with a logic probe. This step is normally listed first because the IC cannot work if it is not properly powered. The J and K inputs as well as $\overline{CLR}$ and $\overline{PRE}$ all must be high for the counter to function. A logic probe will quickly detect a faulty level and isolate a malfunction. *Note:* If power supply problems are indicated in any manner, use a voltmeter to check V_{CC}. Power supply fluctuations may not be detected by a logic probe, but they can cause erratic operation and many headaches for a technician.

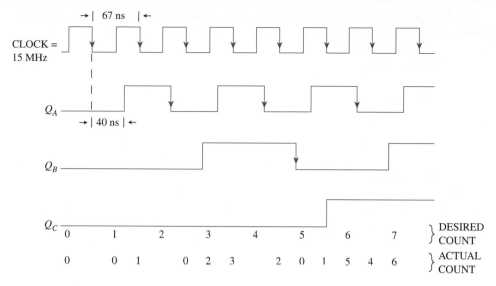

FIGURE 7–130   MOD-8 asynchronous counter waveforms at clock = 15 MHz.

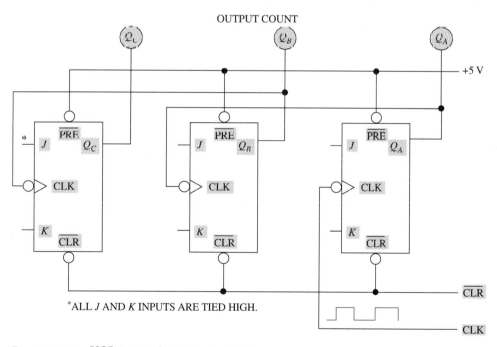

*ALL *J* AND *K* INPUTS ARE TIED HIGH.

FIGURE 7–131   MOD-8 asynchronous up-counter.

A precaution to observe when using the probe: (1) If a trace or pin on an IC is supposed to be *low*, the *green light* on the logic probe should illuminate when the probe tip is placed on the trace/pin. (2) If a trace or pin on an IC is supposed to be *high*, the *red light* on the logic probe should illuminate when the probe tip is placed on the trace/pin. You should always suspect a fault if neither light on the probe illuminates when a trace or IC pin is touched. Once all of the static levels of the flip-flops have been checked and are good, it is time to check the clock inputs.

A good clock input to flip-flop A in the counter in Fig. 7–131 should cause the *amber light* on the probe to flicker. If the clock pulse to flip-flop A is good, use the probe to check Q_A, the clock input to flip-flop B, Q_B, the clock input to flip-flop C, and finally Q_C.

If the lights on the logic probe do not illuminate as described, more than likely the probe is pointing at the problem. Don't forget to isolate the load from the defective output if there is a load connected to it.

Let's assume the clock input to flip-flop A in Fig. 7–131 is good, but there is no pulsing amber light at the Q_A output. If the static checks have already been made, the flip-flop is powered and is not being held in a state due to an active asynchronous input ($\overline{\text{PRE}}$ or $\overline{\text{CLR}}$). It is also known to be set for TOGGLE operation because J and K checked high. Therefore, all that is left to check is the load. If there is an LED or decoder/driver input connected to Q_A, disconnect it. If the problem is persistent, disconnect the clock input to flip-flop B. If the problem persists after all of the load connections have been removed, the flip-flop itself is probably defective.

Keep in mind that synchronous counters can be used to replace asynchronous counters if glitch problems are present. The propagation delays in synchronous counters are not additive so they can be used at higher operating frequencies.

Another problem that might be encountered in counters is **loading.** A MOD-10 74LS163 synchronous counter is shown in Fig. 7–132(a). LEDs connected to the Q outputs of this counter can draw enough current to load down the output levels and prevent the truncating gate from short counting the counter.

Another example of loading is shown in the MOD-16 7493 asynchronous counter in Fig. 7–132(b). If the LEDs draw too much current, the signal at CLK B from the Q_A output may be too weak to trigger the internal B flip-flop. If this happens, the Q_A output will toggle but the remaining outputs will stay low.

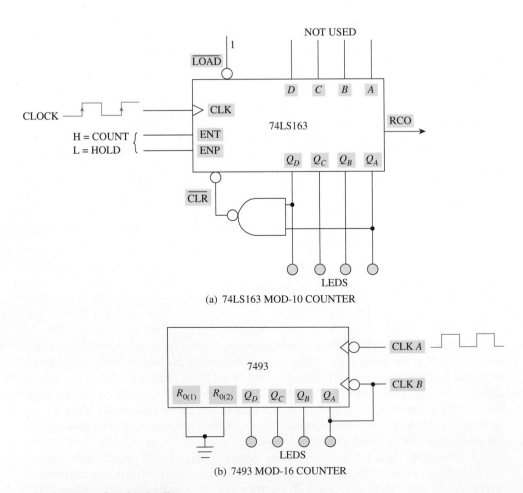

(a) 74LS163 MOD-10 COUNTER

(b) 7493 MOD-16 COUNTER

Figure 7–132 **Counter loading.**

It is always a good idea to use **current-limiting resistors** between the counter outputs and LEDs used to monitor the count. This is also true when the outputs of BCD-to-seven-segment decoder/drivers are used to control seven-segment indicators. The resistors in this case should be placed between the decoder/driver outputs and the input pins of the indicators. Failure to use the current-limiting resistors will normally result in problems.

The 7493 MOD-16 IC counter in Fig. 7–132(b) can be troubleshot in the same manner a counter using discrete flip-flops can be troubleshot. A logic probe can be used to ensure V_{CC} is high, and the ground and reset pins ($R_{0(1)}$ and $R_{0(2)}$) are low. Pulsing signals should be present at CLK A and CLK B and all Q outputs. Troubleshooting analysis will be expanded in the End of Chapter Problems.

CHAPTER 7: SUMMARY

Counters provide two main functions in digital circuits. They are used to sequentially count (up or down) or count in some specified code such as BCD or gray code. They are also used to divide frequencies.

Counters fall into two main categories—**asynchronous** or **synchronous.** The category is determined by how the counter is clocked.

An **asynchronous counter** has its clock input applied to a flip-flop, and that flip-flop's output clocks the next cascaded flip-flop. This clocking arrangement is why these counters are often called **ripple counters.** The flip-flops in asynchronous counters are connected in the **TOGGLE mode of operation.** This produces an output frequency from the flip-flop that is half of its input clock frequency. When several toggling flip-flops are cascaded, each one divides its input clock frequency by two.

The maximum number of states that a counter can assume during its count sequence is its **modulus (MOD).** The MOD of a full-count counter can be calculated by raising 2 to the power of the number of flip-flops (2^n). The **divide-by capability** at a counter's MSB output is its MOD number.

A **truncating gate** is required to **short-count** a counter. The rule of truncation when designing asynchronous counters is to tie the bits that are high on the desired MOD number to the truncating gate. There are exceptions to this rule for some IC counters.

The 7493/74293/74393-series ICs are among the more popular TTL IC counters commercially available. A data book is necessary to acquire pin out information and operating characteristics/descriptions of the ICs.

Counters can be **cascaded** to increase their MOD. The MOD number of cascaded counters is equal to the product of the individual MOD numbers.

The primary disadvantage of asynchronous counters is their internal propagation delays are additive. This reduces the maximum operating frequency of the counters when compared to similar synchronous counters. Short-duration miscounts (glitches) are common with asynchronous counters.

Synchronous counters have the input clock pulses applied to all flip-flops at the same time from one common clock line. This clocking scheme does away with the short-duration miscounts so prevalent in asynchronous counters. The clocking scheme also does away with the additive propagation delays of asynchronous counters, and this increases their maximum operating frequency.

The flip-flops within a synchronous counter cannot be wired to toggle on every active clock transition. Therefore, control circuitry is required to determine when a flip-flop should toggle. This makes synchronous counters a bit more complex, and a little more difficult to design.

MOD-10 counters, regardless of their category, produce ten different states. These counters can count 0000 to 1001 or 1001 to 0000 in BCD. They are sometimes called **decade** counters.

Many **programmable counters'** MOD numbers are controlled by asynchronously preloading a number into the counter. Some of these counters are reversible (up or down). Some have terminal count outputs, which can be used to repetitively preload the counter or to cascade the counter with other counters. Some of the more versatile CMOS counters incorporate all of these features in addition to being programmable for binary or decade counting.

Synchronous counter design follows a few basic steps. The first step is to complete a **present state (Q^n) table.** This table shows the counting sequence desired of the counter. The second step requires a **next state (Q^{n+1}) table.** A **design table** can be used to incorporate the present state data, next state data, and **transition data.** The transition data shows what levels of J and K or D are required on a flip-flop's input(s) to produce the next flip-flop state.

The final design step requires transferring the J and K or D levels from the design table to K-maps. Simplified counter equations are derived from the K-maps. These equations show how the synchronous counter can be implemented.

Hybrid counters employ both synchronous and asynchronous clocking schemes. These counters can produce **symmetrical divide-by-n outputs.** The more popular TTL hybrid ICs produce divide-by-12 and divide-by-6 or divide-by-10 and divide-by-5 outputs.

A counter's output must be **decoded** to make a specific output count usable for some subsequent clock or trigger input. **Active-high decoding** is accomplished with **AND gates; active-low** with **NAND gates.** Decoding circuits can be **strobed** to prevent the decoding of short-duration miscounts.

There are two popular types of shift-register counters-Ring and Johnson. The **Ring counter** must be pre-loaded with a high, and the high is shifted one flip-flop at a time through the counter. The MOD number of a Ring counter is equal to the number of flip-flops in the counter. **Johnson counters** are self-loading, and when clocked they fill up with 1s and then fill up with 0s. The MOD number of a Johnson counter is equal to two times the number of flip-flops in the counter.

Troubleshooting counters is a methodical process that relies on a thorough knowledge of proper operation. The sequential action of a counter makes signal checking from point to point simple.

CHAPTER 7: END OF CHAPTER QUESTIONS/PROBLEMS

Answers are given in the Instructor's Manual.

The use of manufacturer's data sheets is recommended to help answer the End of Chapter Questions/Problems.

SECTION 7-1

1. Define modulus.
2. Identify the counter shown in Fig. 7–133. The identity should include MOD #, type (synchronous or asynchronous) and up- or down-counter.
3. What is the count of the counter in Fig. 7–133 after 4 clock pulses?
4. What is the count of the counter in Fig. 7–133 after 30 clock pulses?
5. What is the frequency of the Q_A output signal of the counter shown in Fig. 7–133?
6. What is the frequency of the Q_B output signal of the counter shown in Fig. 7–133?

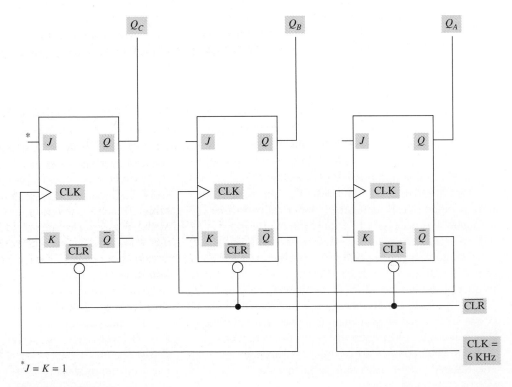

$^*J = K = 1$

FIGURE 7-133

7. What is the frequency of the Q_C output signal of the counter shown in Fig. 7–133?
8. Identify the counter shown in Fig. 7–134.
9. Identify the counter shown in Fig. 7–135.

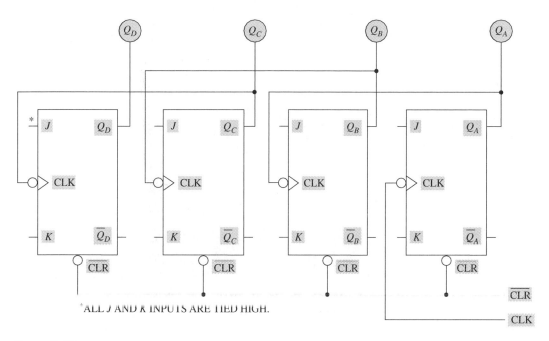

*ALL J AND K INPUTS ARE TIED HIGH.

FIGURE 7–134

10. What is the count of the counter in Fig. 7–135 after 3 clock pulses? Start with the counter in the CLEAR state prior to the application of the first clock pulse.
11. What is the frequency of the Q_C output signal of the counter in Fig. 7–135?
12. How many flip-flops are required to implement a MOD-128 counter?

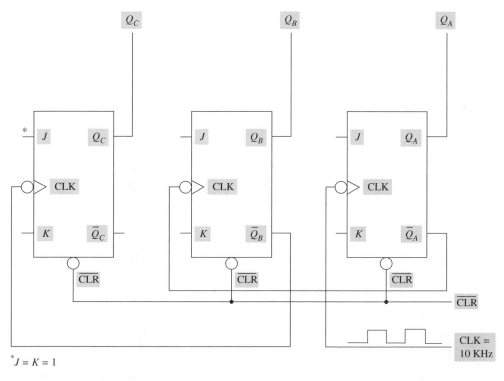

*$J = K = 1$

FIGURE 7–135

13. Identify the circuit shown in Fig. 7–136.
14. What is the frequency of the Q_C output signal of the circuit in Fig. 7–136?
15. What is the maximum count of the circuit in Fig. 7–136?

CT 16. Design a MOD-25 asynchronous up counter using NGT-triggered J-K flip-flops. Draw the circuit.

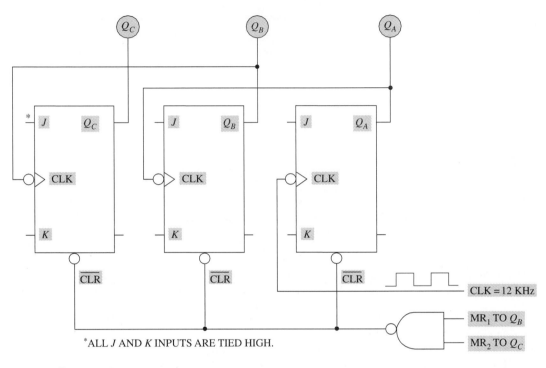

*ALL J AND K INPUTS ARE TIED HIGH.

FIGURE 7-136

17. What is the MOD # of the IC counter in Fig. 7–137?

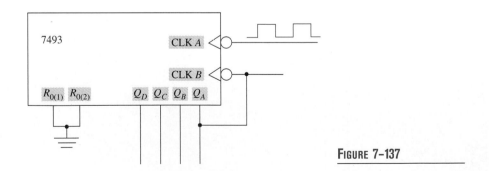

FIGURE 7-137

18. What is the MOD # of the IC counter in Fig. 7–138?
19. What is the overall MOD # of the cascaded counters in Fig. 7–139?
20. What is the frequency at the Q_D output of the #1 counter in Fig. 7–139?
21. What is the frequency at the Q_D output of the #2 counter in Fig. 7–139?
22. What is the maximum clock frequency of a MOD-16 asynchronous up-counter that uses J-K flip-flops that have 25 ns per flip-flop propagation delay?

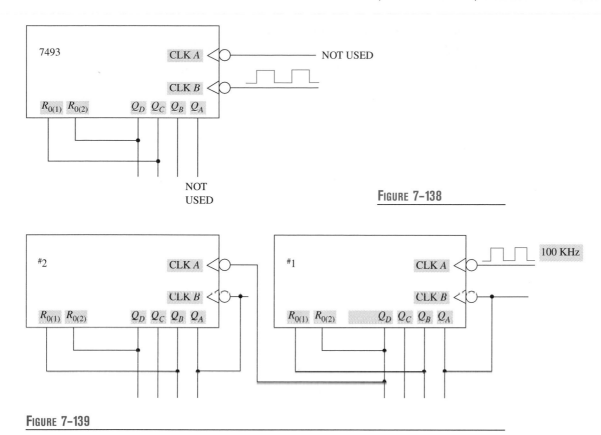

FIGURE 7-138

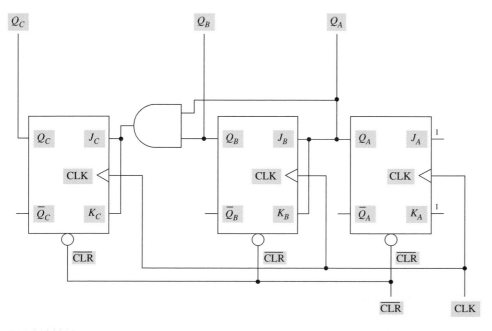

FIGURE 7-139

SECTION 7-3

23. Identify the counter shown in Fig. 7–140.
24. What is the maximum count of the counter in Fig. 7–140?
25. What is the Q_C output frequency of the counter in Fig. 7–140 if the CLK input frequency is 4 KHz?
26. What is the purpose of the AND gate in Fig. 7–140?

FIGURE 7-140

27. Identify the counter shown in Fig. 7–141.
28. What is the output frequency of the counter in Fig. 7–141 at
    a. Q_A?
    b. Q_B?
    c. Q_C?

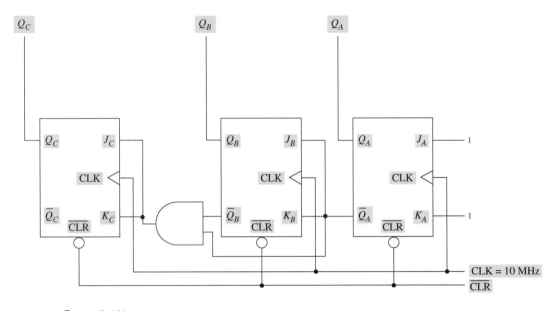

FIGURE 7-141

29. Identify the counter shown in Fig. 7–142.
CT 30. Draw the Q output waveforms for the counter shown in Fig. 7–142.

SECTION 7-4

CT 31. Draw the Q output waveforms for the counter shown in Fig. 7–143.

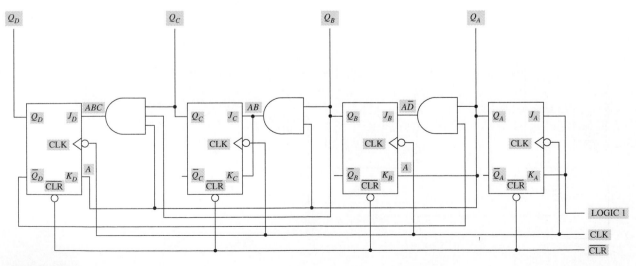

FIGURE 7-142

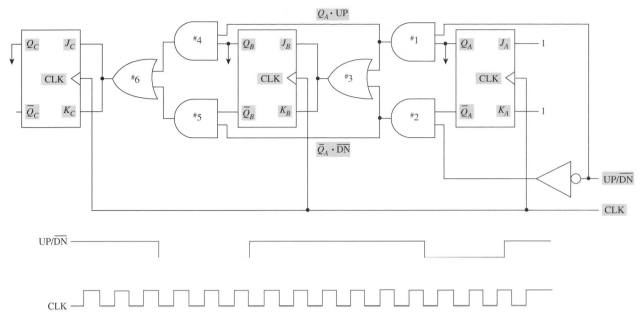

FIGURE 7-143

SECTION 7-5

32. What is the MOD number of the synchronous programmable down-counter in Fig. 7–144?
33. Draw the logic symbol for a 74193 IC and make the required connections to implement a MOD-5 up-counter using the Preload inputs.
34. Repeat question 33 to make the counter a MOD-5 down-counter.

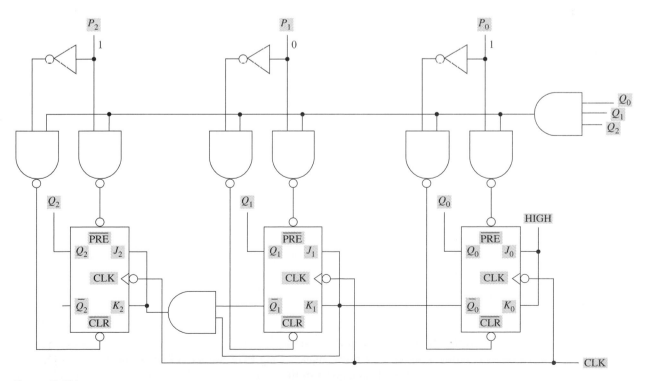

FIGURE 7-144

35. Draw the logic symbol for a 74LS163 IC and make the required connections to make the counter count up to $1111_{(2)}$.

36. Redraw the 74LS163 IC counter to make it a MOD-10 counter. (*Note:* The clearing action of the 74LS163 is synchronous—see data sheet or data book.)

37. The 74LS190 BCD counter in Fig. 7–145 will count *down* on the PGT of the clock pulse when
$\overline{\text{LOAD}}$ = _____
$\overline{\text{CTEN}}$ = _____
$\text{D}/\overline{\text{U}}$ = _____

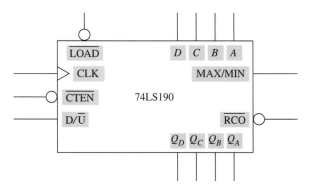

FIGURE 7–145

CT 38. Draw a cascaded MOD-100 BCD up-counter using 74LS190 ICs. Show all connections and use parallel clocking.

SECTION 7-6

CT 39. Design a 3-bit synchronous gray code counter using J-K flip-flops.

SECTION 7-7

40. Define a hybrid counter.

41. Draw the logic symbol for a 74LS92 hybrid counter and make the required connections to make the IC a MOD-6 counter.

42. Connect the 74LS92 of question 41 for MOD-12 operation.

43. Draw the logic symbol for a 74LS90 IC and make the required connections to make the counter a
   a. MOD-5 counter
   b. MOD-10 counter
   c. MOD-10 bi-quinary counter

SECTION 7-8

44. Define/explain counter decoding.

45. Draw a 3-input NAND gate connected to decode a MOD-8 counter's 010 output (outputs are Q_C to Q_A).

46. Refer to question 45. Is the NAND gate's output low or high on the decoded count?

SECTION 7-9

47. Identify the counter in Fig. 7–146.

48. Draw a MOD-8 Johnson counter using PGT-triggered D-type flip-flops.

49. Determine the MOD number of the circuit shown in Fig. 7–147. (*Note:* Two flip-flops are set and two are cleared at the beginning of the cycle.)

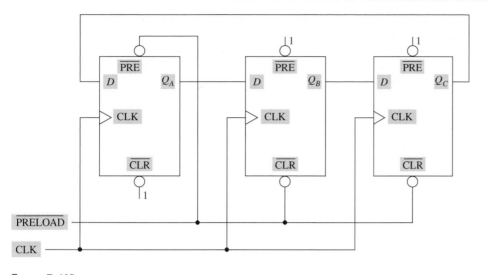

FIGURE 7–148

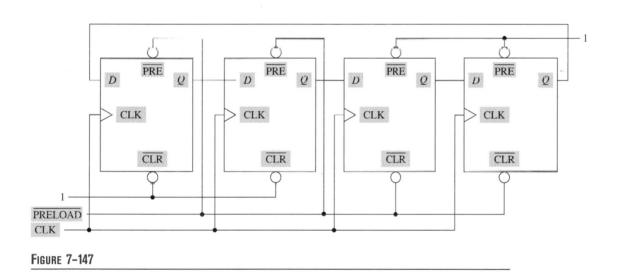

FIGURE 7–147

SECTIONS 7–10 AND 7–11

50. Identify the circuit shown in Fig. 7–148.
51. What is the purpose of the AND gate in Fig. 7–148?
52. What is the frequency of the Q_C output in Fig. 7–148?

CT 53. The counter shown in Fig. 7–148 has the following count sequence:

Q_C	Q_B	Q_A
0	0	0
0	0	1
0	1	0
0	1	1
0	0	0
0	0	1
0	1	0
0	1	1
0	0	0

List the most likely cause(s).

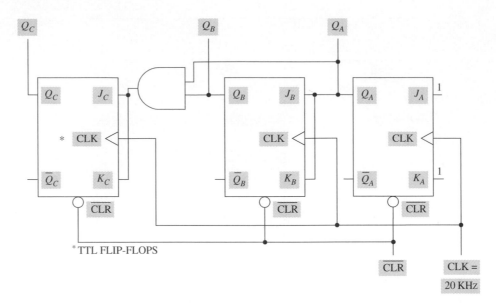

FIGURE 7–148

CT 54. The counter shown in Fig. 7–148 has the following count sequence:

Q_C	Q_B	Q_A
0	0	0
1	0	1
0	1	0
1	1	1
0	0	0
1	0	1
0	1	0
1	1	1

List the most likely cause(s).

Answers to Review Questions

Section 7-1

A. False
B. True
C. False
D. False
E. True
F. True
G. False
H. Increase
I. $n = \log 64 \div \log 2 = 6$
J. $1011_{(2)}$

Section 7-2

A. Short-count a counter.
B. Tie the bits that are high on the desired MOD number to the truncating circuitry.

C. $\log 100 \div \log 2 = 6.6 = $ **7 flip–flops**
D. $6_{(10)} = 110_{(2)}$
E. True
F. Q_D, Q_C, and Q_A
G. MOD-120

H. $f_{max} = \dfrac{1}{4 \times 25 \text{ ns}} = $ **10 MHz**

Section 7-3

A. MOD-8 synchronous down-counter
B. 101 $(Q_C–Q_A)$
C. $Q_A = 5$ MHz; $Q_B = 2.5$ MHz; $Q_C = 1.25$ MHz
D. All clock inputs are connected to one common line.
E. No short-duration miscounts
   Input clock frequency (f_{max}) is higher.

SECTION 7-4

A. $\overline{Q}$
B. MOD 4
C. up
D. 00

SECTION 7-5

A. True
B. True

C. $101 = \overline{\text{LOAD}} = 0/1$
$110 = t_1$
$111 = t_2$
$000 = t_3$
$001 = t_4$
$010 = t_5$
D. Refer to Fig. 7–149.
E. Refer to Fig. 7–150.
F. Refer to Fig. 7–151.

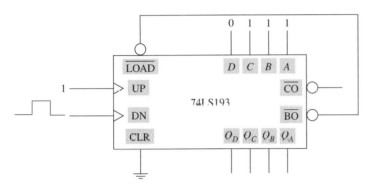

FIGURE 7-149

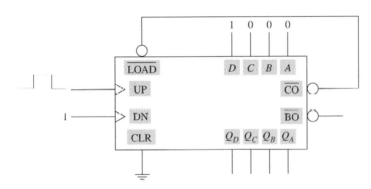

FIGURE 7-150

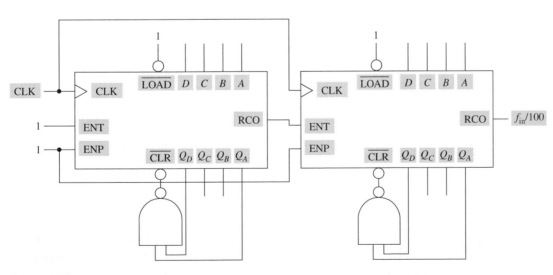

FIGURE 7-151

Note: The RCO output of the left counter can be connected to the ENP input of the second counter and ENT can be tied high. Another configuration includes omitting the NAND gates and preloading 0110 and connecting each RCO through an inverter back to its $\overline{LOAD}$ input as depicted in Fig. 7–56.

G. $\overline{LOAD}$ = 1
$\overline{CTEN}$ = 0
$D/\overline{U}$ = 1
CLK = Pulsing

H. active-high

I. 1001

J. PE = 0
Bin/Decade = 1
Up/$\overline{Dn}$ = 0
Clock = Pulsing

Section 7-6

No review questions on synchronous counter design.

Section 7-7

A. Hybrid counters can produce symmetrical outputs.

B. True
C. See Fig. 7–152.
D. See Fig. 7–153.

Section 7-8

A. True
B. low
C. True
D. See Fig. 7–154.
E. High
F. See Fig. 7–155.
G. Low

Section 7-9

A. Ring counter
B. Yes
C. No
D. Refer to Fig. 7–156
E. 111
F. MOD-6

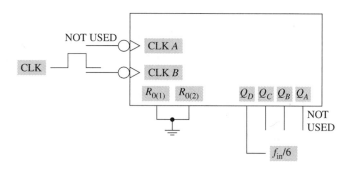

Figure 7-152

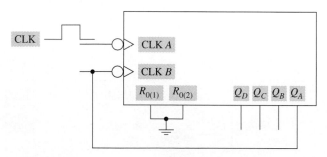

Figure 7-153

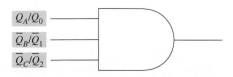

Figure 7-154

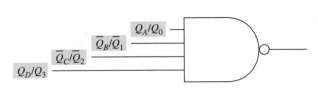

Figure 7-155

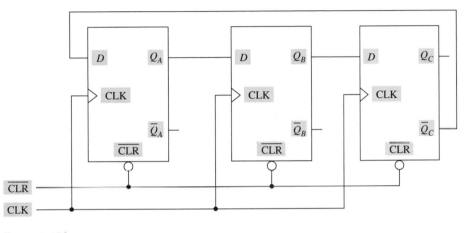

Figure 7-156

Answers to Internal Summary Questions

Sections 7-1 and 7-2

1. b	9. a
2. a	10. b
3. d	11. b
4. b	12. c
5. a	13. b
6. c	14. c
7. d	15. b
8. c	16. b

Sections 7-3 through 7-5

1. a	5. c
2. b	6. d
3. a	7. a
4. c	8. d

9. d	15. d
10. c	16. a
11. b	17. a
12. d	18. b
13. b	19. c
14. c	20. a

Sections 7-7 through 7-9

1. c	9. a
2. d	10. b
3. a	11. a
4. d	12. b
5. b	13. b
6. c	14. b
7. a	15. c
8. c	

8 REGISTERS

Topics Covered in this Chapter

Introduction

A **register** is a group of latches or flip-flops used to store, transfer, or shift data. The requirement for data storage to meet memory requirements in digital systems is self-evident. In some registers, data can be shifted (moved one flip-flop) to the left and/or the right with each clock pulse. Some of the registers presented in this chapter are primarily used to store information. It is this capability to store Logic 0s and Logic 1s that gives digital systems their memory capability. These applications will be detailed in Section 8–3.

Some groundwork regarding register operation and principles has already been accomplished in the text. The general principles of loading data in series and in parallel were presented in Section 5–4. Also, operation of a serial-shift register was covered in Section 6–7. Some of these principles will be reviewed and expanded upon in this chapter.

Registers are classified by the method with which the binary data is input to (loaded) and retrieved from the register. Data may be input in serial or parallel format, and it may be retrieved in either format. Data movement in *serial* format means nothing more than moving data one bit at a time during one clock cycle. Data movement in *parallel* format means moving several bits at a time during one clock cycle.

Chapter Objectives

1. Define the term "register" and state some of the applications of registers in digital circuits.

2. Identify registers by classification as serial-in/serial-out, serial-in/parallel-out, parallel-in/serial-out, parallel-in/parallel-out, or universal.

3. Given the logic diagram or symbol of a register and combinations of input data and control levels, determine the output after X clock input pulses.

4. Given a description of a problem in a register or observed results, identify the most likely output results or malfunction.

The registers presented in this chapter are classified as **serial-input registers** (serial-in/serial-out and serial-in/parallel-out) and **parallel-input registers** (parallel-in/serial-out and parallel-in/parallel-out). Figure 8–1 represents these classifications in block diagram format.

Figure 8–1(a) shows serial data input on the left side of a 4-bit register. The data output is taken from the right side of the register. This serial-in/serial-out register is usually referred to as a **shift register.** In this case, the data is shifted to the right, so this register is a shift-right register. If the input and output lines were reversed, the block diagram would represent a shift-left register. Bidirectional shift registers are very common in digital systems.

Figure 8–1(b) shows the block diagram of a serial-in/parallel-out shift register. The data is input in series and retrieved from the register in parallel format. From previous discussions you should remember that it takes four clock pulses to serially load four bits of data into this 4-bit register. Once loaded, the data is available at the Q_A, Q_B, Q_C, and Q_D outputs. If desired, four additional clock pulses could be applied, and the output taken serially at Q_D. This would allow operation as a serial-in/serial-out register.

449

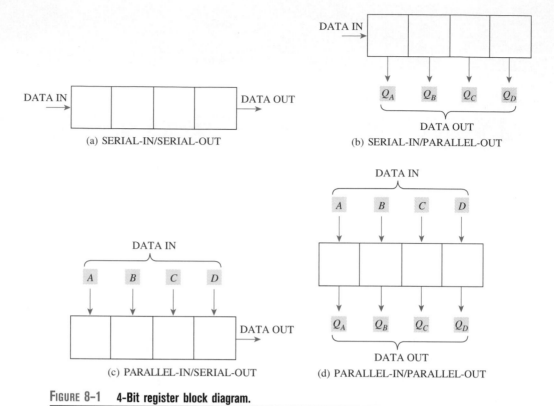

FIGURE 8–1   **4-Bit register block diagram.**

Figure 8–1(c) illustrates parallel-in/serial-out data movement. Some registers of this type load the parallel data asynchronously when a load input pulse is applied. Other types of parallel-in/serial-out registers load the parallel data synchronously with the system clock pulse. The loaded data must be clocked out of the register one bit at a time.

A parallel-in/parallel-out register block diagram is shown in Fig. 8–1(d). The data is loaded from the *A*, *B*, *C*, and *D* inputs synchronously and is immediately available at the register outputs. This class of register is normally employed in memory systems.

Some IC registers are **universal.** In other words, data can be loaded into the register in serial or parallel format, and it may be retrieved from the register in either format.

In addition to memory capabilities, registers offer the ability to change the data format from serial-to-parallel or parallel-to-serial. Section 8–3 presents registers used as code detectors such as those devices used to activate or deactivate a security system. Registers used to multiply and divide by powers of 2 are also presented in Section 8–3, as well as uses for Ring and Johnson counters.

Section 8-1: Serial-Input Registers

Objectives

1. Identify serial-input registers.
2. Given the serial data input levels of a serial-input register, determine the output levels after *X* clock pulses.

Serial-in/Serial-out Shift Register

Figure 8–2(a) shows a 4-bit serial-in/serial-out shift register implemented with D-type flip-flops. The key point to keep in mind during the following analysis is that a flip-flop will respond to the data that is on its input immediately prior to and during the active

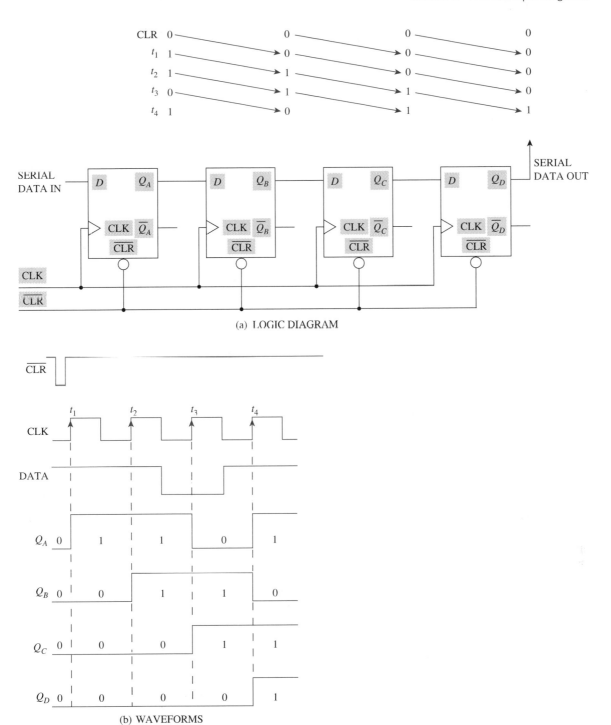

(a) LOGIC DIAGRAM

(b) WAVEFORMS

FIGURE 8–2   **Serial-in/serial-out shift register.**

clock transition. Note that all of the flip-flops in the register in Fig. 8–2(a) are clocked simultaneously from a common clock line.

The serial data input to the register in this example is 1011 as shown in Fig. 8–2(b). The loading of this data is complete after four clock pulses:

t_1:   *The serial data input at t_1 is high* (Fig. 8–2b). This data is clocked into flip-flop A at t_1. Immediately prior to the arrival of the t_1 clock pulse's PGT, the data inputs to the B, C, and D flip-flops are low. Therefore, the 0s are clocked into these flip-flops at t_1. The data in the register after this first clock pulse is 1000 and is shown above the logic diagram in Fig. 8–2(a).

t_2: *The serial data input at t_2 is still high.* This data is clocked into flip-flop A at the t_2 clock pulse PGT. The data outputs of all of the flip-flops are shifted one flip-flop to the right at t_2. Q_A data is shifted into flip-flop B, Q_B data is shifted into flip-flop C, and Q_C data is shifted into flip-flop D. The data in the register after t_2 is 1100.

t_3: *The serial data input at t_3 is low.* This data is clocked into flip-flop A, and the remaining data are shifted one flip-flop to the right. The data in the register after t_3 is 0110.

t_4: *The input data is again high* upon arrival of the t_4 clock pulse PGT. The shift-right process at t_4 results in register data of 1011.

Notice in this serial-in/serial-out shift register the data are loaded one bit at a time. To transfer the loaded data out of this register to another digital device, it must be clocked out one bit at a time also because the only data output pin is Q_D.

The logic diagram in Fig. 8–3 shows two registers connected together. One is called a **source register** (Register A) and the other a **destination register** (Register B). The serial data input for this example is high for the first four clock pulses (t_1-t_4) and then taken low for the second four clock pulses (t_5-t_8).

The basic idea here is relatively simple. Four bits of data are referred to as a **nibble** of data. The first nibble of data (1111) is loaded into Register A with the first four clock pulses (t_1-t_4) as previously explained. Four additional clock pulses (t_5-t_8) are required to transfer the data from Register A to Register B while the new data (0000) is being loaded into Register A.

A couple of conclusions can be drawn from this example. First, it takes the same number of clock pulses as flip-flops in a register to *load* a serial-in register. It also takes the same number of clock pulses as flip-flops to serially *transfer* the data out of a loaded register to another device. In this example, it takes four clock pulses to load Register A and four more clock pulses to transfer that data to Register B. Second, the transfer of data out of Register A destroys that data when it is transferred out. This is because the second nibble of data (0000) is written over the original nibble (1111) during the transfer operation.

7491 8-Bit Shift Register.

Figure 8–4 shows the logic diagram and symbol of this commercially available serial-in/serial-out shift register. The register consists of eight active-high SET–CLEAR flip-flops, an input control gate, and a clock driver. The input control NAND gate can be inhibited or enabled by an input control signal at A or B to control serial data loading of the register. A low into input A or B inhibits the control gate and causes the first flip-flop's inputs to be $S = 0$ and $C = 1$. Thus, the first flip-flop will clear on the PGT of the next clock pulse. A high at input A or B enables the input control gate so that the data on the other input can be clocked into the first flip-flop. The inverter at the output of the NAND gate ensures complementary data inputs to the flip-flops. The **clock driver circuit** is used to invert and condition the clock signal so that it can drive all eight flip-flops. The clock driver causes the flip-flops to shift information on the PGT of the input clock pulse. The register shown in Fig. 8–4(a) can also be implemented with D-type flip-flops or J-K flip-flops.

Serial-In/Parallel-Out Shift Register

A 4-bit serial-in/parallel-out shift register using D-type flip-flops is shown in Fig. 8–5. Like its serial-in/serial-out counterpart, all of the flip-flops are clocked simultaneously. This register can be used to convert data from serial-to-parallel format. The necessity of changing format should be evident from the parity material presented in Chapter 5. Data movement within a system is normally accomplished in parallel format due to faster speed.

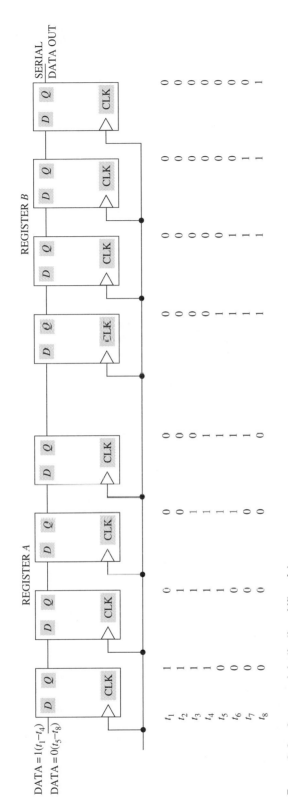

REGISTER A

REGISTER B

SERIAL
DATA OUT

DATA = 1 ($t_1 - t_4$)
DATA = 0 ($t_5 - t_8$)

FIGURE 8-3 Source and destination shift registers

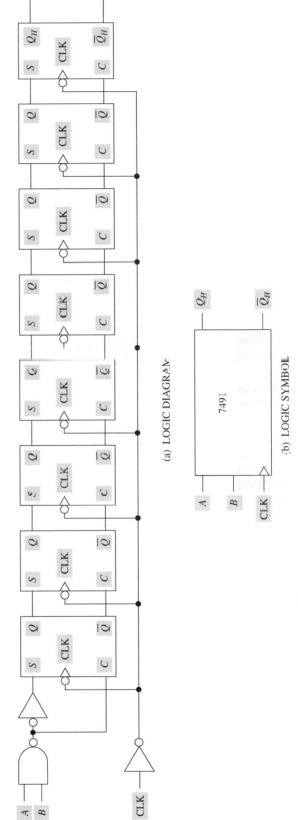

(a) LOGIC DIAGRAM

7491

(b) LOGIC SYMBOL

FIGURE 8-4 7491 8-Bit serial-in/serial-out shift register.

453

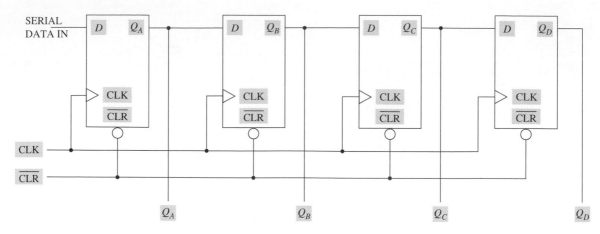

FIGURE 8–5 **Serial-in/parallel-out shift register logic diagram.**

However, when the data is to be transmitted to another system some distance away, it is converted to serial format and transmitted over a telephone line. Naturally, the data needs to be converted back to the parallel format upon receipt at the receiving system.

It takes four clock pulses to load the register in Fig. 8–5 because the data are clocked in serially. The loading action is as described for the serial-in/serial-out shift register in the previous section. Once loaded, the output data is available at Q_A, Q_B, Q_C, and Q_D. From here the data may be clocked into another digital device with a parallel input capability by using only one clock pulse.

74164 8-Bit Parallel-Out Serial Shift Register.

The logic diagram and symbol of this serial-in/parallel-out shift register are shown in Fig. 8–6. The logic diagram of this register is remarkably similar to the 7491 register in Fig. 8–4(a). Other than a $\overline{CLR}$ input, the only difference is the Q output of each flip-flop is brought to an output pin on the IC. The S and C inputs have been reversed in the diagram, and the $\overline{CLR}$ input is shown at the top of each flip-flop to simplify the drawing.

This register also contains eight active-high SET-CLEAR flip-flops, an input control gate, and a clock driver. Their functions are the same as those discussed for the 7491 register.

The ANSI/IEEE symbol for the 74164 register is shown in Fig. 8–7. The symbol contains a common-control block and eight sections representing the eight flip-flops in the register. The SRG8 notation in the common-control block identifies this symbol as representative of an 8-bit shift register. Keep in mind that the inputs to the common-control block represent an input to each element (flip-flop in this case) below the block. The RESET input (R) is an active-low input. The $C_1/\longrightarrow$ at the clock input indicates the clock controls the data shift of 1D into flip-flop A. The arrow ($\longrightarrow$) indicates the data shifts right ($Q_A \longrightarrow Q_B \longrightarrow \cdots \longrightarrow Q_H$) with each PGT of the clock pulse. The AND symbol (&) at the input of flip-flop A represents the gated-input circuit that enables or inhibits the data at flip-flop A's input.

Section 8-1: Review Questions

Answers are given at the end of the chapter.

A. Define the term "register."
B. State the two classifications of serial-input registers.
C. How many clock pulses are required to load an 8-bit serial-input register?
D. Once an 8-bit serial-in/serial-out shift register has been loaded, how many clock pulses are required to transfer the data out of the register?

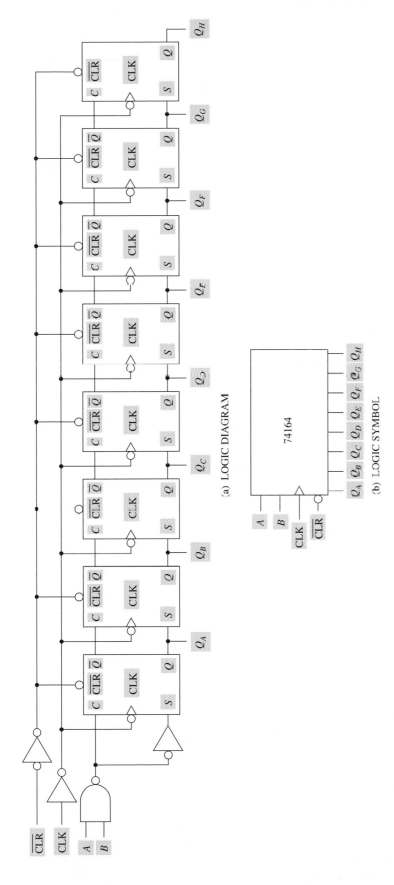

FIGURE 8-6 74164 8-Bit serial-in/parallel-out shift register.

455

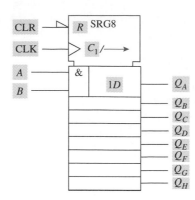

FIGURE 8–7   74164 register ANSI/IEEE symbol.

E. Data loaded in a serial-in/serial-out register are normally destroyed when the data are transferred out.
   (1) True
   (2) False

F. The notation SRG16 on an ANSI/IEEE symbol indicates _____.

SECTION 8-2: PARALLEL-INPUT AND UNIVERSAL REGISTERS

OBJECTIVES

1. Identify parallel-input and universal registers.
2. Given the logic diagram of a universal register and its inputs, determine the outputs after X clock pulses.

Parallel-in/Serial-out Shift Register

A parallel-in/serial-out shift register may be used to convert data from parallel-to-serial format. A 4-bit parallel-in/serial-out shift register implemented with D-type flip-flops is shown in Fig. 8–8. The small boxes labeled DS at the input of each flip-flop represent

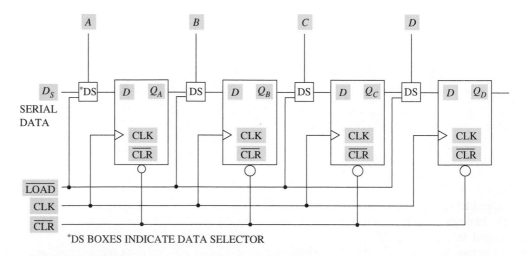

FIGURE 8–8   Parallel-in/serial-out shift register logic diagram.

data selectors. The data selectors are used to select the parallel input data (A, B, C, and D) when $\overline{\text{LOAD}} = 0$ and the serial input data (D_S) when $\overline{\text{LOAD}} = 1$. Data selector circuits will be presented in detail in Chapter 10.

The parallel input data at A–D are clocked into the flip-flops on the PGT of the clock pulse when the $\overline{\text{LOAD}}$ input is activated. Parallel loading is sometimes referred to as **broadside** loading. This type of parallel loading is synchronous. Some parallel-in/serial-out registers employ asynchronous loading which is not clock dependent. In either case, serial data movement is inhibited during the parallel load operation.

The data in the register are serially shifted to the right once the $\overline{\text{LOAD}}$ input is taken high providing clock pulses are present. The loaded data of the register is destroyed when it is transferred out to another digital circuit. This register, like most of the commercially available parallel-in/serial-out shift registers, also has a serial data input (D_S). Therefore, it can be used as a serial-in/serial-out shift register.

74165 8-Bit Parallel Load Shift Register.

The logic diagram and symbol for this register are shown in Fig. 8–9. Probably the easiest method of analyzing the **load circuitry** is to consider only one flip-flop and its associated logic gates as shown in Fig. 8–10. All eight flip-flops in the 74165 use identical load circuitry.

The load circuitry in Fig. 8–10 is enabled when $\overline{\text{LOAD}} = 0$. The inverted $\overline{\text{LOAD}}$ signal enables both NAND gates for asynchronous loading of the flip-flop. If the $\overline{\text{LOAD}}$ input is activated and the parallel data input bit A is *high,* the #1 NAND gate's output goes low. This low to the flip-flop's $\overline{\text{PRE}}$ input asynchronously sets the flip-flop. The low output from the #1 NAND gate also inhibits the #2 NAND gate so that its output ($\overline{\text{CLR}}$) stays at an inactive level.

If the $\overline{\text{LOAD}}$ input is taken low (active) and data input bit A is *low,* the #1 NAND gate's output stays high. This high output along with the high $\overline{\text{LOAD}}$ input from the NOT gate to NAND gate #2 activates the $\overline{\text{CLR}}$ input and clears the flip-flop.

The outputs of both NAND gates are high when the $\overline{\text{LOAD}}$ input is high. This keeps the $\overline{\text{PRE}}$ and $\overline{\text{CLR}}$ inputs inactive. The cascaded arrangement of the two NAND gates in Fig. 8–10 ensures only one flip-flop asynchronous input is activated at a time while the other remains inactive.

The logic diagram in Fig. 8–9(a) shows parallel loaded data are shifted one flip-flop to the right with the application of each clock pulse PGT. If one byte of data is parallel loaded, shifted out with eight clock pulses, and then another byte of data is parallel loaded, the serial data (D_S) input pin may be tied to ground. Notice the only data output pins for this register are Q_H and $\overline{Q}_H$. This allows data in serial format only at the register's output. The input clock pulse may be inhibited at any time by putting a high logic level on the clock inhibit (CLK INH) input pin.

Parallel-in/Parallel-out Register

A parallel-in/parallel-out register may be used to store binary data. Data are normally transferred within a computer in parallel because the parallel format allows high-speed data transfer. The data are usually stored in parallel format because this allows high-speed, nondestructive *read operations* to be performed. Thus, parallel-in/parallel-out register circuits make excellent memory circuits.

A 4-bit parallel-in/parallel-out register implemented with D-type flip-flops is shown in Fig. 8–11. This register loads the input data in parallel. Only one clock pulse is required to load the data after it has been placed on the input lines (A–D). The data are available at the Q outputs of the register after application of the clock pulse. The data loaded in the register may be transferred (read) without being destroyed. Remember, when data is shifted out of a serial-in/serial-out or parallel-in/serial-out register, the data are destroyed.

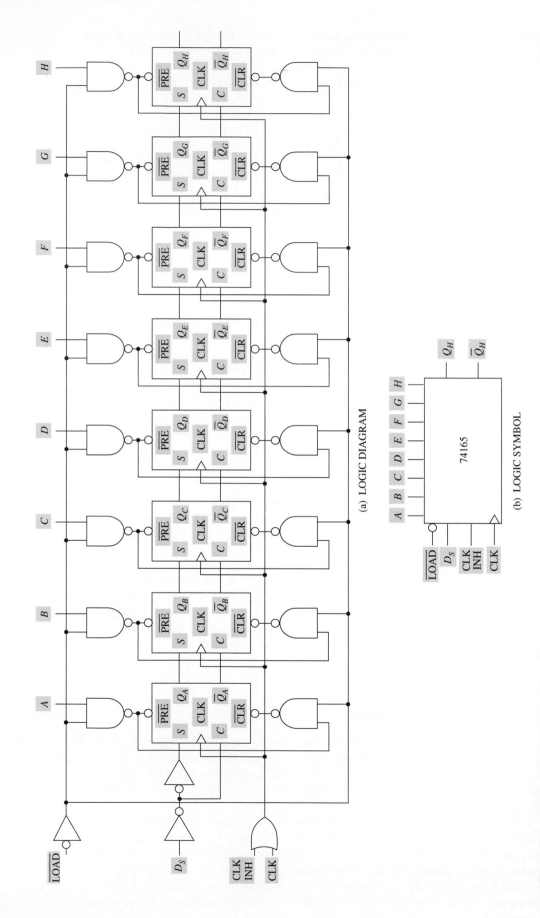

(a) LOGIC DIAGRAM

(b) LOGIC SYMBOL

Figure 8–9 74165 8-Bit parallel load shift register.

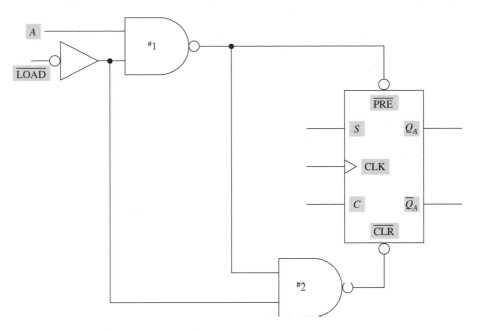

FIGURE 8-10   74165 Register–asynchronous load circuitry.

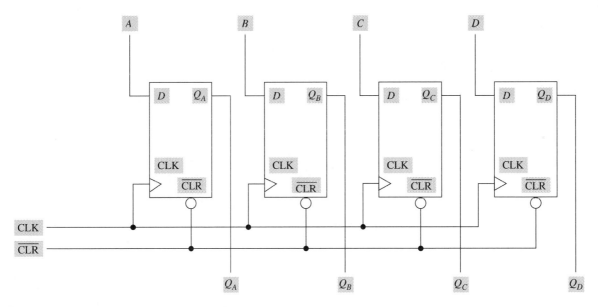

FIGURE 8-11   Parallel-in/parallel-out register logic diagram.

74178 4-Bit Parallel-Access Shift Register.

Figure 8–12 shows the logic diagram, function table, and logic symbol for this register. There are three modes of operation: LOAD, SHIFT, and HOLD.

LOAD function. The register will parallel load data when LOAD = 1 and SHIFT = 0 as shown on the function table in Fig. 8–12(b). There are three AND control gates and a NOR gate at the input of each flip-flop in Fig. 8–12(a). The AND gates are used to control the mode of operation of the register. With LOAD = 1 and SHIFT = 0, the middle AND gate in each set of gates is enabled. The other two AND gates in each set are inhibited with this LOAD/SHIFT combination of inputs. The enabled gate will allow the parallel input data at *A*, *B*, *C*, and *D* to pass to the flip-flop's inputs. However, a clock pulse must

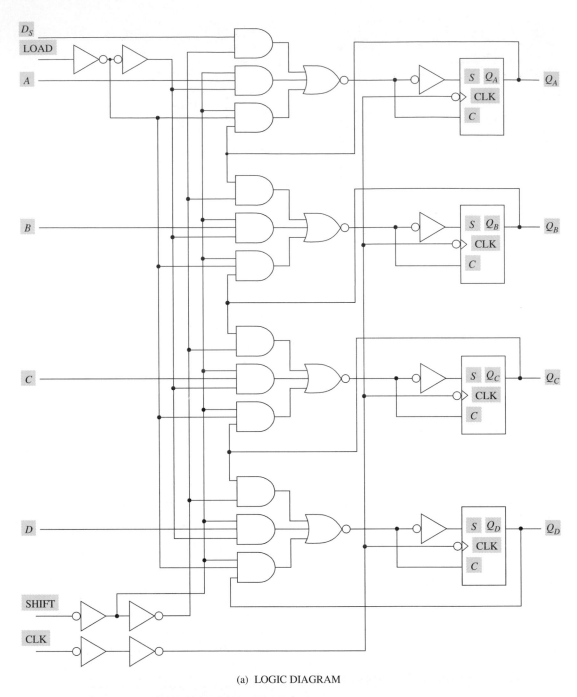

(a) LOGIC DIAGRAM

FIGURE 8–12   74178 4-Bit parallel access shift register.

be applied to load the parallel input data into the flip-flops. This synchronous loading is shown in the function table by the NGT arrow in the clock column. If this register loaded asynchronously the same as the 74165 register just presented, there would be an X in the clock column. The parallel input data is available at the parallel output pins of the IC after application of the clock pulse.

SHIFT function.   Once loaded, the register can serially shift data when SHIFT = 1. The logic diagram in Fig. 8–12(a) shows the serial data movement is through the top AND gate in each set of AND gates. This AND gate is the only one of the three gates in each set that is enabled during the SHIFT operation. The serial data input is connected to flip-flop A's

SHIFT	LOAD	D_S	CLOCK	OUTPUTS		
1	X	1/0	↓	SHIFT	$D_S \longrightarrow Q_A/Q_A \longrightarrow Q_B/$ $Q_B \longrightarrow Q_C/Q_C \longrightarrow Q_D$	
0	1	X	↓	LOAD	$A \longrightarrow Q_A/B \longrightarrow Q_B/$ $C \longrightarrow Q_C/D \longrightarrow Q_D$	
0	0	X	↓	HOLD		

(b) FUNCTION TABLE

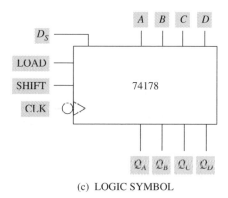

(c) LOGIC SYMBOL

FIGURE 8–12   Continued

input. Q_A is connected to flip-flop B's input, and so on when SHIFT = 1. The data will shift to the next flip-flop on the NGT of each clock pulse.

HOLD function.   The register can be placed in the HOLD mode of operation in two different ways. First, if clock transitions are not applied, the register will retain the loaded data. Second, when SHIFT = 0 and LOAD = 0, the bottom AND gate in each set of gates in Fig. 8–12(a) is enabled. This allows the Q_A output to be fed back to flip-flop A's input. This allows recirculation of the data if clock pulses remain applied to the register. In this mode of operation, the output of each flip-flop is fed back to its own input.

This register, whose logic symbol appears in Fig. 8–12(c), has been classified as a parallel-in/parallel-out register. However, it is *universal* because it can be used with serial data inputs and outputs. To expand on the universal register theory, a 4-bit universal register is presented.

Universal Register

The universal register typically can be used as a serial-in/serial-out, serial-in/parallel-out, parallel-in/serial-out, and parallel-in/parallel-out register. These registers are very popular with circuit designers because of their multiple capabilities.

74194 4-Bit Bidirectional Universal Shift Register.

This register incorporates in one IC most of the features previously discussed. The register can be loaded serially or with parallel data. In addition, data may be transferred out in either format—serial or parallel. The data internal to the register may be shifted to the left or right. The mode of operation is controlled by two select inputs (S_1 and S_0). The logic diagram, function table, and logic symbol are shown in Fig. 8–13.

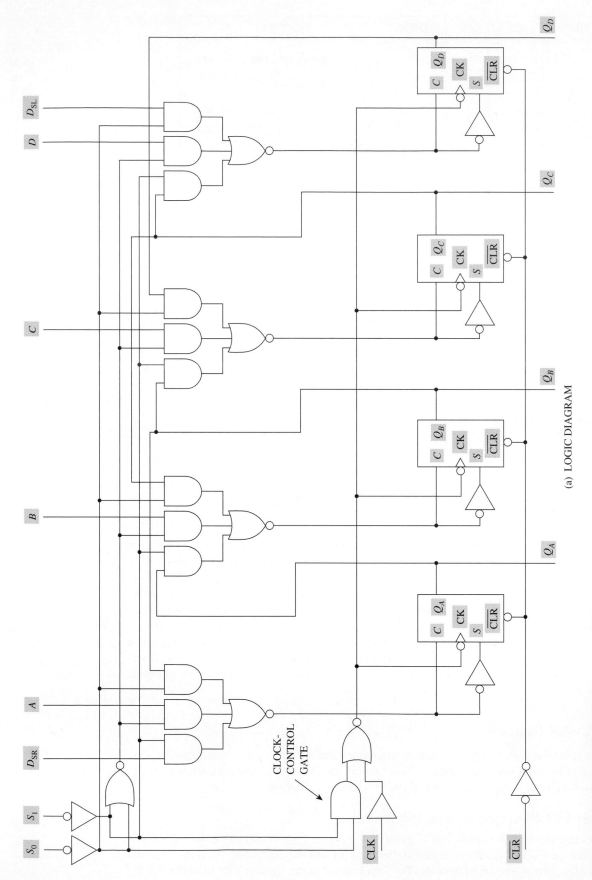

Figure 8–13 74194 4-Bit bidirectional universal shift register.

(a) LOGIC DIAGRAM

462

The 74194's mode of operation, like that of the 74178 just discussed, is controlled by a set of AND gates. These gates are in turn controlled by the select inputs and produce operational modes as follows:

S_1	S_0	Mode of Operation
1	1	PARALLEL LOAD
1	0	SHIFT LEFT
0	1	SHIFT RIGHT
0	0	HOLD

These modes of operation are shown in the function table in Fig. 8–13(b). The function table shows the $\overline{\text{CLEAR}}$ input is active low, and it overrides all other register inputs. It must be left inactive (high) for the register to operate in any of its four normal modes of operation.

$\overline{\text{CLEAR}}$	MODE		CLOCK	SERIAL		OUTPUTS
	S_1	S_0		LEFT	RIGHT	
0	X	X	X	X	X	CLEAR
1	1	1	↑	X	X	PARALLEL LOAD $A \rightarrow Q_A/B \rightarrow Q_B$ / $C \rightarrow Q_C/D \rightarrow Q_D$
1	1	0	↑	1	X	SHIFT LEFT $1 \rightarrow Q_D/Q_D \rightarrow Q_C$ / $Q_C \rightarrow Q_B/Q_B \rightarrow Q_A$
1	1	0	↑	0	X	SHIFT LEFT $0 \rightarrow Q_D/Q_D \rightarrow Q_C$ / $Q_C \rightarrow Q_B/Q_B \rightarrow Q_A$
1	0	1	↑	X	1	SHIFT RIGHT $1 \rightarrow Q_A/Q_A \rightarrow Q_B$ / $Q_B \rightarrow Q_C/Q_C \rightarrow Q_D$
1	0	1	↑	X	0	SHIFT RIGHT $0 \rightarrow Q_A/Q_A \rightarrow Q_B$ / $Q_B \rightarrow Q_C/Q_C \rightarrow Q_D$
1	0	0	X	X	X	HOLD

(b) FUNCTION TABLE

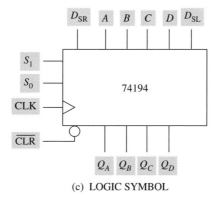

(c) LOGIC SYMBOL

FIGURE 8-13   Continued

LOAD function.   The register in Fig. 8–13(a) *synchronously* loads the parallel input data at A–D when the mode select inputs are both high. The middle gate of each set of AND gates is enabled when $S_1 = S_0 = 1$. The other two AND gates in each set are inhibited with this select input combination. The data appear at the flip-flop outputs after the PGT of the clock pulse. Serial data movement is inhibited during the loading operation.

SHIFT LEFT function.   The register will shift data to the left when $S_1 = 1$ and $S_0 = 0$. This mode of operation utilizes the right-most AND gate in each set of control gates. New serial data is entered into the register at the shift-left serial data input (D_{SL}) as shown in the logic diagram in Fig. 8–13(a). The new serial data is clocked into flip-flop D on the PGT of the clock pulse. Simultaneously, the Q_D output is shifted left into flip-flop C. The function table in Fig. 8–13(b) shows the movement of all of the data through the register during shift left operation. There are two line entries on this table—one for $D_{SL} = 1$ and one for $D_{SL} = 0$.

SHIFT RIGHT function.   The register performs shift right operations when $S_1 = 0$ and $S_0 = 1$. The new serial data to be input to the register is applied to the shift right serial data input (D_{SR}) pin. If the S_1 and S_0 logic levels are traced into the register, you can see the left-most AND gate in each control gate set is enabled, and the other two gates are inhibited.

There are also two line entries on the function table for shift right operation—one for $D_{SR} = 1$ and one for $D_{SR} = 0$. The movement of data within the register for this mode of operation is specified in the function table in Fig. 8–13(b).

HOLD function.   The register goes to the HOLD mode of operation when both mode select inputs are taken low. The clock pulse (CLK) is applied through a **buffer** to a NOR gate in the register. This NOR gate is inhibited by the output of the clock-control AND gate when $S_1 = S_0 = 0$. Therefore, no data movement can be accomplished because clocking of the register is inhibited. This NOR gate is enabled and allows clocking of the flip-flops in all other modes of operation. As long as one of the select inputs is high, a low into the clock-control AND gate produces a low output to enable the NOR gate.

The buffer referred to in the preceding paragraph is a circuit used to isolate the input and output while increasing the drive capability of a signal. The buffer produces no phase inversion and is used entirely to condition the signal. The symbol, as shown in Fig. 8–13(a) at the CLK input, is the same symbol as a NOT gate without the inverting bubble. The $\overline{CLR}$ input of this circuit is also applied through a buffer to the $\overline{CLR}$ inputs of the flip-flops. The bubbles at the input and output of this symbol are used as state indicators. They indicate an active-low input and an active-low output. The standard logic symbol for this universal register is shown in Fig. 8–13(c).

Section 8–2: Review Questions

Answers are given at the end of the chapter.

A. State the two classifications of parallel-input registers.

B. How many clock pulses are required to load an 8-bit parallel-input register?

C. Data transferred out of a parallel-in/parallel-out register are normally destroyed.
    (1) True
    (2) False

D. Data may be loaded into a parallel-input register synchronously or asynchronously.
    (1) True
    (2) False

E. Universal registers can normally be loaded serially or broadside and data can be transferred out in series or parallel format.
    (1) True
    (2) False

SECTIONS 8-1 AND 8-2: INTERNAL SUMMARY

A group of latches or flip-flops is a register. A register is used to store, transfer, or shift data. Data may be input and retrieved from a register in serial or parallel format. The type of register determines the data format.

The two broad classifications of registers are **serial-input** and **parallel-input.** The serial-input register may allow retrieval of data in serial- or parallel-out format. The parallel-input register may also allow serial- or parallel-output data.

Serial-in/serial-out registers can be used to store data or shift it left or right. The number of clock pulses required to load a serial-in/serial-out register is equal to the number of flip-flops in the register. The same is true regarding transferring data serially out of a loaded register.

Serial-in/parallel-out registers are normally used to change data formats. When data is received in a system from another system, it is usually received in serial format from a single input line. The serial data is converted in this type of register back to parallel format to increase effective system speed. **Parallel-in/serial-out** registers reverse the process. They take the parallel data and convert it to serial format.

Parallel-in/parallel-out registers are used to store or transfer data. They do not have a shift capability for processing data. They are often used as memory devices.

Many IC registers provide for loading and transferring data in either serial or parallel format. Some of these **universal registers** allow **bidirectional** shifting of data.

SECTIONS 8-1 AND 8-2: INTERNAL SUMMARY QUESTIONS

Answers are given at the end of the chapter.

Note: The use of manufacturer's data sheets is recommended to assist you in answering these summary questions.

1. A register is a group of latches or flip-flops used to transfer, store, or shift data.
   a. True
   b. False

2. What type of register would be used to take parallel data from within a digital system and convert it to a format for transmission over a single line?
   a. Serial-in/serial-out
   b. Serial-in/parallel-out
   c. Parallel-in/serial-out
   d. Parallel-in/parallel-out

3. How many clock pulses are required to load an 8-bit serial-in/parallel-out shift register?
   a. 1   c. 8
   b. 4   d. 16

4. How many clock pulses are required to read data from a loaded 8-bit serial-in/parallel-out shift register?
   a. 1   c. 8
   b. 4   d. 16

5. Identify the symbol shown in Fig. 8–14.
   a. Serial-in/serial-out register
   b. Serial-in/parallel-out register
   c. Parallel-in/serial-out register
   d. Parallel-in/parallel-out register

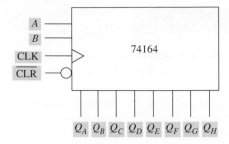

FIGURE 8-14

6. The register shown in Fig. 8–15 can be loaded with series or parallel data.
   a. True
   b. False

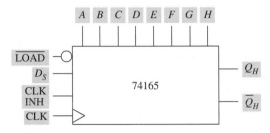

FIGURE 8-15

7. The register shown in Fig. 8–15 can transfer data out in serial or parallel format.
   a. True
   b. False

8. What inputs are required for the 74178 IC in Fig. 8–16 to load parallel data?
   a. SHIFT = 1; LOAD = 1
   b. SHIFT = 1; LOAD = 0
   c. SHIFT = 0; LOAD = 1
   d. SHIFT = 0; LOAD = 0

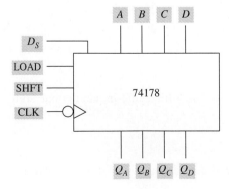

FIGURE 8-16

9. What inputs are required to the 74178 in Fig. 8–16 to shift data?
   a. SHIFT = 1; LOAD = 1
   b. SHIFT = 1; LOAD = 0
   c. SHIFT = 0; LOAD = 1
   d. a or b

10. Identify the register in Fig. 8–17.
    a. Serial-in/serial-out
    b. Parallel-in/serial-out
    c. Parallel-in/parallel-out
    d. Universal

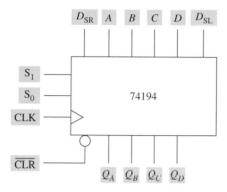

FIGURE 8-17

11. What is the mode of operation of the register in Fig. 8–17 when $\overline{CLR} = 1$; $S_1 = 0$; $S_0 = 1$; CLK = PGT?
    a. HOLD   c. SHIFT RIGHT
    b. SHIFT LEFT   d. PARALLEL LOAD

12. What data are contained in the register in Fig. 8–17 if

    STEP 1: $\overline{CLR} = 0$ and then is taken high; $S_1 = 1$; $S_0 = 1$; $D_{SR} = 0$; $D_{SL} = 1$; $A = B = C = D = 1$; CLK = PGT.

    STEP 2: $S_1 = 0$; $S_0 = 1$; CLK = PGT (one pulse).

    a. 0110   c. 1111
    b. 0111   d. 1110

SECTION 8-3: PRACTICAL APPLICATIONS OF REGISTERS

OBJECTIVES

1. Given the logic diagram of a Ring or Johnson counter implemented with ICs, determine the output after X clock pulses as well as the MOD number.
2. Given the logic diagram and input levels of a bidirectional shift register, determine the product or quotient in the register after X clock pulses.

Code Detector

A practical application of a serial-in/serial-out shift register is the **code-detector** circuit shown in Fig. 8–18. The circuit is used to detect a specified code (set of sequential numbers) and produce a pulse out when the proper code has been entered. The output code pulse may be used to arm or disarm a security system.

A keypad is used to enter the code into the circuit. The exact code, in correct sequence, must be entered for the proper output code pulse. Each time a key is pressed and released a positive pulse is generated. If the 5 key is pressed and released, a PGT is gen-

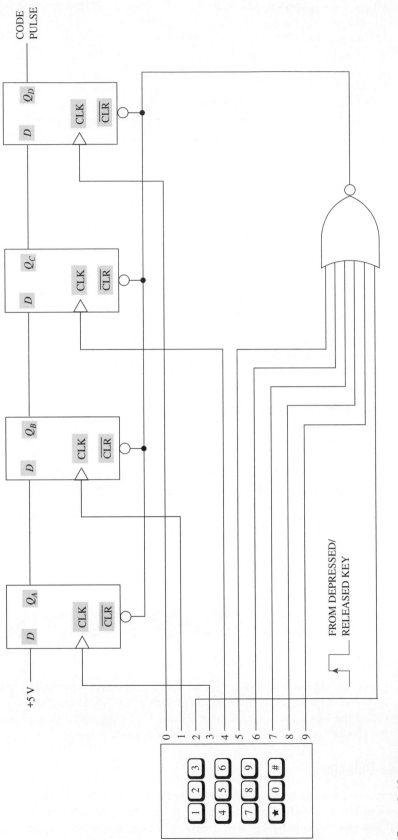

FIGURE 8-18 Code-detector circuit.

erated and the key pulse returns to its normally low level when the key is released. Note in Fig. 8–18 that depressing the number 2 key sends a Logic 1 to the NOR gate. The low output from the NOR gate asynchronously clears all of the flip-flops. This same action will occur if the number 5, 6, 7, 8, or 9 key is depressed and then released.

The ultimate objective of this circuit is to produce a high at the code pulse output for use by a security system. This high output can be achieved only by entering the code 3140 in this particular detector. Any other code will fail to produce the proper output.

Note in the logic diagram that the D input of flip-flop A is tied to +5 V. When the 3 key is depressed, the PGT of the generated pulse clocks flip-flop A and its output (Q_A) goes high. This places a high on the input of flip-flop B. If the 2 key is accidently pressed next, the low output of the NOR gate will clear the flip-flops, and the process must start over.

It is easy to see that the code must be entered sequentially. If it is not entered properly, the high logic level will not be shifted through the register. After the 3 key was depressed, flip-flop A was set. Flip-flop B will set if the 1 key is depressed next. The C flip-flop will set next if the 4 key is pressed. Finally, flip-flop D will set if the 0 key is depressed. The security system is armed, or disarmed, when flip flop D sets.

Shift Register with Data Recirculate

It was mentioned in Section 8–1 that data transfer operations in a serial-in/serial-out shift register normally destroy the data in the register. However, a little ingenuity and logic circuitry can be used to solve this data destruction problem. The circuit shown in Fig. 8–19 is used to recirculate the data during a read operation.

A 7491 8-bit register is used to serially shift the data. The logic gates are used to control READ and WRITE operations. The READ operation automatically provides data recirculation.

READ/$\overline{\text{WRITE}}$ = 0 and $\overline{\text{CLK INH}}$ = 1: This input combination allows writing data into the register. The high $\overline{\text{CLK INH}}$ (Clock Inhibit) signal enables the lower AND gate and allows the clock pulses to clock the register. The low $\overline{\text{WRITE}}$ signal is inverted and applied as an enabler to the middle AND gate. This allows serial data (D_S) to be loaded into the shift register. A byte of serial data is loaded after eight clock pulses. The $\overline{\text{CLK INH}}$ signal may now be taken low, and the data left stored for any period of time.

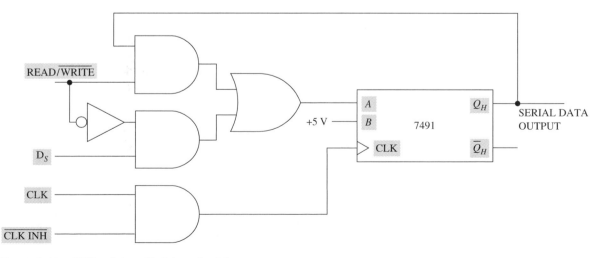

FIGURE 8-19   **Shift register with data recirculate.**

READ/$\overline{\text{WRITE}}$ = 1 and $\overline{\text{CLK INH}}$ = 1: This input combination allows reading the data that is stored in the register. The high READ/$\overline{\text{WRITE}}$ signal enables the top AND gate. This allows the data shifted out of the register to be clocked back into the register at the same time it is transmitted down the serial data output line. This simple circuit prevents the READ operation from being a destructive operation.

Ring/Johnson Counters

The **Ring counter** was presented in Section 7–9. The counter is a shift register that has its output connected back to its own input. IC implementation of a Ring counter is shown in Fig. 8–20(a) using a 74178 4-bit shift register. You may want to review the logic diagram and function table in Fig. 8–12 prior to the following discussion.

The Ring counter must be loaded with a one for proper operation. This is accomplished in the 74178 when LOAD = 1 and SHIFT = 0 (Fig. 8–20a). Since the preload inputs are $A = 1$ and B, C and $D = 0$, the A flip-flop is set and the remaining flip-flops are cleared on the NGT of the clock pulse.

The register will shift the loaded 1 to the right one flip-flop per clock pulse when the shift input is taken high as shown in the figure. The LOAD input is a don't care input when SHIFT = 1. The Q_D output is connected to the serial data input (D_S) in the recirculate method just discussed. The signal available at any one of the outputs is $f_{\text{in}}/4$ (MOD-4 counter).

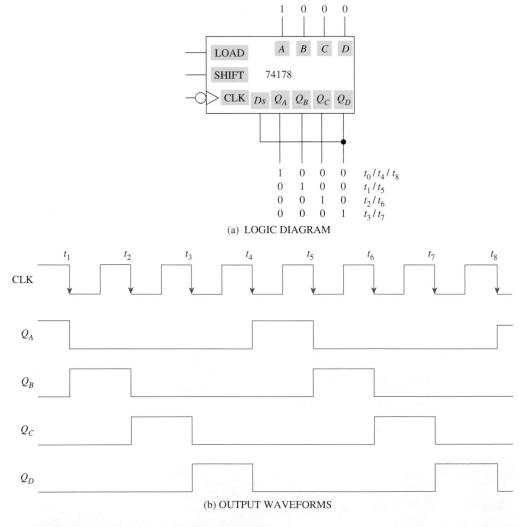

(a) LOGIC DIAGRAM

(b) OUTPUT WAVEFORMS

FIGURE 8-20   Ring counter–74178 implementation.

The Ring counter is normally used to control events (circuits) by triggering them on and then off in sequence. Sometimes the Ring counter is called a **sequencer** because of its equally spaced output pulses, which provide the ability to control a sequence of events. The MOD-4 ring counter output waveforms shown in Fig. 8–20(b) illustrate that only one of the four outputs is high at any given time. This is true because an NGT of the input CLK pulse sets an output high and the next CLK NGT clears that output to a low. Therefore, if four devices (events) were controlled by the Ring counter, one device would be on and the other three off at any given time.

Another Ring counter implementation is shown in Fig. 8–21. This circuit is a **self-loading Ring counter.** The addition of the NOR gate provides the self-loading feature. Regardless of the initial register condition upon application of power, the circuit will end up with one high and three lows for proper operation. Also note the connection from the Q_D output back to the circuit's input flip-flop has been eliminated.

Let's assume upon the application of power the initial circuit condition is 1100 (Q_A–Q_D). The highs from Q_A and Q_B cause the NOR gate to produce a low output. This loads the A flip-flop with a 0 on the next NGT of the clock (t_1), and the outputs are 0110 after the shift right operation. The highs from Q_B and Q_C again cause the NOR gate to produce a low output. This loads another 0 into flip-flop A on the next active transition of the clock (t_2). The remaining data are shifted right to produce 0011 in the register. The high from the Q_C output loads another 0 into flip-flop A via the D_S input (t_3). The register output is now 0001. The three 0s into the NOR gate result in a high output, which sets flip-flop A on the next active clock transition (t_4). The output of the register is now 1000.

The NOR gate will allow only one Logic 1 in the counter after a maximum of three clock pulses. This is true regardless of the initial register condition.

Two other methods of *self-loading* a Ring counter are shown in Fig. 8–22(a) and (b). The Ring counter in Fig. 8–22(a) uses an AND gate for loading the D-type flip-flops. This circuit does not require the output connection back to the shift register's input. The reason will soon become evident. The AND gate will produce a high output to set the D flip-flop only when the counter's output is X000 (Q_A–Q_D). This is because $\overline{Q}_B$, $\overline{Q}_C$, and $\overline{Q}_D$ are connected to the AND gate. This arrangement automatically loads a high into flip-flop D

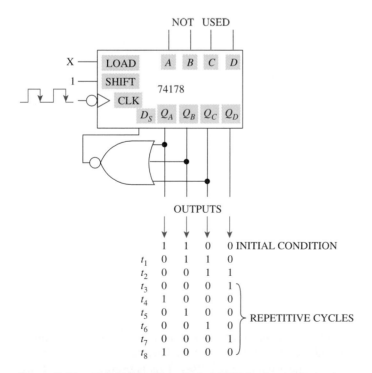

FIGURE 8–21   Self-loading Ring counter–74178 implementation.

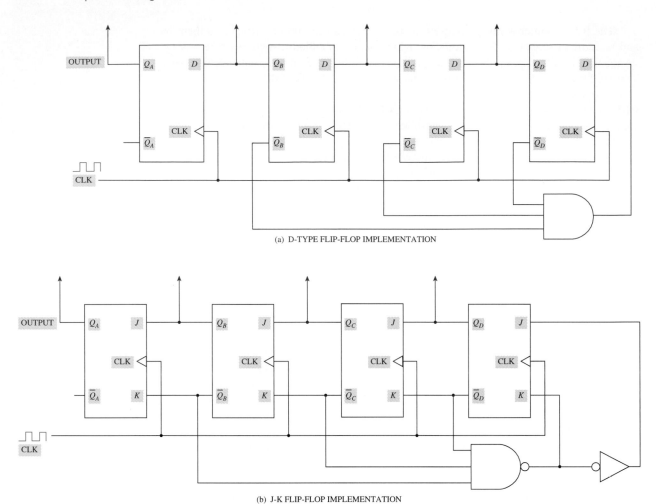

(a) D-TYPE FLIP-FLOP IMPLEMENTATION

(b) J-K FLIP-FLOP IMPLEMENTATION

FIGURE 8–22 **Self-loading Ring counters.**

every time a high is shifted out of flip-flop A. Another version of a self-loading Ring counter using J-K flip-flops is shown in Fig. 8–22(b). Operational analysis of this circuit will be left to the student.

The Johnson counter in Fig. 8–23(a) has been implemented with a 74178 4-bit shift register. The modification used to change a Ring counter to a Johnson counter was presented in Chapter 7. The $\overline{Q}$ output instead of the Q output of the MSB flip-flop is connected back to the input flip-flop through the D_S input pin. A NOT gate is used to invert Q_D because a $\overline{Q}_D$ output pin is not available on the IC.

The outputs of this MOD-8 Johnson counter are

Q_A	Q_B	Q_C	Q_D
0	0	0	0
1	0	0	0
1	1	0	0
1	1	1	0
1	1	1	1
0	1	1	1
0	0	1	1
0	0	0	1

The MOD-8 Johnson counter output waveforms in Fig. 8–23(b) show that any given output is high for four cycles and then low for four cycles. Therefore, a device (circuit) can be turned on and left on for the next three CLK pulses. This is useful when a device needs to be turned on and left on until all devices are turned on.

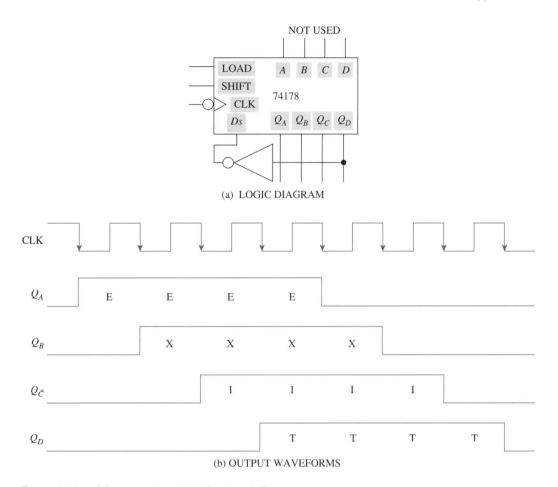

FIGURE 8–23   **Johnson counter–74178 implementation.**

A good example of using the Johnson counter of Fig. 8–23 is illuminating an EXIT sign. The waveforms show the sign would illuminate as follows:

E
EX
EXI
EXIT
XIT
IT
T

Multiply/Divide Register

Shift registers are often used to perform multiplication and division operations. A properly executed shift left operation multiplies a binary number by a factor of 2^n. The n represents the number of shift left operations. A shift right operation divides a binary number by a factor of 2^n.

Let's use the 74198 bidirectional shift register shown in Fig. 8–24 to perform these operations. The 74198 shift register is an 8-bit version of the 74194 register presented in the last section. The mode-control inputs are the same as the 74194. The steps to perform the arithmetic operations are

1. $S_1 = 1$; $S_0 = 1$: Parallel load the number to be multiplied or divided.
2. $S_1 = 1$; $S_0 = 0$: Shift left to multiply.
3. $S_1 = 0$; $S_0 = 1$: Shift right to divide.
4. $S_1 = 0$; $S_0 = 0$: Hold (store) the data after the arithmetic operation.

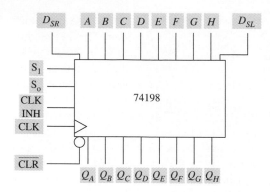

FIGURE 8-24   **74198 8-Bit bidirectional universal shift register.**

The 74198 internal flip-flops have been drawn in block diagrams in Fig. 8–25 for the following examples.

Look at Fig. 8–25(a). A binary 1 has been parallel loaded into the 74198 register as the initial condition. The register is then set up with $S_1 = 1$ and $S_0 = 0$ for shift left operation. If the binary 1 is to be multiplied by 32 (2^5), five clock pulses must be applied to the register.

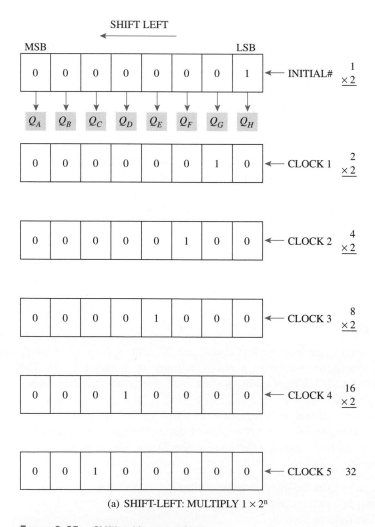

(a) SHIFT-LEFT: MULTIPLY 1×2^n

FIGURE 8-25   **Shifting binary numbers.**

This entire operation, one step at a time, is depicted in Fig. 8–25(a). The first clock pulse multiplies the initial binary 1 (LSB) by 2 to produce a 2 output. The 74198 shift left serial data input (D_{SL}) must be low so that a 0 is loaded into the LSB flip-flop. The second clock pulse shifts the 1 from flip-flop G to flip-flop F. This shift left operation results in a 4 ($00000100_{(2)}$) in the register. Each successive clock pulse shifts the high bit one flip-flop to the left, multiplying the existing binary number by 2 each shift. The register contents after five clock pulses are $00100000_{(2)} = 32$. The 74198 register may now be placed in the HOLD mode to await a READ operation.

Another example of multiplication is shown in Fig. 8–25(b). This example incorporates a binary point between the second and third bits in the register. The initial number is $11.11_{(2)} = 3.75$. Each clock pulse loads a 0 into the LSB position and shifts the register data one flip-flop to the left. Four clock pulses have been applied to the register as shown in the example. Since $2^4 = 16$, the original binary number has been multiplied by 16 ($16 \times 3.75 = 60$).

Notice that a fifth input clock pulse would shift the MSB out of the register. This would render the product invalid. Therefore, registers used for multiplication are often cascaded to insure the product can be accommodated by the circuit. Examples of dividing binary numbers using shift operations are shown in Fig. 8–25(c) and (d).

The first example shows a 96 has been parallel loaded into the register. The first clock pulse shifts the binary number one flip-flop to the right. Note that a 0 is shifted into the MSB position with each clock pulse. Thus, the 74198 shift right serial data input (D_{SR}) must be low during this operation. The shift produces a quotient of 48. Repeated shift right operations result in repeated division by 2. The total number of shift right operations

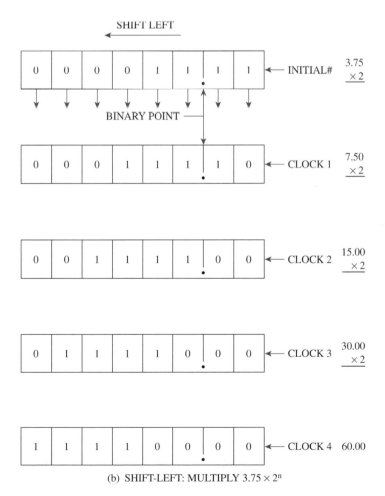

(b) SHIFT-LEFT: MULTIPLY 3.75×2^n

FIGURE 8–25   Continued

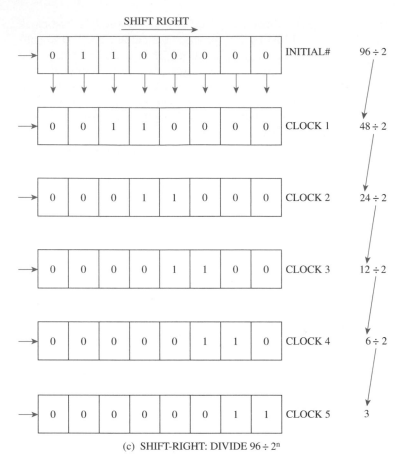

(c) SHIFT-RIGHT: DIVIDE $96 \div 2^n$

FIGURE 8–25   **Continued**

produces a divide-by capability equal to 2^n. In this example, $2^n = 2^5 = 32$. The end re-sult in this example is $96 \div 32 = 3$.

The second example of binary division by 2^n includes a binary point for illustration. This is shown in Fig. 8–25(d). Here the number 80 in binary is shifted right five posi-tions. The quotient produced by the register is $80 \div 32 = 2.5$.

These examples prove that a binary number can be multiplied or divided by a factor of 2^n by shifting that number left or right. Data shifting to produce products or quotients is very popular in digital systems because it is simple and fast.

The shifting process is used in memory systems in many computers to obtain an ac-tual memory address. For example, the contents of a shift register are multiplied by 16 when the data is shifted four positions to the left. These four shifted positions represent shifting one hexadecimal digit (4 bits) of the address to the left because $16_{(10)} = 10_{(16)}$. The shifted address is the segment address portion of an actual address. The segment address is then added to an offset address to obtain the actual memory address. Memory addressing is presented in Chapter 12.

Section 8–3: Review Questions

Answers are given at the end of the chapter.

A. Data transfer (READ) operations using a serial-in/serial-out shift register are always destructive—even if data recirculation is employed.
    (1) True
    (2) False

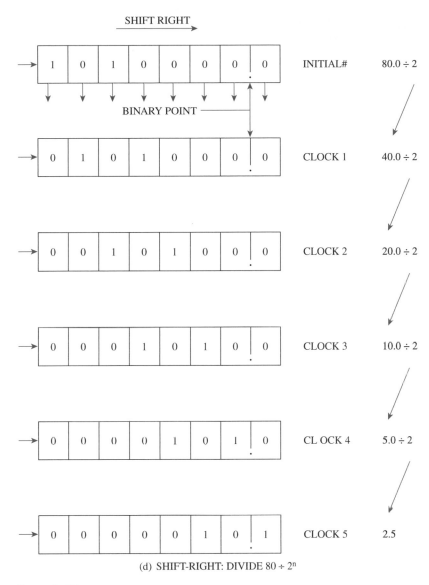

(d) SHIFT-RIGHT: DIVIDE $80 \div 2^n$

FIGURE 8-25   Continued

B. A Ring counter can be implemented using a serial-in/serial-out shift register IC by connecting the Q output back to the serial data input of the IC and loading a high into the register.
   (1) True
   (2) False

C. A Johnson counter can be implemented using a serial-in/serial-out shift register IC by connecting the $\overline{Q}$ output back to the serial data input of the IC.
   (1) True
   (2) False

D. What are the contents of the circuit in Fig. 8–20 after 6 clock pulses? Assume the LOAD input is asserted while the SHIFT input $= 0$ and then SHIFT is taken high.

E. What is the MOD number of the circuit in Fig. 8–20?

F. Shift left register operations _____ a binary number by a factor of 2^n.
   (1) divide
   (2) multiply

G. How many shift operations are required to divide the number 20 by 4?

H. In which direction must the shift be in the preceding question?

SECTION 8-4: TROUBLESHOOTING REGISTERS

OBJECTIVES

1. Given the observed troubleshooting results of a defective register circuit, determine the possible fault(s).
2. Given the malfunction in a register circuit, determine the most probable symptoms.

Since registers are nothing more than groups of latches or flip-flops, the basic trouble-shooting steps presented in Chapter 6 for flip-flops apply in this section. Always keep in mind that to troubleshoot a circuit you must know what it should do when it is operating properly. Initially, visually check for common faults such as proper power connections. Many troubles have been fixed by connecting the power plug! This may seem ridiculous, but talk to some experienced technicians, and they will tell you this is true. In addition, check for blown fuses, broken wires or traces, and poor connections. After your sense of sight has been used, use your sense of feeling. If the circuit contains ICs and has power applied, laying your hands on top of the ICs can locate an IC that is extremely warm. This condition usually indicates a problem area.

Another point—remember from previous studies that the load connected to a circuit can cause the driving circuit to appear faulty when it in fact is good. Whenever an output appears erroneous and it is connected to a load, disconnect the load to isolate the problem.

Let's start our troubleshooting studies with a discrete-component register. The register in Fig. 8–26 is a serial-in/serial-out shift register consisting of TTL flip-flops. Analyze the following:

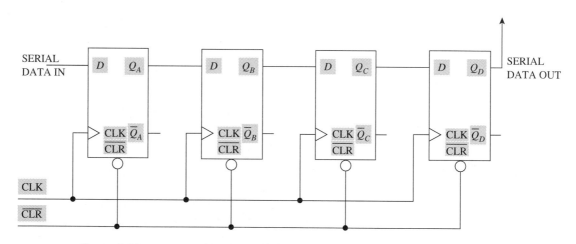

FIGURE 8-26   Serial-in/serial-out register for Case Study 1.

Case Study 1.

Assume the serial data input to the circuit in Fig. 8–26 is tied high. The register is cleared and produces the observed output conditions shown below as it is clocked. The desired output conditions of a properly operating circuit are also shown.

Observed				Desired				
Q_A	Q_B	Q_C	Q_D	Q_A	Q_B	Q_C	Q_D	
0	0	0	0	0	0	0	0	CLR
1	0	1	0	1	0	0	0	t_1
1	1	1	1	1	1	0	0	t_2
1	1	1	1	1	1	1	0	t_3
1	1	1	1	1	1	1	1	t_4

Observation of the output allows us to conclude that the Q_C flip-flop is malfunctioning. This is based on the fact that its input is low prior to the t_1 PGT, yet the flip-flop's output went high at t_1. The desired results show Q_C should not go high until t_3.

Diagnosis.   The most probable cause of this malfunction is an open at the D input of flip-flop C. The open could be the wire (trace) connecting flip-flops B and C, or it could be internal. Isolating the problem to the trace or flip-flop could easily be done with a logic probe. Remember, an open (floating) input to a TTL flip-flop acts as a high. Thus, flip-flop C sets on the first clock pulse.

Case Study 2.

The circuit shown in Fig. 8–27 is a parallel-in/serial-out shift register. Initial troubleshooting steps reveal $\overline{\text{LOAD}} = 1$, $D_S = 0$, CLK INH = 0, and CLK = pulsing. However, application of numerous clock pulses produce a constant $Q_H = 1$ and $\overline{Q}_H = 0$.

Changing the A through H parallel inputs to all low and pulsing $\overline{\text{LOAD}}$ low changes Q_H to 0 and $\overline{Q}_H$ to 1. If D_S is set high and the register is clocked eight times, Q_H should go high, but application of numerous clock pulses continues to produce a constant $Q_H = 0$ and $\overline{Q}_H = 1$.

When the parallel inputs $A–H$ are returned to the levels shown in Fig. 8–27 and $\overline{\text{LOAD}}$ is pulsed low, Q_H returns to 1 and $\overline{Q}_H$ returns to 0. Again, repetitive clock pulses produce no change in the $Q_H = 1$ and $\overline{Q}_H = 0$ outputs.

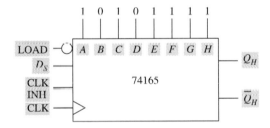

FIGURE 8–27   Parallel-in/serial-out register for Case Study 2.

Diagnosis.   The 74165 does not have output pins available at Q_A through Q_G. This leaves the technician with no way to monitor the internal data as it is clocked within the register.

We can surmise that the parallel load function of the register is working properly. This is based on the fact that Q_H follows the H preload input data when $\overline{\text{LOAD}}$ is activated.

It is evident that the data within the register is not shifting to the right as it should. If it were, the Q_H output would go low on the t_4 PGT of the clock with the parallel input data shown in the figure.

The most probable fault is an internal open in the CLK/CLK INH circuit. An internal open on either input of the OR gate will inhibit the gate and prevent the clock pulse from being applied to the flip-flops within the register. *Note:* The parallel load function in this register is *asynchronous*. Therefore, this mode of operation functions properly since parallel loading is not clock dependent.

Case Study 3.

The logic diagram in Fig. 8–28 shows two 74178 ICs cascaded to form a MOD–16 Johnson counter. Proper operation of the counter will produce the following outputs:

Q_A	Q_B	Q_C	Q_D	Q_E	Q_F	Q_G	Q_H	
0	0	0	0	0	0	0	0	CLR
1	0	0	0	0	0	0	0	t_1
1	1	0	0	0	0	0	0	t_2
1	1	1	0	0	0	0	0	t_3
1	1	1	1	0	0	0	0	t_4
1	1	1	1	1	0	0	0	t_5
1	1	1	1	1	1	0	0	t_6
1	1	1	1	1	1	1	0	t_7
1	1	1	1	1	1	1	1	t_8
0	1	1	1	1	1	1	1	t_9
0	0	1	1	1	1	1	1	t_{10}
0	0	0	1	1	1	1	1	t_{11}
0	0	0	0	1	1	1	1	t_{12}
0	0	0	0	0	1	1	1	t_{13}
0	0	0	0	0	0	1	1	t_{14}
0	0	0	0	0	0	0	1	t_{15}
0	0	0	0	0	0	0	0	t_{16}

This analysis commences with LOAD = 1 and SHIFT = 0. This *synchronously* clears all flip-flops in the register by preloading all 0s. Next, SHIFT is taken high, and this produces the following:

Q_A	Q_B	Q_C	Q_D	Q_E	Q_F	Q_G	Q_H	
0	0	0	0	0	0	0	0	CLR
1	0	0	0	0	0	0	0	t_1
1	1	0	0	0	0	0	0	t_2
1	1	1	0	0	0	0	0	t_3
1	1	1	1	0	0	0	0	t_4
1	1	1	1	1	0	0	0	t_5
1	1	1	1	1	1	0	0	t_6
1	1	1	1	1	1	1	0	t_7
1	1	1	1	1	1	1	1	t_8
1	1	1	1	1	1	1	1	t_9

The outputs all remain high with repetitive clock pulses after t_8. The problem encountered is when Q_H goes high, $\overline{Q}_H$ from the NOT gate should go low, and the counter should fill up with 0s as shown in the proper operation chart. This is not happening.

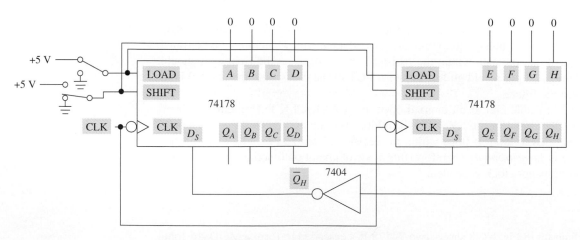

FIGURE 8-28   MOD-16 Johnson counter for Case Study 3.

Diagnosis.   If the D_S input of the first register $(Q_A–Q_D)$ is open, the floating input will always load 1s into the register and cause this problem. The problem could be the connection between the NOT gate and the D_S input. This would also act as a floating input to the register. An open before the NOT gate would not produce the same symptoms.

What would happen if the input to the NOT gate were open? A floating input here would cause the output of the NOT gate to stay low all of the time. Thus, when the counter is cleared, only 0s will be loaded into it during shift operations. The outputs would stay low all of the time if this problem existed.

What could possibly produce the following output conditions from the Johnson counter in Fig. 8–28?

Q_A	Q_B	Q_C	Q_D	Q_E	Q_F	Q_G	Q_H	
0	0	0	0	0	0	0	0	CLR
1	0	0	0	1	0	0	0	t_1
1	1	0	0	1	1	0	0	t_2
1	1	1	0	1	1	1	0	t_3
1	1	1	1	1	1	1	1	t_4
0	1	1	1	1	1	1	1	t_5
0	0	1	1	1	1	1	1	t_6
0	0	0	1	1	1	1	1	t_7
0	0	0	0	1	1	1	1	t_8
0	0	0	0	1	1	1	1	t_9

The outputs remain as shown after t_8 with the application of repetitive clock pulses. The counter should load a 1 starting at Q_A and then fill up with 1s with several successive clock pulses until all of the outputs are high. Then it should fill up with 0s until reaching the initial state. However, Q_E sets high on the first clock pulse. What could cause this to happen? The first 74178's $(Q_A–Q_D)$ output appears normal through t_8. The problem has to be at the input to the second 74178. These results would be produced if the trace/connection between the first register's Q_D output and the second register's D_S input were bad, or if the second 74178's D_S input were open internally.

Case Study 4.

The circuit shown in Fig. 8–29 is a self-loading Ring counter. Application of power and clock pulses should produce the following bit pattern at the output:

	Desired			
Q_A	Q_B	Q_C	Q_D	
0	0	0	1	t_1
0	0	1	0	t_2
0	1	0	0	t_3
1	0	0	0	t_4
0	0	0	1	t_5

Note: The initial circuit condition upon application of power is immaterial. If more than one flip-flop sets, the circuit will rid itself of the excess 1s. If the circuit initializes with all 0s, a 1 will be loaded into the D flip-flop at t_1.

The observed conditions of the defective circuit are

	Observed			
Q_A	Q_B	Q_C	Q_D	
0	0	0	1	t_1
0	0	1	0	t_2
0	1	0	1	t_3
1	0	1	0	t_4
0	1	0	1	t_5

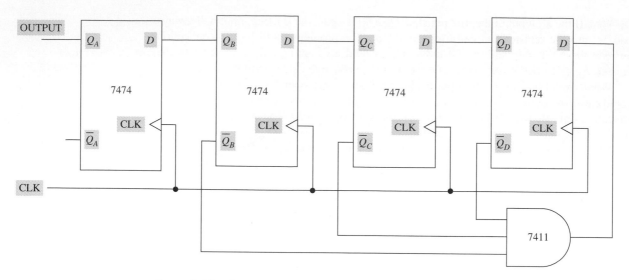

FIGURE 8-29   Self-loading Ring counter for Case Study 4.

Checking the observed output in respect to the desired output condition reveals a disparity at t_3. The counter output should be 0100 after t_3 if we assume both sets of output conditions started at the same initial condition as shown in the desired and observed output condition tables. The observed output shows the malfunctioning output after t_3 is 0101. The problem is the high at the Q_D output (010$\underline{1}$). The only way to synchronously set flip-flop D is to have a high at the D input when the PGT of the clock pulse arrives.

An open at the D input of flip-flop D would not cause this problem. This is evidenced by the fact that flip-flop D's output went low at t_2. This wouldn't happen if the D input were open.

Diagnosis. Troubleshooting the defective circuit reveals the high loaded into flip-flop D at t_3 is produced by the AND gate. As you well know, this gate can output high only when all of its inputs are high. The inputs to this gate after t_2 should be $\overline{Q_B} = 1$, $\overline{Q_C} = 0$, and $\overline{Q_D} = 1$. This information was derived from the output of the circuit after t_2, which is 0010 (Q_A–Q_D). After these levels have been verified as normal with a logic probe, the only conclusion to reach is that the $\overline{Q_C}$ input to the AND gate must be internally open.

Use this same line of thinking and come up with the output conditions and circuit symptoms for this circuit if the $\overline{Q_B}$ output trace to the AND gate is open. The problem here would be easier to discern when using a logic probe. Theoretically, the following output conditions should result from the problem:

Q_A	Q_B	Q_C	Q_D	
0	0	0	1	t_1
0	0	1	0	t_2
0	1	0	0	t_3
1	0	0	1	t_4
0	0	1	0	t_5
0	1	0	0	t_6
1	0	0	1	t_7

The malfunction occurs at t_4. If the counter were functioning properly, the output would be 1000 (Q_A–Q_D). The key is to analyze the input conditions to the AND gate after t_3 and prior to t_4. The conditions are $\overline{Q_B} = 0$, $\overline{Q_C} = 1$, and $\overline{Q_D} = 1$. A logic probe could be used to verify these levels as good at the flip-flop's outputs, but the probe would show no valid logic level at the $\overline{Q_B}$ input of the AND gate because the trace to the gate is open.

Section 8-4: Review Questions

Answers are given at the end of the chapter.

A. Complete the observed output table (t_2–t_8) below for the self-loading ring counter in Fig. 8–29 if the $\overline{Q}_D$ input to the AND gate is internally open.

Observed				Desired				
Q_A	Q_B	Q_C	Q_D	Q_A	Q_B	Q_C	Q_D	
0	0	0	1	0	0	0	1	t_1
				0	0	1	0	t_2
				0	1	0	0	t_3
				1	0	0	0	t_4
				0	0	0	1	t_5
				0	0	1	0	t_6
				0	1	0	0	t_7
				1	0	0	0	t_8

B. Complete the observed output table (t_1–t_5) below for the counter in Fig. 8–29 if the D input to flip-flop A is open. The desired outputs are shown in question (A).

Observed				
Q_A	Q_B	Q_C	Q_D	
0	0	0	0	CLEAR
				t_1
				t_2
				t_3
				l_4
				t_5

C. How can you prove an incorrect AND gate output in Fig. 8–29 is actually caused by the AND gate?

D. It is very important to determine what a properly operating digital circuit should do prior to troubleshooting the circuit.
   (1) True
   (2) False

SECTIONS 8-3 AND 8-4: INTERNAL SUMMARY

Analysis of circuits with flip-flops or registers depends on a thorough understanding of basic flip-flop operation. The addition of logic gates as control circuits enhances the operational capabilities of many registers. Code detection, data recirculation, self-loading counters, and multiplication and division of binary numbers were all explored in Section 8–3. This section serves merely as an introduction to the various uses of registers. More applications will be presented as the need arises.

Troubleshooting concepts applicable to registers were presented in Section 8–4. Although troubleshooting techniques are developed and refined with experience, some of the basics are presented in almost every chapter to lay a foundation for this development.

The initial point in troubleshooting is to know what you are looking for. This can only be done if you know what the circuit should do during normal operation.

1. Check visually for obvious problems.
2. Check for overheated ICs. *Be careful. Don't put your hand near power supply terminals or circuits that have high voltages. Remember, 115 Vac is extremely dangerous.*
3. Compare the observed output results with the desired output results to isolate malfunctions. The best technicians are those who can analyze theoretically and practically.

SECTIONS 8-3 AND 8-4: INTERNAL SUMMARY QUESTIONS

Answers are given at the end of the chapter.

Note: The use of manufacturer's data sheets is recommended to assist you in answering these summary questions.

1. What operation will be performed by a device with a READ/$\overline{\text{WRITE}}$ input when the input is low?
   a. READ
   b. WRITE

2. The 74178 register shown in Fig. 8–30 is connected as a/an
   a. up-counter   c. code detector
   b. Ring counter   d. Johnson counter

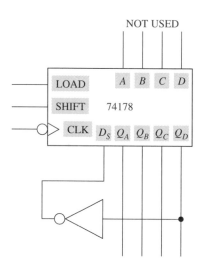

FIGURE 8-30

3. The 74178 register shown in Fig. 8–31 is connected as a/an
   a. up-counter   c. code detector
   b. Ring counter   d. Johnson counter

4. What is the modulus (MOD number) of the circuit in Fig. 8–31?
   a. 4   c. 12
   b. 8   d. 16

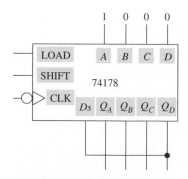

FIGURE 8-31

5. The circuit shown in Fig. 8–31 is self-loading.
   a. True
   b. False

6. The circuit in Fig. 8–32 is a self-loading Ring counter. If the circuit initializes with 1111 (Q_A–Q_D), how many clock pulses will it take before the counter contains only one Logic 1 output?
   a. 1   b. 2   c. 3   d. 4

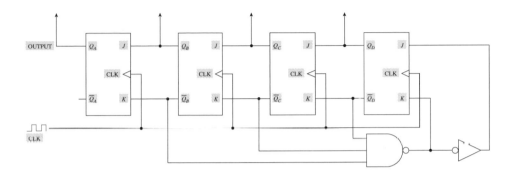

FIGURE 8–32

7. The observed results from the shift register in Fig. 8–33 are shown below. Assume the serial data input is tied high and select the most likely malfunction.

	Observed		
Q_A	Q_B	Q_C	Q_D
0	0	0	0	CLR
1	0	0	1	t_1
1	1	0	1	t_2
1	1	1	1	t_3
1	1	1	1	t_4

   a. D input to flip-flop A is open internally.
   b. D input to flip-flop D is open internally.
   c. $\overline{CLR}$ input to flip-flop A is open internally.
   d. CLK input to flip-flop D is open internally.

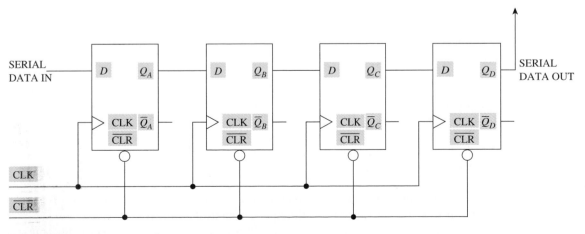

FIGURE 8–33

8. Assume the circuit in Fig. 8–34 initializes (starts up) with 1011 (Q_A–Q_D). What are its outputs after t_2?

	Q_A	Q_B	Q_C	Q_D
a.	0	0	0	0
b.	0	1	0	1
c.	0	0	1	0
d.	1	1	0	0

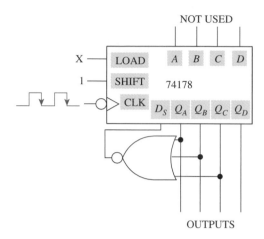

FIGURE 8–34

9. Identify the observed output of the self-loading Ring counter in Fig. 8–34 after four clock pulses if the Q_C input to the NOR gate is open.

	Q_A	Q_B	Q_C	Q_D
a.	0	0	0	0
b.	1	0	0	0
c.	0	0	0	1
d.	1	0	1	0

10. The circuit shown in Fig. 8–34 will operate normally as a Ring counter with SHIFT = 1 and LOAD = 0. What would happen to its operational mode if SHIFT = 0 (shorted to ground) and LOAD = 0?
   a. HOLD.
   b. Preload A–D data.
   c. Outputs would all go low.
   d. Outputs would all go high.

CHAPTER 8: SUMMARY

Groups of latches or flip-flops used to store, transfer, or shift data are called **registers.** Many IC registers are currently commercially available.

Various classifications of registers provide circuit designers with opportunities to store, shift, rotate, and transfer data as well as to change its format—serial to parallel or parallel to serial.

Serial-input and parallel-input registers are used in abundance in digital systems. Computers contain numerous registers, many of which are integrated in high-density ICs such as microprocessors and math coprocessors.

Serial-input shift registers are loaded one bit at a time. Although somewhat slower in moving data, this format is often dictated by system needs. The output data may be in serial or parallel format.

Parallel-input registers are used in digital systems because of their high speed. Only one clock pulse is required to load the data, and if the register has parallel output pins available, the data can be read in parallel format as soon as it is loaded.

Universal registers provide users with the capability to load or read data in serial or parallel format.

These registers normally have a serial and parallel load input mode of operation. Once loaded, they allow shifting of the data within the register to the left or to the right. They also have a HOLD mode of operation for temporarily storing data. Although their operation is a bit more complex, the use of a data book greatly simplifies the complexity.

Troubleshooting registers is based on understanding theory of their operation. Comparing what logic levels are present at the register's output(s) with what should be there will generally guide troubleshooting.

CHAPTER 8: END OF CHAPTER QUESTIONS/PROBLEMS

Answers are given in the Instructor's Manual.

The use of manufacturer's data sheets is recommended to help answer the End of Chapter questions/problems.

SECTION 8-1

1. Define a register.

2. What type of register (classification) is shown in Fig. 8–35?

3. How many clock pulses are required to load the register in Fig. 8–35?

4. If the register in Fig. 8–35 were loaded, how many clock pulses would it take to transfer the loaded data to another register?

5. What is the purpose of the NAND gate at the data input of the register in Fig. 8–35?

6. What type of register is shown in Fig. 8–36?

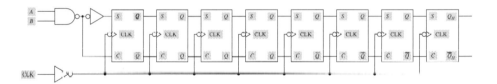

FIGURE 8–35

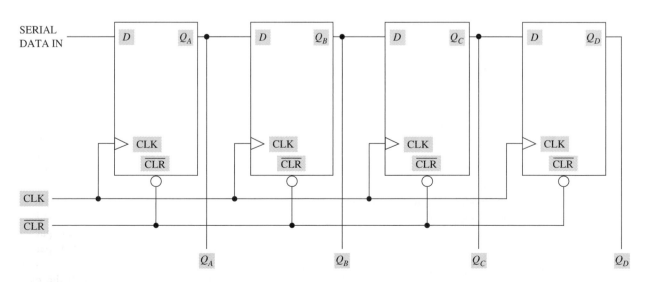

FIGURE 8–36

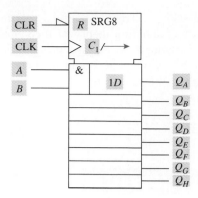

FIGURE 8-37

7. The symbol shown in Fig. 8–37 is of an 8-bit shift register—serial-in/parallel-out. What level must the CLR input be to reset the register?

8. What does $C_1/ \rightarrow$ in Fig. 8–37 indicate?

9. What does the & symbol in Fig. 8–37 indicate?

10. Data loaded in a serial-in/serial-out register are normally destroyed when the data are transferred out.
    (a) True
    (b) False

Section 8-2

11. The logic diagram and symbol in Fig. 8–38 represent a 74165 8-Bit Parallel Load Shift Register. What function is performed by the register when $\overline{\text{LOAD}} = 0$?

12. Can serial data be loaded into the register in Fig. 8–38?

13. What is the purpose of the OR gate in Fig. 8–38?

14. What type of register is shown in Fig. 8–39?

15. How many clock pulses does it take to load data into the register in Fig. 8–39?

16. Data loaded in the register in Fig. 8–39 would normally be destroyed when a READ operation is performed.
    a. True
    b. False

17. What function is performed by the universal register in Fig. 8–40 when LOAD = 1, SHIFT = 0?

18. What function is performed by the universal register in Fig. 8–40 when LOAD = X, SHIFT = 1?

19. What function is performed by the universal register in Fig. 8–40 when LOAD = 0, SHIFT = 0?

Section 8-3

CT 20. The logic circuit shown in Fig. 8–41 represents a self-loading Ring counter. Show the Q outputs after each clock pulse assuming the initial condition shown at t_0.

	Q_A	Q_B	Q_C	Q_D
t_0	1	0	1	0
t_1				
t_2				
t_3				
t_4				
t_5				

Section 8-4

CT 21. The following output conditions are observed from the circuit in Fig. 8–42. What is the most probable malfunction?

	Q_A	Q_B	Q_C	Q_D
t_1	0	0	0	1
t_2	0	0	1	0
t_3	0	1	0	1
t_4	1	0	1	0
t_5	0	1	0	1

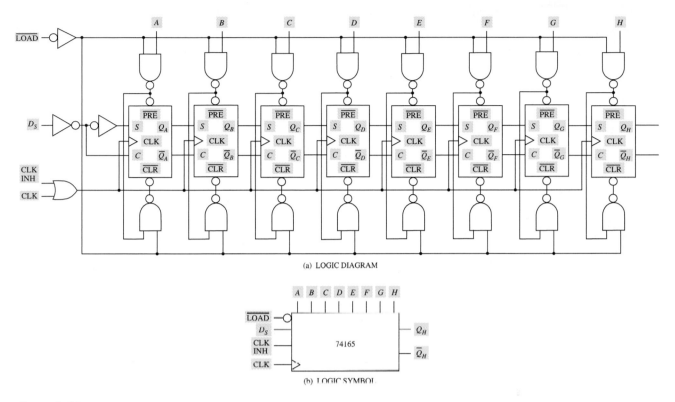

(a) LOGIC DIAGRAM

(b) LOGIC SYMBOL

FIGURE 8–38

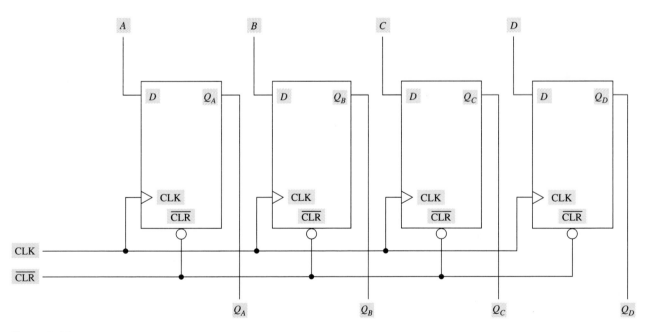

FIGURE 8–39

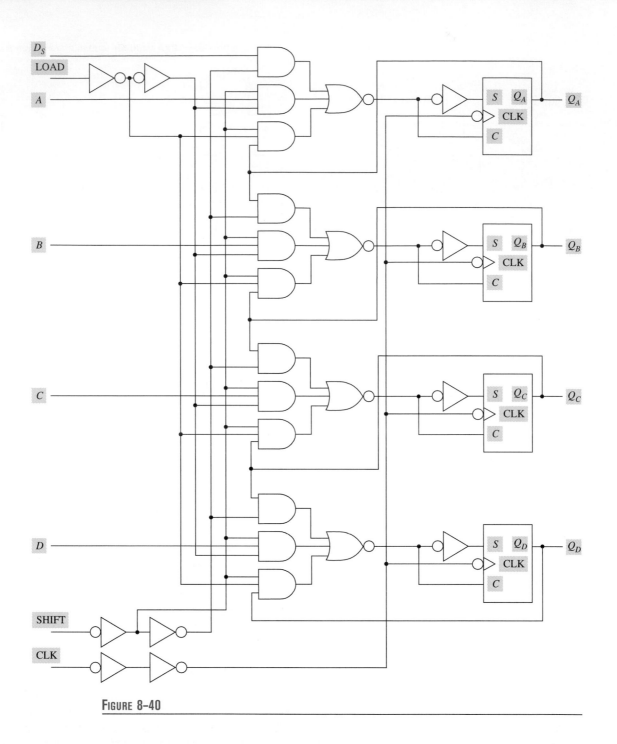

Figure 8-40

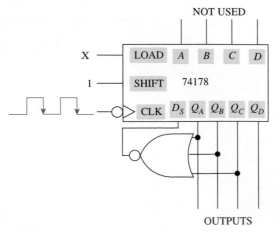

Figure 8-41

OUTPUTS

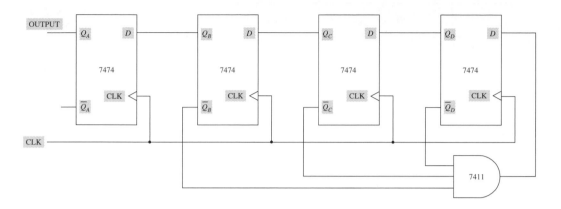

FIGURE 8-42

ANSWERS TO REVIEW QUESTIONS

SECTION 8-1

A. A register is a group of latches or flip-flops used to transfer, store, or shift data.
B. Serial-in/serial-out
   Serial-in/parallel-out
C. 8
D. 8
E. True
F. 16-bit shift register

SECTION 8-2

A. Parallel-in/serial-out
   Parallel-in/parallel-out
B. 1
C. False
D. True
E. True

SECTION 8-3

A. False
B. True
C. True
D. 0010
E. MOD 4
F. multiply
G. 2
H. shift right

SECTION 8-4

A.
Observed				
Q_A	Q_B	Q_C	Q_D	
0	0	0	1	t_1
0	0	1	1	t_2
0	1	1	0	t_3
1	1	0	0	t_4
1	0	0	0	t_5
0	0	0	1	t_6
0	0	1	1	t_7
0	1	1	0	t_8

B.
Observed				
Q_A	Q_B	Q_C	Q_D	
0	0	0	0	CLR
1	0	0	1	t_1
1	0	1	0	t_2
1	1	0	0	t_3
1	0	0	0	t_4
1	0	0	1	t_5

C. Isolate the problem to the AND gate or the load by disconnecting the gate's output from the data input of the D-type flip-flop.
D. True

ANSWERS TO INTERNAL SUMMARY QUESTIONS

SECTIONS 8-1 AND 8-2

1. a
2. c
3. c
4. a
5. b
6. a
7. b
8. c
9. d
10. d
11. c
12. b

SECTIONS 8-3 AND 8-4

1. b
2. d
3. b
4. a
5. b
6. c
7. b
8. c
9. a
10. a

9 DIGITAL ARITHMETIC AND CIRCUITS

Topics Covered in this Chapter

Introduction

Digital systems perform most of their arithmetic using **adder** circuits. This chapter presents a brief review of decimal arithmetic to introduce the concept of complementing decimal numbers. This is done to introduce subtraction by adding complemented decimal numbers.

The review of decimal is a natural progression into binary arithmetic. It takes us from the known to the unknown. Addition, subtraction, multiplication, and division of binary numbers are presented in detail as the chapter progresses. Once these tasks have been mastered, the process of adding Binary-Coded Decimal (BCD) numbers is discussed.

The operation of several arithmetic circuits is explained in the latter part of the chapter. **Half-adders, full-adders, and adder/subtracter** circuits are covered in detail. Binary adders are then connected to error detection/correction circuits to enable BCD addition.

Finally, a detailed discussion of an Arithmetic Logic Unit (ALU) is put forth. The ALU is a multipurpose circuit used to accomplish various arithmetic functions and to perform many logical operations.

Chapter Objectives

1. Ones (1s) and Twos (2s) complement binary numbers.

2. Add, subtract, multiply, and divide binary numbers.

3. Add BCD numbers.

4. Identify binary arithmetic circuits—half-adder, full-adder, adder/subtracter, BCD adder, and ALU.

5. Given the logic diagram or symbol of an arithmetic circuit and a data book, determine the circuit's outputs when various input combinations are applied.

SECTION 9-1: DECIMAL/BINARY ARITHMETIC

OBJECTIVES

1. State the rules for 9s and 10s complementing decimal numbers and 1s and 2s complementing binary numbers.
2. Add, subtract, multiply, and divide binary numbers.

A review of **decimal** arithmetic is in order to orient our thinking toward the binary arithmetic circuits that are presented in this chapter. The four basic arithmetic operations—addition, subtraction, multiplication, and division—allow vast mathematical calculations.

The addition process is stressed in this section because it is by far the most important mathematical operation: Subtraction, multiplication, and division can be performed using addition. The process of subtraction can be accomplished by adding a negative number to a positive number. Multiplication and division can be accomplished by repetitive addition and subtraction respectively. Many computers perform all basic mathematical operations almost entirely with **adder** circuits.

If subtraction can be performed by adding a negative number to a positive number, a method must be employed to differentiate negative numbers from positive numbers. This is accomplished by using **signed numbers.** For now we will use the conventional signs to represent positive and negative numbers in decimal. However, **sign bits** will be used in the binary arithmetic section to identify positive and negative numbers.

A Logic 1 is used to represent a negative sign and a Logic 0 is used to represent a positive sign. A sign bit has a designated position in a binary number. If the sign bit is in the MSB position of a multibit number, the remaining bits are **magnitude bits.**

$$1 \underbrace{0011010}_{\text{Magnitude bits}}{}_{(2)} = -26_{(10)}$$

$$\underbrace{\text{Sign bit}}_{} \quad 0 \underbrace{0011010}_{}{}_{(2)} = +26_{(10)}$$

Let's examine some decimal procedures that will work equally well in binary when binary arithmetic circuits are presented.

Nines (9s) Complement

Deriving the 9s complement of a decimal number allows the subtraction process to be accomplished using addition. A positive number is added to a negative number to perform the subtraction procedure if 9s complement arithmetic is used.

First let's look at how a 9s complement number is derived and then how it is used in a subtraction problem. The 9s complement of a decimal number is derived by subtracting that number from 9. For example, the 9s complement of $-3 = 6$. The signed negative number -3 is represented by the unsigned 9s complement number 6. Theses two numbers, -3 and 6, are complementary, equivalent numbers.

Complementary indicates the numbers can be changed to an opposite condition. Simply put, if the 9s complement of -3 is 6, then the 9s complement of 6 is -3. The 9s complement of -3 is used here to represent the negative number.

A multidigit decimal number can be converted to 9s complement format in the manner previously described. The procedure requires subtracting each digit in the decimal number from 9:

$$2,619_{(10)} = \mathbf{7,380}_{\textbf{(9s comp)}}$$

$$\begin{array}{r} 9,999 \\ -2,619 \\ \hline 7,380 \end{array}$$

The procedure for subtracting by adding a negative decimal number is as follows:

1. Nines complement the negative number.
2. Add the 9s complement to the positive number.
3. Add the carry generated in step 2 to the sum to obtain the difference of the two numbers.

EXAMPLE PROBLEM

$$\begin{array}{r} 8 \\ -4 \\ \hline \end{array}$$

Step 1:
$$\begin{array}{r} 9 \\ -4 \\ \hline 5 \end{array} = (\text{9s complement of } -4)$$

Step 2:
$$\begin{array}{r} 8 \\ +5 \\ \hline ①\ 3 \end{array} = (\text{Initial sum})$$

Step 3: $+1 = (\text{End-around carry})$

   4 = (Difference)

Note: The procedure in step 3 requires adding the carry generated in step 2 to the initial sum. This process is called the **end-around carry.**

EXAMPLE PROBLEM

$$\begin{array}{r} 36 \\ -12 \\ \hline \end{array}$$

Step 1:
$$\begin{array}{r} 99 \\ -12 \\ \hline 87 \end{array} = (\text{9s complement of } -12)$$

Step 2:
$$\begin{array}{r} 36 \\ +87 \\ \hline ①\ 23 \end{array} = (\text{Initial sum})$$

Step 3: $+1 = (\text{End-around carry})$

   24 = (Difference)

EXAMPLE PROBLEM

$$\begin{array}{r} 8,550 \\ -7,431 \\ \hline \end{array}$$

$$\begin{array}{r} 9,999 \\ -7,431 \\ \hline 2,568_{\text{(9s comp)}} \end{array}$$

$$\begin{array}{r} 8,550 \\ +2,568 \\ \hline ①\ 1,118 \end{array}$$

   $+\ \ \ \ 1$

   1,119 = (Difference)

If 2,568 is the 9s complement of X, then X can be determined in the manner used to 9s complement the number originally. In other words, when you take the 9s complement twice, you obtain the original number.

Original number $= X$

$$
\begin{array}{r}
9,999 \\
-2,568 = (\text{9s complement}) \\
\hline
-7,431 = (\text{Original number}, X)
\end{array}
$$

The difference calculated in the preceding example problems was positive, and in each case an end-around carry was generated. *The high end-around carry indicates a positive result.*

A zero end-around carry is generated when the difference is negative. *When the carry is zero (0), the difference is an unsigned 9s complement number.* This difference must be recomplemented to obtain the actual signed value.

EXAMPLE PROBLEM

$$
\begin{array}{r}
1 \\
-6 \\
\hline
\end{array}
$$

Step 1:
$$
\begin{array}{r}
9 \\
-6 \\
\hline
3_{(\text{9s comp})}
\end{array}
$$

Step 2:
$$
\begin{array}{r}
1 \\
+3 \\
\hline
\text{⓪} \, 4
\end{array}
$$

Step 3: $\quad \longrightarrow \underline{+0}$ (End-around carry)

$\qquad\qquad\qquad 4$ (Difference in 9s complement form)

Signed answer $= \mathbf{-5}$

The complement of 4 is 5; thus, the signed value is -5.

EXAMPLE PROBLEM

$$
\begin{array}{r}
47 \\
-56 \\
\hline
\end{array}
$$

Step 1:
$$
\begin{array}{r}
99 \\
-56 \\
\hline
43_{(\text{9s comp})}
\end{array}
$$

Step 2:
$$
\begin{array}{r}
47 \\
+43 \\
\hline
\text{⓪} \, 90
\end{array}
$$

Step 3: $\quad \longrightarrow \underline{+0}$

$\qquad\qquad\qquad 90$ (Difference in 9s complement form)

Signed answer $= \mathbf{-9}$

The complement of 90 is 09; therefore, the signed value of the difference is -9.

EXAMPLE PROBLEM

$$\begin{array}{r} 5 \\ -5 \\ \hline \end{array}$$

Step 1:
$$\begin{array}{r} 9 \\ -5 \\ \hline 4 _{\text{(9s comp)}} \end{array}$$

Step 2:
$$\begin{array}{r} 5 \\ +4 \\ \hline \textcircled{0}\ 9 \end{array}$$

Step 3:
$$\begin{array}{r} +0 \\ \hline 9 \end{array} \text{ (Difference in 9s complement form)}$$

Signed answer $= -\mathbf{0} = \mathbf{0}$

Since the complement of a 9s complement 9 is -0, the 0 represents the signed answer. The sign is not required for a zero value.

Tens (10s) Complement

The 10s complement of a decimal number is nothing more than the 9s complement plus 1. The 10s complement of a decimal number may be used to represent a negative number in the same manner as the 9s complement. This indicates subtraction may be performed by adding the 10s complement of the negative number in a manner similar to that accomplished using 9s complement arithmetic.

The 10s complement of a decimal number is obtained by subtracting that number from 9 and adding 1. For example, the 10s complement of -3 is

$$\begin{array}{r} 9 \\ -3 \text{ (signed number)} \\ \hline 6 \text{ (9s complement of } -3) \\ +1 \\ \hline 7 \text{ (10s complement of } -3) \end{array}$$

The 10s complement of -648 is

$$\begin{array}{r} 999 \\ -648 \text{ (signed number)} \\ \hline 351 \text{ (9s complement of } -648) \\ +1 \\ \hline 352 \text{ (10s complement of } -648) \end{array}$$

Notice in the preceding examples that a signed negative number is represented by an unsigned 10s complement number. The rules for subtracting by adding a 10s complement number are as follows:

1. Tens complement the negative number.
2. Add the 10s complement to the positive number.
3. Ignore the carry except to determine whether the difference is positive or negative. A *1 carry* indicates a positive difference. A *0 carry* indicates the difference is negative and is represented in unsigned complement form.

EXAMPLE PROBLEM

8
−4

Step 1:   9
           −4
            5 (9s complement of −4)
           +1
            6 (10s complement of −4)

Step 2:   8
           +6

Step 3: ①4
Carry ◄──┘

The 1 carry indicates the difference (4) is positive.

EXAMPLE PROBLEM

36
−12

Step 1:   99
          −12
            87 (9s complement of −12)
           +1
            88 (10s complement of −12)

Step 2:   36
          +88

Step 3: ① 24
Carry ◄──┘

EXAMPLE PROBLEM

8,550
−7,431

Step 1:   9,999
          −7,431
            2,568 (9s complement)
          +   1
            2,569 (10s complement)

Step 2:   8,550
          +2,569

Step 3: ① 1,119
Carry ◄──┘

The preceding three example problems are identical to those worked using the 9s complement subtraction procedure except the end-around carry is not added when using 10s complement arithmetic. It is used only to indicate a positive or negative result. If the result is negative (0 carry), it must be recomplemented to obtain its actual signed value. This was also true of a 0 carry in 9s complement arithmetic. Let's work a couple of problems that generate negative results in 10s complement arithmetic.

EXAMPLE PROBLEM

$$\begin{array}{r} 1 \\ -6 \\ \hline \end{array}$$

Step 1: $\begin{array}{r} 9 \\ -6 \\ \hline 3 \end{array}$ (9s complement of −6)

$\begin{array}{r} +1 \\ \hline 4 \end{array}$ (10s complement of −6)

Step 2: $\begin{array}{r} 1 \\ +4 \\ \hline \end{array}$

Step 3: ⓪5
Carry ◄──┘

The 0 carry indicates an unsigned 10s complement negative result. The difference must be recomplemented to derive the signed value.

$\begin{array}{r} 9 \\ -5 \\ \hline 4 \end{array}$ (Unsigned 10s complement difference)

$\begin{array}{r} +1 \\ \hline 5 \end{array}$ = **−5** (**Signed value**)

EXAMPLE PROBLEM

$$\begin{array}{r} 47 \\ -56 \\ \hline \end{array}$$

Step 1: $\begin{array}{r} 99 \\ -56 \\ \hline 43 \end{array}$ (9s complement of −56)

$\begin{array}{r} +1 \\ \hline 44 \end{array}$ (10s complement of −56)

Step 2: $\begin{array}{r} 47 \\ +44 \\ \hline \end{array}$

Step 3: ⓪91
Carry ◄──┘

$\begin{array}{r} 99 \\ -91 \\ \hline 08 \end{array}$

$\begin{array}{r} +1 \\ \hline 09 \end{array}$ = −9 (Signed value)

Binary Arithmetic

The rules for binary addition were presented in Chapter 2. They are presented here as a brief review:

$$
\begin{array}{cccc}
0 & 0 & 1 & 1 \\
+0 & +1 & +0 & +1 \\
\hline
0 & 1 & 1 & 10
\end{array}
$$

Binary adder circuits have the ability to add two bits plus a carry input. The rules for adding three bits are the same if two bits are added and then the sum of the two bits is added to the third bit as shown:

$$
\begin{array}{cc}
 & 1 \\
1 & 1 \\
1 = & +1 \\
+1 & \overline{10} \\
\overline{11} & +1 \\
 & \overline{11}
\end{array}
$$

EXAMPLE PROBLEMS

$$
\begin{array}{l}
11 \longleftarrow \text{Carry} \\
1100 \\
+0101 \\
\hline
10001
\end{array}
$$

$$
\begin{array}{l}
1 \longleftarrow \text{Carry} \\
1000 \\
+1010 \\
\hline
10010
\end{array}
$$

$$
\begin{array}{l}
1111 \longleftarrow \text{Carry} \\
1111 \\
+1001 \\
\hline
11000
\end{array}
$$

$$
\begin{array}{l}
11 \longleftarrow \text{Carry} \\
10100111 \\
+00110100 \\
\hline
11011011
\end{array}
$$

The *rules for binary subtraction* are straightforward:

$$
\begin{array}{cccc}
0 & 1 & 1 & 10 \\
-0 & -1 & -0 & -1 \\
\hline
0 & 0 & 1 & 1
\end{array}
$$

In the final example, a 1 was borrowed from the next higher column to produce the $10_{(2)}$ in the column being subtracted.

EXAMPLE PROBLEMS

$$
\begin{array}{l}
1010 \\
-0110 \\
\hline
0100
\end{array}
$$

$$
\begin{array}{l}
1100 \\
-0001 \\
\hline
1011
\end{array}
$$

The borrow for the LSB column in the preceding example had to be taken from two columns away from that column.

Since a digital circuit normally uses addition to perform the subtraction function, let's investigate the methods that can be used to accomplish this task with adders.

Ones (1s) Complement

The 1s complement in binary is comparable to the 9s complement in decimal. The rules to implement 1s complement subtraction are similar to those used in 9s complement subtraction.

The 1s complement of a binary number is derived by subtracting each bit in the number to be complemented from 1. To 1s complement the binary number 1100, subtract each bit from 1 as shown.

$$
\begin{array}{l}
1111 \\
\underline{-1100} \text{ (Number to be complemented)} \\
0011 \text{ (1s complement)}
\end{array}
$$

It can easily be seen in this example that the 1s complemented number is the inverse (NOT) of each bit in the number that was complemented. Therefore, a simpler method than that previously stated can be used to 1s complement a number—change each 1 to a 0 and each 0 to a 1.

EXAMPLE PROBLEMS

$1011_{(2)} = \mathbf{0100}_{(1s\ comp)}$

$11010110_{(2)} = \mathbf{00101001}_{(1s\ comp)}$

The steps for subtracting a binary number using 1s complement addition are as follows:

1. Ones complement the negative number.
2. Add the 1s complemented number to the positive number.
3. Add the end-around carry to the sum to obtain the difference of the two numbers. *Note:* A 1 end-around carry indicates a positive difference. A 0 end-around carry indicates the difference is negative. In addition, a zero carry also indicates the difference is an unsigned 1s complement number. Therefore when the carry is zero, the difference must be recomplemented to obtain the signed value.

EXAMPLE BINARY SUBTRACTION PROBLEMS

$$
\begin{array}{l}
1110 \\
\underline{-0010}
\end{array}
$$

Step 1: $-0010 = 1101_{(1s\ comp)}$

Step 2: 1110

$\underline{+1101}_{(1s\ comp)}$

$\textcircled{1}\ 1011$

Step 3: $\underline{+\ \ \ 1}$ (End-around carry)

$\mathbf{1100}$ **(Difference)**

$$0110$$
$$-0011$$

$$-0011 = 1100_{(1s\ comp)}$$

$$\begin{array}{r} 0110 \\ +1100 \\ \hline ①0010 \\ +\ \ 1 \\ \hline \mathbf{0011}\ (\textbf{Difference}) \end{array}$$

$$11101000$$
$$-01100100$$

$$-01100100 = 10011011_{(1s\ comp)}$$

$$\begin{array}{r} 11101000 \\ +10011011 \\ \hline ①10000011 \\ +\qquad 1 \\ \hline \mathbf{10000100}\ (\textbf{Difference}) \end{array}$$

A high end-around carry was generated in each of the preceding example problems. This indicates the difference is positive and in its true form. The following example problems generate low end-around carries.

EXAMPLE PROBLEMS

$$0111$$
$$-1001$$

Step 1: $-1001 = 0110_{(1s\ comp)}$

Step 2:
$$\begin{array}{r} 0111 \\ +0110 \\ \hline ⓪1101 \end{array}$$

Step 3:
$$\begin{array}{r} +\ \ 0 \\ \hline 1101\ (\text{Difference in 1s complement form}) \end{array}$$

$$1101 = \mathbf{-0010}$$

$$0100$$
$$-1110$$

$$-1110 = 0001_{(1s\ comp)}$$

$$\begin{array}{r} 0100 \\ +0001 \\ \hline ⓪0101 \\ +\ \ 0 \\ \hline 0101 \end{array}$$

$$0101 = \mathbf{-1010}$$

$$01000111$$
$$-01100100$$

$$-01100100 = 10011011_{(1s\ comp)}$$

$$\begin{array}{r} 01000111 \\ +10011011 \\ \hline ⓪ 11100010 \\ +\qquad 0 \\ \hline 11100010_{(1s\ comp)} \end{array}$$

$$11100010_{(1s\ comp)} = \mathbf{-00011101}$$

A 0 end-around carry was generated in each of the three previous examples. The difference in each of these problems is negative and had to be recomplemented to obtain the signed value.

Twos (2s) Complement

The 2s complement of a binary number is the 1s complement plus 1. Like the 10s (tens) complement in decimal, the 2s complement allows subtraction by adding the 2s complement of the negative number without performing an end-around carry. The carry is ignored except to determine whether the difference is positive or negative.

The 2s complement of a binary number is derived by adding 1 to the 1s complement number. For example, the 2s complement of $-1101_{(2)}$ is

$$
\begin{array}{r}
-1101 = 0010_{(1s\ comp)} \\
+\quad 1 \\
\hline
0011_{(2s\ comp)}
\end{array}
$$

The 2s complement of $-11001000_{(2)}$ is

$$
\begin{array}{r}
-11001000 = 00110111_{(1s\ comp)} \\
+\qquad 1 \\
\hline
00111000_{(2s\ comp)}
\end{array}
$$

There is a short cut that can be used to 2s complement binary numbers.

STEP 1:   Write the bits from the LSB up to and including the first binary 1.

STEP 2:   Write the complement of the remaining bits.

EXAMPLE 2s COMPLEMENT PROBLEMS

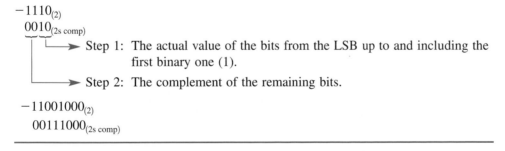

$-1110_{(2)}$

$0010_{(2s\ comp)}$

Step 1: The actual value of the bits from the LSB up to and including the first binary one (1).

Step 2: The complement of the remaining bits.

$-11001000_{(2)}$

$00111000_{(2s\ comp)}$

The rules for adding an unsigned 2s complement number to perform subtraction are as follows:

1. Twos complement the negative number.
2. Add the 2s complement to the positive number.
3. Do not add the end-around carry. A 1 carry indicates a positive difference. A 0 carry indicates the difference is negative and is in its unsigned complement form. The unsigned complement number must be recomplemented to obtain its signed value.

EXAMPLE BINARY SUBTRACTION PROBLEMS

$$
\begin{array}{r}
1011 \\
-0101 \\
\end{array}
$$

Step 1: $-0101 = 1010_{(1s\ comp)}$
$$
\begin{array}{r}
+\ \ \ 1 \\
\hline
1011_{(2s\ comp)} \\
\end{array}
$$

Step 2: $\begin{array}{r} 1011 \\ +1011 \\ \hline \end{array}$

Step 3: ① **0110 (Difference)**
Carry ◄──┘

$$
\begin{array}{r}
1100 \\
-0001 \\
\end{array}
$$

$-0001 = 1110_{(1s\ comp)}$
$$
\begin{array}{r}
+\ \ \ 1 \\
\hline
1111_{(2s\ comp)} \\
\end{array}
$$

$$
\begin{array}{r}
1100 \\
+1111 \\
\hline
\end{array}
$$
① **1011**
Carry ◄──┘

$$
\begin{array}{r}
10100111 \\
-10000110 \\
\end{array}
$$

$-10000110 = 01111001_{(1s\ comp)}$
$$
\begin{array}{r}
+\qquad 1 \\
\hline
01111010_{(2s\ comp)} \\
\end{array}
$$

$$
\begin{array}{r}
10100111 \\
+01111010 \\
\hline
\end{array}
$$
① **00100001**
Carry ◄──┘

$$
\begin{array}{r}
1000 \\
-1100 \\
\end{array}
$$

$-1100 = 0011_{(1s\ comp)}$
$$
\begin{array}{r}
+\ \ \ 1 \\
\hline
0100_{(2s\ comp)} \\
\end{array}
$$

$$
\begin{array}{r}
1000 \\
+0100 \\
\hline
\end{array}
$$
⓪ 1100
Carry ◄──┘

The 0 carry indicates the answer is an unsigned 2s complement negative result. The answer must be recomplemented to produce the signed value.

$1100 = \quad 0011_{(1s\ comp)}$
$$
\begin{array}{r}
+\ \ 1 \\
\hline
\end{array}
$$
−0100 (Signed value)

$$
\begin{array}{r}
10011001 \\
-11110100 \\
\end{array}
$$

$-11110100 = 00001011_{(1s\ comp)}$
$$
\begin{array}{r}
+\qquad 1 \\
\hline
00001100_{(2s\ comp)} \\
\end{array}
$$

$$
\begin{array}{r}
10011001 \\
+\ 00001100 \\
\hline
\end{array}
$$
⓪0100101
Carry ◄──┘

$$10100101 = \quad 01011010$$
$$\underline{+ \qquad 1}$$
$$-\mathbf{01011011} \text{ (Signed value)}$$

Multiplication/Division

Multiplication can be accomplished by adding a number to itself a designated number of times. The number of times is specified by the **multiplier** while the number itself is called the **multiplicand.** The sum produced by the repeated addition is called the **product.**

$$
\begin{array}{r}
8 \\
8 \text{ (Multiplicand)} \qquad 8 \\
\underline{\times 5} \text{ (Multiplier)} \quad = \quad 8 \\
40 \text{ (Product)} \qquad 8 \\
\underline{+8} \\
40
\end{array}
$$

Binary multiplication can also be done in a like manner. The preceding example (8×5) is shown here in binary:

$$
\begin{array}{r}
1000 \\
\underline{+1000} \\
10000 \\
1000 \qquad +1000 \\
\underline{\times 101} \quad = \quad 11000 \\
\underline{+1000} \\
100000 \\
\underline{+1000} \\
101000_{(2)} = 40_{(10)}
\end{array}
$$

Binary multiplication can also be accomplished by shifting and adding the multiplicand.

$$
\begin{array}{r}
1000 \\
\underline{\times 101} \\
1000 \\
0000 \\
\underline{1000} \\
101000
\end{array}
$$

This procedure for multiplying binary numbers is just as easy to implement with digital circuits as repeated addition. Shift and storage registers can perform the proper shift operations and storage of accumulated sums while adders perform the successive additions required to implement the procedure.

Division by longhand is shown here:

$$
\begin{array}{r}
5 \text{ (Quotient)} \\
8 \overline{)40} \text{ (Dividend)} \\
\underline{40} \\
0 \text{ (Remainder)} \\
\text{(Divisor)}
\end{array}
$$

Division may also be accomplished by repetitive subtraction. The **divisor** is subtracted from the **dividend** until the remainder is zero or less than the divisor. The **quotient** is equal to the number of subtractions performed.

$$
\begin{array}{r}
40 \\
\underline{-8} \quad \text{(1st subtraction)} \\
32 \\
\underline{-8} \quad \text{(2nd subtraction)} \\
24 \\
\underline{-8} \quad \text{(3rd subtraction)} \\
16 \\
\underline{-8} \quad \text{(4th subtraction)} \\
8 \\
\underline{-8} \quad \text{(5th subtraction)} \quad \textbf{Quotient = 5} \\
0
\end{array}
$$

This shows division can be accomplished by subtraction, and we know subtraction can be performed by adding. Therefore, division can be performed with adder circuits.

The repetitive subtraction method can be used by employing 1s or 2s complement addition. Again the divisor is subtracted (2s complement added) from the dividend until the remainder is zero. Since the divisor is to be 2s complemented in the following example, it must be extended to 6 bits before it is 2s complemented.

$$
\begin{array}{rl}
\text{Dividend} & = 101000 \\
\text{Divisor} & = 001000 \\
\text{1s complemented divisor} & = 110111 \\
& \underline{+ \quad 1} \\
\text{2s complemented divisor} & = 111000
\end{array}
$$

$$
\begin{array}{l}
\ \ 101000 \\
\underline{+111000} \quad \text{(1st subtraction = 2s complement addition)} \\
1\ \ \ 100000 \\
\underline{+111000} \quad \text{(2nd subtraction)} \\
1\ \ \ 011000 \\
\underline{+111000} \quad \text{(3rd subtraction)} \\
1\ \ \ 010000 \\
\underline{+111000} \quad \text{(4th subtraction)} \\
1\ \ \ 001000 \\
\underline{+111000} \quad \text{(5th subtraction)} \\
1\ \ \ 000000
\end{array}
$$

The quotient is 5 because 5 subtractions were performed before reducing the remainder to 0. The carries shown separated to the left of the sums are ignored.

Brief Overview

This section has thus far covered a lot of ground in a few pages. However, the many rules provided, especially for subtraction, all have similarities that make them easily understood.

The basis of the subtraction procedures allows subtraction to be performed by adding negative numbers represented by some complemented value.

The **9s complement** procedure allows subtraction by addition to be accomplished in the following manner:

1. Nines complement the negative number.
2. Add the 9s complement to the positive number.
3. Add the end-around carry. In addition, a high end-around carry indicates a positive result. A low end-around carry indicates a negative result, and that result must be recomplemented to obtain the signed value.

The **10s complement** procedure uses similar rules as the 9s complement except the carry is not added. It is only used to indicate a positive or negative result.

The **1s complement** in binary is almost identical to the 9s complement in decimal. In fact if the "9s complement" in the decimal rules is replaced with "1s complement," the rules apply in binary.

The **2s complement** is implemented in binary in the same manner as the 10s complement is implemented in decimal.

Section 9-1: Review Questions

Answers are given at the end of the chapter.

A. Nines complement the following decimal numbers.
   (1) −1
   (2) −5
   (3) −9
   (4) −28
   (5) −791

B. Tens complement the numbers in question (A).

C. Ones complement the following binary numbers.
   (1) −0011
   (2) −1000
   (3) −11000100
   (4) −01001000

D. Twos complement the numbers in question (C).

E. Add the following binary numbers.
   (1)   0100
        +0111
   (2)   1000
        +0111
   (3)   1100
        +1100
   (4)   1111
        +0101
   (5)   10101110
        +00101010

F. Subtract the following binary numbers using 2s complement. Show your work.
   (1)   1010
        −0010
   (2)   1111
        −0111
   (3)   0101
        −0100
   (4)   1010
        −1100
   (5)   0011
        −0101

OBJECTIVES

1. Encode a decimal digit to BCD or XS3 code.
2. State the two rules for BCD adjust of an invalid/incorrect BCD sum.
3. Add BCD numbers.

Binary-Coded Decimal (BCD) Addition

Binary-coded decimal, sometimes referred to as the 8–4–2–1 code, was presented in Chapter 2. BCD numbers consist of a group of 4 bits used to encode a single decimal digit. BCD numbers may be added using the rules of binary addition. For example, here is how to add 51 + 38:

$$
\begin{array}{rl}
51_{(10)} = & 0101\ 0001_{(BCD)} \\
+38_{(10)} = & +0011\ 1000_{(BCD)} \\
\hline
89_{(10)} = & 1000\ 1001_{(BCD)}
\end{array}
$$

Six *invalid BCD numbers* can appear in BCD arithmetic: 1010, 1011, 1100, 1101, 1110, and 1111. We know these numbers are incorrect because a BCD number $10_{(10)}$ is equal to $0001\ 0000_{(BCD)}$, $11_{(10)} = 0001\ 0001_{(BCD)}$, and so on.

Keep in mind during the following discussion that a digital system uses adder circuits to add the numbers it processes. The system cannot differentiate between binary numbers and BCD numbers. This means that the BCD numbers are added in the exact same manner as binary numbers are added.

When a digital adder sums two binary numbers, the result is pure binary.

PROBLEM 1

$$
\begin{array}{r}
1000 \\
+0010 \\
\hline
1010
\end{array}
$$

If this binary problem were BCD, the sum would be invalid.

PROBLEM 2

Now consider this problem in binary (8 + 9):

$$
\begin{array}{r}
1000 \\
+1001 \\
\hline
10001
\end{array}
$$

What would the result be if the preceding example were BCD? Let's take a closer look at this problem.

$$
\begin{array}{rl}
1 \longleftarrow \text{Carry} & \\
0000\ 1000_{(BCD)} = & 08_{(10)} \\
+0000\ 1001_{(BCD)} = & +09_{(10)} \\
\hline
0001\ 0001_{(BCD)} = & 11_{(10)} \quad \textbf{Wrong answer}
\end{array}
$$

This problem indicates the sum of BCD 1000 + 1001 is $0001\ 0001_{(BCD)}$, which is a decimal 11. The correct BCD sum should be $0001\ 0111_{(BCD)}$, which is a decimal 17.

The preceding examples point out a couple of problems that are often encountered in BCD arithmetic circuits. Invalid sums regularly appear in these circuits. In the last example, an incorrect BCD answer was produced due to the carry. Both of these problems produce invalid BCD answers. Since these invalid answers occur, a BCD adjust by adding 0110 ($6_{(10)}$) is necessary.

Both of the previous problems could have been corrected by adding 0110 to the answer. The first problem can be corrected as follows:

$$
\begin{array}{rll}
1000_{(2)} = & 0000 \quad 1000_{(BCD)} & = \quad 08_{(10)} \\
+0010_{(2)} = & +0000 \quad 0010_{(BCD)} & = \quad +02_{(10)} \\
\hline
1010_{(2)} & 0000 \quad 1010 \text{ (Invalid)} \\
& + \quad 1 \quad 0110 \\
\hline
& 0001 \quad 0000_{(BCD)} & = \quad 10_{(10)}
\end{array}
$$

In this problem a 0110 was added because the original sum (1010) obtained was invalid. After the 0110 is added, the sum is correct.

The second example problem can also be corrected by adding 0110:

$$
\begin{array}{rll}
& \quad\quad 1* \\
1000_{(2)} = & 0000 \quad 1000_{(BCD)} = & 08_{(10)} \\
+1001_{(2)} = & +0000 \quad 1001_{(BCD)} = & +09_{(10)} \\
\hline
10001_{(2)} = & 0001 \quad 0001 \\
& + \quad\quad 0110 \\
\hline
& 0001 \quad 0111_{(BCD)} = & 17_{(10)}
\end{array}
$$

---

*Note the 1 carry to the upper nibble.

The two previous problems show two distinct cases where BCD adjust by adding 0110 is required. The rules for BCD adjust are to add 0110 (1) when an invalid sum is generated (1010 through 1111); or (2) when a carry-out is generated out of the sum when adding the two BCD numbers in the original problem.

EXAMPLE PROBLEMS

$$
\begin{array}{ll}
0011 & 1000_{(BCD)} \\
+0100 & 0001_{(BCD)} \\
\hline
0111 & 1001_{(BCD)}
\end{array}
$$

No BCD adjust required.

$$
\begin{array}{ll}
\quad 1 \\
0011 & 1001_{(BCD)} \\
+0010 & 1001_{(BCD)} \\
\hline
0110 & 0010 \\
+ & 0110 \\
\hline
0110 & 1000_{(BCD)}
\end{array}
$$

BCD adjust required due to carry-out from the original problem. The only carries shown out of the original nibble addition in these examples are ones that produce a BCD adjust.

$$
\begin{array}{ll}
\quad 1 \\
0111 & 0111_{(BCD)} \\
+0001 & 1001_{(BCD)} \\
\hline
1001 & 0000 \\
+ & 0110 \\
\hline
1001 & 0110_{(BCD)}
\end{array}
$$

BCD adjust required due to carry-out from the original problem.

$$
\begin{array}{r}
0010 \quad 0111_{(BCD)} \\
+0101 \quad 0111_{(BCD)} \\
\hline
0111 \quad 1110 \\
+ \quad 1 \quad 0110 \\
\hline
1000 \quad 0100_{(BCD)}
\end{array}
$$

BCD adjust required due to invalid sum (1110).

$$
\begin{array}{r}
1 \\
0001 \quad 1000 \quad 0101_{(BCD)} \\
+0001 \quad 1000 \quad 0110_{(BCD)} \\
\hline
0010 \quad 0001 \quad 1011 \\
+ \qquad 0110 \quad 0110 \\
\hline
0011 \quad 0111 \quad 0001_{(BCD)}
\end{array}
$$

Both cases of BCD adjust were used in this example. The lower nibble produced an invalid sum (1011). The sum of the middle nibbles in the original problem produced a carry.

$$
\begin{array}{r}
1 \\
0000 \quad 0110 \quad 0111_{(BCD)} \\
+0000 \quad 0100 \quad 1001_{(BCD)} \\
\hline
0000 \quad 1011 \quad 0000 \\
+ \quad 1 \quad 0110 \quad 0110 \\
\hline
0001 \quad 0001 \quad 0110_{(BCD)}
\end{array}
$$

Both cases of BCD adjust were used here again. The lower nibble produced a carry, and the sum of the middle nibbles is invalid.

The circuitry required to implement BCD addition is more complex than that required for binary addition. The error detection and correction circuitry necessary to implement BCD addition is presented in Section 9–4.

Excess-3 (XS3) Code

The XS3 code of a decimal number is derived by adding 3 to each decimal digit and then coding the sum in a group of 4 bits. This unweighted code is shown in Table 9–1. The following 4-bit groups are not used in XS3:

TABLE 9-1

XS3 Code

Decimal #	BCD	XS3 Code
0	0000	0011
1	0001	0100
2	0010	0101
3	0011	0110
4	0100	0111
5	0101	1000
6	0110	1001
7	0111	1010
8	1000	1011
9	1001	1100

0000   0001   0010
1101   1110   1111

The XS3 code for multidigit numbers is derived in the same manner. Add 3 to *each digit* and code the sum in a group of 4 bits:

$$29_{(10)} = \begin{array}{cc} 2 & 9 \\ +3 & +3 \\ \hline 5 & 12 \end{array} = 0101 \quad 1100_{(XS3)}$$

The advantages of XS3 are twofold. First, the 0000 code is not used. Elimination of this code reduces the possibility of errors in a digital system. A 0000 input combination implies no signal is present at any of the 4 inputs. Faulty equipment sometimes produces this code and the errors that go along with it. Second, the XS3 code is **self-complementing.**

The self-complementing feature is provided because the 1s complement of the XS3 code of a decimal number is equal to the XS3 code of the 9s complement of that decimal number. Although this statement may seem confusing, let's see what it means. The decimal number 7 will be used to explain the process. The 9s complement of 7 is 2. The statement indicates the 1s complement of $1010_{(XS3 \text{ of decimal } 7)}$ is equal to the XS3 code representing the 9s complement of 7.

$$7_{(10)} = 1010_{(XS3)} = 0101_{(1s \text{ comp of XS3 of } 7)}$$
$$2_{(9s \text{ comp})} = 5_{(9s \text{ comp} + 3)} = 0101_{(XS3 \text{ of 9s comp of } 7)}$$

Table 9–2 shows this equality for each decimal digit.

TABLE 9-2	Decimal Number	XS3	1s Complement	9s Complement	9s Complement + 3
XS3 Code and 9s Complement	0	0011	1100	9	12
	1	0100	1011	8	11
	2	0101	1010	7	10
	3	0110	1001	6	9
	4	0111	1000	5	8
	5	1000	0111	4	7
	6	1001	0110	3	6
	7	1010	0101	2	5
	8	1011	0100	1	4
	9	1100	0011	0	3

Section 9-2: Review Questions

Answers are given at the end of the chapter.

A. State the two rules for BCD adjust of an invalid/incorrect BCD sum.
B. Code the following decimal numbers in BCD.
   (1) 6
   (2) 29
   (3) 140
   (4) 1000

C. *Add* the following *BCD numbers*. Show your work.

$$(1) \quad \begin{array}{cc} 0011 & 0001_{(BCD)} \\ +0010 & 0010_{(BCD)} \end{array}$$

(2)   0101   0110$_{(BCD)}$
     $+0100$ 0011$_{(BCD)}$

(3)   0110   1000   0001 $_{(BCD)}$
     $+0010$ 0001   0111 $_{(BCD)}$

(4)   0111   1000$_{(BCD)}$
     $+0001$ 1001$_{(BCD)}$

(5)   0100   0101 $_{(BCD)}$
     $+0011$ 0110 $_{(BCD)}$

(6)   0111   0111 $_{(BCD)}$
     $+0011$ 1001 $_{(BCD)}$

D.   Code the following decimal numbers in XS3.
   (1) 3
   (2) 9
   (3) 38
   (4) 194

SECTIONS 9-1 AND 9-2: INTERNAL SUMMARY

The procedure of 9s or 10s complementing a negative decimal number allows the subtraction process to be performed through addition. The **9s complement** of a decimal number is derived by subtracting that number from 9. The **10s complement** of a decimal number is the 9s complement plus 1.

A 9s or 10s complemented number can be used to obtain the difference of two decimal numbers by adding.

The 9s complement method employs the **end-around carry.** The procedure is

1. Nines complement the negative number.
2. Add the 9s complement to the positive number.
3. Add the carry generated in step 2 to the sum to obtain the difference.

EXAMPLE

   7
 -3

Step 1:   9
             -3
              6 (9s comp)

Step 2:   7
             $+6$
            ①3
Step 3: $+1$
             4 = Difference (positive result)

The difference obtained is an unsigned 9s complement number if a zero end-around carry is generated. In this case, the difference must be recomplemented to obtain the signed value of the number.

The 10s complement procedure is basically the same without the end-around carry. The carry is used only to indicate a positive difference (Carry = 1) or a negative difference in unsigned complement form (Carry = 0).

The 1s complement in binary is similar to the 9s complement in decimal. Likewise, the 2s complement in binary is similar to the 10s complement in decimal. The steps used to subtract by adding binary numbers are the same as those used in decimal subtraction.

Multiplication and division can be accomplished by repetitive addition or repetitive subtraction. Multiplication is accomplished by adding a number to itself by the number of times specified by the multiplier. The sum produced by this addition is the product. Division is performed by subtracting the divisor from the dividend until the remainder is zero or less than the divisor. The quotient is the number of subtractions performed.

BCD addition is performed using standard binary adders. The six invalid BCD numbers that often appear in BCD sum outputs and the carries generated when the sum exceeds $15_{(10)}$ pose some unique problems for BCD adder circuits. These conditions must be detected and corrected to produce valid sum outputs. The BCD arithmetic circuit must add 0110 when an invalid sum is generated or when a carry-out is generated out of the sum when adding two BCD numbers. This correction is called a **BCD adjust.**

Excess-3 codes are obtained by adding 3 to each decimal digit and encoding the sum to a 4-bit group of numbers. The XS3 code is self-complementing.

SECTIONS 9–1 AND 9–2: INTERNAL SUMMARY QUESTIONS

Answers are given at the end of the chapter.

1. Nines complement the following decimal numbers.
   a. −2
   b. −10
   c. −66
   d. −382

2. Tens complement the following decimal numbers.
   a. −3
   b. −12
   c. −48
   d. −159

3. Ones complement the following binary numbers.
   a. −01
   b. −1001
   c. −10000111
   d. −00110011

4. Twos complement the following binary numbers.
   a. 0011
   b. 1111
   c. 10001101
   d. 01011100

5. Add the following binary numbers.
   a. 1000
       +1000
   b. 0101
       +0101
   c. 1111
       +1111
   d. 10101000
       +10010111

6. Subtract the following binary numbers. (*Note:* Use 2s complement and show your work.)
   a. 1111
       −0111
   b. 1011
       −1000
   c. 1000
       −0101
   d. 0101
       −1000

7. Code the following decimal numbers in BCD.
   a. 3
   b. 124

8. Code the following decimal numbers in XS3.
   a. 10
   b. 297

9. Add the following BCD numbers. Show your work.

a. $\begin{array}{r} 0110 \quad 0100_{(BCD)} \\ +0011 \quad 0001_{(BCD)} \\ \hline \end{array}$ c. $\begin{array}{r} 0001 \quad 1000_{(BCD)} \\ +0001 \quad 0110_{(BCD)} \\ \hline \end{array}$

b. $\begin{array}{r} 0001 \quad 0010 \quad 0111_{(BCD)} \\ +0001 \quad 0100 \quad 0001_{(BCD)} \\ \hline \end{array}$ d. $\begin{array}{r} 0001 \quad 1000_{(BCD)} \\ +0011 \quad 1000_{(BCD)} \\ \hline \end{array}$

SECTION 9-3: BINARY ADDERS

OBJECTIVES

1. Identify the logic diagram of a binary half-adder and full-adder.
2. Given an adder or adder/subtracter logic diagram or symbol and input levels, identify the circuit and determine its outputs.

Binary adders play a very important role in digital computers. The first section of this chapter made this point evident. This section shows how these adders are implemented.

Half-Adder

Our study of arithmetic circuits will start with one called a **half-adder.** As discussed in Chapter 5, a half-adder is a combinational logic circuit that can add 2 bits and produce both a sum (Σ) and a carry output. The truth table for a circuit that will implement this requirement is shown in Table 9–3.

TABLE 9-3	A	$+$	B	Sum	$C_{(out)}$
Half-Adder Truth Table	0		0	0	0
	0		1	1	0
	1		0	1	0
	1		1	0	1

The sum output expression ($\overline{A}B + A\overline{B}$) is that of an X-OR gate; the carry expression (AB) is produced by an AND gate. The circuit is shown in Fig. 9–1.

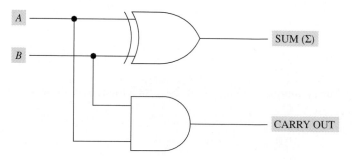

FIGURE 9-1   **Half-adder circuit.**

Full-Adder

A **full-adder** has the ability to add 3 bits—$A + B +$ Carry In (C_{in}). Its outputs, like those of the half-adder, are a **sum bit** and a **carry bit** (C_{out}). The truth table for a logic circuit that will implement this addition is shown in Table 9–4. The data from this truth table has been transferred to K-maps in Fig. 9–2.

TABLE 9-4		A	+	B	+	$C_{(in)}$	Sum	$C_{(out)}$
Full-Adder Truth Table		0		0		0	0	0
		0		0		1	1	0
		0		1		0	1	0
		0		1		1	0	1
		1		0		0	1	0
		1		0		1	0	1
		1		1		0	0	1
		1		1		1	1	1

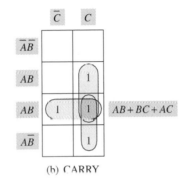

FIGURE 9-2   **K-maps for full-adder circuit outputs.**

(a) SUM   (b) CARRY

The expression for the **sum output** cannot be simplified because no loops can be made on the map. Thus, the expression is

$$\text{Sum} = \overline{A}\,\overline{B}C + \overline{A}B\overline{C} + ABC + A\overline{B}\,\overline{C}$$

The expression for the **carry output** from the truth table is

$$C_{\text{out}} = \overline{A}BC + A\overline{B}C + AB\overline{C} + ABC$$

Once the three pairs on the carry K-map are looped, this expression simplifies to

$$C_{\text{out}} = AB + BC + AC$$

The circuit is shown in Fig. 9–3.

Implementation of the full-adder can be noticeably simplified if it is implemented in the half-adder configuration shown in Fig. 9–4. Two half-adders are cascaded in this figure in a manner to add A, B, and Carry In (C_{in}). (*Note:* The SOP AND and OR gates in the figure can be replaced with NAND gates as discussed in Section 4–9). The expression for each gate output is annotated on the illustration. The equality of this circuit implementation to that of the circuit in Fig. 9–3 is proven by simplifying the expressions in Fig. 9–4.

The sum output expression is simplified as follows (note that $C_{\text{in}} = C$):

$$(\overline{\overline{A}B + A\overline{B}}\overset{x}{})C + (\overline{A}B + A\overline{B})\,\overline{C}$$
$$(\overline{\overline{A}B}\ \overset{x}{}\overline{A\overline{B}}\overset{x}{})C + (\overline{A}B + A\overline{B})\,\overline{C}$$
$$[(A + \overline{B})\,(\overline{A} + B)]C + (\overline{A}B + A\overline{B})\,\overline{C}$$
$$[(A\overline{A} + AB + \overline{A}\,\overline{B} + B\overline{B})\,C] + (\overline{A}B + A\overline{B})\,\overline{C}$$
$$(0 + AB + \overline{A}\,\overline{B} + 0)\,C + (\overline{A}B + A\overline{B})\,\overline{C}$$
$$(AB + \overline{A}\,\overline{B})\,C + (\overline{A}B + A\overline{B})\,\overline{C}$$
$$ABC + \overline{A}\,\overline{B}C + \overline{A}B\overline{C} + A\overline{B}\,\overline{C}$$

*(The X on the overbar means the overbar has been broken and the logical sign under the break has been changed.)

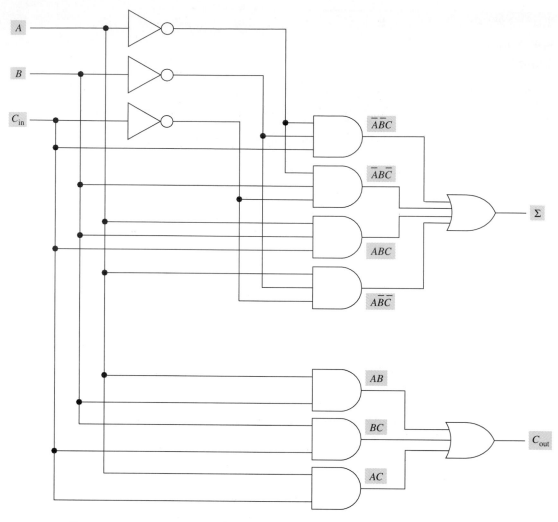

FIGURE 9–3   **Full-adder logic diagram.**

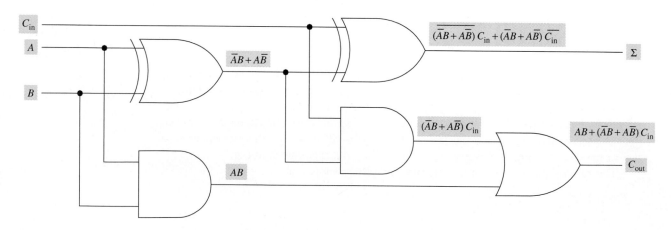

FIGURE 9–4   **Full-adder logic diagram–simplified implementation.**

The carry output expression is simplified as follows:

$$AB + (\overline{A}B + A\overline{B})\, C$$
$$AB + \overline{A}BC + A\overline{B}C$$

This expression can be taken to a K-map. Remember, AB must be plotted on the map as ABC and $AB\overline{C}$ as explained in Chapter 4.

$$AB + \overline{A}BC + A\overline{B}C$$
$$AB + BC + AC$$

The full-adder presented in Fig. 9–4 is a **single-bit adder** because it is adding only one bit from each input binary number. Most adders can add multibit binary numbers such as $1010 + 0011$. The circuits used to accomplish multibit binary addition are called **parallel binary adders.**

Two 4-bit binary numbers are shown in the following addition. It would require a *4-bit parallel binary adder* to sum these two 4-bit binary numbers.

$$\begin{array}{r} 1010 = A_3 A_2 A_1 A_0 \\ +\,0011 = B_3 B_2 B_1 B_0 \\ \hline \end{array}$$

7483—4-Bit Binary Full-Adder.

The 7483 logic symbol is shown in Fig. 9–5. The IC contains eight input pins for the two 4-bit binary numbers to be added in addition to a Carry Input (C_0). It has four output pins that produce the 4-bit sum and a Carry Out pin (C_4).

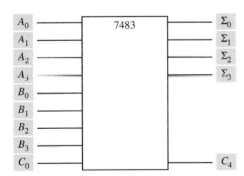

FIGURE 9–5   **7483 4-bit binary full-adder logic symbol.**

Full-adders may be cascaded to expand the number of bits that can be added. An 8-bit parallel binary adder is shown in Fig. 9–6. Notice in the figure the Carry Input to the lower-nibble adder is grounded. Also, the Carry Output of the lower-nibble adder is

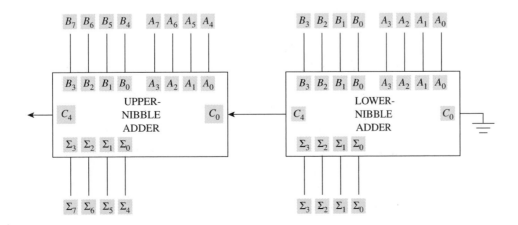

FIGURE 9–6   **8-bit parallel binary adder.**

connected to the Carry Input of the upper-nibble adder. The circuit produces an 8-bit sum. If two numbers were added that produced a 9-bit sum, the C_4 output of the upper-nibble adder would go high. For example,

$$
\begin{array}{r}
1000 \quad 0110 \\
+\ 1000 \quad 0010 \\
\hline
1\ 0000 \quad 1000
\end{array}
$$

The high Carry Output in this case is used to generate an error signal. The error signal indicates the number of sum bits exceeds the circuit's sum capability. This is what happens in your calculator when the sum is too large to be indicated.

The 7483 full-adder features internal **look-ahead carry** across all four bits. Look-ahead carry was discussed in Chapter 7.

The Carry Output (C_4) is generated in this adder prior to generation of the sum outputs. The 74LS83 generates a Carry Output for two BCD inputs in approximately 12 ns, whereas it typically takes 15 ns to generate the sum. This increases speed of operation when several ICs are cascaded to add multibit numbers.

Adder/Subtracter

Section 9–1 illustrated subtraction by adding an unsigned 2s complement number to a positive number. A 7483 full-adder connected as an adder/subtracter is shown in Fig. 9–7. The control input $\overline{\text{ADD}}/\text{SUB}$ indicates the circuit will add when the control input is low and subtract when it is high.

ADD Operation—$\overline{\text{ADD}}/\text{SUB} = 0$.

The Carry In (C_0) input to the 7483 in Fig. 9–7 is low and the B_3–B_0 input bits are routed through the X-OR gates with no phase inversion. The adder sums the true magnitude

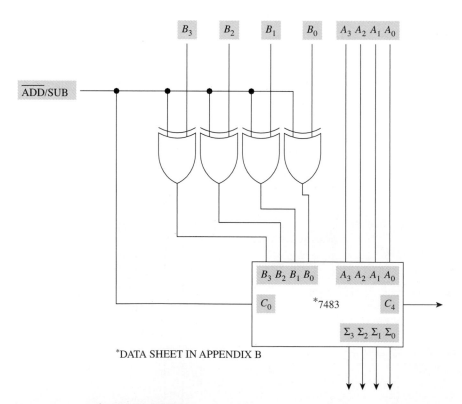

*DATA SHEET IN APPENDIX B

FIGURE 9–7   7483 adder/subtracter.

A_3–A_0 inputs with the true magnitude B_3–B_0 inputs to produce the sum of the two input nibbles.

SUBTRACT Operation—$\overline{\text{ADD}}$/SUB = 1.

The X-OR gate was presented in Chapter 5. It should be remembered from your study of exclusive gates that the X-OR gate functions as a NOT gate when one of its inputs is tied high. Since the X-OR gates invert the B inputs, these inputs to the 7483 are the 1s complement of the B_3–B_0 inputs to the circuit.

The $\overline{\text{ADD}}$/SUB high input also places a Logic 1 on the Carry In (C_0) input to the full-adder. This adds 1 to the 1s complement B_3–B_0 adder inputs, which results in the 2s complement of the negative number. Normal binary addition of the 2s complement number can now be accomplished to produce the difference.

EXAMPLE WITH $\overline{\text{ADD}}$/SUB = 0

```
 0101    A inputs
+0011    B inputs
 1000    Sum
```

EXAMPLE WITH $\overline{\text{ADD}}$/SUB = 1

```
 0101    A inputs
-0011    B inputs =   1100(1s comp)
                    +    1
                      1101(2s comp)
```

```
 0101
+1101
① 0010
```
→ Indicates positive result

Section 9–3: Review Questions

Answers are given at the end of the chapter.

A. What are the sum and carry outputs of the 7483 in Fig. 9–5 when $C_0 = 1$, A_3–$A_0 = 1100$, and B_3–$B_0 = 0100$?

B. What are the sum and carry outputs of the 7483 in Fig. 9–5 when $C_0 = 0$, A_3–$A_0 = 1111$, and B_3–$B_0 = 1111$?

C. What are the sum and carry outputs of the 8-bit parallel binary adder in Fig. 9–6 when $C_0 = 0$ (ground), A_3–$A_0 = 0111$, A_7–$A_4 = 1000$, B_3–$B_0 = 0110$, and B_7–$B_4 = 1000$?

D. What are the sigma (Σ) and C_4 outputs of the 7483 in Fig. 9–7 when $\overline{\text{ADD}}$/SUB = 0, A_3–$A_0 = 1110$, and B_3–$B_0 = 0100$?

E. What are the sigma (Σ) and C_4 outputs of the 7483 in Fig. 9–7 when $\overline{\text{ADD}}$/SUB = 1, A_3–$A_0 = 1110$, and B_3–$B_0 = 0100$?

SECTION 9-4: BCD ADDER

OBJECTIVES

1. Identify the logic diagram of a BCD adder.
2. Given the logic diagram of a BCD adder and its BCD input numbers, determine the sum and carry outputs.

It was mentioned in Section 9–2 that a digital system adds BCD numbers in the same manner it adds binary numbers. However, a **BCD adjust** is required when an invalid sum is present or when a carry is generated out of the original BCD problem.

The BCD adjust circuit must detect invalid sums or carries and add 0110 to correct the result. A truth table identifying the six invalid BCD sums of a 4-bit full-adder is shown in Table 9–5 and was briefly discussed in Chapter 5. The Boolean expression produced by this truth table is $\overline{AB}\overline{C}D + A\overline{B}\overline{C}D + \overline{A}\,\overline{B}CD + A\overline{B}CD + \overline{A}BCD + ABCD$. The ex-

TABLE 9-5	Σ_3 Σ_2 Σ_1 Σ_0				
	D	C	B	A	Out
BCD Adder Sum Outputs Truth Table (Invalid Sum Detector)	0	0	0	0	0
	0	0	0	1	0
	0	0	1	0	0
	0	0	1	1	0
	0	1	0	0	0
	0	1	0	1	0
	0	1	1	0	0
	0	1	1	1	0
	1	0	0	0	0
	1	0	0	1	0
	1	0	1	0	1
	1	0	1	1	1
	1	1	0	0	1
	1	1	0	1	1
	1	1	1	0	1
	1	1	1	1	1

pression, when implemented, will detect all of the invalid sum outputs of a full-adder. The expression is K-mapped in Fig. 9–8. The two quad loops on the map simplify the invalid sum expression to $BD + CD$, which is the same as $\Sigma_1\Sigma_3 + \Sigma_2\Sigma_3$.

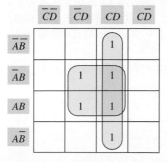

FIGURE 9-8   K-map for BCD sum outputs of a 4-bit full-adder.

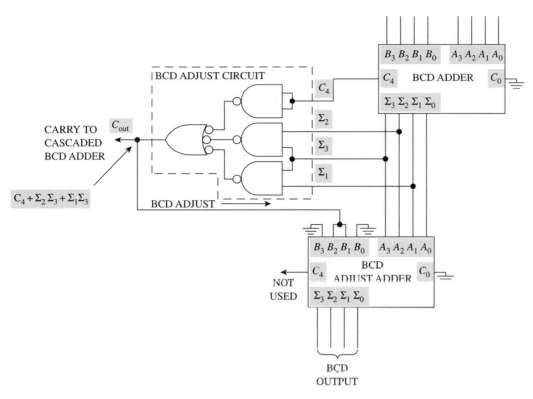

CARRY TO
CASCADED
BCD ADDER

C_{out}

$C_4 + \Sigma_2 \Sigma_3 + \Sigma_1 \Sigma_3$

BCD
OUTPUT

FIGURE 9-9   **BCD adder.**

Three example BCD addition problems are now discussed. The discussion relates to the BCD adder in Fig. 9–9.

PROBLEM 1

$$\begin{array}{r} 0101_{(BCD)} \\ +\,0010_{(BCD)} \\ \hline 0111_{(BCD\ sum)} \end{array}$$

The sum outputs of the BCD adder are applied to the BCD adjust circuit and to the BCD adjust adder. The Boolean expression for the output of the BCD adjust circuit is $C_4 + \Sigma_3 \Sigma_2 + \Sigma_3 \Sigma_1$, and the output is low for this problem. The low BCD adjust signal is applied to inputs B_2 and B_1 of the BCD adjust adder. Since B_0 and B_3 are connected to ground, the B_3 to B_0 inputs to the BCD adjust adder are 0000. This BCD number (0000) is added to the 0111 $(A_3–A_0)$ inputs to the adjust adder. The final output is $0111_{(BCD)}$.

PROBLEM 2

$$\begin{array}{r} 1000_{(BCD)} \\ +\,0111_{(BCD)} \\ \hline 1111 \end{array}$$

The sum generated out of the BCD adder is invalid in Problem 2. This invalid sum is applied to the BCD adjust circuit and to the BCD adjust adder. The BCD adjust circuit produces a high output as a result of the invalid sum. The high BCD adjust signal is applied to B_2 and B_1 of the BCD adjust adder. This adds 0110 to the 1111 sum out of the BCD adder to correct the invalid sum.

$$1000_{(BCD)}$$
$$+0111_{(BCD)}$$
$$\overline{1111}$$
$$+0110$$
$$\overline{① \ 0101_{(BCD)}}$$

→ Not used

Since $8 + 7 = 15_{(10)}$, the BCD LSD nibble is correct. The BCD adjust high signal that produced the add 0110 is also sent as a carry to the next-higher-digit BCD adder, which is not shown in Fig. 9–9. A carry is always generated out of the BCD adjust adder when adding 0110 to an invalid sum. This carry is not used.

PROBLEM 3

$$1000 \ _{(BCD)}$$
$$+1000 \ _{(BCD)}$$
$$\overline{1 \ 0000}$$

The sum produced in this problem is incorrect because a high Carry Out (C_4) signal is produced at the output of the BCD adder. This produces a high BCD adjust signal that is applied to the next BCD adder as well as the BCD adjust adder. This adds 0110 to the incorrect sum to correct the LSD nibble:

$$1000 \ _{(BCD)}$$
$$+1000 \ _{(BCD)}$$
$$\overline{1 \ 0000}$$
$$+0110$$
$$\overline{0110 \ _{(BCD)}}$$

Figure 9–10 shows a **parallel BCD adder** designed to add two 2-digit BCD numbers. The inputs annotated on the figure show the following:

$$A_{input} = \quad 0011 \quad 0111 \ _{(BCD)} = \quad 37_{(10)}$$
$$B_{input} = +0001 \quad 0100 \ _{(BCD)} = +14_{(10)}$$
$$\overline{\quad 0100 \quad 1011}$$
$$\underline{+ \quad 1 \quad 0110}$$
$$\overline{\quad 0101 \quad 0001 \ _{(BCD)} = \quad 51_{(10)}}$$

The lower-nibble BCD adjust circuit produces a high output that is used as the Carry In (C_0) to the upper-nibble BCD adder. The high output also adds 0110 to the invalid sum to correct the LSD nibble.

Section 9–4: Review Questions

Answers are given at the end of the chapter.

A. Determine the BCD and carry out (BCD adjust) outputs of the BCD adder in Fig. 9–9 when $C_0 = 0$, A_3–$A_0 = 0101$, and B_3–$B_0 = 0011$.
B. Determine the BCD and carry out outputs of the BCD adder in Fig. 9–9 when $C_0 = 0$, A_3–$A_0 = 0101$, and B_3–$B_0 = 0111$.
C. Determine the BCD and carry out outputs of the BCD adder in Fig. 9–9 when $C_0 = 0$, A_3–$A_0 = 1000$, and B_3–$B_0 = 1001$.

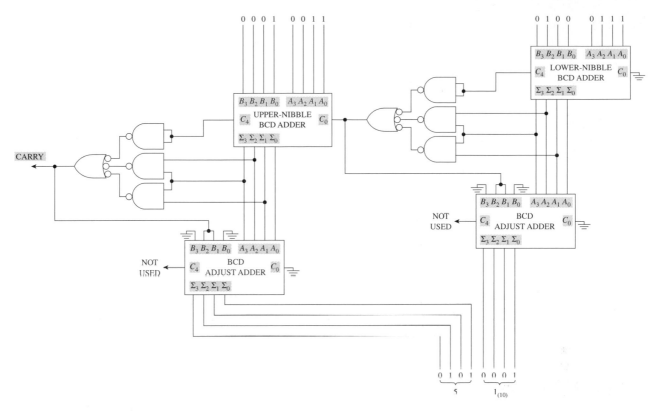

FIGURE 9-10   **Parallel BCD adder.**

SECTION 9-5: ARITHMETIC LOGIC UNIT (ALU)

OBJECTIVES

1. State the functions provided by an arithmetic logic unit.
2. Given a data book and a logic symbol of an ALU annotated with input levels, determine the ALU's mode of operation and output signal levels.

A digital system must perform numerous arithmetic and logic operations. A system would be quite large if all of these operations had to be performed using "glue chips." Instead, many systems use an arithmetic logic unit (ALU) to perform both arithmetic and logic operations.

Arithmetic logic units combine a multitude of operations on one IC. A typical ALU can **compare, double, increment, decrement, shift, add, subtract, multiply,** and **divide** binary numbers. It can also perform the following logical operations: NOT, AND, OR, NAND, NOR, X-OR, and X-NOR.

ALUs operate on two 4-, 8-, 16- or 32-bit binary numbers and are usually incorporated in the microprocessor chip in computers. Several IC versions are commercially available.

74181—4-Bit Arithmetic Logic Unit.

The 74181 IC logic symbol is shown in Fig. 9–11. The 74181 can perform 16 different arithmetic operations and 16 different logical functions. The IC has 8 input pins (A_0–A_3 and B_0–B_3) for application of the two 4-bit words and 4 output function pins (F_0–F_3).

The **mode control input** (M) determines whether arithmetic operations ($M = 0$) or logical functions ($M = 1$) are performed. The unit's internal carries are disabled during logical operations.

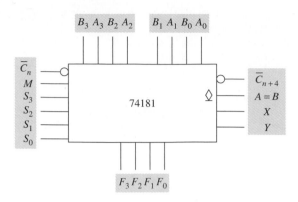

FIGURE 9-11   74181 4-bit arithmetic logic unit logic symbol.

Four **select inputs** $(S_3–S_0)$ determine which of the 16 arithmetic operations or 16 logical functions is selected. Table 9–6 defines the functions of the 74181 using active-high inputs and producing active-high outputs. The table distinguishes between logical operations and mathematical operations by spelling out the latter. The 74181 IC can also be used with active-low inputs and outputs. A separate function table is used for active-low operation.

TABLE 9-6	Selection $S_3\ S_2\ S_1\ S_0$	Logic Functions F Output $M = 1$	Arithmetic $M = 0$	
74181 ALU Active-High Input/Output Function Table			$\overline{C}_n = 1$	$\overline{C}_n = 0$
	0 0 0 0	$\overline{A}$	A	A PLUS 1
	0 0 0 1	$\overline{A+B}$	$A+B$	$(A+B)$ PLUS 1
	0 0 1 0	$\overline{A}B$	$A+\overline{B}$	$(A+\overline{B})$ PLUS 1
	0 0 1 1	0	MINUS 1 (2s comp)	Zero
	0 1 0 0	$\overline{A\ B}$	A PLUS $A\overline{B}$	A PLUS $A\overline{B}$ PLUS 1
	0 1 0 1	$\overline{B}$	$(A+B)$ PLUS $A\overline{B}$	$(A+B)$ PLUS $A\overline{B}$ PLUS 1
	0 1 1 0	$A\oplus B$	A MINUS B MINUS 1	A MINUS B
	0 1 1 1	$A\overline{B}$	$A\overline{B}$ MINUS 1	$A\overline{B}$
	1 0 0 0	$\overline{A}+B$	A PLUS AB	A PLUS AB PLUS 1
	1 0 0 1	$\overline{A\oplus B}$	A PLUS B	A PLUS B PLUS 1
	1 0 1 0	B	$(A+\overline{B})$ PLUS AB	$(A+\overline{B})$ PLUS AB PLUS 1
	1 0 1 1	AB	AB MINUS 1	AB
	1 1 0 0	1	A PLUS $A*$	A PLUS A PLUS 1
	1 1 0 1	$A+\overline{B}$	$(A+B)$ PLUS A	$(A+B)$ PLUS A PLUS 1
	1 1 1 0	$A+B$	$(A+\overline{B})$ PLUS A	$(A+\overline{B})$ PLUS A PLUS 1
	1 1 1 1	A	A MINUS 1	A

* Each bit is shifted to next more significant bit position.
$\overline{C}_n = 1 =$ No carry.
$\overline{C}_n = 0 = 1$ carry.

The **compare** operation of the 74181 produces a high output when $A = B$ provided the open-collector output is connected to a pull-up resistor. This is the only output shown in Fig. 9–11 marked with the open-collector output symbol. *Note:* Open-collector logic is presented in Chapter 11. The $A = B$ output is designed so that it can be wire-ANDed with other $A = B$ outputs. This is required when comparing words of more than four bits. The ALU must be in the subtract mode $(S_3–S_0 = 0110)$ with $\overline{C}_n = 1$ (no carry) to compare input values. The $A = B$ output will go high only when the two input words $(B_0–B_3$ and $A_0–A_3)$ are of equal magnitude. The $A = B$ output can be used with the $\overline{C}_{n+4}$ signal to identify $A < B$ or $A > B$.

The logical operations are performed by the 74181 as four distinct 2-input functions. For instance when ANDing in this ALU, A_0 is ANDed with B_0, and the product appears at F_0; A_1 is ANDed with B_1, and the product appears at F_1; and so on. Remember, no internal carries are generated during logical operations.

LOGICAL OPERATION EXAMPLES

A inputs $= 1010$
B inputs $= 0011$

$AB \qquad = 0010 = $ **AND**

A inputs $= 1010$
B inputs $= 0011$

$\overline{AB} \qquad = 1101 = $ **NAND**

A inputs $= 1010$
B inputs $= 0011$

$A + B \quad = 1011 = $ **OR**

A inputs $= 1010$
B inputs $= 0011$

$\overline{A + B} \quad = 0100 = $ **NOR**

A inputs $= 1010$
B inputs $= 0011$

$A \oplus B \quad = 1001 = $ **X-OR**

A inputs $= 1010$
B inputs $= 0011$

$\overline{A \oplus B} \quad = 0110 = $ **X-NOR**

Table 9–6 indicates a variety of logical operations can be performed by the 74181. In addition to standard logical operations, the circuit performs such functions as $\overline{AB}, A\overline{B}, \overline{A} + B$, and $A + \overline{B}$. Since these functions are an integral part of the 74181 operation, the IC is referred to as an ALU/function generator in some data books.

The $\overline{C}_{n+4}$ output is a standard **ripple–carry output** signal. It can be connected to the $\overline{C}_n$ input of a cascaded 74181 to increase word length when high-speed operation isn't required. The **Carry Propagate** (P) and **Carry Generate** (G) signals are used for look-ahead carry between ICs. These pins are called X and Y in the active-high configuration (Fig. 9–11) and $\overline{P}$ and $\overline{G}$ in the active-low configuration. The X and Y signals can be connected to a look-ahead carry generator when high-speed operation is required. The generator anticipates a carry across four binary adders or across a group of adders.

74181 ALU EXAMPLE OPERATIONS (REFER TO TABLE 9-6):

1. Determine the F and $\overline{C}_{n+4}$ outputs of the 74181 in Fig. 9–11 when $M = 0$, $\overline{C}_n = 1$, $S_3–S_0 = 1001$, $A_3–A_0 = 0101$, and $B_3–B_0 = 1100$

Answer: The circuit is in the arithmetic mode with no carry-in. **Addition** is performed on the A and B inputs as shown in Table 9–6. The 1 carry out of the addition problem shown is inverted to a low $\overline{C}_{n+4}$ output.

$$
\begin{array}{r}
0101 = A \text{ inputs} \\
+\,1100 = B \text{ inputs} \\
\hline
①\,0001 = F_3–F_0 \text{ sum outputs}
\end{array}
$$

$\overline{C}_{n+4} = 0 = 1$ carry

2. Change the select inputs to 1100 and determine the F and $\overline{C}_{n+4}$ outputs.

Answer: The circuit is still in the arithmetic mode with no carry in. This selection produces A PLUS A, which indicates each bit is shifted to the next more significant position.

$$
\begin{array}{r}
0101 = A \text{ inputs} \\
+\,0101 = A \text{ inputs} \\
\hline
\text{⓪}\,1010 = F_3\text{–}F_0 \text{ sum outputs} = A \text{ PLUS } A \\
\overline{C}_{n+4} = 1 = \text{No carry}
\end{array}
$$

3. Determine the F and $\overline{C}_{n+4}$ outputs of the 74181 when $M = 0$, $\overline{C}_n = 0$, $S_3\text{–}S_0 = 0110$, $A_3\text{–}A_0 = 1110$, and $B_3\text{–}B_0 = 0001$.

Answer: The difference is obtained in the ALU by 2s complement addition.

$$
\begin{array}{r}
1110 \\
-\,0001 = 1110_{(1\text{s comp})} \\
+\quad 1 \\
\hline
1111_{(2\text{s comp})}
\end{array}
$$

$$
\begin{array}{r}
1110 = A \text{ inputs} \\
+\,1111 = B \text{ inputs (2s complement)} \\
\hline
\text{①}\,1101 = A \text{ MINUS } B \\
\overline{C}_{n+4} = 0 = 1 \text{ carry} = \text{Positive result}
\end{array}
$$

4. Determine the F outputs of a 74181 ALU when $M = 1$, $S_3\text{–}S_0 = 0100$, $A_3\text{–}A_0 = 1001$, and $B_3\text{–}B_0 = 0111$.

Answer: The carries ($\overline{C}_n$ and $\overline{C}_{n+4}$) are disabled because the ALU is in the logic mode. The circuit ANDs and then NOTs (NANDs) each AB input.

$$
\begin{array}{l}
1001 = A \text{ inputs} \\
0111 = B \text{ inputs} \\
\hline
0001 = AB \text{ outputs} \\
1110 = \overline{AB} \text{ outputs}
\end{array}
$$

5. Change the select inputs to 0110 and determine the F outputs.

Answer: The output is the X-OR function of each A and B input.

$$
\begin{array}{l}
1001 = A \text{ inputs} \\
0111 = B \text{ inputs} \\
\hline
1110 = A \oplus B
\end{array}
$$

Section 9-5: Review Questions

Answers are given at the end of the chapter.

A. List 4 arithmetic functions performed by an ALU.
B. List 6 logic functions performed by an ALU.
C. Determine the F and $\overline{C}_{n+4}$ outputs of the 74181 ALU in Fig. 9–11 when: $M = 0$, $\overline{C}_n = 0$, $S_3\text{–}S_0 = 1111$, $A_3\text{–}A_0 = 1010$, and $B_3\text{–}B_0 = 0101$.

D.   Change the $\overline{C}_n$ input in question (C) to $\overline{C}_n = 1$ and determine the F and $\overline{C}_{n+4}$ outputs.
E.   Determine the F outputs of a 74181 ALU when $M = 1$, S_3–$S_0 = 1011$, A_3–$A_0 = 1000$, and B_3–$B_0 = 1110$.
F.   Change the S_3–S_0 inputs in question (E) to 0001 and determine the F outputs of the ALU.

SECTIONS 9-3 THROUGH 9-5: INTERNAL SUMMARY

Binary adders are used to add two bits. A **full-adder** can add a carry in in addition to the two input bits. Multibit addition is accomplished by parallel binary adders.

The 7483 IC is a 4-bit binary full-adder with the ability to add two 4-bit words plus a carry in. Subtraction can be performed by 2s complementing the negative number and adding it to the positive number.

BCD addition is accomplished by connecting an error detection circuit to the output of a standard binary full-adder. The circuit must detect an invalid BCD sum at the adder sigma outputs or a high carry out bit. Either condition will result in adding 0110 (BCD adjust) to the answer.

An **arithmetic logic unit** is used to perform binary arithmetic and several logic operations. The 74181 ALU can perform 16 arithmetic operations and 16 logical functions. Specific operation is controlled by the input levels applied to the mode control, select, and carry in inputs. A function table in a data book is a necessity when studying an ALU.

SECTIONS 9-3 THROUGH 9-5: INTERNAL SUMMARY QUESTIONS

Answers are given at the end of the chapter.

Note: The use of manufacturer's data sheets is recommended to assist you in answering these summary questions.

1. Identify the circuit shown in Fig. 9–12.
   a.   Half-adder
   b.   Full-adder
   c.   BCD adder
   d.   Adder/subtracter

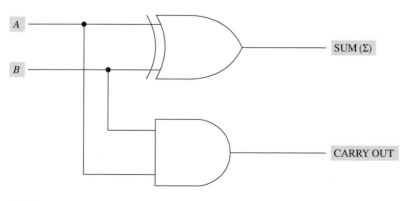

FIGURE 9-12

2. The outputs of the circuit in Fig. 9–12 are _____ when $A = 0$ and $B = 1$.
   a. $\Sigma = 0$ and $C_{out} = 0$
   b. $\Sigma = 0$ and $C_{out} = 1$
   c. $\Sigma = 1$ and $C_{out} = 0$
   d. $\Sigma = 1$ and $C_{out} = 1$

3. The outputs of the circuit in Fig. 9–12 are _____ when $A = 1$ and $B = 1$.
   a. $\Sigma = 0$ and $C_{out} = 0$
   b. $\Sigma = 0$ and $C_{out} = 1$
   c. $\Sigma = 1$ and $C_{out} = 0$
   d. $\Sigma = 1$ and $C_{out} = 1$

4. What are the outputs of the full-adder in Fig. 9–13 when A_3–$A_0 = 1010$, B_3–$B_0 = 0101$, and $C_0 = 0$?
   a. Σ_3–$\Sigma_0 = 0101$ and $C_4 = 0$.
   b. Σ_3–$\Sigma_0 = 0101$ and $C_4 = 1$.
   c. Σ_3–$\Sigma_0 = 1111$ and $C_4 = 0$.
   d. Σ_3–$\Sigma_0 = 1111$ and $C_4 = 1$.

5. What are the outputs of the full-adder in Fig. 9–13 when A_3–$A_0 = 1100$, B_3–$B_0 = 0100$, and $C_0 = 0$?
   a. Σ_3–$\Sigma_0 = 0000$ and $C_4 = 0$.
   b. Σ_3–$\Sigma_0 = 0000$ and $C_4 = 1$.
   c. Σ_3–$\Sigma_0 = 1000$ and $C_4 = 0$.
   d. Σ_3–$\Sigma_0 = 1000$ and $C_4 = 1$.

6. What are the outputs of the full-adder in Fig. 9–13 when A_3–$A_0 = 0100$, B_3–$B_0 = 0111$, and $C_0 = 1$?
   a. Σ_3–$\Sigma_0 = 0011$ and $C_4 = 0$.
   b. Σ_3–$\Sigma_0 = 1100$ and $C_4 = 0$.
   c. Σ_3–$\Sigma_0 = 1101$ and $C_4 = 0$.
   d. Σ_3–$\Sigma_0 = 1101$ and $C_4 = 1$.

7. What are the outputs of the 8-bit adder in Fig. 9–14 when A_3–$A_0 = 0110$, B_3–$B_0 = 1001$, A_7–$A_4 = 0110$, and B_7–$B_4 = 0011$?
   a. Σ_7–$\Sigma_0 = 1001\ 1111$ and upper-nibble $C_4 = 0$.
   b. Σ_7–$\Sigma_0 = 1001\ 1111$ and upper-nibble $C_4 = 1$.
   c. Σ_7–$\Sigma_0 = 1100\ 1100$ and upper-nibble $C_4 = 0$.
   d. Σ_7–$\Sigma_0 = 1100\ 1100$ and upper-nibble $C_4 = 1$.

8. What are the outputs of the 8-bit adder in Fig. 9–14 when A_3–$A_0 = 1000$, B_3–$B_0 = 1100$, A_7–$A_4 = 1001$, and B_7–$B_4 = 0111$?
   a. Σ_7–$\Sigma_0 = 0100\ 0001$ and upper-nibble $C_4 = 0$.
   b. Σ_7–$\Sigma_0 = 0100\ 0001$ and upper-nibble $C_4 = 1$.
   c. Σ_7–$\Sigma_0 = 0001\ 0100$ and upper-nibble $C_4 = 0$.
   d. Σ_7–$\Sigma_0 = 0001\ 0100$ and upper-nibble $C_4 = 1$.

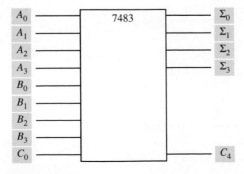

FIGURE 9–13

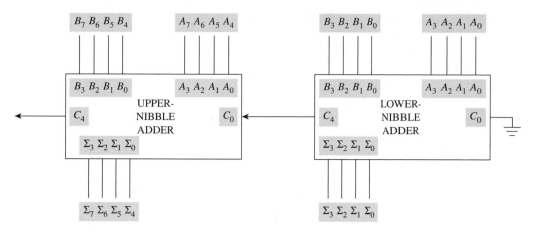

FIGURE 9–14

9. What is the level of the BCD adjust signal to the B_2 and B_1 inputs of the BCD adjust adder in Fig. 9–15 when $A_3\text{–}A_0 = 1000$ and $B_3\text{–}B_0 = 1001$?
    a. Low
    b. High

10. What are the BCD and carry outputs of the circuit in Fig. 9–15 when $A_3\text{–}A_0 = 0111$ and $B_3\text{–}B_0 = 1001$?
    a. $\Sigma_3\text{–}\Sigma_0 = 0000$ and $C_{out} = 0$.
    b. $\Sigma_3\text{–}\Sigma_0 = 0000$ and $C_{out} = 1$.
    c. $\Sigma_3\text{–}\Sigma_0 = 0110$ and $C_{out} = 0$.
    d. $\Sigma_3\text{–}\Sigma_0 = 0110$ and $C_{out} = 1$.

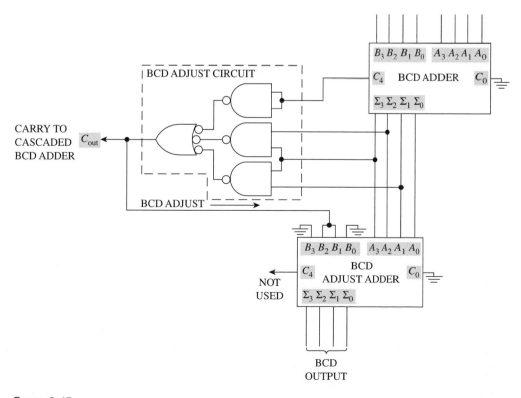

FIGURE 9–15

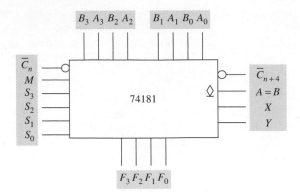

FIGURE 9–16

---

11. What does the symbol next to the compare output ($A = B$) in Fig. 9–16 indicate?
    a. Tristate output
    b. Active-low output
    c. Open-collector output
    d. Schmitt-trigger output

12. What are the outputs of the ALU in Fig. 9–16 if $M = 0$, $\overline{C}_n = 0$, S_3–$S_0 = 0000$, A_3–$A_0 = 1000$, and B_3–$B_0 = 0001$?
    a. F_3–$F_0 = 0001$.
    b. F_3–$F_0 = 0010$.
    c. F_3–$F_0 = 1000$.
    d. F_3–$F_0 = 1001$.

CHAPTER 9: SUMMARY

Block diagrams and truth tables for a **half-adder (HA)** and a **full-adder (FA)** are presented in Fig. 9–17. Review the truth tables to ensure you have a thorough understanding of binary addition.

Keep in mind that 1s complementing a negative number allows subtraction by addition with **end-around carry.** The 2s complement provides the same operation without the end-around carry. Actually, the

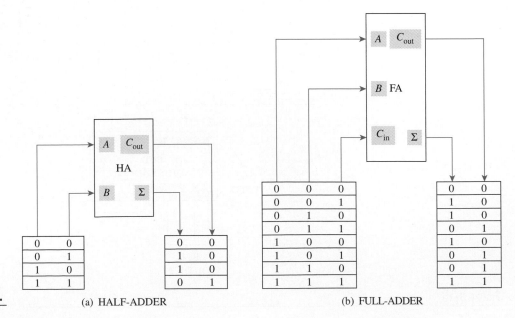

FIGURE 9–17 Block diagram and truth table.

(a) HALF-ADDER   (b) FULL-ADDER

end-around carry is taken care of when a 1s complemented number is 2s complemented.

BCD arithmetic requires error detection and error correction circuitry to validate BCD sums. The extra circuitry required is one of the disadvantages of BCD arithmetic that was discussed in Chapter 2.

Binary **full-adders** can be implemented with X-OR, AND, and OR gates by cascading two half-adders and ORing the carry output of each. Multibit IC adders can be cascaded by connecting the carry output to the carry input of the next-more-significant bit adder.

Subtraction can be performed by placing a high on the carry input of an adder and 1s complementing (in-verting) the negative number. The high carry input produces the desired 2s complement number. The 1s complement was achieved by using X-OR gates as inverters in Section 9–3. It could also be accomplished by loading the negative number into a register and 1s complementing that number by taking it from the $\overline{Q}$ outputs of the register. Again, the carry input must be high to 2s complement the 1s complement number.

An **ALU** is used to accomplish multiple arithmetic and logic operations. The arithmetic functions of an ALU were presented in this chapter; the logic functions were explained in earlier chapters.

CHAPTER 9: END OF CHAPTER QUESTIONS/PROBLEMS

Answers are given in the Instructor's Manual.

Note: The use of manufacturer's data sheets is recommended to assist you in answering these questions.

SECTION 9-1

1. The 9s complement of -12 is _____.
2. The 10s complement of -148 is _____.
3. Which logic level (low or high) is used as the sign bit to represent a positive number?
4. The 9s complement of some unknown number (X) is 270. The signed value of X is _____.
5. Ones complement $1001_{(2)}$.
6. Twos complement $1001_{(2)}$.
7. Add. $\quad 1001_{(2)}$
   $\quad +1011_{(2)}$
8. Add. $\quad 10011011_{(2)}$
   $\quad +00110011_{(2)}$
9. Subtract. $\quad 1111_{(2)}$
   $\quad -0011_{(2)}$
10. Subtract. $\quad 1011_{(2)}$
    $\quad -1100_{(2)}$

11. What must be done with the end-around carry in 1s complement subtraction to obtain the difference of two binary numbers?
12. Multiply. $\quad 1000_{(2)}$
    $\quad \times 11_{(2)}$

SECTION 9-2

13. Add. $\quad 0101 \quad 0111_{(BCD)}$
    $\quad +0100 \quad 0001_{(BCD)}$
14. Add. $\quad 0111 \quad 1000_{(BCD)}$
    $\quad +0111 \quad 1001_{(BCD)}$
15. Code $631_{(10)}$ to XS3.

SECTION 9-3

16. What are the outputs (Σ and C_{out}) of the full-adder in Fig. 9–18 if $A = 1, B = 1$, and $C_{in} = 0$?
17. What are the outputs of the full-adder in Fig. 9–18 if $A = 0, B = 0$, and $C_{in} = 1$?
18. What is the output of the circuit in Fig. 9–19?

SECTION 9-4

19. What is the output of the circuit in Fig. 9–20?

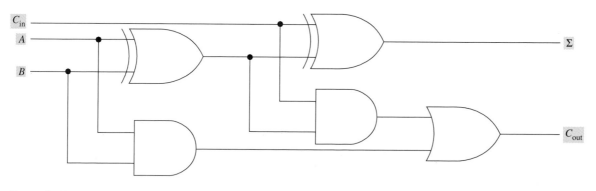

FIGURE 9-18

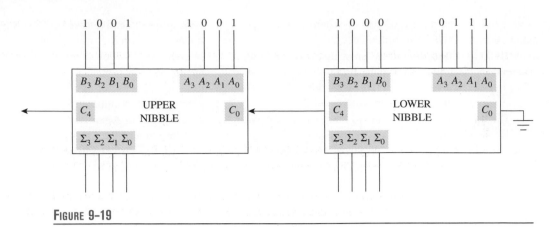

FIGURE 9–19

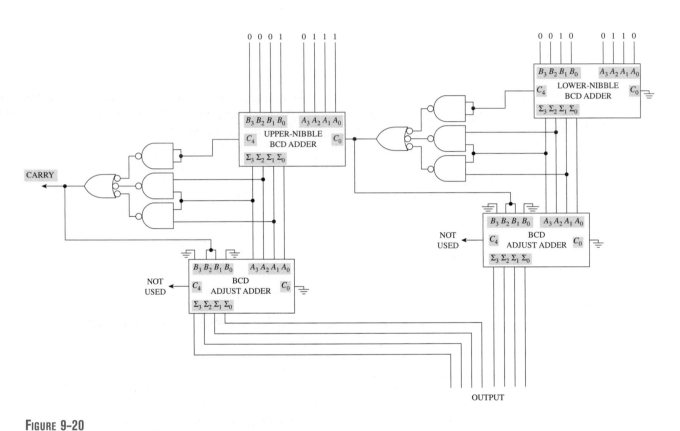

FIGURE 9–20

SECTION 9–5

20. What function does the 74181 ALU in Fig. 9–21 perform when $M = 1$?

CT 21. Determine the F outputs of the 74181 ALU in Fig. 9–21 when $M = 1$, S_3–$S_0 = 1110$, A_3–$A_0 = 1000$, and B_3–$B_0 = 0001$.

CT 22. Determine the F outputs of the 74181 ALU in Fig. 9–21 when $M = 0$, $\overline{C}_n = 0$, S_3–$S_0 = 0110$, A_3–$A_0 = 1100$, and B_3–$B_0 = 0001$.

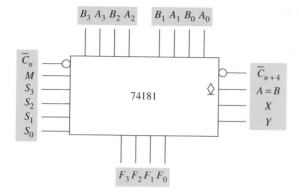

FIGURE 9-21

ANSWERS TO REVIEW QUESTIONS

SECTION 9-1

A. (1) 8
    (2) 4
    (3) 0
    (4) 71
    (5) 208

B. (1) 9
    (2) 5
    (3) 1
    (4) 72
    (5) 209

C. (1) 1100
    (2) 0111
    (3) 00111011
    (4) 10110111

D. (1) 1101
    (2) 1000
    (3) 00111100
    (4) 10111000

E. (1) 1011
    (2) 1111
    (3) 11000
    (4) 10100
    (5) 11011000

F. (1)   1010
          $-$0010 $-$0010 $= 1101_{(1s\ comp)}$
                                     $+\ \ 1$
                                 $\overline{\ 1110_{(2s\ comp)}}$

           1010
         $+$1110
        ①**1000**
        └──► Positive result

(2)   1111
      $-$0111 $-$0111 $= 1000_{(1s\ comp)}$
                                 $+\ \ 1$
                             $\overline{\ 1001_{(2s\ comp)}}$

       1111
     $+$1001
    ①**1000**
    └──► Positive result

(3)   0101
      $-$0100 $-$0100 $= 1011_{(1s\ comp)}$
                                 $+\ \ 1$
                             $\overline{\ 1100_{(2s\ comp)}}$

       0101
     $+$1100
    ①**0001**
    └──► Positive result

(4)   1010
      $-$1100 $-$1100 $= 0011_{(1s\ comp)}$
                                 $+\ \ 1$
                             $\overline{\ 0100_{(2s\ comp)}}$

       1010
     $+$0100
    ⓪**1110**
    └──► Negative result in 2s
         complement

    1110 $= 0001$
          $+\ \ 1$
     $\overline{-\textbf{0010}}$

(5)   0011
      −0101 $-0101 = 1010_{(1s\ comp)}$
      ‾‾‾‾ $+\ \ \ 1$
                            ‾‾‾‾‾‾
                         $1011_{(2s\ comp)}$

                  0011
               + 1011
               ‾‾‾‾‾‾
               ⓪ 1110
                   └──► Negative result in 2s
                        complement

            1110 =   0001
                    +   1
                    ‾‾‾‾‾‾
                    −0010

Section 9-2

A. (1) Add 0110 when an invalid sum is generated.
    (2) Add 0110 when a carry out is generated in the sum when adding two BCD numbers in the original problem.

B. (1) $0110_{(BCD)}$
    (2) $0010\ \ 1001_{(BCD)}$
    (3) $0001\ \ 0100\ \ 0000_{(BCD)}$
    (4) $0001\ \ 0000\ \ 0000\ \ 0000_{(BCD)}$

C. (1) $0011\ \ \ 0001_{(BCD)}$
          $+0010\ \ \ 0010_{(BCD)}$
          ‾‾‾‾‾‾‾‾‾‾‾‾‾‾
           $0101\ \ \ 0011_{(BCD)} = 53_{(10)}$

    (2) $0101\ \ \ 0110_{(BCD)}$
          $+0100\ \ \ 0011_{(BCD)}$
          ‾‾‾‾‾‾‾‾‾‾‾‾‾‾
           $1001\ \ \ 1001_{(BCD)} = 99_{(10)}$

    (3) $0110\ \ 1000\ \ 0001_{(BCD)}$
          $+0010\ \ 0001\ \ 0111_{(BCD)}$
          ‾‾‾‾‾‾‾‾‾‾‾‾‾‾‾‾‾‾‾‾
           $1000\ \ 1001\ \ 1000_{(BCD)} = 898_{(10)}$

    (4)   1
                      ↖
                  $0111\ \ 1000_{(BCD)}$
               $+0001\ \ \backslash\ 1001_{(BCD)}$
               ‾‾‾‾‾‾‾‾‾‾‾‾
                $1001\ \ 0001$
              $+\ \ \ \ \ \ \ 0110$
              ‾‾‾‾‾‾‾‾‾‾‾‾
               $1001\ \ 0111_{(BCD)} = 97_{(10)}$

    (5) $0100\ \ 0101_{(BCD)}$
          $+0011\ \ 0110_{(BCD)}$
          ‾‾‾‾‾‾‾‾‾‾‾‾
           $0111\ \ 1011$
         $+\ \ \ 1\ \ \ 0110$
         ‾‾‾‾‾‾‾‾‾‾‾‾
          $1000\ \ 0001_{(BCD)} = 81_{(10)}$

(6)   1
                  ↖
           $0000\ \ 0111\ \ 0111_{(BCD)}$
        $+0000\ \ 0011\ \backslash\ 1001_{(BCD)}$
        ‾‾‾‾‾‾‾‾‾‾‾‾‾‾‾‾‾‾
         $0000\ \ 1011\ \ 0000$
       $+\ \ \ 1\ \ \ 0110\ \ 0110$
       ‾‾‾‾‾‾‾‾‾‾‾‾‾‾‾‾‾‾
        $0001\ \ 0001\ \ 0110_{(BCD)} = 116_{(10)}$

D. (1) $0110_{(XS3)}$
    (2) $1100_{(XS3)}$
    (3) $0110\ \ 1011_{(XS3)}$
    (4) $0100\ \ 1100\ \ 0111_{(XS3)}$

Section 9-3

A. $\Sigma_3 - \Sigma_0 = 0001$
    $C_4 = 1$

B. $\Sigma_3 - \Sigma_0 = 1110$
    $C_4 = 1$

C. $\Sigma_7 - \Sigma_0 = 0000\ \ \ 1101$
    Lower-nibble adder $C_4 = 0$
    Upper-nibble adder $C_4 = 1$

D. $\Sigma_3 - \Sigma_0 = 0010$
    $C_4 = 1$

E. $\Sigma_3 - \Sigma_0 = 1010$
    $C_4 = 1$

Section 9-4

A. $\Sigma_3 - \Sigma_0 = 1000$
    $C_{out} = 0$

B. $\Sigma_3 - \Sigma_0 = 0010$
    $C_{out} = 1$

C. $\Sigma_3 - \Sigma_0 = 0111$
    $C_{out} = 1$

Section 9-5

A. Add, subtract, multiply, divide, increment, decrement, compare, double

B. NOT; AND; OR; NAND; NOR; X-OR; X-NOR

C. $F_3 - F_0 = 1010 = A$
    $\overline{C}_{n+4} = 0 = 1$ carry (positive result)

D. $F_3 - F_0 = 1001 = A$ MINUS 1
    $\overline{C}_{n+4} = 0 = 1$ carry (positive result)

E. $F_3 - F_0 = 1000 = AB$

F. $F_3 - F_0 = 0001 = \overline{A + B}$

Answers to Internal Summary Questions

Sections 9-1 and 9-2

1. a. 7
    b. 89
    c. 33
    d. 617

2. a. 7
    b. 88
    c. 52
    d. 841

3. a. 10
    b. 0110
    c. 01111000
    d. 11001100

4. a. 1101
    b. 0001
    c. 01110011
    d. 10100100

5. a. 10000
    b. 1010
    c. 11110
    d. 100111111

6. a.
$$\begin{array}{r} 1111 \\ -0111 = 1000 \\ \hline +1 \\ \hline 1001 \end{array} \qquad \begin{array}{r} 1111 \\ +1001 \\ \hline ①\,1000 \end{array}$$
    →Positive result

    b.
$$\begin{array}{r} 1011 \\ -1000 = 0111 \\ \hline +1 \\ \hline 1000 \end{array} \qquad \begin{array}{r} 1011 \\ +1000 \\ \hline ①\,0011 \end{array}$$
    →Positive result

    c.
$$\begin{array}{r} 1000 \\ -0101 = 1010 \\ \hline +1 \\ \hline 1011 \end{array} \qquad \begin{array}{r} 1000 \\ +1011 \\ \hline ①\,0011 \end{array}$$
    →Positive result

    d.
$$\begin{array}{r} 0101 \\ -1000 = 0111 \\ \hline +1 \\ \hline 1000 \end{array} \qquad \begin{array}{r} 0101 \\ +1000 \\ \hline ⓪\,1101 \end{array}$$
    → 2s complement answer

$$\begin{array}{r} 1101 = 0010 \\ +1 \\ \hline \mathbf{-0011} \end{array}$$

7. a. $0011_{(BCD)}$
    b. 0001   0010 $0100_{(BCD)}$

8. a. 0100 $0011_{(XS3)}$
    b. 0101   1100 $1010_{(XS3)}$

9. a.
$$\begin{array}{r} 0110 \quad 0100_{(BCD)} \\ +0011 \quad 0001_{(BCD)} \\ \hline 1001 \quad 0101_{(BCD)} \end{array}$$

    b.
$$\begin{array}{r} 0001 \quad 0010 \quad 0111_{(BCD)} \\ +0001 \quad 0100 \quad 0001_{(BCD)} \\ \hline 0010 \quad 0110 \quad 1000_{(BCD)} \end{array}$$

    c.
$$\begin{array}{r} 0001 \quad 1000_{(BCD)} \\ +0001 \quad 0110_{(BCD)} \\ \hline 0010 \quad 1110 \\ |\quad 1\quad 0110 \\ \hline 0011 \quad 0100_{(BCD)} \end{array}$$

    d.
$$\begin{array}{r} 1 \\ 0001 \quad 1000_{(BCD)} \\ +0011 \quad 1000_{(BCD)} \\ \hline 0101 \quad 0000 \\ +\quad\quad 0110 \\ \hline 0101 \quad 0110_{(BCD)} \end{array}$$

Sections 9-3 through 9-5

1. a		7. a	
2. c		8. d	
3. b		9. b	
4. c		10. d	
5. b		11. c	
6. b		12. d	

10 MSI Digital Circuits

Introduction

Medium-scale integration (MSI) ICs were defined in Chapter 1 as ICs that contain 12 to 99 logic gates or circuitry of basically the same complexity. Some MSI digital circuits have already been presented in the text. Many of the counter and register ICs presented in Chapters 7 and 8 fall in this category. This chapter has been set aside to include those MSI digital circuits that do not fall under the main topic of one of the other chapters.

A **decoder** circuit is used to **detect** a binary code or number at the circuit's inputs and indicate that code or number by activating one of its output lines. The activated output line is representative of one of several possible input combinations (codes). Decoders are also used to convert an input code or number to some other usable format. They are used in digital systems for detection and data routing. They are also extremely popular in memory address circuits.

The logic symbol in Fig. 10–1(a) represents a 2-line to 4-line decoder. This decoder is sometimes called a 1-of-4 decoder because it can detect any one of four possible input combinations—00, 01, 10, or 11. Only one of the four output lines will be activated for a given input code. If the input is $I_1 = 1$ and $I_0 = 0$, the O_2 output line will be activated.

Code Converter	Leading Zero Suppression	n-bit Encoder
Decoder		1-of-n Decoder
Demultiplexer	Magnitude Comparator	Priority Encoder
Encoder	Multiplexer	Ripple Blanking

Chapter Objectives

1. Define decoding, encoding, multiplexing, demultiplexing, and magnitude comparing.

2. Identify a decoder, encoder, multiplexer, demultiplexer, or magnitude comparator using a logic diagram or logic symbol.

3. Given the logic diagram or logic symbol of a decoder, encoder, multiplexer, demultiplexer, and/or magnitude comparator, determine the circuit's output when various input combinations are specified.

4. Identify the main use of a circuit containing one or more MSI ICs.

5. Given a description of a problem or a stated malfunction in a circuit containing one or more MSI ICs, identify the most likely output results.

6. Given the improper output result(s) of a defective circuit containing one or more MSI ICs, determine the most likely malfunction.

Other types of decoders are used to convert a binary code into some usable format other than a single activated output line. For example, 7-segment indicators are often used in digital systems. If the system is using binary-coded decimal (BCD), the binary codes (numbers) representing the counts of 0 through 9 must be converted to 7-segment codes that will illuminate the proper digit on the 7-segment indicator.

Encoders perform the opposite function of their decoder counterparts. An encoder produces a distinct binary code for each of its inputs. The output code produced depends on the input line that is activated. A 4-line to 2-line encoder logic symbol is shown in

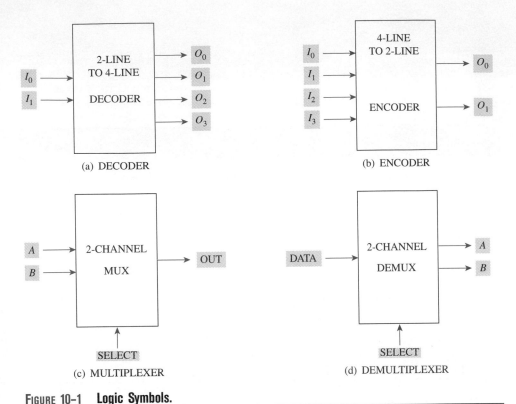

FIGURE 10–1   **Logic Symbols.**

Fig. 10–1(b). The output would be 11 if the I_3 input line were activated. Simply put, the output binary number will be equal to the activated input line number.

Multiplexers are used in digital systems as **data selectors.** They are used to select certain input data and route it to the output. A multiplexer can be thought of as a multiposition switch whose contact arm routes data from different input positions to one single output line. A 2-channel multiplexer logic symbol is shown in Fig. 10–1(c). The Select input is used to select the data at input A or B and route it to the output.

Demultiplexers perform the exact opposite function provided by multiplexers. The demultiplexer distributes the data from one input line over several output lines. Demultiplexers are often referred to as **data distributors.** The Select input on the demultiplexer logic symbol in Fig. 10–1(d) is used to distribute the data input to the A or B output line. Applications of some of these MSI circuits are presented in Section 10–6.

SECTION 10-1:   DECODERS

OBJECTIVES

1. Define decoding and state the function of decoder circuits.
2. Given a decoder logic diagram or logic symbol annotated with input levels, identify the circuit and determine its outputs.

The process of detecting and identifying a binary number or code is called **decoding.** Two basically different types of decoders will be presented in this section. The first type detects the binary code or number applied to its input lines/pins and activates one of its output lines/pins. The one activated output indicates which binary code or number has been applied to the inputs. The second type of decoder detects a binary code or number and converts it to another code such as a 7-segment code to drive a 7-segment indicator. This type can be called a **code converter.**

2-Line to 4-Line Decoders

Figure 10–2 shows the logic diagram and symbol for a **2-line to 4-line decoder.** A 2-bit binary number or code is applied to the AB inputs of the decoder. The A input is the LSB of the number/code. The outputs, Y_0 through Y_3, are used to indicate which 2-bit code has been applied to the AB inputs. The logic symbol in Fig. 10–2(b) indicates the outputs of the device are active low.

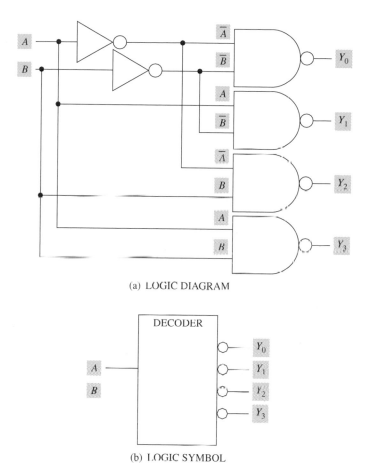

(a) LOGIC DIAGRAM

DECODER

(b) LOGIC SYMBOL

FIGURE 10–2   **Decoder: 2-line to 4-line.**

If the input combination is $AB = 00$, the Y_0 output line goes low and the three remaining output lines stay at the high (inactive) level. The input combinations and resulting outputs are shown in Table 10–1. The circuit is often called a 1-of-4 decoder because only one line goes active during each of the four possible input combinations.

The logic diagram of the 1-of-4 decoder is shown in Fig. 10–2(a). Tracing logic levels through the diagram will prove the function table (Table 10–1) entries to be correct. This is a good one-time exercise to prove the validity of the function table and logic circuitry used to implement the decoder function.

TABLE 10–1		Inputs		Outputs			
1-of-4 Decoder Function Table		**B**	**A**	Y_3	Y_2	Y_1	Y_0
		0	0	1	1	1	0
		0	1	1	1	0	1
		1	0	1	0	1	1
		1	1	0	1	1	1

B = 0 and A = 0.

The inputs to the top NAND gate have both been NOTed to high logic levels. Therefore, the output (Y_0) of that gate goes low. This activated output indicates the binary number 00 has been applied to the inputs. The remaining NAND gates all have at least one low input, which keeps their outputs high (inactive).

Take time to complete this one-time exercise for the other three input combinations. This exercise will not be necessary for any of the other 1-of-n decoders because their operation is similar to this basic decoder. Also, it is best now to begin to rely entirely on the logic symbols for this type of circuit because that is what you will see in system/subsystem digital diagrams.

74LS139.

The logic diagram and symbol for this dual 2-line to 4-line decoder are shown in Fig. 10–3. The logic diagram of this IC in Fig. 10–3(a) shows remarkable similarity to the generic 2-line to 4-line decoder in Fig. 10–2(a).

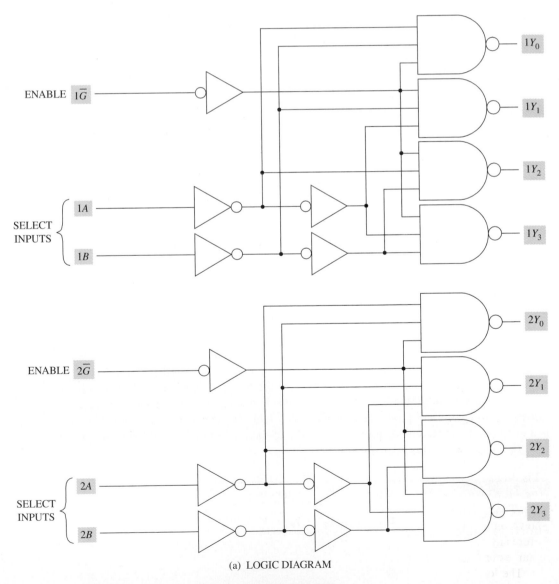

(a) LOGIC DIAGRAM

FIGURE 10-3   **Decoder: 74LS139–dual 2-line to 4-line.**

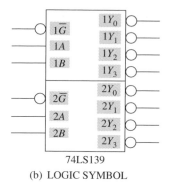

(b) LOGIC SYMBOL

FIGURE 10–3   Continued

One notable difference between the 74LS139 decoder in Fig. 10–3(a) and the generic decoder previously presented is the **enable input** $(\overline{G})$. This active-low input must be brought low for the decoder to function. If the input is held high, all of the NAND gates are inhibited, and their outputs are inactive (high). This is illustrated in the function table in Table 10–2. Note the select inputs are don't care inputs when the enable input is inactive. The enable input to this decoder can also be used as a data input in demultiplexing (data distribution) applications. The 74LS139's use as a demultiplexer will be presented in Section 10–4.

TABLE 10-2	Inputs			Outputs			
	Enable	Select					
74LS139 Decoder **Function Table**	$\overline{G}$	B	A	Y_3	Y_2	Y_1	Y_0
	1	X	X	1	1	1	1
	0	0	0	1	1	1	0
	0	0	1	1	1	0	1
	0	1	0	1	0	1	1
	0	1	1	0	1	1	1

The 74LS139 circuit is two separate 2-line to 4-line decoders in one IC. Each decoder has its own enable input. The logic symbol in Fig. 10–3(b) shows two select inputs (1A and 1B or 2A and 2B), which produce $2^2 = 4$ outputs for each decoder, only one of which will be activated at any given time.

3-Line to 8-Line Decoder

These decoders decode 1-of-8 output lines during each of the eight possible input combinations. The input combinations are produced by three select inputs.

74LS138.

This 3-line to 8-line decoder's logic diagram and logic symbol are depicted in Fig. 10–4. The 74LS138 contains two active-low ($\overline{G}_2A$ and $\overline{G}_2B$) and one active-high (G_1) enable inputs. The enable inputs must all be activated for the decoder to function. Both $\overline{G}$ inputs shown on the logic symbol in Fig. 10–4(b) must be low and the G_1 input must be high before any output line can go active. These three enable inputs allow expansion with minimum external control circuitry.

The logic diagram in Fig. 10–4(a) shows the enable inputs are internally applied to a NOR gate. If all of the inputs to the NOR gate are low, its output is high and the out-

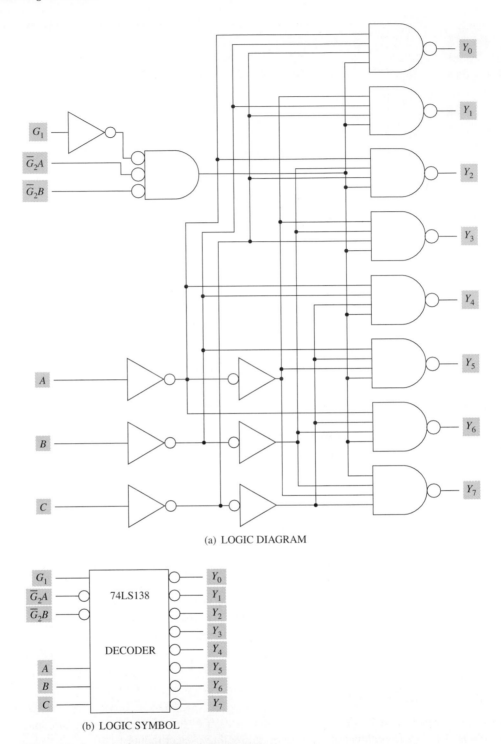

(a) LOGIC DIAGRAM

(b) LOGIC SYMBOL

FIGURE 10-4   Decoder: 74LS138–3-line to 8-line.

put NAND gates are enabled. Any high input to the NOR gate will inhibit the output NAND gates. Note the G_1 input is NOTed prior to its application to the NOR gate. The decoder will activate one line corresponding to the select inputs (CBA) once it is enabled. The function table (Table 10–3) verifies the decoder's operation. This type of decoder is known as a **1-of-n decoder.** It should be evident that the name is derived from the fact that the decoder activates only one of its output lines for a given input number or code.

The circuit shown in Fig. 10–5 consists of two 74LS138s connected as a 1-of-16 decoder. No external control circuitry is required. The circuit could be further expanded to

TABLE 10-3	Inputs						Outputs							
74LS138 Decoder Function Table	Enable			Select										
	G_1	$\overline{G_2}A$	$\overline{G_2}B$	C	B	A	Y_7	Y_6	Y_5	Y_4	Y_3	Y_2	Y_1	Y_0
	0	X	X	X	X	X	1	1	1	1	1	1	1	1
	X	1	X	X	X	X	1	1	1	1	1	1	1	1
	X	X	1	X	X	X	1	1	1	1	1	1	1	1
	1	0	0	0	0	0	1	1	1	1	1	1	1	0
	1	0	0	0	0	1	1	1	1	1	1	1	0	1
	1	0	0	0	1	0	1	1	1	1	1	0	1	1
	1	0	0	0	1	1	1	1	1	1	0	1	1	1
	1	0	0	1	0	0	1	1	1	0	1	1	1	1
	1	0	0	1	0	1	1	1	0	1	1	1	1	1
	1	0	0	1	1	0	1	0	1	1	1	1	1	1
	1	0	0	1	1	1	0	1	1	1	1	1	1	1

a 1-of-24 decoder without additional circuitry. A 1-of-32 decoder expansion would require one NOT gate.

The count sequence required to cycle the decoder through all sixteen input combinations is

D	C	B	A
0	0	0	0
0	0	0	1
0	0	1	0
0	0	1	1
0	1	0	0
0	1	0	1
0	1	1	0
0	1	1	1
1	0	0	0
1	0	0	1
1	0	1	0
1	0	1	1
1	1	0	0
1	1	0	1
1	1	1	0
1	1	1	1

The D input is used to enable the proper decoder. The counts of 0000 through 0111 (D–A) result in the lower-order decoder being enabled while the higher-order decoder is disabled. The reverse condition is true for the counts of 1000 through 1111.

If the input to the circuit is $D = 1$, $C = 1$, $B = 0$, and $A = 0$, which decoder is enabled and which output line is active? The $D = 1$ input enables the higher-order decoder because this input is connected to the G_1 enable input and $\overline{G_2}A$ and $\overline{G_2}B$ are both tied to ground. The $\overline{G_2}A$ (grounded) and G_1 (tied high) enable input pins are permanently enabled by hard-wire connections on the lower-order decoder. However, the $D = 1$ input disables the lower-order decoder because it is connected to the $\overline{G_2}B$ input pin.

The higher-order decoder will activate the Y_4 output line because the decoder is enabled, and the select input lines are 100 (C–A). This would illuminate LED 12 because its cathode is taken low by the decoder's output. All of the other decoder output lines will stay high and their corresponding LEDs will be off. An input of 0101 to the 1-of-16 decoder in Fig. 10–5 will disable the higher-order decoder, enable the lower-order decoder, and illuminate LED 5.

Each LED in Fig. 10–5 is in series with a **current-limiting resistor.** The resistors are required to protect the LEDs. An LED typically drops about 2 V when it is illuminated.

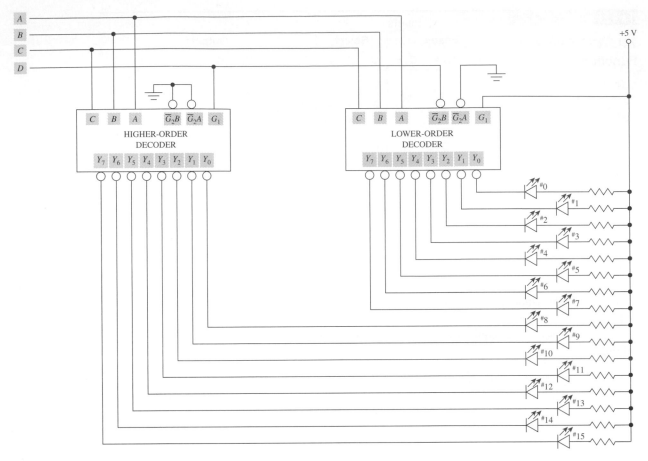

FIGURE 10–5   Decoder: 1-of-16 implemented with 74LS138 ICs.

The current limiting resistor must drop 2.7 V if the active-low output level from the decoders is assumed to be approximately 0.3 V. The maximum forward current for an LED is typically 25 to 30 mA. Therefore, the circuit can be designed safely for an LED current of 10 to 20 mA.

$$R_S = 2.7\ \text{V}/10\ \text{mA} = 270\ \Omega\ @\ 10\ \text{mA}$$
$$R_S = 2.7\ \text{V}/20\ \text{mA} = 135\ \Omega\ @\ 20\ \text{mA}$$

Current-limiting resistors of any of the following sizes will suffice and limit current to a safe value through the LEDs: 150 Ω, 180 Ω, 220 Ω, or 270 Ω. The two decoders shown in Fig. 10–5 are available in a single IC package—74154. The 74154 is a 4-line to 16-line decoder and is very popular in memory decoding circuits.

4-Line to 10-Line Decoder

Decoders in this group are called **BCD-to-decimal decoders.** This 1-of-10 decoder works in the same manner as those previously discussed. A BCD number $(0000–1001_{(BCD)})$ applied to the circuit's input will cause one output line to be activated. The activated line indicates which one of the ten possible input BCD numbers was applied to the input.

7442.

A 4-line BCD to 10-line decimal decoder is shown in Fig. 10–6. The logic diagram in Fig. 10–6(a) shows four input lines $(D–A)$ and ten active-low output lines $(O_9–O_0)$. It

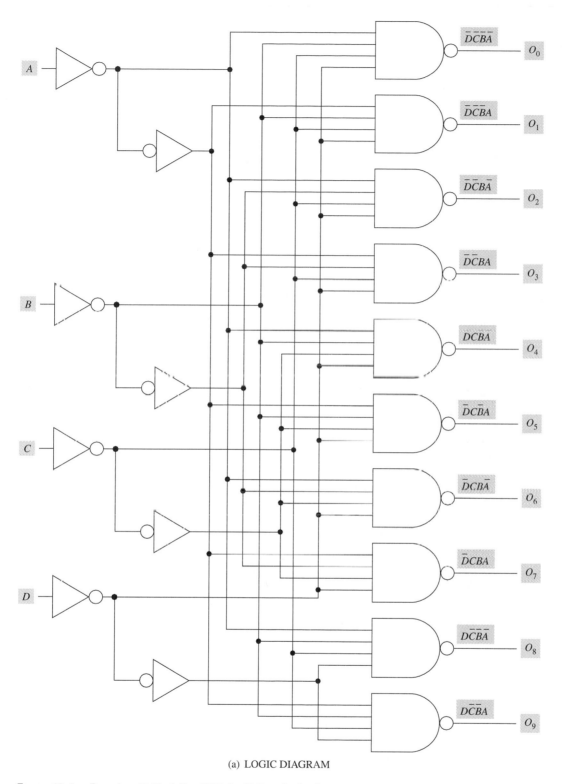

(a) LOGIC DIAGRAM

FIGURE 10–6   Decoder: 7442–4-line BCD to 10-line decimal.

would not be unusual to see the outputs labeled $\overline{O}_9$ through $\overline{O}_0$ because they are active-low outputs. The standard symbol in Fig. 10–6(b) uses the bubbles on the output lines; the ANSI/IEEE symbol in Fig. 10–6(c) uses the active-low qualifying symbols at the output.

If we assume an input BCD number 0111, tracing logic levels through the logic diagram in Fig. 10–6(a) will show output line 7 (O_7) goes active low. This exercise will

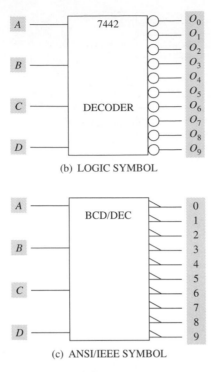

(b) LOGIC SYMBOL

(c) ANSI/IEEE SYMBOL

Figure 10-6 Continued

also show all of the other output NAND gates are inhibited, and their outputs are inactive (high). This information is shown in Table 10–4.

The function table also shows that no output lines are activated if the decoder input is an invalid number (1010–1111). For example, if the invalid number 1111 is applied to the input pins, tracing logic levels through the logic diagram proves this point. Every NAND gate in the decoder is inhibited by low inputs from the input NOT gates. Therefore, all of the decoder outputs will remain inactive (high).

		BCD Input							Outputs						
TABLE 10-4	#	D	C	B	A	O_9	O_8	O_7	O_6	O_5	O_4	O_3	O_2	O_1	O_0
7442 Decoder Function Table	0	0	0	0	0	1	1	1	1	1	1	1	1	1	0
	1	0	0	0	1	1	1	1	1	1	1	1	1	0	1
	2	0	0	1	0	1	1	1	1	1	1	1	0	1	1
	3	0	0	1	1	1	1	1	1	1	1	0	1	1	1
	4	0	1	0	0	1	1	1	1	1	0	1	1	1	1
	5	0	1	0	1	1	1	1	1	0	1	1	1	1	1
	6	0	1	1	0	1	1	1	0	1	1	1	1	1	1
	7	0	1	1	1	1	1	0	1	1	1	1	1	1	1
	8	1	0	0	0	1	0	1	1	1	1	1	1	1	1
	9	1	0	0	1	0	1	1	1	1	1	1	1	1	1
	I	1	0	1	0	1	1	1	1	1	1	1	1	1	1
	N	1	0	1	1	1	1	1	1	1	1	1	1	1	1
	V	1	1	0	0	1	1	1	1	1	1	1	1	1	1
	A	1	1	0	1	1	1	1	1	1	1	1	1	1	1
	L	1	1	1	0	1	1	1	1	1	1	1	1	1	1
	I	1	1	1	1	1	1	1	1	1	1	1	1	1	1
	D														

BCD-to-7-Segment Decoders

The requirement for decoders with capabilities other than the 1-of-n type comes from devices such as numeric LED displays. These displays come in two basic types, which mandates at least two different types of decoders. The basic requirements of these indicators will be presented prior to the decoder circuits that drive them. These numeric LED displays are commonly called 7-segment indicators. The outline drawing in Fig. 10-7 shows the seven segments. The letter designations for the segments are standard and easy to commit to memory.

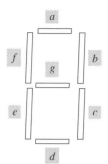

FIGURE 10-7   Seven-segment indicator outline drawing.

Common-Anode 7-Segment Indicator.

Each of the seven segments in this display unit is an LED. The common-anode 7-segment indicator is shown in Fig. 10-8(a). Each segment (a–g) is shown in the diagram. The type of indicator, **common-anode,** means the anodes of the LEDs are all connected to V_{CC}. There are usually two pins on a common-anode 7-segment indicator that require external connections to V_{CC}. The segments are illustrated again in the alternate diagram in Fig. 10-8(b). The anodes of all of the segments are internally connected together.

The indicator displays for each input BCD combination to the decoder are shown in Fig. 10-8(c). The standard BCD-to-7-segment decoder will generate outputs for invalid BCD inputs as shown in the figure. Sometimes these symbols for invalid inputs are used when testing or troubleshooting a system.

A logic circuit containing a BCD counter, BCD-to-7-segment decoder/driver, and a common-anode 7-segment indicator is shown in Fig. 10-9. The decoder circuit must produce active-low outputs to illuminate segments in the 7-segment indicator because this indicator is a common-anode type.

If the count of the BCD counter is 0001 as illustrated in Fig. 10-9, the b and c segments of the indicator will illuminate. This can be accomplished if the b and c outputs of the decoder are low and the remaining outputs are high.

The common-anode display 7-segment codes are shown in Table 10-5. The previous example indicated the b and c outputs of the decoder must be low to illuminate a 1 on the indicator. This can be verified in the table. For any given BCD count in the table, a low will illuminate the indicated segment and a high will keep the segment off.

7446.

This BCD-to-7-segment decoder/driver's logic diagram is shown in Fig. 10-10. The 7446 is designed to drive common-anode LEDs or incandescent indicators. The decoder will produce the display patterns of Fig. 10-8(c) on the 7-segment indicator.

The function table for the 7446 is shown in Table 10-6. The table has been expanded when compared to Table 10-5. The 7446 table includes invalid BCD input numbers as well as some special-function inputs. Operation of the decoder is basic and similar to those previously discussed except more than one output line must be taken low to illuminate multiple segments. The generic name **code converter** for these devices was derived from this change of one code (BCD) to another (segment code). The special function inputs/output of the decoder are discussed in the following paragraphs.

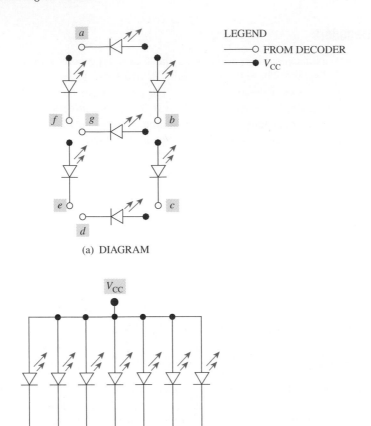

(a) DIAGRAM

LEGEND

——○ FROM DECODER

——● V_{CC}

(b) ALTERNATE DIAGRAM

(c) DISPLAYS

FIGURE 10-8   **Common-anode 7-segment indicator.**

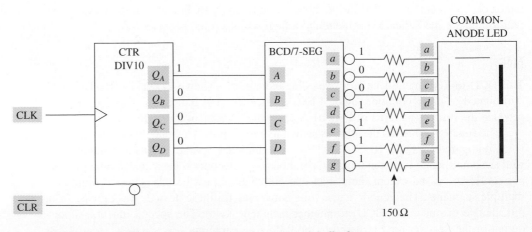

FIGURE 10-9   **BCD counter, decoder, and 7-segment indicator.**

TABLE 10-5		Outputs						
Seven-Segment Codes for Common-Anode Display	Digit	a	b	c	d	e	f	g
	0	0	0	0	0	0	0	1
	1	1	0	0	1	1	1	1
	2	0	0	1	0	0	1	0
	3	0	0	0	0	1	1	0
	4	1	0	0	1	1	0	0
	5	0	1	0	0	1	0	0
	6	0	1	0	0	0	0	0
	7	0	0	0	1	1	1	1
	8	0	0	0	0	0	0	0
	9	0	0	0	0	1	0	0

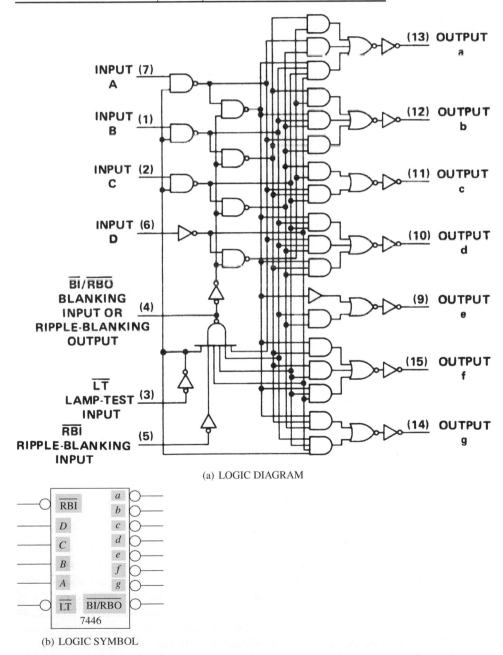

(a) LOGIC DIAGRAM

(b) LOGIC SYMBOL

FIGURE 10–10   Decoder: 7446—BCD-to-7-segment for common-anode displays. *(a) Reprinted by Permission of Texas Instruments.*

The following legend will aid understanding of the discussion of the 7446.

$\overline{BI}$ = Active-low *blanking input*
$\overline{RBI}$ = Active-low *ripple blanking input*
$\overline{RBO}$ = Active-low *ripple blanking output*
$\overline{LT}$ = Active-low *lamp test*

The function table (Table 10–6) shows all segment indicators will be off when the blanking input ($\overline{BI}$) is activated.

Ripple blanking signals are used in the display unit to suppress leading zeros when several decoders are used to drive multiple 7-segment indicators. **Leading zero suppression** is a method employed to blank out (suppress) the leading (most-significant) zero (zeros) in a number. If six decoders are connected to drive six displays and the display

TABLE 10-6 **7446 Decoder Function Table**	Decimal or Function	$\overline{LT}$	$\overline{RBI}$	D	C	B	A	$\overline{BI}$ / $\overline{RBO}$	a	b	c	d	e	f	g
	0	1	1	0	0	0	0	1	ON	ON	ON	ON	ON	ON	OFF
	1	1	X	0	0	0	1	1	OFF	ON	ON	OFF	OFF	OFF	OFF
	2	1	X	0	0	1	0	1	ON	ON	OFF	ON	ON	OFF	ON
	3	1	X	0	0	1	1	1	ON	ON	ON	ON	OFF	OFF	ON
	4	1	X	0	1	0	0	1	OFF	ON	ON	OFF	OFF	ON	ON
	5	1	X	0	1	0	1	1	ON	OFF	ON	ON	OFF	ON	ON
	6	1	X	0	1	1	0	1	ON	OFF	ON	ON	ON	ON	ON
	7	1	X	0	1	1	1	1	ON	ON	ON	OFF	OFF	OFF	OFF
	8	1	X	1	0	0	0	1	ON	ON	ON	ON	ON	ON	ON
	9	1	X	1	0	0	1	1	ON	ON	ON	ON	OFF	ON	ON
	10	1	X	1	0	1	0	1	OFF	OFF	OFF	ON	ON	OFF	ON
	11	1	X	1	0	1	1	1	OFF	OFF	ON	ON	OFF	OFF	ON
	12	1	X	1	1	0	0	1	OFF	ON	OFF	OFF	OFF	ON	ON
	13	1	X	1	1	0	1	1	ON	OFF	OFF	ON	OFF	ON	ON
	14	1	X	1	1	1	0	1	OFF	OFF	OFF	ON	ON	ON	ON
	15	1	X	1	1	1	1	1	OFF	OFF	OFF	OFF	OFF	OFF	OFF
	BI	X	X	X	X	X	X	0	OFF	OFF	OFF	OFF	OFF	OFF	OFF
	RBI	1	0	0	0	0	0	0	OFF	OFF	OFF	OFF	OFF	OFF	OFF
	LT	0	X	X	X	X	X	1	ON	ON	ON	ON	ON	ON	ON

number is 011146, the actual display is 11146 when the leading zero is suppressed. In the same manner, if the 6-digit display number is 001146, the display is actually 1146 with both leading zeros suppressed. To accomplish this suppression, $\overline{RBI}$ = 0, $\overline{LT}$ = 1, and $DCBA$ = 0000 to the MSD decoder when the count input is 011146. Table 10–6 shows when the preceding input conditions exist, all MSD segments will be off and $\overline{RBO}$ will go low. The zero will not be blanked (suppressed) if $\overline{RBI}$ is high or if it is left disconnected.

Leading zero suppression is accomplished in multidigit displays by connecting $\overline{RBI}$ of the MSD display decoder low and connecting its $\overline{RBO}$ to the $\overline{RBI}$ of the next lesser significant digit. Continuous suppression of leading zeros can be accomplished by connecting $\overline{RBO}$ to $\overline{RBI}$ of the next lesser significant digit decoder, down to the LSD.

If the blanking input ($\overline{BI}$) to a decoder is low, all display segments are off. If it is left open or has a high applied, the segments in the indicator are functional. If the $\overline{BI}$ is open or high and lamp test ($\overline{\text{LT}}$) is taken low, all display segments will illuminate. The $\overline{\text{LT}}$ input is used when testing or troubleshooting to identify defective display segments. The 7446 features **open-collector outputs.** This feature will be discussed in detail in Chapter 11.

Common-Cathode 7-Segment Indicator.

The common-cathode 7-segment indicator is shown in Fig. 10–11(a). The legend in this figure shows the cathodes of all seven segments are grounded. There are normally two pins on the indicator unit that require connections to circuit ground. The cathodes are internally connected together as depicted in the alternate diagram in Fig. 10–11(b).

The 7-segment code from the decoder is connected to the anodes of the segments (a–g). This indicates the outputs of the decoder must go high to illuminate a segment. Since the decoder outputs must go high instead of low as they did for the common-anode display, the 7-segment codes (a–g) are exactly opposite those shown in Table 10–5.

7448:

Figure 10–12 illustrates the 7448 BCD-to-7-segment decoder/driver's logic diagram and logic symbol. This IC features active-high outputs. The outputs can be used to drive common-cathode LEDs.

The 7446 decoder function table presented in Table 10–6 is applicable to the 7448. Keep in mind that the 7446 produces a low output to illuminate a certain segment of the common-anode display. On the other hand, the 7448 produces a high output to illuminate a certain segment of the common-cathode display. The reason the function table can be

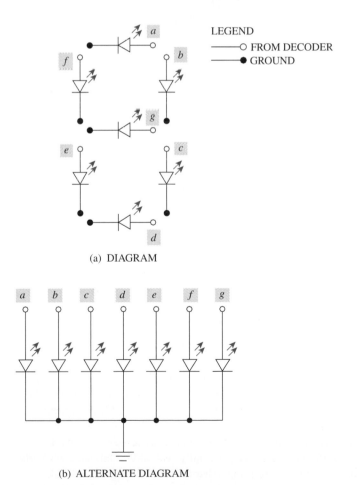

LEGEND
————○ FROM DECODER
————● GROUND

(a) DIAGRAM

(b) ALTERNATE DIAGRAM

FIGURE 10–11   **Common-cathode 7-segment indicator.**

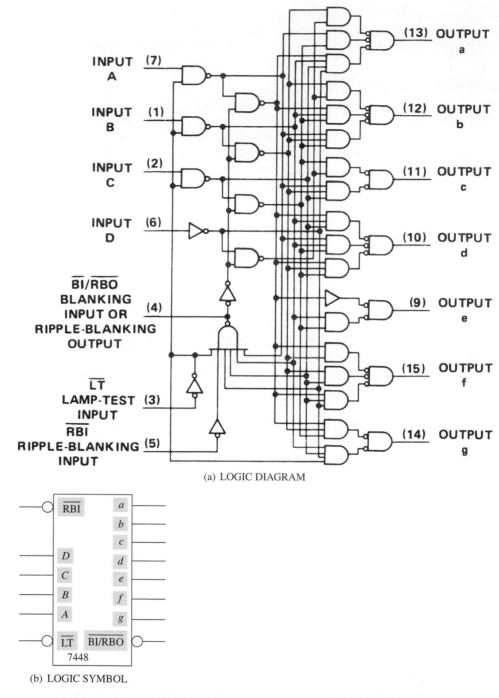

(a) LOGIC DIAGRAM

(b) LOGIC SYMBOL

FIGURE 10-12 **Decoder: 7448–BCD-to-7-segment for common-cathode displays. *(a) Reprinted by Permission of Texas Instruments.***

used for either decoder is that the *a* through *g* outputs in Table 10–6 are put in terms of a specific segment (*a–g*) being off or on. The special functions discussed for the 7446 decoder/driver are also incorporated in the 7448. These functions are **blanking, leading zero suppression,** and **lamp testing.**

Decoders with the capability to produce outputs to display all hexadecimal digits from a MOD-16 counter or other source are also commercially available. One such IC is the 9368 7-segment decoder/driver/latch. The numerical displays of 0 through 9 and A through F are shown in Fig. 10–13. Note the difference between these displays and the ones presented for the 7446 and 7448 in Fig. 10–8(c). The hexadecimal digits *b* and *d* are dis-

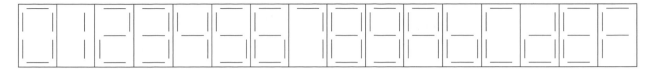

FIGURE 10-13   **Hexadecimal display–7-segment indicator.**

played in lowercase letters to differentiate them from the numbers 8 and 0 respectively. The remaining hex digits are displayed in uppercase letters.

The decoders presented in this section included two basic types. The first type detected a binary number/code and indicated the detected input by activating one of its output lines. The second type converted an input number to a code suitable for driving 7-segment indicators. There are also decoders that can be used to drive liquid crystal displays as well as produce characters for printers and cathode-ray tubes (CRTs).

Section 10-1: Review Questions

Answers are given at the end of the chapter.

A. Describe the two basic types of decoders.
B. If the input to a 1 of 4 decoder is $11_{(2)}$, which output is activated?
C. Which output line of a 74LS138 decoder (Fig. 10–4) is activated if its inputs are $A = B = C = \overline{G_2A} = \overline{G_2B} = 0$ and $G_1 = 1$?
D. Which output line of a 74LS138 decoder (Fig. 10–4) is activated if its inputs are $A = B = C = 1$ and $\overline{G_2A} = \overline{G_2B} = G_1 = 0$?
E. A BCD-to-decimal decoder is a _____-line to _____-line decoder.
F. What logic level output is required from a 7-segment decoder to illuminate a segment of a common-anode 7-segment indicator?
G. Define leading zero suppression.

SECTION 10-2:   ENCODERS

OBJECTIVES

1. Define encoding and state the function of encoder circuits.
2. Given an encoder logic diagram or logic symbol annotated with input levels, identify the circuit and determine its outputs.

Encoding is the process of producing a specific binary code or number. An **encoder** performs the exact opposite of the function accomplished by a basic decoder. Where a decoder detects a binary number and converts it to a single active output, an encoder detects an active-input line and converts it to a binary number or code at the output. A prime example of encoding is the use of an encoder with a keyboard to produce a specific binary code for each key depressed. The alphanumeric American Standard Code for Information Interchange (ASCII) presented in Chapter 2 is formed by an encoder from a 7-bit scan code when a key on a PC keyboard is depressed.

4-Line to 2-Line Encoder

A 4-line to 2-line encoder logic diagram and logic symbol are shown in Fig. 10–14. The logic symbol shows the inputs (I_0–I_3) are active-high inputs in this encoder. The principle of encoding is to activate an input line and convert that particular input to a code or bi-

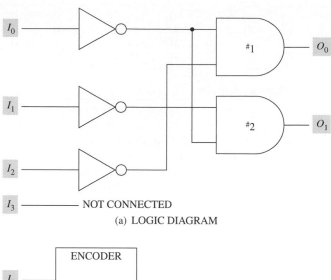

(a) LOGIC DIAGRAM

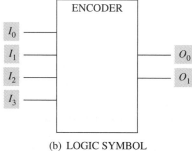

(b) LOGIC SYMBOL

FIGURE 10–14   Encoder: 4-line to 2-line.

nary number. If the I_1 input line in the figure is brought high while the remaining inputs are low, the output will be 01 (O_1O_0). This output binary number corresponds to the activated input. There are only four possible input combinations to this circuit provided only one input is activated at a time. The input combinations are shown in Table 10–7.

The logic diagram in Fig. 10–14(a) reveals some peculiarities about this particular encoder. The first one is that the I_3 input pin is not connected internally to the logic gates. This indicates the I_3 input is a don't care input to the circuit. A look at the logic diagram in Fig. 10–14(a) will reveal why I_3 is a don't care input. First, if the I_0 input is brought high, AND gates #1 and #2 are inhibited, and they produce output 00. If I_1 is activated, the I_1 input is NOTed and applied to AND gate #2. This inhibits the #2 gate. However, the NOTed I_0 and I_2 inputs produce a high out of AND gate #1. This results in the proper 01 output. When I_2 is brought high, the NOTed I_2 input inhibits AND gate #1, but the NOTed I_0 and I_1 inputs produce a high output from AND gate #2. This input combination produces the proper 10 output from the encoder. If I_3 is activated, the low inputs from I_0, I_1, and I_2 are inverted by the NOT gates, and these high inputs produce a 11 output. The results of this analysis are summarized in Table 10–7.

TABLE 10-7	Inputs				Outputs	
Four-Line to 2-Line Encoder Function Table	I_3	I_2	I_1	I_0	O_1	O_0
	0	0	0	1	0	0
	0	0	1	0	0	1
	0	1	0	0	1	0
	1	0	0	0	1	1

Another peculiarity about this encoder can be discerned from the logic diagram in Fig. 10–14(a). If more than one input is brought high (active) at the same time, the encoder's output is invalid. See if you can prove the encoder's output is 00 when $I_0 = I_1 = 1$. Remember, this condition represents two active inputs. The encoder produces a 00 output when I_1 and I_2 are activated simultaneously.

This problem is overcome with **priority encoders.** A priority encoder allows bringing more than one input active at the same time. The priority encoder responds only to the highest-order number input when more than one input is activated at the same time. This allows prioritizing inputs in a system. For example, if a #4 interrupt request in a digital system was in process and the higher priority #6 interrupt request input was also activated, the priority encoder's output would switch to the higher priority interrupt request.

The problem with priority encoders is they are a bit more difficult to design. Although off-the-shelf priority encoders are readily available at a cheap price, let's take a look at how they were designed to better understand how they work.

4-Line to 2-Line Priority Encoder

The 4-line to 2-line encoder previously discussed was designed with don't care outputs for all possible input combinations except 0001, 0010, 0100, and 1000. That active-high input encoder's outputs were invalid for all other input combinations.

The **priority encoder** must look at all possible input combinations and respond only to the highest numbered activated input. The function table for a 4-line to 2-line priority encoder is shown in Table 10–8. The table shows the desired output condition for both O_0 and O_1 for each of the 16 possible input combinations. The circuit outputs are based on active-low inputs to the encoder.

The first eight input combinations (0000–0111) in Table 10–8 all have an active-low I_3 input. Thus, the encoder should output 11 for all of these input combinations because that is the highest numbered active input. The next four sequential counts (1000–1011) dictate a 10 output because the I_2 input is the highest numbered active input for each of these states. The next two sequential counts (1100 and 1101) require a 01 output because the I_1 input is the highest activated input when 1100 is applied and the only activated input when 1101 is applied. The input combination of 1110 requires a 00 output because I_0 is the only active input. The final input combination (1111) outputs are don't care outputs because none of the encoders inputs is active.

TABLE 10–8	Inputs				Outputs	
Four-Line to 2-Line Priority Encoder Function Table	I_3	I_2	I_1	I_0	O_1	O_0
	0	0	0	0	1	1
	0	0	0	1	1	1
	0	0	1	0	1	1
	0	0	1	1	1	1
	0	1	0	0	1	1
	0	1	0	1	1	1
	0	1	1	0	1	1
	0	1	1	1	1	1
	1	0	0	0	1	0
	1	0	0	1	1	0
	1	0	1	0	1	0
	1	0	1	1	1	0
	1	1	0	0	0	1
	1	1	0	1	0	1
	1	1	1	0	0	0
	1	1	1	1	X	X

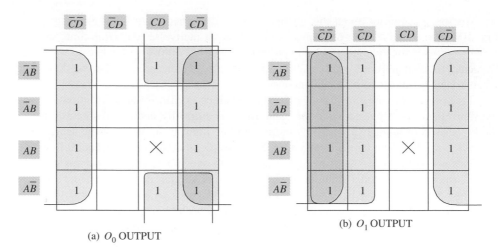

FIGURE 10–15   K-maps for 4-line to 2-line priority encoder.

It is relatively simple to design an encoder if Karnaugh maps are used to simplify the input combinations producing high outputs at O_0 and O_1. Two K-maps are shown in Fig. 10–15. One K-map is used to simplify the O_0 output, and the other is used to simplify the O_1 output. The standard K-map format of Chapter 4 is used.

The O_0 output in Fig. 10–15(a) simplifies to $\overline{B}C + \overline{D}$. The octet circled on the map eliminates the complementary As, Bs, and Cs, and leaves the $\overline{D}$ expression. The quad looped from the lower-right pair to the upper-right pair yields $\overline{B}C$.

The O_1 output in Fig. 10–15(b) reduces to $\overline{C} + \overline{D}$. This expression is a result of the two octet loops on the map. The equations derived from the two maps are

$$O_o = \overline{B}C + \overline{D}$$
$$O_1 = \overline{C} + \overline{D}$$

Implementation of these two simplified expressions is shown in Fig. 10–16(a). The circuit is redrawn in Fig. 10–16(b) using the I_0–I_3 input mnemonics instead of the A–D inputs of the K-maps. Since input A was not part of either output expression, the I_0 input to the priority encoder is not connected to the logic gates.

8-Line to 3-Line Priority Encoder

An 8-line to 3-line priority encoder is used for priority decoding of the highest-order active input of the eight input lines. Once the highest-order input is detected, the data are output on three lines (4–2–1) in Binary-Coded Octal (BCO).

The primary difference between the generic encoders just presented and some of the commercially available ICs is the enable input and output pins, which allow cascading of the encoders without external circuitry. These encoders are often referred to as **n-bit encoders** because they have the ability to take one active-input line and encode it to n bits.

74148.

This 8-line to 3-line priority encoder logic symbol is illustrated in Fig. 10–17. The logic diagram is available in most data books, but it is unnecessary for the following discussion. The symbol shows the inputs to the encoder are active-low and the binary-coded-octal output appears in inverted form.

The function table for this priority encoder appears in Table 10–9. Notice when the encoder is enabled ($\overline{EI} = 0$) and the $\overline{7}$ input line is active, the output is 000 ($\overline{A}_2 - \overline{A}_0$). This is the *complement* of the desired 111 BCO output. Also note when the $\overline{7}$ input is activated that all of the remaining inputs ($\overline{6}-\overline{0}$) are don't care inputs because $\overline{7}$ is the highest-

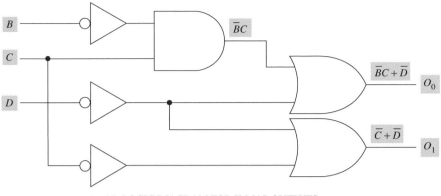

(a) LOGIC DIAGRAM FOR K-MAP OUTPUTS

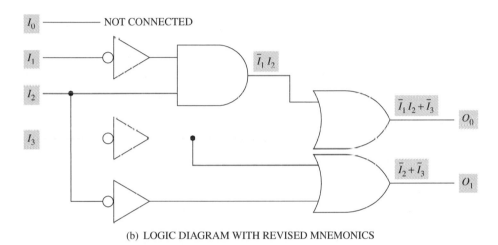

(b) LOGIC DIAGRAM WITH REVISED MNEMONICS

FIGURE 10–16   Priority encoder: 4 line to 2 line.

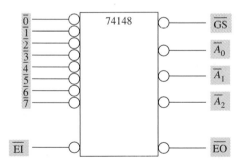

FIGURE 10–17   Priority encoder: 74148–8-line to 3-line logic symbol.

order input line. If the $\overline{5}$ input is activated, the $\overline{6}$ and $\overline{7}$ inputs must be inactive for the output to be an inverted $101_{(BCO)}$.

The table, as well as the logic symbol in Fig. 10–17, shows an enable input $(\overline{EI})$ that must be low for the encoder to function. In addition to the inverted BCO output at $\overline{A}_2$, $\overline{A}_1$, and $\overline{A}_0$, two other outputs are available from the 74148. The $\overline{EO}$ output is active (low) when the encoder is enabled and *all* input lines are inactive. The $\overline{GS}$ output is active when the encoder is enabled and *any* of the inputs is active. An active $\overline{GS}$ output indicates a valid input/output code.

A 4-bit encoder (16-line to 4-line) is implemented in Fig. 10–18 by cascading two 74148 ICs. The active-low $\overline{EO}$ and $\overline{GS}$ pins of the encoders allow expansion with no external circuitry. The lower-order encoder is permanently enabled because $\overline{EI}$ is grounded. The higher-

TABLE 10-9	Inputs									Outputs				
74148 Priority Encoder Function Table	$\overline{EI}$	$\overline{0}$	$\overline{1}$	$\overline{2}$	$\overline{3}$	$\overline{4}$	$\overline{5}$	$\overline{6}$	$\overline{7}$	$\overline{A}_2$	$\overline{A}_1$	$\overline{A}_0$	$\overline{GS}$	$\overline{EO}$
	1	X	X	X	X	X	X	X	X	1	1	1	1	1
	0	1	1	1	1	1	1	1	1	1	1	1	1	0
	0	X	X	X	X	X	X	X	0	0	0	0	0	1
	0	X	X	X	X	X	X	0	1	0	0	1	0	1
	0	X	X	X	X	X	0	1	1	0	1	0	0	1
	0	X	X	X	X	0	1	1	1	0	1	1	0	1
	0	X	X	X	0	1	1	1	1	1	0	0	0	1
	0	X	X	0	1	1	1	1	1	1	0	1	0	1
	0	X	0	1	1	1	1	1	1	1	1	0	0	1
	0	0	1	1	1	1	1	1	1	1	1	1	0	1

order encoder is disabled if any of the input pins of the lower-order encoder are active. For example if the $\overline{6}$ input is activated, the lower-order encoder will produce a 001 inverted BCO output at $\overline{A}_2-\overline{A}_0$. The $\overline{EO}$ pin will be high when this occurs. The $\overline{EO} = 1$ disables the higher-order encoder. The $\overline{GS}$ output of the higher-order encoder will be high when this encoder is

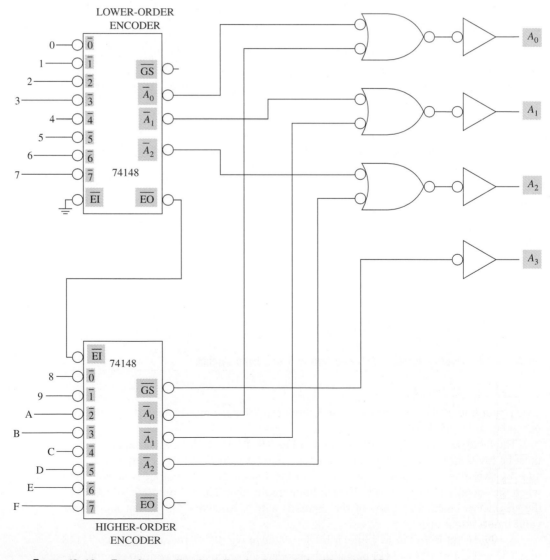

FIGURE 10–18   Encoder: 16-line to 4-line implemented with 74148 ICs.

disabled. Thus, the circuit output will be 1001. This output indicates the $\overline{6}$ line of the lower-order encoder is activated. The output becomes 0110 out of the NOT gates.

10-Line to 4-Line Priority Encoder

A 10-line decimal to 4-line BCD encoder is used for priority encoding of an input, typically from a keypad, to binary-coded decimal (BCD). This circuit may be referred to as a 4-bit encoder.

74147.

Figure 10–19 shows the logic symbols of this decimal-to-BCD encoder. The 74147 IC is similar to the 74148 without the enable input and outputs for cascading. The standard symbol in Fig. 10–19(a) shows nine active-low inputs. There is no input line for the $\overline{0}$ input. This condition is implied because zero is encoded when all inputs are inactive.

The 74147 produces the *complement* of the desired BCD output. This is shown in Table 10–10. If the $\overline{6}$ input line is the highest active input, the output is 1001. This output is the complement of $0110_{(BCD)}$. If none of the input lines is active, the output is 1111—complement of 0000.

The ANSI/IEEE symbol is shown in Fig. 10–19(b). HPRI represents the highest-priority input. Therefore, HPRI/BCD indicates the highest-priority active input will be encoded to an active-low BCD output—$\overline{BCD}$.

An application of a decimal-to-BCD encoder (74147) and a BCD-to-decimal decoder (7442) is illustrated in Fig. 10–20. This circuit allows transmission of data from the transmitting encoder to the receiving decoder using only four data lines. The pull-up resistors R_P are used to ensure the logic levels applied to the encoder are high ($+5$ V) when the switches are open. The note indicates a pull-up resistor must be used on each input line to the encoder.

If the #7 switch in Fig. 10–20 is depressed, a low on that encoder input will result in a 1000 output from the encoder. This data is inverted (0111), and the BCD number is

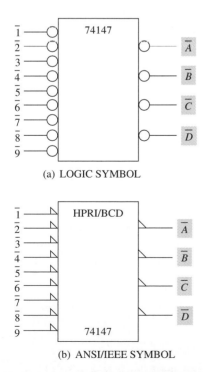

(a) LOGIC SYMBOL

(b) ANSI/IEEE SYMBOL

Figure 10-19   Priority encoder: 74147–10-line to 4-line.

TABLE 10-10	Inputs									Outputs			
74147 Priority Encoder Function Table	$\overline{1}$	$\overline{2}$	$\overline{3}$	$\overline{4}$	$\overline{5}$	$\overline{6}$	$\overline{7}$	$\overline{8}$	$\overline{9}$	$\overline{D}$	$\overline{C}$	$\overline{B}$	$\overline{A}$
	1	1	1	1	1	1	1	1	1	1	1	1	1
	X	X	X	X	X	X	X	X	0	0	1	1	0
	X	X	X	X	X	X	X	0	1	0	1	1	1
	X	X	X	X	X	X	0	1	1	1	0	0	0
	X	X	X	X	X	0	1	1	1	1	0	0	1
	X	X	X	X	0	1	1	1	1	1	0	1	0
	X	X	X	0	1	1	1	1	1	1	0	1	1
	X	X	0	1	1	1	1	1	1	1	1	0	0
	X	0	1	1	1	1	1	1	1	1	1	0	1
	0	1	1	1	1	1	1	1	1	1	1	1	0

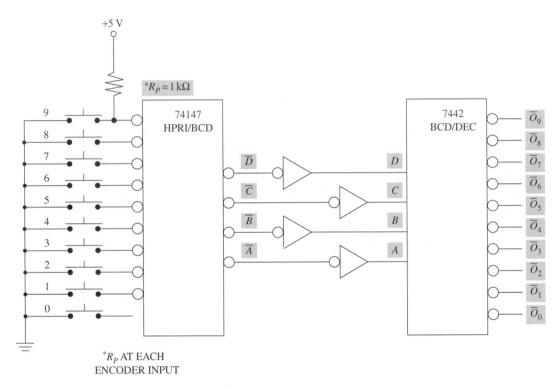

*R_p AT EACH
ENCODER INPUT

FIGURE 10-20   Data transfer via encoder and decoder.

applied to the decoder. The decoder activates the $\overline{O}_7$ output line. This active output indicates the #7 switch was closed at the input of the circuit.

Since the encoder and decoder perform exact opposite functions, the logic of this circuit should be evident. The #7 switch (key) was depressed, bringing a low to the input of the circuit. The end result is that low is routed to the $\overline{O}_7$ output of the decoder. However, the circuit reduced the number of lines required to transmit the data from ten to four.

Section 10-2: Review Questions

Answers are given at the end of the chapter.

A. State the function of an encoder.
B . What is the output of the encoder in Fig. 10–14(a) if the I_3 input is activated?

C. A priority encoder responds to the _____-order input.
   (1) lowest
   (2) highest

D. What is the output of the 74148 priority encoder in Fig. 10–17 if $\overline{EI} = 0$ and $\overline{7} = 0$?

E. What is the level of the $\overline{EO}$ of a 74148 encoder with the input conditions stated in question (D)?

F. The 74147 decimal-to-BCD encoder has nine input lines, yet there are ten valid decimal inputs (0–9). Explain why?

SECTIONS 10-1 AND 10-2: INTERNAL SUMMARY

The decoder and encoder ICs presented in the first two sections of this chapter fall in the **medium-scale integration (MSI)** IC classification.

One type of **decoder** is used to detect a binary code/number at its input and activate one corresponding output. Another type of decoder is the **code converter.** This circuit detects a binary code or number at its input and converts it to a different code or format.

The first type of decoder is often called a **1-of-n decoder** because of the nature of its operation. The output line of this decoder corresponding to a binary number input is activated when the number is applied. The output lines of these decoders are usually **active low.**

Most of the decoder ICs contain one or more enable inputs for use in circuit expansion. These inputs eliminate or reduce the control circuitry required to cascade two or more decoders.

BCD-to-decimal (1-of-10) decoders and **BCD-to-7-segment decoders** are popular in circuits employing binary-coded decimal. Care must be taken when selecting a BCD-to-7-segment decoder/driver circuit. One type of decoder/driver produces active-low outputs to illuminate segments of a **common-anode 7-segment indicator.** The other type produces active-high outputs to illuminate segments of a **common-cathode 7-segment indicator.** BCD-to-7-segment decoder ICs often offer special features such as **blanking, lamp testing,** and **leading zero suppression.**

Encoders are used to detect an active-input line and convert it to a binary number or code at the output pins. They are often used to convert a numeric keypad entry to a binary number or code. Encoders perform the opposite function performed by their decoder counterparts.

Priority encoders respond to the highest activated input number when more than one input is activated simultaneously. An 8-line to 3-line priority encoder can encode an active input to a binary-coded-octal output. Outputs of some of the available encoder ICs appear in **inverted** form. Some of the encoder ICs provide an enable input and output pins for cascading. The use of a data book function table will always aid your understanding of the operation of these circuits.

SECTIONS 10-1 AND 10-2: INTERNAL SUMMARY QUESTIONS

Answers are given at the end of the chapter.

Note: The use of manufacturer's data sheets is recommended to assist you in answering these summary questions.

1. An IC that contains fifty (50) logic gates or circuitry of the same complexity is in the _____ category.
   a. SSI   c. LSI
   b. MSI   d. VLSI

2. Decoders are used to
   a. Select specific input data and route it to the output.
   b. Distribute input data from one input line to several output lines.
   c. Detect a binary code/number at the input and convert it to one active output.
   d. Detect an active-input line and convert it to a binary code/number at the outputs.

3. Which output line of the 74LS138 shown in Fig. 10–21 is active if the inputs are
   $A = 0, B = 1, C = 1, G_1 = 1, \overline{G_2A} = 0$, and $\overline{G_2B} = 0$?
   a. 0   c. 6
   b. 3   d. None

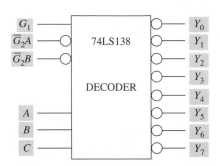

FIGURE 10-21

4. Which output line of the 74LS138 shown in Fig. 10–21 is active if the inputs are
   $A = 0, B = 0, C = 0, G_1 = 0, \overline{G_2A} = 0$, and $\overline{G_2B} = 0$?
   a. 0   c. 6
   b. 3   d. None

5. Which output line of the 74LS138 shown in Fig. 10–21 is active if the inputs are
   $A = 0, B = 0, C = 0, G_1 = 1, \overline{G_2A} = 0$, and $\overline{G_2B} = 0$?
   a. 0   c. 6
   b. 4   d. None

6. Which output line(s) of the 74LS139 in Fig. 10–22 is (are) active if the inputs are
   $1\overline{G} = 0, 1A = 0, 1B = 0, 2\overline{G} = 1, 2A = 0$, and $2B = 0$?
   a. $1Y_0$ c. $2Y_0$
   b. $1Y_2$ d. $1Y_0$ and $2Y_0$

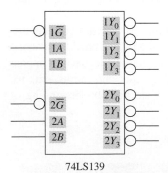

74LS139

FIGURE 10-22

7. The circuit shown in Fig. 10–23 has active- _____ outputs.
   a. low
   b. high

8. What is the output of the circuit in Fig. 10–23 if $A = 1, B = 1, C = 1$, and $D = 0$?
   a. All outputs $= 0$.
   b. All outputs $= 1$.
   c. $O_7 = 0$; all other outputs $= 1$.
   d. $O_7 = 1$; all other outputs $= 0$.

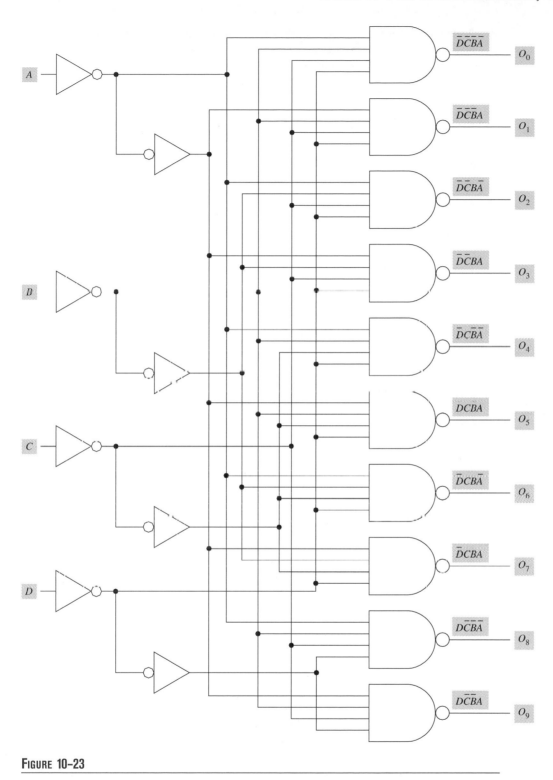

FIGURE 10–23

9. What is the output of the circuit in Fig. 10–23 if
   $A = 0, B = 1, C = 0,$ and $D = 1$?
   a. All outputs = 0.
   b. All outputs = 1.
   c. $O_5 = O$; all other outputs = 1.
   d. $O_5 = 1$; all other outputs = 0.

10. A BCD-to-7-segment decoder with active-low outputs is designed to drive a common-_____ 7-segment indicator.
    a. anode   b. cathode

11. What is the condition of the segments in a 7-segment indicator if the decoder blanking input is active?
    a. All segments are on.
    b. All segments are off.
    c. All zeros are suppressed.
    d. Leading zeros are suppressed.

12. An encoder is used to
    a. Select specific input data and route it to the output.
    b. Distribute input data from one input line to several output lines.
    c. Detect a binary code/number at the input and convert it to one active output.
    d. Detect an active-input line and convert it to a binary code/number at the output.

13. What is the output of the circuit shown in Fig. 10–24 if
    $I_0 = 1, I_2 = 0, I_3 = 0$, and $I_4 = 0$?
    a. 00   c. 10
    b. 01   d. 11

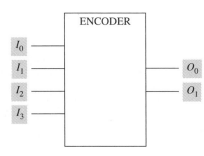

FIGURE 10-24

14. A priority encoder responds to
    a. no activated inputs
    b. all activated inputs
    c. lowest-order active input
    d. highest-order active input

15. What is the output of the circuit shown in Fig. 10–25 if
    $\overline{EI} = 0, \overline{5} = 0$, and all other inputs $= 1$?

	$\overline{A_2}$	$\overline{A_1}$	$\overline{A_0}$
a.	0	0	0
b.	0	1	0
c.	1	0	1
d.	1	1	1

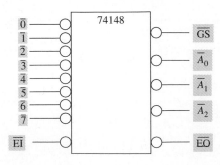

FIGURE 10-25

16. What is the output of the circuit shown in Fig. 10–25 if $\overline{EI} = 0, \overline{5} = 0, \overline{7} = 0$, and all other inputs $= 1$?

$\overline{A_2}$	$\overline{A_1}$	$\overline{A_0}$
a. 0	0	0
b. 1	1	1
c. 0	1	0
d. 1	0	1

17. What are the levels of $\overline{EO}$ and $\overline{GS}$ of the circuit shown in Fig. 10–25 if $\overline{EI} = 0, \overline{0} = 0$, and $\overline{1} - \overline{7} = 1$?

$\overline{EO}$	$\overline{GS}$
a. 0	0
b. 0	1
c. 1	0
d. 1	1

18. The circuit shown in Fig. 10–26 is a _____.
    a. BCD-to-decimal decoder
    b. BCD-to-decimal encoder
    c. decimal-to-BCD decoder
    d. decimal-to-BCD encoder

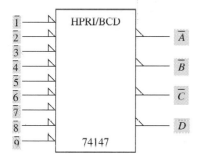

FIGURE 10–26

SECTION 10–3: MULTIPLEXERS

OBJECTIVES

1. Define multiplexing and identify multiplexer logic circuits and logic symbols.
2. Given a multiplexer logic diagram or logic symbol and input conditions, determine the output.

Multiplexing is the process of selecting data. Multiplexers are usually called **data selectors.** This generic name comes from the function these ICs perform in digital systems. Data selection is accomplished in Fig. 10–27 with a two-position switch. Input data A is fed to the output in the figure. Input data B is applied to the output if the switch

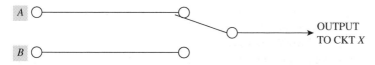

FIGURE 10–27   Multiplexing analogy.

is toggled to the other position. A multiplexer accomplishes this selecting task electronically.

A multiplexer **(MUX)** can be used to expand data input points. The one-line input to circuit X in Fig. 10–27 has been expanded to a two-input capability with the switch (MUX). Multiplexing is often used in PCs to address random-access memory (RAM) chips. Memory locations are laid out within chips in rows and columns. Multiplexers are used to extract the row address from the address bus and then the column address. A detailed example of this process is provided in Chapter 12.

Multiplexing is also used in telephone circuitry to select transmitted data so one line can be used for multiple conversations. Multiplexing in this sense may be thought of as "time-sharing." In addition, a multiplexer can be used to generate logic functions in a sum-of-product form. Logic function generation using a MUX is explained in Section 10–6. Finally, multiplexing can be used for **data conversion.** A MUX can be used to convert parallel data to serial data.

2-Line to 1-Line Multiplexers

Figure 10–28 illustrates a basic 2-line to 1-line multiplexer. The circuit is nothing more than a sum-of-products circuit. The **select input** (S) controls which of the AND gates is enabled.

If $S = 0$, the data at input A is routed to the MUX output. The lower AND gate in Fig. 10–28 is inhibited when $S = 0$. The output of this inhibited gate enables the OR gate. If $S = 0$ and $A = 1$, the two high inputs to the top AND gate produce a high output that is routed to the output of the MUX. If $S = 1$, the data at input B is routed to the MUX output. Regardless of the level of the select input, one of the data inputs is applied as true data (no phase inversion) to the MUX output. The output expression for this MUX is $A\overline{S} + BS$.

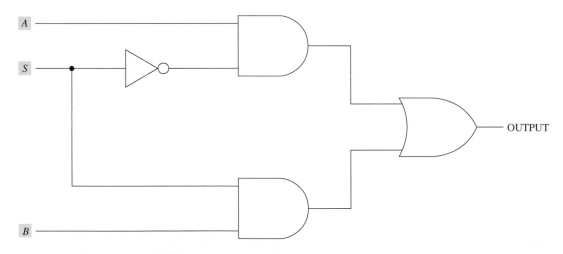

FIGURE 10–28   Multiplexer: 2-line to 1-line logic diagram.

74LS157.

The 74LS157 is a quadruple 2-line to 1-line data selector/multiplexer. The logic diagram and logic symbols for the MUX are shown in Fig. 10–29. The circuit is merely four of the 2-line to 1-line data selectors already discussed.

The logic diagram in Fig. 10–29(a) shows a **strobe input** $(\overline{G})$. This input is used to enable the MUX when the input is low and disable the MUX when it is high. The MUX output stays low when the circuit is not enabled. The $\overline{G}$ input allows for cascading.

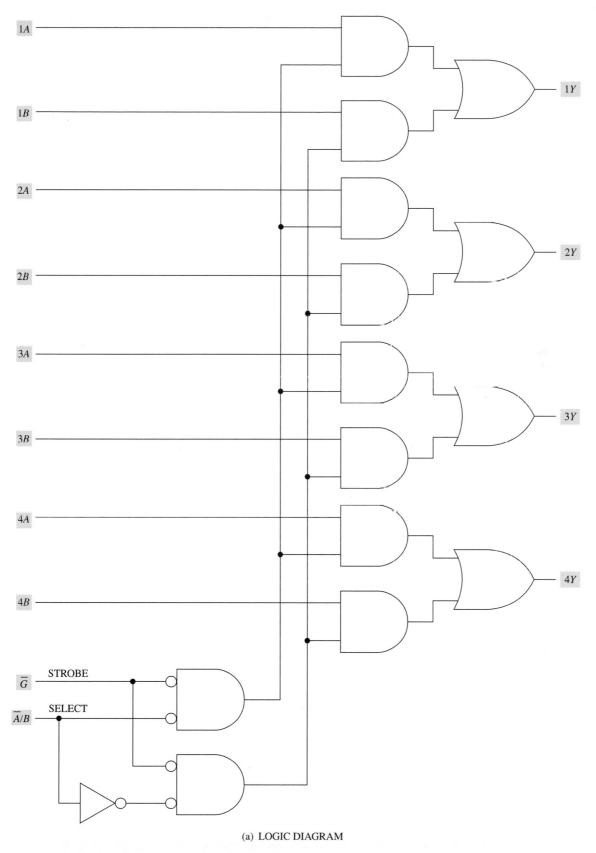

(a) LOGIC DIAGRAM

FIGURE 10–29   **Multiplexer: 74LS157–quad 2-line to 1-line.**

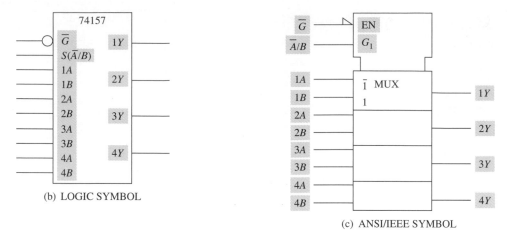

(b) LOGIC SYMBOL

(c) ANSI/IEEE SYMBOL

FIGURE 10-29 Continued

The function table for the 74LS157 is presented in Table 10–11. The table verifies true data are selected and routed to the MUX output when the circuit is enabled. The input data at the A inputs are selected when the select input is low. The B input data are selected when the select input is high.

The logic symbols for the 74LS157 are shown in Fig. 10–29(b) and (c). The ANSI/IEEE symbol in Fig. 10–29(c) shows an active-low enable (EN) input in the common-control block. In addition, G_1 indicates AND dependency of the select input with the indicated 1 and $\overline{1}$ in the MUX. In other words, G_1 indicates the select input $(\overline{A}/B)$ is ANDed with input $1B$ and the complement of the select input $(\overline{A}/B)$ is ANDed with $1A$. These AND functions are true also for the other three pairs of data inputs in the MUX.

TABLE 10-11	Inputs				Output
74LS157	**Strobe** $\overline{G}$	**Select** $\overline{A}/B$	**A**	**B**	**Y**
Multiplexer					
Function Table	1	X	X	X	0
	0	0	0	X	0
	0	0	1	X	1
	0	1	X	0	0
	0	1	X	1	1

4-Line to 1-Line Multiplexers

A 4-line to 1-line multiplexer selects data from one of four inputs and routes the data to the output. Two select input lines are required to perform this operation because there are four (2^2) possible inputs from which to select.

A basic 4-line to 1-line MUX is shown in Fig. 10–30. When the S_1S_0 inputs are 00, the A data is selected and sent to the Y output. $S_1S_0 = 01$ selects B data, $S_1S_0 = 10$ selects C data, and $S_1S_0 = 11$ selects D data.

74153.

This IC is a dual 4-line to 1-line MUX. The logic symbol is shown in Fig. 10–31. The symbol contains the select inputs (S_1S_0) and two enable inputs ($1\overline{G}$ and $2\overline{G}$). The two enable inputs allow the multiplexers to operate independently. All of the input and output labels with a 1 prefix are for one multiplexer. The inputs and output with a 2 prefix represent the other multiplexer. The select inputs are common to both sections.

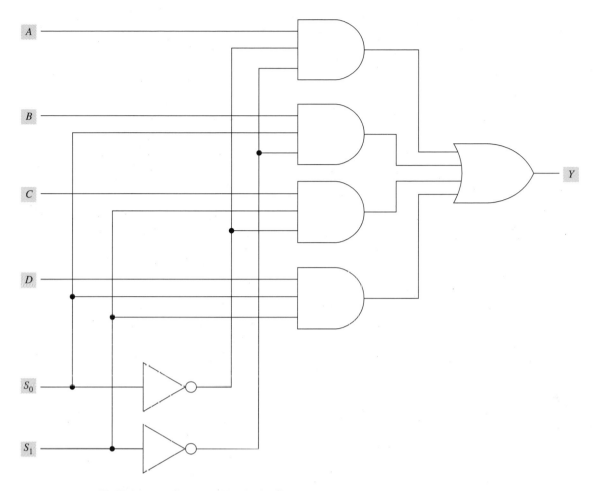

FIGURE 10–30   **Multiplexer: 4-line to 1-line logic diagram.**

The function table for the 74153 MUX is presented in Table 10–12. The table shows the Y output remains low when the strobe input $(\overline{G})$ is high. The remainder of the table indicates selected input data in its true form is applied to the Y output when the MUX is enabled. For example, if $1\overline{G} = 0$, $2\overline{G} = 1$, $S_1 = 1$, $S_0 = 0$, and $1C_2$ data input $= 1$, the $1Y$ output $= 1$. The remaining 1C data inputs are don't care inputs because they are not selected. Also keep in mind that the 2Y output will be low because the $2\overline{G}$ input $= 1$. This means the #2 section of the MUX is not enabled and all of the 2C data inputs are don't care inputs. The 74153 MUX can function as an 8-line to 2-line multiplexer if both strobe inputs are low.

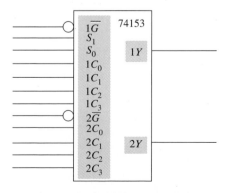

FIGURE 10–31   **Multiplexer: 74153–dual 4-line to 1-line logic symbol.**

TABLE 10-12	Select Inputs		Data		Inputs		Strobe	Output
74153 Multiplexer Function Table	S_1	S_0	C_3	C_2	C_1	C_0	$\overline{G}$	Y
	X	X	X	X	X	X	1	0
	0	0	X	X	X	0	0	0
	0	0	X	X	X	1	0	1
	0	1	X	X	0	X	0	0
	0	1	X	X	1	X	0	1
	1	0	X	0	X	X	0	0
	1	0	X	1	X	X	0	1
	1	1	0	X	X	X	0	0
	1	1	1	X	X	X	0	1

8-Line to 1-Line Multiplexer

74151.

The 74151 circuit is an 8-line to 1-line data selector. The logic diagram and symbol are shown in Fig. 10–32. Three select input lines (CBA) are required to select data from one of eight data inputs (D_0-D_7). The strobe input ($\overline{G}$) must be low for the MUX to function. The address buffers shown in Fig. 10–32(a) have symmetrical propagation delay times. This reduces the possibility of erroneous outputs occurring when the select inputs are changed.

The 74151 MUX provides a data output (Y) and a complementary data output (W) for design convenience. This is shown in Fig. 10–32(b). The function table for the 74151 appears in Table 10–13. The table shows the Y output is low when the MUX is not enabled. The output is determined by the levels of C, B, and A when $\overline{G} = 0$. True selected input data is routed to the Y output, and complementary selected input data is routed to the W output during the selection process.

The circuit shown in Fig. 10–33 consists of two 74151 multiplexers cascaded to form a 16-line to 1-line data selector. The most significant bit (S_3) of the select inputs is used to control which MUX is enabled. Select inputs 0000 through 0111—MSB is low—enable the lower-order MUX at the top in Fig. 10–33. The S_2-S_0 inputs select input data from I_0-I_7 when the lower-order MUX is enabled.

The higher-order MUX is disabled when the select inputs are as listed above—0000 through 0111. The function table (Table 10–13) shows the Y output is low when the strobe input ($\overline{G}$) to the MUX is high as is the case here. The low Y output from the higher-order MUX is used to enable the OR gate. This allows true selected data from the lower-order MUX to be routed to the circuit's output.

Select inputs 1000 through 1111—MSB is high—enable the higher-order MUX and disable the lower-order MUX. The S_2-S_0 inputs now select input data from I_8-I_{15}. This data, in its true form, is passed through the enabled OR gate to the circuit's output.

TABLE 10-13	Inputs				Output	
74151 Multiplexer Function Table	Select			Strobe		
	C	B	A	$\overline{G}$	Y	W
	X	X	X	1	0	1
	0	0	0	0	D_0	$\overline{D}_0$
	0	0	1	0	D_1	$\overline{D}_1$
	0	1	0	0	D_2	$\overline{D}_2$
	0	1	1	0	D_3	$\overline{D}_3$
	1	0	0	0	D_4	$\overline{D}_4$
	1	0	1	0	D_5	$\overline{D}_5$
	1	1	0	0	D_6	$\overline{D}_6$
	1	1	1	0	D_7	$\overline{D}_7$

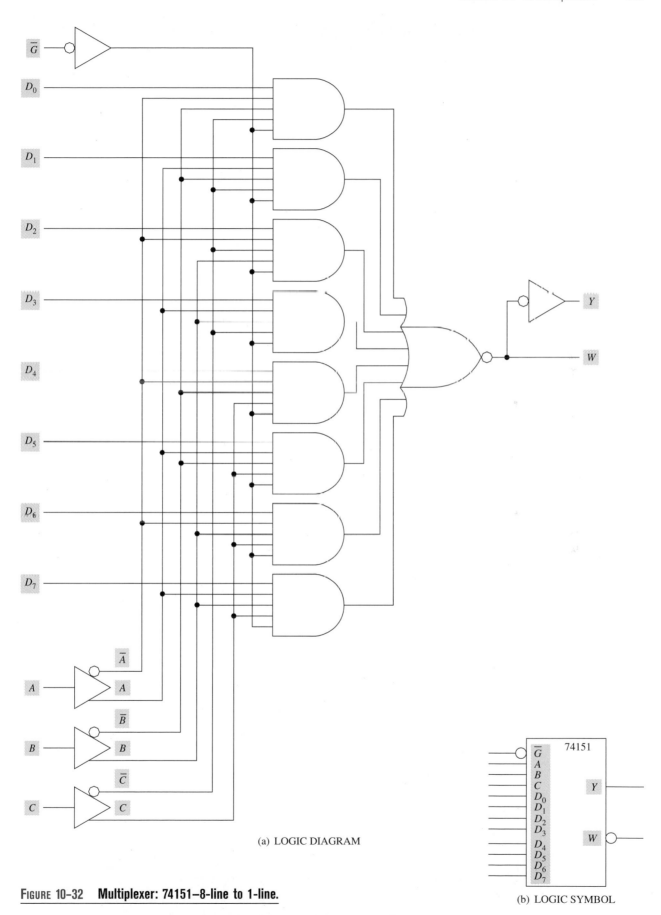

(a) LOGIC DIAGRAM

FIGURE 10–32   Multiplexer: 74151–8-line to 1-line.

(b) LOGIC SYMBOL

The circuit shown in Fig. 10–33 can be implemented with a single 74150 IC. The 24-pin 74150 MUX is a 16-line to 1-line data selector. The IC has 1 enable input pin, 4 select input pins, and 16 data input pins. The output data from the MUX is the *complement* of the selected input data. There is no output pin available on this IC for true data.

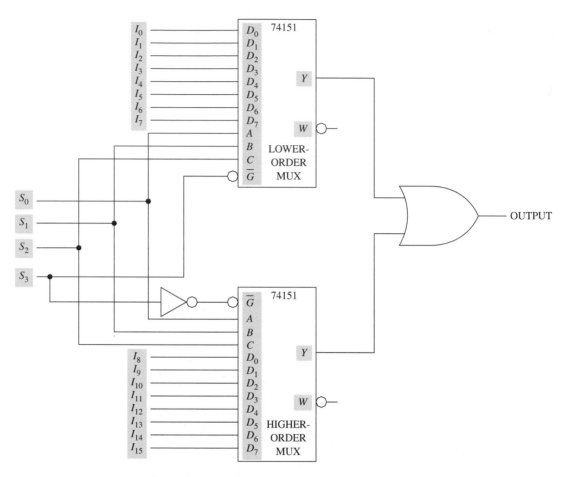

FIGURE 10–33   **Multiplexer: 16-line to 1-line implemented with 74151 ICs.**

Section 10–3: Review Questions

Answers are given at the end of the chapter.

A. Define multiplexing.
B. What is the output of the circuit in Fig. 10–28 when $S = 1, A = 0$, and $B = 1$?
C. What is the purpose of the circuit in Fig. 10–28?
D. What is the output of the circuit in Fig. 10–29(a) if $\overline{G} = 0, \overline{A}/B$ (Select) = 0, all A input data = 1, and all B input data = 0?
E. What is the output of the circuit in Fig. 10–30 if $S_1 = 1, S_0 = 0, A = B = 0$, and $C = D = 1$?
F. What is the Y output of the circuit in Fig. 10–32(a) if $\overline{G} = 0, A = B = C = 1$, $D_0 = D_1 = D_2 = D_3 = 1$, and $D_4 = D_5 = D_6 = D_7 = 0$?
G. What is the Y output of the circuit in Fig. 10–32(a) if $\overline{G} = 1, A = B = C = 0$, $D_0 = D_1 = D_2 = D_3 = 1$, and $D_4 = D_5 = D_6 = D_7 = 0$?

Section 10-4: Demultiplexers

Objectives

1. Define demultiplexing and identify demultiplexer logic circuits and logic symbols.
2. Given a demultiplexer logic diagram or logic symbol annotated with input conditions, determine the output.

Demultiplexing is the process of **data distribution.** Demultiplexers are used to perform this function in digital systems. A demultiplexer (DEMUX) performs the opposite function performed by a multiplexer. Where a MUX selects data from the input, the DEMUX takes the input data from a single input line and distributes it to several output lines.

Data distribution is performed in Fig. 10-34 with a four-position switch. The input data is distributed by the switch to the W, X, Y, or Z output. Demultiplexers perform this function electronically and are often used to convert serial data to parallel data. The decoders presented in Section 10-1 can be wired as demultiplexers. In fact, the two ICs presented in this section are listed in data books as **decoder/demultiplexers.**

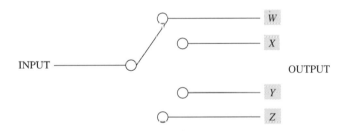

Figure 10-34   Demultiplexing analogy.

1-Line to 4-Line Demultiplexers

A basic 1-line to 4-line DEMUX is shown in Fig. 10-35. The similarities between this DEMUX logic diagram and a basic decoder with an enable input line quickly reveal why the decoder is a multifunctional device. The enable input line of a decoder is used for the data input in a DEMUX. Before we delve into this subject any deeper, let's analyze the DEMUX in Fig. 10-35.

The select inputs $(S_1 S_0)$ of the DEMUX are used to select which AND gate is enabled. When $S_1 S_0 = 00$, the top AND gate is enabled. The remaining AND gates are inhibited. This allows the data input (D_I) to be distributed to the Y_0 output. The expression for the output of this gate is $Y_0 = D_I \overline{A} \, \overline{B}$. If the data input is low, Y_0 will be low. If the data input is high, Y_0 will be high. Thus, true selected data is passed to the output.

74LS139.

One half of the 74LS139 decoder/demultiplexer logic diagram is shown in Fig. 10-36. The circuit functions as a demultiplexer when data is input to the enable input.

Operation of the circuit is identical to that just described. The top NAND gate is enabled, and the remaining NAND gates are inhibited when $S_1 S_0 = 00$. If the D_I is low, it is applied as a high after inversion to the enabled NAND gate. The gate outputs low to produce true data at Y_0. If the D_I is high, the DEMUX is disabled, and the Y_0 output is high.

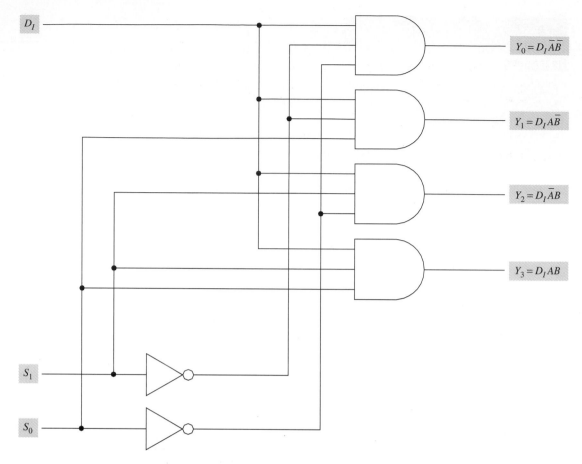

FIGURE 10–35 Demultiplexer: 1-line to 4-line logic diagram.

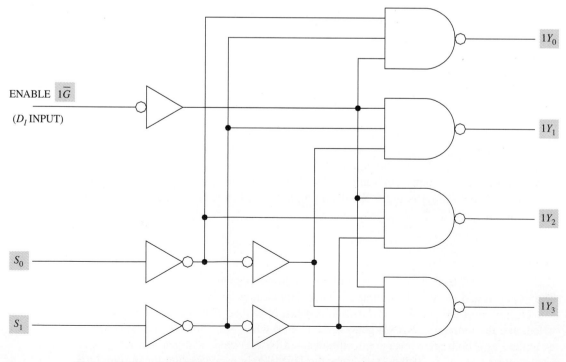

FIGURE 10–36 Demultiplexer: 74LS139–1-line to 4-line logic diagram.

1-line to 8-line Demultiplexer

The 74LS138 decoder/demultiplexer was discussed in Section 10–1. The logic symbol is shown in Fig. 10–37 wired as a DEMUX. Data is applied to one of the active-low enable inputs. The serial data input applied to $\overline{G_2}A$ can be converted to parallel data by using the select inputs (CBA) to distribute the data to the output pins.

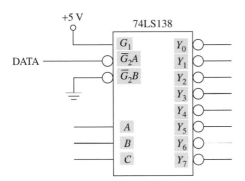

FIGURE 10–37   Demultiplexer: 74LS138–1-line to 8-line logic symbol.

An example of multiplexing data and then demultiplexing it is shown in Fig. 10–38. The reason for multiplexing and then demultiplexing is evident if you assume the input data to the MUX must be routed to a remote control center some distance away

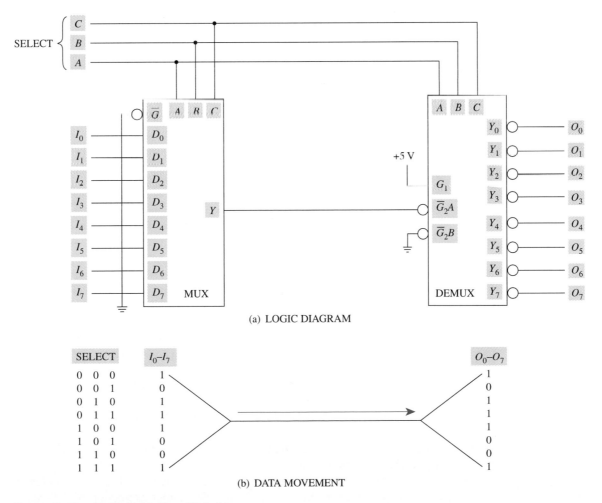

(a) LOGIC DIAGRAM

(b) DATA MOVEMENT

FIGURE 10–38   Multiplexing/demultiplexing.

The select inputs (CBA) are used to select data at the MUX input and transmit it down the single line to the data input of the DEMUX at $\overline{G}_2\,A$. That transmitted data is then distributed across the output lines of the DEMUX. In simple terms, parallel data at the input is converted to serial data, transmitted some distance, and then converted back to its original parallel format. This action is similiar to the action of a PC when it converts data for transmission via a modem. Data movement is simulated in Fig. 10–38(b). A 74154 decoder/demultiplexer can be used to distribute data from 1 line to 16 lines. Operation of this circuit is not presented here because it is similar to that of the other demultiplexers previously presented.

Section 10–4: Review Questions

Answers are given at the end of the chapter.

A. Define demultiplexing.
B. A demultiplexer can be used to convert parallel data to the serial data format.
    (1) True
    (2) False

C. A decoder IC can be used as a demultiplexer.
    (1) True
    (2) False

D. What is the output of the decoder/demultiplexer in Fig. 10–37 if $\overline{G}_2A = 0$ and $A = B = C = 1$?

SECTION 10–5: MAGNITUDE COMPARATORS

OBJECTIVES

1. State the function of a magnitude comparator.
2. Given a function table and a circuit containing two cascaded 4-bit magnitude comparators, determine the circuit's outputs when operating under stated input conditions.

Magnitude is defined as a number assigned to a member of a set to form a basis for comparison with other numbers of the same set. Fortunately in digital circuits we are only concerned with two numbers—0 and 1. Therefore, comparisons are made in digital systems between two input quantities to see which of the quantities, if either, is larger. These comparisons are made using MSI combinational logic circuits called **magnitude comparators.**

The simplest way to gain an insight into magnitude comparator operation is to compare two bits (B_0 and A_0). The circuit used to make this comparison is considered to be a 1-bit comparator because it is comparing only 0 subscripted bits. To avoid confusion, keep in mind that a 2-bit magnitude comparator compares B_1B_0 with A_1A_0, a 3-bit comparator compares $B_2B_1B_0$ to $A_2A_1A_0$, and so on. A magnitude comparator must identify all three possible magnitude conditions. These conditions are A less than B ($A < B$), A equal to B ($A = B$), and A greater than B ($A > B$). Commercially available magnitude comparators contain three output pins for this purpose.

The function table for a 1-bit magnitude comparator is shown in Table 10–14. Active-high outputs are used to represent the specific output condition. The circuit must output high at the $A = B_{\text{out}}$ output when $B_0A_0 = 00$ and $B_0A_0 = 11$. The logic diagram for the 1-bit magnitude comparator is presented in Fig. 10–39.

The top AND gate in the figure will output a high when $A < B$ ($B_0\overline{A}_0$). The lower AND gate will output a high when $A > B$ ($\overline{B}_0A_0$). The output expression of the exclu-

TABLE 10–14	Inputs	Outputs		
1-Bit Magnitude Comparator Function Table	B_0 A_0	$A > B_{out}$	$A = B_{out}$	$A < B_{out}$
	0 0	0	1	0
	0 1	1	0	0
	1 0	0	0	1
	1 1	0	1	0

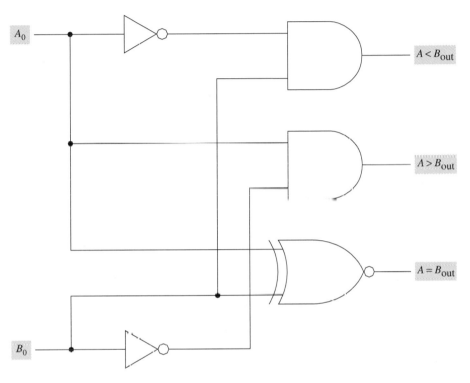

FIGURE 10–39   1-Bit magnitude comparator logic diagram.

sive-NOR (X-NOR) gate is $\overline{A \oplus B}$. This expression is the same as $B_0 A_0 + \overline{B_0}\,\overline{A_0}$. The expression indicates the X-NOR gate will output a high when the inputs are equal. Think back to the short logic of the X-NOR gate—any complementary input = 0 output. This logic indicates equal inputs will produce a high output.

7485.

This 16-pin IC is a 4-bit magnitude comparator. It is used to compare two 4-bit binary or binary-coded decimal numbers. Remember, the 4-bit name is derived from the size of the numbers being compared.

The logic symbol in Fig. 10–40 shows three decoded outputs regarding the two 4-bit "words" are available—$A > B_{out}$, $A = B_{out}$, and $A < B_{out}$. Larger words can be compared by cascading 7485s using the $A > B_{in}$, $A = B_{in}$, and $A < B_{in}$ input pins. These input pins were incorporated in the 7485 to allow expansion without the use of external control logic.

The function table for the 7485 4-bit magnitude comparator is shown in Table 10–15. The uniqueness of this table deserves a detailed look. The first 8 lines of Table 10–15 are relatively self-explanatory. Let's assume we are comparing the following sets of numbers:

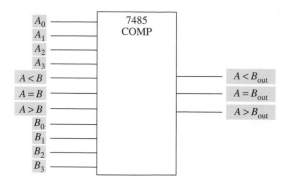

FIGURE 10-40   4-Bit magnitude comparator: 7485 logic symbol.

	B_3	B_2	B_1	B_0		A_3	A_2	A_1	A_0
Set 1	1	1	0	0		0	1	1	1
Set 2	1	1	0	0		1	1	1	1
Set 3	1	1	0	0		1	1	0	0

Set 1: $A_3 < B_3$ in set 1. The second row of the $A_3 B_3$ Comparing Inputs columns in the table indicates the $A < B$ output is active (high) (see Outputs column) and the remaining 7485 outputs are inactive with this set of inputs.

Set 2: In this set of inputs $A_3 = B_3, A_2 = B_2, A_1 > B_1$, and $A_0 > B_0$. Go down the Comparing Inputs columns until this set of input conditions is located. It is located in the fifth row down. The $A_0 B_0$ input in this set of inputs is "don't care" and the $A > B$ output is active.

Set 3: This set of input numbers has all values equal. Rows 9 through 13 of Table 10–15 under the Comparing Inputs columns include this set of input conditions. Therefore, the cascading inputs must now be considered in most circumstances.

The data book specifies the 7485 magnitude comparator handling the least-significant bits (LSBs) must have a high applied to the $A = B$ input. The $A > B$ and $A < B$ cascading inputs to the LSB stage are don't care inputs if only one 7485 is being used for comparison.

TABLE 10-15

7485 4-Bit Magnitude Comparator Function Table

	Comparing Inputs							Cascading Inputs			Outputs		
A_3	B_3	A_2	B_2	A_1	B_1	A_0	B_0	$A > B$	$A < B$	$A = B$	$A > B$	$A < B$	$A = B$
$A_3 > B_3$		X		X		X		X	X	X	1	0	0
$A_3 < B_3$		X		X		X		X	X	X	0	1	0
$A_3 = B_3$		$A_2 > B_2$		X		X		X	X	X	1	0	0
$A_3 = B_3$		$A_2 < B_2$		X		X		X	X	X	0	1	0
$A_3 = B_3$		$A_2 = B_2$		$A_1 > B_1$		X		X	X	X	1	0	0
$A_3 = B_3$		$A_2 = B_2$		$A_1 < B_1$		X		X	X	X	0	1	0
$A_3 = B_3$		$A_2 = B_2$		$A_1 = B_1$		$A_0 > B_0$		X	X	X	1	0	0
$A_3 = B_3$		$A_2 = B_2$		$A_1 = B_1$		$A_0 < B_0$		X	X	X	0	1	0
$A_3 = B_3$		$A_2 = B_2$		$A_1 = B_1$		$A_0 = B_0$		1	0	0	1	0	0
$A_3 = B_3$		$A_2 = B_2$		$A_1 = B_1$		$A_0 = B_0$		0	1	0	0	1	0
$A_3 = B_3$		$A_2 = B_2$		$A_1 = B_1$		$A_0 = B_0$		X	X	1	0	0	1
$A_3 = B_3$		$A_2 = B_2$		$A_1 = B_1$		$A_0 = B_0$		1	1	0	0	0	0
$A_3 = B_3$		$A_2 = B_2$		$A_1 = B_1$		$A_0 = B_0$		0	0	0	1	1	0

However, the $A = B$ input must be tied high. The eleventh row contains the condition of the set 3 inputs. The $A = B$ output is activated for this set of inputs.

An 8-bit magnitude comparator is shown in Fig. 10–41. The circuit consists of two 7485s wired to accomplish the expansion from 4-bit to 8-bit comparison. The $A = B$ input to the lower-order (LSB) comparator is tied to +5 V per the data book. The outputs of the lower-order comparator are connected to the corresponding cascade inputs of the higher-order comparator.

Comparison of the two 8-bit numbers shown is analyzed below:

	B_7 B_6 B_5 B_4	B_3 B_2 B_1 B_0	A_7 A_6 A_5 A_4	A_3 A_2 A_1 A_0
Set 4	1 0 0 0	0 0 0 0	1 0 0 1	0 0 0 0

The inputs to the lower-order comparator are all equal, and the $A = B$ input to this comparator is tied high as shown in Fig. 10–41. This produces $A = B_{out}$ high, $A > B_{out}$ low, and $A < B_{out}$ low outputs from this comparator.

The higher-order comparator has the following comparisons: $A_4 > B_4, A_5 = B_5,$ $A_6 = B_6$, and $A_7 = B_7$. (*Note:* For the higher-order input bit numbers, A_7–A_4 and B_7–B_4 must be found using A_3–A_0 and B_3–B_0 In the Comparing Inputs columns in Table 10–15. Row 7 shows the $A > B_{out}$ output is activated to indicate the "A" 8-bit number is greater than the "B" 8-bit number for set 4.

One more example will illustrate the significance of the cascading inputs.

	B_7 B_6 B_5 B_4	B_3 B_2 B_1 B_0	A_7 A_6 A_5 A_4	A_3 A_2 A_1 A_0
Set 5	1 0 0 0	0 0 0 0	1 0 0 0	1 0 1 0

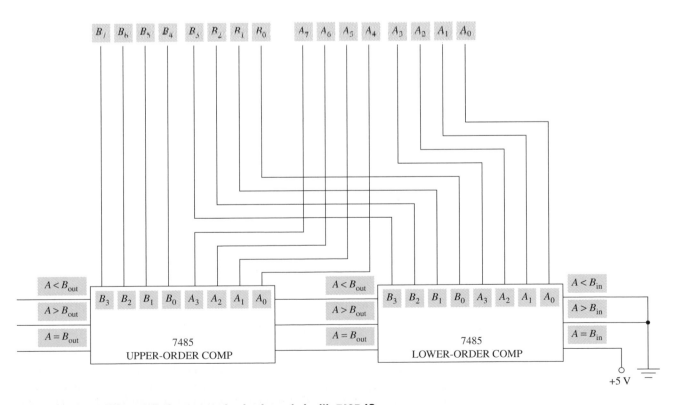

FIGURE 10–41   8-Bit magnitude comparator implemented with 7485 ICs.

The inputs to the lower-order comparator are $A_0 = B_0$, $A_1 > B_1$, $A_2 = B_2$, and $A_3 > B_3$. Since $A_3 > B_3$, the comparator produces $A > B_{out} = 1$, $A < B_{out} = 0$, and $A = B_{out} = 0$.

The A and B inputs to the higher-order comparator are all equal. The cascade input $A > B$ is high. The table shows the circuit produces a high $A > B_{out}$ signal from this comparator. One final note about the function table—if all cascading inputs are low, the circuit produces an active-low output at $A = B$. This is shown as the bottom line entry in Table 10–15.

Section 10–5: Review Questions

Answers are given at the end of the chapter.

A. What is the function of a magnitude comparator?
B. What are the typical outputs that are available on a magnitude comparator IC and what are their active levels?
C. Give the three output levels of the 7485 in Fig. 10–40 when

B_3	B_2	B_1	B_0	A_3	A_2	A_1	A_0
1	0	0	0	0	1	1	1

D. Give the three output levels of the 7485 in Fig. 10–40 when

B_3	B_2	B_1	B_0	A_3	A_2	A_1	A_0
1	1	1	0	1	1	1	1

E. Give the three output levels of the 7485 in Fig. 10–40 when

B_3	B_2	B_1	B_0	A_3	A_2	A_1	A_0	$A < B_{in}$	$A > B_{in}$	$A = B_{in}$
1	1	0	0	1	1	0	0	1	0	0

SECTIONS 10–3 THROUGH 10–5: INTERNAL SUMMARY

These three sections deal with important processes in digital circuits—**multiplexing, demultiplexing,** and **magnitude comparing.**

Multiplexers are used in digital as **data selectors.** The process can be thought of in terms of **time-sharing.** A MUX is often used to convert data from its parallel format to the serial format.

Multiplexer ICs can be purchased to multiplex 2, 4, 8, and 16 lines to 1 line. They select one line's data input and send that selected data to the output. A multiplexer IC usually has an enable input pin in addition to its select and data input pins. The enable input allows for cascading MUX ICs without the necessity of designing additional control circuitry.

The process of demultiplexing a digital signal is the inverse of multiplexing it. A **DEMUX** takes data from a single input line and distributes it across several output lines. Demultiplexers can be used to convert serial data to parallel data.

Demultiplexer ICs are not manufactured. Instead, decoder ICs are used to demultiplex a signal. This is accomplished by bringing serial data in on an enable input pin of a decoder.

Magnitude comparators are used to compare binary quantities to determine whether the quantities are equal. If they are not equal, the comparator reveals which one of the unequal quantities is larger or smaller.

The **7485** 4-bit magnitude comparator is used to compare two 4-bit binary or BCD numbers. In this case the binary quantity is called a word. Multiple 7485s can be cascaded to compare larger words. Many computer programs are written with commands that indicate the program should jump to a specific address if the compared quantities are equal, or if A is greater than B or vice versa.

SECTIONS 10-3 THROUGH 10-5: INTERNAL SUMMARY QUESTIONS

Answers are given at the end of the chapter.

Note: The use of manufacturer's data sheets is recommended to assist you in answering these summary questions.

1. Multiplexers are used to _____.
   a. select data
   b. distribute data
   c. produce a distinct code/number for each of its inputs
   d. indicate a selected code/number by asserting one output

2. Demultiplexers are used to _____.
   a. select data
   b. distribute data
   c. produce a distinct code/number for each of its inputs
   d. indicate a selected code/number by asserting one output

3. Which of the following digital circuits can be used to convert data from the serial to the parallel format?
   a. Decoder   c. Multiplexer
   b. Encoder   d. Demultiplexer

4. Identify the circuit shown in Fig. 10–42.
   a. Dual 4-line to 1-line multiplexer
   b. Quad 2-line to 1-line multiplexer
   c. Dual 4-line to 1-line demultiplexer
   d. Quad 2-line to 1-line demultiplexer

5. What is the output of the circuit shown in Fig. 10–42 if $\overline{G} = 0$, $\overline{A}/B = 1$, all A input data $= 1$, and all B input data $= 0$?
   a. $1Y$ through $4Y = 0$.
   b. $1Y$ through $4Y = 1$.
   c. $1Y = 2Y = 0$ and $3Y = 4Y = 1$.
   d. $1Y = 2Y = 1$ and $3Y = 4Y = 0$.

6. What is the output of the circuit shown in Fig. 10–42 if $\overline{G} = 1$, $\overline{A}/B = 0$, all A input data $= 1$, and all B input data $= 0$?
   a. $1Y$ through $4Y = 0$.
   b. $1Y$ through $4Y = 1$.
   c. $1Y = 2Y = 0$ and $3Y = 4Y = 1$.
   d. $1Y = 2Y = 1$ and $3Y = 4Y = 0$.

7. What is the output of the dual 4-line to 1-line MUX shown in Fig. 10–43 if $1\overline{G} = 2\overline{G} = 0$, $S_1 = S_0 = 1$, $1C_3 - 1C_0 = 0$, and $2C_3 - 2C_0 = 1$?
   a. $1Y = 2Y = 1$.
   b. $1Y = 2Y = 0$.
   c. $1Y = 0$ and $2Y = 1$.
   d. $1Y = 1$ and $2Y = 0$.

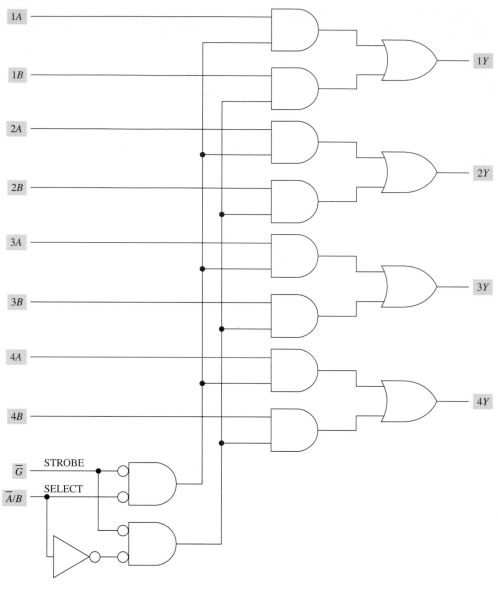

FIGURE 10–42

8. What is the output of the dual 4-line to 1-line MUX shown in Fig. 10–43 if $1\overline{G} = 2\overline{G} = 0$, $S_1 = 0$, $S_0 = 1$, $1C_3$–$1C_0 = 0101$, and $2C_3$–$2C_0 = 1010$?
   a. $1Y = 2Y = 1$.
   b. $1Y = 2Y = 0$.
   c. $1Y = 0$ and $2Y = 1$.
   d. $1Y = 1$ and $2Y = 0$.

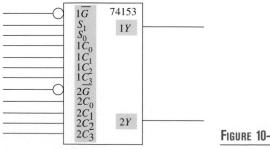

FIGURE 10–43

9. What is the output of the 74151 shown in Fig. 10–44 when $\overline{G} = 0, C = B = A = 1$, and $D_7 - D_0 = 10100011$?
   a. $Y = 0$ and $W = 0$.   c. $Y = 1$ and $W = 0$.
   b. $Y = 0$ and $W = 1$.   d. $Y = 1$ and $W = 1$.

10. What is the output of the 74151 shown in Fig. 10–44 when $\overline{G} = 1, C = B = A = 0$, and $D_7 - D_0 = 10100011$?
    a. $Y = 0$ and $W = 0$.   c. $Y = 1$ and $W = 0$.
    b. $Y = 0$ and $W = 1$.   d. $Y = 1$ and $W = 1$.

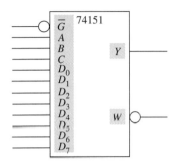

FIGURE 10-44

11. The 74LS138 shown in Fig. 10–45 is wired as a _____.
    a. decoder   c. multiplexer
    b. encoder   d. demultiplexer

12. The 74LS138 shown in Fig. 10–45 can be used to convert _____.
    a. serial data to parallel data
    b. parallel data to serial data

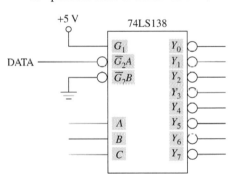

FIGURE 10-45

13. Give the outputs of the 7485 magnitude comparator in Fig. 10–46 if the inputs are

B_3	B_2	B_1	B_0	A_3	A_2	A_1	A_0	$A > B_{in}$	$A < B_{in}$	$A = B_{in}$
1	0	0	0	0	1	0	0	X	X	X

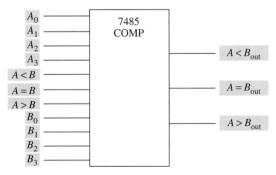

FIGURE 10-46

   a. $A > B_{out} = 1$, $A < B_{out} = 1$, and $A = B_{out} = 1$.
   b. $A > B_{out} = 0$, $A < B_{out} = 0$, and $A = B_{out} = 1$.
   c. $A > B_{out} = 1$, $A < B_{out} = 0$, and $A = B_{out} = 0$.
   d. $A > B_{out} = 0$, $A < B_{out} = 1$, and $A = B_{out} = 0$.

14. Give the outputs of the 7485 in Fig. 10–46 if the inputs are

B_3	B_2	B_1	B_0	A_3	A_2	A_1	A_0	$A > B_{in}$	$A < B_{in}$	$A = B_{in}$
1	0	0	0	1	0	0	0	X	X	1

a. $A > B_{out} = 1$, $A < B_{out} = 1$, and $A = B_{out} = 1$.
b. $A > B_{out} = 0$, $A < B_{out} = 0$, and $A = B_{out} = 1$.
c. $A > B_{out} = 1$, $A < B_{out} = 0$, and $A = B_{out} = 0$.
d. $A > B_{out} = 0$, $A < B_{out} = 1$, and $A = B_{out} = 0$.

15. Give the outputs of the 7485 in Fig. 10–46 if the inputs are

B_3	B_2	B_1	B_0	A_3	A_2	A_1	A_0	$A > B_{in}$	$A < B_{in}$	$A = B_{in}$
1	0	0	0	1	0	0	0	0	1	0

a. $A > B_{out} = 1$, $A < B_{out} = 1$, and $A = B_{out} = 1$.
b. $A > B_{out} = 0$, $A < B_{out} = 0$, and $A = B_{out} = 1$.
c. $A > B_{out} = 1$, $A < B_{out} = 0$, and $A = B_{out} = 0$.
d. $A > B_{out} = 0$, $A < B_{out} = 1$, and $A = B_{out} = 0$.

SECTION 10-6: APPLICATIONS AND TROUBLESHOOTING

OBJECTIVES

1. Given a circuit containing decoders, encoders, multiplexers, demultiplexers, and/or magnitude comparators, determine the circuit's output.
2. State the main purpose for or application of various circuits containing decoders, encoders, multiplexers, demultiplexers, and/or magnitude comparators.
3. Given the observed troubleshooting results of a defective circuit containing MSI ICs, determine the possible fault.
4. Given a malfunction in a circuit containing MSI ICs, determine the most probable symptom(s).

The purpose of this section is not to teach theory of computer operation or computer terminology. The circuits have been selected because they are actual circuits used in digital systems. The important points in this section are to understand the MSI circuits, what they accomplish, and what happens if they or their associated circuitry malfunctions. The terminology used is common computer terminology. The terminology is not the point of the lesson—circuit operation is.

Most MSI ICs are relatively simple to troubleshoot. They work if they are properly powered, or else they are defective. The primary consideration if they appear defective is to *isolate their load* to ensure it isn't the cause of the problem. Since our problem is to locate the defective IC in a circuit, troubleshooting in this chapter is incorporated into the applications section to provide a more real-world approach to troubleshooting. After all, the most important part of troubleshooting is to know what the circuit should do when it is operating properly. Once a problem has been isolated to a suspect IC, the troubleshooting information presented earlier in the text can be used to ensure the IC is in fact defective.

Address Decoders

Our first brief discussion of **read-only memory** (ROM) was in Section 3–10. **ROM address decoding** was presented in Section 5–5. The decoding circuit presented in Section 5–5 was shown implemented with logic gates. That circuit is actually a 74LS138 decoder. This can be seen by comparing the 74LS138 decoder logic diagram in Fig. 10–4(a) with the decoding circuitry of Fig. 5–36. The ROM address decoder is redrawn with the 74LS138 logic symbol in Fig. 10–47. A quick overview of the address decoder operation is presented here. The details may be reviewed in Chapter 5 if desired.

There are two active-low enable inputs ($\overline{G_2A}$ and $\overline{G_2B}$) on the 74LS138 decoder in Fig. 10–47. Address bits A_{19} through A_{16} must be high for the NAND gate to produce a low at $\overline{G_2A}$. This low output from the NAND gate is called the **ROM page** ($\overline{ROMPG}$) signal because it is used to select a page of memory from ROM if it is activated during a memory–read operation. The **memory–read** ($\overline{MEMR}$) signal from the microprocessor must also be low to enable the 74LS138 decoder. Note the active-high enable input (G_1) on the decoder is wired high.

An active-low output from the decoder will select one of the ROM chips. The selection is based on the level of the address bits A_{15}, A_{14}, and A_{13} that are applied to the decoder *CBA* inputs. If the select inputs to the decoder are $A_{15} = 0, A_{14} = 1$, and $A_{13} = 1$, the $\overline{Y_3}$ output is active. This low to the **chip select** ($\overline{CS}$) input of ROM chip #2 enables this chip.

The direction of data movement through the **transceiver** is controlled by the $\overline{ROMPG}$ and $\overline{MEMR}$ signals. The transceiver was presented briefly in Chapter 5. It is a transmitter/receiver that allows data movement in one direction or the opposite direction. It is evident in Fig. 10–47 that data must move from left to right through the transceiver if the microprocessor is to read the data from the ROM chip. The active-high **ROM address select** (ROM ADDRSEL) signal is activated when $\overline{MEMR}$ and $\overline{ROMPG}$ to NOR gate #1 are active. This signal inhibits NOR gate #3 and sets the direction (DIR) pin on the transceiver low. Data movement is from left to right through the transceiver when the DIR pin is low. This allows ROM data to be read by the microprocessor.

The ROM address decoder in Fig. 10–47 is based on the 8088 microprocessor system. A page layout of the system's memory is shown in Table 10–16. Memory in the 8088 microprocessor-based computer is allocated in **64 KB pages.**

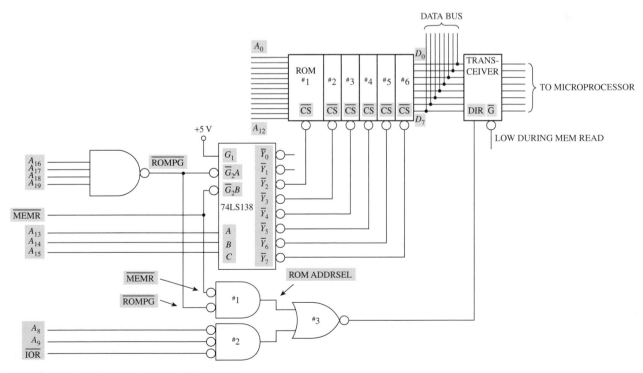

FIGURE 10–47   **ROM address decoder circuit.**

TABLE 10-16	A_{19}	A_{18}	A_{17}	A_{16}	
8088 System Page Layout	0	0	0	0	Page 0 = RAM
	0	0	0	1	Page 1 = RAM
	0	0	1	0	Page 2 = RAM
	0	0	1	1	Page 3 = RAM
	0	1	0	0	Page 4 = RAM
	0	1	0	1	Page 5 = RAM
	0	1	1	0	Page 6 = RAM
	0	1	1	1	Page 7 = RAM
	1	0	0	0	Page 8 = RAM
	1	0	0	1	Page 9 = RAM
	1	0	1	0	Page A = RESERVED
	1	0	1	1	Page B = VIDEO/GRAPHICS
	1	1	0	0	Page C = RESERVED
	1	1	0	1	Page D = RESERVED
	1	1	1	0	Page E = ROM BASIC
	1	1	1	1	Page F = ROM

Sixteen pages of memory can be addressed in this system. These 16 pages of memory equal a total memory capability of 1,048,576 memory locations. Each memory location within a memory chip in this system contains a byte (8 bits) of data.

The page layout in Table 10–16 shows the system uses 20 address bits. The upper nibble (4 bits) of the address (A_{19}–A_{16}) forms the **page address.** The ROM chips are addressed only when (A_{19}–A_{16}) = 1111. This is page F in the page layout table. Note hex page addresses 0 through E are used for other purposes. A ROM address is specified here:

$$A_{19}\ A_{18}\ A_{17}\ A_{16}\ A_{15}\ A_{14}\ A_{13}\ A_{12}\ A_{11}\ A_{10}\ A_9\ A_8\ A_7\ A_6\ A_5\ A_4\ A_3\ A_2\ A_1\ A_0$$
$$\underbrace{1\ \ 1\ \ 1\ \ 1}_{F}\ \underbrace{X\ \ X\ \ X\ \ X}_{X}\ \underbrace{X\ \ X\ \ X\ \ X}_{X}\ \underbrace{X\ \ X\ \ X\ \ X}_{X}\ \underbrace{X\ \ X\ \ X\ \ X}_{X}{}_{(2)}$$
$$(16)$$

The address is shown in both binary and hex. The only part of the address that must be considered here is the upper nibble. Thus, the remaining address bits are don't care inputs for this discussion. The upper four bits, as we have already seen, select the ROM page. Figure 10–47 shows any one of the six ROM ICs can be selected when the page address is F and $\overline{MEMR}$ = 0. Address bits $A_{15}, A_{14},$ and A_{13} determine which one of the six ROM chips is actually selected. Address bits A_{12} through A_0 read data from a specific memory location within the selected ROM chip. The data retrieved from ROM provides information for power-on initialization. It configures the system and initializes the support chips required for normal operation. This initialization process is called **boot-up.**

What symptoms would occur if the $\overline{G}_2A$ or $\overline{G}_2B$ pin on the 74LS138 decoder in Fig. 10–47 were internally open? The outputs of the 74LS138 would never go active because the decoder is not enabled. This would prevent selecting a ROM chip during boot-up. Since the system receives its initial instructions from ROM, it would not initialize. This would be referred to as a dead system. These same results would be evident if the decoder were defective or +5 V was not tied to G_1 or to the V_{CC} pin.

What symptoms would appear if the A_{19} input pin to the NAND gate were open? This floating input would act as a high to the $\overline{ROMPG}$ NAND gate. This would not affect the selection of a ROM chip during normal boot-up operation. However, if RAM page 7 (page address 0111 in Table 10–16) was addressed for a read operation, the ROM decoder circuit would interpret this input address as a request for ROM data. The result would be the application of ROM and RAM data to the system data bus simultaneously. This bus contention problem would cause the system to "crash."

What would be the result if the DIR input pin to the transceiver opened? The DIR pin is low during ROM read and I/O read operations. Data cannot be read by the microprocessor if the DIR pin is not low. Thus, the system would not initialize—dead system.

A **programmable address decoder** is shown in Fig. 10–48. The circuit incorporates two 7485 4-bit magnitude comparators. This circuit allows the user to program the desired address of a device with DIP switches. This circuit could be used for decoding the address of an input/output (I/O) port in a computer.

The circuit in Fig. 10–48 is set to enable a device when, and only when, the address from the microprocessor is $B0_{(16)}$. The lower-order comparator $A = B_{\text{in}}$ input is tied high. The IC compares the lower-order nibble of the microprocessor's address with the lower four DIP switch settings. If the lower address nibble is $0000_{(2)}$, the output of this comparator is $A = B_{\text{out}} = 1$ (see Table 10–15). The $A > B_{\text{out}}$ and $A < B_{\text{out}}$ outputs are not used. If the upper-address nibble from the microprocessor is $1011_{(2)}$ which is a hex B, the output of the upper-order comparator is also $A = B_{\text{out}} = 1$.

The output of the upper-order comparator is NANDed with an input-output/memory signal ($\overline{\text{IO/M}}$) to generate an active-low select signal ($\overline{\text{SEL}}$) when the microprocessor is addressing this I/O port. The $\overline{\text{SEL}}$ signal must be ORed with the microprocessor READ

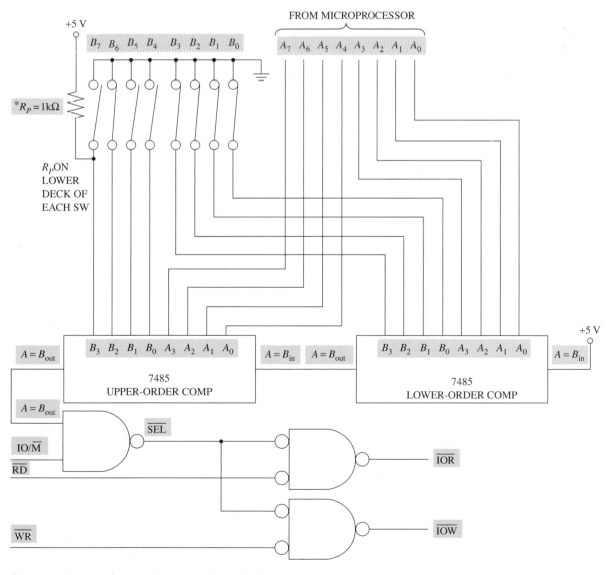

FIGURE 10–48   Programmable address decoder circuit.

$(\overline{RD})$ and WRITE $(\overline{WR})$ signals to activate the proper I/O read $(\overline{IOR})$ or I/O write $(\overline{IOW})$ command. $\overline{IOR}$ will be low when the microprocessor indicates a read cycle.

A quick overview of this operation will put it in the proper perspective. The microprocessor places an address on the address bus and signifies a read operation. The address is that of a specific I/O port. This port's address has been designated by the settings of the DIP switches. The setup port address is compared in the 7485 ICs with the microprocessor address. The $A = B_{out}$ signal of the upper-order comparator is high when the two addresses are equal. A combinational logic circuit is then used to indicate a read operation of the port by activating $\overline{IOR}$.

Troubleshooting this circuit is pretty straightforward, although its design is less than desirable. The 7485's outputs will both be high at the $A = B_{out}$ output pins if the address from the microprocessor matches the port address set up with the DIP switches. If the $A = B_{out}$ input to the NAND gate is open, the $\overline{SEL}$ signal will go low during an I/O read or write operation. This is true even for nonmatching addresses. If any of the OR gate inputs are open, the OR gate will be inhibited, and its output cannot be activated.

Display Decoders

The circuit in Fig. 10–49 contains a MOD-1000 counter—maximum count is 999. The counts of each MOD-10 counter are decoded by a 7446 BCD-to-7-segment decoder/driver. The 7446 converts the BCD counts to 7-segment codes for driving the common-anode 7-segment indicators.

Leading zero suppression is employed to blank out the leading zero (zeros) in the multidigit display. This is accomplished by connecting the ripple blanking signals as shown in Fig. 10–49. Note the $\overline{RBI}$ of the MSD decoder/driver is tied low and $\overline{LT}$ is tied high. When the MSD counter's output is $0000_{(BCD)}$, all segments of the MSD indicator will be off and $\overline{RBO}$ will be low. This is illustrated in the 3-digit display for the number 55 in the figure. The MSD 7-segment display is blank.

If the counter's output is 005 ($0000\ 0000\ 0101_{(BCD)}$), both leading zeros will be suppressed. This ripple blanking is accomplished by connecting the MSD decoder's $\overline{RBO}$ to the $\overline{RBI}$ of the next lower significant digit decoder. Keep in mind the MSD decoder's $\overline{RBO}$ goes low only when its count is $0000_{(BCD)}$.

Leading zero suppression can be accomplished for any number of 7-segment displays by connecting the ripple blanking signals as shown—RBO to $\overline{RBI}$ of the next lesser significant digit decoder/driver, down to and including the LSD.

Counter troubleshooting was discussed at length in Chapter 7. The easy part of troubleshooting the circuit in Fig. 10–49 is that its operation is sequential. The LSD counter's Q_D output clocks the next higher digit counter. That counter's Q_D output clocks the MSD counter. If the problem is no count or a miscount from a counter, disconnect the 7446 from the suspect counter to ensure the decoder is not the problem. Also, make sure there is a good clock input to the suspect counter. Always remember to check V_{CC} and ground. Take the lamp test $(\overline{LT})$ input low if an indicator produces no count (blank). This will test all of the segments in the display.

Check the outputs of the decoder if the counter driving the decoder counts normal, and the lamp test indicates the segments of the display unit are good. The decoder outputs an active-low signal to turn on a segment of the common-anode 7-segment indicator. Therefore, you can apply a specific count from a counter and determine which decoder outputs should be active. For example, pulse the counter with a logic probe until a count of $0111_{(BCD)}$ is output to the suspected faulty decoder. The a, b, and c segments should illuminate, and the remaining segments should be off. The outputs of the decoder should be $a = b = c = 0$ and $d = e = f = g = 1$. Replace the suspect decoder if applicable.

What if the circuit in Fig. 10–49 has just been built, and it is not functioning properly? The observed results along with the desired results of the LSD 7-segment indicator are shown in Fig. 10–50. The other two 7-segment indicators are functioning properly. Determine the probable cause of the erroneous indications.

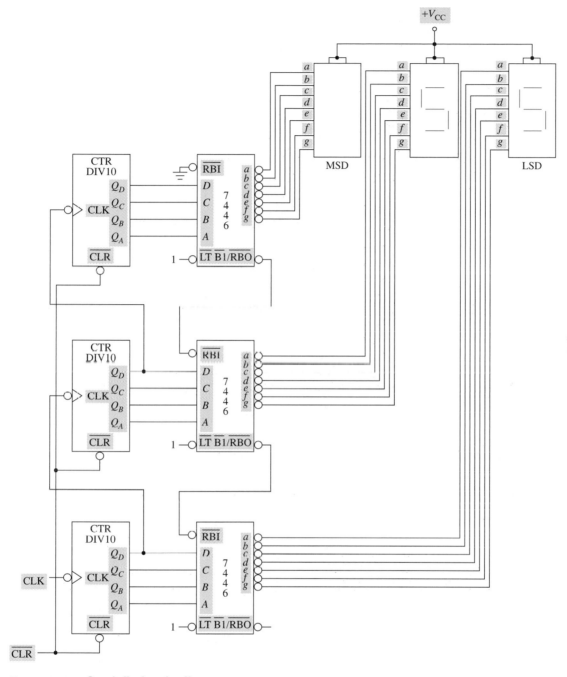

FIGURE 10-49 **Count display circuit.**

The first look at Fig. 10–50 shows the 1, 4, 5, 6, 7, 8, and 9 counts appear normal. The 0 count shows the *f* segment off (should be on) and the *g* segment on (should be off). This first clue indicates the *f* and *g* outputs of the decoder or the *f* and *g* inputs to the 7-segment indicator have been reversed. The 2 count and the 3 count indications verify this conclusion.

Logic Function Generator

Our knowledge of truth tables and logic functions from Chapter 4 will be beneficial in analysis of the logic function generator in Fig. 10–51(a). The Boolean expression for the truth table in Fig. 10–51(b) is $A\overline{BC} + A\overline{BC} + \overline{ABC}$. The circuit required to implement

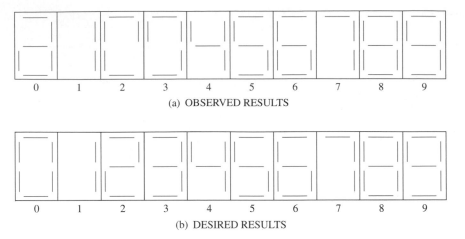

(a) OBSERVED RESULTS

(b) DESIRED RESULTS

FIGURE 10-50 **Count display circuit–troubleshooting indications.**

this expression should be in the sum-of-products (SOP) configuration. A multiplexer can be used to implement the SOP expression because the MUX is an SOP circuit. The data input lines to the 74151 MUX have been connected low or high as indicated by the Y output on the truth table. Since $D_3, D_5,$ and D_6 input lines are connected high, the MUX will output high only when the select inputs are $CBA = 011, 101,$ or 110. The Y output is low for the remaining input combinations. This is true because the input lines selected by the remaining input combinations are tied low.

Determine the output expression for the circuit shown in Fig. 10–52(a). Note the select input conditions on the truth table in Fig. 10–52(b) that require a high output. The conditions are $\overline{A}\,\overline{B}\,C, \overline{A}\,B\overline{C},$ and $\overline{A}BC$. Therefore, the circuit produces $\overline{A}\,\overline{B}\,C + \overline{A}\,B\overline{C} + \overline{A}BC$ at its Y output. If the output is taken from the complementary output pin (W) on the MUX, the expression is $\overline{\overline{A}\,\overline{B}\,C + \overline{A}\,B\overline{C} + \overline{A}BC}$.

Although the Y output expression $\overline{A}\,\overline{B}\,C + \overline{A}\,B\overline{C} + \overline{A}BC$ can be simplified to $\overline{A}\,B + \overline{A}C$, its implementation would still require two 2-input AND gates, two NOT gates, and a 2-input OR gate—three ICs. The circuit would require five 2-input NAND gates if it were implemented with only NAND gates—two ICs. Thus, it would be beneficial to implement this circuit with one MUX IC.

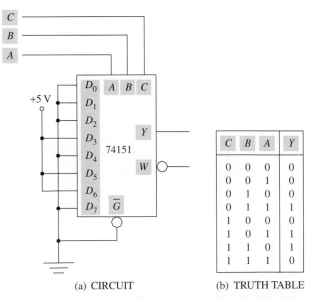

C	B	A	Y
0	0	0	0
0	0	1	0
0	1	0	0
0	1	1	1
1	0	0	0
1	0	1	1
1	1	0	1
1	1	1	0

(a) CIRCUIT   (b) TRUTH TABLE

FIGURE 10-51 **Logic function generator–$AB\overline{C} + A\overline{B}C + \overline{A}BC$.**

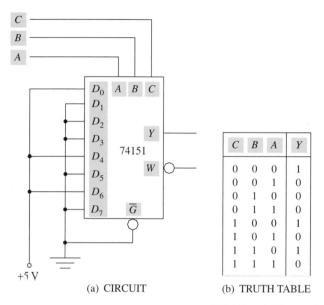

C	B	A	Y
0	0	0	1
0	0	1	0
0	1	0	0
0	1	1	0
1	0	0	1
1	0	1	0
1	1	0	1
1	1	1	0

(a) CIRCUIT   (b) TRUTH TABLE

FIGURE 10-52   Logic function generator–$\overline{A}\,\overline{B}\,\overline{C} + \overline{A}\,\overline{B}C + A\overline{B}C$.

The circuits in Fig. 10–51 and 10–52 produced a Boolean function that consisted of three variables. This coincides with the number of select inputs. Design of the circuits was accomplished by simply connecting the data inputs of the MUX low or high per the truth table.

The 74151 MUX can also be connected to produce a four-input variable function. The circuit is shown in Fig. 10–53(a). The circuit implements the truth table function in Fig. 10–53(b). The expression is $\overline{A}\,\overline{B}\,\overline{C}\,\overline{D} + \overline{A}\,\overline{B}\,CD + AB\overline{C}D + ABCD$. The MSB of the $DCBA$ input combination of the truth table is connected to the D_3 and D_7 inputs of the MUX. The CBA inputs select the input data to be routed to the Y output in exactly the same manner previously described.

The truth table shows the CBA inputs cycle through from 000 to 111 two times during the 16 input combinations. The MSB is low during the first cycle, and it is high during the recycle. Refer to Fig. 10–53 for the in-depth look at operation.

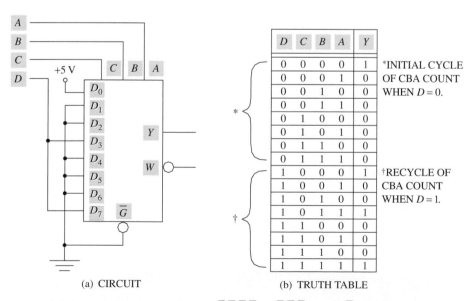

D	C	B	A	Y	
0	0	0	0	1	*INITIAL CYCLE
0	0	0	1	0	OF CBA COUNT
0	0	1	0	0	WHEN $D = 0$.
0	0	1	1	0	
0	1	0	0	0	
0	1	0	1	0	
0	1	1	0	0	
0	1	1	1	0	
1	0	0	0	1	†RECYCLE OF
1	0	0	1	0	CBA COUNT
1	0	1	0	0	WHEN $D = 1$.
1	0	1	1	1	
1	1	0	0	0	
1	1	0	1	0	
1	1	1	0	0	
1	1	1	1	1	

(a) CIRCUIT   (b) TRUTH TABLE

FIGURE 10-53   Logic function generator–$\overline{A}\,\overline{B}\,\overline{C}\,\overline{D} + \overline{A}\,\overline{B}\,CD + AB\overline{C}D + ABCD$.

$DCBA = 0000-Y$ is high because D_0 is tied high.

$DCBA = 0001$ and $0010-Y$ is low because D_1 and D_2 are tied low.

$DCBA = 0011-Y$ is low because D_3 is selected and input D is low.

$DCBA = 0100-0110-Y$ is low because D_4, D_5 and D_6 are tied low.

$DCBA = 0111-Y$ is low because D_7 is selected and input D is low.

Note: At this point in the select input cycle the *CBA* inputs start over (recycle) at 000. This time the *D* input is high.

$DCBA = 1000-Y$ is high because D_0 is high.

$DCBA = 1001$ and $1010-Y$ is low because D_1 and D_2 are tied low.

$DCBA = 1011-Y$ is high because D_3 is selected and input D is high.

$DCBA = 1100-1110-Y$ is low because D_4, D_5 and D_6 are tied low.

$DCBA = 1111-Y$ is high because D_7 is selected and input D is high.

Another example of implementing a 4-variable logic function with a 74151 MUX follows shortly. First, let's look at how these circuits are designed. The key in this analysis is to compare the *CBA* counts during the initial cycle when *D* is low ($\underline{0000}-\underline{0111}$) and during the recycle when *D* is high ($\underline{1000}-\underline{1111}$).

Design/Implementation Rules

Refer to Fig. 10–54 for the explanations that accompany the design rules.

Rule 1. Connect a Logic 0 to the selected data input line if the desired Y output is low during the initial and recycle counts of *CBA*.

The desired low outputs occur at $0\underline{001}/1\underline{001}$, $0\underline{100}/1\underline{100}$, $0\underline{101}/1\underline{101}$, and $0\underline{110}/1\underline{110}$. The *CBA* inputs have been underscored because they control the selected input data. The circuit shows D_1, D_4, D_5, and D_6 are all connected to ground per Rule 1.

Rule 2. Connect a Logic 1 to the selected data input line if the desired *Y* output is high during the initial and recycle counts of *CBA*.

No condition in this circuit meets this requirement. However, check back to the previous circuit in Fig. 10–53(a), and you will see D_0 is tied high to comply with this rule.

Rule 3. Connect *D* to the selected data input line if the desired *Y* output is different during the initial cycle and

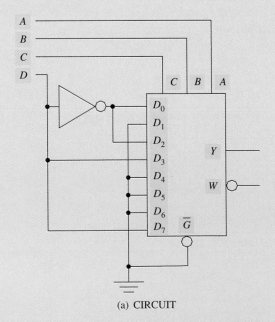

(a) CIRCUIT

D	C	B	A	Y
0	0	0	0	1
0	0	0	1	0
0	0	1	0	1
0	0	1	1	0
0	1	0	0	0
0	1	0	1	0
0	1	1	0	0
0	1	1	1	0
1	0	0	0	0
1	0	0	1	0
1	0	1	0	0
1	0	1	1	1
1	1	0	0	0
1	1	0	1	0
1	1	1	0	0
1	1	1	1	1

(b) TRUTH TABLE

FIGURE 10-54 **Logic function generator—$\overline{A}\,B\,\overline{C}\,\overline{D} + \overline{ABC}\,\overline{D} + AB\overline{C}D + ABCD$.**

the recycle counts of *CBA*, and if the desired *Y* output level equals the level of *D*.

The *CBA* counts of $\overline{0011}/1011$ and $\overline{0111}/1111$ comply with this rule. On the count of $\overline{0011}$, the desired *Y* output is low and on the count of $10\overline{11}$, the desired *Y* output is high. Thus, the two counts require different output levels and those levels are equal to the level of the *D* input. The same is true for $\overline{0111}/1111$.

Rule 4.   Connect $\overline{D}$ to the selected data input line if the desired *Y* output is different during the initial cycle and the recycle counts of *CBA*, and if the desired *Y* output level equals the complement of *D* ($\overline{D}$).

The *CBA* counts of $\overline{0000}/1000$ and $\overline{0010}/1010$ comply with this rule. The desired *Y* output for the count of $\overline{0000}$ is high. The desired *Y* output for the count of 1000 is low. The two counts require different output levels, but these desired outputs are the complement of the *D* input. Note in the circuit in Fig. 10–54(a) that $\overline{D}$ is connected to the D_0 and D_2 inputs.

The circuit implemented in the figure produces the following functional expression: $\overline{A}\,\overline{B}\,C\,\overline{D} + \overline{A}BC\,\overline{D} + AB\overline{C}D + ABCD$. This expression can be simplified. However, the MUX and one NOT gate is the simplest method of implementation.   ∎

Data Conversion

Data conversion is accomplished by the circuit shown in Fig. 10–55. The parallel input data (I_0–I_7) is converted to serial data in the 74151 MUX. The data is transmitted in serial format to the 74LS138 where it is demultiplexed. Once returned to its original parallel format, the data is available at the circuit's output lines (O_0–O_7). Data $63_{(16)}$ is being transmitted to the demultiplexer. The data is transmitted serially as $01100011_{(2)}$ (D_7–D_0) from the MUX IC to the DEMUX IC. The 7493 counter is wired as a MOD-8 counter, and it is used to synchronize the transmitting MUX with the receiving DEMUX.

Troubleshooting in this circuit should be accomplished by isolation to a stage. If the data is not available at any of the output lines, ensure it is present at the input lines. Next, make sure the select input levels to the MUX are present. These CBA inputs can be checked

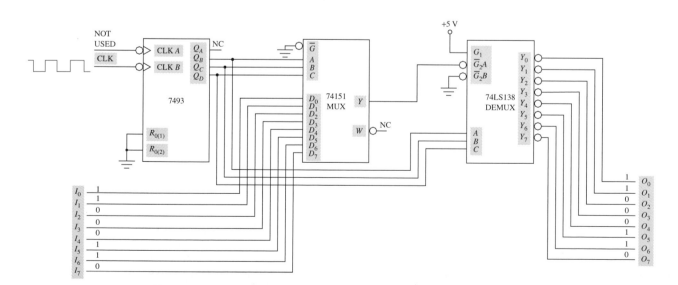

FIGURE 10-55   Data transmission circuit.

with a logic probe to ensure they are pulsing. If they are not pulsing, troubleshoot the counter as described in Chapter 7. If the select inputs are pulsing, assume the counter is functioning properly and check the MUX output. The use of a logic probe at the MUX output will only indicate whether the line is pulsing or not. An oscilloscope may be used to view the data serially at the Y output as it appears on the input lines. If the MUX output is not pulsing, check the IC's V_{CC}, ground, and enable pins for proper voltage levels. If they are good, disconnect the load and check the MUX output again. If it is still not pulsing, replace the MUX.

If the MUX Y output pin is pulsing, check the DEMUX data input at $\overline{G_2}A$. If this signal is normal, check the select input pins. Remember to check the DEMUX $\overline{G_1}$, $\overline{G_2}B$, V_{CC}, and ground pins before deciding to replace it.

This section is summarized in the chapter summary. The end of chapter troubleshooting questions and problems will expand your knowledge in this area. In addition, they will enhance your understanding of the circuits' operations.

CHAPTER 10: SUMMARY

A representative sampling of medium-scale integrated circuits was presented in this chapter. Other MSI circuits are interspersed throughout the text in their appropriate chapters.

Some of the **decoders** in this chapter are used to detect codes or numbers and activate a corresponding output line to indicate that input code or number. These circuits are often called **1-of-n decoders.**

Other types of decoders convert a binary code or number to another format. A prime example of this type of **code converter** is the BCD-to-7-segment decoder. This decoder detects a specific BCD number at the input and converts it to a 7-segment code to illuminate the proper segments of a 7-segment indicator.

There are two basic types of 7-segment indicators—common-anode and common-cathode. Each requires its own particular type of decoder. Many of the decoders improve displays with special features such as **blanking** and **leading zero suppression.** They also usually incorporate a **lamp test** input to aid in troubleshooting.

Encoders are decoders. In other words, they perform the opposite function of their decoder counterparts. Encoders detect an active-input line and convert it to a binary code or number at the output.

Priority encoders respond to the highest-order activated input number when more than one input is activated simultaneously. This capability allows prioritizing inputs, and at the same time does away with the invalid outputs produced by nonpriority encoders when two inputs are activated.

Multiplexers are called **data selectors.** A MUX is used to expand the data input points of a circuit. This is done by taking 2-line, 4-line, 8-line, or 16-line inputs and routing the selected input data to the single output line. This multiplexing action is often used to convert **parallel data to serial data.** The process can be thought of in terms of time-sharing. In this sense, the multiple-input lines are time-shared with a single output line.

Demultiplexers are multiplexers. The DEMUX performs the opposite function of a MUX. A MUX selects data from the input, where the DEMUX takes the data from one input line and distributes that data to several output lines. Thus, they are referred to as **data distributors.**

Decoders are used as demultiplexers when one of their enable input pins is used to input data to the IC. A DEMUX circuit can be used to convert data from its **serial format to the parallel format.**

Most of the decoder, encoder, and multiplexer ICs have enable input pins. These pins allow cascading of the ICs with minimal or no extra control circuitry. Cascading allows expansion of the circuits so they can handle larger binary numbers.

Magnitude comparators are used to determine which of two input quantities is larger. These ICs have three available output pins—$A > B, A = B, A < B$. The 7485 IC is a 4-bit magnitude comparator. It is used to compare two 4-bit binary or BCD numbers. Multiple magnitude comparators can be cascaded to compare larger binary words.

CHAPTER 10: END OF CHAPTER QUESTIONS/PROBLEMS

Answers are given in the Instructor's Manual.

Note: The use of manufacturers data sheets is recommended to assist you in answering these questions.

SECTION 10-1

1. What is the purpose of a decoder?
2. What are the Y output levels of the 74LS139 decoder in Fig. 10–56 when $1\overline{G} = 1, 2\overline{G} = 0$, $1A = 0, 1B = 0, 2A = 1,$ and $2B = 0$?

3. What are the Y output levels of the 74LS138 decoder in Fig. 10–57 when $G_1 = 0, \overline{G_2}A = 0, \overline{G_2}B = 0$, and $CBA = 000$?
4. What are the Y output levels of the 74LS138 decoder in Fig. 10–57 when $G_1 = 1, \overline{G_2}A = 0, \overline{G_2}B = 0$, and $CBA = 011$?
5. Which decoder in Fig. 10–58 is enabled when $DCBA = 0101$?
6. Which LED in Fig. 10–58 is illuminated when $DCBA = 1100$?

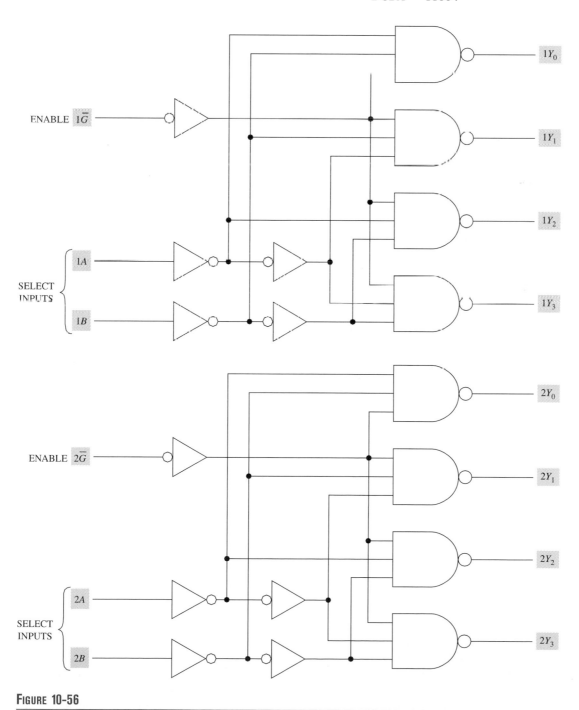

FIGURE 10-56

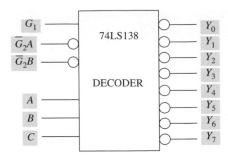

FIGURE 10-57

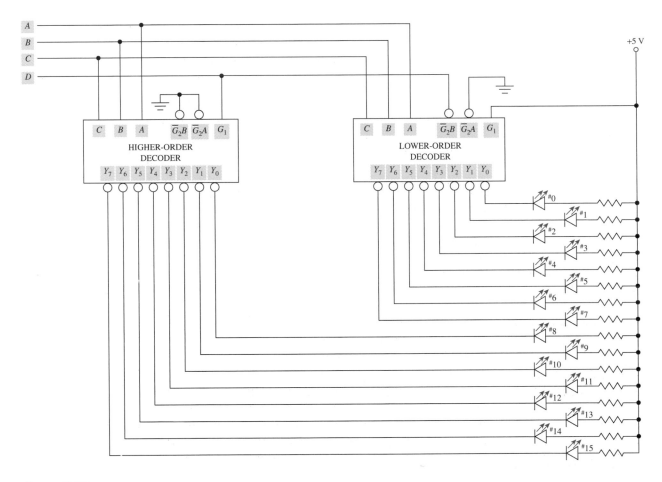

FIGURE 10-58

7. How many decoder outputs are required to decode inputs $00000–11111_{(2)}$?

8. A BCD-to-7-segment decoder with active-low outputs is designed to drive common-_____ 7-segment indicators.

9. The 7-segment indicator in Fig. 10–59 will indicate _____ when $\overline{LT} = 0$, $\overline{BI/RBO} = 1$, and DCBA = 0000.

10. The 7-segment indicator in Fig. 10–59 will indicate _____ when $\overline{LT} = 1$, $\overline{BI/RBO} = 0$, $\overline{RBI} = 0$, AND DCBA = 0000.

11. Implement a 1-of-24 decoder using only 74LS138 ICs.

12. Which output line of the decoder designed in question 11 is activated with an input of $EDCBA = 10111$?

CT 13. Draw the Y_0 output of the circuit in Fig. 10–60. *Note:* Draw the output in relation to the select input waveforms (ABC).

CT 14. Draw the Y_5 output of Fig. 10–60 in relation to the select input waveforms.

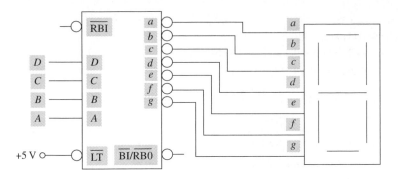

Figure 10-59

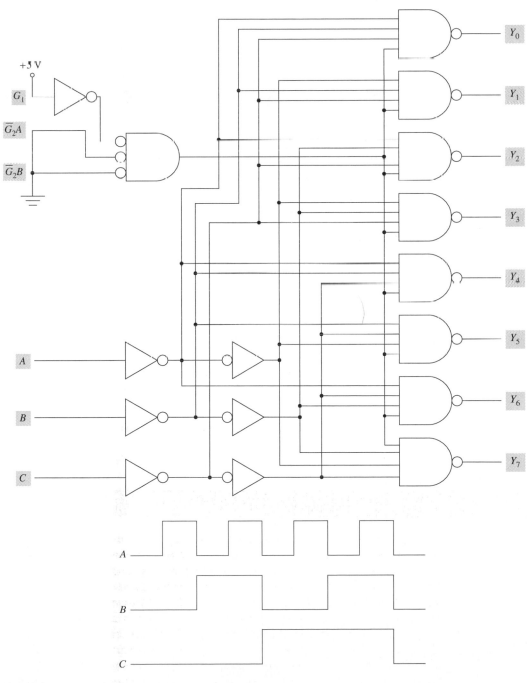

Figure 10-60

Section 10-2

15. What is the purpose of an encoder?
16. What are the outputs of the encoder in Fig. 10–61 when $I_3 = I_2 = 1$ and $I_1 = I_0 = 0$?
17. If more than one input of a priority encoder is activated, which active input is responded to?
18. What is the output of the 74147 priority encoder in Fig. 10–62 when $\bar{6} = \bar{2} = 0$ and the remaining inputs are high?
19. What is the output of the encoder in Fig. 10–62 when none of the input lines is active?
20. What are the A_3–A_0 output levels of the circuit in Fig. 10–63 when the A input line is activated and the remaining inputs are inactive?

CT 21. What are the A_3–A_0 output levels of the circuit in Fig. 10–63 when the 2 and 6 inputs are activated, and the remaining inputs are inactive? Explain your answer.

CT 22. What are the A_3–A_0 output levels of the circuit in Fig. 10–63 when the 7 and A inputs are activated, and the remaining inputs are inactive? Explain your answer.

Section 10-3

23. What is the function of a multiplexer in a digital circuit?
24. Which MUX in Fig. 10–64 is enabled when the select inputs are S_3–$S_0 = 0011$?
25. What is the level of the output data in the circuit in Fig. 10–64 with the data inputs shown and S_3–$S_0 = 0011$?
26. What is the level of the output data in the circuit in Fig. 10–64 with the data inputs shown and S_3–$S_0 = 1101$?
27. The circuit shown in Fig. 10–65 is a _____.

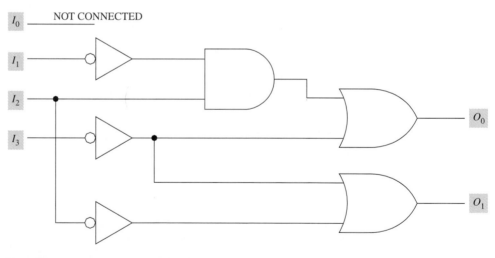

Figure 10-61

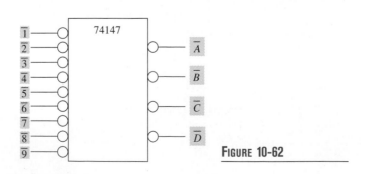

Figure 10-62

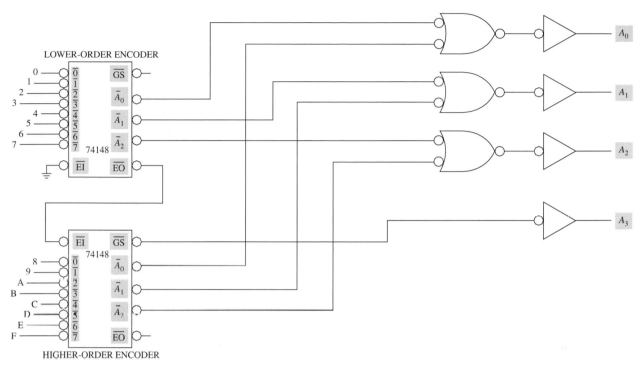

FIGURE 10-63

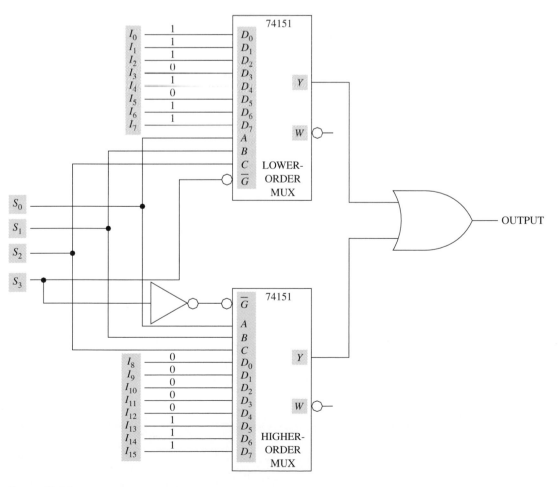

FIGURE 10-64

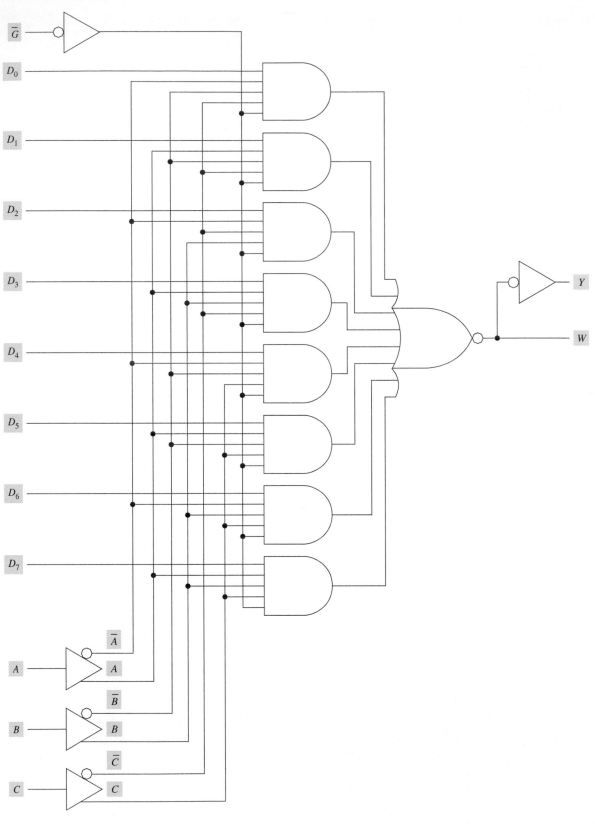

Figure 10-65

28. What is the Y output of the circuit shown in Fig.10–65 if $\overline{G} = 0$, $CBA = 100$, $D_7 - D_4 = 1$, and $D_3 - D_0 = 0$?

CT 29. Draw the Y output of the circuit in Fig. 10–66 in relation to the input waveforms.

CT 30. Draw the Y and W outputs of the 74151 MUX in Fig. 10–67 in relation to the select input waveforms (ABC).

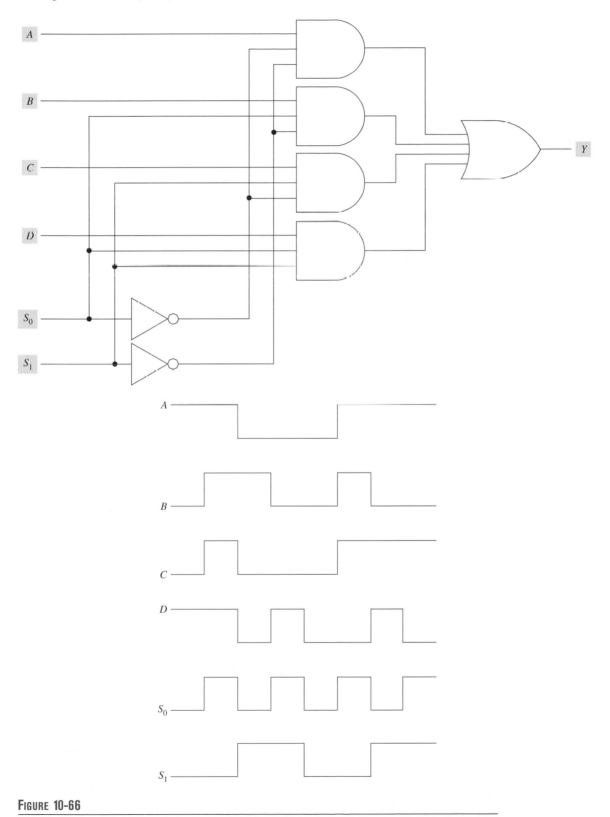

FIGURE 10-66

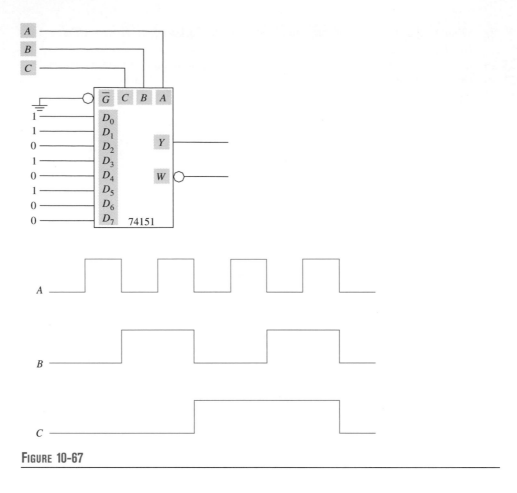

Figure 10-67

Section 10-4

31. What is the function of a demultiplexer in a digital circuit?
32. A decoder circuit that has data connected to its enable input functions as a
    _____ .
33. What data is on each output line of the circuit in Fig. 10–68 after the select inputs
    have cycled through 00–11 in synchronism with the data input?

Section 10-5

34. The circuit shown in Fig. 10–69 is used to compare two _____ -bit numbers.
35. Give the three output logic levels from the lower-order comparator in Fig. 10–69
    when

B_7	B_6	B_5	B_4	B_3	B_2	B_1	B_0	A_7	A_6	A_5	A_4	A_3	A_2	A_1	A_0
1	1	1	1	1	1	0	0	1	1	1	1	1	1	0	1

36. What are the three output logic levels from the upper-order comparator in Fig. 10–69
    under the same input conditions specified in question 35?

Section 10-6

37. What input signal conditions in Fig. 10–70 (on p. 604) are required to enable the
    74LS138 decoder?
    $A_{19}-A_{16} = $ _____
    $\overline{\text{MEMR}} = $ _____
38. What signal levels are required in Fig. 10–70 to activate the ROM ADDRSEL
    signal?

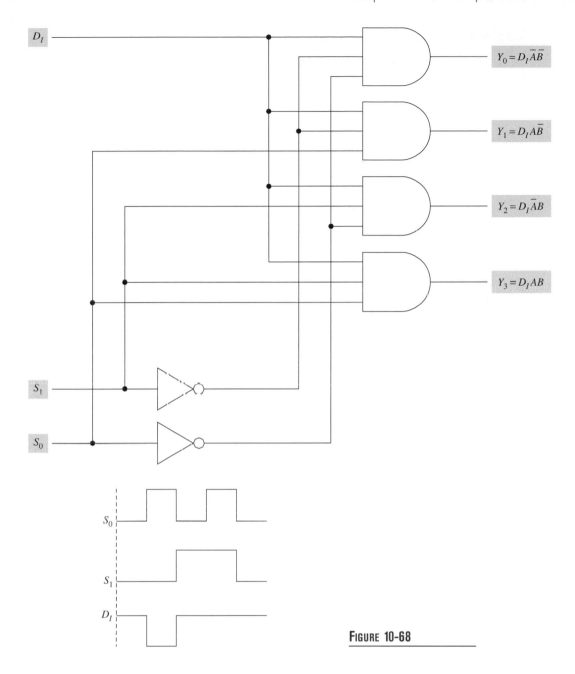

$$Y_0 = D_I \overline{A}\,\overline{B}$$

$$Y_1 = D_I A \overline{B}$$

$$Y_2 = D_I \overline{A} B$$

$$Y_3 = D_I A B$$

FIGURE 10-68

CT 39. Determine if ROM in Fig. 10–70 is being addressed when $\overline{\text{MEMR}} = 1$ and

A_{19}	A_{18}	A_{17}	A_{16}	A_{15}	A_{14}	A_{13}	A_{12}	A_{11}	A_{10}	A_9	A_8	A_7	A_6	A_5	A_4	A_3	A_2	A_1	A_0
1	1	1	1	0	0	0	0	X	X	X	X	X	X	X	X	X	X	X	X

If ROM is being addressed, which ROM chip?

CT 40. Determine if ROM in Fig. 10–70 is being addressed when $\overline{\text{MEMR}} = 0$ and

A_{19}	A_{18}	A_{17}	A_{16}	A_{15}	A_{14}	A_{13}	A_{12}	A_{11}	A_{10}	A_9	A_8	A_7	A_6	A_5	A_4	A_3	A_2	A_1	A_0
1	1	1	1	1	1	1	1	X	X	X	X	X	X	X	X	X	X	X	X

If ROM is being addressed, which ROM chip?

CT 41. Determine if ROM in Fig. 10–70 is being addressed when $\overline{\text{MEMR}} = 0$ and

A_{19}	A_{18}	A_{17}	A_{16}	A_{15}	A_{14}	A_{13}	A_{12}	A_{11}	A_{10}	A_9	A_8	A_7	A_6	A_5	A_4	A_3	A_2	A_1	A_0
1	1	1	0	0	0	0	0	X	X	X	X	X	X	X	X	X	X	X	X

If ROM is being addressed, which ROM chip?

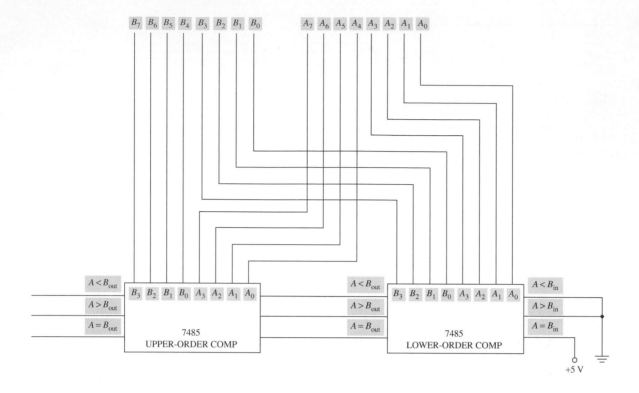

FIGURE 10-69

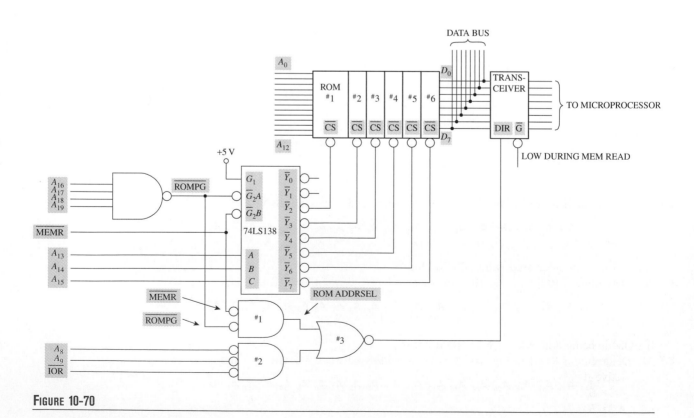

FIGURE 10-70

CT 42. What would be the result if the *A* input to the 74LS138 decoder in Fig. 10–70 stuck low?

43. What is the direction of data flow through the transceiver in Fig. 10–70 during a read ROM operation?

CT 44. The observed results and desired results of the LSD 7-segment indicator are shown in Fig. 10–71. What is the most probable malfunction in the circuit?

ANSWERS TO REVIEW QUESTIONS

SECTION 10-1

A. (1) Detects an input code/number and activates one of its outputs
   (2) Code converter

B. $Y_3(O_3)$

C. Y_0 (line 0)

D. No output line is activated because $G_1 = 0$ and the decoder is not enabled.

E. 4, 10

F. low (Logic 0)

G. A method employed to blank out the leading zero or zeros in a displayed number.

SECTION 10-2

A. To detect an active input and convert it to a binary number or code.

B. 11

C. highest

D. 000

E. high (1)

F. There is no input line for a 0 input. This input is implied when all of the encoder input lines are inactive.

SECTION 10-3

A. The process of selecting data

B. 1

C. Data selection

D. $1Y = 2Y = 3Y = 4Y = 1$ (all output data = 1).

E. 1 (*C* input data is selected.)

F. 0 (D_7 input data is selected.)

G. 0 (output is 0 because the MUX is not enabled.)

SECTION 10-4

A. The process of data distribution

B. False

C. True

D. $Y_7 = 0$; all other outputs = 1.

SECTION 10-5

A. Compare two quantities to see which of the quantities, if either, is larger.

B. $A > B$, $A < B$, and $A = B$.
   active-high

C. $A > B = 0$, $A < B = 1$, and $A = B = 0$.

D. $A > B = 1$, $A < B = 0$, and $A - B = 0$.

E. $A > B = 0$, $A < B = 1$, and $A = B = 0$.

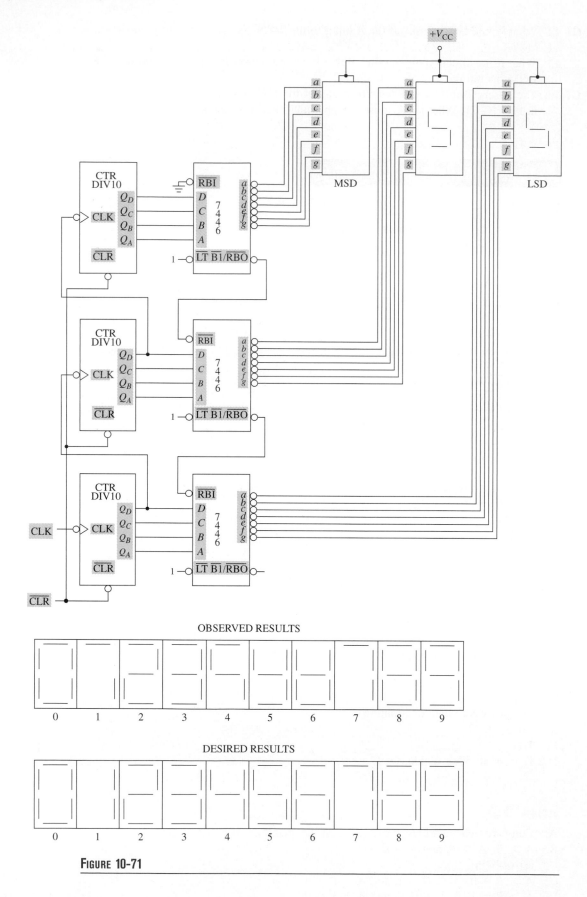

FIGURE 10-71

ANSWERS TO INTERNAL SUMMARY QUESTIONS

SECTIONS 10-1 AND 10-2

1.	b	10.	a
2.	c	11.	b
3.	c	12.	d
4.	d	13.	a
5.	a	14.	d
6.	a	15.	b
7.	a	16.	a
8.	c	17.	c
9.	b	18.	d

SECTIONS 10-3 THROUGH 10-5

1.	a	9.	c
2.	b	10.	b
3.	d	11.	d
4.	b	12.	a
5.	a	13.	d
6.	a	14.	b
7.	c	15.	d
8.	c		

INTERFACING AND DATA CONVERSION

Topics Covered in this Chapter

Introduction

Digital ICs are available in several different logic technologies (families). The most popular, transistor-to-transistor logic (TTL) and complementary MOS (CMOS) are presented in this chapter. A look at internal operation of these two families is presented in Appendix A. Several important characteristics of logic families are discussed in this chapter. Among them are voltage levels, current requirements, current sourcing and sinking, power dissipation, propagation delay, and fan-out.

The **interface** of logic devices sometimes requires special circuitry to condition the logic signal for compatibility between different logic circuits or families. This may be required when two different logic families are connected together. Analysis of interface requirements is accomplished in Section 11–2.

Wire-ANDing gate outputs with ICs that have **open-collector outputs** is presented in detail in this chapter. The necessity for this special-purpose group of ICs is discussed and their operation is covered. In addition, a section has been set aside to discuss **tristate logic.** Tristate (3-state) logic circuits/ICs have an output state that produces an extremely high resistance between the logic circuit and its output pin. This state is called the **high-**

Analog-to-Digital Conversion

Analog-to-Digital Converter (ADC)

Buffer

Bus Contention

Bus Transceiver

CMOS Technology

Current Sinking

Current Sourcing

Differential Nonlinearity

Digital-to-Analog Conversion

Digital-to-Analog Converter (DAC)

Drive Gate

Fan-In

Fan-Out

Flash ADC

High-Impedance (Hi-Z) State

Interface

Load

Load Gate

Monotonic

Open-Collector Output

Resolution

R–2R Ladder

Simultaneous ADC

Sink Current

Source Current

Successive-Approximation ADC

Switched Current-Source DAC

Totem-Pole Output

Tristate

TTL Technology

Unit Load

Wire-ANDing

Chapter Objectives

1. Match terms associated with interfacing and data conversion with their definitions.

2. Given the logic diagram or logic symbol of an open-collector or tristate output circuit, identify the circuit and determine its output under stated input conditions.

3. Determine the interface requirements of a circuit.

4. State the principles of digital-to-analog and analog-to-digital conversion.

5. Identify specified types of data conversion circuits—R–2R ladder, successive-approximation, and flash—and state the principles of each.

6. Troubleshoot circuits containing bus transceivers, DACs, or ADCs.

impedance state. Buffers, bus transceivers, and registers with 3-state outputs are presented in Section 11–3.

Finally, Section 11–4 presents methods and ICs used to convert digital signals to analog signals using **digital-to analog converters (DACs)** and to convert analog-to-digital signals using converters called **ADCs.**

Section 11-1: Integrated Circuit Technologies

Objective

Describe the families and the basic components used in TTL, CMOS, and BiCMOS devices.

Transistor-to-transistor logic and CMOS technology were introduced in Chapter 3. They are discussed again in this chapter and will receive a final look in Appendix A.

Transistor-to-Transistor Logic (TTL)

Transistor-to-transistor logic (TTL) technology is characterized by its use of bipolar transistors. The logic gates, flip-flops, counters, registers, arithmetic, and MSI circuits presented in this textbook are all available in TTL technology. The popularity of TTL technology stems from its medium to high speed, low cost, and excellent drive capability. However, with advances in MOS technology and the requirement for fewer glue chips in systems, the popularity of TTL devices is fading somewhat.

The original TTL logic circuits are referred to as **standard TTL** ICs. They are numbered 54XX or 74XX. The XX represents the numbers of the specific function of the circuit in the IC. For example, the 7400 is a standard quadruple 2-input NAND gate. The XX may be three digits, as is evidenced in any TTL data book. The 54XX-series ICs are designed for military use over a temperature range of −55°C to 125°C. The 74XX commercial series IC's temperature range is 0°C to 70°C.

Detailed operation of the 7400 NAND gate circuit and a detailed explanation of the data sheets for that gate are presented in Appendix A. Various modifications were made to the original standard TTL circuit to improve operating speed and reduce power consumption. The first change involved varying the size of the resistors in the circuit. Decreasing resistor values improved speed but increased power consumption. The opposite resulted when the resistor values were increased. The ICs employing these modifications are no longer available because later modifications resulted in improved performance.

The Schottky TTL family employs Schottky transistors to achieve higher operating speeds. The 74SXX family was improved via slight circuit modification to form the low-power Schottky (74LS version) and the advanced low-power Schottky (74ALS version). Operating characteristics of the Schottky family may be compared to those of other families in Table 11–1. Propagation delays are listed as low-to-high transition (propagation)

TABLE 11-1		74XX	74SXX	74LSXX	74FXX	74ACXX	74ACTXX	74HCXX
TTL/CMOS Typical Characteristics	t_{PLH}	11 ns	3 ns	9 ns	6 ns	1.5–7.4 ns	1.5–12.3 ns	28 ns t_{PD}
	t_{PHL}	7 ns	3 ns	10 ns	5.3 ns	1.5–6.8 ns	1.5–8.8 ns	
	I_{CCH}	4 mA	10 mA	0.8 mA	2.8 mA	40 μA max I_{CC}	40 μA max I_{CC}	20 μA max I_{CC}
	I_{CCL}	12 mA	20 mA	2.4 mA	10.2 mA			
	PD	40 mW	75 mW	8 mW	32.5 mW	200 μW	200 μW	100 μW
	NM	0.4 V	0.3 V	0.3 V	0.3 V	0.95 V	0.85 V	0.69 V

time (t_{PLH}) and high-to-low time (t_{PHL}) on most data sheets. Some sheets list total propagation delay time (t_{PD}). Power dissipation is derived by averaging the power supply currents for both high-and low-level outputs.

$$I_{CC(avg)} = \frac{I_{CCH} + I_{CCL}}{2}$$

The product of this average current and V_{CC} is the average Power Dissipation (PD) as shown in Table 11-1. Noise Margin (NM) is explained in the next section.

FAST (F) technology is a high-performance TTL family (74FXX). FAST devices offer comparable performance to the advanced Schottky technology ICs but FAST device power consumption is reduced by 25 to 50%. Speed and performance characteristics of the FAST family are also presented in Table 11-1 for comparison to other logic families.

Complementary MOS (CMOS) Technology

CMOS technology is characterized by its use of unipolar transistors (FETs). The high density of CMOS technology allows many more transistors to be diffused in an IC than bipolar technology allows. In addition, CMOS power consumption is very low. CMOS circuits use p-channel and n channel enhancement-mode MOSFETs. Most of the gates, flip-flops, and MSI circuits presented in this text are available in CMOS technology. The CMOS family of ICs consists of several subfamilies. The 74C version is a TTL pin-compatible CMOS IC that has been outdated by technological improvements.

The 74HC and 74HCT series are high-speed, improved versions of the 74C. One of the primary disadvantages of the 74HC series is the minimum input high voltage level is 3.5 V. This makes it incompatible with valid TTL high output voltages of 2.4 V to 3.5 V. This problem was resolved in the 74HCT series when the minimum input high voltage level was decreased to 2.0 V.

The advanced CMOS series includes the 74AC and 74ACT ICs. The 74AC series is faster than the 74HC series. The 74ACT series ICs are modified 74AC technology, which makes the ACT series TTL compatible. A detailed explanation of the internal circuit operation of CMOS gates is presented in Appendix A.

BiCMOS Technology

An integration of CMOS and bipolar technologies produced the BiCMOS bus interface family. The 74BCT family provides high-speed switching characteristics and high drive currents. Latches, flip-flops, buffers, drivers, and transceivers are available in 74BCT technology.

Section 11-1: Review Questions

Answers are given at the end of the chapter.

A. What type of transistor is used in TTL devices?
B. What type of transistor is used in CMOS devices?
C. What type of MOSFET is used in CMOS technology?
   (1) Depletion-mode
   (2) Enhancement-mode
D. What types of transistors are used in BiCMOS devices?

SECTION 11-2: VOLTAGE/CURRENT COMPATIBILITY

OBJECTIVES

1. Define source current, sink current, fan-out, and noise immunity/margin.
2. Identify drive and load circuits when gates are interconnected and determine whether they are sourcing or sinking current when the inputs are specified.
3. Calculate fan-out.
4. Determine the interface requirements of a circuit.
5. Identify and state the purpose of open-collector output devices.

All sections or subsections of a digital circuit such as the one shown in Fig. 11–1 must be compatible to make the circuit function properly. A **drive circuit** is a logic gate or device used to provide the input(s) to another logic gate(s) or device(s). A **load circuit** consists of the logic gate(s) or device(s) connected to a drive gate's output. The inputs to the load circuits must have enough, but not too much, voltage amplitude and current to ensure proper operation.

If the load circuits' inputs are not proper, an **interface circuit** must be designed to condition the output of the drive circuit so that it will produce proper operation in the load circuits. To interface these circuits is to connect them together with the extra components/circuits necessary to make the connection compatible.

Manufacturer's data sheets can be used to determine compatibility of a drive output to load (driven) input requirements. Minimum and maximum levels of voltage and current must be checked on the data sheet when designing a circuit. For example if the minimum high-level output voltage of a drive gate is 2.4 V, and the minimum high-level input voltage required to drive a load gate is 3.5 V, there is a compatibility problem. The problem is that the drive gate's minimum high-level output voltage (2.4 V) will be below the load gate's minimum input voltage (3.5 V) requirement.

Interface circuits are not always necessary between drive and load circuits. For example, if the output of a TTL NAND gate is used to drive the inputs of several more TTL gates, interface circuitry may not be necessary because these gates are usually compatible. However, interface is often necessary when a TTL device is used to drive the inputs of CMOS devices or vice versa.

Prior to delving into specific interface problems and circuits, an understanding of **current sourcing** and **current sinking** is necessary. Appendix A provides a detailed explanation of electrical and switching characteristics of standard TTL gates. A review of the appendix is beneficial, though not an absolute necessity, at this point to aid understanding of the concepts presented in this section. The voltage and current levels used in the following examples were taken from the SN7400 standard NAND gate data sheet (Appendix B) that is discussed in Appendix A. The terms used for these levels are:

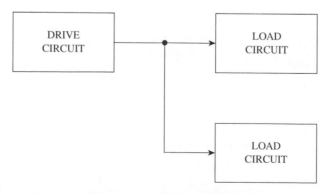

FIGURE 11-1   Drive/Load circuit block diagram.

V_{OL}:   Low-level output voltage of a gate.

V_{OH}:   High-level output voltage of a gate.

V_{IL}:   Low-level input voltage to a gate.

V_{IH}:   High-level input voltage to a gate.

I_{OL}:   Low-level output current into an output terminal when the output voltage is low (V_{OL}).

I_{OH}:   High-level output current from an output terminal when the output voltage is high (V_{OH}).

I_{IL}:   Low-level input current flowing out of an input terminal when a low-level voltage (V_{IL}) is applied to that input.

I_{IH}:   High-level input current into an input terminal when a high-level voltage (V_{IH}) is applied to that input.

Current flowing *toward* a gate's input/output pin is designated *positive*; current flowing *away* is designated *negative*. Current direction arrows are normally drawn toward a gate's input/output pins (positive designation) in data books.

The problem with this current flow convention is illustrated in Fig. 11–2. The arrows for I_{OH} and I_{IH} are drawn in accordance with the positive convention. However, the source of the current is not evident using this convention. This problem is resolved when looking at the data sheet for a given IC. The I_{OH} current specified for the 7400 NAND gate is −0.4 mA. *The minus sign indicates the current is actually flowing out of the output pin of the drive gate even though convention has it drawn in the opposite direction.* If the I_{OH} current arrow is reversed (as should be done in your mind when you see the minus sign on the current value), it is easy to see the current source of this circuit originates in the drive gate.

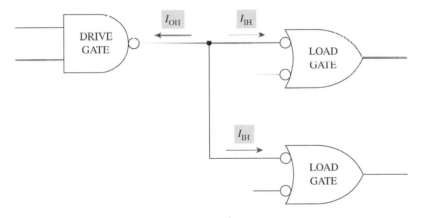

FIGURE 11-2   **Current convention.**

Figure 11–3 presents the two input current conditions (I_{IH} and I_{IL}) and the two output current conditions (I_{OH} and I_{OL}) for a NAND gate. The current values discussed here are from the SN7400 data sheet.

Figure 11–3(a) shows the *current flow into an input* of a NAND gate when a high is applied to that input. The data sheet specifies $I_{IH(MAX)}$ is 40 μA.

Figure 11–3(b) shows the *current flow out of an input* of a NAND gate when a low is applied to that input. Although the positive convention for drawing I_{IL} is used in the figure, the data sheet specifies $I_{IL(MAX)}$ is −1.6 mA. Thus, the current is actually flowing out of the input terminal of the gate as shown in the figure.

Figure 11–3(c) shows *current flowing out of the output pin* of the NAND gate when its output voltage is high (V_{OH}). The output current direction is again determined by the fact that $I_{OH(MAX)}$ is specified as −0.4 mA on the data sheet.

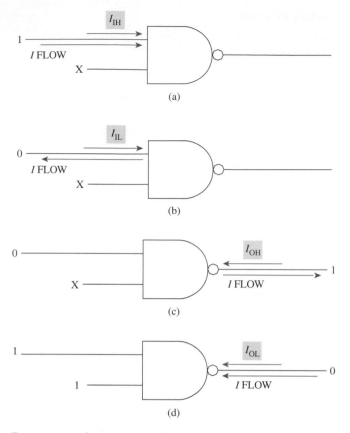

FIGURE 11-3 Current convention.

Figure 11–3(d) shows *current flowing into the output pin* of the NAND gate when its output voltage is low (V_{OL}). The NAND gate data sheet specifies $I_{OL(MAX)}$ is 16 mA.

Figure 11–3 leads us directly to the subject of current sourcing and current sinking.

Current Sourcing and Current Sinking

All devices used to drive TTL gates must source and sink current. **Source current** is the current that flows *out* of a logic circuit. The source current may flow out of an input or output pin. **Sink current** is the current that flows *into* a logic circuit on an input or output pin. Conventional current flow is used in the following analysis.

Figure 11–4 illustrates **current sourcing.** The low-level input voltage to the drive gate produces a high output voltage (V_{OH}). The high-level output current (I_{OH}) that flows

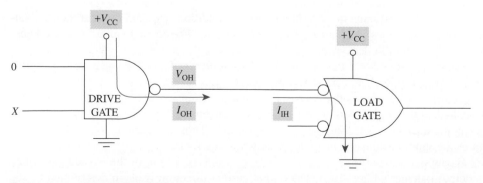

FIGURE 11-4 Current sourcing.

TABLE 11-2		
Standard TTL Gate Current Specifications	I_{IH}	40 μA*
	I_{IL}	−1.6 mA
	I_{OH}	−0.4 ma
	I_{OL}	16 mA

* All values are maximum.

out of the output pin is specified on the data sheet as −0.4 mA. The input and output current specifications for standard TTL gates are summarized in Table 11–2. As previously stated, the minus sign before the current value indicates current out of the gate; therefore, the implication of that minus sign is that this is a source current. This source current originates at V_{CC} in the drive gate and flows to ground (sink current) in the load gate.

Current sinking is illustrated in Fig. 11–5. The two high-level inputs to the drive gate produce a low output signal (V_{OL}). The low-level output current (I_{OL}) that flows into the output pin is 16 mA maximum (Table 11–2). This current is sink current because it is flowing into the output pin of the drive gate. The sink current originates at V_{CC} in the load gate and flows to ground (sunk) in the drive gate.

A point to keep in mind about this discussion is *the drive gate is the reference gate* when determining whether a circuit is sourcing or sinking current. The example of current sourcing in Fig. 11–4 shows the source of current in the circuit is the drive gate. However, that source current (I_{OH}) is sink current (I_{IH}) in the load gate. Since the drive gate is the reference point, the circuit is said to be current sourcing. In addition, the source current (I_{OH}) and sink current (I_{IH}) in Fig. 11–4 must have opposite signs. The same is true of the source current (I_{IL}) and sink current (I_{OL}) in the current sinking circuit of Fig. 11–5. It should also be clear at this point that a logic output sources current when its output is high (V_{OH}) and sinks current when its output is low (V_{OL}).

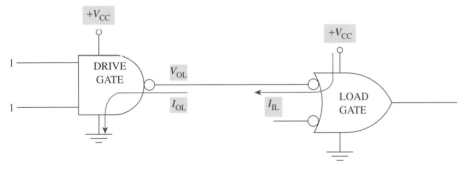

FIGURE 11–5   Current sinking.

The preceding discussion should bring another point to mind regarding source current and sink current. Figure 11–6 illustrates this point with two load gates. The maximum value of the sink current (I_{OL}) is specified on the data sheet as 16 mA. The maximum value of the sum of the two source currents in Fig. 11–6 (I_{IL_1} and I_{IL_2}) is 3.2 mA. This presents no problem because the sink current value of the drive gate (16 mA maximum) is high enough to sink the source currents of both load gates.

However, this indicates there is a point beyond which a circuit designer should not venture. If too many load gates are connected to the output of a TTL gate, the sum of the source currents from the load gates will exceed the sink current capability of the drive gate. The following describes how to determine how many load gates are too many.

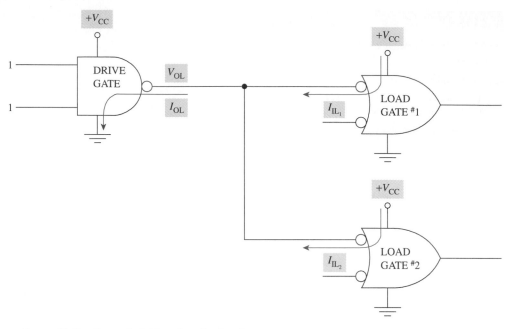

FIGURE 11-6   Current sinking—two load gates.

Fan-out and Fan-in

Fan-out is the maximum number of inputs that the output of a gate can reliably drive. Table 11–3 presents the input and output voltage specifications for standard TTL gates. Figure 11–7 graphically displays these specifications. We already know that TTL logic low levels range from 0 V to 0.8 V and logic high levels range from 2 V to 5 V. The data in Fig. 11–7 corresponds to this information, but it includes all of the minimum, maximum, and typical voltage values for the following discussion.

Figure 11–8 illustrates a drive gate sinking current from ten load gates when its output is in the low state. Since the output of the drive gate (I_{OL}) can sink a maximum of 16 mA, the maximum number of inputs (I_{IL}) that the output can drive is 10.

$$\frac{I_{OL}}{I_{IL}} = \frac{16\,\text{mA}}{1.6\,\text{mA}} = 10$$

Each load gate can source up to 1.6 mA (I_{IL}). Each of these currents represents 1 Unit Load (UL) on the output of the drive gate. Therefore, the fan-out (low) is 10 inputs or 10 ULs. A **unit load** is the measure of the input/output current of a load gate. One UL for the standard TTL family is equal to 1.6 mA in the low state. A unit load is sometimes referred to as **fan-in** because it connects to a drive gate's output which is called fan-out.

What happens if the fan-out is exceeded? If 12 load gates were connected to the output of the drive gate in Fig. 11–8, the I_{OL} (sum of all I_{IL}s) would be equal to 19.2 mA maximum. The data sheet specifies 16 mA maximum for I_{OL}.

TABLE 11-3		
Standard TTL Gate Voltage Specifications	V_{IH}	2 V min
	V_{IL}	0.8 V max
	V_{OH}	2.4 V min/3.4 V typ
	V_{OL}	0.4 V max/0.2 V typ

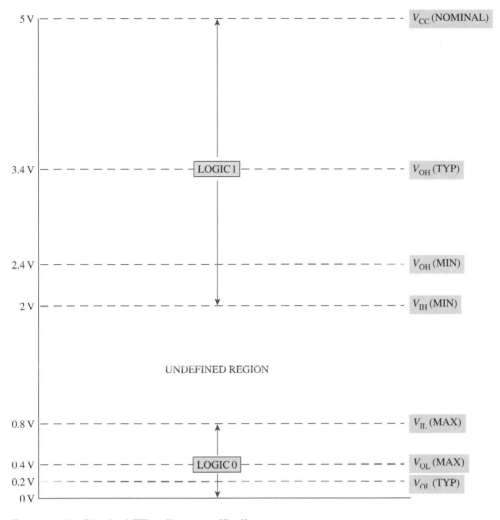

FIGURE 11–7   Standard TTL voltage specifications.

If we take this one step further, the output low voltage (V_{OL}) is specified as 0.4 V (max) in Table 11–3. This voltage is developed across a saturated transistor in the output circuit of the drive gate. The low output voltage will exceed its maximum value (0.4 V) when the drive gate is required to sink more than 16 mA. The circuit will not operate properly if the sink current drives V_{OL} above 0.8 V. The undefined logic levels for TTL gates range from 0.8 V to 2 V. If V_{OL} is driven greater than 0.4 V but not in excess of 0.8 V, the noise margin of the circuit is reduced. Noise margin is the next topic.

Figure 11–9 depicts a drive gate sourcing current to 10 load gates when its output is in the high state. Table 11–2 specifies $I_{OH} = -0.4$ mA (400 μA) and $I_{IH} = 40$ μA. This indicates the drive gate can provide (source) current for 10 inputs.

Each of the load gate's input currents (I_{IH}) represents 1 UL on the output of the drive gate. Thus, 1 UL = 40 μA in the high state.

$$\frac{I_{OH}}{I_{IH}} = \frac{400 \text{ μA}}{40 \text{ μA}} = 10$$

These numbers show the fan-out (high) for a standard TTL gate is also 10 inputs or 10 ULs. If the fan-out is exceeded by connecting more than 10 inputs to the drive gate, the output voltage (V_{OH}) of the drive gate will decrease below $V_{OH(MIN)}$, and the circuit will

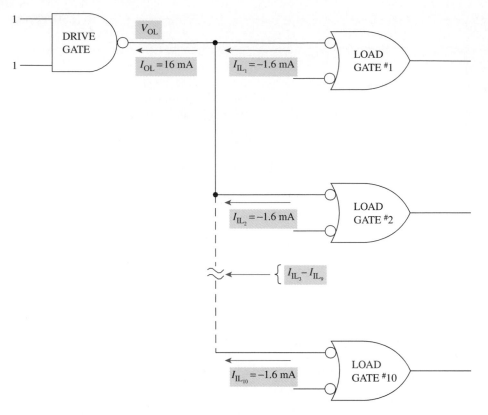

FIGURE 11-8   **Current sinking—ten load gates.**

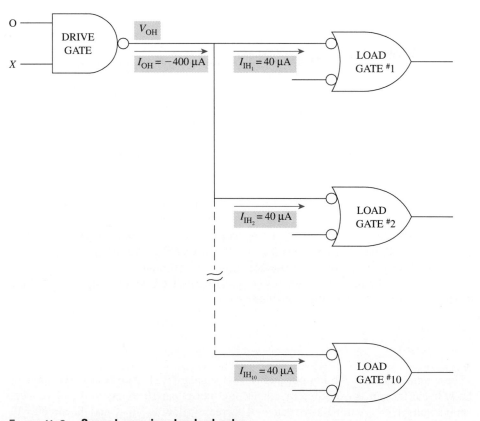

FIGURE 11-9   **Current sourcing—ten load gates.**

not function properly. The noise margin of the circuit is affected when V_{OH} is between 2.4 V and 2.0 V.

Fan-out for most TTL families ranges from 10 to 20. CMOS devices have a very high input resistance; thus, they sink or source very little current from/to the drive gate. This allows fan-out to be up in the range of 50 when operating at low frequencies. On the other hand, the input capacitance of the CMOS load gates causes switching time to deteriorate at high frequencies (MHz range) and fan-out decreases at these frequencies.

Noise Margin/Immunity

Noise immunity is a circuit's ability to tolerate noise signals on its inputs and still maintain reliable operation. **Noise margin** is a measure of noise immunity. Noise margin is calculated by taking the difference of the worst case output and the worst case input voltages. Figure 11–10(a) shows a noise signal riding a high logic level and a low logic level.

Low-State Noise Margin.

The TTL voltage levels annotated on Fig. 11–10(a) show the maximum low-input logic level ($V_{IL(MAX)}$) is 0.8 V. The figure also shows the maximum low-output logic level ($V_{OL(MAX)}$) is 0.4 V. This $V_{OL(MAX)}$ value represents a fully loaded, worst case condition for the low output of the drive gate.

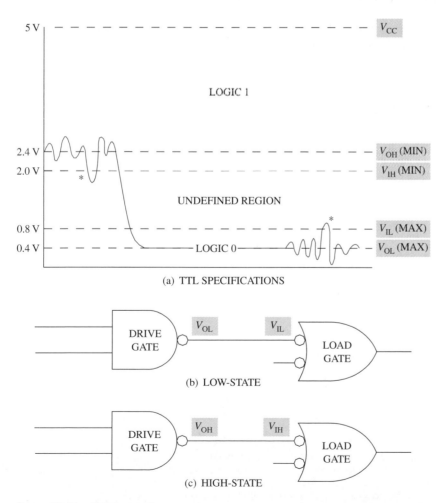

(a) TTL SPECIFICATIONS

(b) LOW-STATE

(c) HIGH-STATE

FIGURE 11–10   Noise margin.

Figure 11–10(b) displays these maximum voltages pictorially in a circuit. If the maximum worst case output (V_{OL}) from the drive gate is 0.4 V and up to a 0.8-V input (V_{IL}) is acceptable at the load gate's input, a margin of 0.4-V exists. This margin is referred to as the noise margin for the low state. Noise levels riding a worst case 0.4-V output can be as high as 0.4 V without having an effect on the load gate. Should the noise amplitude exceed the 0.4-V noise margin, the input to the load gate would be greater than 0.8 V. This puts the input to the load gate in the undefined logic region, and unreliable operation might be expected.

High-State Noise Margin.

The voltage levels of Fig. 11–10(a) indicate $V_{OH(MIN)}$ is 2.4 V and $V_{IH(MIN)}$ is 2.0 V. These voltages are labeled on the logic diagram in Fig. 11–10(c). The output of the drive gate is at least 0.4 V greater than the minimum acceptable input voltage required for proper operation of the load gate. Again, a margin of 0.4 V exists in the high-state condition. Noise up to 0.4 V riding the worst case 2.4-V output of the drive gate is acceptable. If the noise exceeds 0.4 V in amplitude, the load gate's input voltage will go below the 2.0-V minimum. This puts that input level in the undefined logic region. An input in the undefined region will produce erratic or unreliable operation of a logic gate.

This section commenced with a brief description of interfacing, progressed to current sourcing and current sinking, took a slight detour to fan-out, and then presented noise margin and noise immunity.

Now it is time to return to the subject of interfacing and apply some of the knowledge we have gained up to this point in the chapter. The purpose of the remainder of this section is to determine compatibility of drive outputs to load inputs. The compatibility check must include the low-state and high-state conditions for both voltage and current. This type of analysis was accomplished earlier in the section. Now, the process will include mixed technologies.

Interface Requirements

Example 1: Standard TTL Gate Driving Standard TTL Gate.

Our first example of drive/load gate interconnecting is the simplest because the technologies are not mixed. This example is included to show the steps necessary to perform this type of analysis. Figure 11–11(a) represents the example using two standard TTL gates—7400 NAND gate and 7432 OR gate.

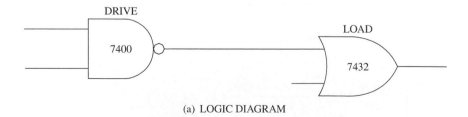

(a) LOGIC DIAGRAM

DRIVE 7400	LOAD 7432
$V_{OH} = 2.4$ V (MIN) $V_{OL} = 0.4$ V (MAX) $I_{OH} = -400\,\mu\text{A}$ (MAX) $I_{OL} = 16\,\text{mA}$ (MAX)	$V_{IH} = 2.0$ V (MIN) $V_{IL} = 0.8$ V (MAX) $I_{IH} = 40\,\mu\text{A}$ (MAX) $I_{IL} = -1.6\,\text{mA}$ (MAX)

(b) VOLTAGE/CURRENT SPECIFICATIONS

FIGURE 11–11   Standard TTL gate driving standard TTL gate.

The methods previously discussed are employed in this analysis—voltage and current levels are checked for compatibility. The output voltage and current levels for the drive gate and the input voltage and current levels for the load gate have been transferred from data sheets to Fig. 11–11(b).

The voltage and current level comparisons reveal

$$V_{OH(\text{Drive MIN})} > V_{IH(\text{Load MIN})}$$
$$V_{OL(\text{Drive MAX})} < V_{IL(\text{Load MAX})}$$
$$I_{OH(\text{Drive MAX})} > I_{IH(\text{Load MAX})}$$
$$I_{OL(\text{Drive MAX})} > I_{IL(\text{Load MAX})}$$

These conditions indicate there are no compatibility problems when a standard TTL gate is used to drive a standard TTL gate. The output high voltage of the driver is more than enough to drive the load gate. Likewise, the output low voltage is below the maximum allowed at the load gate input. The drive gate is sourcing current ($-400 \mu A$) when its output is high. It provides enough current to drive 10 load gates. The drive gate sinks current when its output is low, and it can sink current from 10 load gates.

Example 2: Standard TTL Gate Driving 74LS TTL Gate.

Figure 11–12(a) is a logic diagram illustrating this example problem. The voltage and current levels for the drive and load gates have been transferred from data sheets to Fig. 11–12(b). The voltage levels for the drive and LS load gates are the same as they were in the preceding example. This indicates the two gates are voltage compatible.

Input current to the load gate (I_{IH}) is not required when its input is at the high logic level. The driving gate can source current up to -0.8 mA when necessary, but the 74LS series will sink a maximum of only 20 μA. Since $I_{OH} > I_{IH}$, there is no interface problem for the high logic level. In the low state the drive gate can sink current for as many as 40 load gates.

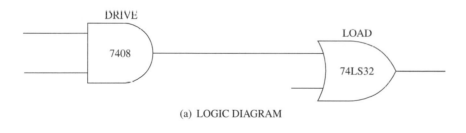

(a) LOGIC DIAGRAM

DRIVE 7408	LOAD 74LS32
$V_{OH} = 2.4$ V (MIN)	$V_{IH} = 2.0$ V (MIN)
$V_{OL} = 0.4$ V (MAX)	$V_{IL} = 0.8$ V (MAX)
$I_{OH} = -0.8$ mA (MAX)	$I_{IH} = 20\mu$ A (MAX)
$I_{OL} = 16$ mA (MAX)	$I_{IL} = -0.4$ mA (MAX)

(b) VOLTAGE/CURRENT SPECIFICATIONS

FIGURE 11–12   Standard TTL gate driving 74LS TTL gate.

Example 3: Standard TTL Gate Driving Advanced CMOS (AC) Gate

A logic diagram of a standard TTL NAND gate driving a 74AC00 (CMOS) NAND gate is shown in Fig. 11–13(a).

The minimum input high voltage (V_{IH}) to the CMOS load gate is 3.85 V @ $V_{CC} = 5.5$ V, and the output of the drive gate may be as low as 2.4 V as depicted in Fig. 11–13(b). These voltage levels are incompatible. The output and input voltage levels in the low state

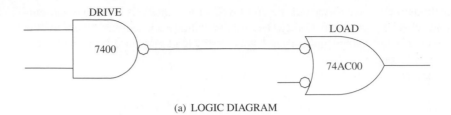

(a) LOGIC DIAGRAM

DRIVE 7400	LOAD 74AC00
$V_{OH} = 2.4$ V (MIN)	$V_{IH} = 3.85$ V (MIN)*
$V_{OL} = 0.4$ V (MAX)	$V_{IL} = 1.65$ V (MAX)*
$I_{OH} = -400\,\mu A$ (MAX)	$I_{IH} = 1.0\,\mu A$ (MAX)
$I_{OL} = 16$ mA (MAX)	$I_{IL} = -1.0\,\mu A$ (MAX)

*$V_{CC} = 5.5$ V

(b) VOLTAGE/CURRENT SPECIFICATIONS

FIGURE 11–13   Standard TTL gate driving advanced CMOS (AC) gate.

---

are compatible. There are no current problems because the drive gate can source or sink the load gate's input current requirements. Since CMOS devices are voltage controlled instead of current controlled like their TTL counterparts, no current problems should be encountered when a TTL gate drives a CMOS gate.

The high-level voltage incompatibility in this example can be resolved with an **interface circuit.** An open-collector high-voltage output buffer such as the one shown in Fig. 11–14 can be used to solve the problem. The 7417 TTL buffer is specifically designed for interfacing with CMOS circuits or for increasing current levels to drive TTL inputs. The 7417 buffer can be connected to V_{CC} values up to 15 V. Another comparable buffer, the 7407, can be connected to V_{CC} values up to 30 V.

This high-voltage incompatibility problem would not have been encountered if a 74ACT00 NAND gate had been used as the load gate instead of the 74AC00. The minimum input voltage for the 74ACT00 gate has been reduced to 2.0 V, which makes it TTL compatible.

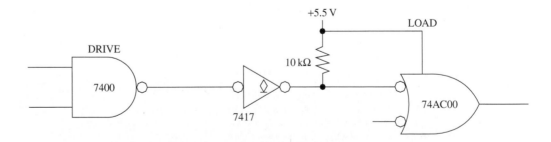

FIGURE 11–14   TTL-to-advanced CMOS interface.

Example 4: Advanced CMOS (AC) Gate Driving Standard TTL Gate.

The voltage and current analysis for the circuit shown in Fig. 11–15 reveals

$$V_{OH(\text{Drive MIN})} > V_{IH(\text{Load MIN})}$$
$$V_{OL(\text{Drive MAX})} < V_{IL(\text{Load MAX})}$$
$$I_{OH(\text{Drive MAX})} > I_{IH(\text{Load MAX})}$$
$$I_{OL(\text{Drive MAX})} > I_{IL(\text{Load MAX})}$$

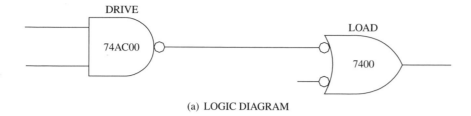

(a) LOGIC DIAGRAM

DRIVE 74AC00	LOAD 7400
$V_{OH} = 4.8$ V (MIN)* $V_{OL} = 0.44$ V (MAX)* $I_{OH} = -24$ mA (MAX)* $I_{OL} = 24$ mA (MAX)*	$V_{IH} = 2.0$ V (MIN) $V_{IL} = 0.8$ V (MAX) $I_{IH} = 40$ µA (MAX) $I_{IL} = -1.6$ mA (MAX)

*$V_{CC} = 5.5$ V

(b) VOLTAGE/CURRENT SPECIFICATIONS

FIGURE 11–15   **Advanced CMOS (AC) gate driving standard TTL gate.**

There are no incompatibility problems in this example of a standard TTL load gate being driven by an advanced CMOS (AC) gate.

Open-Collector ICs.

An open-collector TTL buffer (7417) was mentioned in Example 3. A brief presentation of **open-collector logic** is in order here to ensure proper use of this type of device. A look inside the IC and more details are presented in Appendix A.

Wire-ANDing Device Outputs.

The logic circuit in Fig. 11–16 ANDs the outputs of two NAND gates. The circuit will output high only when the inputs to the AND gate are both high. The expression $\overline{AB} \cdot \overline{CD}$ equals $(\overline{A} + \overline{B})(\overline{C} + \overline{D}) = \overline{A}\,\overline{C} + \overline{A}\,\overline{D} + \overline{B}\,\overline{C} + \overline{B}\,\overline{D}$. The latter equivalent expression indicates the circuit will output high when there is at least one low input to each NAND gate. These low inputs produce the required NAND gate high outputs to the AND gate.

If the circuit in Fig. 11–16 is modified as shown in Fig. 11–17(a), a wire-ANDed circuit results. Note in Fig. 11–17(a) the output AND gate has been removed, and the outputs of the two NAND gates are connected together. This is called wire-ANDing outputs. The principle, as depicted in the truth table in Fig. 11–17(b), is when the outputs of both NAND gates are low, the output of the wire-ANDed connection is low. When the outputs

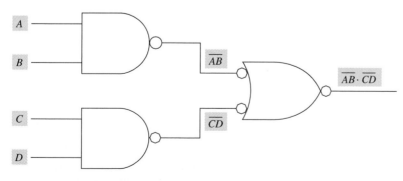

FIGURE 11–16   **NAND/AND circuit logic.**

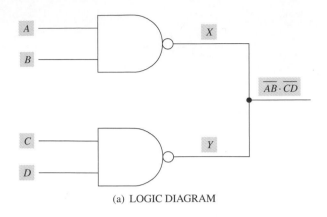

(a) LOGIC DIAGRAM

$X = \overline{AB}$	$Y = \overline{CD}$	OUTPUT
0	0	0
0	1	0
1	0	0
1	1	1

(b) TRUTH TABLE

FIGURE 11–17   Wire-ANDed circuit.

of both NAND gates are high, the output of the wire-ANDed connection is high. Finally, when one NAND gate output is low and one is high, the output of the wire-ANDed connection is pulled low. The truth table in Fig. 11–17(b) is representative of the AND function. Anytime the outputs of logic gates are connected together as shown in this example, the wire-ANDing is represented as shown in Fig. 11–18.

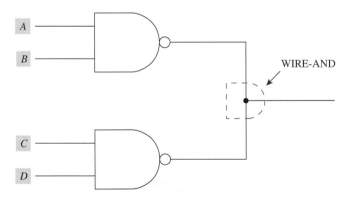

FIGURE 11–18   Wire-ANDing symbology.

Now let's look at some of the ramifications of wire-ANDing outputs. There is a rule that states, "*Do not wire-AND totem-pole outputs.*" The reason for this rule will soon be evident. Figure 11–19 is a simplified schematic diagram of the TTL totem-pole output circuit used in many TTL gates. The output voltage of a logic gate is developed across Q_4. In TTL gates other than the tristate type (Section 11–3), when Q_3 is on, Q_4 is off and vice versa. This type of output arrangement is popular because it keeps power consumption reasonable in TTL devices. The output signal is low when Q_4 is on and high when Q_4 is off. This should make it easy to understand that the output circuit shown in Fig. 11–19 is sinking current when Q_4 is on (Q_3 is off) and sourcing current when Q_4 is off (Q_3 is on).

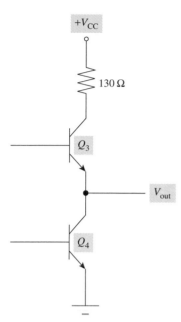

FIGURE 11–19   TTL totem-pole output circuit–simplified
schematic diagram.

Figure 11–20(a) shows two TTL NAND gates with their outputs wire-ANDed. Only the totem-pole output transistors of the NAND gate's internal circuitry are shown in the figure.

High Output.   Low inputs at *A* or *B and* at *C* or *D* will produce high outputs from both gates in Fig. 11–20(a). Q_4 in each gate is off, and Q_3 is on. The NAND gates are current sourcing to the load, and no problems are encountered in this example.

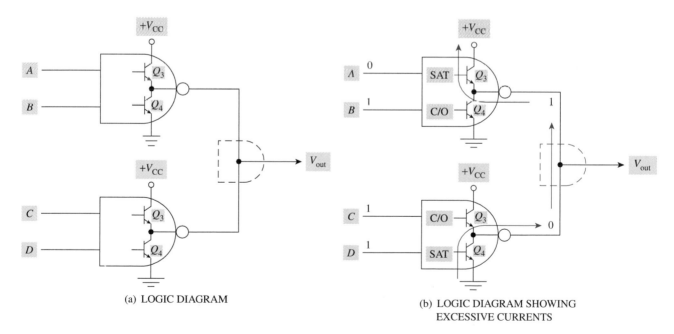

(a) LOGIC DIAGRAM

(b) LOGIC DIAGRAM SHOWING
EXCESSIVE CURRENTS

FIGURE 11–20   Wire-ANDing TTL totem-pole outputs.

Low Output.   High inputs at *A*, *B*, *C*, and *D* will produce low outputs from both gates. In the example, Q_4 in each gate is on, and Q_3 is off. The NAND gates' function is to sink current from the load(s) gates. Again, no problems are encountered in this example.

High/Low Output.   Figure 11–20(b) has been annotated $A = 0$ and $B = C = D = 1$. These input combinations produce a high from the top NAND gate and a low from the bottom NAND gate. The totem-pole transistors Q_3 in the top gate and Q_4 in the bottom gate are saturated. Although a 130-Ω resistor in the totem-pole circuit of each gate is not shown, it is evident that even with the resistor in the circuit a high current would be flowing. The excessive current, which can range up to -55 mA, can destroy the output transistors. This last condition clearly indicates *the totem-pole outputs of TTL devices should never be wire-ANDed.*

Open-Collector Logic.   TTL gates with totem-pole outputs were modified by removing the 130-Ω resistor, Q_3, and a diode (see Fig. 11–19—the diode is not shown in the simplified schematic but is shown in Appendix A). This modification produced a type of TTL device with open-collector outputs, which allows wire-ANDing of the outputs. However, the modification removed V_{CC} from the gate's output and Q_4. This prevents operation of the gate unless a **pull-up resistor** is used to replace the removed components. The pull-up resistor must be connected to the gate's output and V_{CC}. The value of the pull-up resistor is determined by the number of open-collector outputs connected to the wire-AND node and the amount of source and sink current in the circuit. An example problem demonstrating how to calculate the size of R_{PU} is included in Appendix A.

The logic diagram in Fig. 11–21 shows the outputs of three open-collector output NAND gates wire-ANDed. Note that these gate outputs consist of Q_4 only. They do not contain R_3, Q_3, and D_3 as shown in the standard NAND gate circuit (Fig. A-1). Each output in Fig. 11–21 is connected through the pull-up resistor (R_{PU}) to V_{CC}. The 5- to 10-kΩ

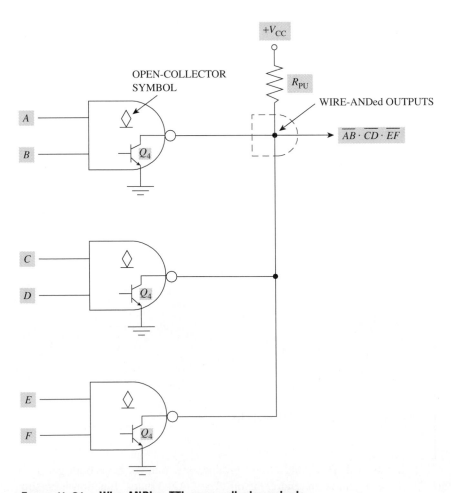

Figure 11–21   **Wire-ANDing TTL open-collector outputs.**

pull-up resistor prevents the excessive currents that would flow if the circuit used the conventional totem-pole output circuit.

The diamond on the base in each gate symbol in Fig. 11–21 is the symbol used to represent an open-collector output device. There are many open-collector TTL devices currently available. Among them are standard logic gates, exclusive gates, buffers, decoders, multiplexers, registers, and bus transceivers.

Section 11–2: Review Questions

Answers are given at the end of the chapter.

A. Special interface circuits are always necessary between a drive gate and load gate(s).
   (1) True
   (2) False

B. Identify the drive gate(s) in Fig. 11–22.
   (1) OR gates
   (2) NAND gate

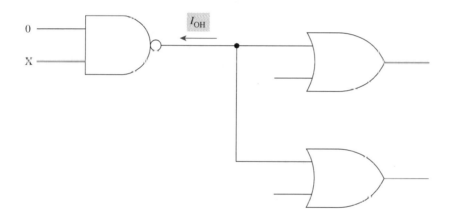

FIGURE 11-22

C. I_{OH} in Fig. 11–22 is $-400 \mu A$. The minus sign indicates the current is actually flowing out of the NAND gate output pin.
   (1) True
   (2) False

D. The drive gate in Fig. 11–22 is current _____.
   (1) sinking
   (2) sourcing

E. The drive gate in Fig. 11–23 is current _____.
   (1) sinking
   (2) sourcing

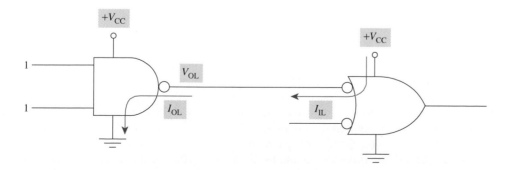

FIGURE 11-23

F. A logic gate _____ current when its output is low (V_{OL}).
   (1) sinks
   (2) sources

G. Define fan-out.

H. A logic circuit may not work properly if the fan-out (low or high) is exceeded.
   (1) True
   (2) False

I. Define noise immunity.

J. Standard TTL gate's outputs *can be wire-ANDed.*
   (1) True
   (2) False

K. Write the Boolean expression for the circuit in Fig. 11–24.

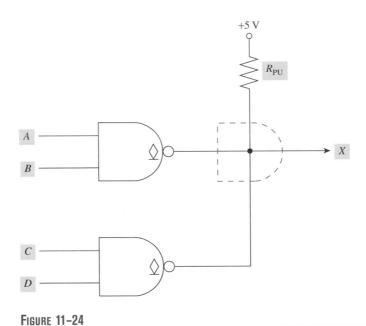

FIGURE 11–24

L. The circuit in Fig. 11–24 will operate properly if R_{PU} is removed.
   (1) True
   (2) False

M. What do the symbols within the NAND gates in Fig. 11–24 represent?

SECTIONS 11-1 AND 11-2: INTERNAL SUMMARY

The most popular integrated circuit technologies in use today are TTL and CMOS. **TTL technology** is popular because of its medium to high speed, low cost, and good drive capability. FAST (F) technology is a high-performance group of TTL devices. **CMOS technology** is popular because of its very high density and low power consumption. **BiCMOS technology** incorporates CMOS and bipolar technologies in one IC.

Digital components must be compatible for proper circuit design. The voltage and current levels of the driving circuit must be compatible with the load circuit. Interface circuits are sometimes required to condition a signal so it is compatible to drive a load.

Compatibility is determined during circuit design by checking input/output current and voltage levels of the drive circuit and the load circuit. Part of the design criteria is to ensure fan-out of a driving circuit is not exceeded. During this part of the design, current sourcing and current sinking are important. If fan-out is exceeded during circuit design, noise margin may be reduced or the voltage levels may be driven into the indeterminate (undefined) range. Proper circuit operation will not be realized should this occur.

Open-collector output devices allow the circuits output to be wire-ANDed with other open-collector devices' outputs. This cannot be done with other types of TTL devices. Open-collector output devices require an external pull-up resistor to operate.

SECTIONS 11–1 AND 11–2: INTERNAL SUMMARY QUESTIONS

Answers are given at the end of the chapter.

1. What type of digital device uses bipolar transistors?
   a. TTL
   b. CMOS

2. Which type of digital device is characterized by medium to high speed, low cost, and excellent drive capability?
   a. TTL
   b. CMOS

3. Which type of digital device is characterized by high density and very low power consumption?
   a. TTL
   b. CMOS

4. Schottky TTL devices can operate at higher clock speeds than standard TTL devices.
   a. True
   b. False

5. Some CMOS ICs are TTL compatible.
   a. True
   b. False

6. The I_{IH} current in Fig. 11–25 is 40 μA. What is the fan-out (high) if I_{OH} is -400 μA?
   a. 1
   b. 2
   c. 10
   d. 20

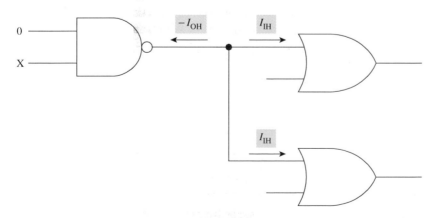

FIGURE 11–25

7. With the currents specified in question 6, are these devices current compatible?
   a. No
   b. Yes

8. The drive gate in Fig. 11–25 is current _____.
   a. sinking
   b. sourcing

9. The minus current sign ($I_{OH} = -400$ μA) indicates the I_{OH} current in Fig. 11–25 is actually flowing out of the output pin of the NAND gate.
   a. True
   b. False

10. The term V_{IH} represents _____.
    a. high-level output voltage of a gate
    b. low-level output voltage of a gate
    c. high-level input voltage to a gate
    d. low-level input voltage to a gate

11. Standard convention current direction arrows are always drawn toward a gate's input/output pins.
    a. True
    b. False

12. The maximum number of inputs that the output of a logic gate can reliably drive is referred to as _____.
    a. noise immunity
    b. noise margin
    c. fan-out
    d. fan-in

13. The drive gate in Fig. 11–26 is current _____.
    a. sinking
    b. sourcing

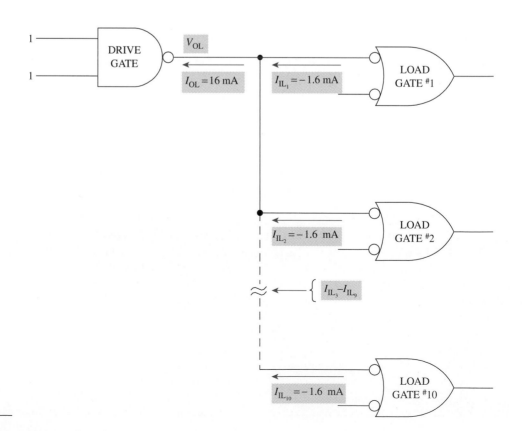

FIGURE 11-26

14. What is the X output of the circuit in Fig. 11–27 if $A = B = C = 1$ and $D = 0$?
    a. Logic 0
    b. Logic 1

15. The diamonds in Fig. 11–27 represent _____.
    a. TTL outputs
    b. CMOS outputs
    c. tristate outputs
    d. open-collector outputs

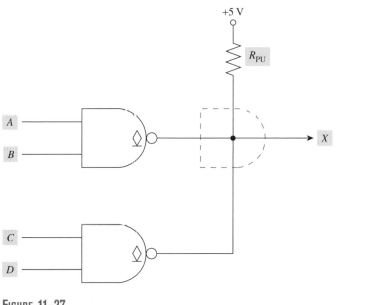

FIGURE 11–27

SECTION 11–3: TRISTATE LOGIC

OBJECTIVES

1. Define tristate, high-impedance state, and bus contention.
2. State the reason for using tristate circuits and how they can be identified.
3. Given the logic diagram/symbol of a device with tristate outputs, identify the device and determine its output(s) when various input conditions are specified.

There are many requirements in digital systems, especially PCs, that mandate the output of a register or some other device be electrically disconnected from a common line (bus).

The circuit in Fig. 11–28 shows the outputs of several buffers connected to a common data bus line (D_0). The bus in this example may be being used by the microprocessor to communicate with memory and peripheral devices. The problem that can occur here is the output of one of the buffers may be high while the outputs of the other buffers are low. This problem, known as **bus contention,** is caused anytime two or more digital devices are allowed on-line simultaneously. The problem may damage the buffers connected to the bus, and it will usually produce an invalid output.

Circuits with tristate (3-state) outputs are used to solve the bus contention problem. The three-position switch in Fig. 11–29 is used to represent the output circuitry of a tristate device. Position 3 of the switch produces a Logic 0 to the data bus. Position 1 produces a Logic 1. The output of a tristate device is electrically disconnected from the

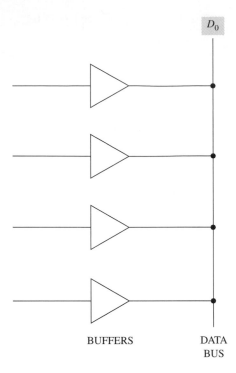

BUFFERS   DATA
                 BUS

FIGURE 11–28   **Buffers connected to data bus.**

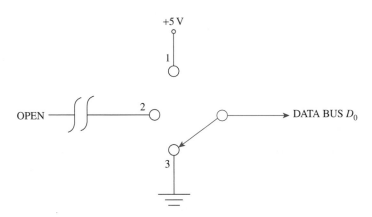

FIGURE 11–29   **Switch representing 3-state output.**

bus when the switch is in position 2. This last condition produces an infinite resistance between the bus and the tristate device's output. This condition is called the **high-impedance state** of a tristate output circuit. The term "tristate" indicates the device's output can be high, low, or high impedance. If the buffers in Fig. 11–28 had tristate outputs, the bus contention problem would be alleviated provided three of the four buffer's outputs were in the high-impedance (Hi-Z) state at any given time. The outputs of tristate circuits are often said to be "floating" when they are in the Hi-Z state.

Many digital devices come with 3-state outputs. **Buffers, bus transceivers,** and **registers** with 3-state outputs are presented in this section. Other available devices with this capability include logic gates, counters, encoders, multiplexers, memory ICs, and microprocessors. All devices that connect to a system bus must have a tristate output capability.

Buffers with 3-State Outputs

74125 Quadruple Bus Buffer with 3-State Outputs.

A **buffer** is a circuit with one data input and one data output. It is used for isolation or to increase current levels to provide a greater fan-out. The logic diagram for one section of the 74125 buffer is shown in Fig. 11–30(a). This symbol represents one of four identical buffers contained in the IC. The $\overline{G}$ active-low input enables the buffer to produce a logic low or logic high output when it is activated (low). The buffer provides additional drive capability when its output is at the high level. This signal conditioning allows driving heavily loaded bus lines without external pull-up resistors.

The buffer in Fig. 11–30(a) is in the high-impedance (Hi-Z) state when $\overline{G}$ is high. This is shown in the truth table in Fig. 11–30(b). The output of this buffer is taken between the totem-pole output transistors discussed in Section 11–2. Both of the totem-pole transistors are turned off when the buffer is not enabled ($\overline{G} = 1$). How this is accomplished is discussed in Appendix A. The two off transistors present a very high impedance to the bus line, which effectively disconnects the buffer from the bus.

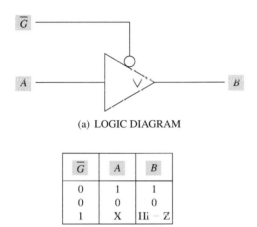

(a) LOGIC DIAGRAM

$\overline{G}$	A	B
0	1	1
0	0	0
1	X	Hi – Z

(b) TRUTH TABLE

FIGURE 11–30   74125 buffer with 3-state output.

74126 Quadruple Bus Buffer with 3-State Outputs.

The logic diagram for one section of the 74126 buffer is shown in Fig. 11–31(a). This IC contains four such buffers, each with its own active-high enable input (G). The active level of the enable input is the only difference between this buffer and the 74125. The truth

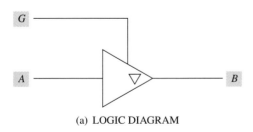

(a) LOGIC DIAGRAM

G	A	B
1	1	1
1	0	0
0	X	Hi – Z

(b) TRUTH TABLE

FIGURE 11–31   74126 buffer with 3-state output.

table in Fig. 11–31(b) shows the input is passed to the output with no phase inversion when the buffer is enabled ($G = 1$). The buffer is in the Hi-Z state when $G = 0$.

The 74125 and 74126 configuration diagrams are shown in Fig. 11–32. The upside-down triangle shown on the diagrams represents tristate outputs. This symbology is used on all ANSI/IEEE symbols to denote a device has 3-state output capability. Standard tristate bus buffer symbols may not have this upside-down triangle, but the devices are readily identifiable by the third input (enable) line.

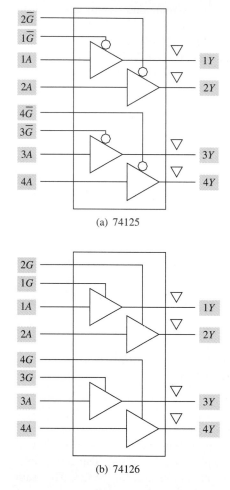

(a) 74125

(b) 74126

FIGURE 11–32   **Buffer configuration diagrams.**

Bus Transceivers

The word "transceiver" implies two-way communication. A **bus transceiver** in a digital system allows asynchronous two-way communications between two data busses. A block diagram of a transceiver is shown in Fig. 11–33. The purpose of the transceiver is to interface data bus A with data bus B.

There are times when data must be transmitted from data bus A to data bus B. At other times, data may need to be transmitted in the opposite direction. Finally, there are times when the two busses must be isolated from each other. In this last condition, all signals must be isolated from the bus.

The G inputs on the block diagram (Fig. 11–33) show direction of data movement. Data are transferred from bus A to bus B when $\overline{GAB}$ and $GBA = 0$. The direction of data flow is reversed when $\overline{GAB}$ and $GBA = 1$. The transceiver is used to isolate the two busses

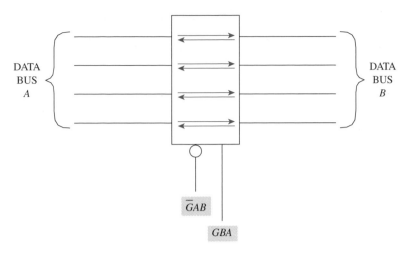

Bus transceiver block diagram.

when $GAB = 1$ and $GBA = 0$. Note by the active-level input indicators (overbar or lack of an overbar over G) that both enable inputs are inactive when this last input combination ($GAB = 1$ and $\overline{GBA} = 0$) is applied. This combination puts the transceiver's outputs in the Hi-Z (floating) state.

74LS243 Quadruple Bus Transceiver with 3-State Outputs.

The logic diagram and function table for this bus transceiver are shown in Fig. 11–34. The first thing to note in the logic diagram in Fig. 11–34(a) is the active levels of the two control signals (GAB and GBA). Data movement, as shown in the function table in Fig. 11–34(b), is from A to B (A_1 to B_1, A_2 to B_2, etc.) when $\overline{GAB}$ is active (low) and GBA is inactive (low). Data movement is from B to A when $\overline{GAB}$ is inactive (high) and GBA is active (high). Inactive levels at both inputs cause the outputs of both buffers in each set of bidirectional buffers to go to the Hi-Z state. This isolates bus A from bus B.

From the previous discussion it should be evident that activating one control input and leaving the other inactive sets the direction of data movement. The direction is as specified by the activated control signal mnemonic. To illustrate, $\overline{GAB} = 0$ allows data movement only from A to B. $GBA = 1$ allows data movement only from B to A. These two examples must be accompanied by an inactive signal on the opposite control input. If neither input signal is activated, the circuit does not allow data movement in either direction. This is the **Hi-Z state,** where the two busses are totally isolated from each other.

What would happen in the circuit if both control inputs were activated simultaneously? This allows both buffers in each set of buffers to become enabled and data movement can be in either direction. In other words, bus A is latched to bus B.

74LS245 Octal Bus Transceiver with 3-State Outputs.

The 74LS245 is basically an 8-bit version of the 4-bit 74LS243. The logic diagram, function table, and logic symbol are presented in Fig. 11–35. Although the control inputs differ slightly, the function table (Fig. 11-35b) clarifies their use. The $\overline{G}$ input is used to enable the device. When $\overline{G}$ is high (inactive), the outputs are in the Hi-Z state. The direction (DIR) input controls the direction of data movement through the transceiver when $\overline{G} = 0$. The buffers contain Schmitt-trigger-input circuits as they were in the 74LS243. This im-

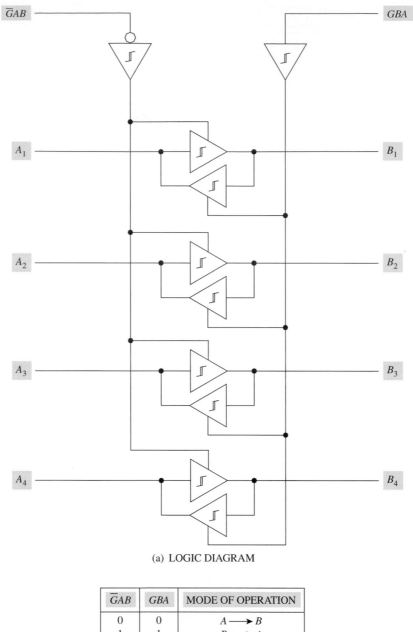

(a) LOGIC DIAGRAM

$\overline{GAB}$	GBA	MODE OF OPERATION
0	0	$A \longrightarrow B$
1	1	$B \longrightarrow A$
1	0	Hi − Z
0	1	LATCH $(A = B)$

(b) FUNCTION TABLE

FIGURE 11–34 **74LS243 quad bus transceiver with 3-state outputs.**

proves the noise margins of the buffers. The 74LS245 IC is discussed again in Section 11–5, where its application in a PC is presented.

Registers with 3-State Outputs

Our discussion of transceivers and bus buffers with 3-state outputs wouldn't be complete without addressing bus buffer registers and storage registers with 3-state outputs. Two different 4-bit registers have been selected for this discussion—74173 and 74295.

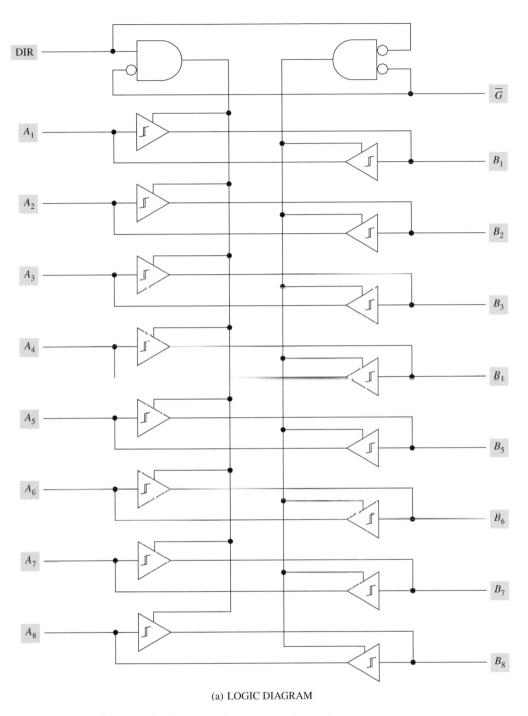

(a) LOGIC DIAGRAM

FIGURE 11-35   74LS245 octal bus transceiver with 3-state outputs.

$\overline{G}$	DIR	FUNCTION
0	0	*B* DATA TO BUS *A*
0	1	*A* DATA TO BUS *B*
1	X	Hi-Z

(b) FUNCTION TABLE

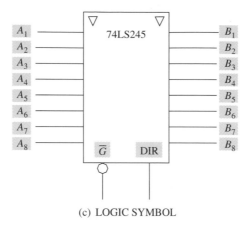

(c) LOGIC SYMBOL

FIGURE 11–35   Continued

74173 4-Bit D-Type Register with 3-State Outputs.

This register can be used as a bus buffer register. An 8-bit version is also commercially available. Figure 11–36 shows the logic diagram and function table for this register. The basic operation and method of analyzing this circuit are similar to the parallel-in/parallel-out registers presented in Section 8–2.

The function table of the 74173 is shown in Fig. 11–36(b). The register in Fig. 11–36(a) will be in the HOLD mode if either of the data enable inputs $(\overline{G}_1$ or $\overline{G}_2)$ is high or if the clock input is held low. The Q data of each flip-flop is recirculated through that flip-flop via the top AND gate of each pair of flip-flop AND gates when $\overline{G}_1 = 1$ or $\overline{G}_2 = 1$ and clock pulse PGTs are applied. The output control inputs $(\overline{M}$ and $\overline{N})$ must be active for the data to be present at the output pins. The tristate output buffers at $1Q$–$4Q$ are in the Hi-Z state if $\overline{M} = 1$ or $\overline{N} = 1$. The note at the bottom of the function table indicates clocked flip-flop operation is unaffected by the output control signals $(\overline{M}$ or $\overline{N})$. This indicates the register can be loaded while its outputs are in the Hi-Z state provided the data enable inputs $(\overline{G}_1$ and $\overline{G}_2)$ are active. The output Qs follow the data input Ds when CLEAR $= \overline{G}_1 = \overline{G}_2 = \overline{M} = \overline{N} = 0$ and a clock PGT is applied to the register.

The standard logic symbol and ANSI/IEEE symbol for the 74173 register are depicted in Fig. 11–37. The gates drawn on the standard symbol are internal to the circuit. The ANSI/IEEE symbol shows the AND of the active-low output control inputs $(\overline{M}$ and $\overline{N})$ enables the outputs. If they are not enabled, the upside-down triangle at $1Q$ ($4Q$ on the standard logic symbol) indicates the outputs are in the Hi-Z state. The remaining AND symbol indicates the active-low data enable inputs $(\overline{G}_1$ and $\overline{G}_2)$, along with the clock PGT, load data to the outputs.

74LS295 4-Bit Right Shift/Left Shift Register with 3-State Outputs.

Operation of this register is similar to operation of the 74194 4-Bit Bidirectional Universal Shift Register presented in Section 8–2. The tristate outputs are the main difference. Other minor differences and a detailed explanation of the function table are presented. The logic diagram and function table appear in Fig. 11–38.

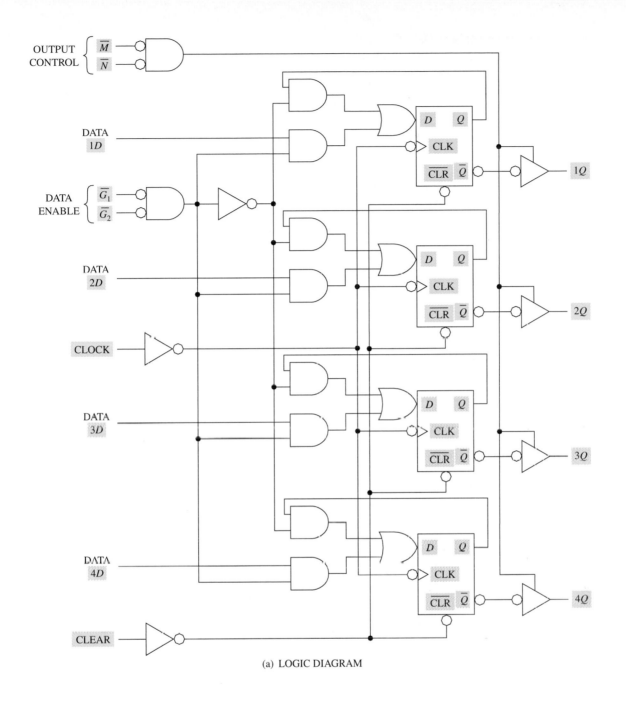

(a) LOGIC DIAGRAM

CLEAR	CLOCK	$\overline{G_1}$	$\overline{G_2}$	$\overline{M}$	$\overline{N}$	DATA	OUTPUT
1	X	X	X	0	0	X	CLEAR
0	0	X	X	0	0	X	HOLD
0	↑	1	X	0	0	X	HOLD
0	↑	X	1	0	0	X	HOLD
0	↑	0	0	0	0	0	$Q = 0$
0	↑	0	0	0	0	1	$Q = 1$
0	* ↑	0	0	1	X	X	Hi-Z
0	* ↑	0	0	X	1	X	Hi-Z

*CLOCKED OPERATION OF THE FLIP-FLOPS IS NOT AFFECTED BY $\overline{M}$ OR $\overline{N}$
WHEN EITHER IS HIGH.

(b) FUNCTION TABLE

FIGURE 11–36   **74173 4-bit D-type register with 3-state outputs.**

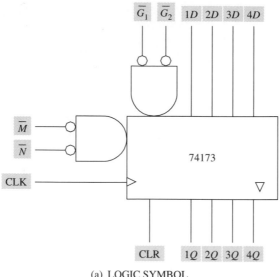

(a) LOGIC SYMBOL

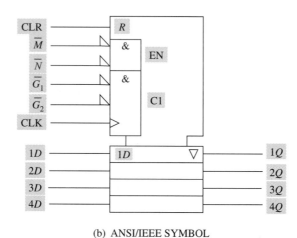

(b) ANSI/IEEE SYMBOL

Figure 11-37   74173 IC.

Hold.   The function table indicates two conditions that will establish the HOLD mode of operation. In both conditions, the CLK is held high which means the flip-flops cannot change states. The outputs for the HOLD mode are shown as Q_{A_0}, Q_{B_0}, Q_{C_0}, and Q_{D_0}. These outputs can be defined as equal to the level of Q_A, Q_B, Q_C, and Q_D respectively immediately prior to the CLK input being taken high and held at that level.

Parallel load.   Parallel loading of A, B, C, and D occurs on the first NGT of the clock pulse after the mode control input (LD/$\overline{SH}$) is taken high. This indicates the loading of the register is synchronous. The serial data input (SER) is inhibited during the loading operation. The Q outputs of the register are shown on the function table in Fig. 11–38(b) as a, b, c, and d. These lowercase letters reflect the levels at inputs A, B, C, and D respectively.

Shift Left.   Shift left operation requires external IC connections. The 74LS295 logic symbol wired for the shift left mode of operation is shown in Fig. 11–39. The figure shows the output of each flip-flop is connected to the parallel input of the previous flip-flop. Serial data is applied to the D input and taken from the Q_A output. The mode control input (LD/$\overline{SH}$) must be high to implement shift left operation.

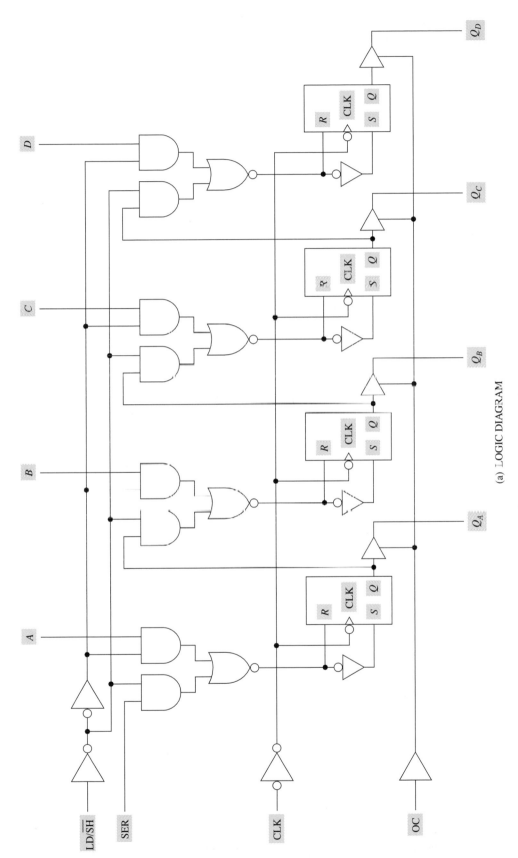

FIGURE 11–38 **74LS295 4-bit right shift/left shift register with 3-state outputs.**

(a) LOGIC DIAGRAM

LD/$\overline{\text{SH}}$	CLK	OC	SER	A B C D PARALLEL	Q_A	Q_B	Q_C	Q_D	OUTPUT
1	1	1	X	X X X X	Q_{A_0}	Q_{B_0}	Q_{C_0}	Q_{D_0}	HOLD
1	↓	1	X	a b c d	a	b	c	d	PARALLEL LOAD
1	↓	1	X	Q_B Q_C Q_D d	Q_{B_n}	Q_{C_n}	Q_{D_n}	d	SHIFT-LEFT†
0	1	1	X	X X X X	Q_{A_0}	Q_{B_0}	Q_{C_0}	Q_{D_0}	HOLD
0	↓	1	1	X X X X	1	Q_{A_n}	Q_{B_n}	Q_{C_n}	SHIFT-RIGHT
0	↓	1	0	X X X X	0	Q_{A_n}	Q_{B_n}	Q_{C_n}	SHIFT-RIGHT
	*	0							Hi-Z

*CLOCKED OPERATION OF THE FLIP-FLOPS IS NOT AFFECTED BY OC.
†SEE TEXT.

(b) FUNCTION TABLE

FIGURE 11–38   Continued

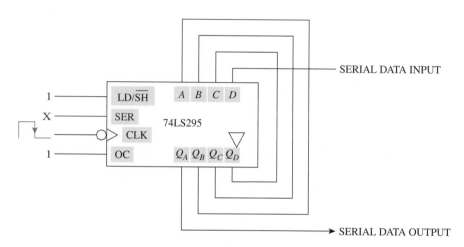

FIGURE 11–39   74LS295 IC logic symbol wired for shift left operation.

The outputs on the function table applicable to the shift left mode show $Q_A = Q_{B_n}$, $Q_B = Q_{C_n}$, $Q_C = Q_{D_n}$, and $Q_D = d$. Let's start with Q_D because data is being shifted to the left. $Q_D = d$ indicates Q_D is equal to the level applied to the D input when the CLK NGT occurred. $Q_C = Q_{D_n}$ shows Q_C is equal to the level at Q_D prior to the CLK NGT. Simply put, the data is shifted left through the register.

Shift Right.   The function table in Fig. 11–38(b) shows the shift right mode of operation occurs when the mode control input (LD/$\overline{\text{SH}}$) is low. The first shift right entry on the table shows serial data (SER) is high. The outputs for this mode when SER = 1 show $Q_A = 1$, $Q_B = Q_{A_n}$, $Q_C = Q_{B_n}$, and $Q_D = Q_{C_n}$ after the active transition of the clock. The level of Q_A changes to a high on the NGT because SER = 1. This can be verified on the logic diagram in Fig. 11–38(a). $Q_B = Q_{A_n}$ indicates that Q_B is equal to the level Q_A was prior to the most recent CLK NGT. This means the data at Q_A is shifted right to Q_B, Q_B to Q_C, and so on. The second shift right entry on the function table has the serial data input low.

High-Impedance State.   The **output control (OC)** signal has been high during the preceding modes of operation. The OC = 1 condition allows the register outputs to drive the bus lines. The outputs are asynchronously taken to the Hi-Z state when OC is taken low. This action occurs because the output buffers are not enabled when OC = 0. The asterisk on this line entry on the function table states, "Clocked operation of the flip-flops is not affected by OC." Thus, sequential operations such as loading and shifting can be accomplished while the outputs are in the Hi-Z state.

Section 11–3: Review Questions

Answers are given at the end of the chapter.

The use of manufacturer's data sheets is recommended to assist you in answering these review questions.

A. Define tristate.
B. Define bus contention.
C. The output of the buffer in Fig. 11–40 is _____ when $\overline{G} = 0$.

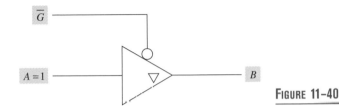

FIGURE 11–40

D. The output of the buffer in Fig. 11–40 is _____ when $\overline{G} - 1$.
E. The upside-down triangle in Fig. 11–40 indicates _____.
F. State the purpose of a bus transceiver in a digital system.
G. What function is performed by the circuit in Fig. 11–41 when $\overline{G} = 0$ and DIR = 0?
H. What function is performed by the circuit in Fig. 11–41 when $\overline{G} = 1$ and DIR $-$ 1?

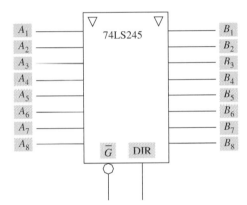

FIGURE 11–41

I. In what mode of operation is the shift register in Fig. 11–42?
J. In what mode of operation is the shift register in Fig. 11–42 if the output control (OC) is taken low?

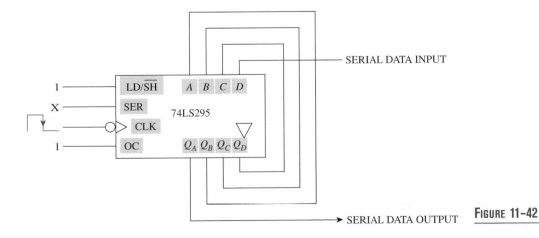

SERIAL DATA OUTPUT **FIGURE 11–42**

Section 11-4: Data Conversion

Objectives

1. Define terms relative to data conversion.
2. State the essential principles of digital-to-analog and analog-to-digital conversion.
3. Calculate the output of an R–$2R$ ladder when the data input levels are given.
4. State the principles of successive-approximation analog-to-digital conversion.
5. State the principles of flash (simultaneous) analog-to-digital conversion and the advantage and disadvantage of using this method.
6. State the purpose of each block in a data sampling system that uses an ADC and a DAC.

Most real-world measurable quantities are analog. Temperature, velocity, acceleration, pressure, position, and intensity signals are a few examples. If the information represented by these analog signals is to be processed by a computer or some other digital processing system, it must be converted to a series of bits of information representing the analog values. This is accomplished by an **analog-to-digital converter (ADC).** The digital signal may be reconverted to analog by a circuit known as a **digital-to-analog converter (DAC).** Changing back and forth from analog to digital and digital to analog is a necessity in order to benefit from today's digital technology.

Many industrial processes are controlled by computers. The digital output of the computer must normally be converted to an analog signal for actual process control. The analog signal's magnitude is proportional to the digital output of the computer after conversion. The conversion is accomplished by a DAC. A DAC may be thought of as the potentiometer in Fig. 11–43 whose output voltage level is controlled by and proportional to a digital input. In actuality the conversion process can be accomplished with a ladder network such as the one discussed on the following page.

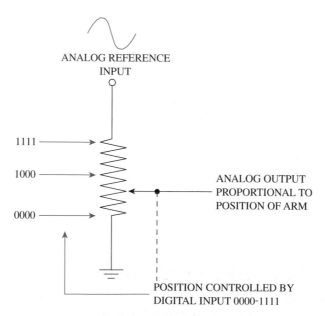

Figure 11–43   **Digital-to-analog representation.**

Digital-to-Analog Conversion

One of the more popular methods of converting digital to analog employs a ladder network. The ladder network shown in Fig. 11–44 is known as an R–$2R$ ladder. The R–$2R$ ladder is presented in the following analysis with inputs 1000, 0100, 0010, and 0001.

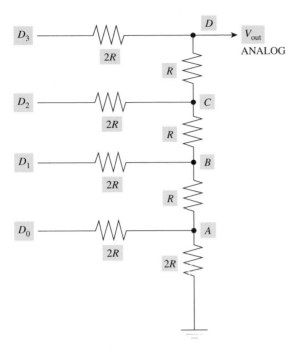

FIGURE 11–44   R-2R ladder.

R-2R Ladder Network.

Input 1000 (D_3 D_0).   The analog output voltage of the ladder in Fig. 11–45(a) is developed from node D to ground. The figure shows the complete ladder network, and the digital input levels (D_3–D_0) have been labeled. The equivalent resistance from node D to ground in this circuit is $2R$. For example, if $R = 10 \text{ k}\Omega$ and $2R = 20 \text{ k}\Omega$, then $R_{eq} = 20 \text{ k}\Omega$.

The equivalent resistance $2R$ can readily be seen if analysis of the ladder network starts at node A and progresses to node D. Since the D_0 input is connected to Logic 0 (ground) in this example, both of the lower $2R$ resistors in Fig. 11–45(a) are in parallel. This results in an equivalent resistance (R_{eq_1}) from node A to ground equal to R. This is illustrated in Fig. 11–45(b). The equivalent resistance (R_{eq_1}) is in series with the resistance connected between nodes B and A, and this will be referred to as R_{BA}. Therefore, the equivalent resistance from node B to ground (R_{eq_2}) is $2R$ and this resistance is in parallel with the $2R$ input resistor connected to D_1 as seen in Fig. 11–45(c). This parallel combination produces R_{eq_3} in series with R_{CB} as shown in Fig. 11–45(d). Continuation of the analysis through Fig. 11–45(e) and (f) reveals the equivalent resistance from node D to ground is $2R$. Figure 11–45(g) shows that D_3 is connected to Logic 1, which is some positive voltage. The voltage level is typically +3.3 V to +5 V at the high data input(s). The output voltage (V_{out}) at node D is equal to $V_{in} \div 2$ when the digital inputs are 1000.

When one data input is high and the remaining data inputs are low, V_{out} can be calculated as follows:

$$V_{out} = V_{IN} \div 2^{N-n}$$

Where N is the number of binary inputs to the ladder and n is the bit position number of the *high* input bit.

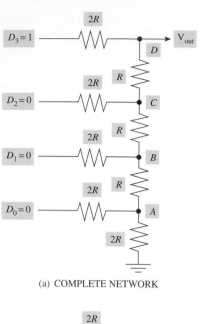

(a) COMPLETE NETWORK

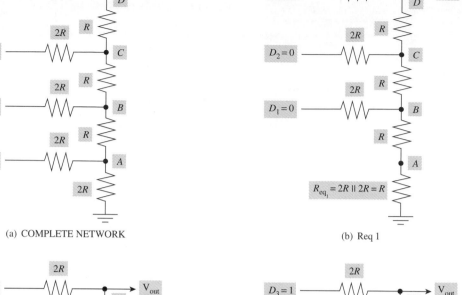

(b) Req 1

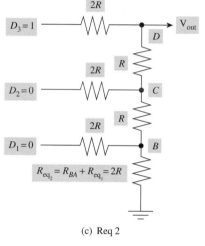

(c) Req 2

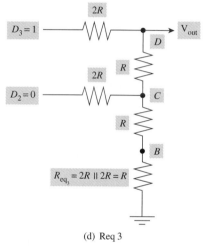

(d) Req 3

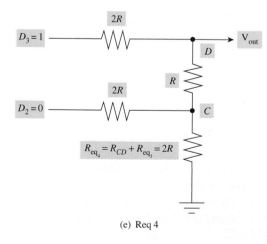

(e) Req 4

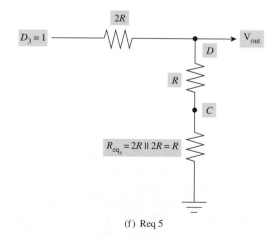

(f) Req 5

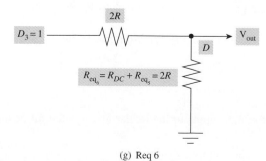

(g) Req 6

FIGURE 11–45 *R–2R*
ladder (1000 inputs).

If the input logic level at the D_3 input in the preceding example is $+4$ V, the output voltage is

$$\begin{aligned}
V_{out} &= +4 \text{ V} \div 2^{4-3} \\
&= +4 \text{ V} \div 2^1 \\
&= +4 \text{ V} \div 2 \\
&= +\mathbf{2} \textbf{ V}
\end{aligned}$$

This example problem proves V_{out} is $V_{IN} \div 2$ when the data inputs are 1000.

Input 0100 (D_3–D_0). The D_2 input is high, and the remaining data inputs are low in this example. The complete ladder network is shown in Fig. 11–46(a). This example assumes $R = 10$ kΩ, thus $2R = 20$ kΩ.

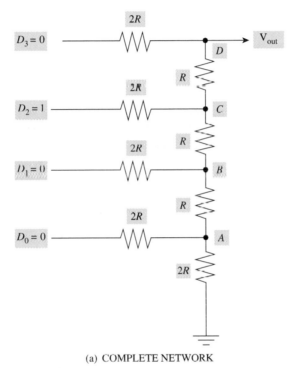

(a) COMPLETE NETWORK

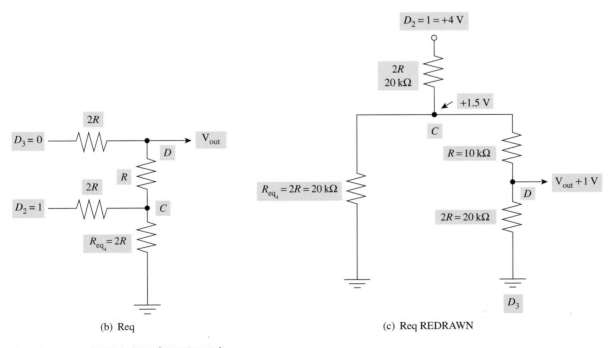

(b) Req

(c) Req REDRAWN

FIGURE 11–46   *R-2R* ladder (0100 inputs).

647

The circuit is drawn showing the equivalent resistance in Fig. 11–46(b). The preceding example showed the equivalent resistance from node C to ground is $2R$ (R_{eq_4} in Fig. 11–45e) if the R_{DC} and $2R$ resistance from node D to D_3 are ignored. However, that resistance is not in parallel with the D_2 $2R$ input resistor because D_2 is connected to a Logic 1 input in this example. In the preceding example the D_2 input was connected to Logic 0 (ground).

The equivalent resistance circuit has been redrawn in Fig. 11–46(c). The D_3 input is shown grounded because $D_3 = 0$. Therefore, R_{eq_4} is in parallel with $R_{DC} + 2R$. This produces an actual resistance from node C to ground of 12 kΩ. This resistance produces a voltage at node C of $+1.5$ V and an output voltage at node D of $+1$V.

The formula provided in the preceding example verifies V_{out}.

$$
\begin{aligned}
V_{out} &= V_{IN} \div 2^{N-n} \\
&= +4\text{ V} \div 2^{4-2} \\
&= +4\text{ V} \div 2^2 \\
&= +4\text{ V} \div 4 \\
&= \mathbf{+1\ V}
\end{aligned}
$$

Input 0010 (D_3–D_0). The ladder network with data inputs 0010 is shown in Fig. 11–47(a). The circuit in Fig. 11–47(b) shows the equivalent resistance from node B to ground is $2R$ (R_{eq_2} in Fig. 11–45c) if the D_3 and D_2 input resistances are ignored.

The circuit is redrawn in Fig. 11–47(c). The D_3 and D_2 $2R$ resistors are shown connected to ground because $D_3 = 0$ and $D_2 = 0$. The total resistance in this circuit from node B to ground is 10.476 kΩ. This produces a node B voltage of $+1.375$ V, node C voltage of $+0.75$ V, and node D output voltage of $+0.5$ V.

$$
\begin{aligned}
V_{out} &= V_{IN} \div 2^{N-n} \\
&= +4\text{ V} \div 2^{4-1} \\
&= +4\text{ V} \div 2^3 \\
&= +4\text{ V} \div 8 \\
&= \mathbf{+0.5\ V}
\end{aligned}
$$

Input 0001 (D_3–D_0). The output voltage of the R–$2R$ ladder is $+0.25$ V.

$$
\begin{aligned}
V_{out} &= +4\text{ V} \div 2^{4-0} \\
&= +4\text{ V} \div 2^4 \\
&= +4\text{ V} \div 16 \\
&= \mathbf{+0.25\ V}
\end{aligned}
$$

A detailed analysis is similar to the preceding examples and is unnecessary at this point. The circuit is shown in Fig. 11–48(a) and (b).

The V_{out} formula proves the output voltage is $V_{IN} \div 2$ when the input is 1000; $V_{IN} \div 4$ when the input is 0100; $V_{IN} \div 8$ when the input is 0010; and $V_{IN} \div 16$ when the input is 0001. The output voltage of the ladder network can be calculated using the **superposition method** when more than one data input is high.

The superposition method of determining V_{out} is shown in Fig. 11–49. The complete ladder network with 1100 inputs is shown in Fig. 11–49(a). This example assumes the data high input levels are $+4$ V and the resistors are 10 kΩ and 20 kΩ. The circuit has been redrawn in Fig. 11–49(b).

Figure 11–49(c) shows the ladder network with the D_2 source shorted using the superposition method. The output voltage is $V_{IN} \div 2 = 2$ V. Figure 11–49(d) on p. 652 illustrates the D_3 source shorted. The output voltage for this circuit is $V_{IN} \div 4 = 1$ V. The voltages resulting from each source are the same polarity so they are added.

$$
\begin{aligned}
V_{out} &= 2\text{ V with } D_2 \text{ source shorted} \\
V_{out} &= 1\text{ V with } D_3 \text{ source shorted} \\
V_{out} &= \mathbf{3\ V}
\end{aligned}
$$

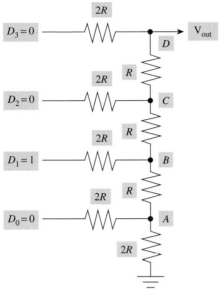

(a) COMPLETE NETWORK

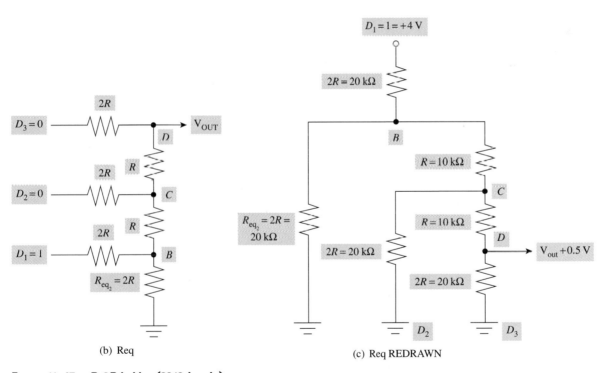

(b) Req

(c) Req REDRAWN

FIGURE 11–47   *R-2R* ladder (0010 inputs).

This example shows the output voltages produced by high data inputs are added to calculate V_{out} using the superposition method.

Calculate the output voltage for a DAC *R-2R* ladder when the data inputs are 1001 and the high data level is 3.5 V:

$$V_{out}(1000) = V_{IN} \div 2 = 1.75 \text{ V}$$
$$V_{out}(0001) = V_{IN} \div 16 = 0.219 \text{ V}$$
$$V_{out} = \textbf{1.969 V}$$

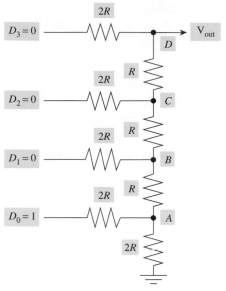

(a) COMPLETE NETWORK

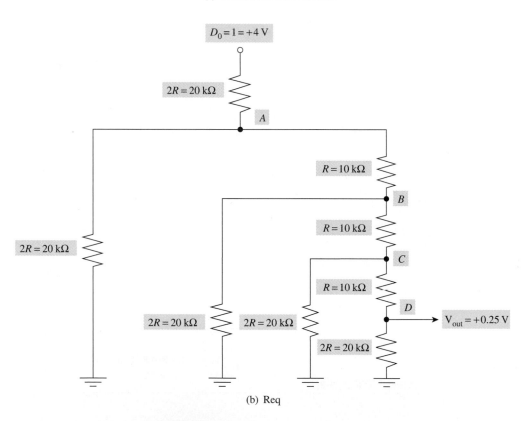

(b) Req

FIGURE 11–48 *R–2R* **ladder (0001 inputs).**

Figure 11–50(a) shows a DAC with four data inputs from a MOD-16 up-counter. The output voltage increases 0.25 V for each count increment. The output voltage waveform from the DAC is shown in Fig. 11–50(b). Each sequential analog output increases by 0.25 V and equates to 1 LSB.

A DAC implemented with an *R–2R* ladder and an operational amplifier (op-amp) is illustrated in Fig. 11–51 (on p. 653). The op-amp has a very high input impedance and a

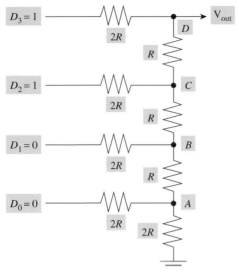

(a) COMPLETE NETWORK

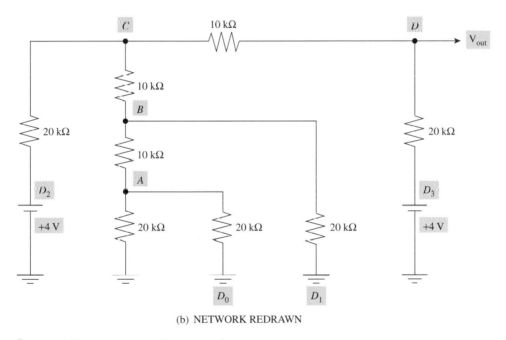

(b) NETWORK REDRAWN

FIGURE 11–49   *R–2R* ladder (1100 inputs).

voltage gain of one in this configuration. The op-amp configuration is referred to as a voltage follower. The high input impedance of the op-amp prevents loading the ladder network and results in accurate conversion.

Switched current-source DAC.

Operation of the *R–2R* ladder DAC is based on summing input voltages to produce an analog output voltage. Many DAC ICs employ current switching to achieve a higher speed and more accurate conversion. The inputs are used to control switches, which in turn furnish current to an op-amp. The switches are actually transistors controlled by the data inputs. The purpose of the DAC in this case is to produce an output current proportional to the data input. The output current can be converted to an output voltage using an op-amp.

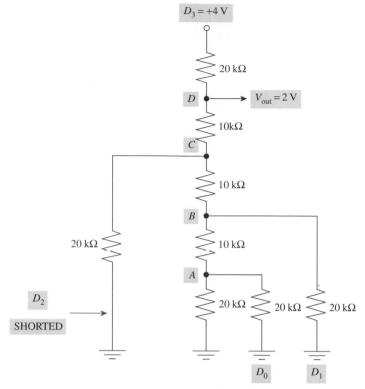

(c) D2 INPUT SOURCE SHORTED

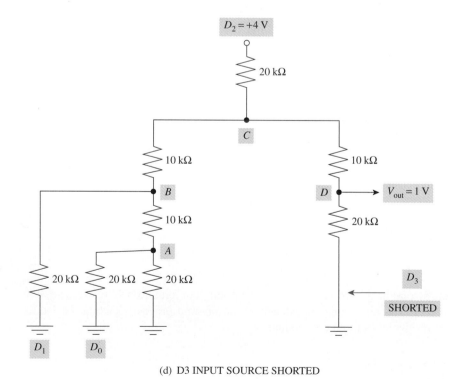

(d) D3 INPUT SOURCE SHORTED

FIGURE 11–49 **Continued**

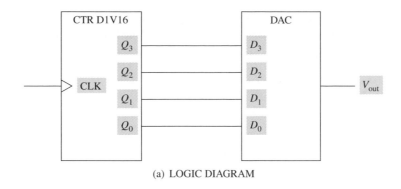

(a) LOGIC DIAGRAM

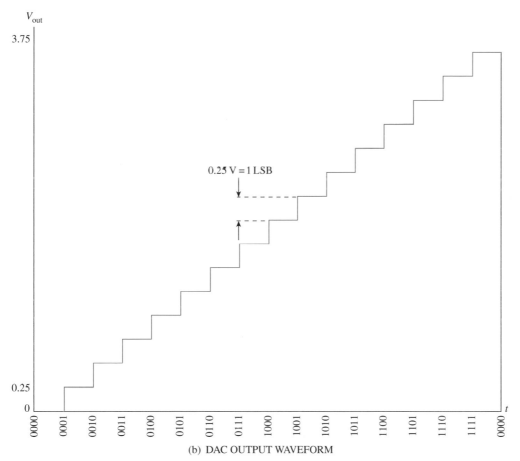

(b) DAC OUTPUT WAVEFORM

FIGURE 11–50   **Digital-to-analog conversion.**

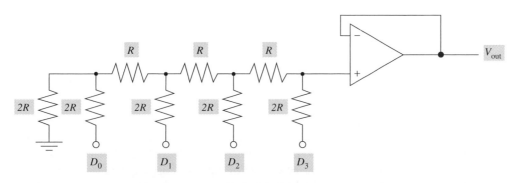

FIGURE 11–51   **R-2R ladder buffered with voltage follower.**

The circuit presented in Fig. 11–52 shows an R–$2R$ ladder network furnishing current(s) proportional to the digital input to an op-amp circuit. The inverting op-amp acts as a buffer. The D_3 input is high, so the I_3 current flows through the op-amp feedback resistor (R_f). The amount of current is set by the voltage reference (V_{REF}). Therefore, slight variations of data input (D_3–D_0) voltage levels have no effect on the conversion process.

The feedback current (I_f) flowing through R_f is equal to the sum of I_3 and I_2 currents if both D_3 and D_2 are high and D_1 and D_0 are low. The output voltage is

$$V_{out} = -I_f R_f$$

Many variations of DACs are commercially available. The operation of most is based on the principles just discussed. Op-amp configurations allow for inverted or noninverted

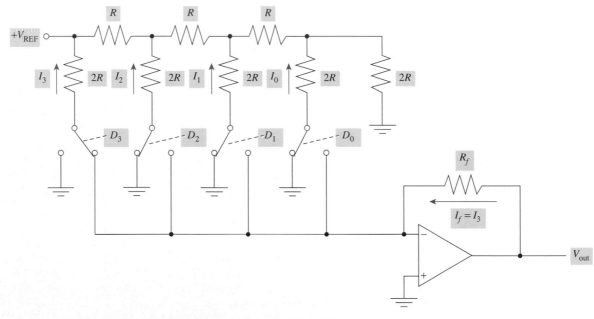

FIGURE 11–52 **Switched current-source DAC.**

output voltages. In addition, the value of the op-amp feedback resistor can be chosen to allow for voltage gain. The symbol for a DAC is shown in Fig. 11–53. This 8-bit DAC is comparable to National Semiconductor's DAC0808.

Data Converter Specifications

Manufacturer's data sheets for DACs and ADCs provide the specifications for data converters. These specifications are defined here. An understanding of the specs will take you from the ideal implementation to real-world circuits.

Resolution is defined as the smallest analog incremental or decremental change in output voltage of a *DAC* that results from a 1-LSB converter code change. Resolution is expressed in bits as 2^n, where n is the number of input bits to a DAC. The DAC0808 is an 8-bit converter. Therefore, this converter maps the output signal into 2^8 (256) analog voltage levels. An LSB in a DAC corresponds to the *height of a step* between sequential analog outputs and represents a DAC's resolution. An LSB is labeled in Fig. 11–50(b). Resolution for *ADCs* is defined as the change in input voltage required to increment or decrement the output of the ADC 1 LSB where the input is incremented or decremented by a value equal to 1 LSB. Resolution of an ADC is the nominal value of the *width of*

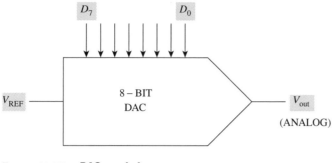

FIGURE 11–53   DAC symbol.

the step. In linear data conversion, the magnitude of the resolution is normally used as the reference unit LSB.

An 8-bit DAC can produce 256 analog output voltages. Each of the 256 output voltage levels corresponds to one binary input combination. If an analog output voltage range of 0 V to +8 V is desired, an LSB = 8 V ÷ 256 = 0.03125 V. This voltage represents 0.39% of full scale. A 12-bit DAC increases resolution significantly by being able to produce 4096 analog output voltages. An LSB = 8 V ÷ 4096 = 0.001953 V for this DAC when a 0-V to +8-V output range is desired. Note the smaller incremental changes per LSB of the 12-bit DAC. This would produce an output analog signal that is much more representative of the ideal output than an 8-bit converter. An ADC with 12-bit resolution can resolve 1 part in 4096 (2^{12}) or 0.0244% of the full-scale range.

Differential nonlinearity is the worst case deviation from the ideal 1-LSB step. This type of nonlinearity may be expressed as a percentage of full scale or in fractional bits. A differential nonlinearity of ±$\frac{1}{2}$ LSB maximum indicates V_{out} for a specific digital input is within an amount equal to the change in voltage that would result by half the LSB change. For example if V_{out} changes 0.02 V for one LSB change, then V_{out} should always be within 0.01 V of the expected voltage level. If the differential nonlinearity of a converter exceeds 1 LSB, there is a possibility that the magnitude of the output will get smaller for an increase in the magnitude of the input. This is referred to as **nonmonotonic** operation. **Monotonic** operation indicates the output has a slope whose sign does not change for an increasing input signal. In other words, the output will consistently increase in proportion to an increasing input but will never decrease (oscillate) in value. The DAC output waveform in Fig. 11–50(b) is monotonic.

Settling time is the interval of time between a DAC's input code change and the time the analog output remains within some specified tolerance of the final value. The tolerance is often specified as "to 0.5 LSB."

Analog-to-Digital Conversion

Computers are often used to control the environment in buildings. Analog signals representing temperature and humidity must be converted to digital signals for use in the computer. The conversion is accomplished in an IC called an analog-to-digital converter (ADC). The more popular methods of analog-to-digital conversion are presented in the remainder of this section.

Successive-Approximation ADC.

Successive-approximation ADCs are extremely popular, as evidenced by the large number of different types commercially available. A successive-approximation ADC consists of an op-amp comparator, successive-approximation register (SAR), and DAC as shown in Fig. 11–54. The SAR is merely a group of flip-flops used to store Logic 0s and 1s.

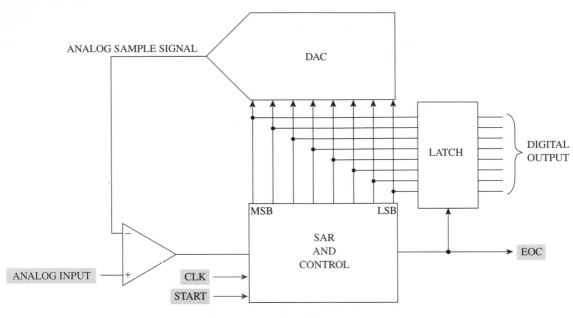

FIGURE 11–54   Successive-approximation ADC.

The SAR starts a successive-approximation cycle with its MSB high and the remaining bits low $(100 \ldots 0)$ when a start conversion signal is input. This digital output from the SAR represents mid-range and is converted to analog in the DAC. The DAC's analog sample output signal is compared in the op-amp with the analog input signal. If the DAC's sample signal is greater than the analog input signal, the op-amp's output is low, the MSB is reset low, and the second MSB is set high. If the DAC's analog sample signal is less than the analog input signal, the op-amp's output goes high, the MSB is left high, and the second MSB is set high. Once the SAR's flip-flops have been set or reset in accordance with the op-amp's output, the sample and adjust procedure is repeated. The process continues down to the LSB if necessary. The contents of the SAR represent the digital equivalent of the analog input upon completion of a successive-approximation cycle. An end-of-conversion (EOC) signal is generated and used to latch the data at the end of the cycle. The EOC signal may also be used to inform another digital device that digital data are currently available from the ADC.

A National Semiconductor ADC0800 8-bit converter block diagram is shown in Fig. 11–55. This successive-approximation converter uses p-channel MOS technology. It contains 256 series resistors (300 Ω each) and 256 analog switches. An unknown analog voltage is compared to the resistor tie point voltages via the analog switches. The conversion cycle is complete when the tie point voltage and the analog input voltage are equal. An 8-bit complementary binary word is latched at the output upon completion of the conversion cycle.

Normal operation of the ADC0800 converter requires a 10-V reference voltage (V_{REF}) be applied across the resistor network. The analog input voltage is compared to the center point of the resistor network at the start of a conversion cycle. If the analog input voltage is greater than 5 V $(0.5\ V_{\text{REF}})$ internal logic changes the switch points and next compares the analog input to 7.5 V $(0.75\ V_{\text{REF}})$. This process is continued until the compared voltages are equal. The latched digital data outputs are enabled when the output enable (OE) = 1. The ADC outputs are in the Hi-Z state when OE = 0.

Each ADC0800 conversion cycle requires 40 clock periods. The EOC signal is low (ADC Busy) during this period. At the end of the cycle, EOC goes high, and 4 additional clock periods must be allowed prior to the arrival of another start conversion pulse. The converter may be set up to free run by connecting the EOC output to the start conversion

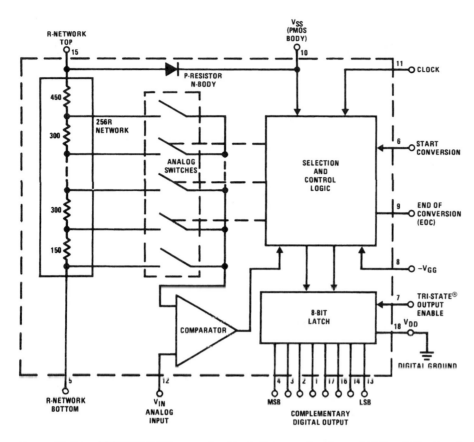

FIGURE 11–55   ADC0800 8-bit converter block diagram. *Courtesy of National Semiconductor.*

input. In this configuration, an external start conversion pulse is required only during circuit initialization (power-up).

Flash (Simultaneous) ADC.

Flash converters employ op-amp comparators as shown in Fig. 11–56. The primary advantage of flash conversion is **high-speed operation.** A reference voltage is applied to the inverting input of each op-amp from a series voltage divider. The analog input signal is applied to the noninverting input of each op-amp. When the analog input signal is less than the reference voltage, the comparator outputs a Logic 0. A Logic 1 output is generated by the comparator when the analog input is greater than the reference voltage.

The outputs of the op-amp comparators are connected to a **priority encoder.** Our study of priority encoders in Chapter 10 revealed two main points that should be kept in mind here: (1) An encoder detects an active-input line (high in this case) and converts it to a binary number or code at the output; and (2) a priority encoder responds only to the highest-order active-input number when more than one input is activated at the same time. If the #1, #2, and #3 inputs to the priority encoder in Fig. 11–56 are high and the remaining inputs are low, the encoder will output 011 (D_2–D_0).

A brief analysis of flash ADC operation reveals the simplicity of the circuit. The voltage divider is connected between ground and $+10$ V (V_{REF}) in this example. The voltages at the resistor tie points are $V_7 = 8.75$ V, $V_6 = 7.5$ V, $V_5 = 6.25$ V, $V_4 = 5$ V, $V_3 = 3.75$ V, $V_2 = 2.5$ V, and $V_1 = 1.25$ V.

An instantaneous analysis with the analog input (V_{in}) at 4.25 V reveals V_{in} is less than V_7, V_6, V_5, and V_4 and greater than V_3, V_2 and V_1. Comparator outputs are low from op-amps #7–#4 because V_{in} is less than V_{REF}. The outputs of op-amps #3–#1 are high. The priority encoder will output 011 because the #3 input is the highest-order active input.

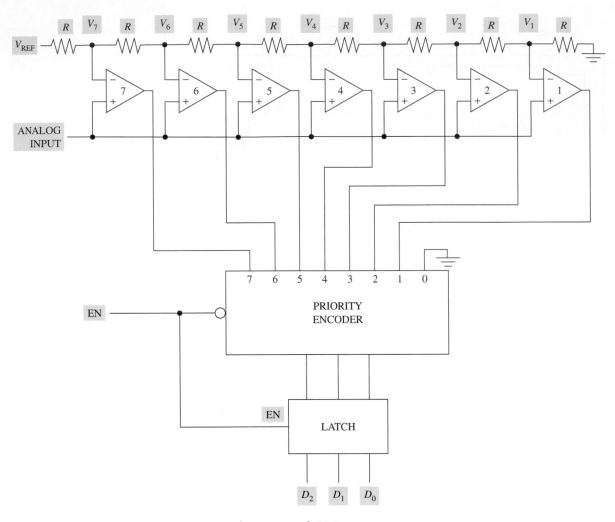

FIGURE 11-56 **Flash (simultaneous) ADC.**

If the analog input signal increases to 5.25 V, the priority encoder outputs 100 (D_3-D_0) because $V_{in} > V_4$, and the #4 encoder input is now active.

It is evident from the preceding analysis the digital output is a rough approximation of the analog input. The circuit would produce a much better approximation if more comparisons were made. However, more comparisons require more op-amps. The number of op-amp comparators required in a flash ADC is $2^n - 1$ to convert to an n-bit binary number.

Most computers utilize an 8-bit or larger data bus. Thus, it would take 255 $(2^n - 1)$ comparators in a flash ADC to encode an analog signal to 8 bits. This would significantly improve resolution, but the disadvantage is evident—the increase in the number of comparators increases both cost and power consumption.

The disadvantage of the large number of comparators required for flash conversion was overcome when **semiflash converters** were introduced. These converters are referred to as **half-flash converters** by some manufacturers. The modified flash ADC uses upper and lower 4-bit sampling comparators to encode 8 bits.

Conversion Accuracy.

Accuracy of a DAC is compared to the linear transfer function in Fig. 11–57. The conversion accuracy of an 8-bit DAC cannot be better than $\pm \frac{1}{2}$ LSB or ± 1 part in 2^{8+1}. This

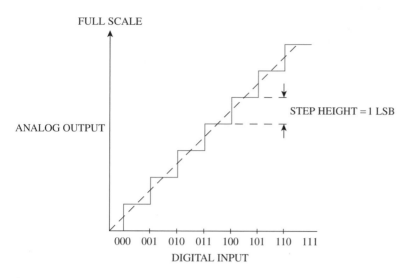

FULL SCALE

ANALOG OUTPUT

STEP HEIGHT = 1 LSB

000   001   010   011   100   101   110   111

DIGITAL INPUT

FIGURE 11–57   **DAC ideal linear transfer function.**

calculates to 0.195% of full scale. This percentage is the best possible accuracy that an 8-bit DAC with no errors can achieve. The accuracy is actually 99.805%, but convention gives the accuracy spec as 0.195%, which is in reality an inaccuracy percentage.

Accuracy of an ADC describes the difference between the analog input voltage and the weighted equivalent of the binary output. For example, if the accuracy of an 8-bit ADC is ± 1 LSB, the output is accurate within 0.39%. An ideal transfer function for an ADC is shown in Fig. 11–58.

Quantization is defined as limiting the possible values of a quantity to a discrete set of values by quantum mechanical rules. Naturally, an ADC must quantize an analog input into a finite number of output codes (bits). This produces a quantization error in all ADCs.

Numerous other errors are encountered in DAC and ADC performance. Some of these errors were discussed in the Data Conversion Specifications section of this chapter.

The Nyquist sampling theorem states that in order to reproduce a periodic signal, it is necessary to sample at a rate greater than twice the sampled signal's highest frequency component:

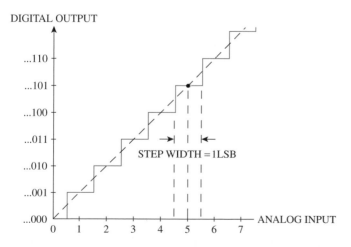

DIGITAL OUTPUT

...110

...101

...100

...011

STEP WIDTH = 1LSB

...010

...001

...000

0   1   2   3   4   5   6   7   ANALOG INPUT

FIGURE 11–58   **ADC ideal linear transfer function.**

$$f_S \geq 2f_{in}$$
where f_S is the sample frequency and f_{in} is the highest frequency to be sampled.

In actuality, the sampling rate may be several times the highest input frequency. Accepted practice is to use a sampling rate that is 2.2 times the sampled signal's highest frequency component. If the sampling rate is not at least twice the sampled signal's (f_S) highest frequency component, distortion is produced at lower frequencies. This distortion is called **aliasing.** It is caused by the difference between the sample frequency (f_S) and the frequency being sampled (f_{in}) where this difference is less than f_{in}. The alias frequencies produce distortion and incorrect ADC outputs.

The effects of aliasing are prevented by routing the input to an ADC through an **antialiasing filter.** This filter is extremely important because aliasing errors are not correctable after the fact. For example in data gathering systems, aliasing errors can neither be identified nor removed from the data stream. The frequency response of an antialiasing filter is selected at half the sample frequency. The filter output should be down at least 40 dB at $f_S/2$. This will prevent difference frequencies below f_{in} and prevent aliasing.

The rapid advancement in digital and computer technology in the past few years has resulted in improved conversion technology. High-density MOS technology has permitted IC manufacturers to incorporate control circuits and ADCs on a single chip. A look at a data acquisition databook reveals many converters designed specifically for computer compatibility with no requirement for interfacing logic.

A complete data sampling system is presented in the block diagram in Fig. 11–59. Most of the circuits in the diagram have been presented in this section. A review of the block diagram will introduced a couple of points and summarize the data conversion processes already presented.

We have seen that the low-pass antialiasing input filter is necessary to ensure proper reproduction of the analog signal at the output of the ADC. The input filter may be an active or passive filter type.

The **sample and hold (S/H)** circuit in Fig. 11–59 is not necessary in all ADC circuits. It is required in high-speed operations where conversion speed of the ADC is insufficient to allow stand-alone operation. To assure conversion accuracy during high-speed operation, the S/H circuit reduces the amplitude uncertainty error. This is accomplished by measuring the analog input in a smaller aperture time than the ADC conversion time aperture and then storing that sample until the next sampling period begins.

The ADC quantizes and codes the analog input signal into a set of discrete digital output levels. **Signal processing,** as the name implies, allows manipulation or conditioning of the digital data. At times this processing may be merely storage of the digital data for reconversion at a later time.

The DAC converts the digital signal data back to an analog signal. Each analog output level is a direct function of its binary weight value. Each sample level is held constant until the next sample level arrives. The **smoothing filter** is used as the output circuit in this sampling system to smooth the analog output.

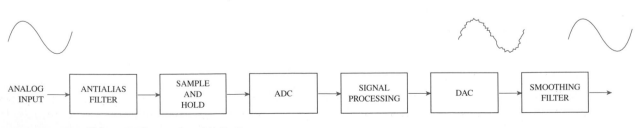

FIGURE 11–59   **Data sampling system block diagram.**

Section 11-4: Review Questions

Answers are given at the end of the chapter.

A. Calculate V_{out} of the ladder network in Fig. 11-44 if $D_3 = 0$ V, $D_2 = +4$ V, $D_1 = 0$ V, and $D_0 = 0$ V.

B. Calculate V_{out} of the ladder network in Fig. 11-44 if $D_3 = +4$ V, $D_2 = +4$ V, $D_1 = 0$ V, and $D_0 = 0$ V.

C. How many analog output voltage levels can be produced by a 10-bit DAC?

D. Calculate the analog step height of 1 LSB for a 10-bit DAC if the desired output voltage range is 0 V to +5 V.

E. What percentage of full scale is the 1 LSB calculated in question (D)?

F. Define monotonic.

G. State the main advantage and disadvantage of flash converters.

H. State the Nyquist sampling theorem.

I. Distortion caused by the lower frequencies generated when the sampling rate of an ADC is too low is called _____.

J. State the purpose of each block in Fig. 11-59.

SECTIONS 11-3 AND 11-4: INTERNAL SUMMARY

Bus contention is caused when two or more devices output data to a data bus simultaneously. The result is unintelligible data. This problem is prevented by having all devices that are connected to the data bus contain **tristate-output** capability. Tristate-output circuits must be enabled by an address specific to one device. Thus, only one device at a time can communicate on the bus. The remaining circuit's outputs are electrically disconnected from the data bus by placing them in the high-impedance (Hi-Z) state.

Many types of gates, buffers, registers, memory devices, and transceivers with 3-state outputs are commercially available. Transceivers with 3-state outputs are often used in computers to assist in the function of bus control.

Data conversion is necessary in many electronic applications. If the speed of an ac motor is to be controlled by a computer, the digital speed signal from the computer must be converted to an analog signal to drive the motor. Often analog feedback is converted back to digital format for updating information in the computer.

The data conversion is accomplished by **digital-to-analog converters (DACs)** and **analog-to-digital converters (ADCs).** A DAC produces an analog output voltage that is proportional to the binary input value. The greater the number of binary inputs to the DAC, the truer the analog output. An ADC produces a digital output code (number) representative of the analog input signal.

SECTIONS 11-3 AND 11-4: INTERNAL SUMMARY QUESTIONS

Answers are given at the end of the chapter.

The use of manufacturer's data sheets is recommended to assist you in answering these summary questions.

1. What are the $1Y$–$4Y$ outputs of the circuit shown in Fig. 11-60 with the inputs shown?

	1Y	2Y	3Y	4Y
a.	1	Hi-Z	1	Hi-Z
b.	1	1	0	0
c.	0	0	1	1
d.	Hi-Z	1	Hi-Z	0

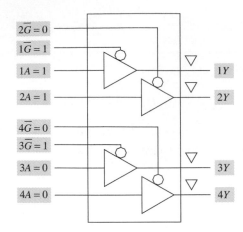

FIGURE 11–60

2. What are the $1Y$–$4Y$ outputs of the circuit shown in Fig. 11–61 with the inputs shown?

	1Y	**2Y**	**3Y**	**4Y**
a.	1	Hi-Z	0	Hi-Z
b.	1	1	0	0
c.	0	0	1	1
d.	Hi-Z	1	Hi-Z	0

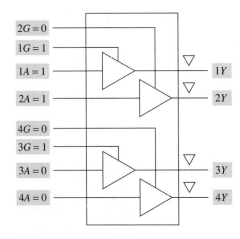

FIGURE 11–61

3. What function is being performed by the 74LS245 in Fig. 11–62 when $\overline{G} = 0$ and DIR = 1?

   a. Hi-Z   c. *B* data to Bus A
   b. Latch $(A = B)$ d. *A* data to Bus B

4. What function is being performed by the 74LS245 in Fig. 11–62 when $\overline{G} = 1$ and DIR = 1?

   a. Hi-Z   c. *B* data to Bus A
   b. Latch $(A = B)$ d. *A* data of Bus B

5. The upside-down triangles in Fig. 11–62 indicate the bus transceiver has open-collector outputs.

   a. True
   b. False

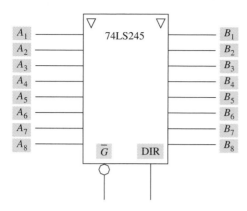

FIGURE 11–62

6. What function is being performed by the 74173 register in Fig. 11–63 when $\overline{\text{CLEAR}} = 1, \overline{G_1} = 0, \overline{G_2} = 0, \overline{M} = 0, \overline{N} = 0,$ Data $= 1,$ and CLK $=$ PGT?
   a. Hi-Z
   b. HOLD
   c. CLEAR
   d. Q Outputs $=$ Data

7. What function is being performed by the 74173 register in Fig. 11–63 when $\overline{\text{CLEAR}} = 0, \overline{G_1} = 1, \overline{G_2} = 1, \overline{M} = 0, \overline{N} = 0,$ Data $= 1,$ and CLK $=$ PGT?
   a. Hi-Z
   b. HOLD
   c. CLEAR
   d. Q Outputs $=$ Data

8. What function is being performed by the 74173 register in Fig. 11–63 when $\overline{\text{CLEAR}} = 0, \overline{G_1} = 0, \overline{G_2} = 0, \overline{M} = 1, \overline{N} = 0,$ Data $= 1,$ and CLK $=$ PGT?
   a. Hi-Z
   b. HOLD
   c. CLEAR
   d. Q Outputs $=$ Data

9. What function is being performed by the 74173 register in Fig. 11–63 when $\overline{\text{CLEAR}} = 0, \overline{G_1} = 0, \overline{G_2} = 0, \overline{M} = 0, \overline{N} = 0,$ Data $= 1,$ and CLK $=$ PGT?
   a. Hi-Z
   b. HOLD
   c. CLEAR
   d. Q Outputs $=$ Data

10. The output of the R–$2R$ ladder in Fig. 11–64 is $V_{\text{IN}} \div 2$ when the input is 1000 $(D_3–D_0)$.
    a. True
    b. False

11. The output of the R–$2R$ ladder in Fig. 11–64 is $V_{\text{IN}} \div 2$ when the input is 0001 $(D_3–D_0)$.
    a. True
    b. False

12. An op-amp voltage follower can be connected to the R–$2R$ ladder network in Fig. 11–64 to prevent loading the ladder.
    a. True
    b. False

13. The circuit in Fig. 11–65 is a _____.
    a. flash ADC
    b. switched current-source DAC
    c. successive-approximation ADC
    d. successive-approximation DAC

14. The circuit in Fig. 11–66 is a _____.
    a. flash ADC
    b. switched current-source DAC
    c. successive-approximation ADC
    d. successive-approximation DAC

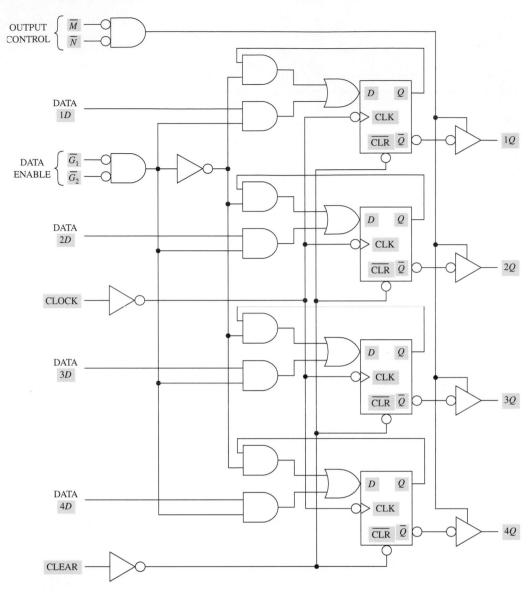

FIGURE 11-63

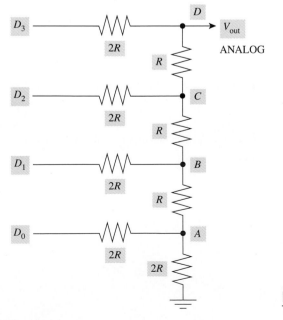

FIGURE 11-64

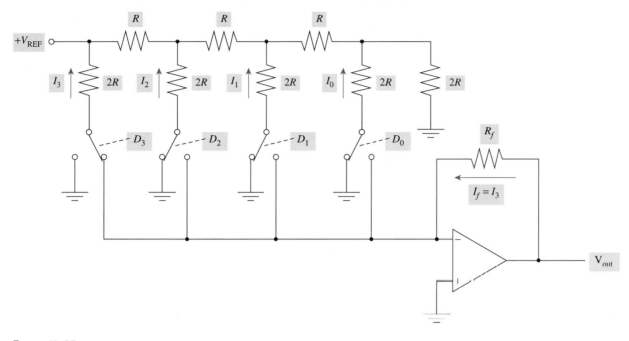

FIGURE 11–65

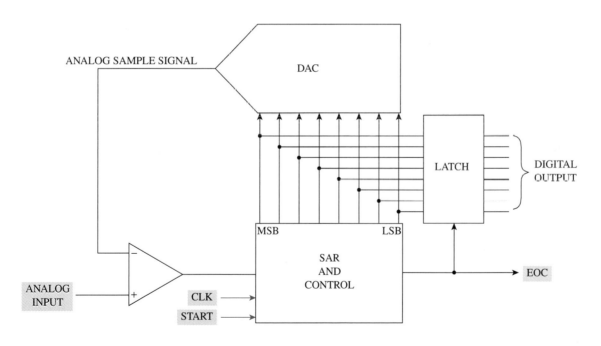

FIGURE 11–66

15. The circuit in Fig. 11–67 is a _____.
    a. flash ADC
    b. switched current-source DAC
    c. successive-approximation ADC
    d. successive-approximation DAC

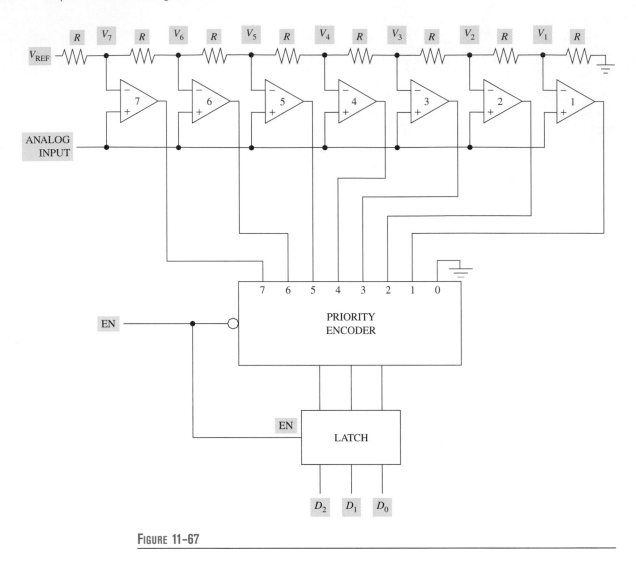

FIGURE 11-67

OBJECTIVES

1. Given a computer circuit containing bus transceivers, DACs, or ADCs, determine the circuit's output(s) when the inputs are specified.
2. Given a malfunction in a circuit containing bus transceivers, DACs, or ADCs, determine the most probable symptom(s).

Troubleshooting mixed-technology (TTL and MOS) logic circuits for compatibility is a static operation. Proper level inputs are applied to the drive gate, and V_{OH} and V_{OL} of the load gate(s) are checked to ensure they meet manufacturer's specifications. If the levels are out of tolerance, the fan-out and compatibility may be checked per specs, but this is seldom a problem except in newly engineered systems. If compatibility and loading are acceptable and problems exist, the device is probably faulty and requires replacement.

The bidirectional data bus and parallel configuration of a computer require bus transceivers to assist in controlling the data bus. The block diagram in Fig. 11–68 illustrates the use of the 74LS245 bus transceiver described in Section 11–3.

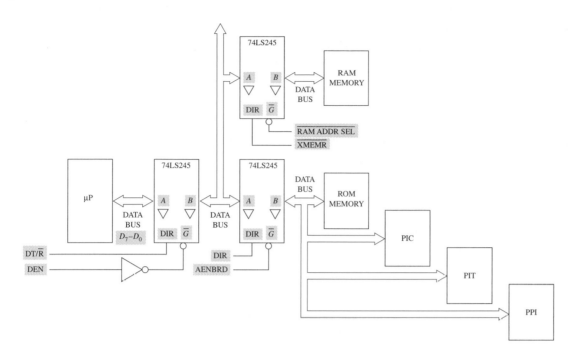

FIGURE 11–68   **Partial computer block diagram showing octal bus transceivers.**

The $\overline{G}$ input of the 74LS245 transceiver is used to enable the device. The IC's outputs are in the Hi-Z state when $\overline{G}$ is high. The direction (DIR) input controls the direction of data movement through the transceiver when it is enabled ($\overline{G} = 0$).

The computer microprocessor is connected to memory (RAM and ROM), to the expansion slots on the system board, and to several peripheral devices called intelligent chips as shown in Fig. 11–68. The peripheral devices include the Programmable Interrupt Controller (PIC), Programmable Interval Timer (PIT), and Programmable Peripheral Interface (PPI) ICs, which are used to support microprocessor operation.

The bus transceiver (74LS245) connected to the microprocessor is enabled when the data enable (DEN) signal from the system bus controller is activated. This high signal (DEN) indicates a data transfer is to take place over the data bus. If the signal is inactive, the outputs of the transceiver are in the Hi-Z state and the system data bus is disconnected from the microprocessor. The data transmit/receive ($DT/\overline{R}$) signal from the bus controller controls the direction of data movement through the transceiver. A microprocessor WRITE operation sets the signal high and a READ operation brings it low. Data movement is from A to B when this signal is high and B to A when it is low.

The RAM (memory) data bus transceiver is enabled by a low RAM address select ($\overline{\text{RAM ADDR SEL}}$) signal. This signal is active only when a valid RAM page address is placed on the address bus. The direction of data movement is controlled by the extended memory read ($\overline{\text{XMEMR}}$) signal. An active $\overline{\text{XMEMR}}$ signal sets data movement from B to A (RAM memory to microprocessor) during a READ operation. Data movement through the transceiver is from A to B (microprocessor to RAM memory) during a write-to-memory operation.

The bus transceiver connected to ROM (memory) is enabled by a low address enable board (AENBRD) signal. This signal goes high and disables this transceiver only during direct-memory access (DMA) operations. The AENBRD signal is also used to disable the bus controller's DEN signal so that the microprocessor's bus transceiver is also disabled.

Disabling both of these transceivers puts them in the Hi-Z state and allows the DMA controller to take over complete control of the system's busses. The DIR input to the $\overline{\text{ROM}}$ transceiver is set low (B to A) during a ROM read operation or an input output read ($\overline{\text{IOR}}$) operation when address bits A_8 and $A_9 = 0$. The circuit controlling B to A data movement was discussed in Section 5–5 under the heading "ROM Address Decoder Logic Circuit."

Troubleshooting a circuit similar to the one in Fig. 11–68 is based on the troubleshooting techniques discussed in previous chapters. Data on the data bus may be checked and verified with a logic analyzer. Pulsing of the amber light on a logic probe indicates activity on the data bus during PC operation. Although this is only a quick check, it verifies the bus is not dead when activity is present. Keep in mind the 74LS245 is a TTL IC; therefore, an open input acts as a high. An open at the $\overline{G}$ input will put the chip in the Hi-Z state. An open at the DIR input will set the direction of data movement from A to B permanently. If the DIR input to the microprocessor bus transceiver were open, the microprocessor would never be able to read memory. This would prevent boot-up (initialization) of the system.

It was mentioned in Section 11–4 that IC converters designed specifically for computer capability are commercially available. One example is National Semiconductor's DAC0830 8-bit microprocessor-compatible, double-buffered DAC. The IC uses an R–$2R$ ladder network to divide the reference current. It employs CMOS current switches and control logic, but it is TTL compatible. A 10-bit version (DAC1000 series) and a 12-bit version (DAC1208–1230) are available from National Semiconductor when higher resolution is required.

A functional diagram of the DAC0830 is provided in Fig. 11–69. Two 8-bit latches are available at the IC input, thus the term "double-buffered." The double latch configuration allows the microprocessor to load the input latch, store the data, and then transfer it to the DAC latch when conversion is desired. This allows fast updating of the DAC output on demand. In addition, numerous DACs can be updated simultaneously with one common strobe signal.

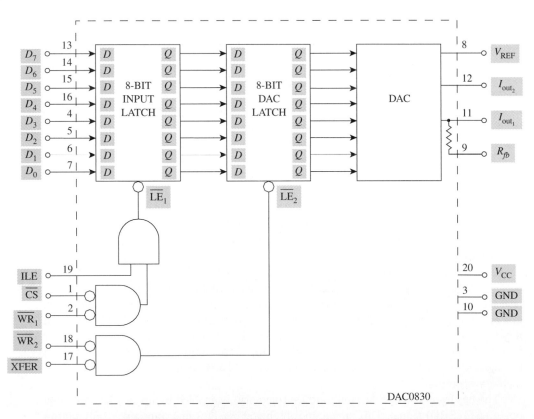

FIGURE 11–69   DAC0830 8-bit microprocessor-compatible, double-buffered DAC–function diagram. *Courtesy of National Semiconductor.*

Figure 11–69 reveals two separate addresses must be decoded for latch operation. The input latch is controlled by the input latch enable (ILE), chip select ($\overline{CS}$), and write 1 ($\overline{WR_1}$) signals. This latch is enabled for loading only when $\overline{CS} = \overline{WR_1} = 0$ and ILE = 1. This input combination places a high on the latch internal latch enable 1 ($\overline{LE_1}$) input, which allows Q to follow D to load the latch. The loaded data are latched when $\overline{LE_1} = 0$. The ILE input expands the addressing scheme of the DAC and should be tied high if not needed. When ILE = 0 the input latch data are stored and new data cannot be loaded in the DAC. The DAC latch is loaded with the data from the input latch when transfer ($\overline{XFER}$) = write 2 ($\overline{WR_2}$) = 0. This input combination places a high on $\overline{LE_2}$ and allows Q to follow D. The DAC latch data are stored when $\overline{LE_2} = 0$.

Figure 11–70 shows two DACs interfaced with a microprocessor-controlled system. More DACs may be connected in parallel in this manner if desired. Additional chip select addresses would have to be decoded if more DACs were added to the circuit.

The two DAC chip select inputs are controlled by separate addresses. The transfer signal ($\overline{XFER}$) inputs to the DACs are connected together in Fig. 11–70 and controlled by one address to allow simultaneous updating of both DACs. The analog output is equal to $-I_{out_1} \times R_{fb}$ and is of the opposite polarity of V_{REF}.

All unused digital inputs to the DAC0830 must be connected to ground or V_{CC}. Floating inputs act like a high, but static-discharge damage may be incurred if the inputs are left open.

The DAC0830 was designed for microprocessor interface compatibility; however, it can be connected to allow the internal latches to continuously convert the digital inputs to analog outputs. This setup is often called **flow-through operation.** The flow-through configuration is achieved by grounding $\overline{CS}$, $\overline{WR_1}$, $\overline{WR_2}$, and $\overline{XFER}$ and connecting ILE to +5 V.

Troubleshooting the circuit in Fig. 11–70 is accomplished by first isolating the malfunction to one DAC or both DACs. If neither DAC produces an analog output, the

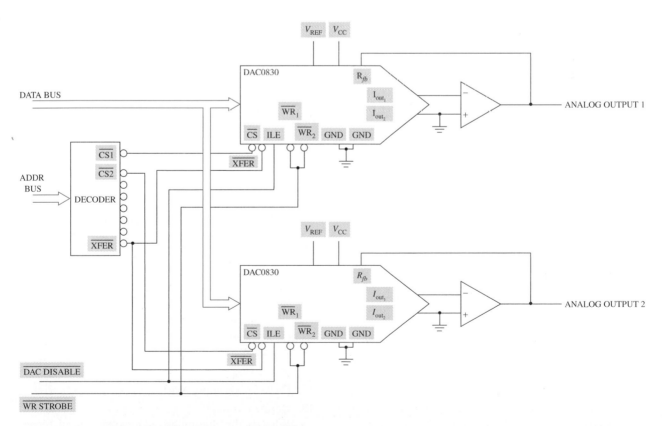

Figure 11–70 **Multiple microprocessor-controlled DACs.**

problem has to be in an area that affects both DACs. The procedure to follow here is:

1. Check V_{CC} and ground to all ICs.
2. Check the DAC reference voltage (V_{REF}) and ensure it is within specs.
3. Ensure the DAC $\overline{CS}$ inputs from the decoder pulse low during circuit operation.
4. Ensure the DAC $\overline{XFER}$ inputs from the decoder pulse low during circuit operation.
5. Ensure the $\overline{WR}$ Strobe input strobes low to enable latch loading.
6. Ensure $\overline{DAC}$ Disable (ILE) is high to enable latch loading.
7. Ensure data are available on the data bus.

If the problem appears at the output of only one DAC channel, check the inputs to that DAC as described and check the op-amp.

CHAPTER 11: SUMMARY

A general introduction to TTL and CMOS technology was presented in the first section of this chapter. Interfacing problems were presented, and the methods used to solve these problems were discussed. Current sourcing, current sinking, and voltage compatibility are important parameters that must be considered when designing logic circuits. Although digital-to-analog and analog-to-digital conversion was set aside in a separate section, these converters are interface circuits used to make the digital and analog world compatible.

Special-purpose **open-collector output** and **tri-state output** circuits were presented by discussing the need for such devices and how that need has been filled.

Digital systems communicate with the world through conversion circuits called DACs and ADCs. Digital data quantities are converted to analog quantities by DACs. The reverse process is accomplished by ADCs.

The quality of the conversion depends on the device's resolution. Analog-to-digital conversion is accomplished using several different methods. Two of the more popular methods—successive-approximation and flash—were covered in detail in this chapter.

The conversion terms defined in the chapter are important for an understanding of data conversion in general—they should be committed to memory.

CHAPTER 11: END OF CHAPTER QUESTIONS/PROBLEMS

Answers are given in the Instructor's Manual

Note: The use of manufacturer's data sheets is recommended to assist you in answering these questions.

SECTION 11-1

1. What type of transistor is used in TTL ICs?
2. What do the letters LS indicate in a 74LS00 part number?
3. What type of transistor is used in CMOS ICs?
4. Which technology, TTL or CMOS, provides the highest packing density?
5. What do the letters BCT indicate in a digital IC part number?

SECTION 11-2

6. What is the purpose of connecting digital drive circuits to load circuits using interface circuits between the two?

7. Figure 11–71 indicates current *sinking/sourcing*. (Select correct answer.)
8. Which direction is I_{OH} (-0.4 mA on the data sheet) flowing in Fig. 11–71?
9. The NAND gate on the left in Fig. 11–71 is the *drive/load* gate. (Select correct answer.)
10. State the meaning of the specification I_{IL} on a data sheet.
11. Current that flows *out* of a logic circuit is *sink/source* current. (Select correct answer.)
12. A logic gate *sinks/sources* current when its output is low (V_{OL}). (Select correct answer.)
13. Define fan-out.
14. What is the minimum input voltage level recognized as a high input in a TTL gate?

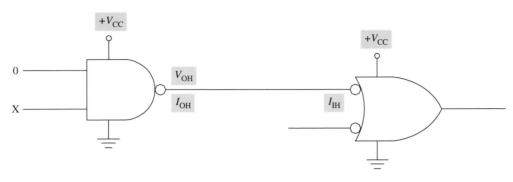

15. What is the maximum input voltage level recognized as a low input in a TTL gate?
16. What is the fan-out of a 74S08 AND gate whose low-level output current (I_{OL}) is 20 mA if the load gate's low-level input current is −2 mA?
17. What happens in a TTL circuit when fan-out is exceeded?
18. Define noise immunity in relation to digital circuits/gates.
19. What does the dashed AND gate symbol in Fig. 11–72 represent?

20. What do the diamond symbols on each logic gate in Fig. 11–72 represent?
21. Standard TTL logic gate outputs *can/cannot* be wire-ANDed. (Select correct answer.)

SECTION 11–3

22. What does the symbol within the buffer in Fig. 11–73 represent?
23. What is the output of the buffer in Fig. 11–73 when.
$\overline{G} = 1$?
$\overline{G} = 0$?

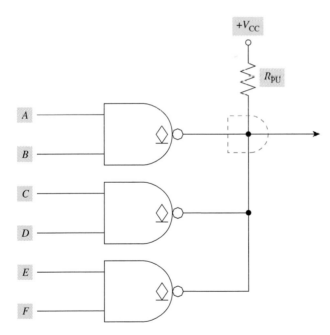

FIGURE 11–72

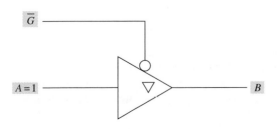

FIGURE 11–73

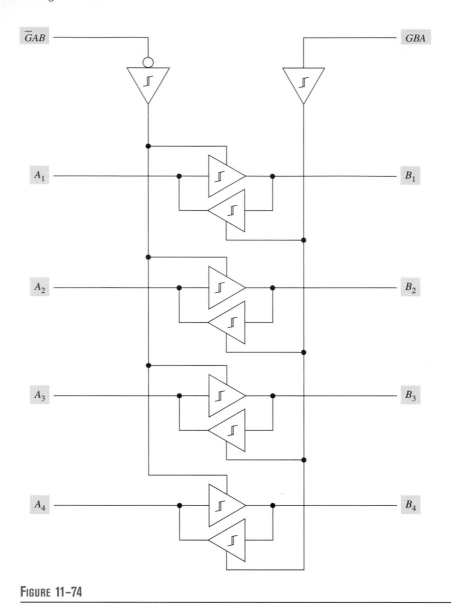

Figure 11-74

CT 24. What is the mode of operation of the 74LS243 Bus Transceiver in Fig. 11–74 when
    a. $\overline{GAB} = 0$ and $GBA = 0$?
    b. $\overline{GAB} = 0$ and $GBA = 1$?
    c. $\overline{GAB} = 1$ and $GBA = 0$?

25. What is the purpose of a bus transceiver in a digital system?

Section 11-4

26. What is the output of the R–$2R$ ladder network in Fig. 11–75 when $D_3 = +4$ V, $D_2 = 0$ V, $D_1 = +4$ V, and $D_0 = 0$ V?

27. What is the output of the R–$2R$ ladder network in Fig. 11–75 when $D_3 = 0$ V, $D_2 = 0$ V, $D_1 = 0$ V, and $D_0 = +4$ V?

28. What type of data conversion is accomplished by the circuit in Fig. 11–76?

CT 29. What is the output voltage of the converter in Fig. 11–76 when $D_3 = +5$ V, $D_2 = 0$ V, $D_1 = 0$ V, and $D_0 = +5$ V?

30. What is the purpose of the op-amp in Fig. 11–76?

31. What is the voltage gain of the op-amp in Fig. 11–76?

32. Define DAC resolution.

33. What type of data conversion is accomplished by the circuit in Fig. 11–77?

34. Identify the circuit in Fig. 11–77.

35. Identify the circuit in Fig. 11–78.

36. What is the advantage of the converter in Fig. 11–78 when compared to other types of converters?

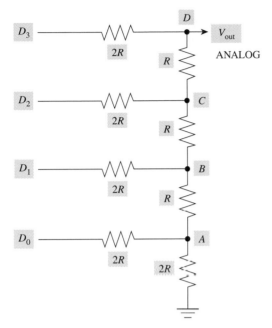

FIGURE 11–75

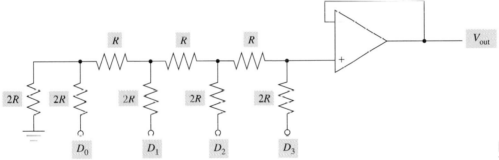

FIGURE 11–76

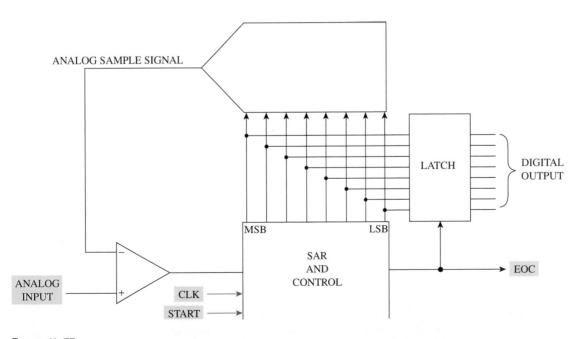

FIGURE 11–77

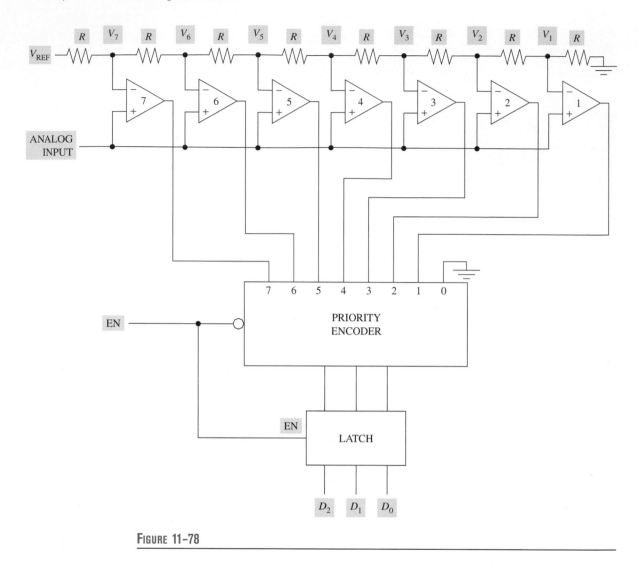

FIGURE 11-78

All Sections: Match each of the following definitions to their term. There are more terms than definitions, so some of the terms will not be used.

DEFINITIONS

_____ 37. A logic gate or device used to provide the input(s) to another logic gate(s) or device(s).

_____ 38. The current that flows into a logic circuit on an input or output pin when its output is low (V_{OL}).

_____ 39. The maximum number of inputs the output of a logic gate can reliably drive.

_____ 40. A circuit used to convert digital to analog.

_____ 41. Low-frequency distortion produced in ADCs that is caused by sampling at too low a frequency.

_____ 42. A type of gate or circuit that must be used when connecting to a data bus line in a digital system.

_____ 43. A type of gate/circuit that can be used for wire-ANDing outputs.

_____ 44. The smallest analog change in output voltage of a DAC that results from a 1-LSB converter code change.

TERMS

a. fan-in
b. fan-out
c. ADC
d. DAC
e. drive circuit
f. load circuit
g. source current
h. sink current
i. aliasing
j. resolution
k. open-collector output
l. tristate output
m. quantization

ANSWERS TO REVIEW QUESTIONS

SECTION 11-1

A. Bipolar
B. Unipolar
C. Enhancement-mode
D. Bipolar and Unipolar

SECTION 11-2

A. False
B. NAND Gate
C. True
D. sourcing
E. sinking
F. sinks
G. The maximum number of inputs that the output of a gate can reliably drive.
H. True
I. A circuit's ability to tolerate noise signals on its inputs and still maintain reliable operation.
J. False
K. $\overline{AB} \cdot \overline{CD}$
L. False
M. Open-collector output gate

SECTION 11-3

A. Three states available at the output of an IC—Logic 0, Logic 1, and Hi-Z.
B. The invalid bus condition that occurs when more than one device is allowed on the bus at the same time.
C. Logic 1
D. floating (Hi-Z state)
E. tristate output
F. A device that allows asynchronous two-way communications between two data busses.
G. B data to Bus A
H. Hi-Z state
I. Shift left
J. Hi-Z state

SECTION 11-4

A. $V_{out} = V_{IN} \div 2^{N-n}$
   $= +4 \text{ V} \div 2^{4-2} = +4 \text{ V} \div 2^2 =$
   $+4 \text{ V} \div 4 = +\mathbf{1V}$
B. $V_{out} = V_{IN} \div 2 = 2 \text{ V}$ with D_2 source shorted
   $V_{out} = V_{IN} \div 4 = 1 \text{ V}$ with D_3 source shorted
   $V_{out} = \mathbf{3V}$
C. $2^{10} = \mathbf{1024}$
D. $\text{LSB} = 5 \text{ V} \div 1024 = \mathbf{0.00488 \text{ V}}$
E. $0.00488 \text{ V} \div 5 \text{ V} = 0.000976 = \mathbf{0.0976\%}$
F. Operation of a data converter when the output increases or decreases in response to a consistent increase or decrease, respectively, of the input signal.
G. Advantage: High-speed operation.
   Disadvantage: Higher resolution requires more op-amp comparators—increases IC cost and power consumption
H. To reproduce a periodic signal, it is necessary to sample at a rate greater than twice the sampled signal's highest frequency component.
I. aliasing
J. Antialias filter: Prevents low-frequency distortion caused by sampling at a rate that is not at least two times the sampled signal's highest frequency component.
   Sample and hold circuit: A circuit used in high-frequency conversion that samples the analog input to an ADC in a shorter time aperture than the conversion time of the ADC.
   ADC: A circuit used to quantize and code the analog input into a digital number/code.
   Signal processing: A digital circuit used to manipulate, condition, or store data.
   DAC: A circuit used to convert the quantized digital signal to an analog signal.
   Smoothing filter: A filter used to smooth the analog output signal.

ANSWERS TO INTERNAL SUMMARY QUESTIONS

SECTIONS 11-1 AND 11-2

1. a	9. a
2. a	10. c
3. b	11. a
4. a	12. c
5. a	13. a
6. c	14. a
7. b	15. d
8. b	

SECTIONS 11-3 AND 11-4

1. d	9. d
2. a	10. a
3. d	11. b
4. a	12. a
5. b	13. b
6. c	14. c
7. b	15. a
8. a	

12 MEMORY

Topics Covered in this Chapter

Introduction

Memory is required in all computers. Memory is used to store the boot-up program for computer initialization; it contains the operating system data and the steps of an applications program; and it holds the data to be processed and the data that has been processed. **Data** (binary 1s and 0s) can be used to represent instructions, numbers, letters, and characters in a computer. There are several types of memory used to accomplish the tasks just listed as well as many other varied tasks in a computer.

The neat thing about memory circuits in digital systems is that they only have to store a Logic 0 or a Logic 1. The flip-flops presented in Chapter 6 had this capability. It is the foundation of Chapter 6 upon which this chapter is built.

One of the first PCs put on the market commercially (1974) contained 256 bytes of memory. It was called the ALTAIR 8800 and was sold by Micro Instrumentation Telemetry Systems. By 1977, 4K and 8K of memory became the norm. The Tandy Corporation TRS-80 PC was also put on the market in 1977. It had 4K Random-Access Memory (RAM). Some computers of the era had a respectable 32K memory.

The IBM PC (released in 1981) and the Commodore 64 (released in 1982) both had an amazing 64K of RAM. By 1983, the IBM PC-XT sported 128K of RAM, a 360K floppy-disk drive, and a 10 MB hard drive. The rest of this story is current history.

IMPORTANT TERMS

Address Bus

Bus

Bus Cycle

Byte

Central Processing
Unit (CPU)

Control Bus

Data Bus

Dynamic RAM (DRAM)

Electrically Erasable
PROM (EEPROM)

Erasable
Programmable ROM
(EPROM)

Firmware

Flash Memory

Hardware

Input/Output (I/O)

Mask ROM (MROM)

Memory Address

Memory Capacity

Memory Cell

Primary Memory

Programmable ROM
(PROM)

Pseudo Static RAM
(PSRAM)

Random-Access
Memory (RAM)

Read Bus Cycle

Read-Only Memory
(ROM)

Refresh

Secondary Memory

Software

Static Ram (SRAM)

Volatile Memory

Write Bus Cycle

Chapter Objectives

1. Match memory terms with their definitions.

2. Calculate the number of address inputs required to access
   memory within a chip of specified capacity.

3. Identify different types of ROM and state how their stored data
   are erased and reprogrammed.

4. State the advantages and disadvantages of various types of
   ROM and of flash memory.

5. Identify different types of RAM.

6. Compare the various types of RAM in terms of advantages and
   disadvantages.

7. State the purpose of refresh.

8. Determine the address range of a specified memory IC and draw
   a memory map.

9. Design a memory decoder that will decode a specific address
   range.

677

During this PC revolution, with memory expanding beyond belief, system speed has changed just as significantly. From a clock speed of 2 MHz for Motorola's 6800 microprocessor, current system operating speeds have steadily increased. All of these advances have dictated the trail of memory development over the last decade or two.

This chapter presents the basics of memory circuits. It includes how memory circuits work and how they are organized and addressed. Specific chips are seldom alluded to in the text because they will surely be obsolete in the near future. Rather, a generic but applicable view of memory devices and circuits is presented. This information will prepare you for the detailed analysis of specific memory circuits required in a microprocessor or computer repair course.

SECTION 12–1: MEMORY FAMILIARIZATION/INTRODUCTORY CONCEPTS

OBJECTIVES

1. Define memory terms.
2. Given the capacity of a memory chip, calculate the number of address lines required to access all memory locations in the chip.

Computer memory falls into two categories—primary and secondary. **Primary memory** may be thought of as **on-board memory.** It is used to store data and programs currently being used by the microprocessor. The data must be readily accessible, and its retrieval must be fast. **Random-Access Memory (RAM)** and **Read-Only Memory (ROM)** fall in the primary memory category, and they are **semiconductor memory.** Access to the data they store is exceptionally fast. The biggest disadvantage of this type of memory is that its capacity is somewhat limited when compared to the needs of modern computer programs.

Secondary memory has the ability to store much more data than primary memory. This category of memory must be used to fill the needs of the memory-hungry applications programs that are so popular today. Its bulk capability is generally so great that it is called **mass memory.** Hard disks, floppy disks, Compact Discs (CDs), and magnetic tapes fall in this memory category. Secondary memory devices are hybrid and as such are as much mechanical as electronic. This makes data retrieval from them slower than from primary memory devices.

Regardless of the type of memory, the data stored as a sequence of instructions in a computer is called a **program.** The programs, which consist of Logic 1s and 0s, are called **software.** Microprocessors and all other ICs, plus a multitude of other components and parts in a computer, are called **hardware.** A hardware component such as a memory chip with software programmed into it is called **firmware.**

Data are constantly being processed, transferred, and stored in a computer. The transfer of data may be to or from memory, or to or from an Input/Output (I/O) device.

A computer uses busses to accomplish these data transfers. A **bus** is a conductor or set of conductors that connects two or more devices in a system. It is the group of lines used by a computer's microprocessor to communicate with memory and input/output devices. These conductors/lines may be wires or traces on a circuit board. A computer contains three busses as shown in Fig. 12–1. The **address bus** is a unidirectional bus used by the microprocessor to address a specific memory or I/O location. The **data bus** is a bidirectional bus used to transfer data to or from the microprocessor. The width (number of lines) of a data bus usually signifies the size of a word in a system. A **word** is defined as the number of bits of data processed simultaneously in a digital system. If a data bus contains 8 lines, the word size is 8 bits (1 byte). If it is 16 lines, the word size is 16 bits, and they are usually referred to as an **upper byte** and a **lower byte.** The **control bus** is a bidirectional bus used to indicate a specific operation such as a **read, write,** or **interrupt.** This bus is also used to monitor status of certain devices or acknowledge an event.

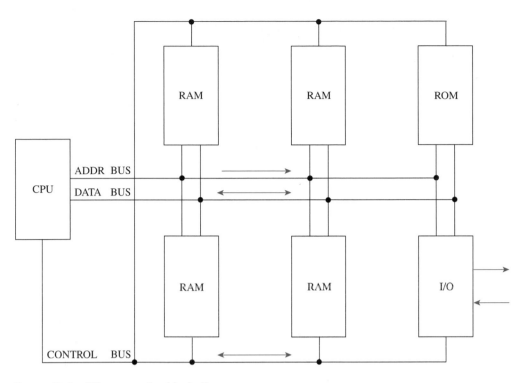

Figure 12-1   Microcomputer block diagram.

The 3-bus architecture of personal computers is shown again in Fig. 12–2. Although the figure looks like a piping diagram, the "pipes" represent multiline busses. The address bus in the figure represents 20 address lines because it is labeled A_{19}–A_0. The data bus represents 8 data lines. It can readily be seen that the microprocessor is the heart of this system. It is also the brain—often called the **Central Processing Unit (CPU)**—that controls all actions in the computer.

A microprocessor **bus cycle** occurs with each transfer of data in the system. A bus cycle is the length of time it takes to execute an instruction, and the bus cycle may take several clock cycles. If the microprocessor is fetching data from memory, a *read* bus cycle is in process. When it is writing data into a memory location, a *write* bus cycle is in process.

The microprocessor generates several different control signals to identify the type of bus cycle it desires.

> **Read:** The $\overline{\text{RD}}$ active-low output signal from a microprocessor indicates it is reading data from a memory or I/O location. This operation is sometimes called a **fetch** operation.

> **Write:** The $\overline{\text{WR}}$ active-low output signal identifies the write bus cycle. It indicates the microprocessor is writing data to a memory or I/O location. This operation is called a **store** operation when memory is involved.

> **Input–output/memory ($\text{IO}/\overline{\text{M}}$):** This microprocessor signal indicates the current bus cycle is a memory access when it is *low* and an I/O access when it is *high*.

A microprocessor must interface with the outside world to be beneficial to the user. Keyboards, monitors, printers, and disk drives are all necessary additions to a computer to make it usable. These devices, among others, are called **input/output (I/O) devices. Input/output ports** in the computer are the channels of interface through which the microprocessor communicates with the I/O devices.

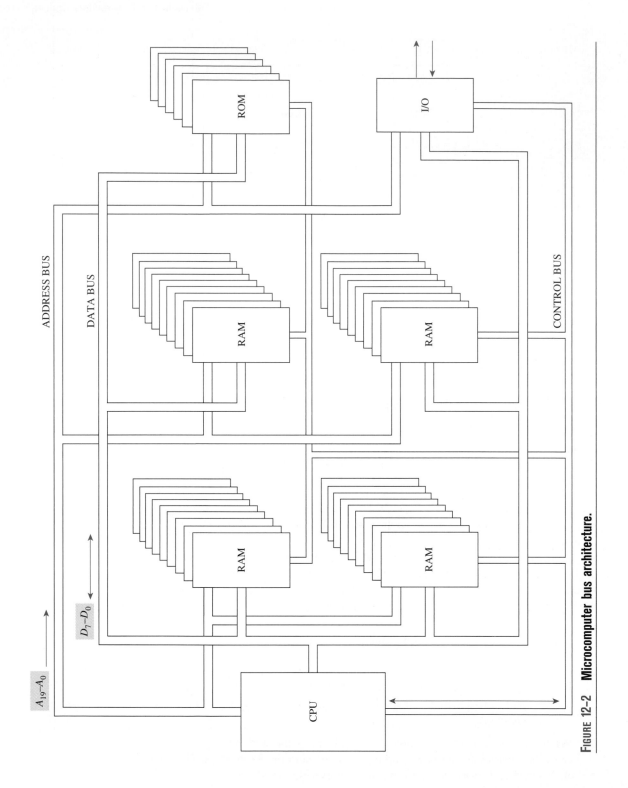

Figure 12–2 Microcomputer bus architecture.

The RD, WR, and IO/M signals are used in some computers to generate memory read (MEMRD), memory write (MEMWR), I/O read (IORD), and I/O write (IOWR) signals. These signals are used to differentiate addresses for memory devices from those for I/O devices. The logic circuit generating these signals was presented in Chapter 5.

Most of the blocks in Fig. 12–1 and Fig. 12–2 are memory blocks. One block is Read-Only Memory (ROM). This ROM chip is programmed by the manufacturer. Its data (program) is **nonvolatile.** In simple terms, the data is permanent—not subject to loss or change under any conditions.

ROM contains much data, but the all-important initialization data in ROM is what gets the computer up and running. This data is called the **bootstrap program.** Once the system has self-tested and initialized, the bootstrap program loads a **disk operating system** program from the hard disk or floppy drive into **Random-Access Memory (RAM).** The disk operating system program performs many computer housekeeping chores.

Once the system is up and running, it is ready for an **applications program.** The program may be word processing, a spreadsheet, or a game. The program is normally loaded into RAM when the selection is made.

When power is applied to a computer, some testing information is displayed on the screen while the computer is ticking away. This information is presented during **boot-up.** The resident portion of the disk operating system is then automatically loaded from the hard drive in most computers. Then the main menu appears on the screen. The menu displays the applications programs available to the user. Although there are variations of this scheme, the process is basic.

Now let's think about what all of this means. Memory is such an integral part of the computing process that it is tested during boot-up. Every RAM memory location in the computer is written to and read back several times during initialization to ensure it is working properly. Several different bit patterns are used during this RAM test. ROM gets the system up and running, and RAM provides the applications program an on-board residence for quick access plus storage for user-generated data.

Many RAM (Random-Access Memory) chips are shown in Fig. 12–2. The term "random-access" relates to how the data in a RAM chip is stored or fetched. All memory locations in RAM are equally accessible. This can be contrasted to the sequential access of a magnetic tape, which is terribly slow. The term RAM always implies data can be written to and read from the memory device. Memory chips with this characteristic are known as **read/write (R/W) memory.** The term "RAM" is actually a misnomer. The data in a ROM chip is read in the same random-access manner it is read in a RAM chip. Nonetheless, *RAM implies read/write memory whereas ROM is read-only memory.*

The drawback to RAM technology is that it is **volatile.** Information in RAM is present as long as power is applied. The data are lost when power is removed. This is due to the nature of the semiconductor memory cell itself. If you think back to flip-flop operation, remember you could not second-guess whether a flip-flop would initially come up in the SET state or the CLEAR state. The same is true of RAM devices. In fact, the microprocessor generates a *RESET signal* early in the boot-up procedure that ensures the memory devices come up in the RESET (CLEAR) state. This signal is automatically generated when power is applied after a power outage. Thus, the data in RAM are lost.

There are ways to prevent this power interruption from being devastating. One method commonly used is **battery backup.** Newer systems also routinely store RAM data into the hard drive automatically. This stores the data permanently, and this built-in feature has prevented many headaches due to power outages.

A few more memory terms need to be presented prior to presentation of the ROM and RAM devices. A **memory cell** is a device designed to store one bit of data. Different methods are employed to accomplish this task, and they will be presented as the need arises.

Memory chips are identified by their storage-handling capability, or **capacity.** An 8K × 1 memory chip can store 8 kilobits of data. This number equals 8192 bits because **1K = 1024.** Each one of the 8192 memory locations within the chip stores one bit of binary data. Since there are many different locations within the device, each location must have its own specific **address.**

A simple analogy here will set some ground rules and teach some addressing concepts. Figure 12–3 shows a circuit with eight memory locations. Each location is used in this example to store one bit. Therefore, each location can be designated a **memory cell.** A 74LS138 decoder is used to decode the 3-bit address applied to the circuit.

Three address bits (A_2-A_0) are required to address eight different memory locations because $2^3 = 8$. This formula actually solves what we have referred to in previous chapters as the modulus of the circuit; however, the MOD formula is used here to raise 2 to a power (n) that equals the number of memory locations to be addressed (MOD $= 2^n = 8$).

The basic problem of determining how many address lines are required to address X memory locations can be solved in the same manner used in Chapter 7 to determine how many flip-flops are needed to implement a specific MOD counter. If n represents the number of address lines, then $2^n = X$, where X is the number of memory locations. Let's make the number of memory locations in this example 32. Thus, $2^n = 32$, and n can be calculated as follows:

$$n = \log \# \text{ memory locations} \div \log 2$$
$$= \log 32 \div \log 2$$
$$= \textbf{5} \text{ (address lines)}$$

Now back to the circuit in Fig. 12–3. The active-low outputs of the decoder are connected to the active-low inputs of the eight memory cells. This active-low input is labeled $\overline{\text{CS}}$ (chip select). It is sometimes called enable $(\overline{\text{E}})$ or chip enable $(\overline{\text{CE}})$, and it is invariably an active-low signal. If this input is not active, the cell is not selected, and the data output line is in the **high-impedance state** discussed in Chapter 11.

Another input to the cells in Fig. 12–3 is the RD/$\overline{\text{WR}}$ input pin. The level on this pin determines whether data is read from (RD/$\overline{\text{WR}} = 1$) or written to (RD/$\overline{\text{WR}} = 0$) the memory cell. The details of how this is accomplished are presented in Section 12–3.

If we assume RD/$\overline{\text{WR}} = 1$ and $A_2-A_0 = 101$, the data bit stored in memory location #5 is available at the output of the #5 cell. This is true because the 74LS138 decoder is permanently enabled, and the address is 101 to the select inputs. These inputs bring the Y_5 output line of the decoder low, which selects the #5 memory cell because $\overline{\text{CS}} = 0$. The remaining seven memory cells are not selected. Therefore, the data output lines of all of these cells are in the Hi-Z (floating) state.

If each memory location in Fig. 12–3 consisted of an 8-bit parallel-in/parallel-out register, the memory circuit would have the ability to store 8 bits of data in each memory location—total storage is now 8 bytes of data. The circuit would then be designated an 8×8 memory circuit. The first 8 represents the number of memory locations within the circuit; the second 8 represents the number of bits stored in each location.

Realistically, memory capacities run in the kilobyte or megabyte range. Using these terms, 1K equals 1024 and 1M equals 1,048,576. For example, a certain RAM chip is classified as 32K $\times$ 8. This indicates there are 32,768 memory locations within this chip because 1K equals 1024. The device can store 32,768 *bytes* of data because each memory location stores one byte (8 bits). The chip can store a total of 262,144 *bits*, although this number has little significance at this point.

How many address pins would you expect to find on this chip?

$$n = \log 32,768 \div \log 2$$
$$= \textbf{15 address pins}$$

This problem must be solved using the actual number of memory locations. The number of memory locations cannot be entered into the calculator as 32K.

The terminology and concepts presented in this section are basic but important to your understanding of memory devices. Familiarity with the terms presented in this section will increase your understanding of the remainder of the chapter. Some of these terms and their definitions are listed below. Additional memory-related terms will be introduced as the need arises.

Bus: conductor or set of conductors that connects two or more devices in a system.

I/O: Input/output.

1K: 1024

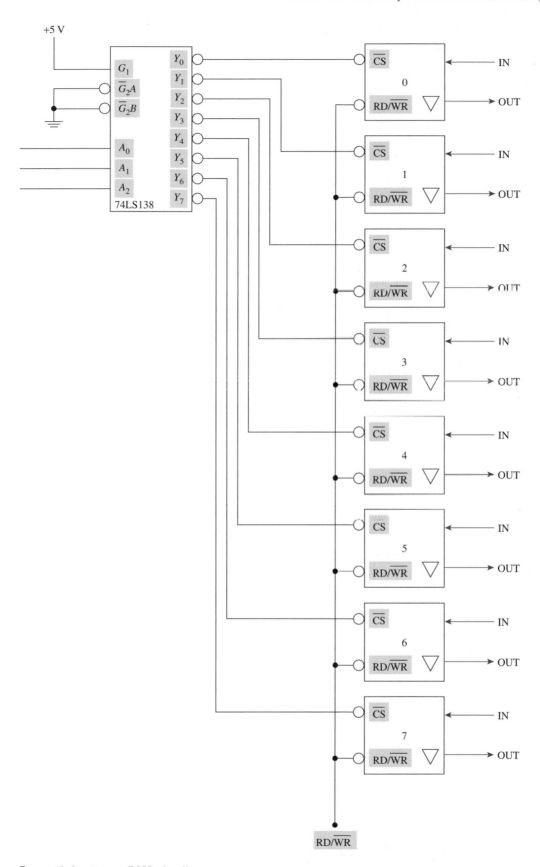

FIGURE 12-3   **8 × 1 RAM circuit.**

1M: 1,048,576

Memory cell: Storage device for one bit of data.

Primary memory: On-board computer memory.

Secondary memory: Mass memory.

ROM: Read-only memory—nonvolatile.

RAM: Random-access memory—volatile.

Section 12–1: Review Questions

Answers are given at the end of the chapter.

A. Match the following terms with their definitions.

(1) Primary memory	a. 1024
(2) Secondary memory	b. 1,048,576
(3) ROM	c. Length of time it takes the microprocessor to execute an instruction.
(4) RAM	d. Volatile memory.
(5) Memory cell	e. Nonvolatile memory.
(6) 1K	f. Input/output.
(7) Bus cycle	g. Storage device for 1 bit.
	h. On-board computer memory.
	i. Mass memory.

B. Define
(1) Volatile memory
(2) Write bus cycle
(3) Byte
(4) Memory capacity

C. A RAM chip is designated 16K × 4. How many memory locations are contained in the chip?

D. How many bits are stored in each memory location in a 16K × 4 RAM chip?

E. How many total bits are stored in a 16K × 4 RAM chip?

F. How many address pins are on a 16K × 4 RAM chip?

G. A RAM chip is designated 1M × 8. How many memory locations are contained in the chip?

H. How many bits are stored in each memory location in a 1M × 8 RAM chip?

I. How many bytes of data are stored in a 1M × 8 RAM chip?

J. How many address pins would you expect to find on a 1M × 8 RAM chip?

SECTION 12–2: READ–ONLY MEMORY (ROM)

OBJECTIVES

1. Define terms relative to read-only memory.
2. Identify conventional types of ROM and state how their memory is erased and reprogrammed.
3. Compare the advantages, disadvantages, and characteristics of MROM, PROM, EPROM, EEPROM, and flash memory.

Read-only memory is nonvolatile. The data in a ROM chip are written to the memory locations within the chip and become permanent. The data are not subject to destruction with the removal or interruption of power.

ROM ICs can be programmed to contain an initialization (boot-up) program. This data is contained in the **ROM BIOS (Basic Input/Output System)** program. The RAM test mentioned previously is part of this program. This test and many others in the BIOS program are called the **Power-On Self-Tests (POST).** The POST program checks the computer for troubles. Once the system has been tested and passed the POST, the BIOS routine looks in drive A for an operating system. If there is a floppy disk in drive A that contains an operating system, the BIOS loads the data into RAM. If no disk is in drive A, the BIOS program looks to the hard disk for an operating system. In this manner, the BIOS turns over control of the PC to the new program.

Additional ROM chips in PCs are used as character generators, data converters, code converters, and look-up tables.

Several types of ROM chips are commercially available. Keep in mind during the following discussion that RAM is random-access read/write memory and ROM is random-access read-only memory.

The 4×4 ROM circuit in Fig. 12–4 uses the **memory matrix** shown in dotted lines. The circuit contains four memory locations, and each location is programmed with four

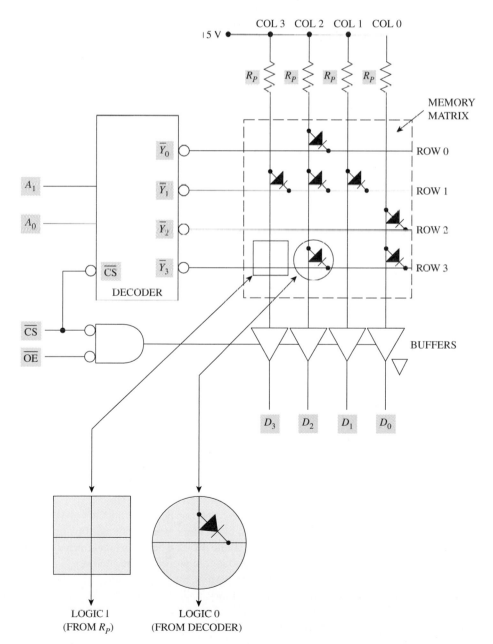

FIGURE 12–4 4×4 ROM circuit.

TABLE 12-1	A_1	A_0	D_3	D_2	D_1	D_0	Hex
4 × 4 ROM Circuit	0	0	1	0	1	1	B
Data for Fig. 12–4	0	1	0	0	0	1	1
	1	0	1	1	1	0	E
	1	1	1	0	1	0	A

bits of data (nibble). There are two enable inputs to the circuit—chip select ($\overline{\text{CS}}$) and output enable ($\overline{\text{OE}}$). Both of these inputs must be activated to read data from the ROM circuit. When $\overline{\text{CS}} = 0$, the 1-of-4 decoder is enabled. The $\overline{Y}_3$ output of this decoder will be low when the input address is $A_1A_0 = 11$.

The highlights on Fig. 12–4 show how Logic 1s and Logic 0s are permanently programmed in memory. The Logic 1 highlight shows no diode in place between the intersect point of Row 3 and Column 3. The Column 3 line is pulled high by the pull-up resistor because there is no connection to the low level on Row 3 at this point. This high is applied to the input of the D_3 output buffer. The Logic 0 highlight shows a diode connected between the intersect point of Row 3 and Column 2. Since $\overline{Y}_3 = 0$, the low on Row 3 from the decoder is applied to the Column 2 line. This pulls the Column 2 line low, and this low is applied to the input of the D_2 output buffer.

The output buffers are enabled when $\overline{\text{CS}}$ and $\overline{\text{OE}}$ are both active. The buffers will be in the Hi-Z state if the memory circuit is not selected ($\overline{\text{CS}} = 1$) or if the output is not enabled ($\overline{\text{OE}} = 1$). The output data for the four memory locations are shown in Table 12–1. Verify the data in the table are correct. Remember, during verification, the decoder will only allow data from a single row to be read for any given address. If there is a diode at an intersect point in the matrix, that point is pulled low when the row line is low. The point will be high if there is no diode.

This ROM circuit is general in nature and operation. It has been discussed to present principles of read-only memory—not details. Four types of ROM chips are presented in this section: **Mask ROM (MROM),** which is normally referred to as ROM; **Programmable ROM (PROM); Erasable Programmable ROM (EPROM),** which is sometimes called UVPROM for the ultraviolet light used during program erasure; and **Electrically Erasable Programmable ROM (EEPROM).**

Mask Read-Only Memory (MROM)

Mask ROMs are programmed by the manufacturer and are not reprogrammable. Logic 1s and Logic 0s are permanently programmed in the chip. This is accomplished by the presence or absence of a diode or transistor at an intersect point in the memory matrix. The presence or absence of an active device at an intersect point is determined by a photo mask. Development of the mask and the masking process are expensive. This makes MROMs costly for limited-production chips. If the setup cost of MROMs is spread across thousands or tens of thousands of chips, it becomes an insignificant cost per IC.

The BIOS programs for some PCs might be contained in an MROM chip. The high-quantity production of these chips keeps their cost down to a minimum. Low-quantity production such as the ROM programs used in research and development projects makes the use of MROMs undesirable.

The use of a 16 × 4 binary–to–gray code converter demonstrates how MROMs can be used as data/code converters or look-up tables. The circuit is shown in Fig. 12–5(a). The binary and gray code data are shown in Table 12–2. The binary data are used to address the MROM decoder, and the gray code data are stored in the memory.

Figure 12–5 (b) and (c) demonstrates the permanent programming of a Logic 1 and a Logic 0 using bipolar transistors in the memory matrix of an MROM IC. This could also be accomplished using unipolar MOSFETs. Every row in this MROM contains four logic intersect points as shown in Fig. 12–5(a).

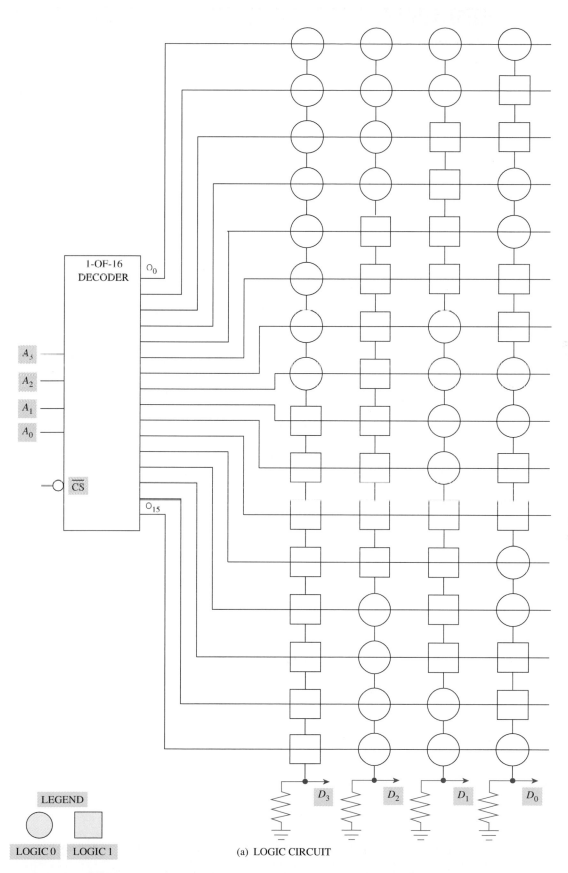

LEGEND

LOGIC 0   LOGIC 1

(a) LOGIC CIRCUIT

FIGURE 12–5   **MROM for binary-to-gray code conversion.**

	Binary	Gray Code
TABLE 12-2		
Binary/Gray Codes	0000	0000
	0001	0001
	0010	0011
	0011	0010
	0100	0110
	0101	0111
	0110	0101
	0111	0100
	1000	1100
	1001	1101
	1010	1111
	1011	1110
	1100	1010
	1101	1011
	1110	1001
	1111	1000

The Logic 1 circuit appears in the box in Fig. 12–5(b). The 1-of-16 decoder produces an active-high output for the specified row address when the chip is selected ($\overline{CS} = 0$). This high is applied to the base circuit of each of the transistors shown in a box on the selected row. The transistor in the box conducts when a high is applied to its base from the decoder. The transistor's path for conduction is through the output resistor in Fig. 12–5(a), which develops the Logic 1 level.

The transistors represented by circles in Fig. 12–5(a) are highlighted in Fig. 12-5(c). The connection from the decoder output line to the base of the transistor has not been made. This is accomplished during the manufacturing process through masking. Since the base in this example is not connected, the transistors in circles will always be off for the selected row. A Logic 0 output is present because no current flows through the output resistor. If the decoder is not enabled, all of its output lines remain inactive (low). This causes all transistors in the matrix to be off.

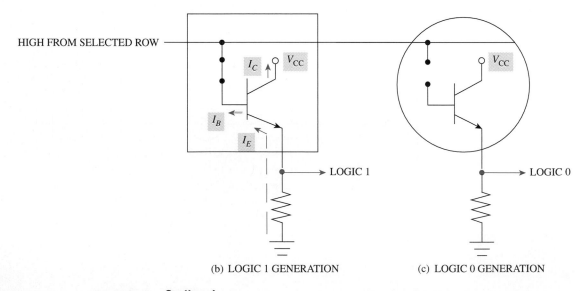

(b) LOGIC 1 GENERATION   (c) LOGIC 0 GENERATION

FIGURE 12–5   Continued

Programmable Read-Only Memory (PROM)

The Mask ROMs just discussed are suitable only for high-volume production. Naturally, requirements exist for low-volume production ROMs in industry. This requirement often exists in experimentation projects requiring fast turnaround. This requirement was filled with Programmable ROMs (PROMs).

Programmability is made possible for the user in a PROM chip by manufactured **fusible links** in the transistor circuit. Figure 12–6 shows one method of producing user programmability. This will be accomplished in a similar manner in programmable logic devices in Chapter 13. Figure 12–6 can be compared operationally to Fig. 12–5(b) and (c). When the fusible link is left intact and the decoder output is high, the transistor conducts and develops a Logic 1 output. When the fusible link is blown, the transistor remains cut off even when the decoder selects the row the transistor is in. Thus, the output is Logic 0.

The user can program these PROM chips with a device called a **PROM programmer.** The programmer is used to blow the selected fuse links in accordance with the data required by the user. Programming a ROM in this manner is called **burning** or **burning-in** a program. In fact, the PROM programmer is usually called a **PROM burner.**

The PROM ICs are shipped with all fuses intact. The user selects those that require blowing. The main problem encountered here is that the effect of blowing a fuse is permanent. This is why PROMs are referred to as **one-time programmable (OTP)** chips. The transistor shown in Fig. 12–6 is only one method employed to establish ROM programmability. In this case, the PROM is shipped from the manufacturer with highs at all memory locations. The act of blowing a fuse establishes a Logic 0.

Some PROMs are designed so they output lows when the fuses are intact. In this case, blowing a fuse establishes a Logic 1. The differences in PROM manufacturing techniques is of little significance to the user/programmer. These manufacturing differences become transparent when you follow the manufacturer's programming instructions in the product data sheets.

Figure 12–7(a) shows the block diagram of a 256 × 4 TTL PROM chip. The logic symbol for this 1024-bit PROM is shown in Fig. 12–7(b). Eight address bits are used to address the 256 memory locations within the IC. Each memory location stores 4 bits. The square memory matrix in this IC is laid out in 32 rows and 32 columns— 32 × 32 = 1024. The memory matrix is shown in Fig. 12–8.

The 1-of-32 decoder enables one row with an active-high output as it did for the circuit in Fig. 12–5(a). The row numbers are shown on the right side of the matrix in Fig. 12–8. The column numbers are labeled across the top of the matrix. The outputs of each group of eight columns are fed to a 1-of-8 multiplexer. This allows selection of four

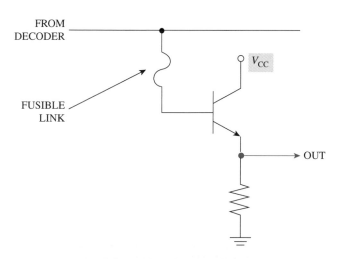

FIGURE 12–6   **PROM transistor with fusible link.**

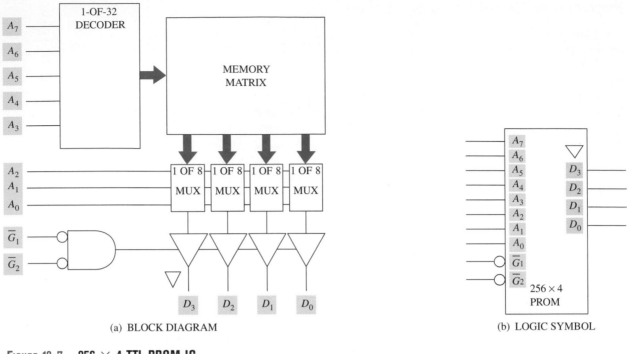

FIGURE 12-7 **256 × 4 TTL PROM IC.**

bits from the selected row. The columns have been renumbered for their specific MUX on the bottom of the figure.

Let's apply address A_7–$A_0 = 11111000_{(2)}$. Bits A_7 through $A_3 = 11111_{(2)} = 31_{(10)}$. These bits are applied to the 1-of-32 decoder and enable Row 31. The lower three address bits (A_2–$A_0 = 000$) set each MUX to select the data at the D_0 input line and pass it to the buffers. These four D_0 intersect points are highlighted by circles in Fig. 12–8. The tri-state buffers in Fig. 12–7(a) are enabled when $\overline{G}_1 = \overline{G}_2 = 0$. The outputs of the buffers are in the Hi-Z state if the enable inputs are not active.

This TTL PROM chip would be programmed by using the following procedure:

STEP 1: Apply the selected address to the PROM chip while it is disabled. Connect $\overline{G}_2$ to ground and leave $\overline{G}_1$ high.
Note: Program only one output bit at a time. Outputs not being programmed may be left open.

STEP 2: Select one output to be programmed high and increase that output pin voltage level and V_{CC} from $+5\,V$ to $+10.5\,V$ at a slew rate between 1 and 10 V/μs.

STEP 3: Enable the chip by taking $\overline{G}_1$ low with a 10-μs pulse.

STEP 4: Verify the desired high-level output has been programmed. Verification should be accomplished with the output loaded within specifications (I_{OL} and I_{OH}), the device enabled, and V_{CC} reduced to $+6$ V and then to $+4$ V. A typical output load would be one TTL gate.

STEP 5: Apply five more programming pulses to the output pin being programmed after verification.

The waveforms for the programming operation are illustrated in Fig. 12–9. The top wave-form represents the address input to the PROM IC. The crossed lines immediately prior to t_1 on this waveform indicate the microprocessor is changing address levels to apply a new address to the IC. The address is stable and valid at t_1.

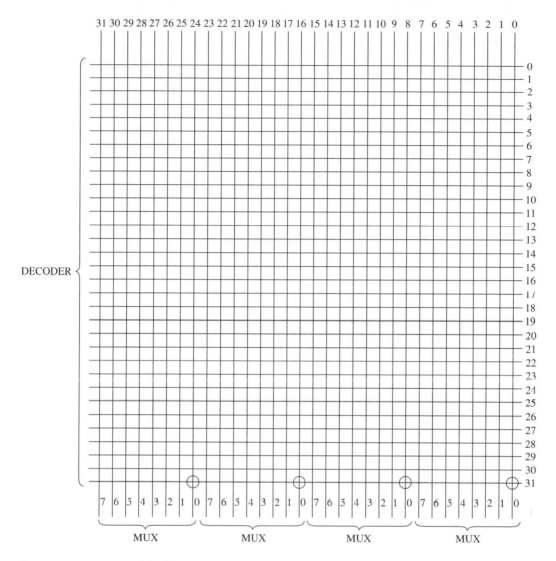

FIGURE 12-8   **256 × 4 PROM memory matrix.**

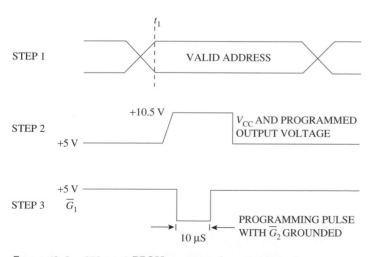

FIGURE 12-9   **256 × 4 PROM programming waveforms.**

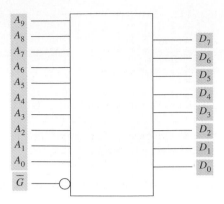

FIGURE 12–10   **ROM chip pin diagram analysis.**

The 256 × 4 designation of this PROM chip implies the word size is four bits (nibble). This may well be true, but keep in mind that another 256 × 4 IC can be connected in parallel with this IC to provide the memory system with a 256 × 8 capability.

Figure 12–10 shows the pin diagram of a memory chip. Several things can be derived by merely looking at this diagram. The lack of a read/write ($RD/\overline{WR}$) or write enable ($\overline{WE}$) input pin on the IC signifies it is a read-only memory chip. Ten address lines (A_9–A_0) indicate there are 1024 (1K) memory locations within the IC. Since there are 8 data output pins, 8 bits are stored in each memory location. Therefore, this is a 1K × 8 ROM IC with a total capacity of 8192 bits. The 8 data pins also imply the system is using 8-bit words and an 8-bit data bus.

Erasable Programmable ROM (EPROM)

The biggest problem encountered by users of PROM chips is their **one-time programmability.** Pattern experimentation is almost impossible, and mistakes in programming are costly because they cannot be corrected using PROM chips.

Chip manufacturers solved these problems with **erasable ROM** ICs. Two methods can be used to erase programs depending upon the type of ROM purchased. Both types will be presented here to round out the discussion of ROM devices. Although these devices can be written to and read from, they are normally programmed and then read. They are often called **read-mostly memories** because they are primarily read from as any other ROM device.

The EPROM is an ultraviolet erasable, electrically reprogrammable ROM. Once the data is programmed into the chip, it is nonvolatile. The EPROM is encapsulated in a package containing a **quartz window** as shown in Fig. 12–11. The data programmed in an EPROM can be erased by exposing the quartz window to ultraviolet (UV) light for a period of about 20 minutes. The window is usually covered with opaque tape to prevent accidental program erasure. Never remove this tape unless the contents of the EPROM are to be erased.

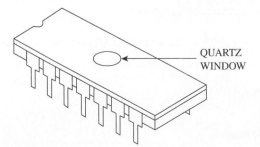

QUARTZ WINDOW

FIGURE 12–11   **EPROM package.**

The memory cells in EPROMs are enhancement-mode MOSFETs (E-MOSFETs) that contain an extra gate. This extra gate, shown in Fig. 12–12(a), is "insulated" and "floating" within the device between the control gate and the channel.

The threshold voltage of an E-MOSFET is the level of voltage required to enhance the channel in the device and turn it on. If there is no charge on the floating gate, the threshold level is much less than 5 V, and the cell conducts when it is selected. Its conduction equates to a Logic 1.

A memory cell is programmed for a Logic 0 output by avalanche injection. This process, also known as hot-electron injection, occurs when the drain voltage is at $+6$ V and a programming pulse of about $+12$ V is applied to the gate as shown in the matrix in Fig. 12–12(b). The pulse causes electrons to penetrate the oxide insulating material and deposit themselves on the floating gate. Once completed, this action increases the threshold voltage required to enhance the channel to 2 to 3 times V_{CC}. The row voltage applied to the gate of the selected memory cell never reaches this high a value, so the cell remains cut off and produces a Logic 0 when read. The two cells marked with an asterisk in Row 2 of Fig. 12–12(b) are programmed with Logic 0s. The row data are 10110. The high voltage used during programming makes in-circuit programming almost impossible.

The EPROM is exposed to ultraviolet light for about 20 minutes to erase the entire contents of the chip. The UV light gives the electrons trapped in the floating gate region enough energy to escape from the region. This returns all memory cells to the conducting Logic 1 state.

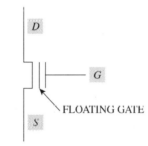

(a) MOSFET WITH FLOATING GATE

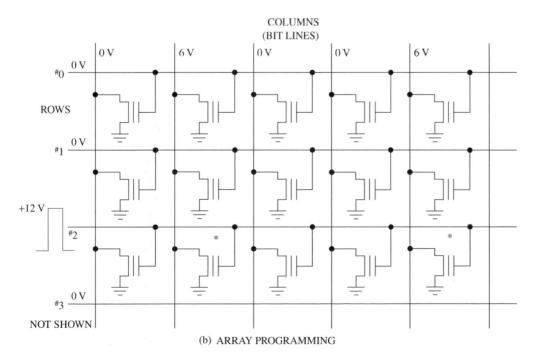

(b) ARRAY PROGRAMMING

Figure 12–12   EPROM.

The primary drawback of the EPROM device is that the entire program is erased when the quartz window is exposed to UV light. This is a distinct disadvantage when only one or two memory locations need data changed. Furthermore, the chip must be removed from the circuit to be erased and reprogrammed.

Electrically Erasable Programmable ROM (EEPROM)

The drawbacks of the EPROM device are partly overcome when **EEPROM** chips are used. The electrically erasable PROM (**E²PROM**) is a nonvolatile, static storage device that is electrically erasable. Modern EEPROM technology allows for bit, byte, or entire data erasures. The chip can be erased and written to rapidly without removal from the circuit. Furthermore, a PROM programmer is not required.

EEPROMs are steadily gaining in popularity. They are being used to replace configuration DIP switches and jumpers in PCs; they are extremely useful in data collection and security systems; they are used in remote control units, cordless phones, radios, and cameras to store the multitude of numbers, channels, stations, and calibration data necessary for proper operation; they are popular for storage of access codes for phones and door openers; and they have found a multitude of other uses in microcontroller-based systems.

An EEPROM uses a variation of the floating-gate technology presented for EPROMs. A basic memory element is illustrated in Fig. 12–13. The device consists of an n-channel FET with an additional floating gate similar to the EPROM. The floating gate is isolated from the rest of the device by silicon dioxide (SiO_2). An extremely thin tunnel oxide layer is used between the floating gate and the drain end of the channel. The charge is transferred to or from the floating gate by a process called Fowler–Nordheim **tunneling.** This quantum mechanical phenomenon occurs only between the floating gate and the drain, which will be referred to as the tunneling region.

Figure 12–14(a) displays setup and operation during a Write-to-Logic 0 operation. It also shows the E²PROM memory cell is a two-transistor cell. This accounts for the somewhat lower densities of these ICs when compared to EPROM devices. The Selected Column is connected to +19 V, and the Select Line is connected to +21 V. The control gate of the memory element (Word Line) is tied to ground, and the source is left floating. Application of these voltages causes the electrons on the floating gate to tunnel into the drain and allow the transistor to conduct. Unlike the EPROM, transistor conductance is read as a Logic 0 at the memory output.

Figure 12–14(b) displays the setup for an Erase operation. The Select Line remains connected to +21 V, and the source is left floating. However, the Selected Column and Word Line inputs are reversed from the Write-to-Logic 0 operation. Here, the Selected Column is grounded and the Word Line is connected to +21 V. These voltage levels cause

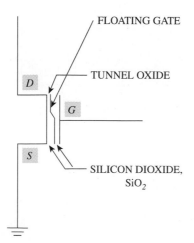

FIGURE 12–13   EEPROM basic memory element.

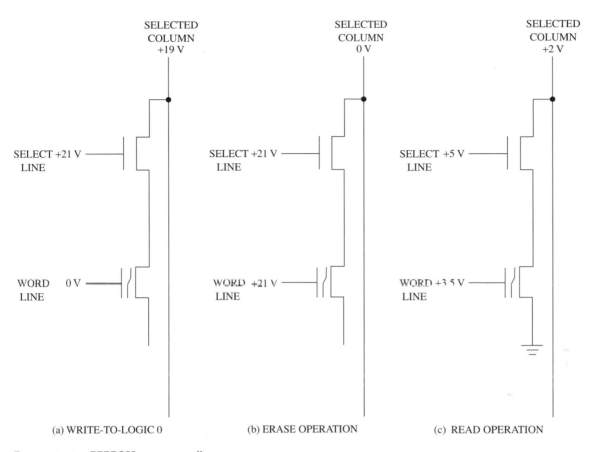

SELECTED COLUMN +19 V

SELECT +21 V LINE

WORD 0 V LINE

(a) WRITE-TO-LOGIC 0

SELECTED COLUMN 0 V

SELECT +21 V LINE

WORD +21 V LINE

(b) ERASE OPERATION

SELECTED COLUMN +2 V

SELECT +5 V LINE

WORD +3.5 V LINE

(c) READ OPERATION

FIGURE 12–14   **EEPROM memory cell.**

electrons to tunnel from the drain into the floating-gate region. This results in FET pinch-off, no drain current, and a Logic 1 at the memory output. This action produces a negative charge that is trapped on the floating gate. This trapped charge typically provides data retention for up to ten years.

A Read operation is illustrated in Fig. 12–14(c). Nominal voltages are shown in the illustration. If the Select Line and Selected Column are both enabled as shown, the +3.5 V applied to the gate of the memory cell will result in a Logic 1 or Logic 0 output. If the floating gate has a negative charge, the transistor is cutoff, and a Logic 1 is present. If the floating gate has a positive charge, the transistor conducts, and the output is Logic 0.

An n-bit serial electrically erasable PROM IC normally contains its own instruction set and the necessary registers, decoders, and control circuits to perform the following functions*:

1. **Read:** A Read command provides serial data on the data output (DO) pin. This pin is also used to monitor status of a programming or erase cycle.
2. **Erase:** This command sets all bits in the addressed register to Logic 1 levels.
3. **Erase/Write Enable:** These memory devices default to Erase/Write Disable when they power up. Thus, all programming commands must be preceded by an Erase/Write Enable instruction.

*This example is taken from National Semiconductor NMC93C06 (256-bit)/NMC93C46 (1024-bit) EEPROMs utilizing 16-bit registers. A block diagram and package outline are shown in **Fig. 12–15(a) and (b).** The serial organization of the chip allows packaging in an 8-pin DIP.

4. **Write:** This instruction to the Instruction Register is followed by 16 data bits clocked (SK pin) into the data input (DI) pin. Chip Select (CS) must be brought low to initiate the write cycle.

5. **Write All:** This command simultaneously programs all 16-bit registers with the data specified in the instruction.

6. **Erase/Write Disable:** Disables all programming and allows read only.

The big advantage of the E^2PROM is the selectivity that can be employed during reprogramming. A single bit or byte of data at a specific address can be erased and rewritten. This saves much time and money when a program needs only slight modification during its development. In addition, the reprogramming can be accomplished with the IC in cir-

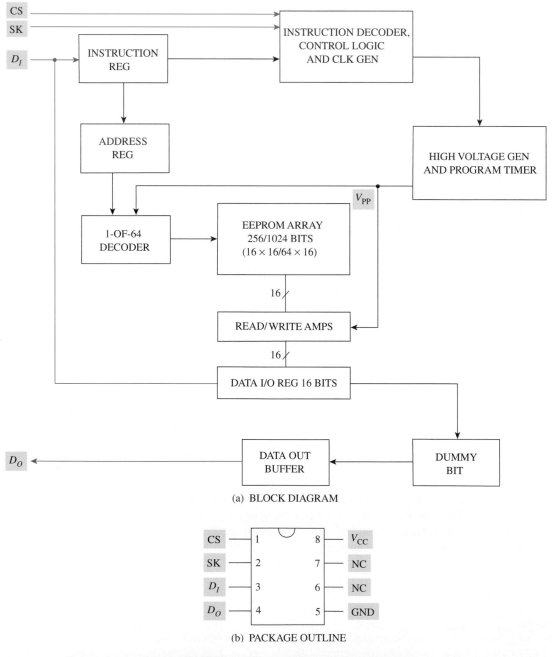

(a) BLOCK DIAGRAM

(b) PACKAGE OUTLINE

FIGURE 12–15 **Serial EEPROM.** *Courtesy of National Semiconductor.*

cuit. Most E²PROMs come with an on-board programming voltage generator that is used to supply the programming voltage. This generator is shown in Fig. 12–15(a). Finally, many E²PROMs can be erased and reprogrammed as many as 40,000 times.

ROM Access Time (t_{ACC})

Memory access time (t_{ACC}) for ROM ICs is the interval of time that elapses between valid address inputs and valid data outputs. The ROM timing diagram in Fig. 12–16 depicts t_{ACC}.

The address inputs to the ROM chip become valid at t_1. Chip Enable ($\overline{CE}$) is activated at t_2. The $\overline{CE}$ to output delay time (t_{CE}) occurs from t_2 to t_4. This is the critical timing factor in ROM.

If the maximum access time (t_{ACC}) is listed on a data sheet as 100 ns, chip enable ($\overline{CE}$) to output delay time (t_{CE}) is normally the same. Therefore, the $\overline{CE}$ input needs to be activated as soon as possible after valid addresses have been established. The output enable ($\overline{OE}$) to output delay time (t_{OE}) is not critical. In fact, $\overline{OE}$ activation can be delayed by an amount equal to $t_{ACC} - t_{OE}$ after $\overline{CE}$ activation. This delay can be as high as 50 to 60% of t_{ACC} without slowing down the read operation.

Memory access time is the interval from valid address to chip enable ($t_1 - t_2$) plus chip enable to output delay ($t_2 - t_4$). One other time is worth noting in Fig. 12-16. Output enable ($\overline{OE}$) high to output float (t_{DF}) is the interval between returning $\overline{OE}$ to the inactive level and returning the data bus to the high-impedance (floating) state.

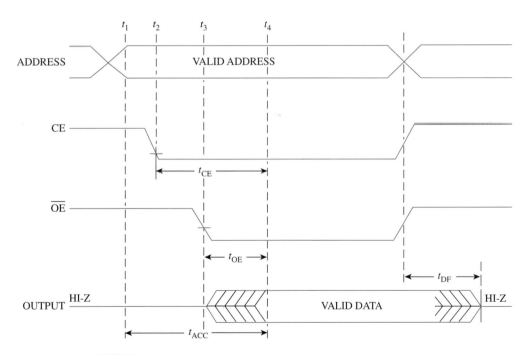

FIGURE 12–16   **ROM timing diagram.**

Flash Memory

Erasable PROMs provide high density (typically 16K–4M bits) and relatively fast access times (typically 150 ns). The out-of-circuit erasure cycle is terribly slow (about 20 minutes) and erases all stored data. Reprogramming must be done with a PROM programmer.

Electrically Erasable PROMs provide in-circuit programmability but at lower densities (typically 256–4K bits) and slower read times because the data are read serially. EEPROMs offer selective bit/byte erasure and reprogramming in the millisecond range.

Flash memory has taken the industry one step closer to the ideal memory device—nonvolatile, high density, fast access time, reprogrammability, and low cost. Flash memory incorporates the EPROM high density and the EEPROM in-circuit programmability in one technology. Flash memory can be reprogrammed (updated) under software control while it is resident on a system board.

Flash memories are currently available in 1M-, 4M-, and 16M- bit ICs. These high densities are achieved because flash memory uses only one transistor per cell. The cell structure is comparable to that used in EPROM technology. However, the oxide thickness between the substrate and the gate has been reduced in flash memory to a point that allows electrical erasure of data.

Flash memory is very suitable for portable applications that require high-density storage but cannot support a disk drive due to power or weight considerations. Power management is one of the key drivers in developing portable PCs. Disk and hard drive motors consume too much power for effective use in battery-powered units. Flash memory offers power consumption around 0.05 watt-hour compared to 1 watt-hour for hard-disk drives. In addition, they offer read access times in the 60-ns range, write times per byte of about 10 μs and block erase times of 1.6 seconds for a 64 KB block.

There are two types of flash memory. The two types, NAND flash and NOR flash, differ primarily in how they are programmed and erased. Both types offer bit-by-bit programmability.

NAND flash memory uses Fowler–Nordheim (F–N) tunneling for both programming and erasing. The F–N tunneling technique is also used in EEPROM devices. This type of flash memory possesses some inherent advantages over NOR flash memory. The low current (microamperes) used during F–N tunneling eases power supply requirements. In addition, the device receives less stress on the gate oxide since the tunneling is from the channel. This produces an extended life of up to 1,000,000 program/erase cycles compared to 100,000 for NOR flash.

NOR flash memory uses the Hot-Electron Injection (HEI) method used for programming EPROMs. This process causes electrons to be injected from the drain to the floating gate. HEI requires a 12-V power source and uses much more current (milliamperes) during programming than the NAND flash. The upside of HEI is programming time is about 10 μs/byte, and individual byte programmability is possible. NOR flash uses F–N tunneling during the erase cycle. The problem here is NOR flash must be programmed to all Logic 1s prior to erasure to prevent over erasure. This extra step slows the erase cycle to about 1 second.

Individual byte erasure is currently not available in flash memory devices. Consequently, block erasure must be accomplished to erase stored data. Some flash memories offer block erasure in blocks as small as 4K bytes.

National Semiconductor's NM29N16 is a 16-Mbit CMOS NAND Flash E²PROM. The basic architecture is shown in Fig. 12–17. Each **page** of memory consists of 256 bytes of storage plus an extra 8 bytes that are allotted to redundancy or error code correction. This results in 264 bytes per page. A **block** of this flash memory consists of 16 pages, and the device contains 512 blocks. Total memory capacity is 264 bytes × 16 pages × 512 blocks = 2,162,688 bytes (17,301,504 bits).

A block of memory (16 pages) stores 4K bytes of data. A block of 16 pages is the smallest unit within this device that can be erased. Both reading and programming can be accomplished for individual pages of 264 bytes. Typical block erase time is 6 ms, sequential read access time is 80 ns, and average program time is 300 μs/page. This indicates quite an improvement over its ROM counterparts.

The NM29N16 is available in a plastic thin small outline package (TSOP). The pin connection diagram is shown in Fig. 12–18. Note there are no dedicated address input pins on the IC. All data and commands utilize the I/O pins. The device supports three modes of operation—read, program (write), and erase. The following in-depth discussion relates to the NM29N16 block diagram in Fig. 12–19.

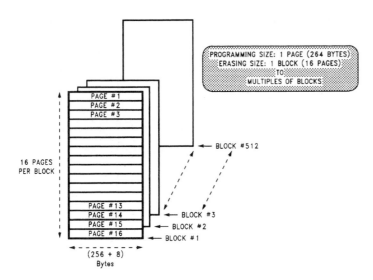

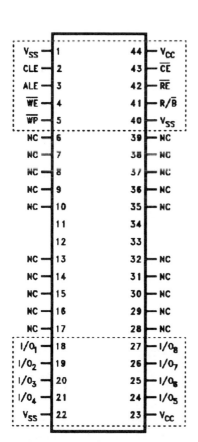

FIGURE 12–17 Flash memory architecture. *Courtesy of National Semiconductor.*

	Pin Assignment	
I/O$_{1-8}$	I/O Port	
$\overline{CE}$	Chip Enable	
$\overline{WE}$	Write Enable	
$\overline{RE}$	Read Enable	
CLE	Command Latch Enable	
ALE	Address Latch Enable	
$\overline{WP}$	Write Protect	
R/$\overline{B}$	Ready/Busy	
V$_{CC}$/V$_{SS}$	Power Supply/Ground	

FIGURE 12–18 NM29N16 flash memory pin connection diagram. *Courtesy of National Semiconductor.*

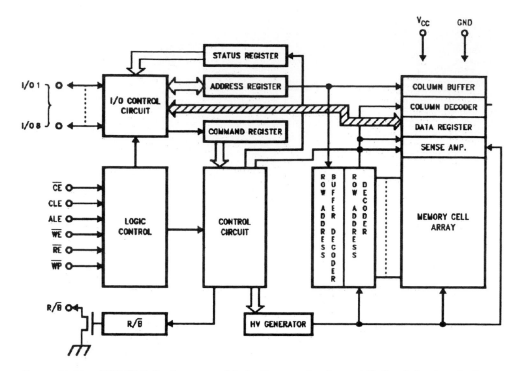

FIGURE 12–19   **NM29N16 flash memory block diagram.** *Courtesy of National Semiconductor.*

IN-DEPTH   Look at Programming

Read modes. The NM29N16 offers five read modes of operation. These and other command modes are shown in Table 12–3. The command codes are written into the command register which is used to control chip operation. The primary modes of operation are summarized next.

Read mode (1) is initiated by writing $00_{(16)}$ to the command register and following it with an address. The address specifies where on a specific page the pointer should be set. The data on the addressed page will be routed to an on-chip buffer for reading when $\overline{WE}$ is taken high. Sequential byte reading is accomplished by consecutive read enable ($\overline{RE}$) pulses. Read mode (2) allows reading of the extra 8 bytes from the addressed page.

Program mode. Automatic programming is accomplished by command code $80_{(16)}$ followed by an address for the specific block and page. Next, the data are written into the I/O ports starting with the lowest byte. This action is followed by command code $10_{(16)}$ to initiate the

actual programming. The **ready/busy** ($R/\overline{B}$) output monitors the device status during an operation. The signal is low (busy) during programming and returns high upon program completion. Once the signal has returned high, the status/read command code $70_{(16)}$ may be input to verify the program.

Erase modes. The chip supports two erase modes — automatic block erase and auto multiblock erase. Two command codes are required to execute an erasure. The command code $60_{(16)}$ followed by an address will initiate the automatic block erase mode. The address must be followed by command code $D0_{(16)}$ to actually execute the erase instruction. The second command code ($D0_{(16)}$) requirement prevents accidental erasure. Auto multiblock erasure is executed in a similar manner with each command code $60_{(16)}$ followed by an address. The execute command code $D0_{(16)}$ is not entered until all of the block addresses have been entered. Erase verification may be obtained by issuing the Status/Read code $70_{(16)}$.   ■

TABLE 12-3	Mode	First Cycle	Second Cycle	Acceptable Command During Busy
NM29N16	Serial data input	80		
Flash Memory	Read mode (1)	00		
Command	Read mode (2)	50		
Modes	Reset	FF		Yes
	Auto program	10		
	Auto block erase	60	D0	
	Auto multiblock			
	erase	60 … 60	D0	
	Suspend in erasing	B0		Yes
	Resume	D0		
	Status read	70		Yes
	Register read	E0		
	ID read	90		

Section 12–2: Review Questions

Answers are given at the end of the chapter.

A. ROM is .
   (1) volatile
   (2) nonvolatile

B. ROM is random-access read-only memory.
   (1) True
   (2) False

C. List the four types of ROM.

D. Which type of ROM is considered to be one-time programmable?

E. Which type of ROM is best suited for high-volume production?

F. Which type of ROM uses a programmer to blow selected fuse links during the programming process?

G. Which type of ROM uses ultraviolet (UV) light to erase the programmed data?

H. Define memory access time.

I. State the meaning/purpose of the following:
   (1) $\overline{CS}$ (2) $\overline{OE}$ (3) $\overline{WE}$ (4) RD/$\overline{WR}$
   (5) One-time programmable (6) PROM burning

J. Which type of memory incorporates the high density of EPROM technology with the in-circuit programmability of EEPROM?

K. Individual bit/byte erasure is available in flash memory.
   (1) True
   (2) False

SECTIONS 12–1 AND 12–2: INTERNAL SUMMARY

Primary memory includes semiconductor ROM and RAM. Access time to this category of memory is fast, yet the memory is somewhat limited in capacity. **Secondary memory** is used to solve the capacity problem of primary memory, but it does so at the expense of sacrificed speed. This type of **mass memory** includes magnetic tapes, hard drives, and disk drives.

Computers use three busses to accomplish the multitude of communications required for them to function. They are the **address, data, and control busses.**

Read-Only Memory (ROM) is random-access read-only memory, and it is non-volatile. **Random-Access Memory** (RAM) is random-access read/write memory, and it is volatile.

Memory chips come in a variety of packages and capacities. A 256 × 8 chip has 256 memory locations containing 8 bits per location. The chip requires 8 address input pins and 8 data output pins.

ROM chips are used for permanent data storage, so they must be nonvolatile. **Mask ROM** chips are programmed when purchased and are not reprogrammable. **Programmable ROM** chips can be programmed by the user with a PROM burner. Once the fuse link is blown and the logic level set, the process is irreversible. For this reason, they are often called **one-time programmable** chips. **Erasable PROM** chips can be programmed by the user. Additionally, the program can be erased by exposing the quartz window on the chip to ultraviolet light. The main problems encountered in erasure are (1) the length of time it takes to erase the program, and (2) the entire program is erased which requires complete reprogramming. **Electrically erasable PROM** chips can be programmed by the user, and stored data can be selectively erased and reprogrammed using programming voltage pulses.

Flash memory provides nonvolatile, high-density, high-speed memory at relatively low cost. The memory is available in NAND flash and NOR flash technologies. The device memory architecture is laid out in pages, and the pages are grouped into blocks. Current technology allows only block erasure.

SECTIONS 12-1 AND 12-2: INTERNAL SUMMARY QUESTIONS

Answers are given at the end of the chapter.

1. RAM and ROM semiconductor memories are memory.
   a. primary
   b. secondary

2. Firmware is .
   a. software
   b. hardware
   c. software in a hardware package

3. The bus is a unidirectional bus.
   a. data   c. address
   b. control   d. hardware

4. $\overline{MEMWR}$ is an active- signal.
   a. low
   b. high

5. RAM is .
   a. volatile
   b. nonvolatile

6. A 2K × 8 memory chip contains address input pins.
   a. 2   c. 10
   b. 8   d. 11

7. A 2K × 8 memory chip contains data output pins.
   a. 2   c. 10
   b. 8   d. 11

8. A 2K × 8 memory chip contains _____ memory locations.
   a. 1024
   b. 2048
   c. 4096
   d. 16,384

9. A 2K × 8 memory chip can store _____ bytes of data.
   a. 1024
   b. 2048
   c. 4096
   d. 16,384

10. How many address bits are required to address a 512K × 1 memory chip?
   a. 1
   b. 16
   c. 19
   d. 512

11. A conductor or set of conductors that connects two or more devices in a system is called a/an _____.
   a. bus
   b. cell
   c. I/O port
   d. bus cycle

12. Which type(s) of ROM is user programmable?
   a. PROM
   b. EPROM
   c. EEPROM
   d. All of the above

13. Mask ROMs are very expensive when used in low-volume production.
   a. True
   b. False

14. The BIOS program used in PCs is best suited for _____ memory.
   a. RAM
   b. ROM

15. Which type of ROM is considered one-time programmable?
   a. PROM
   b. EPROM
   c. UVPROM
   d. EEPROM

16. Is the memory chip shown in Fig. 12–20 a RAM or ROM chip?
   a. RAM
   b. ROM

17. How many memory locations does the memory chip in Fig. 12–20 contain?
   a. 128
   b. 256
   c. 512
   d. 1024

18. How many bits are stored in each memory location in the chip in Fig. 12–20?
   a. 8
   b. 9
   c. 16
   d. 512

19. What is the capacity of the chip in Fig. 12–20 in bytes?
   a. 9
   b. 256
   c. 512
   d. 1024

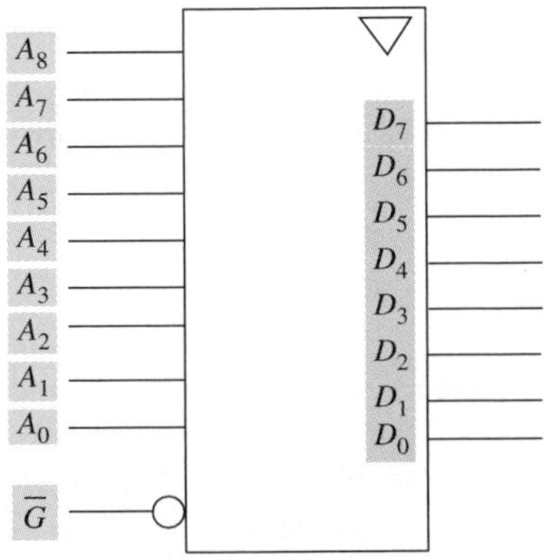

Figure 12–20

20. What is the capacity of the chip in Fig. 12–20 in bits?
    a. 9   c. 1024
    b. 512   d. 4096

21. How many bits are stored in a memory cell?
    a. 1   c. 4
    b. 2   d. 8

22. Which of the following types of memory offers the shortest erase time?
    a. PROM
    b. EPROM
    c. Flash

23. Flash memory incorporates the in-circuit programmability of what ROM technology?
    a. MROM   c. EPROM
    b. PROM   d. EEPROM

24. Flash memory incorporates the high density of which of the following ROM technologies?
    a. MROM   c. EPROM
    b. PROM   d. EEPROM

SECTION 12-3: RANDOM-ACCESS MEMORY (RAM)

OBJECTIVES

1. Identify different types of RAM.
2. Define terms relative to random-access memory.
3. State the main advantages and disadvantages, and compare characteristics of SRAM, DRAM, and PSRAM. Compare these technologies to ROM chip technology as well as to each other.
4. State the purpose of DRAM refresh.

The evolution of **Random-Access Memory** has paralleled the development of the PC. In fact, RAM's evolution has been an integral part of PC development. Industry has met the demands for high-density, high-speed semiconductor memory devices in a remarkable manner. Today, RAMs with 1M, 4M, and 16M storage are readily available. These RAMs, volatile in nature, are used for short-term data storage. They may be thought of as scratch-pad memory.

Three major types of RAM devices are presented in this section—**Static RAM (SRAM), Dynamic RAM (DRAM),** and **Pseudo-Static RAM (PSRAM).** The advantages and disadvantages of each type are discussed and usage tradeoffs are presented in this section.

Static RAM

Static RAM (SRAM) may be defined as random-access, volatile memory that retains stored data as long as power is applied. A basic SRAM IC uses a flip-flop as a memory cell. Static RAM is available in several logic families—NMOS, CMOS, BICMOS, HMOS, TTL, and ECL. Access times generally vary from about 10 ns up to 150 ns.

A six-transistor static RAM memory cell is shown in Fig. 12–21. NMOS transistors Q_1 and Q_2 form the flip-flop storage device, transistors Q_3 and Q_4 are gating transistors

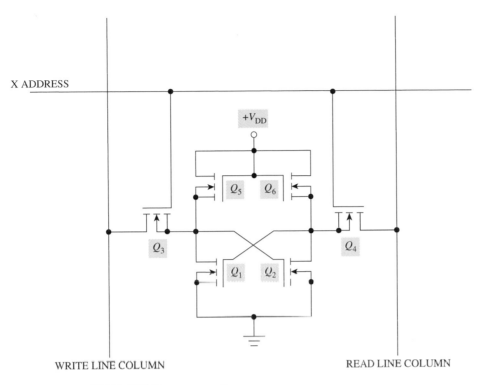

X ADDRESS

$+V_{DD}$

Q_5　Q_6

Q_3

Q_4

Q_1　Q_2

WRITE LINE COLUMN　　　　　　READ LINE COLUMN

FIGURE 12–21　**NMOS SRAM memory cell.**

for read/write functions, and transistors Q_5 and Q_6 function as loads. The cell's purpose is to store a Logic 0 or a Logic 1. An in depth look at the cell's operation appears at the bottom of the page.

The memory cell of Fig. 12–21 is redrawn with its associated column read/write circuitry in Fig. 12–22. Conventional MOSFET logic symbols are used for the NMOS devices in the figure.

A high on the row input line turns on its Q_3 and Q_4 and allows the memory cell to interface with the external read/write circuitry. A high at the write input allows the data input to set or clear the flip-flop for data storage provided the column (Col) input is high. Keep in mind the column and row inputs are controlled by the address inputs applied to the memory device. Both Q_7 and Q_8 are allowed to conduct when the column input is high. When WRITE = 1, the READ input = 0 and Q_{10} is off. On the other hand, a high on the column and read inputs allows the stored complementary data to be read. There are many ways to accomplish the storage just described. Despite the technology of the static RAM, most of the storage methods are similar to the one presented in Fig. 12–22.

IN-DEPTH　Look at Operation

The flip-flop in the memory cell in Fig. 12–21 is isolated until the X address (row) line is brought high. Q_3 and Q_4 turn on when this occurs. Assume the initial condition of the flip-flop is Q_1 on and Q_2 off. Now bring the Write Line high with the application of a Logic 1. This high is passed through gating transistor Q_3 and coupled to the gate of Q_2. This high logic level turns Q_2 on. The drain voltage (V_{DS}) of Q_2 immediately decreases. This decreasing voltage level is coupled to the gate of Q_1 and decreases its conduction.

This causes the (V_{DS}) of Q_1 to increase, and this increase is coupled back to the gate of Q_2. Therefore, the cross-coupled feedback forces Q_2 to saturate and Q_1 to turn off.

The Read Line ($\overline{\text{data}}$ output) is connected through gating transistor Q_4 to the drain of Q_2. Note the output logic level is low because Q_2 is saturated. This represents the complement of the desired logic level written into the memory cell. Thus, the output logic level of the cell must be inverted to produce the true stored logic.　■

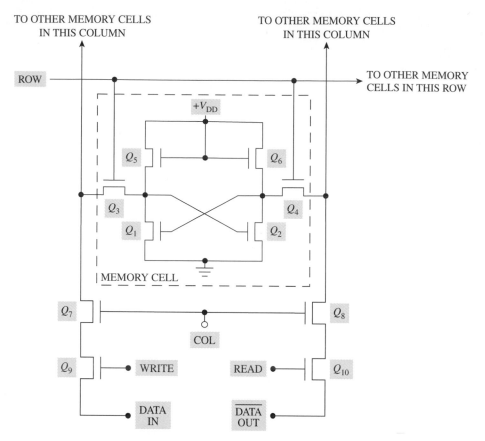

TO OTHER MEMORY CELLS
IN THIS COLUMN

TO OTHER MEMORY CELLS
IN THIS COLUMN

ROW

TO OTHER MEMORY
CELLS IN THIS ROW

$+V_{DD}$

Q_5 Q_6

Q_3 Q_4

Q_1 Q_2

MEMORY CELL

Q_7 Q_8

COL

Q_9 WRITE   READ Q_{10}

DATA
IN

DATA
OUT

FIGURE 12-22   **NMOS SRAM memory cell with read/write circuitry.**

256K × 1 SRAM

A block diagram and logic symbol of a 256K × 1 SRAM IC are shown in Fig. 12–23. Eighteen address input pins are provided on the IC (2^{18} = 262,144, and 262,144 ÷ 1024 = 256K). The memory matrix is laid out in 256 rows of 1024 bits. The 1-of-256 row decoder is used to enable one row of memory at a time. The column decoder and I/O circuit are used to enable the memory cell that intersects the row and column address. Only one memory cell at a time is addressed since this is a 256K × 1 chip.

The row address consists of bits A_0–A_4 and A_{15}–A_{17} (8 bits). The column address consists of bits A_5–A_{14} (10 bits). Table 12–4 provides a representative sample of addresses for this memory IC. The lower 5 address bits are used to enable one of the lower 31 rows of memory. Row addresses for rows 32, 64, 128, and 255 are also shown in the table. The column addresses are shown as don't care conditions in the table.

The don't care column addresses range from 0 to $1023_{(10)}$ in Row 0 and increase up to 261,120 to $262,143_{(10)}$ in Row 255. Determining specific memory locations within the chip and its total address range—lowest and highest addresses—will be studied in Section 12–4.

The memory chip represented in Fig. 12–23(a) contains two dual-controlled **data buffers.** If $\overline{CS}$ = 1, the low level out of the NOT gate is applied to one active-high enable input of the input buffer (D_{IN}) and one active-high enable input of the output buffer (D_{OUT}). Since the input and output data buffers are not enabled, the chip is isolated from the data bus and is in the Hi-Z state. Most memory ICs revert to standby operation when they are not selected. This drastically reduces operating power supply current (I_{CC}), which results in minimum power dissipation when the chips are not in use.

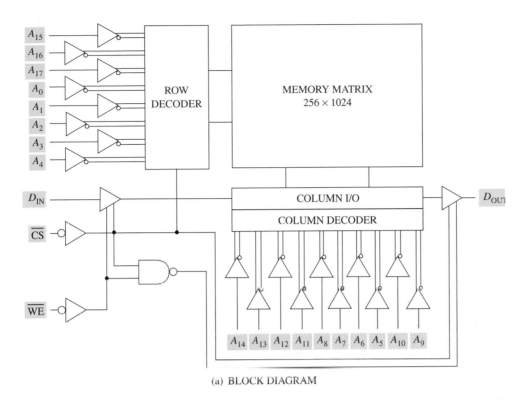

(a) BLOCK DIAGRAM

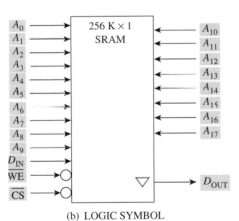

(b) LOGIC SYMBOL

Figure 12–23   256K × 1 SRAM.

The chip in Fig. 12–23(a) can be written to or read from when $\overline{CS}$ = 0. This differentiates it from the read-only memory chip presented in the previous section. Data can be written into a specific address location when $\overline{CS}$ = 0 and write enable ($\overline{WE}$) = 0. Note the inverted $\overline{CS}$ signal enables the row decoder when it is active. The D_{IN} buffer is also enabled when $\overline{CS}$ and $\overline{WE}$ are activated. In addition, these two signals produce a low out of the NAND gate which disables the output buffer during the write operation.

When $\overline{CS}$ = 0 and $\overline{WE}$ = 1, true data can be read from the D_{OUT} buffer. This buffer is enabled by the inverted $\overline{CS}$ signal and the high output from the NAND gate. A summary of this chip's operation is shown in the function table (Table 12–5).

Another standard memory IC is shown in Fig. 12–24. This chip is a 1K × 4 SRAM IC. The configuration shares common data input and data output (I/O) pins. The chip size is kept nominal by reducing the number of data pins required when I/O pins are used.

When $\overline{CS}$ and $\overline{WE}$ are activated during a **write cycle,** the input buffers routing data to the input data circuit are enabled. The output buffers are disabled at the same time. The **read cycle** occurs when $\overline{CS}$ = 0 and $\overline{WE}$ = 1. This set of inputs enables the output buffers and disables the input buffers.

TABLE 12-4

256K × 1 SRAM Sample Addresses

A_{17}	A_{16}	A_{15}	A_{14}	A_{13}	A_{12}	A_{11}	A_{10}	A_9	A_8	A_7	A_6	A_5	A_4	A_3	A_2	A_1	A_0		
0	0	0	X	X	X	X	X	X	X	X	X	X	0	0	0	0	0	= Row	0
0	0	0	X	X	X	X	X	X	X	X	X	X	0	0	0	0	1	= Row	1
0	0	0	X	X	X	X	X	X	X	X	X	X	1	0	0	0	0	= Row	16
0	0	0	X	X	X	X	X	X	X	X	X	X	1	1	1	1	1	= Row	31
0	0	1	X	X	X	X	X	X	X	X	X	X	0	0	0	0	0	= Row	32
0	1	0	X	X	X	X	X	X	X	X	X	X	0	0	0	0	0	= Row	64
1	0	0	X	X	X	X	X	X	X	X	X	X	0	0	0	0	0	= Row	128
1	1	1	X	X	X	X	X	X	X	X	X	X	1	1	1	1	1	= Row	255

The columns A_{17}–A_{15} and A_4–A_0 are labeled **Row Addresses**. The columns A_{14}–A_5 are labeled **Column Addresses**.

Column Addresses		
1,023 0		In Row 0
2,047 1,024		In Row 1
17,407 16,384		In Row 16
32,767 31,744		In Row 31
33,791 32,768		In Row 32
66,559 65,536		In Row 64
132,095 131,072		In Row 128
262,143 261,120		In Row 255

TABLE 12-5

256K × 1 SRAM Function Table

$\overline{CS}$	$\overline{WE}$	Data Out	Function
1	X	Hi-Z	Chip not selected
0	1	D_{OUT}	Read cycle
0	0	Hi-Z	Write cycle

SRAM Timing

Read-cycle timing (t_{RC}) waveforms are shown in Fig. 12–25(a) and (b). The two sets of waveforms represent two different read-cycle conditions.

Read Cycle—Device Continuously Selected.

Figure 12–25(a) shows read-cycle timing waveforms when the memory chip is continuously selected. This set of waveforms could also be used to represent the condition of $\overline{CS}$ being activated prior to the application of a valid address. The output hold time from address change (t_{OH}) represents data from the previous read cycle. Address access time (t_{ACC}) indicates the time interval from valid address to valid data. The hatched data-out pattern represents changing data. There is no Hi-Z state shown in this figure because it is assumed the IC is continuously selected. The Hi-Z state only occurs when the chip is deselected or the output buffers are not enabled with an $\overline{OE} = 0$ input. This timing waveform set does not use an $\overline{OE}$ input to the IC.

Read Cycle—Address Valid Prior to or Coincident with $\overline{CS}$ Activation.

Figure 12–25(b) shows read-cycle timing waveforms when a valid address is applied prior to an active $\overline{CS}$ input. The chip is in the Hi-Z state when $\overline{CS}$ is initially activated. Chip select to output in the low-Z state (t_{LZ}) is the interval required to enable the output buffers. The actual chip select access time (t_{ACS}) represents the time interval from $\overline{CS}$ activation

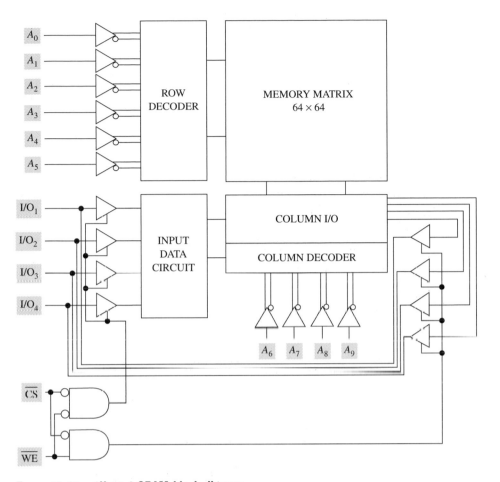

FIGURE 12–24   1K × 4 SRAM block diagram.

to valid output data. Once the chip is deselected ($\overline{\text{CS}} = 1$), the time interval to return to the Hi-Z state is represented by t_{HZ}.

The times involved with placing the chip in standby to reduce power consumption are shown in the I_{CC} supply current waveform in Fig. 12–25(b). Chip select to power-up time is shown as t_{PU}, and chip deselect to power-down time is shown as t_{PD}.

The various times discussed above are dependent upon the technology of the chip. Write-cycle timing (t_{WC}) waveforms are provided in Fig. 12–26(a) and (b). One of the sets of waveforms is primarily controlled by activation of chip select while the other set is primarily controlled by activation of write enable.

Write Cycle–Chip-Select Controlled.

Figure 12–26(a) shows write-cycle waveforms when the chip select input is activated *after* the write enable input has been activated. The Q (data out) output remains in the Hi-Z state during the entire write cycle. A write cycle occurs during the overlap of $\overline{\text{CS}} = 0$ and $\overline{\text{WE}} = 0$.

Write Cycle–Write-Enable Controlled.

Figure 12–26(b) shows write-cycle waveforms when the chip select input is activated *before* the write enable input has been activated. The acronyms used in Fig. 12–25 and 12–26 are summarized in Table 12–6. Some example times for SRAM ICs are also included in the table.

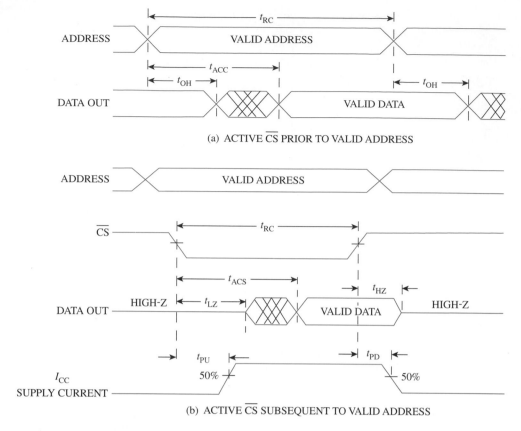

(a) ACTIVE $\overline{\text{CS}}$ PRIOR TO VALID ADDRESS

(b) ACTIVE $\overline{\text{CS}}$ SUBSEQUENT TO VALID ADDRESS

FIGURE 12-25   **SRAM read–cycle timing waveforms.**

TABLE 12-6	Cycle	Acronym	Parameter	Typical Time
Read/Write Cycle Timing Data	Read	t_{RC}	Read cycle time	35 ns min
		t_{ACC}	Address access time	35 ns max
		t_{OH}	Output hold from address change	4 ns min
		t_{ACS}	Enable access time	35 ns max
		t_{LZ}	Output enable low to output active	0 ns min
		t_{HZ}	Output enable high to output Hi-Z	10 ns max
		t_{PU}	Power-up time	0 ns min
		t_{PD}	Power-down time	35 ns max
	Write	t_{WC}	Write cycle time	35 ns min
		t_{AW}	Address valid to end of write	20 ns min
		t_{AS}	Address setup time	0 ns min
		t_{CW}	Enable to end of write	15 ns min
		t_{WR}	Write recovery time	0 ns min
		t_{WP}	Write pulse width	20 ns min
		t_{DW}	Data valid to end of write	10 ns min
		t_{DH}	Data hold time	0 ns min
		t_{WZ}	Write low to output Hi-Z	10 ns max
		t_{OW}	Write high to output active	4 ns min

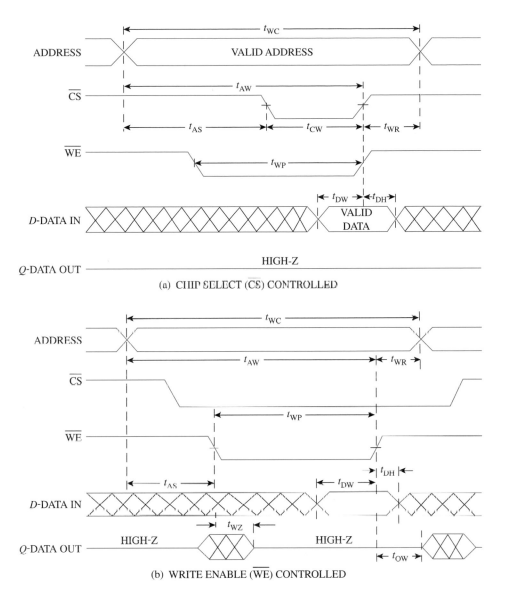

FIGURE 12-26   **SRAM write-cycle timing waveforms.**

Dynamic RAM

Dynamic RAM (DRAM) consists of only one transistor and one capacitor per memory cell. This simplicity in cell design produces very high density and the lowest cost per bit of any type of memory currently available.

A DRAM cell is shown in Fig. 12–27. The capacitor is used to store a Logic 0 or Logic 1 bit. The transistor in the cell functions as a switch to control the capacitor's charge and discharge. All capacitors are leaky, and they lose their charge over a period of time. Unfortunately, that time is a few milliseconds for a DRAM cell. There is only a small difference between a Logic 0 and a Logic 1 charge on a DRAM capacitor. Thus, the capacitors in the cells must be periodically recharged to maintain Logic 1 levels. The periodic recharging is called **refresh.**

Refresh is accomplished by accessing each row of cells in the memory array. Sometimes, two rows are accessed simultaneously to accomplish refresh. When a row is accessed (Fig. 12–27), each capacitor in the row is recharged to its initial value. This process consists of connecting the addressed row (Word Line) to the word-line driver. This connection enables

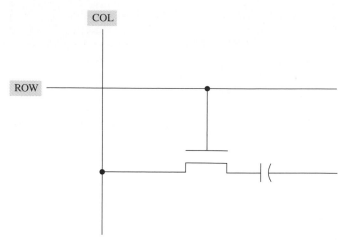

FIGURE 12-27   **DRAM cell.**

the transfer switches for the cells on the addressed row and connects the DRAM capacitors to the columns (bit lines). In this manner, the stored logic levels are detected by the sense amplifiers, and the high stored logic levels are recharged (refreshed). The maximum refresh time interval for a memory device depends on the technology and the number of cells in the device. Refresh times range from 2 ms up to 32 ms.

The disadvantages of refresh are self-evident. First, external circuitry must be implemented to accomplish the task. Second, all time consumed by refresh operations lessens the time available for normal memory read/write operations.

The advantage of very high density per chip results in another disadvantage. When a DRAM chip has a 4M × 1 density, 22 address lines are necessary to address the chip. A look at the outline drawing in Fig. 12–28 for a 20-lead 4M × 1 DRAM package will shed some light on the addressing problem.

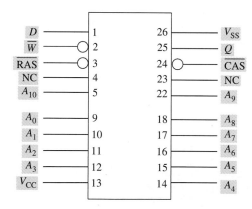

DRAM Pins

D	Data Input
Q	Data Output
$\overline{W}$	Read/Write Input
$\overline{RAS}$	Row Address Strobe
$\overline{CAS}$	Column Address Strobe
A_0–A_{10}	Address Inputs
V_{CC}	Power Supply
V_{SS}	Ground
NC	No Connection

FIGURE 12-28   **4M × 1 DRAM package outline.**

All DRAM chips require the following pins (minimum): (1) Data Input, (2) Data Output, (3) Read/Write Input, (4 and 5) $\overline{RAS}$ and $\overline{CAS}$ (used for refresh operations), (6) V_{CC}, and (7) V_{SS} (ground). This means the IC must have at least 7 pins, in addition to the address pins. If there were 22 address pins in addition to the mandatory pins, the IC would need 29 pins to interface with the outside world. This would make the DRAM chips bulky, and they would take up a lot of system board real estate.

To solve this problem, address inputs to DRAM chips are time-shared through **multiplexing.** The address is broken up into a **row address** and a **column address.** This allows the DRAM designer to use half the required number of address pins to access the memory locations on the chip. Note the total number of pins actually used in Fig. 12–28 is the required 7 plus 11 address pins. The process of address multiplexing will be discussed in detail shortly.

The addressing components of a 64K $\times$ 1 DRAM circuit are illustrated in Fig. 12–29. The refresh circuitry has been omitted for this introductory explanation.

The memory matrix in Fig. 12–29 is laid out in 256 rows and 256 columns. This arrangement provides 65,536 memory locations. Since $2^{16} = 65,536$, sixteen address inputs must be applied to the circuit to access all memory locations. This is accomplished by using a 2-line to 1-line multiplexer (MUX). The system microprocessor places the entire 16-bit address on the address bus to start the cycle. Address bits A_0 to A_7 represent the row address and A_8 to A_{15} represent the column address. The timing waveforms of a read cycle for this simplified DRAM circuit are shown in Fig. 12–30. The select input to the MUX (not shown in this example) is set to select the A_0–A_7 row address bits when the Row Address Strobe ($\overline{RAS}$) signal is active. As shown in Fig. 12–29, these bits are latched into the row latch at $\overline{RAS} = 0$ and decoded to enable one memory row in the matrix. Then the MUX select input changes to a high level, and the MUX selects the A_8–A_{15} column address bits when the column address strobe ($\overline{CAS}$) signal is active. These bits

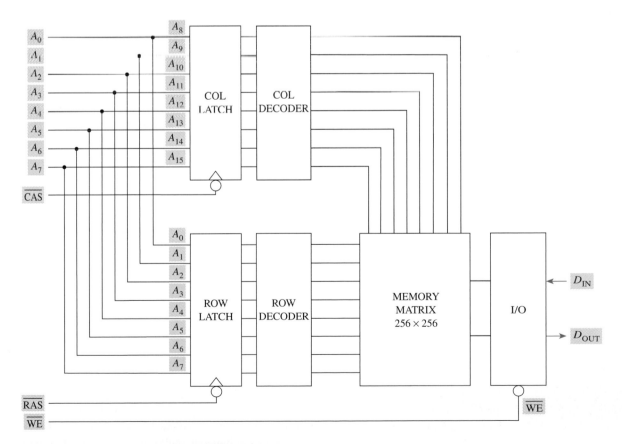

Figure 12-29   64K $\times$ 1 DRAM simplified block diagram.

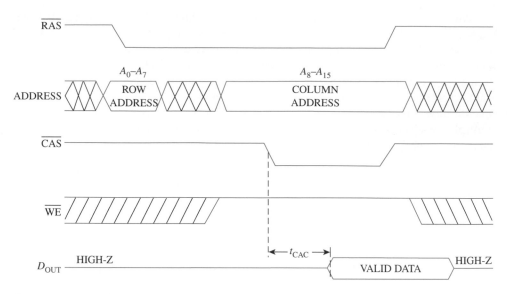

FIGURE 12-30   DRAM read cycle—simplified timing waveforms.

are latched into the column latch at $\overline{CAS}$ = 0 and decoded to enable one memory column in the matrix. It can be seen from this explanation that DRAM locations are accessed with row and column address strobe signals.

A memory location is enabled once the row and column address bits have been decoded. The timing waveforms in Fig. 12–30 indicate this is a read cycle because $\overline{WE}$ is high during the time the memory location is enabled. $\overline{WE}$ can remain high, and usually does, for the entire read cycle. Valid data appears at the output of the circuit by a delay interval equal to t_{CAC} (access time from $\overline{CAS}$). The output of the circuit remains in the Hi-Z state except when valid data are available.

1M × 1 CMOS DRAM/Timing

The block diagram and logic symbol of a 1M × 1 CMOS DRAM IC are shown in Fig. 12–31(a) and (b). The memory array is laid out in 1024 rows and 1024 columns. This layout provides 1,048,576 (1M) memory locations. To calculate the required address inputs: log 1,048,576 ÷ log 2 = 20 address inputs. The 10 address input pins are time multiplexed by $\overline{RAS}$ and $\overline{CAS}$. Bits A_0 to A_9 are used to address a row, and bits A_{10} to A_{19} are used to address a column.

Read Cycle.

Read cycle timing waveforms for the 1M × 1 DRAM chip are presented in Fig. 12–32. $\overline{RAS}$ is activated prior to $\overline{CAS}$ in all DRAM read and write cycles. The interval between activating $\overline{RAS}$ and $\overline{CAS}$ is called the **multiplex window.** The timing diagram in Fig. 12–32 shows both t_{RAS} ($\overline{RAS}$ pulse width) and t_{CAS} ($\overline{CAS}$ pulse width). The pulse width of these signals can typically be up to 10,000 ns. With this much time variance, system designers have plenty of flexibility in setting up and multiplexing address inputs to DRAM. The typical read cycle time (t_{RC}) for these DRAMs runs in the 100- to 150-ns (minimum) range.

The $\overline{WE}$ signal must be inactive (high) before activating $\overline{CAS}$ to enable the read cycle. The minimum time for this action is shown as t_{RCS} (read command setup time) in Fig. 12–32. However, in most cases this time is specified in data sheets as 0 ns. This means as long as $\overline{WE}$ is brought inactive prior to or simultaneous with $\overline{CAS}$ activation there will be no problems.

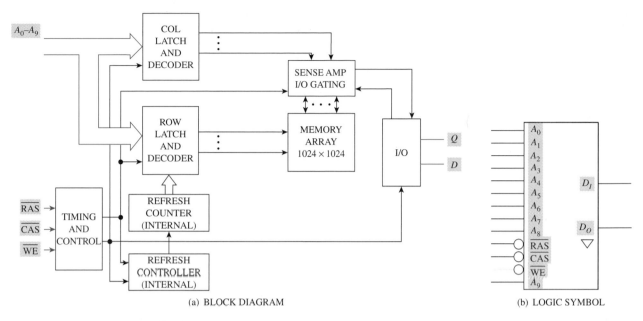

(a) BLOCK DIAGRAM   (b) LOGIC SYMBOL

FIGURE 12–31   1M × 1 CMOS DRAM.

There are three access times to valid data shown in the read cycle timing waveforms. The times are (1) t_{RAC}, access time from RAS; (2) t_{CAA}, access time from column address; and (3) t_{CAC}, access time from $\overline{CAS}$. Access times for the Motorola 1M × 1 CMOS DRAM (PN MCM511000A–70) are as follows: (1) $t_{RAC} = 70$ ns maximum, (2) $t_{CAA} = 35$ ns maximum, and (3) $t_{CAC} = 20$ ns maximum. Since $\overline{RAS}$ must be activated first during a read or write cycle, t_{RAC} (access time from $\overline{RAS}$) is the longest time, and it must be considered the access time for a read cycle for this IC. The acronyms used in the timing diagram in Fig. 12–32 are summarized in Table 12–7. The times provided in the table are for a Motorola PN MCM511000A–70 1M × 1 DRAM.

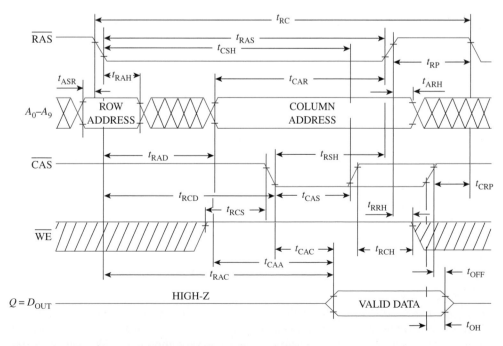

FIGURE 12–32   1M × 1 DRAM read-cycle timing waveforms.

	Acronym	Parameter	Time
TABLE 12-7 **DRAM Read-Cycle Timing Data**	t_{RC}	Read-cycle time	130 ns
	t_{RAS}	$\overline{RAS}$ pulse width	70–10,000 ns
	t_{CAS}	$\overline{CAS}$ pulse width	20–10,000 ns
	t_{CSH}	$\overline{CAS}$ hold time	70 ns
	t_{RSH}	$\overline{RAS}$ hold time (read cycle)	20 ns
	t_{RP}	$\overline{RAS}$ precharge time	50 ns
	t_{CRP}	$\overline{CAS}$ to $\overline{RAS}$ precharge time	5 ns
	t_{RAH}	Row address hold time	10 ns
	t_{ARH}	Column address hold time from $\overline{RAS}$	10 ns
	t_{RCH}	Read command hold time (reference to $\overline{CAS}$)	0 ns
	t_{ASR}	Row address setup time	0 ns
	t_{CAR}	Column address to $\overline{RAS}$ setup time	–
	t_{RAD}	$\overline{RAS}$ to column address delay time	15–35 ns
	t_{RCD}	$\overline{RAS}$ to $\overline{CAS}$ delay	20–50 ns
	t_{RCS}	Read command setup time	0 ns
	t_{RRH}	Read command hold time (reference to $\overline{RAS}$)	0 ns
	t_{CAC}	Access time from $\overline{CAS}$	20 ns max
	t_{CAA}	Access time from column address	35 ns max
	t_{RAC}	Access time from $\overline{RAS}$	70 ns max
	t_{OFF}	Output buffer turn-off delay	0–20 ns
	t_{OH}	Data hold time from $\overline{CAS}$	–

Note: All times are minimum intervals unless otherwise stated.

Write Cycle.

It is evident from the read-cycle waveforms in Fig. 12–32 that timing in DRAM ICs is critical. However to avoid redundancy, the write-cycle waveforms will not be presented. These waveforms are available in any DRAM data book.

The only major difference between a read cycle and a write cycle is the level of the $\overline{WE}$ pulse. The write cycle is initiated when $\overline{WE}$ is activated. The write-cycle sequence is

1. $\overline{RAS} = 0$ with valid address.
2. $\overline{CAS} = 0$ with valid address.
3. $\overline{WE} = 0$ (late write cycle).
4. Data written.
5. Data out will be Hi-Z.

The minimum pulse widths (t_{RAS} and t_{CAS}) discussed in the read cycle also apply to the write cycle.

Modern DRAM ICs offer various types of write cycles. An **early write cycle** occurs when $\overline{WE}$ is activated prior to the $\overline{CAS}$ active transition. A **late write cycle** occurs if $\overline{WE}$ is activated after the $\overline{CAS}$ active transition.

Read–Write Cycle.

A read–write cycle is, as the name implies, a read cycle followed by a write cycle at the same address. $\overline{WE}$ must remain high for a minimum $\overline{CAS}$-to-write delay time after $\overline{CAS}$ activation to ensure valid read data prior to the write operation.

Page Mode Cycle.

The page mode cycle offers fast successive read or write operations within the DRAM. This allows reading or writing blocks of data on a row while changing only the column address. This is accomplished by activating $\overline{RAS}$ and toggling $\overline{CAS}$. This latches a specific row address and the $\overline{CAS}$ high-to-low transitions select successive columns on the addressed row. Read access time during page mode operation is typically less than half t_{RAC} (access time from $\overline{RAS}$). Read, write, and read–write operations can be accomplished in page mode cycles.

Refresh Operation

DRAM refresh is accomplished by cycling through all row addresses in sequence. This must be accomplished within the required refresh time of the DRAM. All bits on a row are refreshed each time the row is addressed. Therefore, a normal read, write, or read–write cycle will refresh all bits on the addressed row. These read or write cycles during normal operation occur in a rather haphazard manner within a DRAM and surely would not refresh *all* rows within the memory array in the allotted time. There are three methods of refresh available in most DRAMs. These methods ensure *all* array rows are refreshed in the allotted refresh time.

$\overline{RAS}$-Only Refresh (ROR).

An external counter is used when the ROR method is employed to refresh the memory rows. The externally generated addresses are usually supplied by a **DRAM controller.** The ROR method of refresh leaves $\overline{CAS}$ inactive while $\overline{RAS}$ is activated to latch a row address for refresh.

$\overline{CAS}$ Before $\overline{RAS}$ Refresh (CBR).

This method of refresh employs the on-board refresh counter to supply the row addresses for refresh. It switches $\overline{CAS}$ low while $\overline{RAS}$ is high, and then takes $\overline{RAS}$ active after the required setup time. External row addresses are ignored during this refresh operation.

Hidden Refresh.

This method is based on the CBR method of refresh. In CBR, $\overline{CAS}$ is activated and then $\overline{RAS}$ is activated. However, $\overline{CAS}$ is held low beyond the normal $\overline{CAS}$ hold time in hidden refresh. $\overline{RAS}$ is then taken high while $\overline{CAS}$ is held low, and then $\overline{RAS}$ is activated again. Valid data can be read while $\overline{CAS}$ is active. The refresh cycle appears hidden since it occurs during a read cycle.

Normal memory operation is suspended during refresh operations. *DRAM controllers* were designed to control refresh by (1) simplifying DRAM support circuitry, (2) ensuring the memory is refreshed in a timely manner, and (3) minimizing the disruptions caused by refresh operations.

A typical refresh controller block diagram is shown in Fig. 12–33. This controller is used with a 1M × 1 DRAM. A MUX circuit provides the normal row and column address inputs to the DRAM. The particular input, row or column, is determined by the level of the row/column enable signal ($R/\overline{C}$) as long as the refresh signal is inactive (low).

The output of the MOD 1024 counter represents a row address. This output is routed through the MUX to the DRAM row latch and decoder when the refresh signal is active. This provides a row enable signal to the memory matrix which refreshes all 1024 memory locations on that row.

Pseudo-Static RAM (PSRAM)

Pseudo-Static RAM (PSRAM) is a type of DRAM. It employs the one-transistor memory cell like its DRAM counterpart. It produces the low cost per bit of DRAM while it provides the simplicity of SRAM. In other words, it provides high density without all of the external refresh circuits required for DRAM. PSRAM offers low power consumption with total internal refresh operations. The internal refresh of this type of memory makes the chip operation static as far as the outside world is concerned. A block diagram and logic symbol of a PSRAM are shown in Fig. 12–34(a) and (b). This PSRAM is a 32-pin 128K × 8 IC.

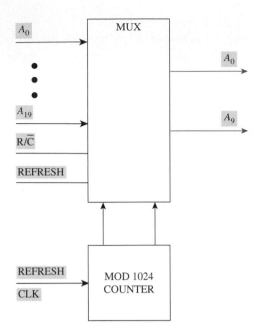

FIGURE12-33 **Refresh controller circuit.**

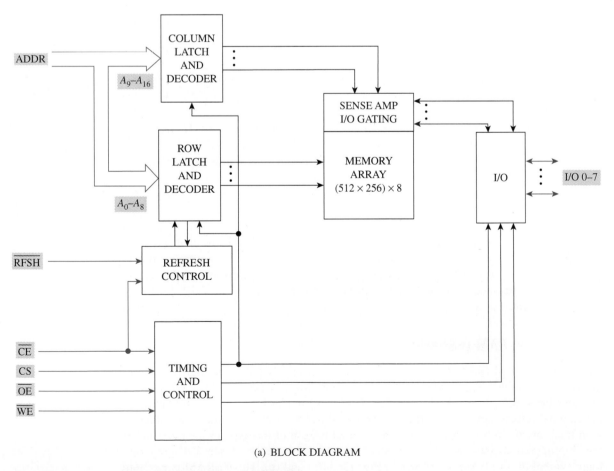

(a) BLOCK DIAGRAM

FIGURE 12-34 **128K × 8 PSRAM.**

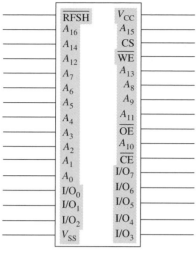

(b) LOGIC SYMBOL

FIGURE 12-34   Continued

Seventeen address pins are required because PSRAM chips do not employ multiplexed address inputs as conventional DRAMs do. This high pin count is a disadvantage for a high-density chip. Nonmultiplexing the address inputs also does away with the $\overline{\text{RAS}}$ and $\overline{\text{CAS}}$ input requirements seen on DRAM ICs. The similarities of the PSRAM block diagram (Fig. 12–34a) to the DRAM block diagram (Fig. 12–31a) are evident. The chips are comparable except as noted. Common memory capacities are 32K × 8, 128K × 8, and 512K × 8. The PSRAM device supports three different refresh options—address refresh, automatic refresh, and self-refresh.

Address Refresh.

This option is similar to that of the $\overline{\text{RAS}}$-Only Refresh of the DRAM chip. The address refresh mode refreshes stored data by inputting row addresses 0–511 through the row address input pins (A_0–A_8).

Automatic Refresh.

Row addresses are generated internally by the refresh control circuit when this option is employed. Automatic refresh is accomplished when the chip is not enabled ($\overline{\text{CE}} = 1$) and $\overline{\text{OE}}$ and $\overline{\text{RFSH}}$ are both active (low).

Self-Refresh.

Refresh request signals and refresh addresses are generated internally during self-refresh. This option is enabled by keeping $\overline{\text{CE}}$ high and activating $\overline{\text{OE}}$ and $\overline{\text{RFSH}}$ for *more than* 8 μs. These same signal levels will put the chip in the automatic refresh option if $\overline{\text{OE}}$ and $\overline{\text{RFSH}}$ are left active for less than 8 μs while $\overline{\text{CE}}$ is high. This mode is most beneficial in low power applications.

Electrostatic Discharge

Semiconductor CMOS memory devices are subject to malfunctions caused by static electricity. Most of these devices have built-in protection circuits to prevent damage due to mishandling and insure reliable operation during normal use; however, the following precautions should be observed:

1. The devices should be stored and shipped in conductive material so the pins on the IC are kept at the same potential. When a circuit board containing memory devices is shipped, the entire board should be enclosed with conductive material.

2. If the device must be handled during inspection or placement on a board, the technician should be grounded. A 1-MΩ resistor should be connected in series with the ground strap to protect against shock should the wearer contact a voltage source. The devices should never be inserted in a board when the power is on.

3. Cotton clothing is preferable for a technician when handling static-sensitive memory devices.

4. Soldering should be done with a low-voltage (12 V or 24 V) iron with its tip grounded.

Section 12–3: Review Questions

Answers are given at the end of the chapter.

A. RAM is _____.
   (1) volatile
   (2) nonvolatile

B. List the three major types of RAM.

C. Which type of RAM provides the highest density?
   (1) SRAM
   (2) DRAM

D. Which type of RAM requires refresh?
   (1) SRAM
   (2) DRAM

E. Which type of memory uses row and column address strobing?
   (1) SRAM
   (2) DRAM

F. What is the purpose of multiplexing address inputs to DRAM ICs?

G. What are the benefits of using PSRAM?

H. PSRAM uses address multiplexing.
   (1) True
   (2) False

SECTION 12–4: MEMORY ADDRESSING

OBJECTIVES

1. Given the lowest or highest address and the capacity of a memory chip, determine the address range.
2. Draw a memory map based on the calculations made in objective 1.
3. Design a decoder circuit to decode a specific address range.

Memory Addresses

Memory addresses are normally generated in a computer by the microprocessor. The address is a group of 0s and 1s placed on the address bus and used to select one memory chip. In addition, the memory address is used to enable one specific memory location within the selected memory chip.

Several memory addressing examples were provided in the ROM and RAM sections of this chapter. The addressing concept is expanded in this section. Some memory locations store only one bit—some store more. If a memory chip has a capacity of 1K × 8, eight bits of data are stored in eight memory cells at each memory location.

Memory addresses are specified in hexadecimal to make them understandable. The process of coding binary numbers was presented in Chapter 2. The conversion starts at the LSB of the binary number, and the bits are marked in groups of four from that point.

$$\underbrace{1111}_{F} \quad \underbrace{1111}_{F} \quad \underbrace{1110}_{E} \quad \underbrace{0000.}_{0}{}_{(2)}$$

$F \quad F \quad E \quad 0_{(16)}$

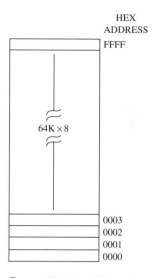

Figure 12–35   64K × 8 memory address range.

A memory address of $FFE0_{(16)}$ is much simpler and easier to relate to than $1111111111100000_{(2)}$. A memory **address range** is the range of addresses from the lowest to the highest address.

If an IC is a 64K × 8 memory, it requires 16 address bits to access all 65,536 locations within the memory. The *lowest*-numbered address lines are always used to access memory locations within the memory chip. As will be seen shortly, the higher-numbered address lines may be used to enable the chip. For example, the lowest address within the 64K × 8 memory is $0000_{(16)}$ and its highest address is $FFFF_{(16)}$ as shown in Fig. 12–35. The hex number FFFF = 65,535. Don't forget the $0000_{(16)}$ address is a memory location. Therefore, the address size (capacity) is 64K = 65,536 = $10000_{(16)}$.

If an IC is a 2K × 8 memory, it requires 11 address bits. Its address range is

A_{10}	A_9	A_8	A_7	A_6	A_5	A_4	A_3	A_2	A_1	A_0	
X	0	0	0	0	0	0	0	0	0	0	$0._{(2)} - 000_{(16)}$
X	1	1	1	1	1	1	1	1	1	1	$1._{(2)} = 7FF_{(16)}$

The entire address range of the memory locations within the IC is $000-7FF_{(16)}$. The X condition in the MSB group is not considered when determining the address range because it is not applied to the chip. In fact, it may not even be generated, depending on the size of the address bus. The highest address in this example is $7FF_{(16)}$, which equals 2047. Since its capacity includes the $000_{(16)}$ address location, its capacity is 2048 = $800_{(16)}$.

If we desire to access memory location $35_{(10)}$ within this 2K × 8 memory, the address is

$$X000\ 0010\ 0011_{(2)} = 023_{(16)}$$

How would the starting (lowest) address of a 48K × 8 ROM memory be calculated if the ending (highest) address were $FFFFF_{(16)}$?

48K = 49,152

$$\begin{array}{r} -\quad 1 \\ \hline 49,151 = BFFF_{(16)}* \end{array}$$

Highest address = $FFFFF_{(16)}$
                    $-0BFFF$
Lowest address   **$F4000_{(16)}$**

Memory Map.

Computers employ both RAM and ROM memory. RAM chips are employed to support the user's applications program and to store user input data. In addition, video RAM is required to support the monitor. Memory maps are designed to specify the address ranges of the different memory involved. If a RAM test program specifies a memory error at location $6F000_{(16)}$, the map identifies the location of the defective RAM chip(s) within the system.

A sample memory map is shown in Fig. 12–36. This generic map was laid out for older IBM-compatible PCs. The memory is mapped using 64K pages. A total of 16 pages supports 1,048,576 (1M) memory locations. Twenty address lines are required for memory access. The 5-digit hex address range of each page is shown on the map.

---

*Total memory locations minus 1.

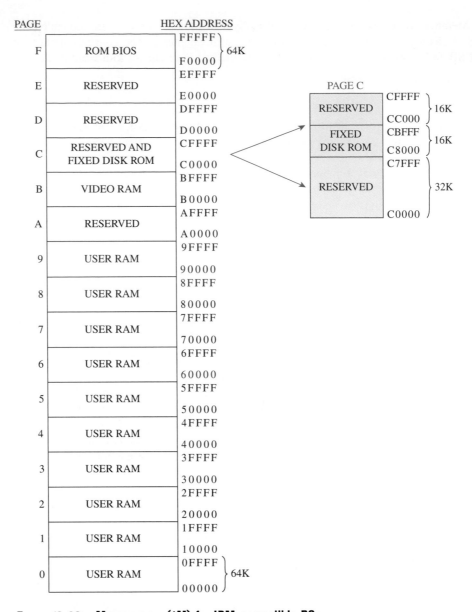

PAGE   HEX ADDRESS

F   ROM BIOS   FFFFF ⎫
                           F0000 ⎬ 64K

E   RESERVED   EFFFF
                           E0000

D   RESERVED   DFFFF
                           D0000

C   RESERVED AND   CFFFF
     FIXED DISK ROM   C0000

B   VIDEO RAM   BFFFF
                           B0000

A   RESERVED   AFFFF
                           A0000

9   USER RAM   9FFFF
                           90000

8   USER RAM   8FFFF
                           80000

7   USER RAM   7FFFF
                           70000

6   USER RAM   6FFFF
                           60000

5   USER RAM   5FFFF
                           50000

4   USER RAM   4FFFF
                           40000

3   USER RAM   3FFFF
                           30000

2   USER RAM   2FFFF
                           20000

1   USER RAM   1FFFF
                           10000

0   USER RAM   0FFFF ⎫
                           00000 ⎬ 64K

PAGE C

RESERVED   CFFFF ⎫ 16K
                 CC000 ⎭

FIXED   CBFFF ⎫ 16K
DISK ROM   C8000 ⎭

RESERVED   C7FFF ⎫ 32K
                 C0000 ⎭

Figure 12–36   Memory map (1M) for IBM-compatible PCs.

Page 0 through Page 9 (10 pages) are dedicated to RAM as shown in Fig. 12–36. This provides 640K user RAM. Pages A, D, and E are reserved. Page C is divided, with the lower 32K reserved, 16K dedicated to fixed-disk ROM, and the upper 16K of the page also reserved.

The memory map of Page C is highlighted in Fig. 12–36. The starting address for Page C is $C0000_{(16)}$ as dictated by its position on the map. The lower half (32K) of Page C is reserved for ROM expansion. The address range for this reserved section is $C0000–C7FFF_{(16)}$. To calculate the address range of this 32K section as well as the remainder of the page, proceed as follows:

$$32K = 32,768$$
$$\underline{-1^*}$$
$$32,767 = 7FFF_{(16)}$$

$$16K = 16,384$$
$$\underline{-1^*}$$
$$16,383 = 3FFF_{(16)}$$

---

*The highest addressable location in a memory chip is the total number of memory locations minus 1. This can be thought of as presented in counter theory, where the highest count of a counter is the total number of states (MOD) minus 1.

Starting address = C0000
 +7FFF (32K − 1)
Highest address C7FFF
 + 1 (Increment to next section)
Lowest address C8000
 +3FFF (16K − 1)
Highest address CBFFF
 + 1 (Increment to next section)
Lowest address CC000
 +3FFF (16K − 1)
Highest address CFFFF

All of the addresses shown in this example are hexadecimal. The hex arithmetic shows the **page address** of Page C is C0000–CFFFF$_{(16)}$. The sectional addresses were calculated to show the usable fixed-disk ROM address range of C8000–CBFFF$_{(16)}$.

Now let's assume we need to map ROM for a system that employes six 8K × 8 ROM ICs whose starting address is F0000$_{(16)}$. We will also assume the lower 16K memory is reserved for future expansion. Thirteen address bits (A_{12}–A_0) are required to access the data within each IC because $2^{13} = 8192$ (8K). Twenty address bits will be used per Fig. 12–36. This ROM circuit was discussed in Chapter 11 and is presented here in Fig. 12–37 for memory mapping.

All of the ROM memory resides on Page F on the memory map. The page address is provided by the microprocessor in the upper nibble (A_{19}–A_{16}) of the 20-bit address. When a memory read cycle is in progress and A_{19}–$A_{16} = 1111_{(2)}$, the $\overline{\text{ROMPG}}$ signal is activated, and the 74LS138 decoder is enabled. One of the decoder's outputs will be activated once the decoder has been enabled. Address bits A_{15}–A_{13} (select inputs to the decoder) determine which decoder output goes low.

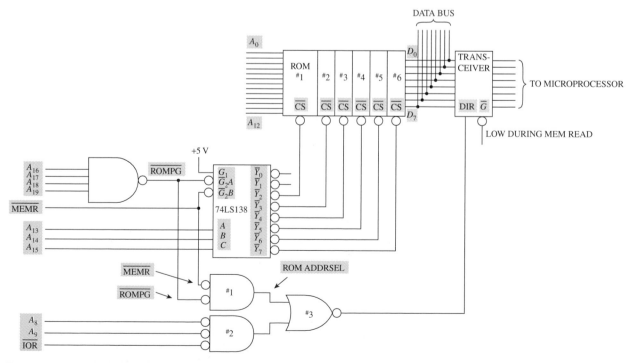

FIGURE 12–37 **ROM address decoder circuit.**

The $\overline{Y}_0$ and $\overline{Y}_1$ decoder outputs are not connected because the lower 16K of memory is reserved. ROM #1 IC is enabled when the $\overline{Y}_2$ decoder output is activated. The complete address for this chip is

A_{19}	A_{18}	A_{17}	A_{16}	A_{15}	A_{14}	A_{13}	A_{12}	A_{11}	A_{10}	A_9	A_8	A_7	A_6	A_5	A_4	A_3	A_2	A_1	A_0
1	1	1	1	0	1	0	X	X	X	X	X	X	X	X	X	X	X	X	X

The ROM page is selected by A_{19}–A_{16}, and the decoder's output is controlled by A_{15}–A_{13} as shown in Fig. 12–37. The remaining 13 address bits (A_{12}–A_0) are used to access a memory location within the selected ROM IC. The address range of the chip is determined by the lowest and highest address of the chip. The remaining bits, A_{12}–A_0, are don't care inputs as far as the address range is concerned because they are not used to select which ROM chip is enabled.

The complete address range for ROM #1 is determined by making all Xs low for the starting address and all Xs high for the highest address in the chip. The starting address may be referred to as the **base address.**

ROM #1 Select:	1111	010X	XXXX	XXXX	XXXX
Starting address:	1111	0100	0000	0000	0000 = F4000$_{(16)}$
Highest address:	1111	0101	1111	1111	1111 = F5FFF$_{(16)}$

The address range for ROM #1 is F4000–F5FFF$_{(16)}$. The unused lower 16K of the memory map starting at F0000$_{(16)}$ is calculated as follows:

$$16K = 16{,}384$$
$$\underline{-\quad 1}$$
$$16{,}383 = 3FFF_{(16)}$$

The specified ROM starting address is F0000$_{(16)}$.

$$\text{Starting address} = F0000$$
$$\underline{+3FFF}$$
$$F3FFF_{(16)}*$$

To summarize:

ROM Unused = F0000–F3FFF$_{(16)}$

ROM #1 = F4000–F5FFF$_{(16)}$

The remaining address ranges are calculated as follows:

$$8K = 8{,}192$$
$$\underline{-\quad 1}$$
$$8{,}191 = 1FFF_{(16)}$$

Highest address =	F5FFF	ROM #1
	$\underline{+\quad 1}$	
Lowest address =	F6000	ROM #2
	$\underline{+1FFF}$	(8K − 1)
Highest address =	F7FFF	ROM #2
	$\underline{+\quad 1}$	
Lowest address =	F8000	ROM #3
	$+1FFF$	

---

*Highest address of the unused ROM.

$$\begin{aligned}
\text{Highest address} &= \text{F9FFF} \quad \text{ROM \#3} \\
&\underline{+\quad 1} \\
\text{Lowest address} &= \text{FA000} \quad \text{ROM \#4} \\
&\underline{+1\text{FFF}} \\
\text{Highest address} &= \text{FBFFF} \quad \text{ROM \#4} \\
&\underline{+\quad 1} \\
\text{Lowest address} &= \text{FC000} \quad \text{ROM \#5} \\
&\underline{+1\text{FFF}} \\
\text{Highest address} &= \text{FDFFF} \quad \text{ROM \#5} \\
&\underline{+\quad 1} \\
\text{Lowest address} &= \text{FE000} \quad \text{ROM \#6} \\
&\underline{+1\text{FFF}} \\
\text{Highest address} &= \text{FFFFF} \quad \text{ROM \#6}
\end{aligned}$$

The memory map for this ROM system is shown in Fig. 12–38.

PAGE F

ROM #6	FFFFF
	FE000
ROM #5	FDFFF
	FC000
ROM #4	FBFFF
	FA000
ROM #3	F9FFF
	F8000
ROM #2	F7FFF
	F6000
ROM #1	F5FFF
	F4000
NOT USED	F3FFF
NOT USED	
	F0000

FIGURE 12–38   ROM memory map.

Address Decoding

Figure 12–37 presented the basics of decoding an address. The page of memory (Page F in this example) was selected by a NAND gate that produced the $\overline{\text{ROMPG}}$ signal. The specific ROM IC within the page was selected by a 74LS138 decoder. The internal memory location within the selected chip was determined by the lower address bits A_{12}–A_0.

A decoder circuit can be designed by laying out the address scheme required to set the base address of the memory. A decoder circuit for the fixed disk ROM addresses highlighted in Fig. 12-36 is designed as follows: The base address per the system memory map is C8000$_{(16)}$. The memory space allocated for this function is 16K. Fourteen address bits are required for internal chip access. The address scheme is

A_{19}	A_{18}	A_{17}	A_{16}	A_{15}	A_{14}	A_{13}	A_{12}	A_{11}	A_{10}	A_9	A_8	A_7	A_6	A_5	A_4	A_3	A_2	A_1	A_0
1	1	0	0	1	0	X	X	X	X	X	X	X	X	X	X	X	X	X	X

Bits A_{19}–A_{16} put the memory at Page C. Bits A_{15}–A_{14} set the memory at a base address of C8000$_{(16)}$. The address increments from this base address 16K − 1 (16,383) times to reach the highest fixed-disk ROM memory location. Bits A_{13}–A_0 control the internal memory address.

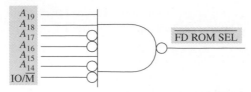

FIGURE 12-39 Fixed-disk ROM decoder (C8000-CBFFF$_{(16)}$).

The range of addresses is determined by making all of the Xs low for the base address and then by making them high for the highest chip address:

Address scheme:	1100	10XX	XXXX	XXXX	XXXX	
Base address:	1100	1000	0000	0000	0000	= C8000$_{(16)}$
Highest address:	1100	1011	1111	1111	1111	= CBFFF$_{(16)}$

The decoder can be implemented with a NAND gate. The base address bits A_{19}, A_{18}, and A_{15} are high, and A_{17}, A_{16}, and A_{14} are low during fixed-disk ROM access. Figure 12–39 shows the implemented decoder. The IO/$\overline{\text{M}}$ read signal ensures the address is applicable to a memory read cycle.

Section 12-4: Review Questions

Answers are given at the end of the chapter.

A. Why are memory addresses specified in hex?
B. Define the memory address range of an IC.
C. How many address bits are required to address a location within an 8K $\times$ 8 ROM IC?
D. If the base address of the 8K $\times$ 8 ROM chip in question (C) is FA000$_{(16)}$, what is its highest address?
E. Draw a memory map for 1M RAM consisting of 128K $\times$ 8 RAM chips starting at address 00000$_{(16)}$. Show the lowest and highest address of each RAM chip.
F. Design a NAND gate decoder to decode the following 32K $\times$ 1 RAM address:

A_{19}	A_{18}	A_{17}	A_{16}	A_{15}	A_{14}	A_{13}	A_{12}	A_{11}	A_{10}	A_9	A_8	A_7	A_6	A_5	A_4	A_3	A_2	A_1	A_0
0	0	1	1	1	X	X	X	X	X	X	X	X	X	X	X	X	X	X	X

SECTIONS 12-3 AND 12-4: INTERNAL SUMMARY

Random-Access Memory is volatile. It retains its data only as long as power is applied.

Static RAM uses a flip-flop as part of the memory cell. This high-speed semiconductor device consists of a 6-transistor cell. Thus, SRAM is not a very high-density device. Unlike their ROM counterparts, SRAM may be written to with the same ease and speed as it is read.

Dynamic RAM uses one transistor and one capacitor to implement a memory cell. This allows very high-density chips to be manufactured. The capacitor within the cell must be periodically recharged to maintain a stored Logic 1 level. This recharge is called refresh. **Refresh** is accomplished by accessing a row in the memory array. The storage capacitors in the row are recharged every time a row is accessed. **DRAM controllers** are often used to control the refresh operations of DRAM ICs.

Address inputs to DRAM are multiplexed to reduce the number of pins on the IC. The row address is routed through the MUX and latched. Then the column address is sent through and latched. A specific memory location within the DRAM is enabled once the latched addresses have been decoded.

Pseudo-static RAM is a DRAM chip with **internal refresh control circuitry.** It provides the high density of DRAM technology without the necessity of external refresh circuits.

Memory addresses are specified in technical literature in hexadecimal. The **address range** of a memory IC is the range of addresses from lowest to highest. The **internal address range** of a memory chip starts at zero and increments up to the highest memory location. The highest memory location may be calculated by raising 2 to the power of the number of address bits minus 1 and converting that number to hex.

Memory maps specify the address range of an IC or address ranges within a memory system. Lowest and highest addresses of a section of memory or an IC may be calculated for a memory map by writing the address scheme. The scheme is written with the higher-order bits that select/enable the desired device(s). All lower-order bits used to internally access memory locations are written as Xs. The total address range is derived by making all Xs low to provide the starting (base) address and then making them high for the highest address.

Decoding circuits use the higher-order device select bits to enable an IC(s). If the address scheme of a device is

A_{19}	A_{18}	A_{17}	A_{16}	A_{15}	A_{14}	A_{13}	A_{12}	A_{11}	A_{10}	A_9	A_8	A_7	A_6	A_5	A_4	A_3	A_2	A_1	A_0
1	0	1	0	1	1	X	X	X	X	X	X	X	X	X	X	X	X	X	X

the base and highest addresses can be calculated as follows:

$$
\begin{aligned}
1010 \quad 11XX \quad XXXX \quad XXXX \quad XXXX_{(2)} \\
1010 \quad 1100 \quad 0000 \quad 0000 \quad 0000_{(2)} &= AC000_{(16)} \text{ (Base address)} \\
1010 \quad 1111 \quad 1111 \quad 1111 \quad 1111_{(2)} &= AFFFF_{(16)} \text{ (Highest address)}
\end{aligned}
$$

The higher-order bits in this example must be used to decode the address. Bits A_{19}, A_{17}, A_{15}, and A_{14} must be tied directly to a NAND gate; bits A_{18} and A_{16} must be routed through NOT gates and then connected to the NAND gate.

SECTIONS 12-3 AND 12-4: INTERNAL SUMMARY QUESTIONS

Answers are given at the end of the chapter.

1. Random-Access Memory is.
   a. volatile
   b. nonvolatile

2. Which type of RAM uses flip-flop technology to store data?
   a. DRAM
   b. SRAM
   c. PSRAM

3. Which type of memory uses one transistor and one capacitor to store data?
   a. DRAM
   b. SRAM

4. What is the capacity of the memory IC in Fig. 12–40?
   a. 128K × 1   c. 256K × 1
   b. 128K × 8   d. 256K × 8

5. What type of memory is shown in Fig. 12–40?
   a. ROM
   b. SRAM
   c. DRAM
   d. PSRAM

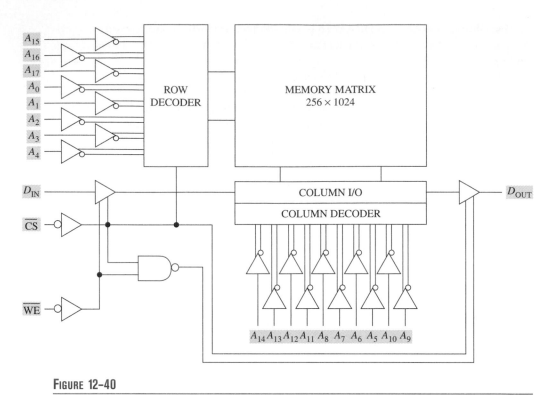

FIGURE 12–40

6. What is D_{OUT} of the memory in Fig. 12–40 when $D_{IN} = 1$, $\overline{CS} = 1$, and $\overline{WE} = 0$?
   a. Logic 0
   b. Logic 1
   c. High-impedance state
   d. Logic level stored in memory

7. What is D_{OUT} of the memory in Fig. 12–40 when $D_{IN} = 1$, $\overline{CS} = 0$, and $\overline{WE} = 0$?
   a. Logic 0
   b. Logic 1
   c. High-impedance state
   d. Logic level stored in memory

8. What is D_{OUT} of the memory in Fig. 12–40 when $D_{IN} = 1$, $\overline{CS} = 0$, and $\overline{WE} = 1$?
   a. Logic 0
   b. Logic 1
   c. High-impedance state
   d. Logic level stored in memory

9. The timing waveforms shown in Fig. 12–41 represent a SRAM _____ cycle.
   a. read
   b. write
   c. erase
   d. refresh

10. What type of memory requires refresh?
    a. DRAM   c. EPROM
    b. SRAM   d. E²PROM

11. Which type of memory requires address multiplexing?
    a. DRAM   c. EPROM
    b. SRAM   d. E²PROM

12. What type of memory is shown in Fig. 12–42?
    a. DRAM   c. EPROM
    b. SRAM   d. E²PROM

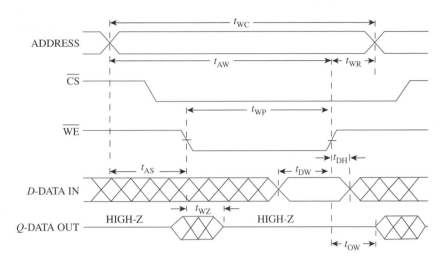

FIGURE 12–41

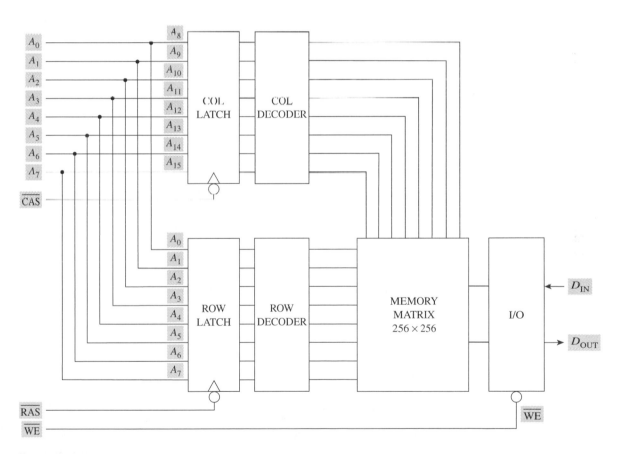

FIGURE 12–42

13. What is the capacity of the memory IC in Fig. 12–42?
    a. 256×1 c. $32K \times 1$
    b. 512×1 d. $64K \times 1$

14. What is the purpose of the $\overline{\text{RAS}}$ input to the memory in Fig. 12–42?
    a. Disable the I/O port
    b. Enable the I/O port
    c. Input the column address
    d. Input the row address

15. $\overline{\text{RAS}}$-only refresh can be accomplished by activating $\overline{\text{RAS}}$ to latch a row address while leaving $\overline{\text{CAS}}$ inactive.
   a. True
   b. False

16. Pseudo-static RAM requires a DRAM controller to accomplish refresh.
   a. True
   b. False

17. What is the address range of a 4K × 8 memory?
   a. 000–7FF$_{(16)}$ c. 0000–1FFF$_{(16)}$
   b. 000–FFF$_{(16)}$ d. 0000–FFFF$_{(16)}$

18. How many bytes of data are stored in a 4K × 8 memory?
   a. 4096   c. 16,384
   b. 8192   d. 32,768

19. 19. What is the address range of a 256K × 1 SRAM?
   a. 000–FFF$_{(16)}$ c. 00000–7FFFF$_{(16)}$
   b. 00000–3FFFF$_{(16)}$ d. 00000–FFFFF$_{(16)}$

20. What is the highest address for a 32K × 1 memory with a base address of D0000$_{(16)}$?
   a. 32, 768$_{(16)}$
   b. D9000$_{(16)}$
   c. D8000$_{(16)}$
   d. D7FFF$_{(16)}$

CHAPTER 12: SUMMARY

Memory is required in all computers for the microprocessor to function. Data must be stored where the microprocessor can access it quickly. This chapter has dealt exclusively with **primary memory—ROM and RAM.**

Read-Only Memory (ROM) is nonvolatile (permanent) memory. The data stored in ROM are always present. Power failures and system turn off and on have no affect on the data.

ROM is available as **Mask ROM (MROM), Programmable ROM (PROM), Erasable Programmable ROM (EPROM),** and **Electrically Erasable Programmable ROM (EEPROM/E²PROM).**

Mask ROMs are not reprogrammable. The MROM initial setup cost is high. Thus, they typically are manufactured only in high volume.

Programmable ROMs can be programmed by the user. Programming is done with a **PROM programmer** which blows selected fuse links and leaves the remaining links intact. Once selected fuses are blown, the program is set and cannot be changed. Therefore, PROMs are **one-time programmable** chips.

Erasable programmable ROMs are also user programmable. In comparison to PROMs, programmed data in an EPROM may be erased by exposing the chip's quartz window to UV light for approximately 20 minutes. A couple of disadvantages are encountered when using EPROMs. First, the erasure time is long, and the erase operation clears all data in memory. Second, the chip must be removed from the circuit for erasure and reprogramming.

Electrically erasable programmable ROMs allow selective in-circuit data erasure and are user programmable. The EEPROM chip supports **read, write (program),** and **erase** modes of operation.

Memory access time is the time interval from valid address on the chip's inputs to valid data on the chip's output(s). **Flash memory** rates three high marks in memory technology—high density, fast access time, and in-circuit programmability. Flash memory works well in portable equipment that requires high density data storage but cannot support a disk drive(s). Flash memory offers block erasure in blocks as small as 16 Kbits but does not offer bit/byte erasure.

Random-Access Memory (RAM) is volatile read/write memory. The data written to RAM is stored only as long as power is applied. The data is cleared when power is removed from the memory device(s). A memory cell stores one bit of data.

The number of address lines (n) required to access X memory locations within a memory IC is calculated by $n = \log X \div \log 2$. The **capacity** of a memory chip is given as the number of memory locations times the number of bits per location. A 64K × 8 chip contains 65,536 memory locations with 8 bits stored in each location. Sixteen address bits are required to access all of the memory locations within this chip.

	SRAM	DRAM	PSRAM	TABLE 12-8
Density	1 Mbit	4 Mbit	4 Mbit	**SRAM, DRAM, PSRAM Comparisons**
Current consumption	1 μA	150 μA	70 μA	
Refresh	Not required	Required (external control)	Required (internal control)	
Data retention voltage	2–5.5 V	4.5–5.5 V	3–5.5 V	
Cell	6 Transistors	1 Transistor 1 Capacitor	1 Transistor 1 Capacitor	
Address	Non-MUX	MUX	Non-MUX	

Static RAMs (SRAMs) employ flip-flop technology for data storage. They do not require refresh, and they use a nonmultiplex approach to addressing. SRAM has a lower density than DRAM because it uses a larger cell area than DRAM. A SRAM cell is organized with four transistors and two resistors or six transistors.

Dynamic RAM (DRAM) memory cells are implemented with one transistor and one capacitor. The capacitor is used as the storage device within the cell. The simplicity of design provides very high density. Capacitors leak. This drawback forces DRAM cells to be periodically recharged (**refreshed**). Refresh is accomplished each time a row of memory is accessed.

The disadvantages of DRAM are threefold: (1) refresh circuitry is required to track and control refresh operations; (2) refresh operations decrease the time available for memory READ and WRITE operations, which can slow a system down; and (3) memory addresses must be multiplexed. **DRAM controllers** are often used to simplify the extra refresh circuitry. Refresh is an integral function of DRAM operations. Manufacturers have increased refresh time requirements and incorporated some of the refresh circuitry on the DRAM chip to simplify this mandatory process.

Pseudo-Static RAM (PSRAM) chips are high-density DRAM chips with internal refresh control circuits. Several methods of refresh are available to the designer of circuits using PSRAM.

The three types of RAM presented in this section are compared in Table 12–8. The 4-Mbit density of the DRAM and PSRAM are achieved using the same technology employed for the 1-Mbit SRAM.

Memory addresses are used to enable/select a specific device(s) and enable one memory location within that device. Memory addresses are specified in hexadecimal.

The **address range** of a chip(s) is the range of addresses from lowest to highest. **Memory maps** are used to specify address ranges of designated memory areas within a system. The system memory map is established during design of a system. It is used to allocate memory locations. Address ranges can be determined by writing the address scheme that will select the device. Once this has been determined, Xs are used for all internal address bits. The total range is determined by first making all Xs low for the base address. Next, all Xs are made high to determine the highest address.

CHAPTER 12: END OF CHAPTER QUESTIONS/PROBLEMS

Answers are given in the Instructor's Manual.

SECTION 12-1

1. Define primary memory.

2. Give an example of computer primary memory.

3. What is the size of a word in a computer that uses a 20-bit address bus and an 8-bit data bus?

4. What type of bus cycle occurs when $IO/\overline{M} = 0$?

5. What type of computer memory contains the bootstrap program that initializes a computer?

6. Which type of computer memory is volatile?

7. A RAM IC has a capacity of 8K × 1. How many memory locations are contained within the chip?

8. How many address bits must be applied to a memory IC with a capacity of 8K × 1?

9. How many memory locations are contained in a 64K × 8 memory IC?

10. How many *bytes* of data are stored in a 64K × 8 memory IC?

11. Define a memory cell.

12. Define nonvolatile memory.

13. Give an example of nonvolatile memory.

14. How many address bits are required to address a 2M × 1 memory IC?

15. What is the capacity of the ROM IC shown in Fig. 12–43?

16. What type of ROM would be most desirable for programming high-quantity BIOS programs for computers?

17. Programmable ROM (PROM) ICs can be reprogrammed.
    a. True
    b. False

18. Erasable Programmable ROM (EPROM) ICs allow for bit, byte, and total data erasures.
    a. True
    b. False

19. Which type of ROM chip utilizes the application of ultraviolet light to erase data?

20. Define ROM access time (t_{ACC}).

21. What are the data outputs of the memory circuit in Fig. 12–44 when $\overline{CS} = 0$, $\overline{OE} = 0$, and $A_1A_0 = 10$?

22. What are the data outputs of the circuit in Fig. 12–44 when $\overline{CS} = 0$, $\overline{OE} = 1$, and $A_1A_0 = 11$?

23. What are the data outputs of the circuit in Fig. 12–44 when $\overline{CS} = 0$, $\overline{OE} = 0$, and $A_1A_0 = 01$?

24. What are the data outputs of the circuit in Fig. 12–44 when $\overline{CS} = 1$, $\overline{OE} = 0$, and $A_1A_0 = 00$?

CT 25. How would a 512×8 (4096-bit) PROM be laid out in terms of the internal decoder and multiplexers? Sketch a block diagram of your circuit. (*Hint:* Use a square memory matrix; refer to the discussion of Fig. 12–7.)

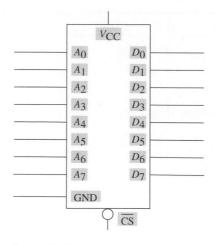

FIGURE 12-43

CT 26. What type of internal decoder and MUX would be required to access memory in a $2K \times 4$ (8192-bit) PROM using a 64×128 memory matrix?

27. Define Static RAM (SRAM).

28. What is the major type of RAM depicted in Fig. 12–45?

29. What is the capacity of the memory IC in Fig. 12–45?

30. What memory operation is performed by the IC in Fig. 12–45 when $\overline{CS} = 1$ and $\overline{WE} = 0$?

31. What memory operation is performed by the IC in Fig. 12–45 when $\overline{CS} = 0$ and $\overline{WE} = 1$?

32. What memory operation is performed by the IC in Fig. 12–45 when $\overline{CS} = 0$ and $\overline{WE} = 0$?

33. What type of memory is illustrated in Fig. 12–46?

34. What is the capacity of the SRAM memory IC shown in Fig. 12–47?

35. What is the capacity of the DRAM memory IC shown in Fig. 12–48?

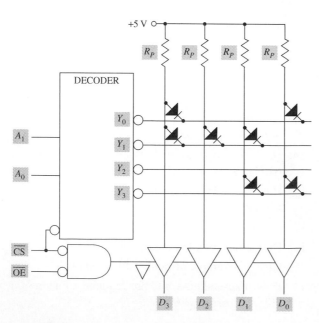

FIGURE 12-44

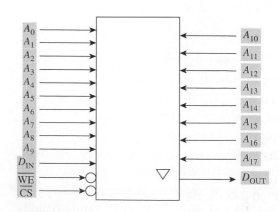

FIGURE 12-45

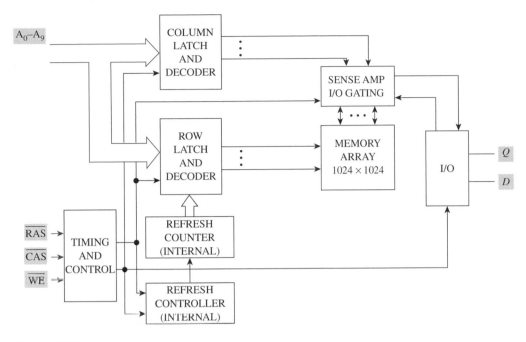

FIGURE 12-46

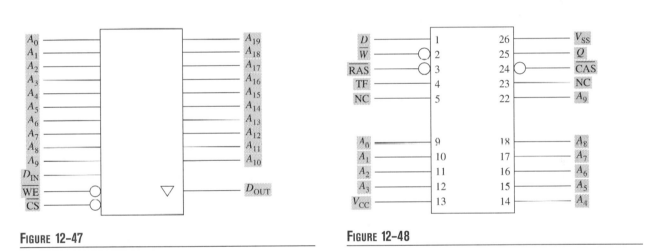

FIGURE 12-47   **FIGURE 12-48**

SECTION 12-4

36. Calculate the highest address of a 132K × 8 memory with a base address of $10000_{(16)}$.

37. What is the address range of a 1M DRAM IC whose base address is $00000_{(16)}$?

38. A memory map of Page B of memory is shown in Fig. 12–49. The base address of the memory is $B0000_{(16)}$ and each section contains 16K. What are the lowest and highest addresses of each section?

39. What is the address range for the following 64K × 1 RAM address?

A_{19}	A_{18}	A_{17}	A_{16}	A_{15}	A_{14}	A_{13}	A_{12}	A_{11}	A_{10}
0	0	0	1	X	X	X	X	X	X

A_9	A_8	A_7	A_6	A_5	A_4	A_3	A_2	A_1	A_0
X	X	X	X	X	X	X	X	X	X

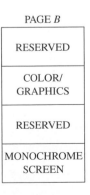

PAGE *B*

RESERVED
COLOR/ GRAPHICS
RESERVED
MONOCHROME SCREEN

FIGURE 12-49

PAGE

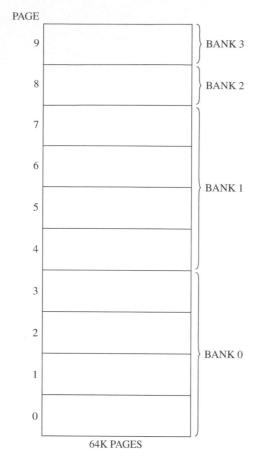

64K PAGES

FIGURE 12-50

CT 40. A RAM memory map is shown in Fig.12–50. Each page (section) of the map represents 64K memory. The RAM is divided into two 256K banks (Bank 0 and Bank 1). Banks 2 and 3 are not used. What are the lowest and highest addresses of Banks 0 and 1 if the base address is $00000_{(16)}$?

41. What are the lowest and highest addresses of each page of memory in Fig. 12–50 starting at address $00000_{(16)}$?

42. What are the base and highest addresses of the following RAM address scheme?

A_{19}	A_{18}	A_{17}	A_{16}	A_{15}	A_{14}	A_{13}	A_{12}	A_{11}	A_{10}
1	1	0	0	1	1	X	X	X	X

A_9	A_8	A_7	A_6	A_5	A_4	A_3	A_2	A_1	A_0
X	X	X	X	X	X	X	X	X	X

CT 43. Design a NAND gate decoder to decode the address scheme of question 42.

CT 44. What is the capacity of the memory decoded in question 43?

CT 45. What range of addresses is decoded by the decoder in Fig.12–51?

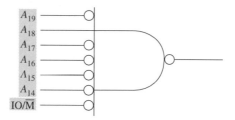

FIGURE 12-51

Answers to Review Questions

Section 12-1

A. (1) h   (5) g
    (2) i   (6) a
    (3) e   (7) c
    (4) d

B. (1) Memory that is lost when power is removed/interrupted.
    (2) A bus cycle that occurs when the microprocessor is writing (storing) data in a memory location or to an I/O port.
    (3) 8 bits of data.
    (4) The number of memory locations within a chip times the number of bits stored per location.

C. 16,384   G. 1,048,576

D. 4   H. 8

E. 65,536   I. 1,048,576

F. 14

J. 20: There are normally 10 address pins on the chip and the address is multiplexed. This information is presented in Section 12–3.

Section 12-2

A. nonvolatile

B. True

C. MROM
    PROM
    EPROM (UVPROM)
    EEPROM (E²PROM)

D. PROM

E. MROM

F. PROM

G. EPROM (UVPROM)

H. The interval of time that elapses between valid address inputs and valid data outputs.

I.   (1) $\overline{CS}$: Chip Select. An active-low signal used to enable a chip.
     (2) $\overline{OE}$: Output Enable. An active-low signal used to enable the output buffers of a chip.
     (3) $\overline{WE}$: Write Enable. An active-low signal used to enable writing to a RAM chip.
     (4) $RD/\overline{WR}$: Read/Write. A signal/line indicating read when it is high and write when it is low.
     (5) One-time programmable: A term related to PROM technology—once a fusible link is blown the program cannot be changed.
     (6) PROM burning: Programming a PROM chip with a PROM programmer.

J.   Flash

K.   False

Section 12-3

A.   Volatile

B.   SRAM, DRAM, PSRAM

C.   DRAM

D.   DRAM

E.   DRAM

F.   Decrease the number of address pins on the IC.

G.   High density and circuit simplicity.

H.   False

Section 12-4

A   Simpler to read.

B.   Lowest to highest address.

C.   13

D.   1111 101X XXXX XXXX XXXX
     1111 1010 0000 0000 0000
     1111 1011 1111 1111 1111
     $= FA000_{(16)} = $ Lowest
     $= FBFFF_{(16)} = $ Highest

or

$$8K = 8{,}192 \qquad\qquad \text{Start} = FA000_{(16)}$$
$$\underline{\quad - \quad 1} \qquad\qquad\qquad \underline{\quad + 1FFF}$$
$$8{,}191 = 1FFF_{(16)} \qquad \text{Highest } FBFFF_{(16)}$$

E.   See Fig. 12–52.

F.   0011 1XXX XXXX XXXX XXXX
     0011 1000 0000 0000 0000 $= 38000_{(16)}$
     0011 1111 1111 1111 1111 $= 3FFFF_{(16)}$

See Fig. 12–53.

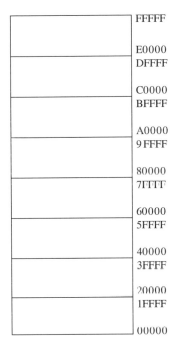

	FFFFF
	E0000
	DFFFF
	C0000
	BFFFF
	A0000
	9FFFF
	80000
	7FFFF
	60000
	5FFFF
	40000
	3FFFF
	20000
	1FFFF
	00000

FIGURE 12–52

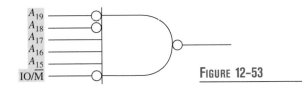

FIGURE 12–53

ANSWERS TO INTERNAL SUMMARY QUESTIONS

SECTIONS 12-1 AND 12-2

1. a		13. a	
2. c		14. b	
3. c		15. a	
4. a		16. b	
5. a		17. c	
6. d		18. a	
7. b		19. c	
8. b		20. d	
9. b		21. a	
10. c		22. c	
11. a		23. d	
12. d		24. c	

SECTIONS 12-3 AND 12-4

1. a		11. a	
2. b		12. a	
3. a		13. d	
4. c		14. d	
5. b		15. a	
6. c		16. b	
7. c		17. b	
8. d		18. a	
9. b		19. b	
10. a		20. d	

13 PROGRAMMABLE LOGIC DEVICES/ARRAYS

Topics Covered in this Chapter

Introduction

Programmable Logic Device (PLD) is a general term that refers to ICs with uncommitted logic arrays. The term is applicable to a multitude of user-programmable ICs that can be programmed to implement digital circuits. PLDs can be used to solve many logic design problems. They contain basic logic gates and flip-flops along with internal connection schemes (interconnects) that can be modified via software and hardware (a programmer) to yield specific logic functions/circuits.

Programming the devices is supported by numerous commercially available programmers and software products. The programmer is normally connected to a PC, and the design software is loaded into that PC. A blank PLD is inserted into the socket of the programmer, and the PLD is programmed by blowing the interconnection fuses or programming the antifuses within the device. These products typically allow the user to describe the logic design with Boolean expressions, truth tables, hardware description, state diagrams, waveforms, or schematics.

There are numerous advantages to using PLDs in lieu of standard SSI logic gates. In many of the ICs presented in this text, four or fewer logic gates were contained on the IC. Many logic functions can be incorporated on one PLD. In fact, some PLDs contain the equivalent of several thousand gates. Therefore, many SSI/MSI functions can be replaced with one PLD. This results in fewer ICs on a circuit board, which saves space.

Antifuse

Application-Specific Integrated Circuit (ASIC)

Field-Programmable Gate Array (FPGA)

Generic Array Logic (GAL*)

Programmable Array Logic (PAL*)

Programmable Logic Array (PLA)

Programmable Logic Device (PLD)

Programmable Logic Element (PLE)

Chapter Objectives

1. Define terms associated with programmable logic.

2. Identify the standard notations used in programmable logic symbology.

3. Classify programmable logic devices as PLEs, PLAs, or PALs.

4. Given a PLD generic logic block or macrocell diagram, determine the type of output and feedback.

5. State the main advantages of using FPGAs with antifuse technology.

Before we progress, keep in mind that the PC revolution has resulted in competition to increase speed and memory capability while making the systems smaller. Computer systems of the past required numerous logic gates and flip-flops to tie the system together so it could function. These ICs, referred to as "glue chips," took up a lot of room on the system board. Thus, the advantage of using less board real estate is a big one. In addition, fewer ICs mean less power consumption and improved reliability. Speed of implementation is another important advantage.

The first PLDs were programmed by blowing fuse links to open an input to a logic array of gates. However, the disadvantage of this technology was that once the fuse link was blown, it could not be reset. These devices are programmable one time only.

Erasable PLDs (EPLDs) use ultraviolet (UV) light to erase the program in the PLD. This allows reprogramming of the device if necessary. These devices, like their EPROM

---

*GAL is a registered trademark of Lattice Semiconductor Corp. PAL is a registered trademark of Advanced Micro Devices, Inc.

counterparts, use stored electrical charges to control the internal connections instead of fuse links. A small window is incorporated in the PLD package through which the UV light can be passed to erase the existing program. This window should be covered with an opaque label except when erasing the program.

Electrically Erasable PLDs (EEPLDs or E²PLDs) can be programmed, erased, and reprogrammed as necessary by applying certain voltage levels to pins on the IC. Like E²PROMs, they use a variation of the EPROM floating-gate technology. These devices use the traditional PLD architecture that will be discussed in this chapter.

The original PLDs functioned and were programmed similar to their Programmable Read-Only-Memory (PROM) counterparts. The PROM's operation was presented in Chapter 12. These PLDs are sometimes referred to as **Programmable Logic Elements (PLEs)** and are used to design logic functions.

Several companies have registered trademark names to their PLDs, which has caused some confusion in the industry. This text classifies PLDs in two broad categories: Programmable Logic Array (PLA)/Programmable Array Logic (PAL^R) devices and Field-Programmable Gate Array (FPGA) devices.

The **Programmable Logic Array (PLA,** developed by Signetics) and **Programmable Array Logic (PAL^R)** devices were designed for programmable **logic applications** whereas the original PROM ICs were designed for **memory applications.** PAL is a registered trademark of Advanced Micro Devices, Inc. The PLE, PLA, and PAL devices allow generation of sum-of-products expressions. The different architectures for these devices will be discussed shortly. These PLDs are available in TTL and CMOS technology.

The **Field-Programmable Gate Array (FPGA)** device category includes the very high-density, multilevel logic, user-programmable devices that are steadily gaining popularity in digital circuits. Some FPGA devices contain circuitry to implement over 100,000 logic gates.

SECTION 13-1: PROGRAMMABLE LOGIC SYMBOLOGY

OBJECTIVES

1. Define programmable logic device, and state the basic logic circuit configuration of PLD circuits.
2. Identify the PLD symbols used to represent intact fuses, blown fuses, and hard-wired connections.

Standard programmable logic devices are programmed by blowing **fuse links** or leaving those links intact. A programmer can be used to apply enough voltage to selected fuses to blow them. A PLD is a device purchased with an uncommitted logic array.

The fuse links are represented by the standard fuse symbol shown in Fig. 13–1. The three fuse links in Fig. 13–1(a) are intact, and the resultant output Boolean expression of the gate is $A \cdot B \cdot C$. On the other hand, the fuse link in line C in Fig. 13–1(b) has been blown. This leaves that input floating. The resultant logic for the gate is $A \cdot B \cdot 1$, which is equal to AB because the 1 in the expression functions as an enabler to the AND gate.

The Sum-of-Products (SOP) configuration shown in Fig. 13–2 allows both AND gates' outputs to be programmable by blowing selected fuse links. The output expression for the circuit is $AB + BC$.

More flexibility can be provided by the circuit shown in Fig. 13–2 if **input buffers** are incorporated to provide both true (A) and complement ($\overline{A}$) inputs to the fuses (fuse links). This flexibility is shown in the circuit in Fig. 13–3. The product term from the upper AND gate is $A\overline{B}\,\overline{C}$, and that from the lower AND gate is $\overline{A}BC$. These product terms are produced by the fuses that are left intact at the AND gate's inputs. The circuit's output is $A\overline{B}\,\overline{C} + \overline{A}BC$. Basic PLDs are laid out in this SOP configuration.

The fuse links depicted in Fig. 13–3 are actually set between the rows and columns as shown in Fig. 13–4. If the fuse link is left intact, the input data on the column is routed

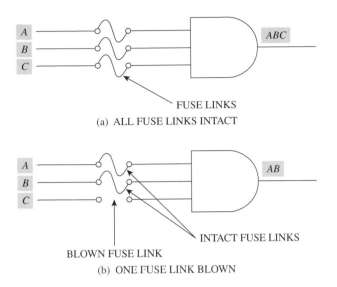

(a) ALL FUSE LINKS INTACT

(b) ONE FUSE LINK BLOWN

FIGURE 13-1   PLD fuse links.

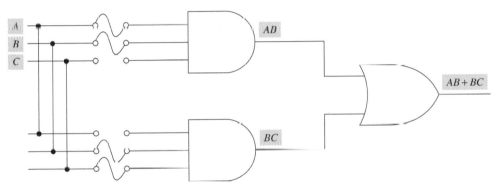

FIGURE 13-2   Programmed sum-of-products circuit.

to the AND gates. If the fuse link is blown, the data on that column is prevented from reaching the logic gates because there is no connection between the column and row.

Some PLDs have 20 or more input pins. If the drawing convention used in Fig. 13–3 were employed for a 20-input PLD, the drawing would be very cluttered with lines and fuses. To simplify matters, a new convention was adopted to simplify the drawings. A 3-input AND gate using this convention is shown in Fig. 13–5. Notice that only one line is shown at the gate's input. This single line is used for simplicity only and it is called the **product line.** The input terms (ABC) are shown as vertical lines. The programmable fuses (fuse links) are located where these vertical input lines intersect the product line.

An X in Fig. 13–5 indicates an *intact fuse*. Therefore, an X makes the input on that line part of the product term from the AND gate. The lack of an X indicates the fuse at that location has been blown. This prevents the data input on that particular line from being part of the product term. A dot at the intersection of any lines indicates a *hard-wired connection*. The dot on the product line at input C indicates C will be part of the output product term, and this variable is not programmable. The output of the AND gate is AC because the A input line has an intact fuse and the C input is hard-wired. The PLD schematic notation is shown again in Fig. 13–6. The notation is the industry standard. Note the two different methods of drawing the input buffers in Fig. 13–6(a).

An example circuit using the symbology of Fig. 13–5 and Fig. 13–6 is shown in Fig. 13–7. The layout provides a 4 × 3 programmable AND array. The product terms of the

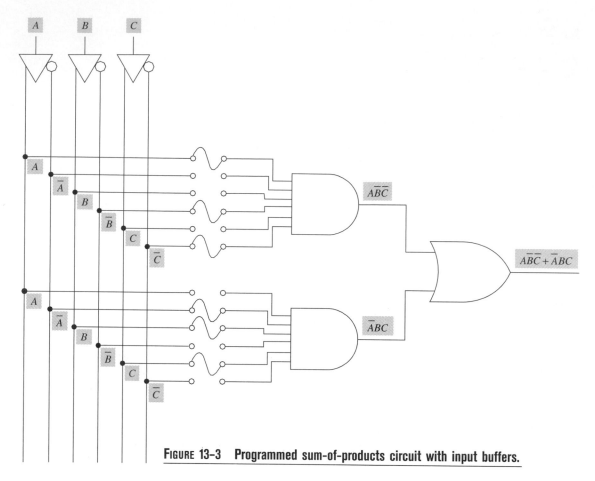

FIGURE 13-3 Programmed sum-of-products circuit with input buffers.

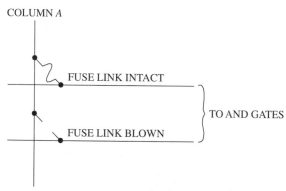

FIGURE 13-4 Fuse links.

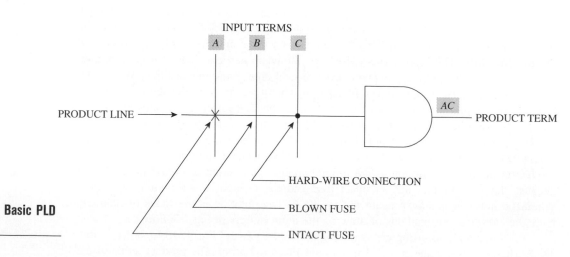

FIGURE 13-5 Basic PLD symbology.

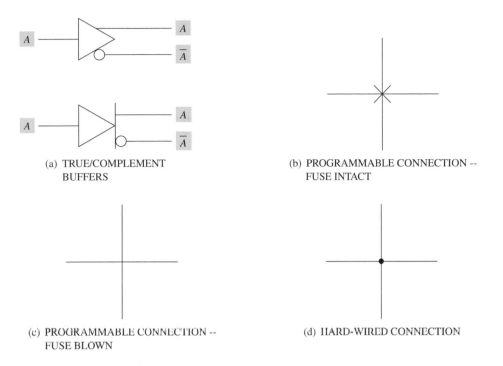

(a) TRUE/COMPLEMENT
   BUFFERS

(b) PROGRAMMABLE CONNECTION --
   FUSE INTACT

(c) PROGRAMMABLE CONNECTION --
   FUSE BLOWN

(d) HARD-WIRED CONNECTION

FIGURE 13–6   PLD notation.

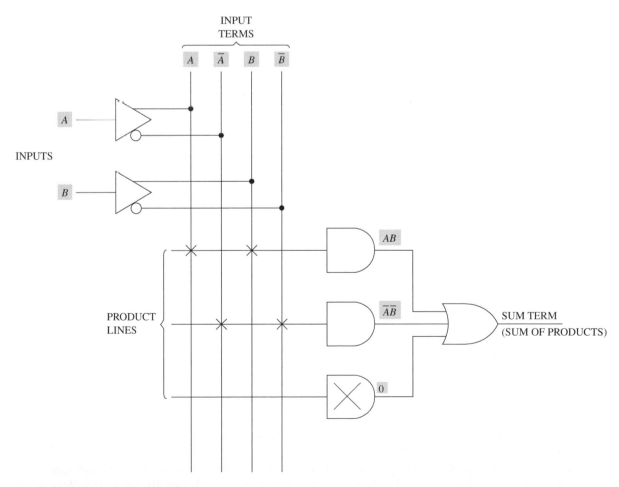

FIGURE 13–7   PLD basic architecture.

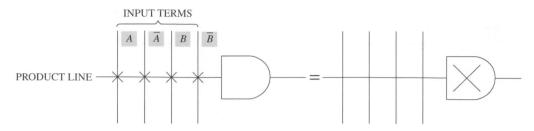

FIGURE 13-8   **Product line: All fuses intact.**

top two AND gates (AB and $\overline{A}\,\overline{B}$) are self-explanatory because the fuses applying these terms to the AND gates are left intact. However, the bottom AND gate contains an X within the symbol. This signifies that all fuse links at the gate's inputs have been left intact. Note that placing the X on the AND gate is exactly the same as placing an X on every programmable input connection to that gate. The equality of this convention is shown in Fig. 13–8. Therefore, this particular product line has no effect on the circuit's output. This is because the inputs to gate X contain two Logic 0s and two Logic 1s from the input buffers. The Logic 0s from the buffers act as inhibitors to the AND gate so its output will always be Logic 0. The Logic 0 functions as an enabler to the OR gate ($A + 0 = A$). The output of the circuit shown in Fig. 13–7 is $AB + \overline{A}\,\overline{B} + 0$, which is equal to $AB + \overline{A}\,\overline{B}$. This expression represents the Exclusive-NOR function.

Think about what would happen to the output of the circuit shown in Fig. 13–7 if the bottom AND gate had all of its input fuse links blown. This action would cause all of the inputs to the gate to be floating. Since these floating inputs act as high inputs, the output of the AND gate would be high, and the OR gate would be inhibited. Thus, the output of the OR gate would be permanently locked to a Logic 1.

Section 13-1: Review Questions

Answers are given at the end of the chapter.

A. Define PLD.
B. What basic logic circuit configuration is used in PLDs?
C. What does an X on a product line in a PLD diagram indicate?
D. What does a dot on a product line in a PLD diagram indicate?
E. What does the lack of an X or dot at the intersection of an input term line and a product line in a PLD diagram indicate?

SECTION 13-2: ARCHITECTURE/SOFTWARE

OBJECTIVES

1. State the three classifications of PLDs.
2. State which array is programmable in each class of PLDs.
3. State how data may be input to program PLDs.
4. Given the part number of a PLD, determine its type, number of inputs/outputs, and configuration.
5. State the meaning of ASIC and list the advantages and disadvantges of ASICs.

There are several programmable logic device architectures currently available. The basic architecture of the PLD employs the **AND/OR (SOP) array.** The programmability of the devices as presented in the following paragraphs determines the type of PLD.

Architecture

The basic architecture of the **Programmable Logic Element (PLE)** is shown in Fig. 13–9(a). This architecture is similar to the programmable read-only memory architecture that was presented in Chapter 12. This architecture consists of an input decoder consisting of AND gates and a programmable OR array on the outputs.

The **Programmable Logic Array (PLA)** architecture is depicted in Fig. 13–9(b). This versatile device offers both a programmable AND array and a programmable OR array.

The basic architecture of the **Programmable Array Logic (PAL)** is shown in Fig. 13–9 (c). This device uses the same AND-OR implementation as the PLE. Note, however, that in the PAL the AND array is programmable and the OR array is fixed. The PAL has been designed with some special features that the PLE does not contain. Some of the outputs of the PAL serve as Input/Output (I/O) ports. This feature allows the outputs to be fed back to the AND-gate array. In addition, the tristate output of the device is controllable by AND logic. Some devices in the PAL family incorporate flip-flops on the outputs. This allows these devices to be configured as counters or registers.

It is easy to see from the foregoing material that the PLDs are classified by the array that is programmable.

1. **PLE:** OR gate array programmable. The PLE is comparable to the PROM technology previously discussed.
2. **PLA:** AND-gate and OR-gate arrays programmable.
3. **PAL:** AND-gate array programmable.

The block diagrams of a PLE, PLA, and PAL in Fig. 13–9 are shown using basic PLD symbology in Fig. 13–10.

Part numbers assigned to PLDs shed light on the specific type of device. Although there are variations, the following format is used by most manufacturers:

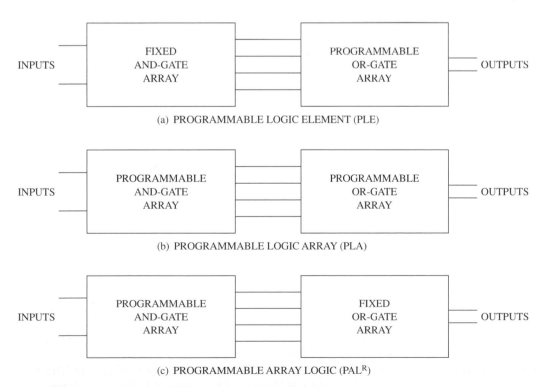

(a) PROGRAMMABLE LOGIC ELEMENT (PLE)

(b) PROGRAMMABLE LOGIC ARRAY (PLA)

(c) PROGRAMMABLE ARRAY LOGIC (PAL^R)

FIGURE 13–9   **Programmable device architecture block diagram.**

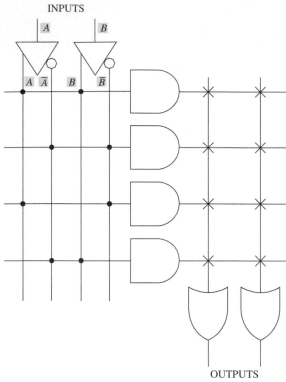

(a) PROGRAMMABLE LOGIC ELEMENT (PLE)

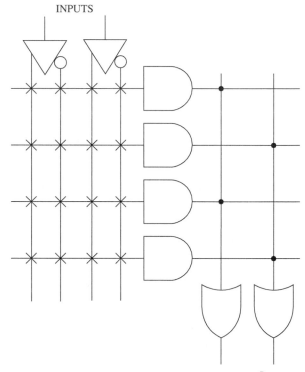

(c) PROGRAMMABLE ARRAY LOGIC (PAL^R)

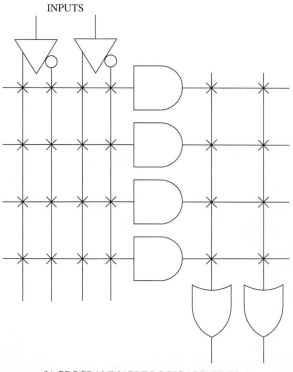

(b) PROGRAMMABLE LOGIC ARRAY (PLA)

FIGURE 13-10 **Programmable logic device logic diagram.**

$$NN_i x NN_o - S$$

where NN_i is the number of inputs
x is the type of configuration
NN_o is the number of outputs
S is the operating speed

An example of a Texas Instruments' (TI) PAL part number is

$$\underset{1}{\underline{TI}}\ \underset{2}{\underline{B}}\ \underset{3}{\underline{PAL}}\ \underset{4}{\underline{20}}\ \underset{5}{\underline{R}}\ \underset{6}{\underline{8}}\ \underset{7}{\underline{-10}}\ \underset{8}{\underline{C}}\ \underset{9}{\underline{N}}$$

1. Manufacturer (Texas Instruments)
2. B: Bipolar; C: CMOS
3. Family/type device
4. Number of array inputs
5. Output configuration
   R: Registered (flip-flops)
   RP: Registered (polarity programmable)
   L: Active-low (no flip-flops)
   V: Versatile (bypassable flip-flops)
   X: X-OR to output
   RX: Registered (X-OR)
6. Number of outputs
7. Speed (ns)
8. Temperature range
   C: Commercial (0°–75°C)
9. Package type

Note that parts 4, 5, 6, and 7 of the TI part number correspond to the basic format used by most manufacturers.

Software

The **software** used to program a PLD uses a Joint Electronic Device Engineering Council (JEDEC) file that represents the desired interconnection fuse map for the device. The JEDEC file is used by the hardware programmer to apply the signals necessary to blow the desired fuses within the PLD.

Design software is written at several different levels. Some programs allow the user to determine which fuses on the interconnection map should be blown and which should remain intact. This low-level fuse-map programming method is very time-consuming and prone to errors. Fortunately, computer software has significantly simplified programming.

More advanced PLD software allows the user to input data via Boolean expressions or truth tables. Logical operators allow AND, OR, NOT, and X-OR operations. Some programs will select the device based on the input design equations—some won't. For devices with active-low outputs, the Boolean expressions input to the PC must be written for active-low outputs. A design example is presented in the next section.

The input equations section of the design software also permits clocked operations, which allow registered outputs on those PLDs with registered configurations.

Schematic capture software allows drawing the desired logic circuit using standard logic symbols on a monitor. The complexity of high-density PLDs and FPGAs requires the use of Computer-Aided Design (CAD) tools. There is an abundance of these support programs commercially available.

A flow chart depicting a CAD system design entry is shown in Fig. 13–11. The design entry might be schematic capture, waveform editor, or hardware description, or it

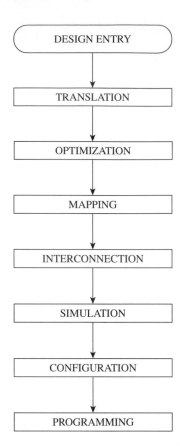

FIGURE 13-11   **CAD system design entry flow chart.**

might be a combination of these and other types. The design entries are translated into Boolean equations and then optimized. The optimization feature simplifies the input equations in much the same manner as you learned to simplify them in Chapter 4. The simplified equations are then mapped to fit the type of logic available in the PLD. Logic is allocated to specific cells, and interconnect wiring is assigned. The design is simulated and mistakes (if applicable) are fixed. Finally, a configuration file is sent to the programmer and the device is programmed. The important part of this is the design software takes care of everything after the initial design entry.

The more advanced software normally includes the optimization feature and final verification of the circuit design. The PLD software translates the Boolean expression, truth table, or schematic inputs in the compiler to a JEDEC file for the hardware programmer.

Application-Specific Integrated Circuit (ASIC)

High-density digital circuit design can be implemented with standard SSI and MSI logic ICs as detailed in the early chapters of this textbook. However, very high-density designs load up board space and are relatively power hungry.

Some designers contract with semiconductor fabrication facilities to manufacture specific high-density circuits on a single IC. These circuits are embedded in **Application-Specific Integrated Circuits (ASICs).**

The ASIC can be thought of as an IC similar to an MROM chip (Section 12–2). However, it is designed specifically for combinational and seqential logic instead of memory. The ASIC, like the MROM, is very expensive and suitable only for high-quantity production. The lead time to design and manufacture ASICs is typically very long.

The ASIC provides numerous advantages over designs utilizing standard SSI and MSI ICs. Board space and power consumption are significantly reduced. In addition, reliability is increased due to the lower number of connections required. The advent of PLDs and FPGAs allows users to design and program their own application-specific circuits and overcome the disadvantages of ASICs.

Section 13–2: Review Questions

Answers are given at the end of the chapter.

A. List the three classifications of PLDs.
B. Which array is programmable in a PLA?
C. Which array is programmable in a PAL?
D. List four methods of inputting data to program a PLD.
E. Determine the number of inputs/outputs and configuration of
    (1) the PAL16R8,
    (2) the PAL20L8, and
    (3) the PAL22V10.
F. State the meaning of ASIC.
G. List the advantages and disadvantages of ASICs.

SECTIONS 13–1 AND 13–2: INTERNAL SUMMARY

Programmable logic devices (PLDs) have revolutionized digital circuit design. The days of implementing circuits with SSI glue chips are gone. Advances in technology have overcome the few drawbacks of early PLDs and have made the devices the product of choice for most designers.

PLD technology, like ROM technology, is available as one-time programmable or reprogrammable. **EPLDs** and **E^{2}PLDs** use the same erasure techniques as their ROM predecessors. The density and versatility of these PLDs have produced their popularity.

The industry standard for PLD notation is depicted in Fig. 13–6. PLDs are classified by their programmable array. (1) **PLE:** fixed AND-gate array and programmable OR-gate array; (2) **PLA:** programmable AND- and OR-gate arrays; (3) **PAL:** programmable AND-gate array and fixed OR-gate array.

Numerous software programs are commercially available to program PLDs. Once the design entry is complete, the remaining parts of the programming procedure are transparent to the user.

SECTIONS 13–1 AND 13–2: INTERNAL SUMMARY QUESTIONS

Answers are given at the end of the chapter.

1. The basic architecture of a PLD is _____ .
    a. sum of sums   c. product of sums
    b. sum of products   d. product of products

2. What does the X at the intersection of the *A* input term line and the product line in Fig. 13–12 indicate?
    a. Blown fuse   c. Floating input
    b. Intact fuse   d. Hard-wired connection

3. What does the "dot" at the intersection of the *C* input term line and the product line in Fig. 13–12 indicate?
    a. Blown fuse   c. Floating input
    b. Intact fuse   d. Hard-wired connection

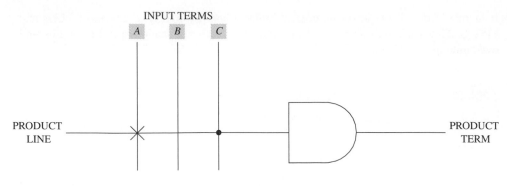

FIGURE   13-12

4. A programmable array logic (PAL) PLD contains _____.
   a. a fixed AND-gate array and a programmable OR-gate array
   b. a programmable AND-gate array and a fixed OR-gate array
   c. programmable AND- and OR-gate arrays

5. A programmable logic array (PLA) PLD contains _____.
   a. a fixed AND-gate array and a programmable OR-gate array
   b. a programmable AND-gate array and a fixed OR-gate array
   c. programmable AND- and OR-gate arrays

6. The PLD in Fig. 13–13 is classified as a _____.
   a. PAL
   b. PLA
   c. PLE

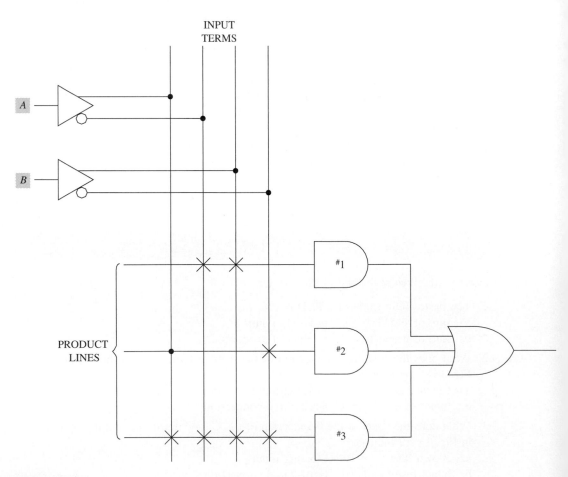

FIGURE 13-13

7. What is the output of the #3 AND gate in Fig. 13–13?
   a. 0
   b. 1
   c. $\overline{A}B$
   d. $\overline{A}\,\overline{B}$

8. When is the output of the #1 AND gate in Fig. 13–13 high?
   a. AB
   b. $A\overline{B}$
   c. $\overline{A}\,\overline{B}$
   d. $\overline{A}B$

9. What is the output sum term from the PLD in Fig. 13–13?
   a. 0
   b. 1
   c. $\overline{A}B + A\overline{B}$
   d. $\overline{A}\,\overline{B} + AB$

10. If all of the fuse links on the #3 AND gate product line in Fig. 13–13 are blown, what is the output sum term from the PLD?
    a. 0
    b. 1
    c. $\overline{A}B + A\overline{B}$
    d. $\overline{A}\,\overline{B} + AB$

11. What is the output of the PAL device in Fig. 13–14?
    a. 0
    b. 1
    c. $\overline{A}\,\overline{B}\,\overline{C} + \overline{A}BC + ABC$
    d. $\overline{A}\,\overline{B}\,\overline{C} + A\overline{B}\,\overline{C} + ABC$

12. What is the output expression of the PAL device in Fig. 13–15?
    a. $\overline{A}B + A\overline{B}$
    b. $\overline{A}B\overline{C} + A\overline{B}\overline{C} + \overline{A}\,\overline{B}\,\overline{C}$
    c. $\overline{A}BC + A\overline{B}C + ABC$
    d. $\overline{A}\,\overline{B}\,\overline{C} + \overline{A}BC + A\overline{B}\,\overline{C}$

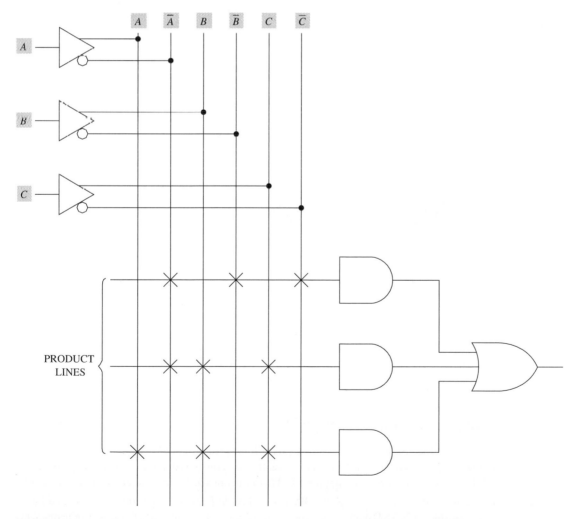

FIGURE 13–14

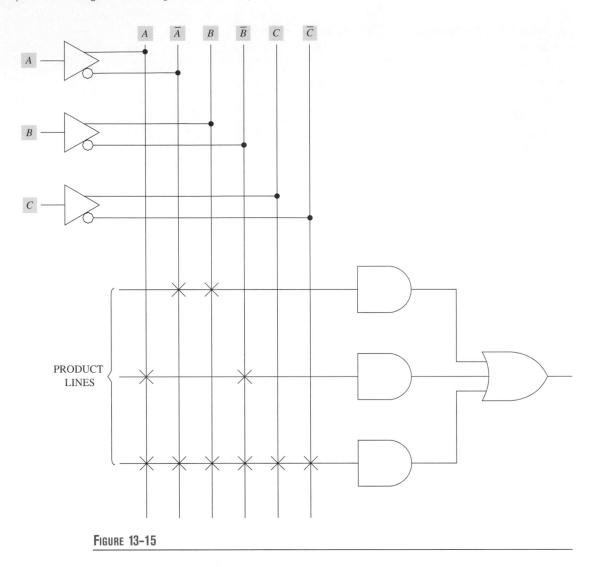

FIGURE 13–15

SECTION 13-3: PROGRAMMABLE LOGIC DEVICES

OBJECTIVES

1. Given the block diagram of a PLD, determine the number of inputs and outputs of the device and its configuration.
2. Given a simplified logic diagram of an output logic macrocell or I/O architecture control unit, determine the type of feedback and output.
3. State the purpose of each main section of a Lattice PLD megablock.
4. Given a generic logic block diagram of a Lattice 1000 family PLD, determine the Q outputs produced by various input combinations.

Programmable Logic Array (PLA)

Programmable Logic Array devices contain both programmable AND-gate and programmable OR-gate arrays. PLA devices actually employ wired logic. The concept of wire-ANDing was presented in Chapter 11. This concept can be expanded to include the OR array. Remember $A + B = \overline{\overline{A}\,\overline{B}}$. Other variations of array structures are also available. For example, a NAND-gate array can be used and its outputs fed back to the array's in-

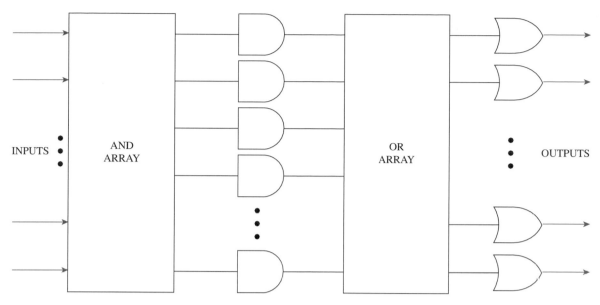

F<small>IGURE</small> 13-16 **Programmable Logic Array (PLA)–logic diagram.**

puts to produce a SOP expression as discussed in Chapter 4. This type of device employs what is sometimes referred to as **folded logic.** Figure 13–16 presents a PLA logic diagram without drawing all of the input and product lines.

Early PLAs inherently had slow operating speeds due to significant propagation delays. This, coupled with high production costs, led manufacturers to design and produce the more popular Programmable Array Logic (PAL) devices. This inexpensive descendant group of PLDs provides input-to-output propagation delays as low as 5 ns. However, technological improvements have revived the popularity of PLAs to some degree. This is especially true when wide OR gates are necessary.

The variety of devices presented in this section is presented to show the different architectures employed by industry to achieve programmable logic. The devices presented show how the addition of flip-flops to the basic SOP programmable devices provides sequential logic capability. All of the PLDs in this section allow implementation of complex circuit designs with minimum effort.

It should be kept in mind the details in this section regarding PLDs are provided only to give a better understanding of device operation. Array fuses are blown, architecture control bits are set, or antifuses are programmed by the hardware programmer/compiler under software control. Implementation of a circuit is virtually transparent to the designer/user.

Programmable Array Logic (PAL^R)

A PAL logic diagram is shown in Fig. 13–17. The programmable wide-input AND-gate array feeds a narrow, fixed OR-gate array. Many of these devices have registered outputs as shown in Fig. 13–18. Most of the PAL devices also incorporate tristate outputs. The type of architecture employed in these PLDs is typically termed **logic rich** due to the high ratio of logic gates to flip-flops.

TIBPAL16L8.

A Texas Instrument PAL block diagram is shown in Fig. 13–19. The TIBPAL16L8 is a bipolar, 16-input PAL with 8 active-low outputs. This PAL produces active-low outputs because of the tristate inverters at the output. The overall layout scheme of the PAL16L8 is AND-OR-NOT. Details of the PAL's programmability and tristate outputs are shown in the logic diagram in Fig. 13–20.

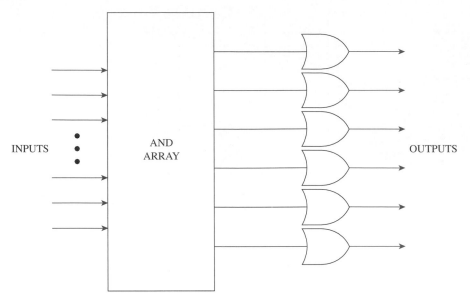

FIGURE 13-17 **Programmable Array Logic (PAL)–logic diagram.**

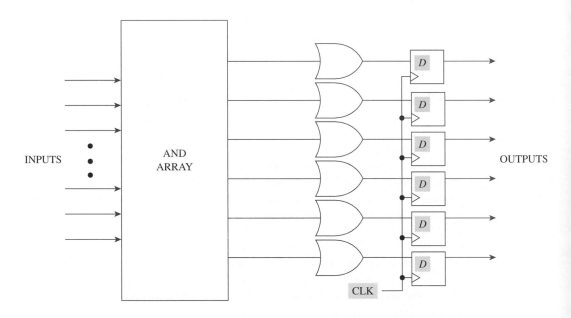

FIGURE 13-18 **Programmable Array Logic (PAL) with registered outputs–logic diagram.**

The PAL16L8 has 10 dedicated input lines and 6 I/O ports. The I/O ports allow feedback into the AND-gate array, or they can be used as input pins. The inputs to the AND-gate array are buffered as indicated by the ▷ symbol on the block diagram in Fig. 13–19. The block diagram also shows there are 32 input lines (16 true inputs and 16 complement inputs provided by the true/complement buffers) and 64 product lines in the AND-gate array (32 × 64). This means each AND gate has 32 inputs. The product terms are obtained by blowing fuses or using some comparable technology in the AND-gate array. The top input line (EN) to each OR gate block in Fig. 13–19 actually bypasses the OR gate and is used to control the 3-state output of the PLD. If this EN input is high, the output buffers are enabled. The remaining 7 input lines are product terms input to the 7-input OR gate.

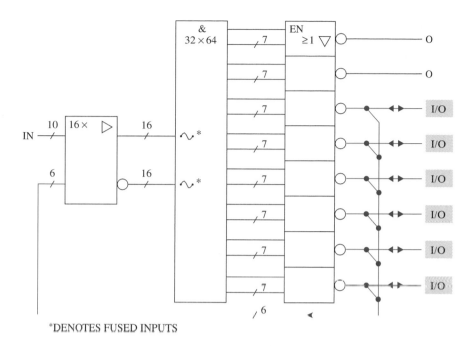

*DENOTES FUSED INPUTS

FIGURE 13-19   **TIBPAL16L8 block diagram.** *Reprinted by Permission of Texas Instruments.*

logic diagram (positive logic)

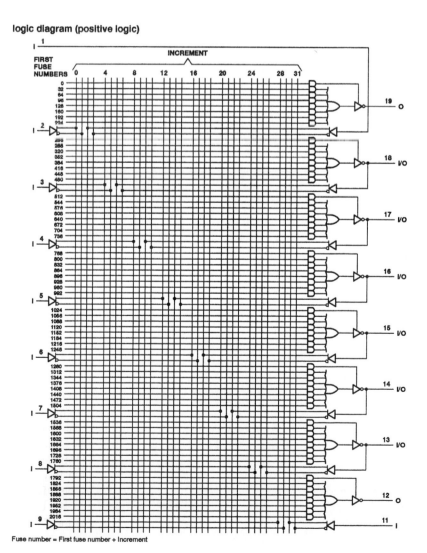

FIGURE 13-20
**TIBPAL16L8 logic
diagram.** *Reprinted by
Permission of Texas
Instruments.*

The PAL logic diagram (Fig. 13–20) uses a different convention than that previously discussed for blown/intact fuses. No Xs are marked on the diagram to represent intact fuses. This modified convention allows the circuit designer to mark the diagram wherever a fuse is to be left intact.

If all fuses on product line 0 are left intact, the Logic 0s from the input buffers inhibit the top AND gate and its low output disables the tristate inverter. This causes the output at pin 19 to be floating (Hi-Z state). This is true of the other PAL outputs if all fuses on product lines 8, 16, 24, 32, 40, 48, and 56 are left intact. Leaving these fuses intact may seem like an undesirable condition since the buffer outputs will remain floating. However, the tristate buffer's outputs must be in the Hi-Z state if I/O pins 13 through 18 are to be used as input pins. Notice even when the I/O pins are used as output pins, the output signal is still fed back into the AND-gate array. If these outputs are programmed back into the circuit, their use is called **folded logic.**

No attempt is made in this text to make you proficient in programming PLDs. There are numerous commercial software packages available to accomplish this task. The programming information provided in this chapter is usually based on general concepts instead of specifics. One exception follows.

IN DEPTH Look at Programming

The following PLD design example uses Boolean equations as the input tool to program a PAL16L8. The design data is entered on a PC keyboard after the design software has been installed. A typical design entry might consists of the following sections:

1. **Documentation Section:** This section should contain the title, author, and date as a minimum.
2. **Declaration Section:** This section starts with the keyword chip. The section should contain the chip name and the type of PLD. The chip name should indicate the type of circuit or design or provide some other indication of the specific PLD use. The next entry is the PLD designation/part number, for example,

   chip EXP 1 PAL16L8

The last entry in the declaration section should contain the pin list. Pin numbers for the TIBPAL16L8 are shown in Fig. 13–21.

The logic functions (gates) that will be implemented in the PLD (Experiment 1) are shown in Fig. 13–22 along with the active-low output equations. Active low is designated by the overbar (NOT sign) in the figure.

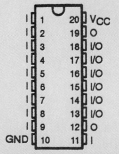

FIGURE 13-21   TIBPAL16L8 pinout diagram. *Reprinted by Permission of Texas Instruments.*

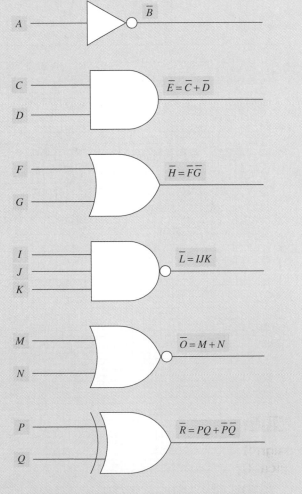

FIGURE 13-22   Logic functions–active-low output expressions.

The first name entered in the pin list shown here is for pin 1, the second name is for pin 2, and so on. The actual PLD pin numbers are included above the pin list names as a comment. Comments begin with a semicolon, extend to the end of the physical line, and may be used anywhere in the input file. The letters Ai, Bo, Ci, and so on are arbitrary. The "i" represents inputs, and the "o" represents outputs. This circuit is used to implement logic functions only.

```
; 1   2   3   4   5   6   7   8   9   10
  Ai  Ci  Di  Fi  Gi  Ii  Ji  Ki  Mi  GND

; 11  12  13  14  15  16  17  18  19  20
  Ni  Ro  Pi  Qi  Oo  Lo  Ho  Eo  Bo  VCC
```

Entries for subexpressions that appear more than one time in the PLD design can be entered in a simplified manner by using **@define**. The format for this entry is

@define LABEL "SUB-EXPRESSION"

An actual entry is shown below. The quotation marks are necessary delimiters in the entry.

@define T1 "AB + C"

3. **Equations Section:** This section starts with the key word *equations*. The Boolean equations for the desired PLD circuit are input in this section. The logical operator symbols used to input the equations in this design example are shown in Table 13–1.

The Boolean equation format is shown below. Keep in mind the equations must be input for active-low outputs for the PAL16L8. (Refer to Fig. 13–22 if necessary.)

SIGNAL-NAME ASSIGNMENT-OPERATOR
LOGIC-EXPRESSION

The **SIGNAL-NAME** is the identifier declared in the pin list (Ai, Bo, etc.). The **ASSIGNMENT-OPERATOR** must be = or :=. The = operator represents unclocked operation and is used for combinational logic circuits. The := operator is used for clocked operation for registered outputs. The **LOGIC-EXPRESSION** is an SOP combination of identifiers and logical operators.

An input file used to program the logic functions of Fig. 13–22 into a PAL16L8 is shown below:

```
TITLE       PLD Experiment 1
AUTHOR      B. Thompson
DATE        1/11/96
chip        EXP 1          PAL16L8
```

```
; 1   2   3   4   5   6   7   8   9   10
  Ai  Ci  Di  Fi  Gi  Ii  Ji  Ki  Mi  GND

; 11  12  13  14  15  16  17  18  19  20
  Ni  Ro  Pi  Qi  Oo  Lo  Ho  Eo  Bo  VCC

@define   T1   "Pi * Qi"    ; PRODUCT TERM
@define   T2   "/Pi * /Qi"  ; PRODUCT TERM

equations

/Bo = Ai              ; INVERTER
/Eo = /Ci + /Di       ; AND GATE
/Ho = /Fi * /Gi       ; OR GATE
/Lo = Ii * Ji * Ki    ; NAND GATE (PRODUCT TERM)
/Oo = Mi + Ni         ; NOR GATE (SUM OF INPUTS)
/Ro = T1 + T2         ; X-OR (SUM-OF-PRODUCTS)
;           END OF INPUT FILE EXP 1
```

The design software should produce two output files—XXXX.DOC and XXXX.JED. The **.DOC** file describes the device pinout along with the names assigned to each pin. The **.JED** file is used by the hardware programmer to "burn" the chip.

Functional equations must be used to control the tristate outputs of the PAL16L8. Functional equations are input in the following format:

SIGNAL-NAME.FUNCTION = LOGIC-EXPRESSION

The **SIGNAL-NAME** is the identifier declared in the pin list, and the **LOGIC-EXPRESSION** is the same as described in the first PLD experiment. The **.FUNCTION** can be used in several ways. The tristate function is our concern in this example. The keyword is *trst* for control of this function. An example of its use is:

OUT1.trst = Pin1 * /Pin2

An input file for PLD Experiment 2 is shown below. The file is similar to Experiment 1 except the tristate outputs of Eo, Ho, Lo, and Oo are defined and controlled by the functional equations.

```
TITLE       PLD Experiment 2
AUTHOR      B. Thompson
DATE        1/11/96
chip        EXP 2          PAL16L8
```

```
; 1   2   3   4   5   6   7   8   9   10
  OE  Ci  Di  Fi  Gi  Ii  Ji  Ki  Mi  GND

; 11  12  13  14  15  16  17  18  19  20
  Ni  Ro  Pi  Qi  Oo  Lo  Ho  Eo  Bo  VCC
```

TABLE 13–1	Symbol	Operation	Example Use	Precedence
Design Software— Logical Operators	/	not	/A	4 (Highest)
	*	and	A * B	3
	+	or	A + B	2
	:+:	XOR	A :+: B	1 (Lowest)

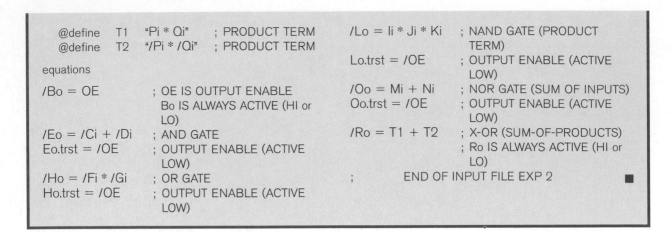

```
    @define  T1   "Pi * Qi"    ; PRODUCT TERM        /Lo = Ii * Ji * Ki    ; NAND GATE (PRODUCT
    @define  T2   "/Pi * /Qi"  ; PRODUCT TERM                                   TERM)
                                                     Lo.trst = /OE        ; OUTPUT ENABLE (ACTIVE
  equations                                                                     LOW)
  /Bo = OE          ; OE IS OUTPUT ENABLE            /Oo = Mi + Ni        ; NOR GATE (SUM OF INPUTS)
                      Bo IS ALWAYS ACTIVE (HI or     Oo.trst = /OE        ; OUTPUT ENABLE (ACTIVE
                      LO)                                                        LOW)
  /Eo = /Ci + /Di   ; AND GATE                       /Ro = T1 + T2        ; X-OR (SUM-OF-PRODUCTS)
  Eo.trst = /OE     ; OUTPUT ENABLE (ACTIVE                               ; Ro IS ALWAYS ACTIVE (HI or
                      LOW)                                                       LO)
  /Ho = /Fi * /Gi   ; OR GATE                        ;          END OF INPUT FILE EXP 2        ■
  Ho.trst = /OE     ; OUTPUT ENABLE (ACTIVE
                      LOW)
```

TIBPAL16R4.

The PAL16L8 just discussed is a combinational logic PLD because it contains no flip-flops. A block diagram of a PAL with registered outputs is shown in Fig. 13–23. The term **registered outputs** is used because a group of flip-flops that is used to transfer or store data is a register, and this PLD contains the necessary flip-flops to form a register. The TIBPAL16R4 is a bipolar, 16-input PAL with 4 D-type flip-flop registered outputs. This information is provided by the part number (16R4) and can be verified by the block diagram.

This PLD has 8 dedicated inputs and 4 I/O ports (pins), and the Q outputs may also be fed back to the AND-gate array. There are 32 input lines and 64 product lines in the AND array. This PLD has an enable input to each of the 4 OR-gate blocks in Fig. 13–23 to control the tristate output of the I/O ports. In addition, it has an Output Enable ($\overline{OE}$) input that must be activated to allow the Q output to be available. A clock (CLK) input allows clocking of the four D-type flip-flops in the PLD. The $\overline{OE}$ and CLK inputs are shown as inputs to the D-type flip-flops—see the flip-flop common control box in the figure. The I = 0 general qualifying symbol indicates the flip-flops power up cleared to

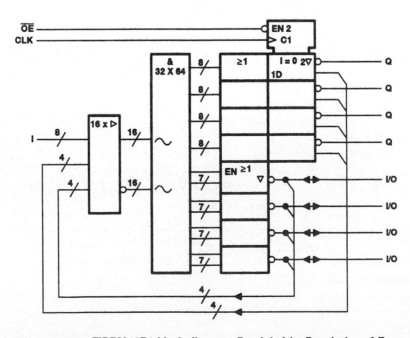

FIGURE 13–23 **TIBPAL16R4 block diagram.** *Reprinted by Permission of Texas Instruments.*

logic diagram (positive logic)

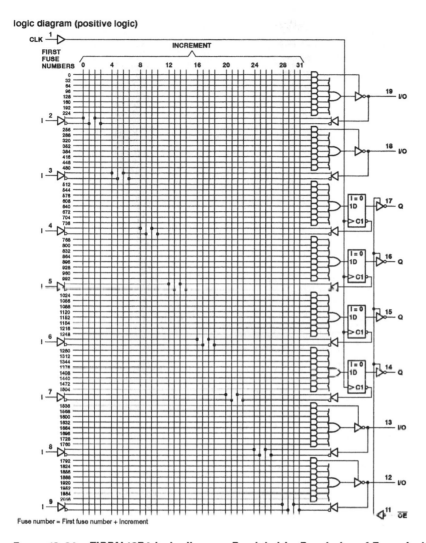

Fuse number = First fuse number + Increment

FIGURE 13–24   **TIBPAL16R4 logic diagram.** *Reprinted by Permission of Texas Instruments.*

the 0 state. The PLD Q outputs are synchronous because all 4 flip-flops are clocked simultaneously. Figure 13–24 shows the logic diagram of a TIBPAL16R4. The D inputs to the four flip-flops are fed by four OR gates in the OR array.

Figure 13–25 presents a view of one section of the PAL16R4. Notice the top AND gate is used to control the tristate inverter at the Q output. Like the PAL16L8, the outputs are active low because of the tristate inverter. The $\overline{Q}$ output of the flip-flop is fed back into the AND-gate array through a true/complement buffer. This folded logic arrangement allows sequential logic circuits such as counters and registers to be implemented when registered-output PLDs are used. This is not possible with combinational logic PLDs that do not contain flip-flops.

Programming sequential circuits is more complex than programming combinational logic circuits due to the timing and storage considerations. One common method of design is to use a **transition table,** which shows the circuit's present state (preclock—usually called Q^n) and the desired next state (post clock—usually called Q^{n+1}). The transition table for synchronous counter design was discussed in Section 7–6. The table incorporates the required level(s) of the D or J and K inputs necessary to achieve the next state condition (Q^{n+1}). The D or J and K data from the transition table are transferred to K-maps and simplified. The simplified expressions determine the type of gating circuits required to implement the sequential circuit.

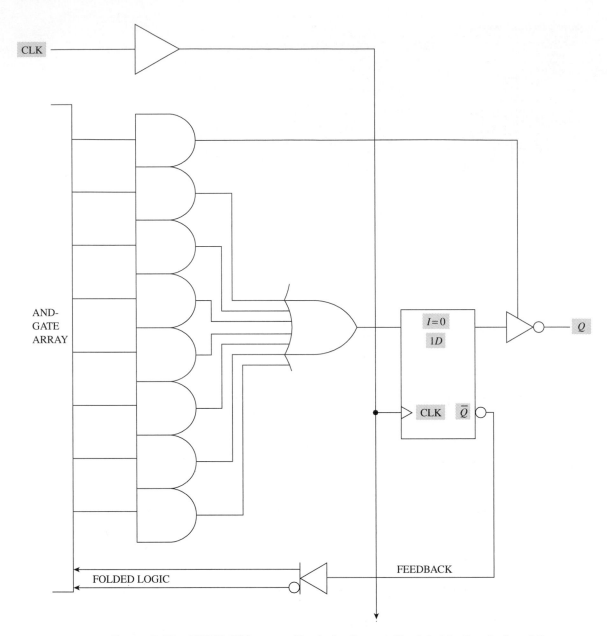

FIGURE 13–25   **TIBPAL16R4 one-section logic diagram.** *Reprinted by Permission of Texas Instruments.*

Another comparable version of the 16R4 is the 16R8. The 16R8 is available from many vendors. They are available with fuse technology and in E²PROM versions. The PAL16R8 has 8 dedicated inputs and 8 dedicated outputs.

TIBPAL22V10.

This PLD is implemented with the basic AND-OR structure of the two PALs previously presented. The PAL22V10 can have up to 22 inputs and 10 outputs. In addition to its basic structure, the 22V10 contains a programmable **output logic macrocell (OLM).** This PLD incorporates the capability to define and program the architecture of each output individually. The outputs are enabled by individual product terms; they may be registered or non-registered; and they may be in true or inverted form. All of these extra features in this versatile (V) PLD are produced by the OLM, which is shown in Fig. 13–26. The macro-

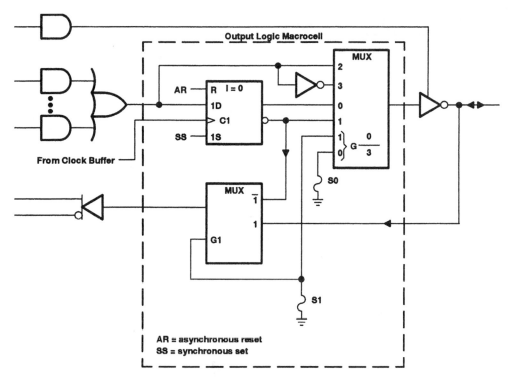

FIGURE 13-26   Output Logic Macrocell (OLM). *Reprinted by Permission of Texas Instruments.*

cell presented here is Texas Instruments' design of the OLM. There are different implementations used by other manufacturers, but the results are similar.

The OLM can be programmed to produce registered or I/O feedback and registered or combinational outputs. The tristate outputs can be in true form (active high) or inverted form (active low). The clock polarity can be selected via the program, so the flip-flops can be clocked on the NGT or the PGT of the input clock signal. In addition, each macrocell flip-flop can be synchronously SET (SS) or asynchronously RESET (AR) by an AND array product term that is written during programming.

The OLM logic diagram of Fig. 13–26 has been simplified in Fig. 13–27. The routing of signals through the OLM is shown for four different configurations as programmed by the macrocell's fuses (S_0 and S_1). A function table (Table 13–2) lists the feedback and output functions for these four OLM configurations. Macrocell components not selected for use in a specific configuration have been omitted in Fig. 13–27 to simplify the illustrations.

There are two programmable fuses (S_0 and S_1) in the macrocell that control the type of feedback and output of the PLD. Table 13–2 indicates an unblown fuse = 0 and a blown fuse = 1. The four different configurations are implemented through the use of multiplexers. Multiplexing (Section 10–3) is nothing more than selecting data, and the selection in the OLM multiplexers is controlled by the programmable fuses. The fuse-controlled configurations are discussed next.

$S_1 = 0$ *and* $S_0 = 0$.   This input combination produces registered feedback and a registered, active-low output. The circuit configuration is shown in Fig. 13–27(a). Both select fuses are left intact (unblown) by the program to achieve this configuration.

The output of the OR array is fed directly to the data input of the flip-flop. The Q output of the flip-flop is applied to data input 0 of MUX #1. *Note:* MUX #1 is a 4-line to 1-line multiplexer. Input data are also applied to data inputs 1, 2, and 3 of MUX #1 as shown in the complete OLM diagram in Fig. 13–26. The signal connections for inputs 1,

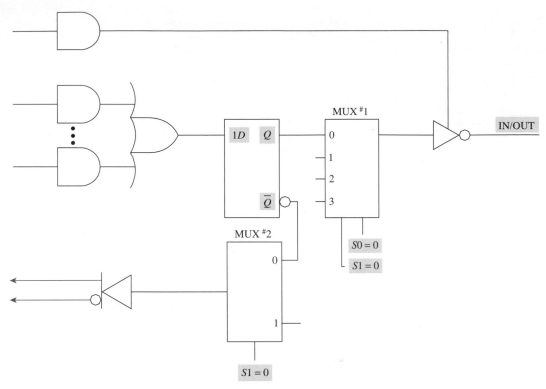

(a) S1 = 0/S0 = 0: REGISTERED FEEDBACK AND REGISTERED/INVERTED (ACTIVE-LOW) OUTPUT

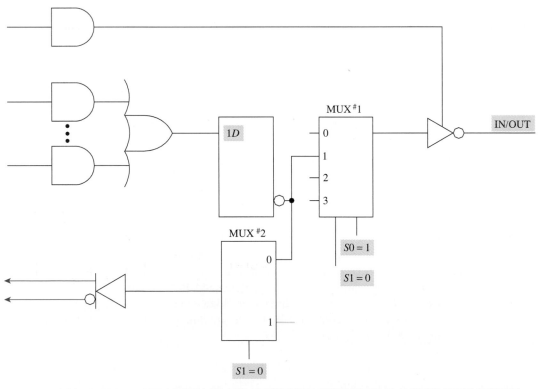

(b) S1 = 0/S0 = 1: REGISTERED FEEDBACK AND REGISTERED/NONINVERTED (ACTIVE-HIGH) OUTPUT

FIGURE 13-27 **Programmed OLM–simplified logic diagram.**

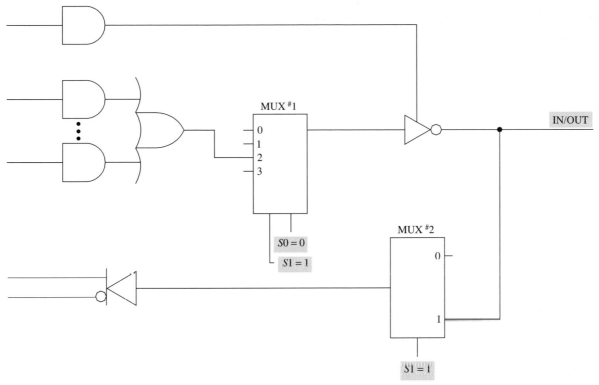

(c) $S1 = 1/S0 = 0$: I/O FEEDBACK AND COMBINATIONAL/INVERTED (ACTIVE-LOW) OUTPUT

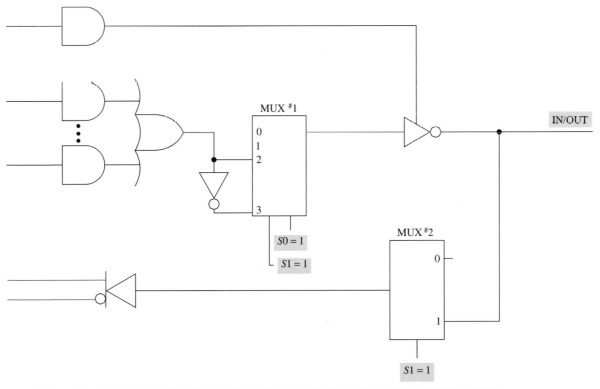

(d) $S1 = 1/S0 = 1$: I/O FEEDBACK AND COMBINATIONAL/NONINVERTED (ACTIVE-HIGH) OUTPUT

FIGURE 13–27   **Continued**

TABLE 13-2	Fuse Select		Data Select		Feedback	Output	
Output Logic Macrocell Function Table	S_1	S_0	MUX #1	MUX #2	Type	Type	Active Level
	0	0	0	0	Register	Registered	Low
	0	1	1	0	Register	Registered	High
	1	0	2	1	I/O	Combinational	Low
	1	1	3	1	I/O	Combinational	High

Unblown fuse = 0; blown fuse = 1.

2, and 3 are not shown in the simplified diagram of Fig. 13–27(a) because these inputs are not selected.

The $S_1 = 0$ and $S_0 = 0$ fuse inputs select the data 0 input and route it to the output of MUX #1. This signal is fed through the tristate inverter when the inverter is enabled by a product term from the AND array. The PLD output is active low due to the inversion produced by the tristate gate. Since the output is taken from the D-type flip-flop, it is a **registered output.**

The $\overline{Q}$ output of the D-type flip-flop is connected to data input 0 of MUX #2. MUX #2 is a 2-line to 1-line MUX. The data at input 0 is selected and routed to the output of the MUX because $S_1 = 0$. Again, the connection of the output data to input 1 of MUX #2 is not shown for simplicity because it is not selected. The feedback to the AND array is taken from the true/complement buffer. This feedback is registered because it is taken from the flip-flop.

$S_1 = 0$ and $S_0 = 1$.   The macrocell's configuration for this fuse combination is shown in Fig. 13–27(b). This configuration produces registered feedback and a registered, active-high output. The only difference in configuration is MUX #1 selects the data 1 input and routes that data to the output. The data 1 input is selected because S_0 is blown and S_1 is left intact during programming. The selected data (MUX #1 input 1) is taken from the $\overline{Q}$ output of the D-type flip-flop. In the first example (Fig. 13–27a), the Q output of the flip-flop was selected. In this example, the $\overline{Q}$ data is selected and then reinverted to its true (active-high) level by the tristate inverter. The $\overline{Q}$ output of the flip-flop is also selected by MUX #2 to produce registered feedback as it was in the previous configuration.

$S_1 = 1$ and $S_0 = 0$.   This input combination is produced when S_1 is blown and S_0 is left intact during programming of the PLD. The circuit configuration is shown in Fig. 13–27(c). This macrocell configuration produces I/O feedback and an active-low, combinational logic output. The output is nonregistered because the macrocell flip-flop is not used.

The output of the OR array, as shown in the macrocell in Fig. 13–26, is connected to pin 2 and through an inverter to pin 3 of MUX #1. Only the data 2 input is shown in Fig. 13–27(c) because the fuse conditions select this input. Therefore, the flip-flop is bypassed in this configuration. The nonregistered combinational output is inverted by the tristate inverter to produce an active-low output when the gate is enabled.

MUX #2 is now set to select data input 1 because fuse S_1 was blown during programming. This allows two possibilities at the I/O port. The output data is fed back to the AND array when the tristate inverter is enabled—folded logic available. The I/O port can be used as an input pin when the tristate inverter is in the Hi-Z state.

$S_1 = 1$ and $S_0 = 1$.   This input combination results when both S_1 and S_0 are blown. The circuit configuration is shown in Fig. 13–27(d). The configuration produces I/O feedback and an active-high, nonregistered, combinational logic output. MUX #1 is set to select the data 3 input. The output of the OR array is inverted prior to its application to input 3 of MUX #1. This inverted data is selected, routed to the MUX output, and reinverted to its true (active-high) level by the tristate inverter.

functional block diagram (positive logic)

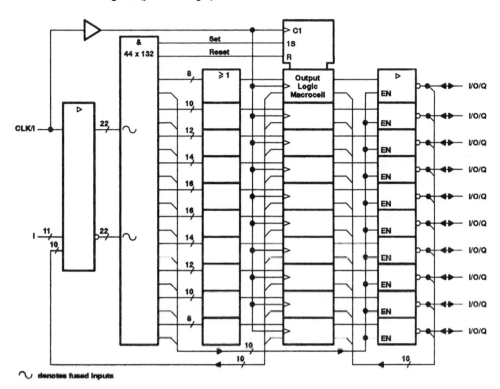

FIGURE 13–28   **TIBPAL22V10 block diagram. *Reprinted by Permission of Texas Instruments.***

MUX #2 operates as described in the preceding example. Its operation makes the I/O port available for an input or allows feedback to the AND array.

The block diagram of the TIBPAL22V10 is shown in Fig. 13–28, and the logic diagram is shown in Fig. 13–29. Compare the block diagram of the 22V10 to that of the 16R4 in Fig. 13–23. The comparison reveals much similarity between the two PLDs. The primary difference is the addition of the OLMs in the 22V10. The logic diagram also reveals the similarity of the two devices. PAL22V10s are available from many different manufacturers in both one-time programmable and reprogrammable versions.

Some 22V10 device macrocells are implemented as shown in Fig. 13–30 on page 766. The multiplexers are typically drawn in the manner shown in the figure for PLD and FPGA MUXs. In many diagrams the select input is not shown since it is under program control.

The output of the OR gate to the macrocell in Fig. 13–30 may be inverted or left in its true form by the X-OR gate. The logic of the X-OR gate as discussed in Chapter 5 is "complementary inputs = 1 out." Thus, the data output of the X-OR gate is inverted if the MUX output is high as illustrated in Fig. 13–31(a). The data output of the X-OR gate is left at its true level if the MUX output is low as shown in Fig. 13–31(b).

Security Bit.   Many PLDs have a **programmable design security bit** that prevents copying or examination of the data stored in the devices. The fuse pattern within the device cannot be copied when the security fuse is blown. Blowing this fuse disables the verification circuitry, and all PLD fuses appear to be open once the fuse has been blown. This is a valuable asset to many designers in this era of reverse engineering.

EP630 Series PLD.

The EP630 series is a 16-macrocell, one-time programmable PLD available from Texas Instruments. The device is capable of implementing over 600 equivalent logic gates in the

logic symbol (positive logic)

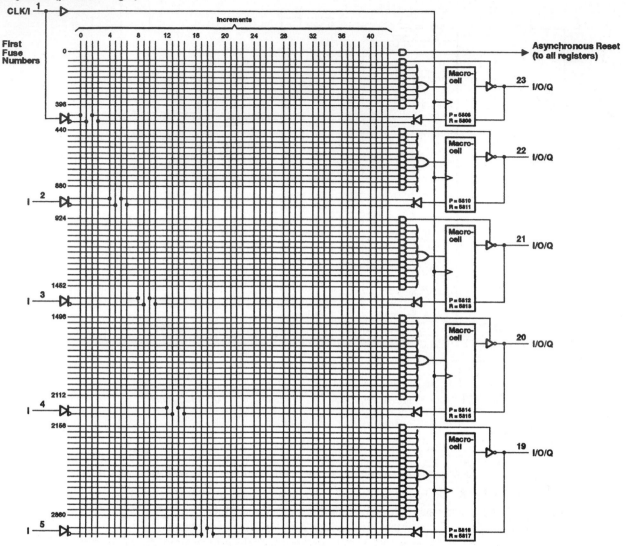

FIGURE 13–29 **TIBPAL22V10 logic diagram.** *Reprinted by Permission of Texas Instruments.*

SOP configuration. The PLD can also produce registered logic functions for sequential operation.

An additional feature of this PLD is the ability to program the flip-flop in the macrocell for D, J-K, S-R, or T (toggle) operation individually. The clock inputs to the flip-flops can be applied separately from any of the input or feedback paths available in the AND array.

The EP630 series PLD has four dedicated input pins and sixteen I/O pins as shown in the block diagram in Fig. 13–32. This allows for up to 40 inputs to the AND array. A **logical false state** is produced at the output of the AND array when both the true and complement forms of an input are applied to the AND array. The input is a don't care to the AND array when the true and complement inputs are both open. A **logical true state** is produced at the output of the AND array when all inputs for the product term are programmed open.

Figure 13–32 also reveals there are two dedicated clock input signals—CLK1 and CLK2. Each of these CLK signals controls a bank of eight flip-flops. The OE/CLK se-

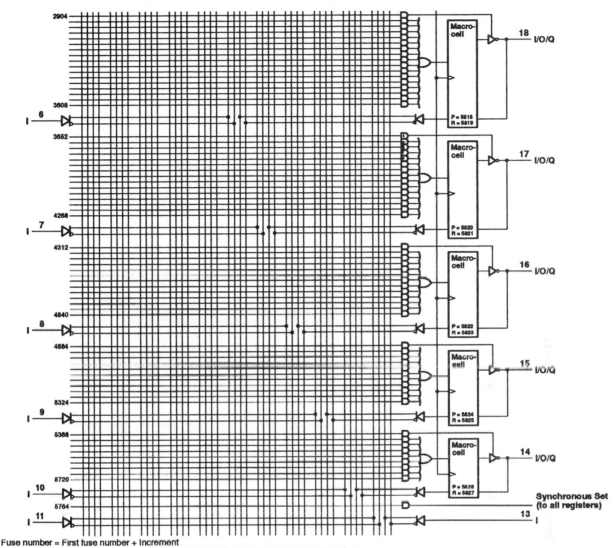

Fuse number = First fuse number + Increment
Inside each MACROCELL the "P" fuse is the polarity fuse and the "R" fuse is the register fuse.

FIGURE 13-29   Continued. *Reprinted by Permission of Texas Instruments.*

lect multiplexer of the EP630 series PLDs is shown in Fig. 13–33. The switches shown are program controlled as indicated by the EPROM MOSFET in the circle directly beneath the switch block.

MODE 0 operation, shown in Fig. 13–33(a), allows the flip-flop to be clocked only by a dedicated clock input—CLK1 or CLK2—which produces banked register clocking. The selected clock signal is common to seven other macrocells for synchronous operation. The tristate output buffer in the figure is controlled by a product (P)–term from the AND array.

Figure 13–33(b) illustrates the I/O architecture for **MODE 1 operation.** This configuration allows for gated clock input signals, which are controlled by a P-term from the AND array. Since the AND array has available both true and complement signals, clocking of the flip-flop in the macrocell may be accomplished on the PGT or NGT of a signal. The tristate buffer is always enabled in MODE 1 operation because the buffer's active-high enable input is connected to V_{CC}.

Figure 13–34 illustrates five I/O configurations of the EP630 series PLDs. Each macrocell may be independently configured during programming of the device. Figure 13–34(a)

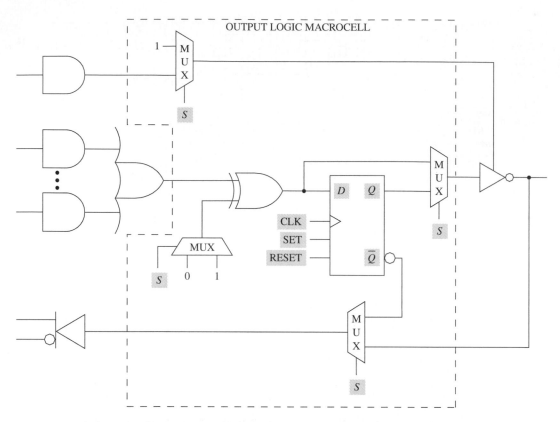

FIGURE 13–30   Output logic macrocell. *Reprinted by Permission of Texas Instruments.*

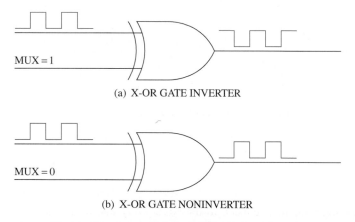

(a) X-OR GATE INVERTER

(b) X-OR GATE NONINVERTER

FIGURE 13–31   Macrocell polarity controller.

shows the basic I/O configuration for a **combinational logic output.** The components not selected for use in the I/O configurations discussed here are not shown in the figures. Note the flip-flop is not shown in the first configuration because the output is combinational, nonregistered logic. The I/O selection table in the figure indicates the combinational logic signal is available at the I/O pin in true (active-high) or inverted (active-low) form. The I/O pin may be used as an input pin when OE = 0 because the output of the tristate buffer is floating. In this case, the I/O Feedback Select switch must be closed.

D-type flip-flop configuration is displayed in Fig. 13–34(b). The flip-flop has an asynchronous RESET (R) input from a dedicated P-term in the AND array. All flip-flops in the EP630 are cleared during power-up. The X-OR gate is programmed to produce true

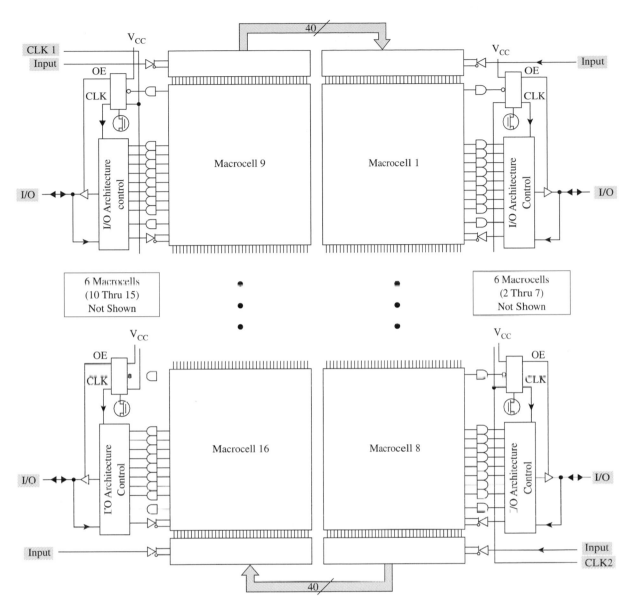

FIGURE 13-32   EP630 Series PLD block diagram. *Reprinted by Permission of Texas Instruments.*

or inverted data at the flip-flop's input. Feedback to the AND array can be left disconnected, or it can be taken from the Q output of the flip-flop or the output buffer. The output pin may be used as an input pin as controlled by the Feedback Select MOSFET.

Figure 13-34(c) illustrates the **TOGGLE (T) flip-flop configuration.** The operation of a T-type flip-flop is relatively simple. The function table in Fig. 13-34(c) shows the flip-flop does not change states when $T = 0$ and a PGT occurs. The output Q_0 for this line entry indicates the output is the level of Q before the indicated steady-state input conditions were established. In other words, the circuit is in the HOLD mode. Note the circuit is also in the HOLD mode in the third line entry in the function table because there is no PGT of the clock pulse. The function table also shows the flip-flop toggles (changes to the opposite level of Q) if it is clocked when $T = 1$. The T-type flip-flop is actually nothing more than a J-K flip-flop with J and K connected together to form one input.

A **J-K flip-flop configuration** is shown in Fig. 13-34(d). Eight product terms are shared between two OR gates in this configuration. Operation of the J-K flip-flop was presented in Section 6-5. The function table line entries represent HOLD, CLEAR, SET,

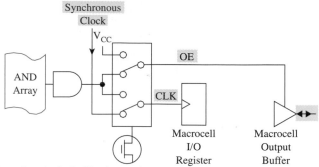

The register is clocked by the synchronous clock signal, which is common to seven other Macrocells. The output is enabled by the logic from the product term.

(a) MODE 0 OPERATION

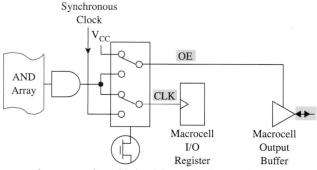

The output is permanently enabled and the register is clocked via the product term. This allows for gated clocks that may be generated from elsewhere in the EP630.

(b) MODE 1 OPERATION

FIGURE 13–33　EP630 Series PLD–OE/CLK select multiplexer. *Reprinted by Permission of Texas Instruments.*

TOGGLE, HOLD (no PGT), and asynchronous CLEAR. Like the previous configurations, the Invert Select EPROM bit sets the output polarity and the Feedback Select MUX enables registered or no feedback to the AND array.

An **S-R flip-flop configuration** is shown in Fig. 13–34(e). Circuit operation is comparable to that of the J-K configuration in the preceding figure. There is one main difference, which was explained in detail in Chapter 6. Since this is an active-high input S-R flip-flop, the output is undefined (invalid) when both inputs are asserted (activated).

Generic Array Logic.

Generic Array Logic (GAL) devices were developed by Lattice Semiconductor Corporation. E^2CMOS technology was combined with PLD technology to form GAL devices. GAL is a registered trademark of Lattice Semiconductor Corp.

There are some distinct advantages to designing with GAL devices instead of standard PLDs:

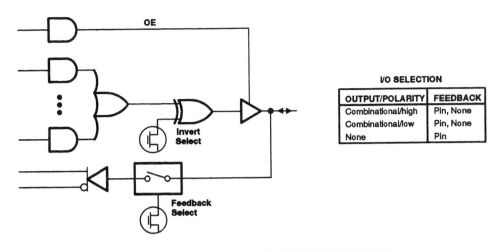

I/O SELECTION	
OUTPUT/POLARITY	FEEDBACK
Combinational/high	Pin, None
Combinational/low	Pin, None
None	Pin

(a) COMBINATIONAL LOGIC OUTPUT

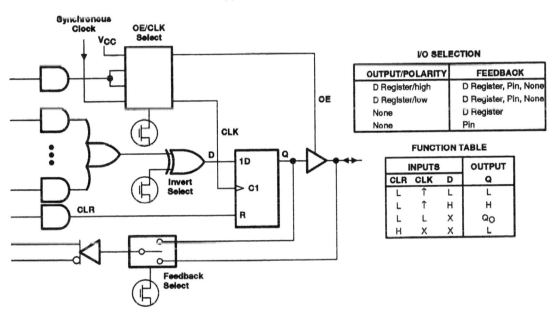

I/O SELECTION	
OUTPUT/POLARITY	FEEDBACK
D Register/high	D Register, Pin, None
D Register/low	D Register, Pin, None
None	D Register
None	Pin

FUNCTION TABLE

INPUTS			OUTPUT
CLR	CLK	D	Q
L	↑	L	L
L	↑	H	H
L	L	X	Q_O
H	X	X	L

(b) D-TYPE FLIP-FLOP

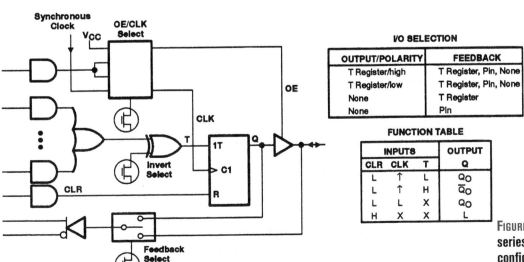

I/O SELECTION	
OUTPUT/POLARITY	FEEDBACK
T Register/high	T Register, Pin, None
T Register/low	T Register, Pin, None
None	T Register
None	Pin

FUNCTION TABLE

INPUTS			OUTPUT
CLR	CLK	T	Q
L	↑	L	Q_O
L	↑	H	$\overline{Q}_O$
L	L	X	Q_O
H	X	X	L

(c) TOGGLE FLIP-FLOP

FIGURE 13-34 **EP630 series PLD – I/O configurations.** *Reprinted by Permission of Texas Instruments.*

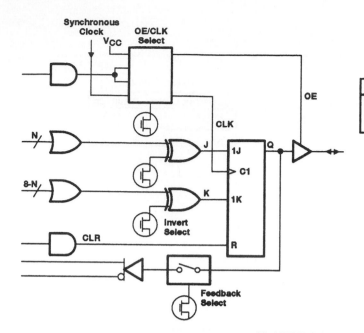

I/O SELECTION

OUTPUT/POLARITY	FEEDBACK
JK Register/high	JK Register, None
JK Register/low	JK Register, None
None	JK Register

FUNCTION TABLE

INPUTS				OUTPUT
CLR	CLK	J	K	Q
L	↑	L	L	Q_O
L	↑	L	H	L
L	↑	H	L	H
L	↑	H	H	$\overline{Q_O}$
L	L	X	X	Q_O
H	X	X	X	L

(d) J-K FLIP-FLOP

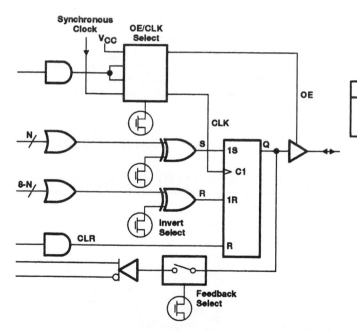

I/O SELECTION

OUTPUT/POLARITY	FEEDBACK
SR Register/high	SR Register, None
SR Register/low	SR Register, None
None	SR Register

FUNCTION TABLE

INPUTS				OUTPUT
CLR	CLK	S	R	Q
L	↑	L	L	Q_O
L	↑	L	H	L
L	↑	H	L	H
L	↑	H	H	Undefined
L	L	X	X	Q_O
H	X	X	X	L

(e) S-R FLIP-FLOP

FIGURE 13-34   **Continued. *Reprinted by Permission of Texas Instruments.***

1. Densities of GAL devices range up to 14,000 gates on an IC.
2. Nonvolatile, in-system programmability.
3. Fast erase and reprogram time.
4. 5-ns (200 MHz) t_{PD} versions available.

GAL16V8.

The block diagram and pin configuration diagram of the Lattice GAL16V8 are shown in Fig. 13–35. The device contains a programmable AND-array (64×32) to produce the normal PLD SOP logic.

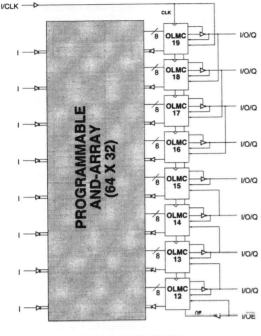

(a) BLOCK DIAGRAM

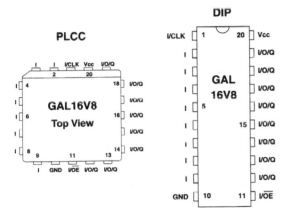

(b) PIN CONFIGURATION DIAGRAMS

FIGURE 13–35 **GAL16V8 PLD.** *Copyright 1994, Lattice Semiconductor Corporation. Used by permission.*

The E^2 floating-gate cell technology of the GAL16V8 allows for reprogramming the devices for reconfiguration. Erasure time is typically less than 100 ms. This PLD contains eight Output Logic Macrocells (OLMCs) as shown in Fig. 13–35(a). The pin configurations for DIP and PLCC packages are shown for reference in Fig. 13–35(b).

The GAL16V8 OLMC can be configured for three modes of operation—**simple mode, complex mode,** and **registered mode** Global bits SYN and AC are used to control the mode of operation.

Simple Mode of Operation.   The OLMCs are configured to produce combinational logic outputs or to act as dedicated inputs in this mode. Figure 13–36 shows three variations for simple mode operation. Figure 13–36(a) shows the configuration required to produce combinational logic at the output pin and output feedback to the AND array. The X-OR architecture control bit determines active-low or active-high output. The outputs in the simple mode have a maximum of eight P-terms. Figure 13–36(b) shows the OLMC configured for combinational logic output with no feedback. Figure 13–36(c) shows the I/O pin dedicated as an input port because the tristate buffer is not enabled. Again, remember the mode/configuration of the OLMCs in the GAL16V8 is controlled by the software via the architecture control bits listed in each figure.

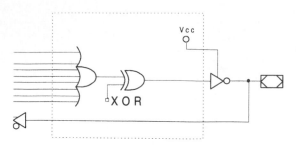

Combinatorial Output with Feedback Configuration for Simple Mode

- SYN=1.
- AC0=0.
- XOR=0 defines Active Low Output.
- XOR=1 defines Active High Output.
- AC1=0 defines this configuration.
- All OLMC **except** pins 15 & 16 can be configured to this function.

(a) COMBINATIONAL OUTPUT WITH FEEDBACK

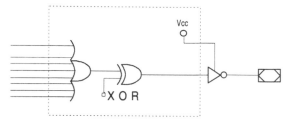

Combinatorial Output Configuration for Simple Mode

- SYN=1.
- AC0=0.
- XOR=0 defines Active Low Output.
- XOR=1 defines Active High Output.
- AC1=0 defines this configuration.
- Pins 15 & 16 are permanently configured to this function.

(b) COMBINATIONAL OUTPUT

FIGURE 13-36
GAL16V8 PLD OLMC—simple mode. *Copyright 1994, Lattice Semiconductor Corporation. Used by permission.*

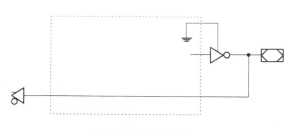

Dedicated Input Configuration for Simple Mode

- SYN=1.
- AC0=0.
- XOR=0 defines Active Low Output.
- XOR=1 defines Active High Output.
- AC1=1 defines this configuration.
- All OLMC **except** pins 15 & 16 can be configured to this function.

(c) DEDICATED OUTPUT

Complex Mode of Operation. The OLMCs are configured as I/O functions or output-only functions in this mode. Figure 13–37(a) and (b) shows the two possible configurations for this mode of operation. Macrocells 12 and 19 cannot be configured as input ports in this mode. Thus, only six I/Os can be configured. Registered mode operation must be used to implement eight I/Os.

Registered Mode of Operation. The macrocells are configured as dedicated registered outputs or as I/O functions in this mode. The registered mode configurations are shown in Fig. 13–38(a) and (b). All OLMCs share common clock and Output Enable (OE) inputs. A

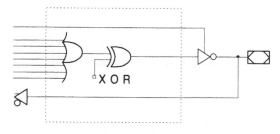

Combinatorial I/O Configuration for Complex Mode

- SYN=1.
- AC0=1.
- XOR=0 defines Active Low Output.
- XOR=1 defines Active High Output.
- AC1=1.
- Pin 13 through Pin 18 are configured to this function.

(a) COMBINATIONAL I/O

FIGURE 13-37
GAL16V8 PLD OLMC—complex mode. *Copyright 1994, Lattice Semiconductor Corporation. Used by permission.*

Combinatorial Output Configuration for Complex Mode

- SYN=1.
- AC0=1.
- XOR=0 defines Active Low Output.
- XOR=1 defines Active High Output.
- AC1=1.
- Pin 12 and Pin 19 are configured to this function.

(b) COMBINATIONAL OUTPUT

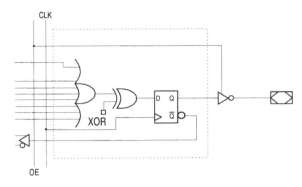

Registered Configuration for Registered Mode

- SYN=0.
- AC0=1.
- XOR=0 defines Active Low Output.
- XOR=1 defines Active High Output.
- AC1=0 defines this output configuration.
- Pin 1 controls common CLK for the registered outputs.
- Pin 11 controls common $\overline{OE}$ for the registered outputs.
- Pin 1 & Pin 11 are permanently configured as CLK & $\overline{OE}$.

(a) REGISTERED CONFIGURATION

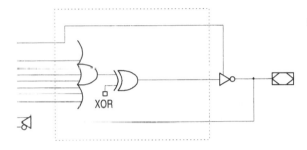

Combinatorial Configuration for Registered Mode

- SYN=0.
- AC0=1.
- XOR=0 defines Active Low Output.
- XOR=1 defines Active High Output.
- AC1=1 defines this output configuration.
- Pin 1 & Pin 11 are permanently configured as CLK & $\overline{OE}$.

(b) COMBINATIONAL CONFIGURATION

FIGURE 13-38   GAL16V8 PLD OLMC—registered mode. *Copyright 1994, Lattice Semiconductor Corporation. Used by permission.*

maximum of eight I/Os or eight registers may be implemented in the registered mode of operation. This mode makes available D, T, or J-K registers.

pLSI[R] and ispLSI™

Lattice Semiconductor **programmable Large-Scale Integration (pLSI)** and **in-system programmable Large-Scale Integration (ispLSI)** families combine PLD performance and simplicity with the high-density, register-rich Field-Programmable Gate Arrays (FPGAs) that will be presented in Section 13–4.

Lattice offers three families of pLSI and ispLSI devices—1000 Family, 2000 Family, and 3000 Family. The families offer different densities (logic capacities) and speed–performance characteristics. The 1000 Family offers 110 MHz system performance with up to 8000 PLD gates. The 2000 Family offers 135 MHz performance with 1000 to 4000 PLD gates. The 3000 Family offers 110 MHz performance with up to 14000 PLD gates.

A block diagram of the Lattice pLSI 1032 is shown in Fig. 13–39. Several unique features of this block diagram are discussed in the following paragraphs. Specifically, the Generic Logic Block (GLB), Output Routing Pool (ORP), and Global Routing Pool (GRP) functions are presented.

Generic Logic Block (GLB).   The 1000 Family PLD GLB is illustrated in Fig. 13–40. The GLB consists of the AND array, which has 18 inputs. Sixteen inputs come from the Global Routing Pool (GRP). These inputs may be feedback from a GLB, or they may be from I/O cells. Inputs 16 and 17 are dedicated inputs. The output of the AND array can produce 20 product-terms (referred to as PTs), which can produce the logical sum of any of the GLB inputs.

The **Product-Term Sharing Array (PTSA)** of the GLB routes the 20 P-terms to the four GLB outputs. Figure 13–40 shows the PTSA in various configurations. The top OR gate in the PTSA has three PTs applied to its inputs. The OR gate next to the bottom in the PTSA has four PTs applied to its inputs. The PTSA is programmed so both of these OR gate's outputs are fed to an X-OR gate. The other input to the X-OR gate is P-term (0). The X-OR gate can be used for logic or to configure the D-type flip-flop to operate

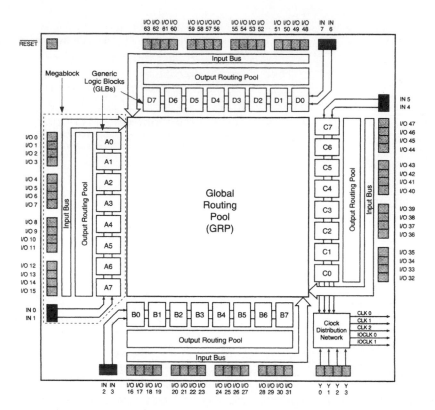

FIGURE 13-39   pLSI 1032 PLD block diagram. *Copyright 1994, Lattice Semiconductor Corporation. Used by permission.*

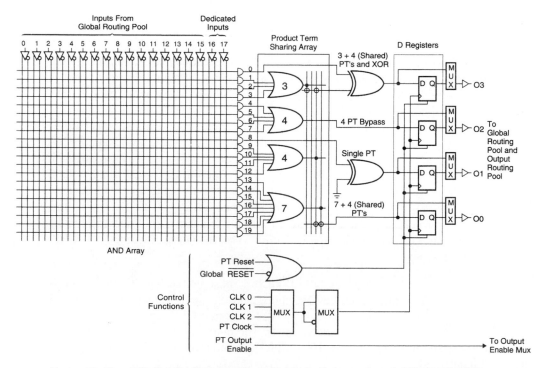

FIGURE 13-40   1000 Family PLD generic logic block diagram. *Copyright 1994, Lattice Semiconductor Corporation. Used by permission.*

as a J-K or T-type flip-flop. The second OR gate in the PTSA is configured to produce a four-product-term bypass directly to the output to increase speed. The bottom OR gate in the PTSA has seven PTs applied to its inputs. This PTSA is programmed so the bottom OR gate's output is shared with the OR gate above it in the figure.

The D-type flip-flops in Fig. 13–40 are reconfigurable. They are bypassed via program control when combinational outputs are desired. The control functions allow selection from four clock inputs to trigger the flip-flops. One of these clocks (PT Clock) is produced by a product term within the GLB. The flip-flops are cleared by PT Reset = 1 or Global $\overline{\text{RESET}}$ = 0.

Megablock.   A Lattice device **Megablock** is shown in Fig. 13–41. The 1000 Family Megablock consists of the Global Routing Pool (GRP), eight GLBs, the Output Routing Pool (ORP), and 16 I/O cells. The 1000 Family pLSI and ispLSI PLDs can contain up to 6 of these Megablocks.

Input signals to the I/O cells are routed directly to the GRP as shown in Fig. 13–41. This allows every GLB in the device access to each I/O cell input. The two dedicated inputs (IN0 and IN1) are connected to all eight GLBs within the Megablock.

The **Output Routing Pool (ORP)** connects signals produced in the GLB to the I/O cells. The cells are configured via programming as input, output, tristate output, or bidirectional (I/O) pins. Routability in the ORP is automatic. A D-type flip-flop within the I/O cell can be configured as a level-sensitive latch or an edge-triggered flip-flop to store input data.

The **Global Routing Pool (GRP),** as the name implies, takes care of the interconnections required within the device. The I/O cell inputs or the GLB outputs can be routed to the inputs of any GLB. Also, GLB outputs can be folded back to the inputs of all other GLBs.

Cross Programming.

The GAL16V8 and GAL20V8 PLDs can be used to replace most standard 20-pin and 24-pin PAL devices. The differences of the GAL and PAL devices' construction prevent programming the GAL device with the same program used for the PAL device. Lattice Semiconductor provides a software utility program that will convert a PAL device JEDEC file to a compatible GAL JEDEC file. This software, called **PALtoGAL,** is just one method available for conversion from existing PAL JEDEC files to GAL JEDEC files.

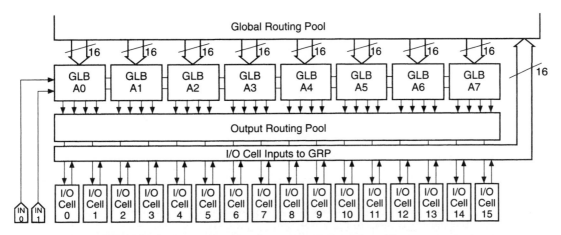

FIGURE 13–41   1000 Family Megablock block diagram. *Copyright 1994, Lattice Semiconductor Corporation. Used by permission.*

Section 13-3: Review Questions

Answers are given at the end of the chapter.

A. PAL devices may produce registered or combinational logic outputs.
   (1) True
   (2) False

B. PAL devices are _____ -rich devices.
   (1) logic
   (2) register

C. The Output Logic Macrocells (OLMs) in a TIBPAL22V10 can be programmed individually.
   (1) True
   (2) False

D. What does the R in part number TIBPAL16R4 indicate?

E. The OLM in Fig. 13–42 is configured to produce _____.
   (1) I/O feedback and combinational/inverted output
   (2) I/O feedback and combinational/noninverted output
   (3) registered feedback and registered/inverted output
   (4) registered feedback and registered/noninverted output

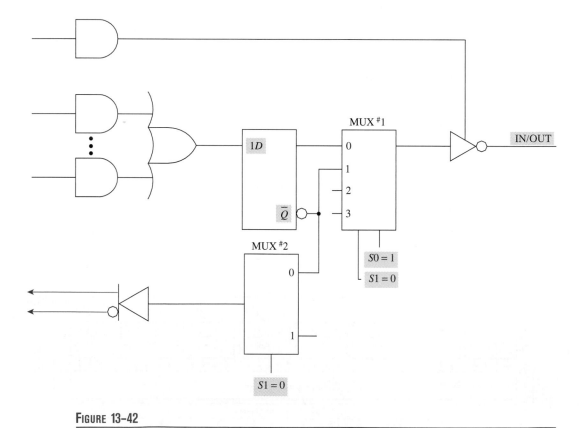

FIGURE 13-42

F. The output data from the OR gate to the macrocell in Fig. 13–43 is _____ if the select (S) input to the X-OR gate control MUX is low.
   (1) inverted
   (2) noniverted

G. What is the purpose of the programmable design security bit in a PLD?

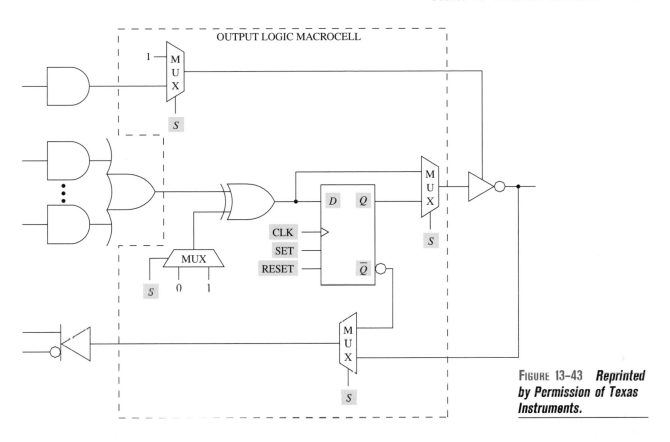

Figure 13–43 *Reprinted by Permission of Texas Instruments.*

H. The EP630 Series PLD I/O configuration in Fig. 13–44 allows _____.
  (1) J-K flip-flop operation
  (2) D-type flip-flop operation
  (3) toggle flip-flop operation
  (4) combinational logic outputs

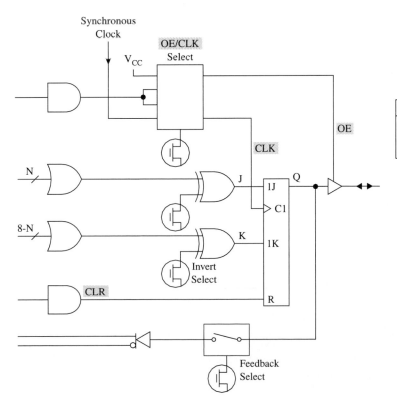

I/O SELECTION	
OUTPUT/POLARITY	FEEDBACK
JK Register/high	JK Register, None
JK Register/low	JK Register, None
None	JK Register

FUNCTION TABLE

INPUTS				OUTPUT
CLR	CLK	J	K	Q
L	↑	L	L	Q_O
L	↑	L	H	L
L	↑	H	L	H
L	↑	H	H	$\overline{Q}_O$
L	L	X	X	Q_O
H	X	X	X	L

Figure 13–44 *Reprinted by Permission of Texas Instruments.*

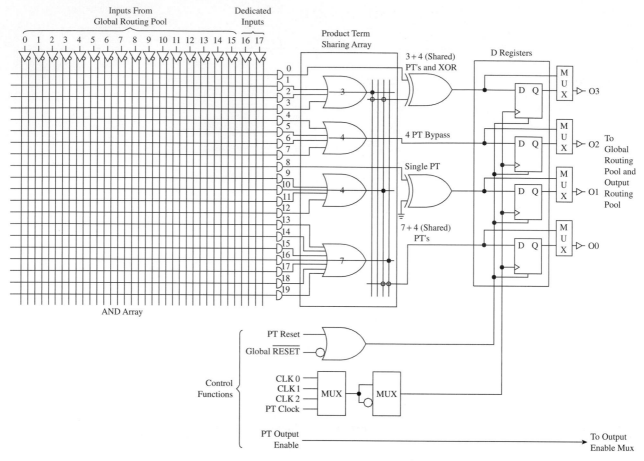

I. The Lattice Family PLD Generic Logic Block (GLB) is shown in Fig. 13–45. The input to the lower flip-flop is _____ shared product-terms.
    (1) 3   (3) 3 + 4
    (2) 4   (4) 7 + 4

J. How can the flip-flops in Fig. 13–45 be cleared?
K. The block diagram in Fig. 13–46 represents one of six _____ in the 1000 Family pLSI or ispLSI PLDs.
L. The function of the Output Routing Pool (ORP) in Fig. 13–46 is to route signals produced in the GLB to the I/O cells.
    (1) True
    (2) False

Section 13-4: Field-Programmable Gate Array

Objectives

1. State the main advantage of using FPGAs.
2. Given a logic diagram of an Actel FPGA logic module, determine the function/circuit implemented by the module.

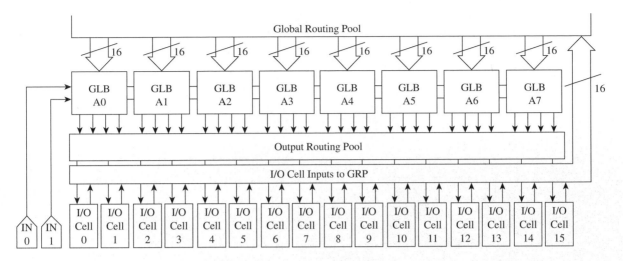

Field-Programmable Gate Array (FPGA) devices provide logic gate capacities with over 100,000 available logic gates. The main advantage of using FPGAs is their density is such that one FPGA can be used to replace hundreds of TTL ICs. The FPGA offers nonvolatile, user programmability, and some offer speeds in the 250 MHz to 350 MHz range.

Field-programmable gate arrays really give the saying "concept-to-silicon in minimum time" true meaning. They are **register rich** when compared to the logic-rich PLDs discussed in the previous section. The FPGA integrates huge amounts of logic into a register rich environment that minimizes power consumption and board space requirements.

Actel Corporation is one of America's leading manufacturers of state-of-the-art FPGA devices. Actel's newest FPGA technology includes the SX and MX families. These families feature very high capacity and high-speed operation.

54SX Family FPGAs

Actel's 54SX family of FPGAs features a revolutionary new **sea-of-modules architecture** that delivers next-generation device performance and integration levels not currently available in other FPGA architectures. The 54SX family of FPGAs provides a gate capacity up to 72,000 available logic gates and the SX-A family gate capacity exceeds 100,000 logic gates. In addition, up to 360 user-programmable input/output (I/O) ports are available. Each I/O can be configured as an input, an output, a tristate output, or a bidirectional pin.

The routing interconnect resources in the SX family are located between the Metal 2 (M2) and Metal 3 (M3) layers of the device as shown in Fig. 13–47. This routing scheme is very efficient because it eliminates the channels of routing and interconnect resources between logic modules used in older FPGA technology that employed channelled array architecture (see Fig. 13–48). This new routing scheme enables the entire floor of the FPGA device to be spanned with an uninterrupted grid of logic modules.

Interconnection between the logic modules is achieved using metal-to-metal *antifuse* interconnect elements patented by Actel. These antifuses are embedded between the M2 and M3 layers (Fig. 13–47). Antifuses are one-time programmable, nonvolatile, two-terminal devices used as interconnect switches in many FPGAs. The antifuse is a normally open switch that exhibits a permanent low-impedance connection when programmed. They offer many advantages when compared to the fuse technology used in PROM devices. One advantage is that it is extremely difficult to distinguish between programmed and

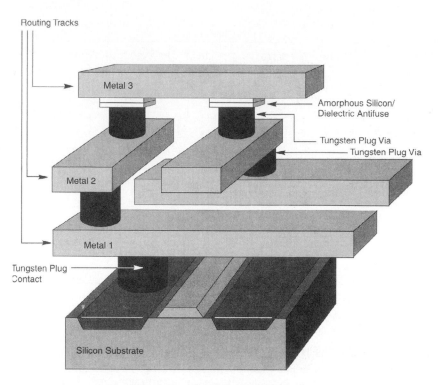

FIGURE 13–47   Actel's SX family interconnect elements. *Courtesy of Actel Corporation.*

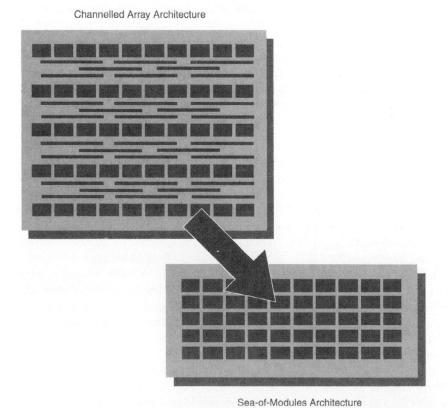

FIGURE 13–48   Channelled array and sea-of-modules architecture. *Courtesy of Actel Corporation.*

unprogrammed antifuses which makes reverse engineering and design theft virtually impossible.

Logic Modules.

The 54SX family of FPGAs provides up to 6036 logic modules. There are two types of logic modules available in this family—the **register cell** and the **combinatorial cell.** There are over 2000 register cells (R-cells) available in some 54SX family devices. Additionally, there are over 4000 combinatorial cells (C-cells) available in some devices in this family. Keep in mind the details of FPGA implementation and architecture are provided only to give you a better understanding of FPGA operation. The architecture of the logic design is software controlled and transparent to the user.

MX Family FPGAs

Actel's 40MX and 42MX families of FPGAs are the newest additions to their Integrator Series of PLDs. These FPGAs provide system logic designers with a high performance, cost-effective ASIC alternative. The MX family of FPGAs provides gate capacity up to 36,000 available logic gates and there are up to 202 user-programmable I/O ports available.

The MX architecture uses vertical and horizontal routing tracks to interconnect the various logic and I/O modules. The tracks are metal interconnects called segments. Varying segment lengths allows the interconnect of over 90% of design tracks to occur with only two antifuse connections. The routing structure of the tracks is shown in Fig. 13–49.

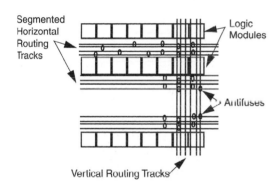

FIGURE 13–49   Actel's MX architecture routing structure. *Courtesy of Actel Corporation.*

40MX Logic Module.

The 40MX logic module is shown in Fig. 13–50. This module is an 8-input, 1-output logic circuit designed to implement a wide range of logic functions with efficient use of interconnect routing resources.

The MUX-based logic module in Fig. 13–50 can be used to implement AND, OR, NAND, NOR, exclusive, and D-latch functions. Latch and flip-flop functions are implemented in the **40MX** device by cross-coupled feedback using the gates in the logic modules in the same manner as was discussed in Chapter 6. The **42MX** family of FPGAs contains flip-flops that greatly simplify design of sequential circuits.

A 3-input AND gate is shown implemented with a 40MX logic module in Fig. 13–51. Input A_0 is selected when SA is low, A_1 is selected when SA is high, B_0 is selected when SB is low, and B_1 is selected when SB is high. The S_1/S_0 inputs control data selection in the logic module output MUX. The Y output is high only when inputs A, B, and C are high in this configuration.

Various other MUX-based implementations are shown in Fig. 13–52. The OR and NOR implementations are straightforward. Note in the exclusive gates in Fig. 13–52(c) and (d) that the A input is applied to both select inputs of the input MUXs. Im-

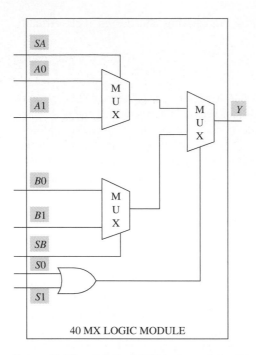

FIGURE 13-50 **Actel's 40MX logic module.** *Courtesy of Actel Corporation.*

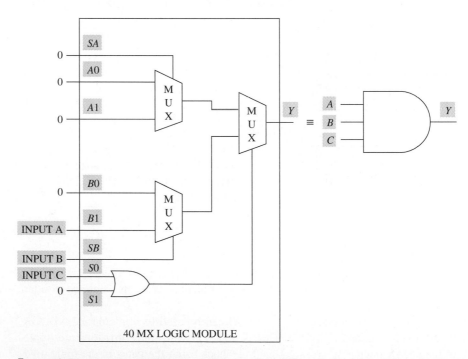

FIGURE 13-51 **Actel's 40MX logic module: 3-input AND gate implementation.** *Courtesy of Actel Corporation.*

plementation of various logic functions is under software control and these examples are provided to aid in a basic understanding of FPGA implementation.

Users normally enter their FPGA designs as schematics built from a hard macro library. Several schematic capture programs are commercially available to accomplish this task. Hard macros are similar to the basic SSI components studied in the early part of this textbook. Soft macros are implemented by combining several hard macros. For example, soft macros are used for counter, adder, and decoder implementation. All Actel FPGAs are fully supported by their Designer Series development tools.

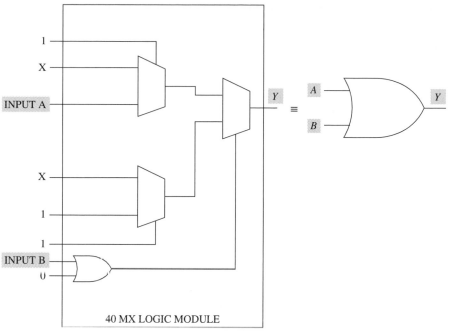

(a) 2-INPUT OR GATE IMPLEMENTATION

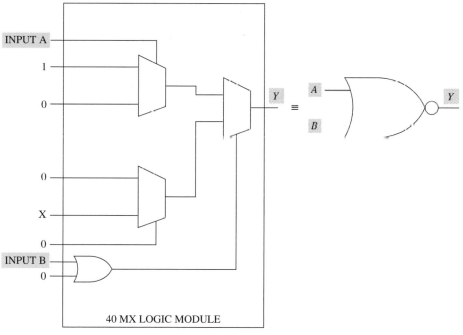

(b) 2-INPUT NOR GATE IMPLEMENTATION

FIGURE 13–52   Actel's 40MX logic module. *Courtesy of Actel Corporation.*

42MX Logic Modules.

The 42MX devices contain three types of logic modules: combinatorial, sequential, and decode.

The 42MX family **combinatorial logic module** (C-module) is shown in Fig. 13–53. Some 42MX devices provide up to 1184 of the C-modules. The logic module in Fig. 13–53 includes a 4-to-1 line MUX. The S_0 select input to the MUX is the AND function (product) of $A_0 B_0$. Select input S_1 is the OR function (sum) of $A_1 + B_1$. The Y output

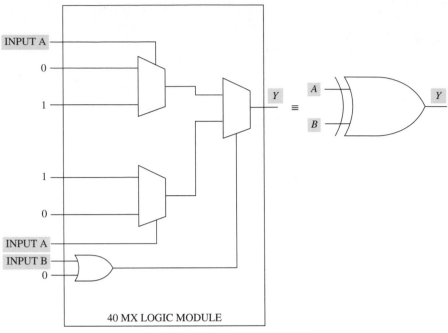

(c) X-OR GATE IMPLEMENTATION

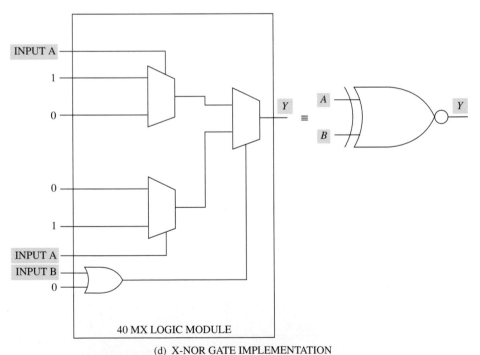

(d) X-NOR GATE IMPLEMENTATION

FIGURE 13-52   **Continued. *Courtesy of Actel Corporation.***

of the C-module is $D_{00}\overline{S_1}\,\overline{S_0} + D_{01}\overline{S_1}S_0 + D_{10}S_1\overline{S_0} + D_{11}S_1S_0$ where $S_0 = A_0B_0$ and $S_1 = A_1 + B_1$. Logic gates with up to five inputs can be implemented with this module.

The 42MX family **sequential logic module** (S-module) configurations are shown in Fig. 13–54. Some 42MX devices contain up to 1230 S-modules. The S-modules shown in Fig. 13–54 are designed to implement high-speed sequential functions. The S-module is nothing more than a C-module with a sequential element (latch or flip-flop). The S-module can be used to implement the Y function of the C-module with one exception—$S_0 = A_0$ (see Fig. 13–53) because the B_0 input is used exclusively to clear the flip-flop. The sequential module can be implemented in four different configurations as shown in Fig. 13–54.

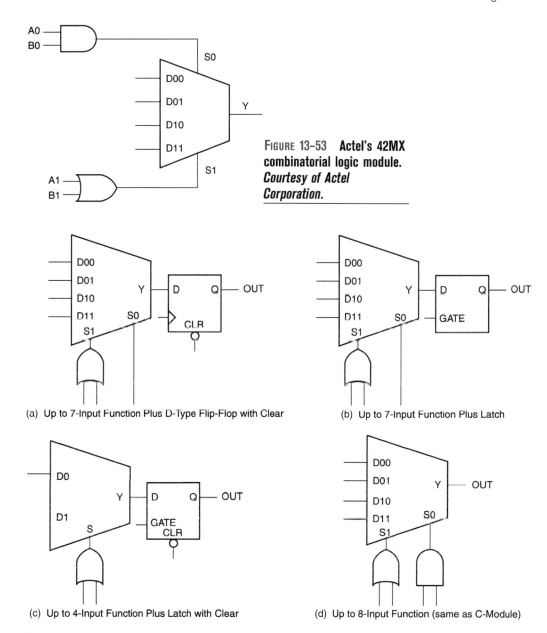

FIGURE 13–53 Actel's 42MX combinatorial logic module. *Courtesy of Actel Corporation.*

(a) Up to 7-Input Function Plus D-Type Flip-Flop with Clear

(b) Up to 7-Input Function Plus Latch

(c) Up to 4-Input Function Plus Latch with Clear

(d) Up to 8-Input Function (same as C-Module)

FIGURE 13–54 Actel's 42MX sequential logic module. *Courtesy of Actel Corporation.*

Some of the 42MX devices contain **decode logic modules** (D-modules). This type of module, shown in Fig. 13-55, contains wide decode circuitry, which provides a fast, wide-input AND function similar to that found in product term sharing arrays. The output of the module can be active low or active high depending upon the logic level programmed to the X-OR gate. Some devices in this family contain as many as 24 D-modules.

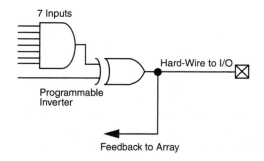

FIGURE 13–55 Actel's 42MX decode logic module. *Courtesy of Actel Corporation.*

Section 13-4: Review Questions

Answers are given at the end of the chapter.

A. FPGAs are _____ rich when compared to PLDs.
   (1) logic
   (2) register

B. Actel's logic modules are based on encoder technology.
   (1) True
   (2) False

C. Identify the type of logic module shown in Fig. 13–55.
D. The 40MX logic module in Fig. 13–56 is an implementation of a/an _____.

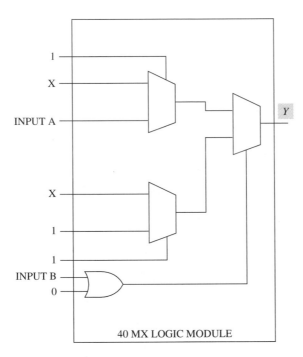

FIGURE 13-56   *Courtesy of Actel Corporation.*

SECTIONS 13-3 AND 13-4: INTERNAL SUMMARY

Programmable Array Logic (PAL[R]) devices contain programmable AND arrays and fixed OR arrays. They are termed **logic-rich** devices due to the high ratio of logic gates to flip-flops.

The PAL16L8 device produces combinational logic outputs. The PAL16R8 device produces registered outputs. The PAL22V10 contains an OLM that allows the user to control the outputs of the device individually. The outputs may be registered or nonregistered, and they may be active high or active low. The feedback in these versatile PLDs may be registered or input/output. The configuration of the devices is established during programming. The program data in the PLD may be protected by a security bit. The Texas Instruments' EP630 series PLDs contain features that allow simple implementation of D-, J-K, S-R, or T-type flip-flops.

Generic Array Logic (GAL) devices from Lattice Semiconductor Corporation use E[2]CMOS technology for fast erasure of data for reconfiguration. These high-density PLDs contain up to 14,000 gates and can be configured for combinational or registered outputs. The pLSI[R] and ispLSI™ Lattice devices contain generic logic blocks with product-term sharing arrays for versatility. The output and global routing pools allow automatic high-speed interconnections within the devices.

Field-Programmable Gate Array (FPGA) devices are register-rich when compared to the logic-rich PLDs. Actel's antifuse allows extreme versatility in routing and interconnect architecture within the FPGAs. The Actel FPGA families employ multiplexer-based, antifuse technology. Multiplexers generate SOP logic expressions from truth tables, which allows generation of desired FPGA output logic expressions.

Sections 13–3 and 13–4: Internal Summary Questions

Answers are given at the end of the chapter

1. PAL devices containing flip-flops at the output are classified as _____ configuration.
   a. registered
   b. active-low
   c. active-high

2. An Output Logic Macrocell (OLM/OLMC) allows outputs to be registered or nonregistered and active low or active high.
   a. True
   b. False

3. The OLM in Fig. 13–57 is configured to produce _____.
   a. I/O feedback and combinational/inverted output
   b. I/O feedback and combinational/noninverted output
   c. registered feedback and registered/inverted output
   d. registered feedback and registered/noninverted output

4. The EP630 series PLD I/O configuration in Fig. 13–58 allows _____.

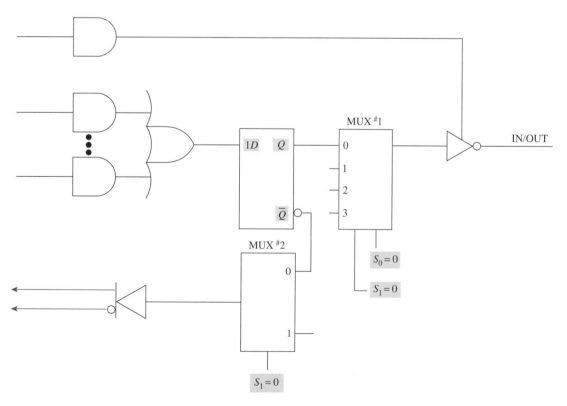

FIGURE 13–57

4. The EP630 series PLD I/O configuration in Fig. 13–58 allows .
   a. J-K flip-flop operation
   b. D-type flip-flop operation
   c. toggle flip-flop operation
   d. combinational logic outputs

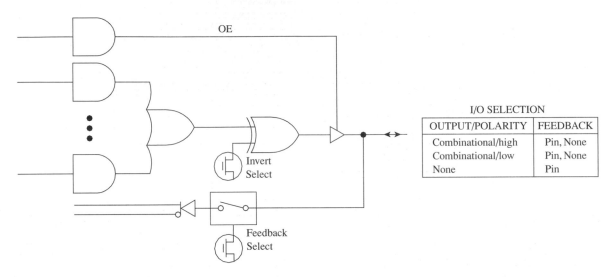

I/O SELECTION	
OUTPUT/POLARITY	FEEDBACK
Combinational/high	Pin, None
Combinational/low	Pin, None
None	Pin

FIGURE 13–58   *Reprinted by Permission of Texas Instruments.*

5. The EP630 series PLD I/O configuration in Fig. 13–59 allows_____.
   a. J-K flip-flop operation
   b. D-type flip-flop operation
   c. toggle flip-flop operation
   d. combinational logic outputs

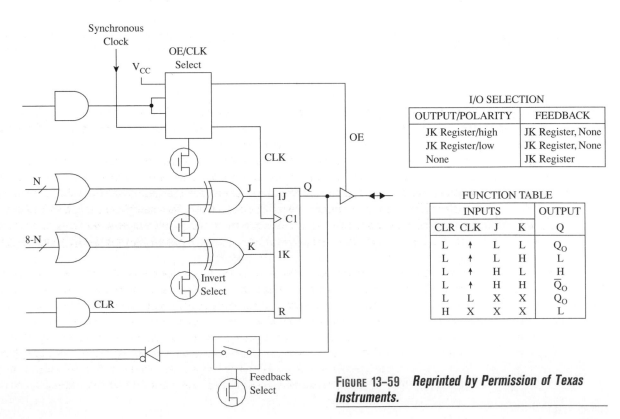

I/O SELECTION	
OUTPUT/POLARITY	FEEDBACK
JK Register/high	JK Register, None
JK Register/low	JK Register, None
None	JK Register

FUNCTION TABLE				
INPUTS				OUTPUT
CLR	CLK	J	K	Q
L	↑	L	L	Q_O
L	↑	L	H	L
L	↑	H	L	H
L	↑	H	H	$\overline{Q_O}$
L	L	X	X	Q_O
H	X	X	X	L

FIGURE 13–59   *Reprinted by Permission of Texas Instruments.*

6. Cross programming is required when a PAL JEDEC file is to be used to program a GAL device.
   a. True
   b. False

7. FPGAs do not have the high density of their PLD counterparts.
   a. True
   b. False

8. FPGAs are register-rich devices.
   a. True
   b. False

9. Antifuses have very high resistance when in the programmed state.
   a. True
   b. False

10. Multiplexers may be used to generate logical expressions in FPGAs.
    a. True
    b. False

CHAPTER 13: SUMMARY

Programmable Logic Device (PLD) is a generic term applicable to ICs with uncommitted logic arrays that are user programmable. They offer advantages of high density (less circuit board real estate), low power consumption, and improved reliability when compared to circuit implementation with SSI/MSI ICs.

PLDs come in one-time programmable, erasable by UV light, or electrically erasable types. They offer SOP output expressions. The schematic notation standard used for PLDs uses an X for an intact fuse, the lack of an X for a blown fuse and a dot for a hard-wired connection.

The type of PLD is dependent upon the device's architecture. A **PLE** contains a fixed AND array and a programmable OR array. A **PLA** contains a programmable AND array and a programmable OR array. A **PAL** contains a programmable AND array and a fixed OR array. PLDs that contain output registers allow sequential logic circuits to be designed with relative simplicity.

PLDs are considered to be **logic-rich** devices, while FPGAs are considered to be **register-rich** devices. The part numbers of PLDs usually identify the maximum number of inputs and outputs in addition to the device configuration.

Programming a PLD may consists of blowing configuration fuses, which sets the device's operation. The device may be set for registered or I/O feedback with registered or combinational logic outputs. The configuration is often set by the output logic macrocell within the device. Some PLDs such as the EP630 series allow internal flip-flops to be configured as D latches, toggle flip-flops, J-K flip-flops, or S-R flip-flops.

Generic Array Logic (GAL) devices by Lattice Semiconductor Corp. offer versatile, nonvolatile in-system programmability with extremely high density. These devices require a software utility program to convert a PAL device JEDEC file to be usable for GAL device programming.

Field Programmable Gate Array (FPGA) devices are register-rich, high-density devices that are quickly replacing many TTL ICs, PLDs, and ASICs in the digital field and in new circuit design. Antifuse technology is used in most FPGAs. The antifuse is the programmable element in an FPGA. Its popularity stems from its small size, which has virtually no effect on circuit density.

Actel Corporation's FPGA logic modules are MUX-based devices. This technology allows implementation of both combinational and sequential logic circuits in minimum time using various computer programs.

CHAPTER 13: END OF CHAPTER QUESTIONS/PROBLEMS

Answers are given in the Instructor's Manual.

SECTION 13-1

1. What is a Programmable Logic Device (PLD)?
2. Write the Boolean expression for the output of the circuit in Fig. 13–60.
3. What does the X on the product line in Fig. 13–61 represent?
4. What does the dot at the intersection of the product and *A* input lines in Fig. 13–61 represent?
5. There is no dot or X at the intersection of the product and *C* input lines in Fig. 13–61. What does this indicate?
6. What is the Boolean expression for the output of the AND gate in Fig. 13–61?
7. Write the output Boolean expression of the PLD in Fig. 13–62.
8. Write the output Boolean expression of the PLD in Fig. 13–63.

SECTION 13-2

9. List the three classifications of PLDs.
10. Which classification of PLD contains both a programmable AND array and a programmable OR array?
11. Which gate array is programmable in a Programmable Array Logic device?
12. How many inputs and outputs does a PAL22V10 device have?
13. List the advantages of using Application-Specific Integrated Circuits (ASICs) in respect to SSI and MSI ICs.

SECTION 13-3

14. What information does the part number PAL16R4 provide the user?
15. The PAL16R4 device in Fig. 13–64 has_____ *dedicated inputs.*

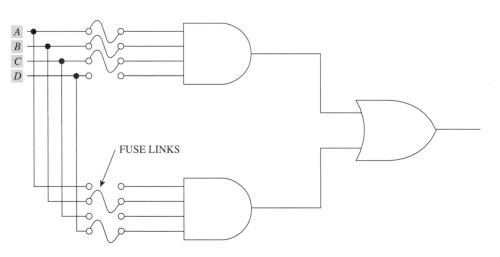

FIGURE 13-60

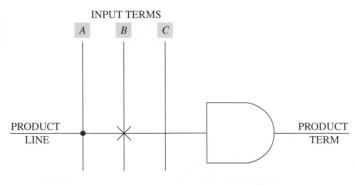

FIGURE 13-61

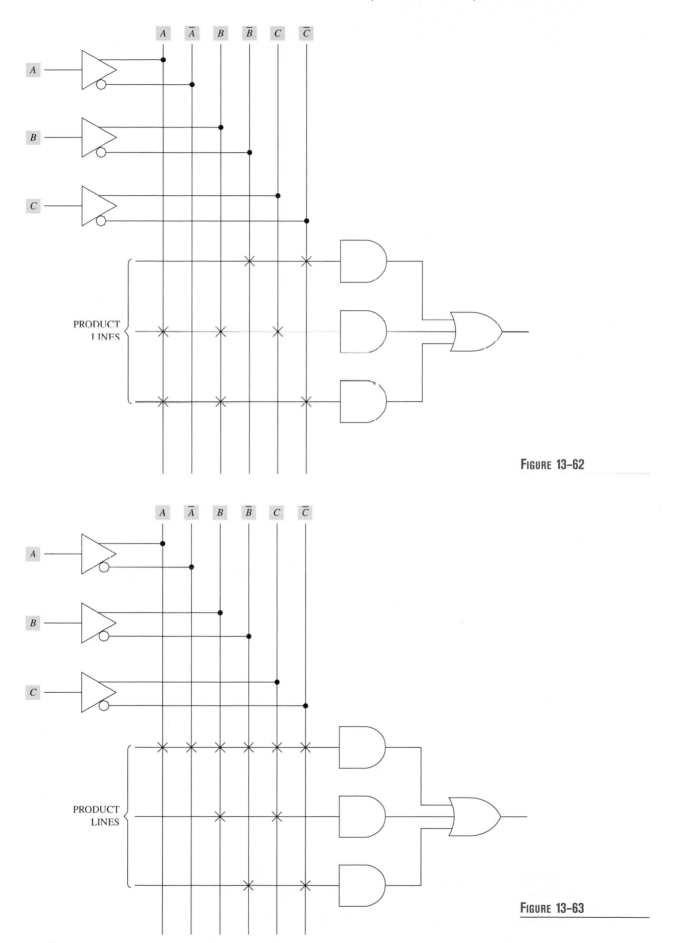

Figure 13-62

Figure 13-63

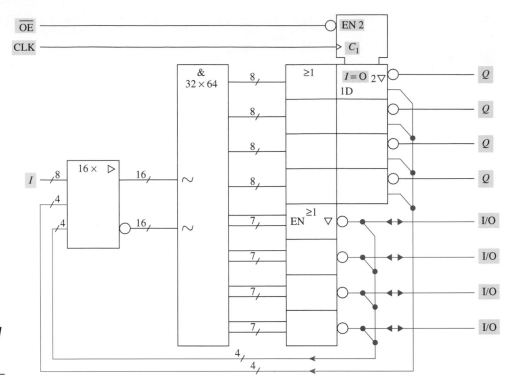

FIGURE 13-64 *Reprinted by Permission of Texas Instruments.*

16. The $\overline{OE}$ input to the Programmable Array Logic device in Fig. 13–64 serves what function?

17. Which gate array is programmable in the PAL16R4 device in Fig. 13–64?

18. Why is an enable (EN) input required to some of the OR-gate blocks in Fig. 13–64?

19. What is the function of the X-OR gate in the Output Logic Macrocell in Fig. 13–65 when its control MUX output is high?

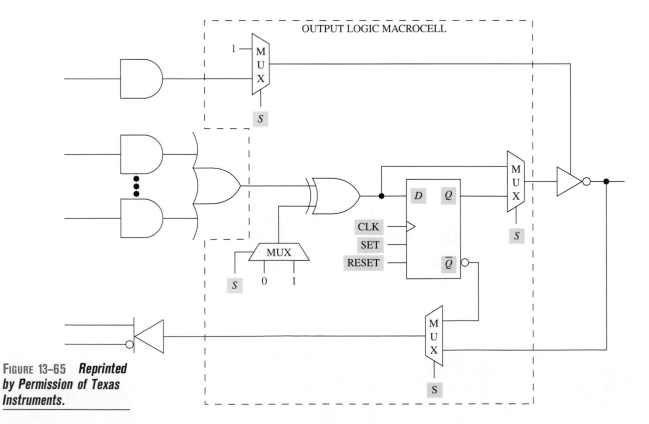

FIGURE 13-65 *Reprinted by Permission of Texas Instruments.*

Section 13-4

20. What do the letters FPGA stand for?
21. State the advantages of using FPGA devices instead of PLD devices.

CT 22. What logic function is implemented by the 40MX Logic Module shown in Fig. 13–66?

CT 23. Write the logical expression for the output of the combinatorial logic module shown in Fig. 13–67.

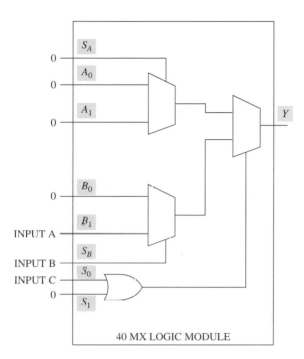

FIGURE 13-66   *Courtesy of Actel Corporation.*

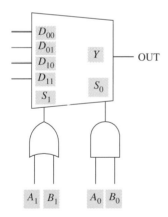

FIGURE 13-67   *Courtesy of Actel Corporation.*

Answers to Review Questions

Section 13-1

A. An IC with an uncommitted logic array
B. SOP
C. Intact fuse—programmable connection
D. Hard-wired connection
E. Blown fuse—programmable connection

Section 13-2

A. PLE, PLA, and PAL
B. AND-gate array and OR-gate array
C. AND-gate array
D. Fuse-map programming
   Boolean expressions
   Truth tables
   Schematic capture
E. (1) Up to 16 inputs and 8 outputs; registered outputs
   (2) Up to 20 inputs and 8 outputs; active-low combinational logic outputs
   (3) Up to 22 inputs and 10 outputs; versatile—combinational or registered outputs

F. Application-Specific Integrated Circuit

G. **Advantages:** High density; reduced board space; less power consumption; improved reliability
   Disadvantages: Expensive; suitable only for high quantity production; long lead time for manufacture

Section 13-3

A. True
B. logic
C. True
D. Registered outputs
E. (4)
F. noninverted
G. Prevents copying/examining stored PLD data
H. (1)
I. (4)
J. PT Reset = 1 or Global $\overline{\text{RESET}}$ = 0
K. Megablocks
L. True

Section 13-4

A. register
B. False
C. Decode logic module
D. 2-input OR

Answers to Internal Summary Questions

Sections 13-1 and 13-2

1. b
2. b
3. d
4. b
5. c
6. a
7. a
8. d
9. c
10. b
11. c
12. a

Sections 13-3 and 13-4

1. a
2. a
3. c
4. d
5. a
6. a
7. b
8. a
9. b
10. a

TECHNOLOGY OF LOGIC FAMILIES

The most popular digital logic families are Transistor-to-Transistor Logic (TTL) and Metal-Oxide Semiconductor (MOS). The MOS family can be subdivided into n-channel MOS (NMOS), p-channel MOS (PMOS), and complementary MOS (CMOS). The discussion of TTL and MOS devices that follows requires an understanding of the operation of bipolar-junction transistors and enhancement-mode (E) MOSFETs.

TTL–STANDARD 2-INPUT NAND GATE

The schematic diagram in Fig. A–1 is of a 7400 standard NAND gate. Diodes D_1 and D_2 serve as protection diodes for transistor Q_1 in case of a negative input signal. A negative input sometimes occurs due to oscillating of the input signal.

The input transistor (Q_1) is a multiple-emitter transistor. Figure A–2 shows the diode equivalent of Q_1. The two diodes on the left represent the emitter–base junctions of this dual-emitter transistor. The diode on the right represents the collector–base junction of the transistor.

Transistor Q_1 in Fig. A–1 controls Q_2. The condition of Q_2, cutoff or conducting, controls the output **totem-pole transistors** (Q_3 and Q_4). The input transistor (Q_1) will be discussed using the diode-equivalent circuit shown in Fig. A–3(a).

Logic 0 Inputs

Figure A–3(a) shows both emitter–base diodes are forward biased and conducting with Logic 0 (V_{IL}) applied to the A and B inputs. This allows current flow (I_{IL}) out of the input emitters of Q_1. The amount of current is controlled by the level of V_{CC} and the 4-kΩ resistance of R_1. The voltage on the base of Q_1 must be greater than $+2.1$V to forward bias the base–collector (BC) junction. However, the forward-biased emitter–base (EB) junctions put the base potential at approximately $+1.0$ V. Therefore, the BC junction is reverse biased, there is no base current path for Q_2, and Q_2 is cut off.

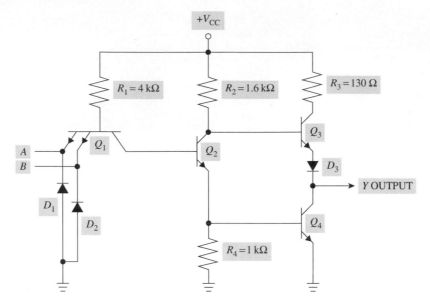

FIGURE A-1   Standard TTL NAND gate–schematic diagram.

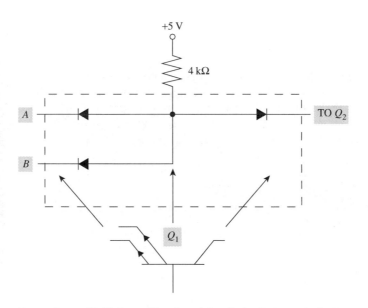

FIGURE A-2   Multiple-emitter transistor (Q₁)–diode equivalent.

The discussion in the preceding paragraph is also applicable when one input to the NAND gate is Logic 0 and one is Logic 1. A Logic 0 applied to one input will forward bias one of the EB junctions and put the base potential at approximately $+1.0$ V. This keeps Q_2 cut off.

The collector voltage of Q_2 is high because Q_2 is cut off. As depicted in Fig. A–3(b), the high on Q_2's collector causes Q_3 to conduct. This action provides a low-resistance path from V_{CC} to the output when Q_3 is saturated. The high output voltage (V_{OH}) is typically $+3.4$ V for the standard TTL NAND gate. V_{OH} can go as low as $+2.4$ V. In addition, because Q_2 is cut off, its emitter and the base of Q_4 are at ground potential. Thus, Q_4 is cut off.

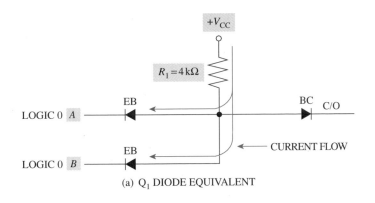

(a) Q_1 DIODE EQUIVALENT

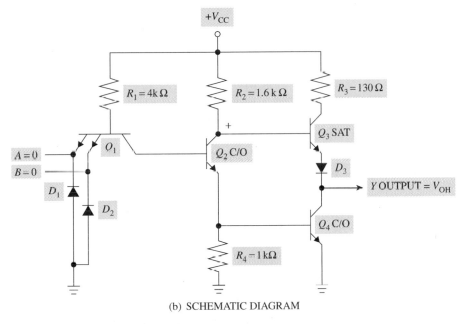

(b) SCHEMATIC DIAGRAM

Figure A-3 **Logic 0 inputs—standard TTL NAND gate.**

Logic 1 Inputs

Figure A–4(a) shows both EB junction diodes reverse biased and cut off when two Logic 1 levels are applied to the NAND gate. This condition forward biases the BC junction diode by placing a high positive potential at its anode. The forward-biased BC diode causes Q_2 to go into saturation.

Figure A–4(b) shows the NAND circuit with two Logic 1 inputs. The electron current flow shown through Q_2 develops a voltage drop across R_4 that forward biases the EB junction of Q_4 and causes it to go into saturation. This provides a sink current path for current from the load and a low output voltage (V_{OL}).

The diagram in Fig. A–4(b) indicates the base voltage of Q_2 is approximately $+1.4$ V. This potential mandates the voltage on the base of Q_1 must be greater than $+2.1$ V to forward bias the BC junction of Q_1 as previously mentioned. Also, the diagram shows Q_3's base voltage as $+0.9$ V when Q_2 is saturated. This amount of base voltage could possibly allow Q_3 to go into conduction if it were not for D_3. With D_3 in the circuit, Q_3's base voltage must be at least $+1.6$ V to allow Q_3 to conduct when Q_4 is conducting. D_3's sole purpose in the circuit is to ensure Q_3 is off when Q_4 is on.

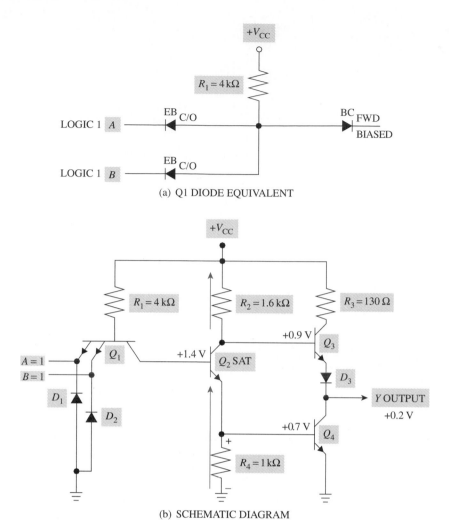

(a) Q1 DIODE EQUIVALENT

(b) SCHEMATIC DIAGRAM

Figure A-4   Logic 1 inputs–standard TTL NAND gate.

TTL–Inverter

Figure A–5 shows the schematic diagram of a TTL NOT gate. The circuit is similar to the NAND gate except it has only one input. This allows the input transistor to be a single-emitter transistor. Operation of the circuit is comparable to that of the NAND gate previously discussed.

Schematic diagrams of other types of TTL logic gates are available in most TTL data books.

TTL–Open-Collector Gates

The NAND gate and the NOT gate just presented both had totem-pole outputs. These gates and other types of logic gates have been modified to incorporate a group of gates referred to as **open-collector output** gates. The use of this group of gates allows wire-ANDing the outputs of several gates together. The subject was presented in Section 11–2 and is presented here to show the schematic diagram of one of the open-collector gates and how to calculate the size of the external pull-up resistor.

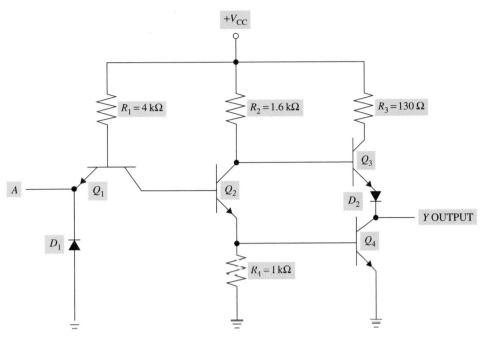

FIGURE A-5   **Standard TTL NOT gate–schematic diagram.**

Figure A–6 is the schematic diagram of a standard TTL NOT gate with open-collector output. Comparison of this NOT gate with the standard inverter in Fig. A–5 reveals Q_3, D_2, and the 130-Ω resistor of the standard inverter have been removed in the open-collector version of the gate. The data sheet for a 7405 open-collector inverter is shown in Appendix B.

A **pull-up resistor** (R_{PU}) must be connected between the tied-together output pins and V_{CC} to wire-AND the outputs of the open-collector gates. This is illustrated in Fig. A–7.

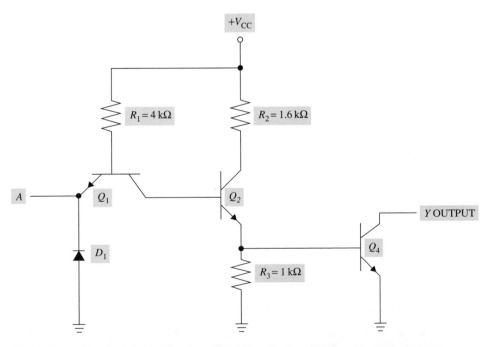

FIGURE A-6   **Standard TTL NOT gate with open-collector outputs–schematic diagram.**

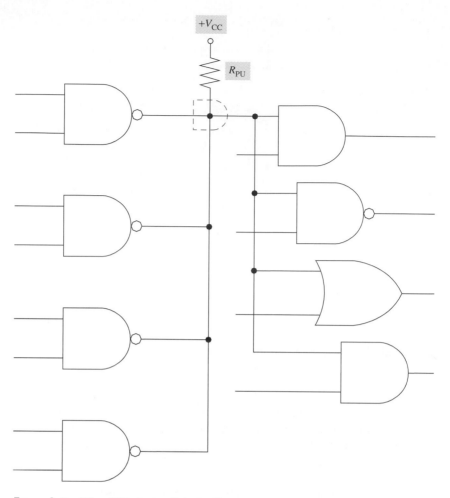

$$\textsc{Figure A-7} \quad \textbf{Wire-ANDed circuit logic diagram.}$$

The value of the pull-up resistor is calculated as follows, using data for the 74LS03 2-input NAND gates with open-collector outputs and using LS family load gates:

$$R_{PU(MIN)} = \frac{V_{CC(MAX)} - V_{OL}}{I_{OL} - N_{2(LOW)} \times 1.6\ \text{mA}}$$

where $N_2 =$ The number of input Unit Loads (ULs) being driven.

The number of unit loads being driven is calculated as follows using the standard 1 UL = 1.6 mA in the low state and 1 UL = 40 μA in the high state.

Input low load factor = I_{IL}/UL = 0.4 mA/1.6 mA = **0.25 UL/gate**

Since there are 4 LS load gates, $N_{2(LOW)}$ = 4 × 0.25 UL = **1 UL**

$$R_{PU(MIN)} = \frac{5.25\ \text{V} - 0.5\ \text{V}}{8\ \text{mA} - 1 \times 1.6\ \text{mA}} = \frac{4.75}{6.4\ \text{mA}} = \textbf{742}\ \boldsymbol{\Omega}$$

$$R_{PU(MAX)} = \frac{V_{CC(MIN)} - V_{OH}}{\#\ \text{Wired outputs} \times I_{OH} + N_{2(HIGH)} \times 40\ \mu\text{A}}$$

Input high load factor = I_{IH}/UL = 20 μA/40 μA = **0.5 UL/gate**

Since there are 4 LS load gates, $N_{2(HIGH)}$ = 4 × 0.5 UL = **2 UL**

$$R_{PU(MAX)} = \frac{4.75\ \text{V} - 2.4\ \text{V}}{4 \times 100\ \mu\text{A} + 2 \times 40\ \mu\text{A}} = \frac{2.35\ \text{V}}{480\ \mu\text{A}} = \textbf{4.89 k}\boldsymbol{\Omega}$$

The value of the pull-up resistor in this example wire-AND problem can be 742 Ω to 4.9 kΩ.

TTL—Tristate Output Gates

Circuits with tristate outputs were discussed in Section 11–3. How the third state—high impedance (Hi-Z)—is achieved can be seen by comparing Fig. A–8(a) and (b).

Figure A–8(a) presents the logic diagram of a standard TTL NOT gate. A high at input A in the figure will cause Q_1's base–collector (BC) junction to conduct as discussed previously in the standard NAND gate's operation. This causes Q_2 to conduct and forward

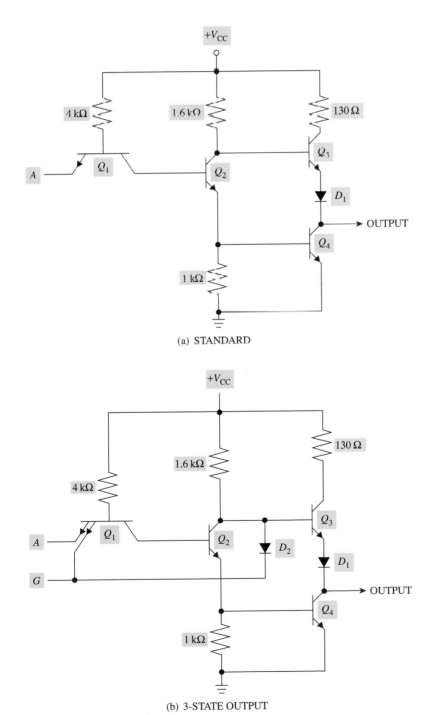

(a) STANDARD

(b) 3-STATE OUTPUT

FIGURE A–8   NOT gate.

bias to be developed across the 1-kΩ resistor. This bias turns on Q_4. Q_4 develops a low output to produce the inverted output signal and insures Q_3 is off.

The NOT gate with 3-state output is shown in Fig. A–8(b). This gate has another input (G) that enables or disables the circuit. The circuit functions as a standard inverter when G is high. The circuit is disabled when G is low. This condition causes both Q_3 and Q_4 to be off, which places the output of the gate in the Hi-Z state. In this state, input A has no effect on the output, and the circuit is electrically disconnected from the load.

$A = 1$ and $G = 1$: Q_1 BC junction is forward biased; Q_2 conducts; and the voltage drop across the 1-kΩ resistor turns on Q_4 to develop a low output. The high at the enable (G) input has no effect on D_2.

$A = 0$ and $G = 1$:   The emitter–base junction at input A is forward-biased and the BC junction of Q_1 is reverse biased, so Q_2 is cut off. The high potential on Q_2's collector causes Q_3 to conduct. Since Q_2 is cut off, no forward-bias voltage is developed across the 1-kΩ resistor, so Q_4 is cut off and produces a high output.

$G = 0$:   The data input is a don't care signal when G is low. Q_2 is off when G is low, and this forces Q_4 to be off. At the same time the low G input places a low through D_2 on the base of Q_3 and forces it to be off. The circuit is in the **Hi-Z state** when both Q_3 and Q_4 are cut off. In this condition, the output of the gate is an open circuit.

TTL Unused Inputs

Unused inputs to logic gates were briefly discussed in Chapter 3. *Unused inputs should not be left floating.* When only two inputs of a 3-input gate are required to accomplish the logic function, the unused input may be jumpered to one of the used inputs. This practice is acceptable for standard TTL gates. One point to keep in mind regarding this subject is the tied-together inputs act as two unit loads when calculating fan-out of a drive gate.

The practice of connecting an unused input to a used input is detrimental to the ac noise immunity of ICs in the LS and FAST families. Therefore, unused inputs of gates in these subfamilies should be tied to an enabler (V_{CC} for AND/NAND gates and 0 V for OR/NOR gates).

TTL Subfamilies

The standard 74-series TTL devices offer relatively high-speed operation and low power dissipation. However, there were users who wanted less power consumption and others who wanted more speed when the family first became available. These user requests quickly resulted in two subfamilies of the standard devices—low-power devices and high-speed devices.

The standard TTL-series ICs were modified in the late 1960s to improve performance. The results produced low-power and high-speed gates. The low-power (74L-series) NAND gate (74L00) appeared as the standard gate shown in Fig. A–1 with one exception—the resistors values were all increased. The increase in resistance produced a reduction in power dissipation.

The high-speed (74H-series) NAND gate (74H00) was also basically the same configuration as the standard NAND gate of Fig. A–1 with a couple of exceptions. First, all of the resistor values in the high-speed version were lower than those of the standard version. Second, the totem-pole transistor Q_3 of the standard version was replaced with a Darlington transistor pair. The main problem encountered with the high-speed version was it used more power.

Subsequent modifications such as those discussed next rendered the 74L- and 74H-series ICs obsolete. The performance ratings of these TTL subfamilies are provided in Table A–1.

TABLE A-1		74-Standard	74L	74H	74S	74LS	74ALS
TTL Subfamilies	t_{PD}(ns)	9	33	6	3	9.5	4
	Power dissipation (mW*)	10	1	23	18.75	2	1.2
	CLK max (MHz)	35	3	50	125	45	70

*Per gate.

The Schottky (74S-series) gates/circuits increased operating speed by using a Schottky diode connected between the collector and base of each transistor as shown in Fig. A–9(a).

The Schottky diode consists of an n-type material joined with metal. This high-frequency, fast-switching diode is characterized by a lower forward voltage drop than a standard pn-diode. The Schottky diode contains no minority-current carriers; thus, it contains no stored charge. The Schottky diode in Fig. A–9(a) diverts excess base current, which prevents the transistor from reaching complete saturation. With practically no stored charges in the transistor or Schottky diode, storage time is minimal and transistor switching time is significantly decreased.

The Schottky diodes are manufactured in the IC when Schottky-family ICs are fabricated. The transistor with the diode in place is called a Schottky transistor, and its symbol is shown in Fig. A–9(b).

A Schottky NAND gate (74S00) diagram is shown in Fig. A–9(c). Transistors Q_3 and Q_5 and their associated resistors were added to the Schottky version of this gate to improve operation by providing more symmetrical switching characteristics. The 74S-series average power dissipation is 18.75 mW per gate, and its typical propagation delay is 3 ns.

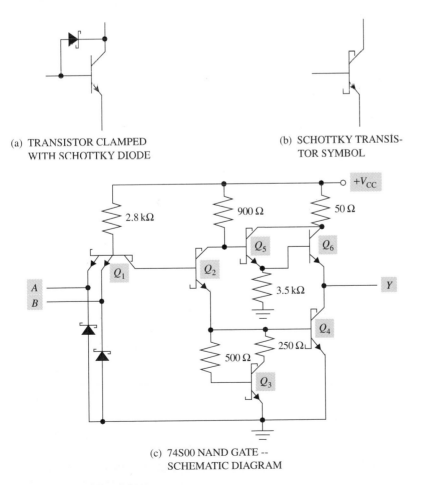

(a) TRANSISTOR CLAMPED
WITH SCHOTTKY DIODE

(b) SCHOTTKY TRANSIS-
TOR SYMBOL

(c) 74S00 NAND GATE --
SCHEMATIC DIAGRAM

FIGURE A-9   **Schottky diode.**

This can be compared to a standard-series TTL gate, which has an average power dissipation of 10 mW per gate and a propagation delay of 9 ns.

Low-power Schottky (74LS-series) gates were modified to use less power than the 74S series, but the modification caused somewhat slower operating speeds. The typical propagation delay of a 74LS gate is 9.5 ns, but power consumption is down to 2 mW per gate.

The 74S-series and the 74LS-series gates were further improved through modification to increase operating speed and decrease power consumption. The modifications resulted in the popular advanced Schottky (74AS) and advanced low-power Schottky (74ALS) series gates.

STANDARD TTL DATA SHEET

Figure A–10 shows the data sheet for a standard TTL NAND gate. *Note:* In addition to the standard 7400 gate, data sheets for the Schottky (74S), low-power Schottky (74LS), advanced Schottky (74AS), and advanced low-power Schottky (74ALS) NAND gates are available in Appendix B. The 7400 data sheet in Fig. A–10 is separated into three sections: (1) recommended operating conditions, (2) electrical characteristics, and (3) switching characteristics.

Recommended Operating Conditions

This section of the data sheet lists the power supply voltage range and the *input* current and voltage requirements. The military version of this IC (SN5400) lists $V_{CC} = 5\ V \pm 10\%$ (4.5 V to 5.5 V). The commercial version (SN7400) lists $V_{CC} = 5\ V \pm 5\%$ (4.75 V to 5.25 V).

Internal gate resistance and loading affect the values of available voltage and current produced by logic gates. Thus, a gate must respond to a range of voltages instead of one specific level to represent a Logic 0 or a Logic 1. The high-level and low-level input voltages are listed as 2 V ($V_{IH\ MIN}$) and 0.8 V ($V_{IL\ MAX}$) on the data sheet. The high-level and low-level output currents (I_{OH} and I_{OL}) are also listed in this section of the data sheet.

Electrical Characteristics

V_{IK} ($-1.5\ V_{max}$) is the input clamp diode voltage, which is typically -0.65 to 0.7 V. The minimum output high-level voltage (V_{OH}) is 2.4 V. Notice this voltage is 0.4 V greater than the required input high voltage level, which is 2.0 V minimum. This ensures V_{OH} is not near the undefined logic region (0.8 to 2.0 V). This is also true for the low logic level because the maximum output low-level voltage (V_{OL}) is 0.4 V—again, this level is 0.4 V less than the maximum 0.8 V for V_{IL} discussed in the recommended operating conditions section.

Power supply current is shown on the data sheet in Fig. A–10 as I_{CCH} and I_{CCL}. These parameters represent the current drawn from the power supply when all of the outputs are high (Logic 1 = I_{CCH}) and when all of the outputs are low (Logic 0 = I_{CCL}). The average power supply current times V_{CC} produces the average power dissipation for the IC. The maximum input currents are listed as I_{IH} and I_{IL}. Negative current values such as $I_{IL} = -1.6$ mA indicate current out of a terminal/pin. Short-circuit output current (I_{OS}) is the current out of the output pin when the output is shorted to ground.

Switching Characteristics

Two switching times are listed on the data sheet—t_{PHL} is the high-to-low switching time and t_{PLH} is the low-to-high switching time. Average propagation delay time t_{pd} is the numerical average of t_{PHL} and t_{PLH}.

SN5400, SN54LS00, SN54S00,
SN7400, SN74LS00, SN74S00
QUADRUPLE 2-INPUT POSITIVE-NAND GATES
DECEMBER 1983–REVISED MARCH 1988

- Package Options Include Plastic "Small Outline" Packages, Ceramic Chip Carriers and Flat Packages, and Plastic and Ceramic DIPs

- Dependable Texas Instruments Quality and Reliability

description

These devices contain four independent 2-input-NAND gates.

The SN5400, SN54LS00, and SN54S00 are characterized for operation over the full military temperature range of −55°C to 125°C. The SN7400, SN74LS00, and SN74S00 are characterized for operation from 0°C to 70°C.

FUNCTION TABLE (each gate)

INPUTS		OUTPUT
A	B	Y
H	H	L
L	X	H
X	L	H

logic symbol†

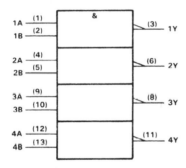

†This symbol is in accordance with ANSI/IEEE Std. 91-1984 and IEC Publication 617-12.
Pin numbers shown are for D, J, and N packages.

SN5400 . . . J PACKAGE
SN54LS00, SN54S00 . . . J OR W PACKAGE
SN7400 . . . N PACKAGE
SN74LS00, SN74S00 . . . D OR N PACKAGE
(TOP VIEW)

```
1A  [1    14]  Vcc
1B  [2    13]  4B
1Y  [3    12]  4A
2A  [4    11]  4Y
2B  [5    10]  3B
2Y  [6     9]  3A
GND [7     8]  3Y
```

SN5400 . . . W PACKAGE
(TOP VIEW)

```
1A  [1    14]  4Y
1B  [2    13]  4B
1Y  [3    12]  4A
Vcc [4    11]  GND
2Y  [5    10]  3B
2A  [6     9]  3A
2B  [7     8]  3Y
```

SN54LS00, SN54S00 . . . FK PACKAGE
(TOP VIEW)

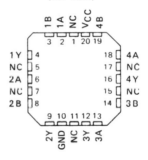

NC – No internal connection

logic diagram (positive logic)

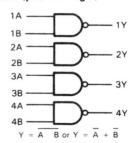

$$Y = \overline{A \cdot B} \text{ or } Y = \overline{A} + \overline{B}$$

FIGURE A-10   SN7400 NAND gate data sheet. *Reprinted by Permission of Texas Instruments.*

SN5400, SN54LS00, SN54S00, SN7400, SN74LS00, SN74S00 QUADRUPLE 2-INPUT POSITIVE-NAND GATES

schematics (each gate)

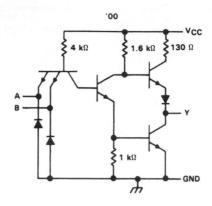

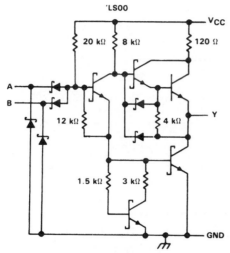

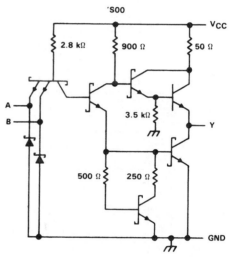

Resistor values shown are nominal.

absolute maximum ratings over operating free-air temperature range (unless otherwise noted)

Supply voltage, V$_{CC}$ (see Note 1) . . . . 7 V
Input voltage: '00, 'S00 . . . . 5.5 V
　　　　　　　'LS00 . . . . 7 V
Operating free-air temperature range: SN54' . . . . −55°C to 125°C
　　　　　　　　　　　　　　　　　　　SN74' . . . . 0°C to 70°C
Storage temperature range . . . . −65°C to 150°C

NOTE 1: Voltage values are with respect to network ground terminal.

FIGURE A–10   Continued. *Reprinted by Permission of Texas Instruments.*

recommended operating conditions

		SN5400			SN7400			UNIT
		MIN	NOM	MAX	MIN	NOM	MAX	
V_{CC}	Supply voltage	4.5	5	5.5	4.75	5	5.25	V
V_{IH}	High-level input voltage	2			2			V
V_{IL}	Low-level input voltage			0.8			0.8	V
I_{OH}	High-level output current			− 0.4			− 0.4	mA
I_{OL}	Low-level output current			16			16	mA
T_A	Operating free-air temperature	− 55		125	0		70	°C

electrical characteristics over recommended operating free-air temperature range (unless otherwise noted)

PARAMETER	TEST CONDITIONS†	SN5400			SN7400			UNIT
		MIN	TYP‡	MAX	MIN	TYP‡	MAX	
V_{IK}	V_{CC} = MIN, I_I = − 12 mA			− 1.5			− 1.5	V
V_{OH}	V_{CC} = MIN, V_{IL} = 0.8 V, I_{OH} = − 0.4 mA	2.4	3.4		2.4	3.4		V
V_{OL}	V_{CC} = MIN, V_{IH} = 2 V, I_{OL} = 16 mA		0.2	0.4		0.2	0.4	V
I_I	V_{CC} = MAX, V_I = 5.5 V			1			1	mA
I_{IH}	V_{CC} = MAX, V_I = 2.4 V			40			40	μA
I_{IL}	V_{CC} = MAX, V_I = 0.4 V			− 1.6			− 1.6	mA
I_{OS}§	V_{CC} = MAX	− 20		− 55	− 18		− 55	mA
I_{CCH}	V_{CC} = MAX, V_I = 0 V		4	8		4	8	mA
I_{CCL}	V_{CC} = MAX, V_I = 4.5 V		12	22		12	22	mA

† For conditions shown as MIN or MAX, use the appropriate value specified under recommended operating conditions.
‡ All typical values are at V_{CC} = 5 V, T_A = 25°C.
§ Not more than one output should be shorted at a time.

switching characteristics, V_{CC} = 5 V, T_A = 25°C (see note 2)

PARAMETER	FROM (INPUT)	TO (OUTPUT)	TEST CONDITIONS	MIN	TYP	MAX	UNIT
t_{PLH}	A or B	Y	R_L = 400 Ω, C_L = 15 pF		11	22	ns
t_{PHL}					7	15	ns

NOTE 2: Load circuits and voltage waveforms are shown in Section 1.

FIGURE A-10   Continued. *Reprinted by Permission of Texas Instruments.*

MOS TECHNOLOGY

Integrated circuits that employ only p-channel E-MOSFETs are referred to as PMOS devices. Those that employ only n-channel E-MOSFETs are known as NMOS devices.

One of the biggest disadvantages of TTL technology is its limited density. This limitation comes, in part, from the resistors that are used in all TTL devices. Resistors consume a lot of chip real estate in ICs.

MOS-type ICs do not use resistors. In some cases they use **load MOSFETs** as resistors, and in other cases the circuits are designed so resistors are unnecessary. If load MOSFETs are used, they do not consume much space compared to standard resistor fabrication on ICs.

The inverter shown in Fig. A–11 is manufactured with n-channel E-MOSFETs. Transistor Q_1's gate is connected to V_{DD}, which means the channel is enhanced, the device is on at all times, and it functions as a load MOSFET. The load MOSFET resistance is approximately 100 kΩ. Transistor Q_2 functions as a switching MOSFET. The

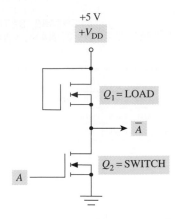

FIGURE A-11   **NMOS inverter (NOT gate).**

switching MOSFET has about 1 kΩ of resistance when it is on (R_{ON}) and approximately 10 GΩ of resistance when it is off (R_{OFF}).

Application of a Logic 1 (+5 V for this example) to the A input enhances the channel of Q_2 and turns it on. With Q_2 on, $R_{ON} \cong 1$ kΩ, and the $\overline{A}$ output = 0.05 V, which is a Logic 0. The circuit is analyzed as two series resistors as illustrated in Fig. A–12(a).

Application of a Logic 0 to the A input of the inverter in Fig. A–11 results in Q_2 being off. Q_2's resistance is now approximately 10 GΩ, so the $\overline{A}$ output = 4.99 V, which is a Logic 1. *Note:* With the very low on resistance of the switching FET and the very high off resistance, the FETs may be thought of as a closed or open switch for purposes of analysis.

Several circuits using FETs are presented in the following paragraphs to illustrate how MOS transistors are configured to produce logic functions. Most digital course objectives (and some employers' entry-level exams) require identification of these circuits in addition to understanding their operation by stating the output when specific input combinations are furnished. You are encouraged to analyze the circuits as described here to identify and determine the output levels and truth tables instead of trying to memorize the circuits. Memorization of circuit identification and truth tables soon fades, and by the time you really need the information, it is gone.

The easiest method of analyzing the following circuits is to construct a truth table and complete that table through simple analysis.

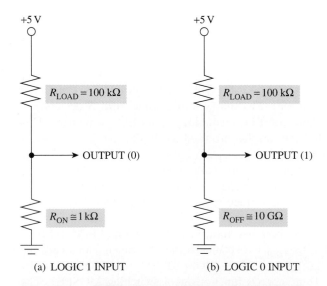

(a) LOGIC 1 INPUT   (b) LOGIC 0 INPUT

FIGURE A-12   **NMOS inverter resistor equivalent circuit.**

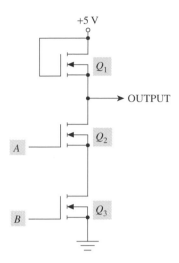

Figure A-13   NMOS NAND gate schematic diagram.

The circuit in Fig. A–13 is an NMOS gate because it contains only n-channel E-MOSFETs. Q_1 is a load MOSFET with 100 kΩ of resistance. Q_2 and Q_3 are both switching MOSFETs with very low on resistance and very high off resistance in respect to the resistance of the load MOSFET.

$A = 0$ **and** $B = 0$ **inputs** leave both N-channel switching FETs off. This produces approximately 20 GΩ of resistance from the output pin to ground in respect to 100 kΩ of load resistance from the load FET. The output is Logic 1.

$A = 0$ **and** $B = 1$ **inputs** cause Q_2 to stay off and Q_3 to turn on. This arrangement produces over 10 GΩ of resistance from the output pin to ground in respect to the 100 kΩ load. The output is Logic 1.

$A = 1$ **and** $B = 0$ **inputs** reverse the condition of Q_2 and Q_3 but leave intact the high resistance between the output and ground that produces the Logic 1 output. During analysis of the remaining gates, only the $A = 0$ and $B = 1$ input combination will be analyzed because the $A = 1$ and $B = 0$ input combination always produces the same output as the $A = 0$ and $B = 1$ combination.

$A = 1$ **and** $B = 1$ **inputs** enhance the channels in both n-channel MOSFETs. This results in about 2 kΩ of resistance to develop the output voltage in respect to the 100-kΩ load resistance. Naturally, the output is Logic 0.

The truth table for this analysis is

A	B	Output
0	0	1
0	1	1
1	0	1
1	1	0

The truth table is representative of the NAND function because "any 0 in equals 1 out." The circuit in Fig. A–13 is identified as an NMOS NAND gate.

The circuit in Fig. A–14 is another NMOS-type gate. A quick analysis allows construction of a truth table which allows gate identification.

$A = 0$ **and** $B = 0$ **inputs** leave the switching n-channel MOSFETs off. The total resistance from V_{OUT} to ground is approximately 5 GΩ, which develops a high logic output.

$A = 0$ **and** $B = 1/A = 1$ **and** $B = 0$ **inputs** turn on one of the switching MOSFETs and leave one of them off. This gives a total resistance of about 1 kΩ from the output to ground, which develops a low output.

$A = 1$ **and** $B = 1$ **inputs** enhance the channel in Q_2 and Q_3, produce a total resistance of about 500 Ω, and produce a low output.

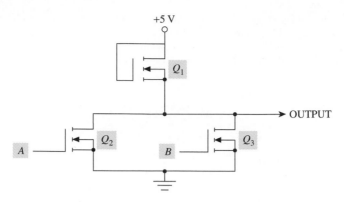

FIGURE A-14   **NMOS NOR gate schematic diagram.**

The following truth table reveals "any 1 in equals a 0 out"—short logic for a NOR gate. Thus, Fig. A–14 is an NMOS NOR gate.

A	B	Output
0	0	1
0	1	0
1	0	0
1	1	0

CMOS technology is used extensively in low-power devices such as digital watches and calculators. CMOS circuits are identified by the fact they use both n-channel and p-channel MOSFETs. This is shown in the CMOS NOT gate in Fig. A–15. The low power consumption characteristic of CMOS gates will soon become evident as circuit analysis is accomplished. The lack of a load MOSFET in this NOT gate is evident because neither E-MOSFET has its gate tied to V_{DD}.

When a Logic 0 is applied to the NOT gate, (1) the p-channel FET (Q_1) turns on because its gate is more negative than its substrate and (2) the n-channel FET (Q_2) is off because its gate voltage does not attract electrons to enhance its channel. This produces the equivalent of a series resistive circuit, with $R_{ON(Q_1)} = 1\ k\Omega$ and $R_{OFF(Q_2)} = 10\ G\Omega =$ Logic 1 output. These conditions are reversed when the input to this gate is Logic 1.

During the remainder of this analysis of CMOS circuits, keep the following in mind: (1) There is always an off MOSFET ($R_{OFF} = 10\ G\Omega$) between ground and V_{DD}, which results in minimal current and power dissipation. (2) Logic inputs to p-channel E-MOSFETs produce the opposite effects of those same inputs to n-channel E-MOSFETs. For example in p-chan-

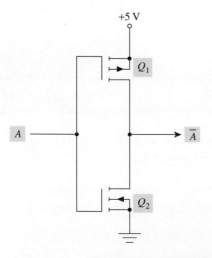

FIGURE A-15   **CMOS inverter (NOT gate).**

nel E-MOSFETs, a *low input* enhances the channel and turns the device *on* whereas a *high input* produces no channel and the device is *off*. The circuit in Fig. A–16 is analyzed next.

A = 0 and B = 0 inputs turn on Q_1 and Q_2 for a total parallel resistance of 500 Ω from the output to V_{DD} and turn off Q_3 and Q_4 for a total resistance of 20 GΩ from the output to ground. This produces a high output.

A = 0 and B = 1 inputs turn on Q_2 and Q_4 and turn off Q_1 and Q_3. This input combination produces a total parallel resistance slightly less than 1 kΩ from the output to V_{DD} and a series resistance of slightly more than 10 GΩ from the output to ground. This resistance combination produces a high output.

A = 1 and B = 1 inputs turn off Q_1 and Q_2 and produce about 5 GΩ of resistance above V_{output}. The input combination turns on Q_3 and Q_4, which develops a low output across about 2 kΩ.

The truth table reveals "any 0 in equals a 1 output" for this CMOS NAND gate whose output is $\overline{AB}$.

If the inverter of Fig. A–15 were connected to the output of the NAND gate in Fig. A–16, a CMOS AND gate would result because $\overline{\overline{AB}} = AB$. The circuit in Fig. A–17 represents a CMOS NOR gate. See if you can prove the short logic "any 1 in equals a 0 output" for this gate.

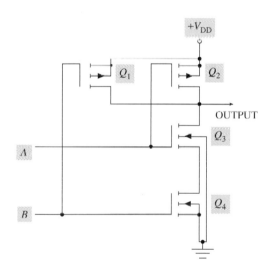

FIGURE A–16   CMOS NAND gate schematic diagram.

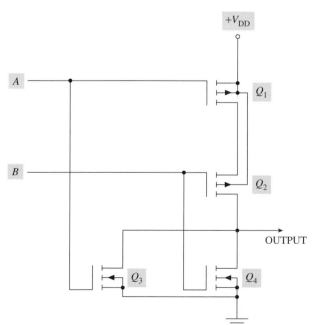

FIGURE A–17   CMOS NOR gate schematic diagram.

EMITTER-COUPLED LOGIC (ECL)

Emitter-coupled logic is another type of bipolar digital logic. ECL gates are used when very high-speed operation is required and other family characteristics are unimportant.

Transistors in the TTL logic family operate in saturation or cutoff. Although this arrangement is simple to control, propagation delay time is important and sometimes significant because the transistor must switch from saturation to cutoff to change output logic levels.

Transistors in the ECL logic family do not operate in saturation and cutoff. Instead, they operate between two voltage levels that are spaced less than 1 V apart. The switching speed is very fast when transistors are operated in this manner. However, this is about the only advantage that can be found for ECL technology. The principle of ECL operation is based on a differential amplifier, which controls current flow through two transistors without letting either of them go into saturation.

Power consumption is high—40 to 50 mW per gate—because the transistors are always conducting. The Logic 0 voltage range (-1.65 V to -1.85 V) and the Logic 1 voltage range (-0.81 V to -0.96 V) make ECL gate voltages incompatible with TTL and CMOS gates. In addition, a negative power source is required. ECL devices have relatively poor noise immunity because the Logic 0 and Logic 1 levels are so close together.

MANUFACTURERS' DATA SHEETS

The data sheets in this appendix are reprinted by permission of Texas Instruments. Several 7400 NAND gate data sheets are shown so that comparisons of electrical and switching characteristics of different families and subfamilies can be made. The standard 2-input NAND gate (7400), Schottky NAND gate (74S00), and low-power Schottky (74LS00) NAND gate are shown in the first set of data sheets. The next set presents the advanced Schottky (74AS00) and the advanced low-power Schottky NAND gates (74ALS00). High-speed CMOS (74HC00/74HCT00) and advanced high-speed CMOS (74AHC00) NAND gate data sheets are presented next. A complete set of data sheets is also presented for the standard, Schottky, and low-power Schottky 7402 NOR gates.

The remaining data sheets (7404–74393) consists of partial sets that normally include circuit description, logic symbol, and logic diagrams. Additional information is available in manufacturers' data books.

7442	4-Line BCD to 10-Line Decimal Decoder
7446	BCD-to-7-Segment Decoder/Driver
7448	BCD-to-7-Segment Decoder/Driver
7474	Dual D-Type Positive-Edge-Triggered Flip-Flops
7475	4-Bit Bistable Latches
7476	Dual J-K Flip-Flops (PGT Triggered)
74LS76	Dual J-K Flip-Flops (NGT Triggered)
7483	4-Bit Binary Full-Adder
7485	4-Bit Magnitude Comparator
7486	Quadruple 2-Input Exclusive-OR Gates
7490	Decade Counter
7491	8-Bit Shift Register
7492	Divide-by-12 Counter
7493	4-Bit Binary Counter
74111	Dual J-K Master–Slave Flip-Flop with Data Lockout
74116	Dual 4-Bit Latches
74125	Quadruple Bus Buffers with 3-State Outputs
74126	Quadruple Bus Buffers with 3-State Outputs
74S135	Quadruple Exclusive-OR/NOR Gates
74LS138	3-Line to 8-Line Decoder/Demultiplexer
74LS139	Dual 2-Line to 4-Line Decoders/Demultiplexers
74147	10-Line Decimal to 4-Line BCD Priority Encoder
74148	8-Line to 3-Line Priority Encoder
74151	Data Selector/Multiplexer
74153	Dual 4-Line to 1-Line Data Selectors/Multiplexers
74157	Quadruple 2-Line to 1-Line Data Selectors/Multiplexers
74163	Synchronous 4-Bit Counter
74164	8-Bit Parallel-Out Serial Shift Register
74165	Parallel-Load 8-Bit Shift Register
74178	4-Bit Parallel-Access Shift Register
74180	9-Bit Odd/Even Parity Generator/Checker
74S181	Arithmetic Logic Unit/Function Generator
74190	Synchronous Up/Down BCD/Binary Counter
74193	Synchronous 4-Bit Up/Down-Counter
74194	4-Bit Bidirectional Universal Shift Register
74198	8-Bit Shift Register
74LS243	Quadruple Bus Transceivers
74LS245	Octal Bus Transceiver with 3-State Outputs
74LS295	4-Bit Right Shift/Left Shift Register with 3-State Output
74393	Dual 4-Bit Decade and Binary Counters

SN5400, SN54LS00, SN54S00,
SN7400, SN74LS00, SN74S00
QUADRUPLE 2-INPUT POSITIVE-NAND GATES
DECEMBER 1983—REVISED MARCH 1988

- Package Options Include Plastic "Small Outline" Packages, Ceramic Chip Carriers and Flat Packages, and Plastic and Ceramic DIPs

- Dependable Texas Instruments Quality and Reliability

description

These devices contain four independent 2-input-NAND gates.

The SN5400, SN54LS00, and SN54S00 are characterized for operation over the full military temperature range of −55°C to 125°C. The SN7400, SN74LS00, and SN74S00 are characterized for operation from 0°C to 70°C.

FUNCTION TABLE (each gate)

INPUTS		OUTPUT
A	**B**	**Y**
H	H	L
L	X	H
X	L	H

logic symbol†

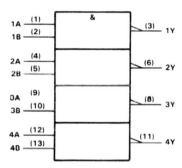

†This symbol is in accordance with ANSI/IEEE Std. 91-1984 and IEC Publication 617-12.

Pin numbers shown are for D, J, and N packages.

SN5400 . . . J PACKAGE
SN54LS00, SN54S00 . . . J OR W PACKAGE
SN7400 . . . N PACKAGE
SN74LS00, SN74S00 . . . D OR N PACKAGE
(TOP VIEW)

1A	1	14 V_{CC}
1B	2	13 4B
1Y	3	12 4A
2A	4	11 4Y
2B	5	10 3B
2Y	6	9 3A
GND	7	8 3Y

SN5400 . . . W PACKAGE
(TOP VIEW)

1A	1	14 4Y
1B	2	13 4B
1Y	3	12 4A
V_{CC}	4	11 GND
2Y	5	10 3B
2A	6	9 3A
2B	7	8 3Y

SN54LS00, SN54S00 . . . FK PACKAGE
(TOP VIEW)

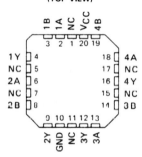

NC – No internal connection

logic diagram (positive logic)

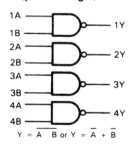

$$Y = \overline{A \cdot B} \text{ or } Y = \overline{A} + \overline{B}$$

FIGURE B-1

SN5400, SN54LS00, SN54S00,
SN7400, SN74LS00, SN74S00
QUADRUPLE 2-INPUT POSITIVE-NAND GATES

schematics (each gate)

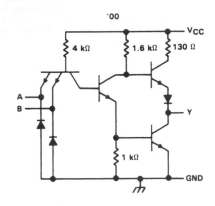

'00

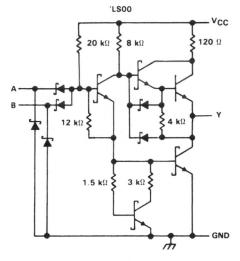

'LS00

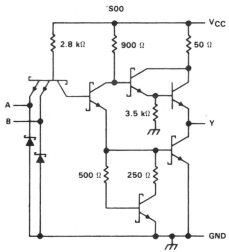

'S00

Resistor values shown are nominal.

absolute maximum ratings over operating free-air temperature range (unless otherwise noted)

Supply voltage, V_{CC} (see Note 1) . . . . 7 V
Input voltage: '00, 'S00 . . . . 5.5 V
           'LS00 . . . . 7 V
Operating free-air temperature range: SN54' . . . . −55°C to 125°C
                             SN74' . . . . 0°C to 70°C
Storage temperature range . . . . −65°C to 150°C

NOTE 1: Voltage values are with respect to network ground terminal.

FIGURE B-2

SN5400, SN7400
QUADRUPLE 2-INPUT POSITIVE-NAND GATES

recommended operating conditions

		SN5400			SN7400			UNIT
		MIN	NOM	MAX	MIN	NOM	MAX	
V_{CC}	Supply voltage	4.5	5	5.5	4.75	5	5.25	V
V_{IH}	High-level input voltage	2			2			V
V_{IL}	Low-level input voltage			0.8			0.8	V
I_{OH}	High-level output current			− 0.4			− 0.4	mA
I_{OL}	Low-level output current			16			16	mA
T_A	Operating free-air temperature	− 55		125	0		70	°C

electrical characteristics over recommended operating free-air temperature range (unless otherwise noted)

PARAMETER	TEST CONDITIONS †			SN5400			SN7400			UNIT
			MIN	TYP‡	MAX	MIN	TYP‡	MAX		
V_{IK}	V_{CC} = MIN,	I_I = − 12 mA			− 1.5			− 1.5		V
V_{OH}	V_{CC} = MIN,	V_{IL} = 0.8 V, I_{OH} = − 0.4 mA	2.4	3.4		2.4	3.4			V
V_{OL}	V_{CC} = MIN,	V_{IH} = 2 V, I_{OL} = 16 mA		0.2	0.4		0.2	0.4		V
I_I	V_{CC} = MAX,	V_I = 5.5 V			1			1		mA
I_{IH}	V_{CC} = MAX,	V_I = 2.4 V			40			40		µA
I_{IL}	V_{CC} = MAX,	V_I = 0.4 V			− 1.6			− 1.6		mA
I_{OS}§	V_{CC} = MAX		− 20		− 55	− 18		− 55		mA
I_{CCH}	V_{CC} = MAX,	V_I = 0 V		4	8		4	8		mA
I_{CCL}	V_{CC} = MAX,	V_I = 4.5 V		12	22		12	22		mA

† For conditions shown as MIN or MAX, use the appropriate value specified under recommended operating conditions.
‡ All typical values are at V_{CC} = 5 V, T_A = 25°C.
§ Not more than one output should be shorted at a time.

switching characteristics, V_{CC} = 5 V, T_A = 25°C (see note 2)

PARAMETER	FROM (INPUT)	TO (OUTPUT)	TEST CONDITIONS	MIN	TYP	MAX	UNIT
t_{PLH}	A or B	Y	R_L = 400 Ω, C_L = 15 pF		11	22	ns
t_{PHL}					7	15	ns

NOTE 2: Load circuits and voltage waveforms are shown in Section 1.

FIGURE B-3

SN54LS00, SN74LS00
QUADRUPLE 2-INPUT POSITIVE-NAND GATES

recommended operating conditions

		SN54LS00			SN74LS00			UNIT
		MIN	NOM	MAX	MIN	NOM	MAX	
V_{CC}	Supply voltage	4.5	5	5.5	4.75	5	5.25	V
V_{IH}	High-level input voltage	2			2			V
V_{IL}	Low-level input voltage			0.7			0.8	V
I_{OH}	High-level output current			−0.4			−0.4	mA
I_{OL}	Low-level output current			4			8	mA
T_A	Operating free-air temperature	−55		125	0		70	°C

electrical characteristics over recommended operating free-air temperature range (unless otherwise noted)

PARAMETER	TEST CONDITIONS †			SN54LS00			SN74LS00			UNIT
				MIN	TYP‡	MAX	MIN	TYP‡	MAX	
V_{IK}	V_{CC} = MIN,	I_I = −18 mA				−1.5			−1.5	V
V_{OH}	V_{CC} = MIN,	V_{IL} = MAX,	I_{OH} = −0.4 mA	2.5	3.4		2.7	3.4		V
V_{OL}	V_{CC} = MIN,	V_{IH} = 2 V,	I_{OL} = 4 mA		0.25	0.4		0.25	0.4	V
	V_{CC} = MIN,	V_{IH} = 2 V,	I_{OL} = 8 mA					0.35	0.5	
I_I	V_{CC} = MAX,	V_I = 7 V				0.1			0.1	mA
I_{IH}	V_{CC} = MAX,	V_I = 2.7 V				20			20	µA
I_{IL}	V_{CC} = MAX,	V_I = 0.4 V				−0.4			−0.4	mA
I_{OS} §	V_{CC} = MAX			−20		−100	−20		−100	mA
I_{CCH}	V_{CC} = MAX,	V_I = 0 V			0.8	1.6		0.8	1.6	mA
I_{CCL}	V_{CC} = MAX,	V_I = 4.5 V			2.4	4.4		2.4	4.4	mA

† For conditions shown as MIN or MAX, use the appropriate value specified under recommended operating conditions.
‡ All typical values are at V_{CC} = 5 V, T_A = 25°C
§ Not more than one output should be shorted at a time, and the duration of the short-circuit should not exceed one second.

switching characteristics, V_{CC} = 5 V, T_A = 25°C (see note 2)

PARAMETER	FROM (INPUT)	TO (OUTPUT)	TEST CONDITIONS		MIN	TYP	MAX	UNIT
t_{PLH}	A or B	Y	R_L = 2 kΩ,	C_L = 15 pF		9	15	ns
t_{PHL}						10	15	ns

NOTE 2: Load circuits and voltage waveforms are shown in Section 1.

FIGURE B-4

SN54S00, SN74S00
QUADRUPLE 2-INPUT POSITIVE-NAND GATES

recommended operating conditions

		SN54S00			SN74S00			UNIT
		MIN	NOM	MAX	MIN	NOM	MAX	
V_{CC}	Supply voltage	4.5	5	5.5	4.75	5	5.25	V
V_{IH}	High-level input voltage	2			2			V
V_{IL}	Low-level input voltage			0.8			0.8	V
I_{OH}	High-level output current			−1			−1	mA
I_{OL}	Low-level output current			20			20	mA
T_A	Operating free-air temperature	−55		125	0		70	°C

electrical characteristics over recommended operating free-air temperature range (unless otherwise noted)

PARAMETER	TEST CONDITIONS †		SN54S00			SN74S00			UNIT
			MIN	TYP‡	MAX	MIN	TYP‡	MAX	
V_{IK}	V_{CC} = MIN,	I_I = −18 mA			−1.2			−1.2	V
V_{OH}	V_{CC} = MIN, V_{IL} = 0.8 V,	I_{OH} = −1 mA	2.5	3.4		2.7	3.4		V
V_{OL}	V_{CC} = MIN, V_{IH} = 2 V,	I_{OL} = 20 mA			0.5			0.5	V
I_I	V_{CC} = MAX,	V_I = 5.5 V			1			1	mA
I_{IH}	V_{CC} = MAX,	V_I = 2.7 V			50			50	µA
I_{IL}	V_{CC} = MAX,	V_I = 0.5 V			−2			−2	mA
I_{OS} §	V_{CC} = MAX		−40		−100	−40		−100	mA
I_{CCH}	V_{CC} = MAX,	V_I = 0 V		10	16		10	16	mA
I_{CCL}	V_{CC} = MAX,	V_I = 4.5 V		20	36		20	36	mA

† For conditions shown as MIN or MAX, use the appropriate value specified under recommended operating conditions.
‡ All typical values are at V_{CC} = 5 V, T_A = 25°C.
§ Not more than one output should be shorted at a time, and the duration of the short-circuit should not exceed one second.

switching characteristics, V_{CC} = 5 V, T_A = 25°C (see note 2)

PARAMETER	FROM (INPUT)	TO (OUTPUT)	TEST CONDITIONS		MIN	TYP	MAX	UNIT
t_{PLH}			R_L = 280 Ω,	C_L = 15 pF		3	4.5	ns
t_{PHL}	A or B	Y				3	5	ns
t_{PLH}			R_L = 280 Ω,	C_L = 50 pF			4.5	ns
t_{PHL}							5	ns

NOTE 2: Load circuits and voltage waveforms are shown in Section 1.

FIGURE B-5

SN54ALS00A, SN54AS00, SN74ALS00A, SN74AS00
QUADRUPLE 2-INPUT POSITIVE-NAND GATES

SDAS187A – APRIL 1982 – REVISED DECEMBER 1994

● **Package Options Include Plastic Small-Outline (D) Packages, Ceramic Chip Carriers (FK), and Standard Plastic (N) and Ceramic (J) 300-mil DIPs**

description

These devices contain four independent 2-input positive-NAND gates. They perform the Boolean functions $Y = \overline{A \cdot B}$ or $Y = \overline{A} + \overline{B}$ in positive logic.

The SN54ALS00A and SN54AS00 are characterized for operation over the full military temperature range of –55°C to 125°C. The SN74ALS00A and SN74AS00 are characterized for operation from 0°C to 70°C.

FUNCTION TABLE
(each gate)

INPUTS		OUTPUT
A	**B**	**Y**
H	H	L
L	X	H
X	L	H

logic symbol†

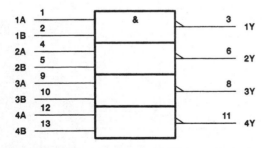

† This symbol is in accordance with ANSI/IEEE Std 91-1984 and IEC Publication 617-12.
Pin numbers shown are for the D, J, and N packages.

FIGURE B-6

SN54ALS00A, SN54AS00 . . . J PACKAGE
SN74ALS00A, SN74AS00 . . . D OR N PACKAGE
(TOP VIEW)

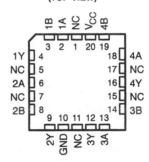

SN54ALS00A, SN54AS00 . . . FK PACKAGE
(TOP VIEW)

NC – No internal connection

SN54ALS00A, SN54AS00, SN74ALS00A, SN74AS00
QUADRUPLE 2-INPUT POSITIVE-NAND GATES

SDAS187A – APRIL 1982 – REVISED DECEMBER 1994

logic diagram (positive logic)

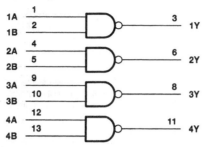

Pin numbers shown are for the D, J, and N packages.

absolute maximum ratings over operating free-air temperature range (unless otherwise noted)†

Supply voltage, V_{CC} .... 7 V
Input voltage, V_I .... 7 V
Operating free-air temperature range, T_A: SN54ALS00A .... −55°C to 125°C
                                            SN74ALS00A .... 0°C to 70°C
Storage temperature range .... −65°C to 150°C

† Stresses beyond those listed under "absolute maximum ratings" may cause permanent damage to the device. These are stress ratings only, and functional operation of the device at these or any other conditions beyond those indicated under "recommended operating conditions" is not implied. Exposure to absolute-maximum-rated conditions for extended periods may affect device reliability.

recommended operating conditions

		SN54ALS00A			SN74ALS00A			UNIT
		MIN	NOM	MAX	MIN	NOM	MAX	
V_{CC}	Supply voltage	4.5	5	5.5	4.5	5	5.5	V
V_{IH}	High-level input voltage	2			2			V
V_{IL}	Low-level input voltage			0.8‡			0.8	V
				0.7§				
I_{OH}	High-level output current			−0.4			−0.4	mA
I_{OL}	Low-level output current			4			8	mA
T_A	Operating free-air temperature	−55		125	0		70	°C

‡ Applies over temperature range −55°C to 70°C
§ Applies over temperature range 70°C to 125°C

FIGURE B-7

SN54ALS00A, SN54AS00, SN74ALS00A, SN74AS00
QUADRUPLE 2-INPUT POSITIVE-NAND GATES

SDAS187A – APRIL 1982 – REVISED DECEMBER 1994

electrical characteristics over recommended operating free-air temperature range (unless otherwise noted)

PARAMETER	TEST CONDITIONS		SN54ALS00A MIN	SN54ALS00A TYP†	SN54ALS00A MAX	SN74ALS00A MIN	SN74ALS00A TYP†	SN74ALS00A MAX	UNIT
V_{IK}	V_{CC} = 4.5 V,	I_I = −18 mA			−1.2			−1.5	V
V_{OH}	V_{CC} = 4.5 V to 5.5 V,	I_{OH} = −0.4 mA	V_{CC} −2			V_{CC} −2			V
V_{OL}	V_{CC} = 4.5 V	I_{OL} = 4 mA		0.25	0.4		0.25	0.4	V
		I_{OL} = 8 mA					0.35	0.5	
I_I	V_{CC} = 5.5 V,	V_I = 7 V			0.1			0.1	mA
I_{IH}	V_{CC} = 5.5 V,	V_I = 2.7 V			20			20	µA
I_{IL}	V_{CC} = 5.5 V,	V_I = 0.4 V			−0.1			−0.1	mA
I_O‡	V_{CC} = 5.5 V,	V_O = 2.25 V	−20		−112	−30		−112	mA
I_{CCH}	V_{CC} = 5.5 V,	V_I = 0		0.5	0.85		0.5	0.85	mA
I_{CCL}	V_{CC} = 5.5 V,	V_I = 4.5 V		1.5	3		1.5	3	mA

† All typical values are at V_{CC} = 5 V, T_A = 25°C.
‡ The output conditions have been chosen to produce a current that closely approximates one half of the true short-circuit output current, I_{OS}.

switching characteristics (see Figure 1)

PARAMETER	FROM (INPUT)	TO (OUTPUT)	V_{CC} = 4.5 V to 5.5 V, C_L = 50 pF, R_L = 500 Ω, T_A = MIN to MAX§				UNIT
			SN54ALS00A MIN	SN54ALS00A MAX	SN74ALS00A MIN	SN74ALS00A MAX	
t_{PLH}	A or B	Y	3	15	3	11	ns
t_{PHL}			2	9	2	8	

§ For conditions shown as MIN or MAX, use the appropriate value specified under recommended operating conditions.

FIGURE B-8

SN54ALS00A, SN54AS00, SN74ALS00A, SN74AS00
QUADRUPLE 2-INPUT POSITIVE-NAND GATES

SDAS187A – APRIL 1982 – REVISED DECEMBER 1994

absolute maximum ratings over operating free-air temperature range (unless otherwise noted)[†]

Supply voltage, V_{CC} .... 7 V
Input voltage, V_I .... 7 V
Operating free-air temperature range, T_A: SN54AS00 .... −55°C to 125°C
                                                SN74AS00 .... 0°C to 70°C
Storage temperature range .... −65°C to 150°C

[†] Stresses beyond those listed under "absolute maximum ratings" may cause permanent damage to the device. These are stress ratings only, and functional operation of the device at these or any other conditions beyond those indicated under "recommended operating conditions" is not implied. Exposure to absolute-maximum-rated conditions for extended periods may affect device reliability.

recommended operating conditions

		SN54AS00			SN74AS00			UNIT
		MIN	NOM	MAX	MIN	NOM	MAX	
V_{CC}	Supply voltage	4.5	5	5.5	4.5	5	5.5	V
V_{IH}	High-level input voltage	2			2			V
V_{IL}	Low-level input voltage			0.8			0.8	V
I_{OH}	High-level output current			−2			−2	mA
I_{OL}	Low-level output current			20			20	mA
T_A	Operating free-air temperature	−55		125	0		70	°C

electrical characteristics over recommended operating free-air temperature range (unless otherwise noted)

PARAMETER	TEST CONDITIONS		SN54AS00			SN74AS00			UNIT
			MIN	TYP[‡]	MAX	MIN	TYP[‡]	MAX	
V_{IK}	V_{CC} = 4.5 V,	I_I = −18 mA			−1.2			−1.2	V
V_{OH}	V_{CC} = 4.5 V to 5.5 V,	I_{OH} = −2 mA	V_{CC} −2			V_{CC} −2			V
V_{OL}	V_{CC} = 4.5 V,	I_{OL} = 20 mA		0.35	0.5		0.35	0.5	V
I_I	V_{CC} = 5.5 V,	V_I = 7 V			0.1			0.1	mA
I_{IH}	V_{CC} = 5.5 V,	V_I = 2.7 V			20			20	µA
I_{IL}	V_{CC} = 5.5 V,	V_I = 0.4 V			−0.5			−0.5	mA
I_O[§]	V_{CC} = 5.5 V,	V_O = 2.25 V	−30		−112	−30		−112	mA
I_{CCH}	V_{CC} = 5.5 V,	V_I = 0		2	3.2		2	3.2	mA
I_{CCL}	V_{CC} = 5.5 V,	V_I = 4.5 V		10.8	17.4		10.8	17.4	mA

[‡] All typical values are at V_{CC} = 5 V, T_A = 25°C.
[§] The output conditions have been chosen to produce a current that closely approximates one half of the true short-circuit output current, I_{OS}.

switching characteristics (see Figure 1)

PARAMETER	FROM (INPUT)	TO (OUTPUT)	V_{CC} = 4.5 V to 5.5 V, C_L = 50 pF, R_L = 500 Ω, T_A = MIN to MAX[¶]				UNIT
			SN54AS00		SN74AS00		
			MIN	MAX	MIN	MAX	
t_{PLH}	A or B	Y	1	5	1	4.5	ns
t_{PHL}			1	5	1	4	

[¶] For conditions shown as MIN or MAX, use the appropriate value specified under recommended operating conditions.

FIGURE B-9

SN54ALS00A, SN54AS00, SN74ALS00A, SN74AS00
QUADRUPLE 2-INPUT POSITIVE-NAND GATES

SDAS187A – APRIL 1982 – REVISED DECEMBER 1994

PARAMETER MEASUREMENT INFORMATION
SERIES 54ALS/74ALS AND 54AS/74AS DEVICES

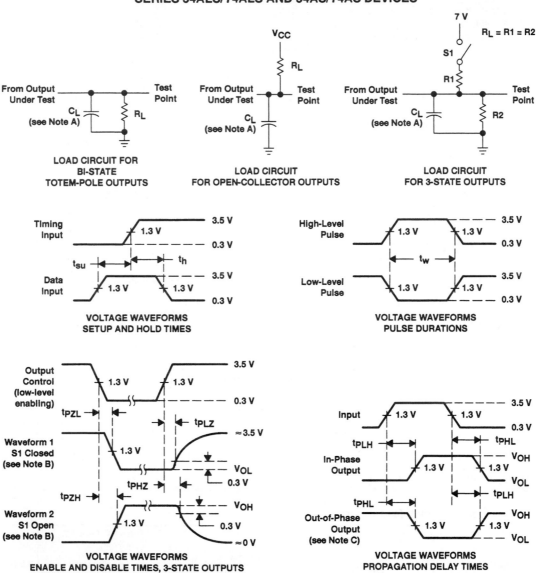

NOTES: A. C_L includes probe and jig capacitance.
B. Waveform 1 is for an output with internal conditions such that the output is low except when disabled by the output control.
Waveform 2 is for an output with internal conditions such that the output is high except when disabled by the output control.
C. When measuring propagation delay items of 3-state outputs, switch S1 is open.
D. All input pulses have the following characteristics: PRR ≤ 1 MHz, t_r = t_f = 2 ns, duty cycle = 50%.
E. The outputs are measured one at a time with one transition per measurement.

Figure 1. Load Circuits and Voltage Waveforms

FIGURE B-10

SN54HC00, SN74HC00
QUADRUPLE 2-INPUT POSITIVE-NAND GATES

SCLS181A – DECEMBER 1982 – REVISED JANUARY 1996

● **Package Options Include Plastic Small-Outline (D), Thin Shrink Small-Outline (PW), and Ceramic Flat (W) Packages, Ceramic Chip Carriers (FK), and Standard Plastic (N) and Ceramic (J) 300-mil DIPs**

description

These devices contain four independent 2-input NAND gates. They perform the Boolean function $Y = \overline{A \cdot B}$ or $Y = \overline{A} + \overline{B}$ in positive logic.

The SN54HC00 is characterized for operation over the full military temperature range of –55°C to 125°C. The SN74HC00 is characterized for operation from –40°C to 85°C.

FUNCTION TABLE
(each gate)

INPUTS		OUTPUT
A	B	Y
H	H	L
L	X	H
X	L	H

SN54HC00 . . . J OR W PACKAGE
SN74HC00 . . . D, N, OR PW PACKAGE
(TOP VIEW)

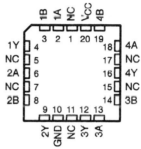

SN54HC00 . . . FK PACKAGE
(TOP VIEW)

NC – No internal connection

logic symbol†

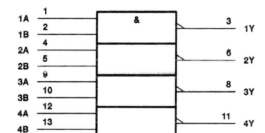

† This symbol is in accordance with ANSI/IEEE Std 91-1984 and IEC Publication 617-12.
Pin numbers shown are for the D, J, N, PW, and W packages.

logic diagram (positive logic)

FIGURE B-11

SN54HC00, SN74HC00
QUADRUPLE 2-INPUT POSITIVE-NAND GATES

SCLS181A – DECEMBER 1982 – REVISED JANUARY 1996

absolute maximum ratings over operating free-air temperature range†

Supply voltage range, V_{CC} .... −0.5 V to 7 V
Input clamp current, I_{IK} ($V_I < 0$ or $V_I > V_{CC}$) (see Note 1) .... ±20 mA
Output clamp current, I_{OK} ($V_O < 0$ or $V_O > V_{CC}$) (see Note 1) .... ±20 mA
Continuous output current, I_O ($V_O = 0$ to V_{CC}) .... ±25 mA
Continuous current through V_{CC} or GND .... ±50 mA
Maximum power dissipation at $T_A = 55°C$ (in still air) (see Note 2): D package .... 1.25 W
                                                                       N package .... 1.1 W
                                                                       PW package .... 0.5 W
Storage temperature range, T_{stg} .... −65°C to 150°C

† Stresses beyond those listed under "absolute maximum ratings" may cause permanent damage to the device. These are stress ratings only, and functional operation of the device at these or any other conditions beyond those indicated under "recommended operating conditions" is not implied. Exposure to absolute-maximum-rated conditions for extended periods may affect device reliability.

NOTES: 1. The input and output voltage ratings may be exceeded if the input and output current ratings are observed.
        2. The maximum package power dissipation is calculated using a junction temperature of 150°C and a board trace length of 750 mils, except for the N package, which has a trace length of zero.

recommended operating conditions

			SN54HC00			SN74HC00			UNIT
			MIN	NOM	MAX	MIN	NOM	MAX	
V_{CC}	Supply voltage		2	5	6	2	5	6	V
V_{IH}	High-level input voltage	$V_{CC} = 2$ V	1.5			1.5			V
		$V_{CC} = 4.5$ V	3.15			3.15			
		$V_{CC} = 6$ V	4.2			4.2			
V_{IL}	Low-level input voltage	$V_{CC} = 2$ V	0		0.5	0		0.5	V
		$V_{CC} = 4.5$ V	0		1.35	0		1.35	
		$V_{CC} = 6$ V	0		1.8	0		1.8	
V_I	Input voltage		0		V_{CC}	0		V_{CC}	V
V_O	Output voltage		0		V_{CC}	0		V_{CC}	V
t_t	Input transition (rise and fall) time	$V_{CC} = 2$ V	0		1000	0		1000	ns
		$V_{CC} = 4.5$ V	0		500	0		500	
		$V_{CC} = 6$ V	0		400	0		400	
T_A	Operating free-air temperature		−55		125	−40		85	°C

Figure B-11   Continued

SN54HC00, SN74HC00
QUADRUPLE 2-INPUT POSITIVE-NAND GATES

SCLS181A – DECEMBER 1982 – REVISED JANUARY 1996

electrical characteristics over recommended operating free-air temperature range (unless otherwise noted)

PARAMETER	TEST CONDITIONS		V_{CC}	T_A = 25°C			SN54HC00		SN74HC00		UNIT
				MIN	TYP	MAX	MIN	MAX	MIN	MAX	
V_{OH}	V_I = V_{IH} or V_{IL}	I_{OH} = −20 µA	2 V	1.9	1.998		1.9		1.9		V
			4.5 V	4.4	4.499		4.4		4.4		
			6 V	5.9	5.999		5.9		5.9		
		I_{OH} = −4 mA	4.5 V	3.98	4.3		3.7		3.84		
		I_{OH} = −5.2 mA	6 V	5.48	5.8		5.2		5.34		
V_{OL}	V_I = V_{IH} or V_{IL}	I_{OL} = 20 µA	2 V		0.002	0.1		0.1		0.1	V
			4.5 V		0.001	0.1		0.1		0.1	
			6 V		0.001	0.1		0.1		0.1	
		I_{OL} = 4 mA	4.5 V		0.17	0.26		0.4		0.33	
		I_{OL} = 5.2 mA	6 V		0.15	0.26		0.4		0.33	
I_I	V_I = V_{CC} or 0		6 V		±0.1	±100		±1000		±1000	nA
I_{CC}	V_I = V_{CC} or 0, I_O = 0		6 V			2		40		20	µA
C_i			2 V to 6 V		3	10		10		10	pF

switching characteristics over recommended operating free-air temperature range, C_L = 50 pF (unless otherwise noted) (see Figure 1)

PARAMETER	FROM (INPUT)	TO (OUTPUT)	V_{CC}	T_A = 25°C			SN54HC00		SN74HC00		UNIT
				MIN	TYP	MAX	MIN	MAX	MIN	MAX	
t_{pd}	A or B	Y	2 V		45	90		135		115	ns
			4.5 V		9	18		27		23	
			6 V		8	15		23		20	
t_t		Y	2 V		38	75		110		95	ns
			4.5 V		8	15		22		19	
			6 V		6	13		19		16	

operating characteristics, T_A = 25°C

PARAMETER		TEST CONDITIONS	TYP	UNIT
C_{pd}	Power dissipation capacitance per gate	No load	20	pF

FIGURE B-11   Continued

SN54HC00, SN74HC00
QUADRUPLE 2-INPUT POSITIVE-NAND GATES

SCLS181A – DECEMBER 1982 – REVISED JANUARY 1996

PARAMETER MEASUREMENT INFORMATION

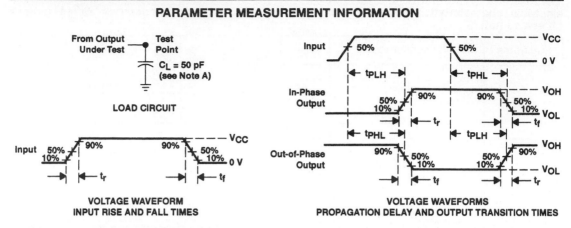

NOTES: A. C_L includes probe and test-fixture capacitance.
        B. Phase relationships between waveforms were chosen arbitrarily. All input pulses are supplied by generators having the following characteristics: PRR ≤ 1 MHz, Z_O = 50 Ω, t_r = 6 ns, t_f = 6 ns.
        C. The outputs are measured one at a time with one input transition per measurement.
        D. t_{PLH} and t_{PHL} are the same as t_{pd}.

Figure 1. Load Circuit and Voltage Waveforms

FIGURE B–11   **Continued**

SN54HCT00, SN74HCT00
QUADRUPLE 2-INPUT POSITIVE-NAND GATES

SCLS062A – NOVEMBER 1988 – REVISED JANUARY 1996

- Inputs Are TTL-Voltage Compatible
- Package Options Include Plastic Small-Outline (D), Thin Shrink Small-Outline (PW), and Ceramic Flat (W) Packages, Ceramic Chip Carriers (FK), and Standard Plastic (N) and Ceramic (J) 300-mil DIPs

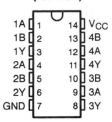

SN54HCT00 . . . J OR W PACKAGE
SN74HCT00 . . . D, N, OR PW PACKAGE
(TOP VIEW)

1A	1	14 Vcc
1B	2	13 4B
1Y	3	12 4A
2A	4	11 4Y
2B	5	10 3B
2Y	6	9 3A
GND	7	8 3Y

description

These devices contain four independent 2-input NAND gates. They perform the Boolean function $Y = \overline{A \cdot B}$ or $Y = \overline{A} + \overline{B}$ in positive logic.

The SN54HCT00 is characterized for operation over the full military temperature range of –55°C to 125°C. The SN74HCT00 is characterized for operation from –40°C to 85°C.

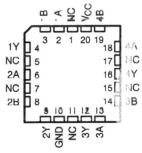

SN54HCT00 . . . FK PACKAGE
(TOP VIEW)

NC – No internal connection

FUNCTION TABLE
(each gate)

INPUTS		OUTPUT
A	B	Y
H	H	L
L	X	H
X	L	H

logic symbol†

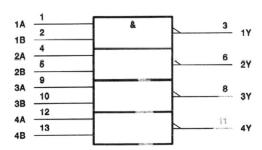

		&		
1A	1		3	1Y
1B	2			
2A	4		6	2Y
2B	5			
3A	9		8	3Y
3B	10			
4A	12		11	4Y
4B	13			

† This symbol is in accordance with ANSI/IEEE Std 91-1984 and IEC Publication 617-12.
Pin numbers shown are for the D, J, N, PW, and W packages.

logic diagram (positive logic)

A
B ———| Y

FIGURE B-11 Continued

SN54HCT00, SN74HCT00
QUADRUPLE 2-INPUT POSITIVE-NAND GATES

SCLS062A – NOVEMBER 1988 – REVISED JANUARY 1996

absolute maximum ratings over operating free-air temperature range[†]

Supply voltage range, V_{CC} .... −0.5 V to 7 V
Input clamp current, I_{IK} ($V_I < 0$ or $V_I > V_{CC}$) (see Note 1) .... ±20 mA
Output clamp current, I_{OK} ($V_O < 0$ or $V_O > V_{CC}$) (see Note 1) .... ±20 mA
Continuous output current, I_O ($V_O = 0$ to V_{CC}) .... ±25 mA
Continuous current through V_{CC} or GND .... ±50 mA
Maximum power dissipation at $T_A = 55°C$ (in still air) (see Note 2): D package .... 1.25 W
    N package .... 1.1 W
    PW package .... 0.5 W
Storage temperature range, T_{stg} .... −65°C to 150°C

[†] Stresses beyond those listed under "absolute maximum ratings" may cause permanent damage to the device. These are stress ratings only, and functional operation of the device at these or any other conditions beyond those indicated under "recommended operating conditions" is not implied. Exposure to absolute-maximum-rated conditions for extended periods may affect device reliability.

NOTES: 1. The input and output voltage ratings may be exceeded if the input and output current ratings are observed.
2. The maximum package power dissipation is calculated using a junction temperature of 150°C and a board trace length of 750 mils, except for the N package, which has a trace length of zero.

recommended operating conditions

			SN54HCT00			SN74HCT00			UNIT
			MIN	NOM	MAX	MIN	NOM	MAX	
V_{CC}	Supply voltage		4.5	5	5.5	4.5	5	5.5	V
V_{IH}	High-level input voltage	$V_{CC} = 4.5$ V to 5.5 V	2			2			V
V_{IL}	Low-level input voltage	$V_{CC} = 4.5$ V to 5.5 V	0		0.8	0		0.8	V
V_I	Input voltage		0		V_{CC}	0		V_{CC}	V
V_O	Output voltage		0		V_{CC}	0		V_{CC}	V
t_t	Input transition (rise and fall) time		0		500	0		500	ns
T_A	Operating free-air temperature		−55		125	−40		85	°C

electrical characteristics over recommended operating free-air temperature range (unless otherwise noted)

PARAMETER	TEST CONDITIONS		V_{CC}	$T_A = 25°C$			SN54HCT00		SN74HCT00		UNIT
				MIN	TYP	MAX	MIN	MAX	MIN	MAX	
V_{OH}	$V_I = V_{IH}$ or V_{IL}	$I_{OH} = -20$ μA	4.5 V	4.4	4.499		4.4		4.4		V
		$I_{OH} = -4$ mA		3.98	4.3		3.7		3.84		
V_{OL}	$V_I = V_{IH}$ or V_{IL}	$I_{OL} = 20$ μA	4.5 V		0.001	0.1		0.1		0.1	V
		$I_{OL} = 4$ mA			0.17	0.26		0.4		0.33	
I_I	$V_I = V_{CC}$ or 0		5.5 V		±0.1	±100		±1000		±1000	nA
I_{CC}	$V_I = V_{CC}$ or 0, $I_O = 0$		5.5 V			2		40		20	μA
ΔI_{CC}[‡]	One input at 0.5 V or 2.4 V, Other inputs at 0 or V_{CC}		5.5 V		1.4	2.4		3		2.9	mA
C_i			4.5 V to 5.5 V		3	10		10		10	pF

[‡] This is the increase in supply current for each input that is at one of the specified TTL voltage levels rather than 0 V or V_{CC}.

FIGURE B–12

SN54HCT00, SN74HCT00
QUADRUPLE 2-INPUT POSITIVE-NAND GATES

SCLS062A – NOVEMBER 1988 – REVISED JANUARY 1996

switching characteristics over recommended operating free-air temperature range, C_L = 50 pF (unless otherwise noted) (see Figure 1)

PARAMETER	FROM (INPUT)	TO (OUTPUT)	V_{CC}	T_A = 25°C			SN54HCT00		SN74HCT00		UNIT
				MIN	TYP	MAX	MIN	MAX	MIN	MAX	
t_{pd}	A or B	Y	4.5 V		11	20		30		25	ns
			5.5 V		10	18		27		22	
t_t		Y	4.5 V		9	15		22		19	ns
			5.5 V		8	14		20		17	

operating characteristics, T_A = 25°C

PARAMETER		TEST CONDITIONS	TYP	UNIT
C_{pd}	Power dissipation capacitance per gate	No load	20	pF

PARAMETER MEASUREMENT INFORMATION

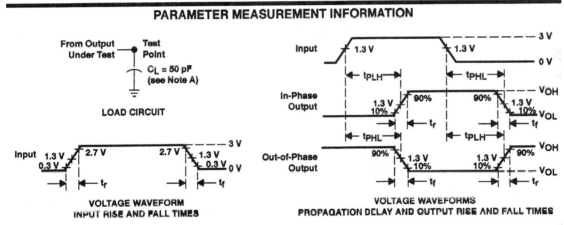

NOTES: A. C_L includes probe and test-fixture capacitance.
B. Phase relationships between waveforms were chosen arbitrarily. All input pulses are supplied by generators having the following characteristics: PRR ≤ 1 MHz, Z_O = 50 Ω, t_r = 6 ns, t_f = 6 ns.
C. The outputs are measured one at a time with one input transition per measurement.
D. t_{PLH} and t_{PHL} are the same as t_{pd}.

Figure 1. Load Circuit and Voltage Waveforms

Figure B-13

SN74AHC00
QUADRUPLE 2-INPUT POSITIVE-NAND GATE

SCLS227 – OCTOBER 1995

- Operating Range: 2-V to 5.5-V V_{CC}
- *EPIC*™ (Enhanced-Performance Implanted CMOS) Process
- High Latch-Up Immunity Exceeds 300 mA Per JEDEC Standard JESD-17
- Package Options Include Plastic Small-Outline (D), Shrink Small-Outline (DB), Thin Shrink Small-Outline (PW) Packages, and Standard Plastic (N) DIPs

D, DB, N, OR PW PACKAGE
(TOP VIEW)

1A	1		14	V_{CC}
1B	2		13	4B
1Y	3		12	4A
2A	4		11	4Y
2B	5		10	3B
2Y	6		9	3A
GND	7		8	3Y

description

The SN74AHC00 performs the Boolean functions $Y = \overline{A \cdot B}$ or $Y = \overline{A} + \overline{B}$ in positive logic.

The SN74AHC00 is characterized for operation from –40°C to 85°C.

FUNCTION TABLE
(each gate)

INPUTS		OUTPUT
A	B	Y
H	H	L
L	X	H
X	L	H

logic symbol†

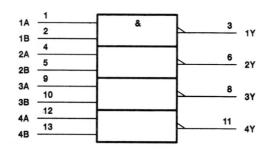

† This symbol is in accordance with ANSI/IEEE Std 91-1984 and IEC Publication 617-12.

logic diagram, each gate (positive logic)

FIGURE B–14

SN74AHC00
QUADRUPLE 2-INPUT POSITIVE-NAND GATE

SCLS227 – OCTOBER 1995

absolute maximum ratings over operating free-air temperature range (unless otherwise noted)†

Supply voltage range, V_{CC} .... −0.5 V to 7 V
Input voltage range, V_I (see Note 1) .... −0.5 V to 7 V
Output voltage range, V_O (see Note 1) .... −0.5 V to V_{CC} + 0.5 V
Input clamp current, I_{IK} ($V_I < 0$) .... −20 mA
Output clamp current, I_{OK} ($V_O < 0$ or $V_O > V_{CC}$) .... ±20 mA
Continuous output current, I_O (V_O = 0 to V_{CC}) .... ±25 mA
Continuous current through V_{CC} or GND .... ±50 mA
Maximum power dissipation at T_A = 55°C (in still air) (see Note 2): D package .... 1.25 W
                                                                    DB or PW package .... 0.5 W
                                                                    N package .... 1.1 W
Storage temperature range, T_{stg} .... −65°C to 150°C

† Stresses beyond those listed under "absolute maximum ratings" may cause permanent damage to the device. These are stress ratings only, and functional operation of the device at these or any other conditions beyond those indicated under "recommended operating conditions" is not implied. Exposure to absolute-maximum-rated conditions for extended periods may affect device reliability.

NOTES: 1. The input and output voltage ratings may be exceeded if the input and output current ratings are observed.
2. The maximum package power dissipation is calculated using a junction temperature of 150°C and a board trace length of 750 mils, except for the N package, which has a trace length of zero.

recommended operating conditions (see Note 3)

			MIN	MAX	UNIT
V_{CC}	Supply voltage		2	5.5	V
V_{IH}	High-level input voltage	V_{CC} = 2 V	1.5		V
		V_{CC} = 3 V	2.1		
		V_{CC} = 5.5 V	3.85		
V_{IL}	Low-level input voltage	V_{CC} = 2 V		0.5	V
		V_{CC} = 3 V		0.9	
		V_{CC} = 5.5 V		1.65	
V_I	Input voltage		0	V_{CC}	V
V_O	Output voltage		0	V_{CC}	V
I_{OH}	High-level output current	V_{CC} = 2 V		50	µA
		V_{CC} = 3.3 V ± 0.3 V		−4	mA
		V_{CC} = 5 V ± 0.5 V		−8	
I_{OL}	Low-level output current	V_{CC} = 2 V		50	µA
		V_{CC} = 3.3 V ± 0.3 V		4	mA
		V_{CC} = 5 V ± 0.5 V		8	
$\Delta t/\Delta v$	Input transition rise or fall rate	V_{CC} = 3.3 V ± 0.3 V		100	ns/V
		V_{CC} = 5 V ± 0.5 V		20	
T_A	Operating free-air temperature		−40	85	°C

NOTE 3: Unused inputs must be held high or low to prevent them from floating.

FIGURE B–15

SN74AHC00
QUADRUPLE 2-INPUT POSITIVE-NAND GATE

SCLS227 – OCTOBER 1995

electrical characteristics over recommended operating free-air temperature range (unless otherwise noted)

PARAMETER		TEST CONDITIONS	V_CC	T_A = 25°C			MIN	MAX	UNIT
				MIN	TYP	MAX			
V_OH		I_OH = −50 µA	2 V	1.9	2		1.9		V
			3 V	2.9	3		2.9		
			4.5 V	4.4	4.5		4.4		
		I_OH = −4 mA	3 V	2.58			2.48		
		I_OH = −8 mA	4.5 V	3.94			3.8		
V_OL		I_OL = 50 µA	2 V			0.1		0.1	V
			3 V			0.1		0.1	
			4.5 V			0.1		0.1	
		I_OL = 4 mA	3 V			0.36		0.44	
		I_OL = 8 mA	4.5 V			0.36		0.44	
I_I	A or B inputs	V_I = V_CC or GND	5.5 V			±0.1		±1	µA
I_CC		V_I = V_CC or GND, I_O = 0	5.5 V			2		20	µA
C_i		V_I = V_CC or GND	5 V		2	10		10	pF

switching characteristics over recommended operating free-air temperature range, V_CC = 3.3 V ± 0.3 V (unless otherwise noted) (see Figure 1)

PARAMETER	FROM (INPUT)	TO (OUTPUT)	LOAD CAPACITANCE	T_A = 25°C			MIN	MAX	UNIT
				MIN	TYP	MAX			
t_PLH	A or B	Y	C_L = 15 pF		5.5	7.9	1	9.5	ns
t_PHL					5.5	7.9	1	9.5	
t_PLH	A or B	Y	C_L = 50 pF		8	11.4	1	13	ns
t_PHL					8	11.4	1	13	

switching characteristics over recommended operating free-air temperature range, V_CC = 5 V ± 0.5 V (unless otherwise noted) (see Figure 1)

PARAMETER	FROM (INPUT)	TO (OUTPUT)	LOAD CAPACITANCE	T_A = 25°C			MIN	MAX	UNIT
				MIN	TYP	MAX			
t_PLH	A or B	Y	C_L = 15 pF		3.7	5.5	1	6.5	ns
t_PHL					3.7	5.5	1	6.5	
t_PLH	A or B	Y	C_L = 50 pF		5.2	7.5	1	8.5	ns
t_PHL					5.2	7.5	1	8.5	

noise characteristics, V_CC = 5 V, C_L = 50 pF, T_A = 25°C (see Note 4)

PARAMETER		MIN	TYP	MAX	UNIT
V_OL(P)	Quiet output, maximum dynamic V_OL		0.3	0.8	V
V_OL(V)	Quiet output, minimum dynamic V_OL		−0.3	−0.8	V
V_OH(V)	Quiet output, minimum dynamic V_OH		4.6		V
V_IH(D)	High-level dynamic input voltage	3.5			V
V_IL(D)	Low-level dynamic input voltage			1.5	V

NOTE 4:   Characteristics are determined during product characterization and ensured by design for surface-mount packages only.

FIGURE B-16

SN74AHC00
QUADRUPLE 2-INPUT POSITIVE-NAND GATE

SCLS227 – OCTOBER 1995

operating characteristics, V_{CC} = 5 V, T_A = 25°C

PARAMETER	TEST CONDITIONS	TYP	UNIT
C_{pd} Power dissipation capacitance	C_L = 50 pF, f = 1 MHz	9.5	pF

PARAMETER MEASUREMENT INFORMATION

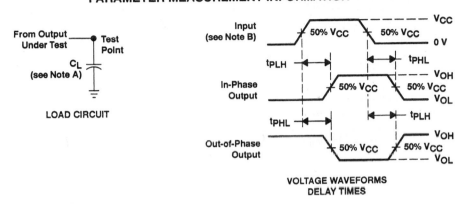

LOAD CIRCUIT

VOLTAGE WAVEFORMS
DELAY TIMES

NOTES: A. C_L includes probe and jig capacitance.
B. All input pulses are supplied by generators having the following characteristics: PRR ≤ 1 MHz, Z_O = 50 Ω, t_r = 3 ns, t_f = 3 ns.
C. The outputs are measured one at a time with one input transition per measurement.

Figure 1. Load Circuit and Voltage Waveforms

FIGURE B-17

SN5402, SN54LS02, SN54S02,
SN7402, SN74LS02, SN74S02
QUADRUPLE 2-INPUT POSITIVE-NOR GATES
DECEMBER 1983—REVISED MARCH 1988

- **Package Options Include Plastic "Small Outline" Packages, Ceramic Chip Carriers and Flat Packages, and Plastic and Ceramic DIPs**

- **Dependable Texas Instruments Quality and Reliability**

description

These devices contain four independent 2-input-NOR gates.

The SN5402, SN54LS02, and SN54S02 are characterized for operation over the full military temperature range of −55°C to 125°C. The SN7402, SN74LS02, and SN74S02 are characterized for operation from 0°C to 70°C.

FUNCTION TABLE (each gate)

INPUTS		OUTPUT
A	B	Y
H	X	L
X	H	L
L	L	H

logic symbol[†]

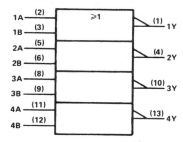

[†]This symbol is in accordance with ANSI/IEEE Std. 91-1984 and IEC Publication 617-12.
Pin numbers shown are for D, J, and N packages.

logic diagram (positive logic)

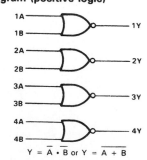

$$Y = \overline{A} \cdot \overline{B} \text{ or } Y = \overline{A + B}$$

SN5402 . . . J PACKAGE
SN54LS02, SN54S02 . . . J OR W PACKAGE
SN7402 . . . N PACKAGE
SN74LS02, SN74S02 . . . D OR N PACKAGE
(TOP VIEW)

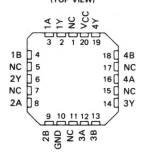

SN5402 . . . W PACKAGE
(TOP VIEW)

1A	1	4Y
1B	2	4B
1Y	3	4A
V_CC	4	GND
2Y	5	3B
2A	6	3A
2B	7	3Y

SN54LS02, SN54S02 . . . FK PACKAGE
(TOP VIEW)

NC – No internal connection

FIGURE B-18

SN5402, SN54LS02, SN54S02, SN7402, SN74LS02, SN74S02
QUADRUPLE 2-INPUT POSITIVE-NOR GATES

schematics (each gate)

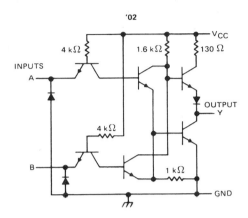

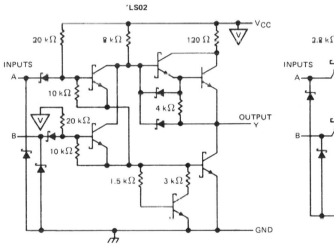

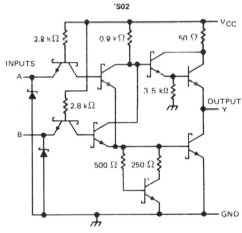

Resistor values shown are nominal.

absolute maximum ratings over operating free-air temperature range (unless otherwise noted)

Supply voltage, V_{CC} (see Note 1) . . . . 7 V
Input voltage: '02, 'S02 . . . . 5.5 V
　　　　　　　　'LS02 . . . . 7 V
Off-state output voltage . . . . 7 V
Operating free-air temperature range: SN54' . . . . −55°C to 125°C
　　　　　　　　　　　　　　　　　　 SN74' . . . . 0°C to 70°C
Storage temperature range . . . . −65°C to 150°C

NOTE 1. Voltage values are with respect to network ground terminal.

FIGURE B-19

SN5402, SN7402
QUADRUPLE 2-INPUT POSITIVE-NOR GATES

recommended operating conditions

		SN5402 MIN	SN5402 NOM	SN5402 MAX	SN7402 MIN	SN7402 NOM	SN7402 MAX	UNIT
V_{CC}	Supply voltage	4.5	5	5.5	4.75	5	5.25	V
V_{IH}	High-level input voltage	2			2			V
V_{IL}	Low-level input voltage			0.8			0.8	V
I_{OH}	High-level output current			− 0.4			− 0.4	mA
I_{OL}	Low-level output current			16			16	mA
T_A	Operating free-air temperature	− 55		125	0		70	°C

electrical characteristics over recommended operating free-air temperature range (unless otherwise noted)

PARAMETER	TEST CONDITIONS†		SN5402 MIN	SN5402 TYP‡	SN5402 MAX	SN7402 MIN	SN7402 TYP‡	SN7402 MAX	UNIT
V_{IK}	V_{CC} = MIN,	I_I = − 12 mA			− 1.5			− 1.5	V
V_{OH}	V_{CC} = MIN, V_{IL} = 0.8 V,	I_{OH} = − 0.4 mA	2.4	3.4		2.4	3.4		V
V_{OL}	V_{CC} = MIN, V_{IH} = 2 V,	I_{OL} = 16 mA		0.2	0.4		0.2	0.4	V
I_I	V_{CC} = MAX,	V_I = 5.5 V			1			1	mA
I_{IH}	V_{CC} = MAX,	V_I = 2.4 V			40			40	μA
I_{IL}	V_{CC} = MAX,	V_I = 0.4 V			− 1.6			− 1.6	mA
I_{OS}§	V_{CC} = MAX		− 20		− 55	− 18		− 55	mA
I_{CCH}	V_{CC} = MAX,	V_I = 0 V		8	16		8	16	mA
I_{CCL}	V_{CC} = MAX,	See Note 2		14	27		14	27	mA

† For conditions shown as MIN or MAX, use the appropriate value specified under recommended operating conditions.
‡ All typical values are at V_{CC} = 5 V, T_A = 25°C.
§ Not more than one output should be shorted at a time.
NOTE 2: One input at 4.5 V, all others at GND.

switching characteristics, V_{CC} = 5 V, T_A = 25°C (see note 3)

PARAMETER	FROM (INPUT)	TO (OUTPUT)	TEST CONDITIONS	MIN	TYP	MAX	UNIT
t_{PLH}	A or B	Y	R_L = 400 Ω, C_L = 15 pF		12	22	ns
t_{PHL}	A or B	Y	R_L = 400 Ω, C_L = 15 pF		8	15	ns

NOTE 3: Load circuits and voltage waveforms are shown in Section 1.

FIGURE B-20

SN54LS02, SN74LS02
QUADRUPLE 2-INPUT POSITIVE-NOR GATES

recommended operating conditions

		SN54LS02			SN74LS02			UNIT
		MIN	NOM	MAX	MIN	NOM	MAX	
V_{CC}	Supply voltage	4.5	5	5.5	4.75	5	5.25	V
V_{IH}	High-level input voltage	2			2			V
V_{IL}	Low-level input voltage			0.7			0.8	V
I_{OH}	High-level output current			-0.4			-0.4	mA
I_{OL}	Low-level output current			4			8	mA
T_A	Operating free-air temperature	-55		125	0		70	$^{\circ}$C

electrical characteristics over recommended operating free-air temperature range (unless otherwise noted)

PARAMETER	TEST CONDITIONS †			SN54LS02			SN74LS02			UNIT
				MIN	TYP‡	MAX	MIN	TYP‡	MAX	
V_{IK}	V_{CC} = MIN,	I_I = -18 mA				-1.5			-1.5	V
V_{OH}	V_{CC} = MIN,	V_{IL} = MAX,	I_{OH} = -0.4 mA	2.5	3.4		2.7	3.4		V
V_{OL}	V_{CC} = MIN,	V_{IH} = 2 V,	I_{OL} = 4 mA		0.25	0.4		0.25	0.4	V
	V_{CC} = MIN,	V_{IH} = 2 V,	I_{OL} = 8 mA					0.35	0.5	
I_I	V_{CC} = MAX,	V_I = 7 V				0.1			0.1	mA
I_{IH}	V_{CC} = MAX,	V_I = 2.7 V				20			20	μA
I_{IL}	V_{CC} = MAX,	V_I = 0.4 V				-0.4			-0.4	mA
I_{OS} §	V_{CC} = MAX			-20		-100	-20		-100	mA
I_{CCH}	V_{CC} = MAX,	V_I = 0 V			1.6	3.2		1.6	3.2	mA
I_{CCL}	V_{CC} = MAX,	See Note 2			2.8	5.4		2.8	5.4	mA

† For conditions shown as MIN or MAX, use the appropriate value specified under recommended operating conditions.
‡ All typical values are at V_{CC} = 5 V, T_A = 25°C
§ Not more than one output should be shorted at a time, and the duration of the short-circuit should not exceed one second.
NOTE 2: One input at 4.5 V all others at GND.

switching characteristics, V_{CC} = 5 V, T_A = 25°C (see note 3)

PARAMETER	FROM (INPUT)	TO (OUTPUT)	TEST CONDITIONS		MIN	TYP	MAX	UNIT
t_{PLH}	A or B	Y	R_L = 2 kΩ,	C_L = 15 pF		10	15	ns
t_{PHL}						10	15	ns

NOTE 3: Load circuits and voltage waveforms are shown in Section 1.

Figure B-21

SN54S02, SN74S02
QUADRUPLE 2-INPUT POSITIVE-NOR GATES

recommended operating conditions

		SN54S02			SN74S02			UNIT
		MIN	NOM	MAX	MIN	NOM	MAX	
V_{CC}	Supply voltage	4.5	5	5.5	4.75	5	5.25	V
V_{IH}	High-level input voltage	2			2			V
V_{IL}	Low-level input voltage			0.8			0.8	V
I_{OH}	High-level output current			−1			−1	mA
I_{OL}	Low-level output current			20			20	mA
T_A	Operating free-air temperature	−55		125	0		70	°C

electrical characteristics over recommended operating free-air temperature range (unless otherwise noted)

PARAMETER	TEST CONDITIONS †	SN54S02			SN74S02			UNIT
		MIN	TYP‡	MAX	MIN	TYP‡	MAX	
V_{IK}	V_{CC} = MIN, I_I = −18 mA			−1.2			−1.2	V
V_{OH}	V_{CC} = MIN, V_{IL} = 0.8 V, I_{OH} = −1 mA	2.5	3.4		2.7	3.4		V
V_{OL}	V_{CC} = MIN, V_{IH} = 2 V, I_{OL} = 20 mA			0.5			0.5	V
I_I	V_{CC} = MAX, V_I = 5.5 V			1			1	mA
I_{IH}	V_{CC} = MAX, V_I = 2.7 V			50			50	μA
I_{IL}	V_{CC} = MAX, V_I = 0.5 V			−2			−2	mA
I_{OS}§	V_{CC} = MAX	−40		−100	−40		−100	mA
I_{CCH}	V_{CC} = MAX, V_I = 0 V		17	29		17	29	mA
I_{CCL}	V_{CC} = MAX, See Note 2		26	45		26	45	mA

† For conditions shown as MIN or MAX, use the appropriate value specified under recommended operating conditions.
‡ All typical values are at V_{CC} = 5 V, T_A = 25°C.
§ Not more than one output should be shorted at a time, and the duration of the short-circuit should not exceed one second.
NOTE 2: One input at 4.5 V, all others at GND.

switching characteristics, V_{CC} = 5 V, T_A = 25°C (see note 3)

PARAMETER	FROM (INPUT)	TO (OUTPUT)	TEST CONDITIONS	MIN	TYP	MAX	UNIT
t_{PLH}	A or B	Y	R_L = 280 Ω, C_L = 15 pF		3.5	5.5	ns
t_{PHL}					3.5	5.5	ns
t_{PLH}			R_L = 280 Ω, C_L = 50 pF		5		ns
t_{PHL}					5		ns

NOTE 3: Load circuits and voltage waveforms are shown in Section 1.

FIGURE B-22

SN5404, SN54LS04, SN54S04,
SN7404, SN74LS04, SN74S04
HEX INVERTERS
DECEMBER 1983 – REVISED MARCH 1988

- Package Options Include Plastic ''Small Outline'' Packages, Ceramic Chip Carriers and Flat Packages, and Plastic and Ceramic DIPs

- Dependable Texas Instruments Quality and Reliability

description

These devices contain six independent inverters.

The SN5404, SN54LS04, and SN54S04 are characterized for operation over the full military temperature range of −55 °C to 125 °C. The SN7404, SN74LS04, and SN74S04 are characterized for operation from 0 °C to 70 °C.

FUNCTION TABLE (each inverter)

INPUTS A	OUTPUT Y
H	L
L	H

logic symbol[†]

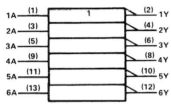

[†]This symbol is in accordance with ANSI/IEEE Std. 91-1984 and IEC Publication 617-12.

Pin numbers shown are for D, J, and N packages.

logic diagram (positive logic)

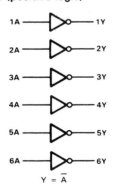

$$Y = \overline{A}$$

SN5404 . . . J PACKAGE
SN54LS04, SN54S04 . . . J OR W PACKAGE
SN7404 . . . N PACKAGE
SN74LS04, SN74S04 . . . D OR N PACKAGE
(TOP VIEW)

1A	1	14 V_CC
1Y	2	13 6A
2A	3	12 6Y
2Y	4	11 5A
3A	5	10 5Y
3Y	6	9 4A
GND	7	8 4Y

SN5404 . . . W PACKAGE
(TOP VIEW)

1A	1	14 1Y
2Y	2	13 6A
2A	3	12 6Y
V_CC	4	11 GND
3A	5	10 5Y
3Y	6	9 5A
4A	7	8 4Y

SN54LS04, SN54S04 . . . FK PACKAGE
(TOP VIEW)

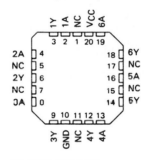

NC - No internal connection

FIGURE B-23

SN5405, SN54LS05, SN54S05,
SN7405, SN74LS05, SN74S05
HEX INVERTERS WITH OPEN-COLLECTOR OUTPUTS
DECEMBER 1983 — REVISED MARCH 1988

- Package Option Includes Plastic "Small Outline" Packages, Ceramic Chip Carriers and Flat Packages, and Plastic and Ceramic DIPs

- Dependable Texas Instruments Quality and Reliability

description

These devices contain six independent inverters. The open-collector outputs require pull-up resistors to perform correctly. They may be connected to other open-collector outputs to implement active-low wired-OR or active-high wired-AND functions. Open-collector devices are often used to generate high V_{OH} levels.

The SN5405, SN54LS05, and SN54S05 are characterized for operation over the full military temperature range of −55°C to 125°C. The SN7405, SN74LS05, and SN74S05 are characterized for operation from 0°C to 70°C.

FUNCTION TABLE (each inverter)

INPUT	OUTPUT
A	Y
H	L
L	H

logic diagram (positive logic)

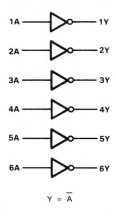

$$Y = \overline{A}$$

SN5405 . . . J PACKAGE
SN54LS05, SN54S05 . . . J OR W PACKAGE
SN7405 . . . N PACKAGE
SN74LS05, SN74S05 . . . D OR N PACKAGE
(TOP VIEW)

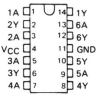

SN5405 . . . W PACKAGE
(TOP VIEW)

SN54LS05, SN54S05 . . . FK PACKAGE
(TOP VIEW)

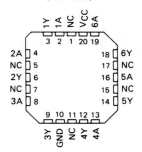

NC—No internal connection

logic symbol[†]

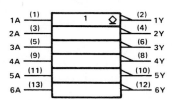

[†] This symbol is in accordance with ANSI/IEEE Std 91-1984 and IEC Publication 617-12.
Pin numbers shown are for D, J, N, and W packages.

Figure B-24

**SN5408, SN54LS08, SN54S08,
SN7408, SN74LS08, SN74S08**
QUADRUPLE 2-INPUT POSITIVE-AND GATES
DECEMBER 1983 — REVISED MARCH 1988

- **Package Options Include Plastic "Small Outline" Packages, Ceramic Chip Carriers and Flat Packages, and Plastic and Ceramic DIPs**

- **Dependable Texas Instruments Quality and Reliability**

description

These devices contain four independent 2-input AND gates.

The SN5408, SN54LS08, and SN54S08 are characterized for operation over the full military temperature range of −55°C to 125°C. The SN7408, SN74LS08 and SN74S08 are characterized for operation from 0° to 70°C.

FUNCTION TABLE (each gate)

INPUTS		OUTPUT
A	B	Y
H	H	H
L	X	L
X	L	L

logic symbol †

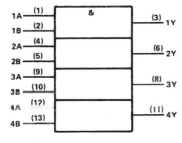

† This symbol is in accordance with ANSI/IEEE Std 91-1984 and IEC Publication 617-12.
Pin numbers shown are for D, J, N, and W packages.

SN5408, SN54LS08, SN54S08 . . . J OR W PACKAGE
SN7408 . . . J OR N PACKAGE
SN74LS08, SN74S08 . . . D, J OR N PACKAGE
(TOP VIEW)

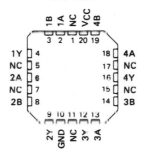

SN54LS08, SN54S08 . . . FK PACKAGE
(TOP VIEW)

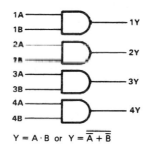

NC—No internal connection

logic diagram (positive logic)

$$Y = A \cdot B \quad \text{or} \quad Y = \overline{\overline{A} + \overline{B}}$$

FIGURE B-25

SN5410, SN54LS10, SN54S10,
SN7410, SN74LS10, SN74S10
TRIPLE 3-INPUT POSITIVE-NAND GATES
DECEMBER 1983—REVISED MARCH 1988

- Package Options Include Plastic "Small Outline" Packages, Ceramic Chip Carriers and Flat Packages, and Plastic and Ceramic DIPs
- Dependable Texas Instruments Quality and Reliability

description

These devices contain three independent 3-input NAND gates.

The SN5410, SN54LS10, and SN54S10 are characterized for operation over the full military temperature range of −55°C to 125°C. The SN7410, SN74LS10, and SN74S10 are characterized for operation from 0°C to 70°C.

FUNCTION TABLE (each gate)

INPUTS			OUTPUT
A	B	C	Y
H	H	H	L
L	X	X	H
X	L	X	H
X	X	L	H

logic symbol[†]

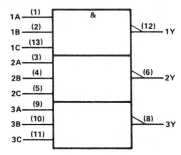

[†]This symbol is in accordance with ANSI/IEEE Std. 91-1984 and IEC Publication 617-12.
Pin numbers shown are for D, J, and N packages.

positive logic

$$Y = \overline{A \cdot B \cdot C} \text{ or } Y = \overline{A} + \overline{B} + \overline{C}$$

SN5410 . . . J PACKAGE
SN54LS10, SN54S10 . . . J OR W PACKAGE
SN7410 . . . N PACKAGE
SN74LS10, SN74S10 . . . D OR N PACKAGE
(TOP VIEW)

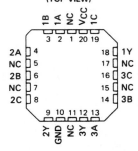

SN54LS10, SN54S10 . . . FK PACKAGE
(TOP VIEW)

NC - No internal connection

logic diagram (positive logic)

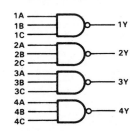

FIGURE B-26

SN54LS11, SN54S11, SN74LS11, SN74S11
TRIPLE 3-INPUT POSITIVE-AND GATES
APRIL 1985 – REVISED MARCH 1988

- Package Options Include Plastic "Small Outline" Packages, Ceramic Chip Carriers and Flat Packages, and Plastic and Ceramic DIPs

- Dependable Texas Instruments Quality and Reliability

description

These devices contain three independent 3-input AND gates.

The SN54LS11 and SN54S11 are characterized for operation over the full military temperature range of −55 °C to 125 °C. The SN74LS11 and SN74S11 are characterized for operation from 0 °C to 70 °C.

FUNCTION TABLE (each gate)

INPUTS			OUTPUT
A	B	C	Y
H	H	H	H
L	X	X	L
X	L	X	L
X	X	L	L

SN54LS11, SN74S11 . . . J OR W PACKAGE
SN74LS11, SN74S11 . . . D OR N PACKAGE
(TOP VIEW)

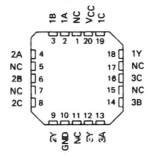

SN54LS11, SN54S11 . . . FK PACKAGE
(TOP VIEW)

NC – No internal connection

logic symbol†

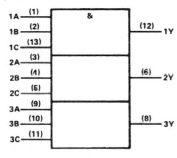

logic diagram (positive logic)

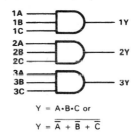

$$Y = A \cdot B \cdot C \text{ or}$$
$$Y = \overline{\overline{A} + \overline{B} + \overline{C}}$$

†This symbol is in accordance with ANSI/IEEE Std. 91-1984 and IEC Publication 617-12.
Pin numbers shown are for D, J, N, and W packages.

FIGURE B-27

APRIL 1985 — REVISED MARCH 1988

SN54LS21, SN74LS21
DUAL 4-INPUT POSITIVE-AND GATES

- **Package Options Include Plastic "Small Outline" Packages, Ceramic Chip Carriers and Flat Packages, and Plastic and Ceramic DIPs**

- **Dependable Texas Instruments Quality and Reliability**

description

These devices contain two independent 4-input AND gates.

The SN54LS21 is characterized for operation over the full military temperature range of −55°C to 125°C. The SN74LS21 is characterized for operation from 0°C to 70°C.

FUNCTION TABLE (each gate)

INPUTS				OUTPUT
A	B	C	D	Y
H	H	H	H	H
L	X	X	X	L
X	L	X	X	L
X	X	L	X	L
X	X	X	L	L

SN54LS21 . . . J OR W PACKAGE
SN74LS21 . . . D OR N PACKAGE
(TOP VIEW)

1A	1	14	V_CC
1B	2	13	2D
NC	3	12	2C
1C	4	11	NC
1D	5	10	2B
1Y	6	9	2A
GND	7	8	2Y

SN54LS21 . . . FK PACKAGE
(TOP VIEW)

NC — No internal connection

logic symbol†

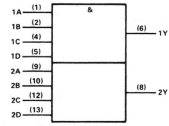

† This symbol is in accordance with ANSI/IEEE Std 91-1984 and IEC Publication 617-12.
Pin numbers shown are for D, J, N, and W packages.

logic diagram

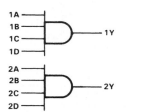

(positive logic) $Y = A \cdot B \cdot C \cdot D$ or $Y = \overline{\overline{A} + \overline{B} + \overline{C} + \overline{D}}$

FIGURE B-28

SN5432, SN54LS32, SN54S32,
SN7432, SN74LS32, SN74S32
QUADRUPLE 2-INPUT POSITIVE-OR GATES

DECEMBER 1983 – REVISED MARCH 1988

- **Package Options Include Plastic "Small Outline" Packages, Ceramic Chip Carriers and Flat Packages, and Plastic and Ceramic DIPs**

- **Dependable Texas Instruments Quality and Reliability**

description

These devices contain four independent 2-input OR gates.

The SN5432, SN54LS32 and SN54S32 are characterized for operation over the full military range of −55°C to 125°C. The SN7432, SN74LS32 and SN74S32 are characterized for operation from 0°C to 70°C.

FUNCTION TABLE (each gate)

INPUTS		OUTPUT
A	B	Y
H	X	H
X	H	H
L	L	L

logic symbol†

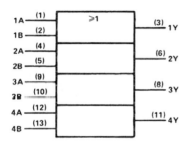

† This symbol is in accordance with ANSI/IEEE Std 91-1984 and IEC Publication 617-12.
Pin numbers shown are for D, J, N, or W packages.

SN5432, SN54LS32, SN54S32 . . . J OR W PACKAGE
SN7432 . . . N PACKAGE
SN74LS32, SN74S32 . . . D OR N PACKAGE
(TOP VIEW)

```
        ____ ____
1A  [ 1      14 ] VCC
1B  [ 2      13 ] 4B
1Y  [ 3      12 ] 4A
2A  [ 4      11 ] 4Y
2B  [ 5      10 ] 3B
2Y  [ 6       9 ] 3A
GND [ 7       8 ] 3Y
```

SN54LS32, SN54S32 . . . FK PACKAGE
(TOP VIEW)

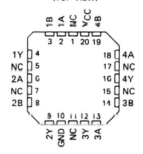

NC – No internal connection

logic diagram

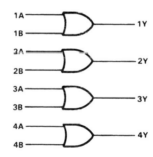

positive logic

$$Y = A + B \text{ or } Y = \overline{\overline{A} \cdot \overline{B}}$$

FIGURE B-29

SN5442A, SN54LS42, SN74442A, SN74LS42
4-LINE BCD TO 10-LINE DECIMAL DECODERS

MARCH 1974—REVISED MARCH 1988

- All Outputs Are High for Invalid Input Conditions

- Also for Application as
  4-Line-to-16-Line Decoders
  3-Line-to-8-Line Decoders

- Diode-Clamped Inputs

TYPES	TYPICAL POWER DISSIPATION	TYPICAL PROPAGATION DELAYS
'42A	140 mW	17 ns
'LS42	35 mW	17 ns

SN5442A, SN54LS42 . . . J OR W PACKAGE
SN7442A . . . N PACKAGE
SN74LS42 . . . D OR N PACKAGE
(TOP VIEW)

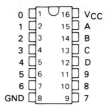

description

These monolithic BCD-to-decimal decoders consist of eight inverters and ten four-input NAND gates. The inverters are connected in pairs to make BCD input data available for decoding by the NAND gates. Full decoding of valid input logic ensures that all outputs remain off for all invalid input conditions.

The '42A and 'LS42 feature inputs and outputs that are compatible for use with most TTL and other saturated low-level logic circuits. DC noise margins are typically one volt.

The SN5442A and SN54LS42 are characterized for operation over the full military temperature range of −55 °C to 125 °C. The SN7442A and SN74LS42 are characterized for operation from 0 °C to 70 °C.

SN54LS42 . . . FK PACKAGE
(TOP VIEW)

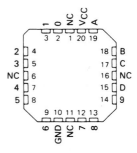

NC - No internal connection

FIGURE B-30

**SN5446A, '47A, '48, SN54LS47, 'LS48, 'LS49,
SN7446A, '47A, '48, SN74LS47, 'LS48, 'LS49
BCD-TO-SEVEN-SEGMENT DECODERS/DRIVERS**

MARCH 1974 — REVISED MARCH 1988

'46A, '47A, 'LS47 feature	'48, 'LS48 feature	'LS49 feature
• Open-Collector Outputs Drive Indicators Directly	• Internal Pull-Ups Eliminate Need for External Resistors	• Open-Collector Outputs
• Lamp-Test Provision	• Lamp-Test Provision	• Blanking Input
• Leading/Trailing Zero Suppression	• Leading/Trailing Zero Suppression	

SN5446A, SN5447A, SN54LS47, SN5448,
SN54LS48 . . . J PACKAGE
SN7446A, SN7447A,
SN7448 . . . N PACKAGE
SN74LS47, SN74LS48 . . . D OR N PACKAGE
(TOP VIEW)

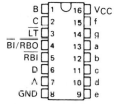

SN54LS47, SN54LS48 . . . FK PACKAGE
(TOP VIEW)

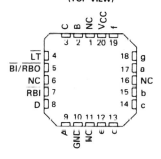

SN54LS49 . . . J OR W PACKAGE
SN74LS49 . . . D OR N PACKAGE
(TOP VIEW)

SN54LS49 . . . FK PACKAGE
(TOP VIEW)

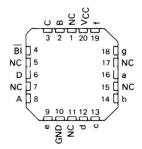

NC — No internal connection

FIGURE B–31

SN5474, SN54LS74A, SN54S74,
SN7474, SN74LS74A, SN74S74
DUAL D-TYPE POSITIVE-EDGE-TRIGGERED FLIP-FLOPS WITH PRESET AND CLEAR
DECEMBER 1983 — REVISED MARCH 1988

- **Package Options Include Plastic "Small Outline" Packages, Ceramic Chip Carriers and Flat Packages, and Plastic and Ceramic DIPs**

- **Dependable Texas Instruments Quality and Reliability**

description

These devices contain two independent D-type positive-edge-triggered flip-flops. A low level at the preset or clear inputs sets or resets the outputs regardless of the levels of the other inputs. When preset and clear are inactive (high), data at the D input meeting the setup time requirements are transferred to the outputs on the positive-going edge of the clock pulse. Clock triggering occurs at a voltage level and is not directly related to the rise time of the clock pulse. Following the hold time interval, data at the D input may be changed without affecting the levels at the outputs.

The SN54' family is characterized for operation over the full military temperature range of −55°C to 125°C. The SN74' family is characterized for operation from 0°C to 70°C.

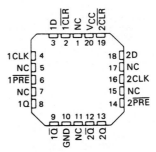

```
SN5474 . . . J PACKAGE
SN54LS74A, SN54S74 . . . J OR W PACKAGE
SN7474 . . . N PACKAGE
SN74LS74A, SN74S74 . . . D OR N PACKAGE
(TOP VIEW)
```

```
SN5474 . . . W PACKAGE
(TOP VIEW)
```

```
SN54LS74A, SN54S74 . . . FK PACKAGE
(TOP VIEW)
```

NC - No internal connection

FUNCTION TABLE

INPUTS				OUTPUTS	
PRE	CLR	CLK	D	Q	Q̄
L	H	X	X	H	L
H	L	X	X	L	H
L	L	X	X	H†	H†
H	H	↑	H	H	L
H	H	↑	L	L	H
H	H	L	X	Q_0	$\overline{Q}_0$

† The output levels in this configuration are not guaranteed to meet the minimum levels in V_{OH} if the lows at preset and clear are near V_{IL} maximum. Furthermore, this configuration is nonstable; that is, it will not persist when either preset or clear returns to its inactive (high) level.

logic symbol‡

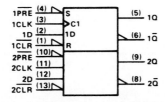

‡This symbol is in accordance with ANSI/IEEE Std 91-1984 and IEC Publication 617-12.
Pin numbers shown are for D, J, N, and W packages.

logic diagram (positive logic)

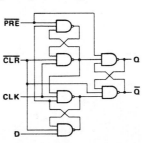

FIGURE B-32

SN5475, SN5477, SN54LS75, SN54LS77,
SN7475, SN74LS75
4-BIT BISTABLE LATCHES
MARCH 1974 — REVISED MARCH 1988

FUNCTION TABLE
(each latch)

INPUTS		OUTPUTS	
D	**C**	**Q**	**$\overline{Q}$**
L	H	L	H
H	H	H	L
X	L	Q_0	$\overline{Q_0}$

H = high level, L = low level, X = irrelevant

Q_0 = the level of Q before the high-to-low transition of G

description

These latches are ideally suited for use as temporary storage for binary information between processing units and input/output or indicator units. Information present at a data (D) input is transferred to the Q output when the enable (C) is high and the Q output will follow the data input as long as the enable remains high. When the enable goes low, the information (that was present at the data input at the time the transition occurred) is retained at the Q output until the enable is permitted to go high

The '75 and 'LS75 feature complementary Q and $\overline{Q}$ outputs from a 4-bit latch, and are available in various 16-pin packages. For higher component density applications, the '77 and 'LS77 4-bit latches are available in 14-pin flat packages.

These circuits are completely compatible with all popular TTL families. All inputs are diode-clamped to minimize transmission-line effects and simplify system design. Series 54 and 54LS devices are characterized for operation over the full military temperature range of −55°C to 125°C; Series 74, and 74LS devices are characterized for operation from 0°C to 70°C.

SN5475, SN54LS75 . . . J OR W PACKAGE
SN7475 . . . N PACKAGE
SN74LS75 . . . D OR N PACKAGE
(TOP VIEW)

```
        ┌───┬─┬───┐
   1Q̄ ┌┤1  └─┘  16├┐ 1Q
   1D ┌┤2       15├┐ 2Q
   2D ┌┤3       14├┐ 2Q̄
3C, 4C ┌┤4       13├┐ 1C, 2C
  VCC ┌┤5       12├┐ GND
   3D ┌┤6       11├┐ 3Q̄
   4D ┌┤7       10├┐ 3Q
   4Q̄ ┌┤8        9├┐ 4Q
        └─────────┘
```

SN5477, SN54LS77 . . . W PACKAGE
(TOP VIEW)

```
        ┌───┬─┬───┐
   1D ┌┤1  └─┘  14├┐ 1Q
   2D ┌┤2       13├┐ 2Q
3C, 4C ┌┤3       12├┐ 1C, 2C
  VCC ┌┤4       11├┐ GND
   3D ┌┤5       10├┐ NC
   4D ┌┤6        9├┐ 3Q
   NC ┌┤7        8├┐ 4Q
        └─────────┘
```

NC - No internal connection

logic symbols†

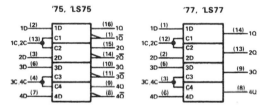

'75, 'LS75   '77, 'LS77

†These symbols are in accordance with ANSI/IEEE Std 91-1984 and IEC Publication 617-12.

absolute maximum ratings over operating free-air temperature range (unless otherwise noted)

Supply voltage, V_{CC} (See Note 1) . . . . 7 V

Input voltage: '75, '77 . . . . 5.5 V

'LS75, 'LS77 . . . . 7 V

Interemitter voltage (see Note 2) . . . . 5.5 V

Operating free-air temperature range: SN54' . . . . −55°C to 125°C

SN74' . . . . 0°C to 70°C

Storage temperature range . . . . −65°C to 150°C

NOTES: 1. Voltage values are with respect to network ground terminal.
2. This is the voltage between two emitters of a multiple-emitter input transistor and is not applicable to the 'LS75 and 'LS77.

FIGURE B-33

SN5476, SN54LS76A, SN7476, SN74LS76A
DUAL J-K FLIP-FLOPS WITH PRESET AND CLEAR
DECEMBER 1983—REVISED MARCH 1988

- Package Options Include Plastic and Ceramic DIPs and Ceramic Flat Packages

- Dependable Texas Instruments Quality and Reliability

description

The '76 contains two independent J-K flip-flops with individual J-K, clock, preset, and clear inputs. The '76 is a positive-edge-triggered flip-flop. J-K input is loaded into the master while the clock is high and transferred to the slave on the high-to-low transition. For these devices the J and K inputs must be stable while the clock is high.

The 'LS76A contain two independent negative-edge-triggered flip-flops. The J and K inputs must be stable one setup time prior to the high-to-low clock transition for predicatble operation. The preset and clear are asynchronous active low inputs. When low they override the clock and data inputs forcing the outputs to the steady state levels as shown in the function table.

The SN5476 and the SN54LS76A are characterized for operation over the full military temperature range of −55°C to 125°C. The SN7476 and the SN74LS76A are characterized for operation from 0°C to 70°C.

SN5476, SN54LS76A . . . J PACKAGE
SN7476 . . . N PACKAGE
SN74LS76A . . . D OR N PACKAGE
(TOP VIEW)

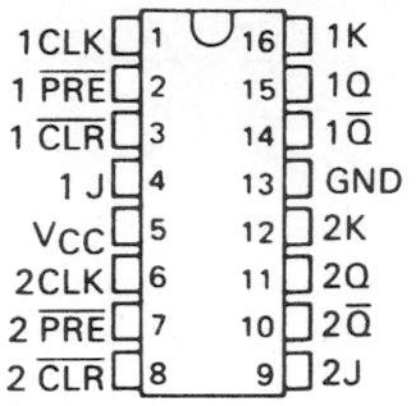

```
1CLK  [ 1   U  16 ]  1K
1 PRE [ 2      15 ]  1Q
1 CLR [ 3      14 ]  1Q̄
  1 J [ 4      13 ]  GND
 VCC  [ 5      12 ]  2K
2CLK  [ 6      11 ]  2Q
2 PRE [ 7      10 ]  2Q̄
2 CLR [ 8       9 ]  2J
```

'76
FUNCTION TABLE

INPUTS					OUTPUTS	
PRE	CLR	CLK	J	K	Q	Q̄
L	H	X	X	X	H	L
H	L	X	X	X	L	H
L	L	X	X	X	H[†]	H[†]
H	H	⊓	L	L	Q_0	$\bar{Q}_0$
H	H	⊓	H	L	H	L
H	H	⊓	L	H	L	H
H	H	⊓	H	H	TOGGLE	

'LS76A
FUNCTION TABLE

INPUTS					OUTPUTS	
PRE	CLR	CLK	J	K	Q	Q̄
L	H	X	X	X	H	L
H	L	X	X	X	L	H
L	L	X	X	X	H[†]	H[†]
H	H	↓	L	L	Q_0	$\bar{Q}_0$
H	H	↓	H	L	H	L
H	H	↓	L	H	L	H
H	H	↓	H	H	TOGGLE	
H	H	H	X	X	Q_0	$\bar{Q}_0$

† This configuration is nonstable; that is, it will not persist when either preset or clear returns to its inactive (high) level.

FIGURE B-34

SN5483A, SN54LS83A, SN7483A, SN74LS83A
4-BIT BINARY FULL ADDDERS WITH FAST CARRY

MARCH 1974 – REVISED MARCH 1988

- Full-Carry Look-Ahead across the Four Bits

- Systems Achieve Partial Look-Ahead Performance with the Economy of Ripple Carry

- SN54283/SN74283 and SN54LS283/SN74LS283 Are Recommended For New Designs as They Feature Supply Voltage and Ground on Corner Pins to Simplify Board Layout

TYPE	TYPICAL ADD TIMES		TYPICAL POWER DISSIPATION PER 4-BIT ADDER
	TWO 8-BIT WORDS	TWO 16-BIT WORDS	
'83A	23 ns	43 ns	310 mW
'LS83A	25 ns	45 ns	95 mW

description

These Improved full adders perform the addition of two 4-bit binary numbers. The sum (Σ) outputs are provided for each bit and the resultant carry (C4) is obtained from the fourth bit. These adders feature full internal look ahead across all four bits generating the carry term in ten nanoseconds typically. This provides the system designer with partial look-ahead performance at the economy and reduced package count of a ripple-carry implementation.

The adder logic, including the carry, is implemented in its true form meaning that the end-around carry can be accomplished without the need for logic or level inversion.

Designed for medium-speed applications, the circuits utilize transistor-transistor logic that is compatible with most other TTL families and other saturated low-level logic families.

Series 54 and 54LS circuits are characterized for operation over the full military temperature range of $-55°C$ to $125°C$, and Series 74 and 74LS circuits are characterized for operation from $0°C$ to $70°C$.

logic symbol[†]

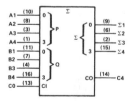

[†]This symbol is in accordance with ANSI/IEEE Std 91-1984 and IEC Publication 617-12.
Pin numbers are for D, J, N, and W packages.

SN5483A,SN54LS83A . . . J OR W PACKAGE
SN7483A . . . N PACKAGE
SN74LS83A . . . D OR N PACKAGE
(TOP VIEW)

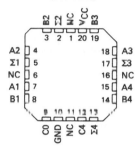

SN54LS83A . . . FK PACKAGE
(TOP VIEW)

NC - No internal connection

FUNCTION TABLE

INPUT				OUTPUT					
				WHEN C0 = L			WHEN C0 = H		
						WHEN C2 = L			WHEN C2 = H
A1 / A2	B1 / B3	A2 / A4	B2 / B4	Σ1 / Σ3	Σ2 / Σ4	C2 / C4	Σ1 / Σ3	Σ2 / Σ4	C2 / C4
L	L	L	L	L	L	L	H	L	L
H	L	L	L	H	L	L	L	H	L
L	H	L	L	H	L	L	L	H	L
H	H	L	L	L	H	L	H	H	L
L	L	H	L	L	H	L	H	H	L
H	L	H	L	H	H	L	L	L	H
L	H	H	L	H	H	L	L	L	H
H	H	H	L	L	L	H	H	L	H
L	L	L	H	L	H	L	H	H	L
H	L	L	H	H	H	L	L	L	H
L	H	L	H	H	H	L	L	L	H
H	H	L	H	L	L	H	H	L	H
L	L	H	H	L	L	H	H	L	H
H	L	H	H	H	L	H	L	H	H
L	H	H	H	H	L	H	L	H	H
H	H	H	H	L	H	H	H	H	H

H = high level, L = low level

NOTE: Input conditions at A1, B1, A2, B2, and C0 are used to determine outputs Σ1 and Σ2 and the value of the internal carry C2. The values at C2, A3, B3, A4, and B4 are then used to determine outputs Σ3, Σ4, and C4.

FIGURE B-35

TYPE	TYPICAL POWER DISSIPATION	TYPICAL DELAY (4-BIT WORDS)
'85	275 mW	23 ns
'LS85	52 mW	24 ns
'S85	365 mW	11 ns

SN5485, SN54LS85, SN54S85 . . . J OR W PACKAGE
SN7485 . . . N PACKAGE
SN74LS85, SN74S85 . . . D OR N PACKAGE
(TOP VIEW)

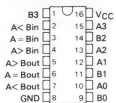

SN54LS85, SN54S85 . . . FK PACKAGE
(TOP VIEW)

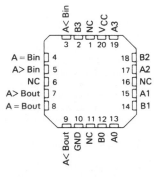

NC - No internal connection

description

These four-bit magnitude comparators perform comparison of straight binary and straight BCD (8-4-2-1) codes. Three fully decoded decisions about two 4-bit words (A, B) are made and are externally available at three outputs. These devices are fully expandable to any number of bits without external gates. Words of greater length may be compared by connecting comparators in cascade. The A > B, A < B, and A = B outputs of a stage handling less-significant bits are connected to the corresponding A > B, A < B, and A = B inputs of the next stage handling more-significant bits. The stage handling the least-significant bits must have a high-level voltage applied to the A = B input. The cascading paths of the '85, 'LS85, and 'S85 are implemented with only a two-gate-level delay to reduce overall comparison times for long words. An alternate method of cascading which further reduces the comparison time is shown in the typical application data.

FUNCTION TABLE

COMPARING INPUTS				CASCADING INPUTS			OUTPUTS		
A3, B3	A2, B2	A1, B1	A0, B0	A > B	A < B	A = B	A > B	A < B	A = B
A3 > B3	X	X	X	X	X	X	H	L	L
A3 < B3	X	X	X	X	X	X	L	H	L
A3 = B3	A2 > B2	X	X	X	X	X	H	L	L
A3 = B3	A2 < B2	X	X	X	X	X	L	H	L
A3 = B2	A2 = B2	A1 > B1	X	X	X	X	H	L	L
A3 = B3	A2 = B2	A1 < B1	X	X	X	X	L	H	L
A2 = B3	A2 = B2	A1 = B1	A0 > B0	X	X	X	H	L	L
A3 = B3	A2 = B2	A1 = B1	A0 < B0	X	X	X	L	H	L
A3 = B3	A2 = B2	A1 = B1	A0 = B0	H	L	L	H	L	L
A3 = B3	A2 = B2	A1 = B1	A0 = B0	L	H	L	L	H	L
A3 = B3	A2 = B2	A1 = B1	A0 = B0	X	X	H	L	L	H
A3 = B3	A2 = B2	A1 = B1	A0 = B0	H	H	L	L	L	L
A3 = B3	A2 = B2	A1 = B1	A0 = B0	L	L	L	H	H	L

FIGURE B-36

SN5486, SN54LS86A, SN54S86,
SN7486, SN74LS86A, SN74S86
QUADRUPLE 2-INPUT EXCLUSIVE-OR GATES
DECEMBER 1972—REVISED MARCH 1988

- Package Options Include Plastic "Small Outline" Packages, Ceramic Chip Carriers and Flat Packages, and Standard Plastic and Ceramic 300-mil DIPs

- Dependable Texas Instruments Quality and Reliability

TYPE	TYPICAL AVERAGE PROPAGATION DELAY TIME	TYPICAL TOTAL POWER DISSIPATION
'86	14 ns	150 mW
'LS86A	10 ns	30.5 mW
'S86	7 ns	250 mW

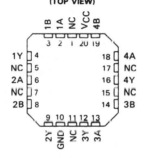

SN5486, SN54LS86A, SN54S86 . . . J OR W PACKAGE
SN7486 . . . N PACKAGE
SN74LS86A, SN74S86 . . . D OR N PACKAGE
(TOP VIEW)

```
      1A [ 1    U  14 ] Vcc
      1B [ 2       13 ] 4B
      1Y [ 3       12 ] 4A
      2A [ 4       11 ] 4Y
      2B [ 5       10 ] 3B
      2Y [ 6        9 ] 3A
     GND [ 7        8 ] 3Y
```

SN54LS86A, SN54S86 . . . FK PACKAGE
(TOP VIEW)

```
        1B  1A  NC  VCC 4B
         3   2   1  20  19
    1Y [ 4              18 ] 4A
    NC [ 5              17 ] NC
    2A [ 6              16 ] 4Y
    NC [ 7              15 ] NC
    2B [ 8              14 ] 3B
         9  10  11  12  13
        2Y GND  NC  3Y  3A
```

NC - No internal connection

description

These devices contain four independent 2-input Exclusive-OR gates. They perform the Boolean functions $Y = A \oplus B = \overline{A}B + A\overline{B}$ in positive logic.

A common application is as a true/complement element. If one of the inputs is low, the other input will be reproduced in true form at the output. If one of the inputs is high, the signal on the other input will be reproduced inverted at the output.

The SN5486, 54LS86A, and the SN54S86 are characterized for operation over the full military temperature range of −55°C to 125°C. The SN7486, SN74LS86A, and the SN74S86 are characterized for operation from 0°C to 70°C.

exclusive-OR logic

An exclusive-OR gate has many applications, some of which can be represented better by alternative logic symbols.

EXCLUSIVE-OR

These are five equivalent Exclusive-OR symbols valid for an '86 or 'LS86A gate in positive logic; negation may be shown at any two ports.

LOGIC IDENTITY ELEMENT	EVEN-PARITY	ODD-PARITY ELEMENT
The output is active (low) if all inputs stand at the same logic level (i.e., A = B).	The output is active (low) if an even number of inputs (i.e., 0 or 2) are active.	The output is active (high) if an odd number of inputs (i.e., only 1 of the 2) are active.

FIGURE B-37

SN5490A, SN5492A, SN5493A, SN54LS90, SN54LS92, SN54LS93, SN7490A, SN7492A, SN7493A, SN74LS90, SN74LS92, SN74LS93
DECADE, DIVIDE-BY-TWELVE AND BINARY COUNTERS
MARCH 1974—REVISED MARCH 1988

'90A, 'LS90 . . . Decade Counters

'92A, 'LS92 . . . Divide By-Twelve Counters

'93A, 'LS93 . . . 4-Bit Binary Counters

TYPES	TYPICAL POWER DISSIPATION
'90A	145 mW
'92A, '93A	130 mW
'LS90, 'LS92, 'LS93	45 mW

description

Each of these monolithic counters contains four master-slave flip-flops and additional gating to provide a divide-by-two counter and a three-stage binary counter for which the count cycle length is divide-by-five for the '90A and 'LS90, divide-by-six for the '92A and 'LS92, and the divide-by-eight for the '93A and 'LS93.

All of these counters have a gated zero reset and the '90A and 'LS90 also have gated set-to-nine inputs for use in BCD nine's complement applications.

To use their maximum count length (decade, divide-by-twelve, or four-bit binary) of these counters, the CKB input is connected to the Q_A output. The input count pulses are applied to CKA input and the outputs are as described in the appropriate function table. A symmetrical divide-by-ten count can be obtained from the '90A or 'LS90 counters by connecting the Q_D output to the CKA input and applying the input count to the CKB input which gives a divide-by-ten square wave at output Q_A.

SN5490A, SN54LS90 . . . J OR W PACKAGE
SN7490A . . . N PACKAGE
SN74LS90 . . . D OR N PACKAGE
(TOP VIEW)

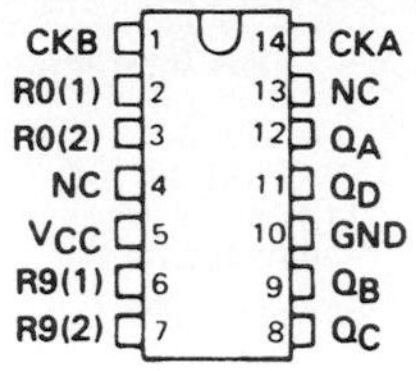

```
       ┌───┬─┐───┐
 CKB  □│1  └─┘ 14│□ CKA
R0(1) □│2      13│□ NC
R0(2) □│3      12│□ Q_A
  NC  □│4      11│□ Q_D
 V_CC □│5      10│□ GND
R9(1) □│6       9│□ Q_B
R9(2) □│7       8│□ Q_C
       └─────────┘
```

SN5492A, SN54LS92 . . . J OR W PACKAGE
SN7492A . . . N PACKAGE
SN74LS92 . . . D OR N PACKAGE
(TOP VIEW)

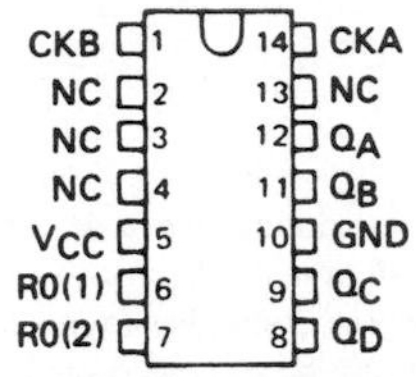

```
       ┌───┬─┐───┐
 CKB  □│1  └─┘ 14│□ CKA
  NC  □│2      13│□ NC
  NC  □│3      12│□ Q_A
  NC  □│4      11│□ Q_B
 V_CC □│5      10│□ GND
R0(1) □│6       9│□ Q_C
R0(2) □│7       8│□ Q_D
       └─────────┘
```

SN5493A, SN54LS93 . . . J OR W PACKAGE
SN7493 . . . N PACKAGE
SN74LS93 . . . D OR N PACKAGE
(TOP VIEW)

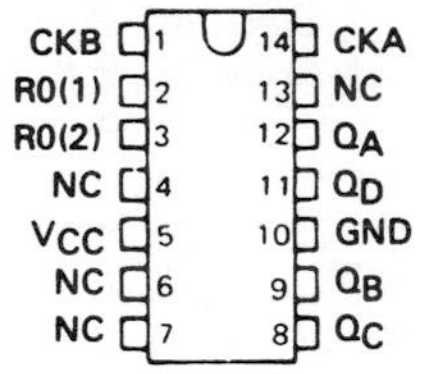

```
       ┌───┬─┐───┐
 CKB  □│1  └─┘ 14│□ CKA
R0(1) □│2      13│□ NC
R0(2) □│3      12│□ Q_A
  NC  □│4      11│□ Q_D
 V_CC □│5      10│□ GND
  NC  □│6       9│□ Q_B
  NC  □│7       8│□ Q_C
       └─────────┘
```

NC—No internal connection

FIGURE B-38

SN5491A, SN54LS91, SN7491A, SN74LS91
8-BIT SHIFT REGISTERS

MARCH 1974 – REVISED MARCH 1988

- **For applications in:**
  - **Digital Computer Systems**
  - **Data-Handling Systems**
  - **Control Systems**

TYPE	TYPICAL MAXIMUM CLOCK FREQUENCY	TYPICAL POWER DISSIPATION
'91A	18 MHz	175 mW
'LS91	18 MHz	60 mW

description

These monolithic serial-in, serial-out, 8-bit shift registers utilize transistor-transistor logic (TTL) circuits and are composed of eight R-S master-slave flip-flops, input gating, and a clock driver. Single-rail data and input control are gated through inputs A and B and an internal inverter to form the complementary inputs to the first bit of the shift register. Drive for the internal common clock line is provided by an inverting clock driver. This clock pulse inverter/driver causes these circuits to shift information one bit on the positive edge of an input clock pulse.

SN5491A, SN54LS91 . . . J PACKAGE
SN7491A . . . N PACKAGE
SN74LS91 . . . D OR N PACKAGE
(TOP VIEW)

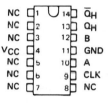

```
NC   [1      14]  Q̄H
NC   [2      13]  QH
NC   [3      12]  A
NC   [4      11]  B
VCC  [5      10]  GND
NC   [6       9]  CLK
NC   [7       8]  NC
```

SN5491A, SN54LS91 . . . W PACKAGE
(TOP VIEW)

```
NC   [1      14]  Q̄H
NC   [2      13]  QH
NC   [3      12]  B
VCC  [4      11]  GND
NC   [5      10]  A
NC   [6       9]  CLK
NC   [7       8]  NC
```

NC - No internal connection

schematics of inputs and outputs

FUNCTION TABLE

INPUTS AT t_n		OUTPUTS AT t_{n+8}	
A	B	Q_H	$\overline{Q}_H$
H	H	H	L
L	X	L	H
X	L	L	H

t_n = Reference bit time, clock low

t_{n+8} = Bit time after 8 low-to-high clock transitions.

logic symbol†

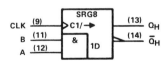

```
CLK (9) ──▷C1/→    SRG8   (13)── QH
 B  (11) ──┐         &   1D (14)── Q̄H
 A  (12) ──┘
```

†This symbol is in accordance with ANSI/IEEE Std 91-1984 and IEC Publication 617-12.

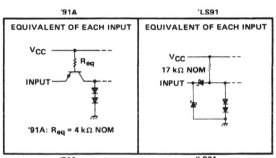

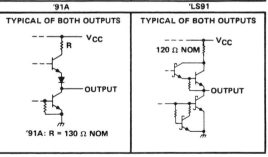

FIGURE B-39

SN54111, SN74111
DUAL J-K MASTER-SLAVE
FLIP-FLOPS WITH DATA LOCKOUT
DECEMBER 1983 — REVISED MARCH 1988

- **Package Options Include Plastic and Ceramic DIPs and Ceramic Flat Packages**
- **Dependable Texas Instruments Quality and Reliability**

description

The SN54111 and SN74111 are d-c coupled, variable-skew, J-K flip-flops which utilize TTL circuitry to obtain 25-MHz performance typically. They are termed "variable-skew" because they allow the maximum clock skew in a system to be a direct function of the clock pulse width. The J and K inputs are enabled to accept data only during a short period (30 nanoseconds maximum hold time) starting with, and immediately following the rising edge of the clock pulse. After this, inputs may be changed while the clock is at the high level without affecting the state of the master. At the threshold level of the falling edge of the clock pulse, the data stored in the master will be transferred to the output. The effective allowable clock skew then is minimum propagation delay time minus hold time, plus clock pulse width. This means that the system designer can set the maximum allowable clock skew needed by varying the clock pulse width. Thus system design is made easier and the requirements for sophisticated clock distribution systems are minimized or, in some cases, entirely eliminated. These flip-flops have an additional feature-the synchronous input has reduced sensitivity to data change while the clock is high because the data need be present for only a short period of time and the system's susceptibility to noise is thereby effectively reduced.

The SN54111 is characterized for operation over the full military temperature range of −55°C to 125°C; the SN74111 is characterized for operation from 0°C to 70°C.

SN54111 . . . J PACKAGE
SN74111 . . . N PACKAGE
(TOP VIEW)

```
       ┌────┐ ┌────┐
  1K  │1     16│ VCC
 1PRE │2     15│ 2K
 1CLR │3     14│ 2PRE
  1J  │4     13│ 2CLR
 1CLK │5     12│ 2J
  1Q̄  │6     11│ 2CLK
  1Q  │7     10│ 2Q̄
 GND  │8      9│ 2Q
       └────────┘
```

logic symbol[†]

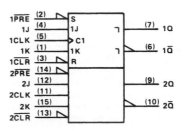

[†]This symbol is in accordance with ANSI/IEEE Std 91-1984 and IEC Publication 617-12.

FUNCTION TABLE

INPUTS					OUTPUTS	
PRE	CLR	CLK	J	K	Q	Q̄
L	H	X	X	X	H	L
H	L	X	X	X	L	H
L	L	X	X	X	H[‡]	H[‡]
H	H	⎍	L	L	Q_0	$\overline{Q}_0$
H	H	⎍	H	L	H	L
H	H	⎍	L	H	L	H
H	H	⎍	H	H	TOGGLE	

[‡]This configuration is non-stable; that is, it will not persist when preset or clear return to their inactive (high) level.

FIGURE B-40

SN54116, SN74116
DUAL 4-BIT LATCHES WITH CLEAR

DECEMBER 1972—REVISED MARCH 1988

- Two Independent 4-Bit Latches in a Single Package
- Separate Clear Inputs Provide One-Step Clearing Operation
- Dual Gated Enable Inputs Simplify Cascading Register Implementations
- Compatible for Use with TTL Circuits
- Input Clamping Diodes Simplify System Design

SN54116 . . . J OR W PACKAGE
SN74116 . . . N PACKAGE
(TOP VIEW)

```
1CLR  [ 1    24 ]  VCC
1C̄1   [ 2    23 ]  2Q4
1C̄2   [ 3    22 ]  2D4
1D1   [ 4    21 ]  2Q3
1Q1   [ 5    20 ]  2D3
1D2   [ 6    19 ]  2Q2
1Q2   [ 7    18 ]  2D2
1D3   [ 8    17 ]  2Q1
1Q3   [ 9    16 ]  2D1
1D4   [ 10   15 ]  2C̄2
1Q4   [ 11   14 ]  2C̄1
GND   [ 12   13 ]  2C̄LR
```

description

These monolithic TTL circuits utilze D-type bistables to implement two independent four-bit latches in a single package. Each four-bit latch has an independent asynchronous clear input and a gated two-input enable circuit. When both enable inputs are low, the output levels will follow the data input levels. When either or both of the enable inputs are taken high, the outputs remain at the last levels setup at the inputs prior to the low-to-high-level transition at the enable input(s). After this, the data inputs are locked out.

The clear input is overriding and when taken low will reset all four outputs low regardless of the levels of the enable inputs.

The SN54116 is characterized for operation over the full military temperature range of −55°C to 125°C; the SN74116 is characterized for operation from 0°C to 70°C.

FUNCTION TABLE
(EACH LATCH)

INPUTS				OUTPUT
CLEAR	ENABLE		DATA	Q
	C̄1	C̄2		
H	L	L	L	L
H	L	L	H	H
H	X	H	X	Q_0
H	H	X	X	Q_0
L	X	X	X	L

H   high level, L – low level, X – irrelevant
Q_0 – the level of Q before these input conditions were established.

logic symbol[†]

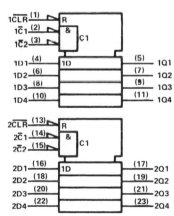

[†]This symbol is in accordance with ANSI/IEEE Std. 91-1984 and IEC Publication 617-12.

FIGURE B-41

**SN54125, SN54126, SN54LS125A, SN54LS126A,
SN74125, SN74126, SN74LS125A, SN74LS126A**
QUADRUPLE BUS BUFFERS WITH 3-STATE OUTPUTS
DECEMBER 1983 — REVISED MARCH 1988

- **Quad Bus Buffers**
- **3-State Outputs**
- **Separate Control for Each Channel**

description

These bus buffers feature three-state outputs that, when enabled, have the low impedance characteristics of a TTL output with additional drive capability at high logic levels to permit driving heavily loaded bus lines without external pull-up resistors, when disabled, both output transistors are turned off presenting a high-impedance state to the bus so the output will act neither as a significant load nor as a driver. The '125 and 'LS125A outputs are disabled when $\overline{G}$ is high. The '126 and 'LS126A outputs are disabled when G is low.

**SN54125, SN54126, SN54LS125A,
SN54LS126A . . . J OR W PACKAGE
SN74125, SN74126 . . . N PACKAGE
SN74LS125A, SN74LS126A . . . D OR N PACKAGE
(TOP VIEW)**

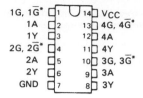

**SN54LS125A, SN54LS126A . . . FK PACKAGE
(TOP VIEW)**

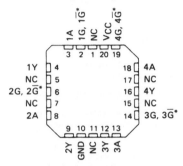

*$\overline{G}$ on '125 and 'LS125A; G on 126 and 'LS126A

NC — No internal connection

logic diagram (each gate)

'125, 'LS125A

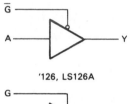

'126, 'LS126A

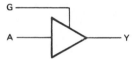

positive logic Y = A

logic symbols †

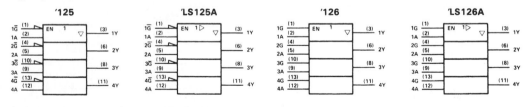

†These symbols are in accordance with ANSI/IEEE Std. 91-1984 and IEC Publication 617-12.
Pin numbers shown are for D, J, N, and W packages.

FIGURE B-42

SN54S135, SN74S135
QUADRUPLE EXCLUSIVE-OR/NOR GATES

DECEMBER 1972 – REVISED MARCH 1988

- **Fully Compatible with Most TTL and TTL MSI Circuits**

- **Fully Schottky Clamping Reduces Delay Times . . . 8 ns Typical**

- **Can Operate as Exclusive-OR Gate (C Input Low) or as Exclusive-NOR Gate (C Input High)**

FUNCTION TABLE

INPUTS			OUTPUT
A	B	C	Y
L	L	L	L
L	H	L	H
H	L	L	H
H	H	L	L
L	L	H	H
L	H	H	L
H	L	H	L
H	H	H	H

H = high level, L = low level

logic diagram (one half)

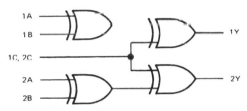

positive logic

$$Y = A \oplus B \oplus C = A\overline{B}\overline{C} + \overline{A}B\overline{C} + \overline{A}\overline{B}C + ABC$$

logic symbol†

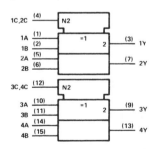

†This symbol is in accordance with ANSI/IEEE Std. 91-1984 and IEC Publication 617-12. Pin numbers are for D, J, N, and W packages.

SN54S135 . . . J OR W PACKAGE
SN74S135 . . . D OR N PACKAGE
(TOP VIEW)

1A	1	16	V_CC
1B	2	15	4B
1Y	3	14	4A
1C,2C	4	13	4Y
2A	5	12	3C,4C
2B	6	11	3B
2Y	7	10	3A
GND	8	9	3Y

SN54S135 . . . FK PACKAGE
(TOP VIEW)

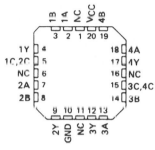

NC – No internal connection

schematics of inputs and outputs

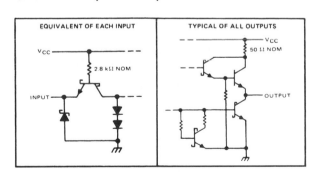

EQUIVALENT OF EACH INPUT	TYPICAL OF ALL OUTPUTS

Resistor values shown are nominal.

FIGURE B-43

SN54LS138, SN54S138, SN74LS138, SN74S138A
3-LINE TO 8-LINE DECODERS/DEMULTIPLEXERS

DECEMBER 1972 – REVISED MARCH 1988

- **Designed Specifically for High-Speed:**
  Memory Decoders
  Data Transmission Systems

- **3 Enable Inputs to Simplify Cascading**
  and/or Data Reception

- **Schottky-Clamped for High Performance**

description

These Schottky-clamped TTL MSI circuits are designed to be used in high-performance memory decoding or data-routing applications requiring very short propagation delay times. In high-performance memory systems, these docoders can be used to minimize the effects of system decoding. When employed with high-speed memories utilizing a fast enable circuit, the delay times of these decoders and the enable time of the memory are usually less than the typical access time of the memory. This means that the effective system delay introduced by the Schottky-clamped system decoder is negligible.

The 'LS138, SN54S138, and SN74S138A decode one of eight lines dependent on the conditions at the three binary select inputs and the three enable inputs. Two active-low and one active-high enable inputs reduce the need for external gates or inverters when expanding. A 24-line decoder can be implemented without external inverters and a 32-line decoder requires only one inverter. An enable input can be used as a data input for demultiplexing applications.

All of these decoder/demultiplexers feature fully buffered inputs, each of which represents only one normalized load to its driving circuit. All inputs are clamped with high-performance Schottky diodes to suppress line-ringing and to simplify system design.

The SN54LS138 and SN54S138 are characterized for operation over the full military temperature range of −55°C to 125°C. The SN74LS138 and SN74S138A are characterized for operation from 0°C to 70°C.

SN54LS138, SN54S138 . . . J OR W PACKAGE
SN74LS138, SN74S138A . . . D OR N PACKAGE
(TOP VIEW)

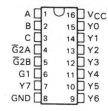

SN54LS138, SN54S138 . . . FK PACKAGE
(TOP VIEW)

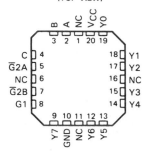

NC – No internal connection

logic symbols[†]

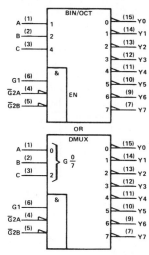

[†]These symbols are in accordance with ANSI/IEEE Std 91-1984 and IEC Publication 617-12.
Pin numbers shown are for D, J, N, and W packages.

FIGURE B-44

SN54LS139A, SN54S139, SN74LS139A, SN74S139A
DUAL 2-LINE TO 4-LINE DECODERS/DEMULTIPLEXERS

DECEMBER 1972 – REVISED MARCH 1988

- **Designed Specifically for High-Speed:**
  Memory Decoders
  Data Transmission Systems

- **Two Fully Independent 2- to 4-Line Decoders/Demultiplexers**

- **Schottky Clamped for High Performance**

description

These Schottky-clamped TTL MSI circuits are designed to be used in high-performance memory-decoding or data-routing applications requiring very short propagation delay times. In high-performance memory systems, these decoders can be used to minimize the effects of system decoding. When employed with high-speed memories utilizing a fast-enable circuit, the delay times of these decoders and the enable time of the memory are usually less than the typical access time of the memory. This means that the effective system delay introduced by the Schottky-clamped system decoder is negligible.

The circuit comprises two individual two-line to four line decoders in a single package. The active-low enable input can be used as a data line in demultiplexing applications.

All of these decoders/demultiplexers feature fully buffered inputs, each of which represents only one normalized load to its driving circuit. All inputs are clamped with high-performance Schottky diodes to suppress line-ringing and to simplify system design. The SN54LS139A and SN54S139 are characterized for operation range of −55°C to 125°C. The SN74LS139A and SN74S139A are characterized for operation from 0°C to 70°C.

SN54LS139A, SN54S139 . . . J OR W PACKAGE
SN74LS139A, SN74S139A . . . D OR N PACKAGE
(TOP VIEW)

```
1G̅  [1   16] VCC
1A  [2   15] 2G̅
1B  [3   14] 2A
1Y0 [4   13] 2B
1Y1 [5   12] 2Y0
1Y2 [6   11] 2Y1
1Y3 [7   10] 2Y2
GND [8    9] 2Y3
```

SN54LS139A, SN54S139 . . . FK PACKAGE
(TOP VIEW)

NC—No internal connection

logic symbols (alternatives)†

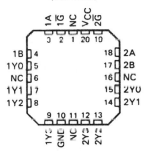

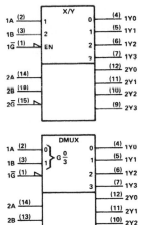

†These symbols are in accordance with ANSI/IEEE Std. 91-1984 and IEC Publication 617-12.
Pin numbers shown are for D, J, N, and W packages.

FUNCTION TABLE

INPUTS			OUTPUTS			
ENABLE	SELECT					
G̅	B	A	Y0	Y1	Y2	Y3
H	X	X	H	H	H	H
L	L	L	L	H	H	H
L	L	H	H	L	H	H
L	H	L	H	H	L	H
L	H	H	H	H	H	L

H = high level, L = low level, X = irrelevant

FIGURE B-45

SN54147, SN54148, SN54LS147, SN54LS148,
SN74147, SN74148 (TIM9907), SN74LS147, SN74LS148
10-LINE TO 4-LINE AND 8-LINE TO 3-LINE PRIORITY ENCODERS
OCTOBER 1976 — REVISED MARCH 1988

'147, 'LS147

- **Encodes 10-Line Decimal to 4-Line BCD**

- **Applications Include:**

  Keyboard Encoding
  Range Selection: '148, 'LS148

- **Encodes 8 Data Lines to 3-Line Binary (Octal)**

- **Applications Include:**

  N-Bit Encoding
  Code Converters and Generators

TYPE	TYPICAL DATA DELAY	TYPICAL POWER DISSIPATION
'147	10 ns	225 mW
'148	10 ns	190 mW
'LS147	15 ns	60 mW
'LS148	15 ns	60 mW

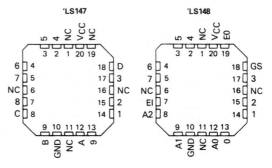

SN54147, SN54LS147,
SN54148, SN54LS148 . . . J OR W PACKAGE
SN74147, SN74148 . . . N PACKAGE
SN74LS147, SN74LS148 . . . D OR N PACKAGE
(TOP VIEW)

SN54LS147, SN54LS148 . . . FK PACKAGE
(TOP VIEW)

NC - No internal connection

description

These TTL encoders feature priority decoding of the inputs to ensure that only the highest-order data line is encoded. The '147 and 'LS147 encode nine data lines to four-line (8-4-2-1) BCD. The implied decimal zero condition requires no input condition as zero is encoded when all nine data lines are at a high logic level. The '148 and 'LS148 encode eight data lines to three-line (4-2-1) binary (octal). Cascading circuitry (enable input EI and enable output EO) has been provided to allow octal expansion without the need for external circuitry. For all types, data inputs and outputs are active at the low logic level. All inputs are buffered to represent one normalized Series 54/74 or 54LS/74LS load, respectively.

'147, 'LS147
FUNCTION TABLE

\multicolumn{9}{c}{INPUTS}									OUTPUTS			
1	2	3	4	5	6	7	8	9	D	C	B	A
H	H	H	H	H	H	H	H	H	H	H	H	H
X	X	X	X	X	X	X	X	L	L	H	H	L
X	X	X	X	X	X	X	L	H	L	H	H	H
X	X	X	X	X	X	L	H	H	H	L	L	L
X	X	X	X	X	L	H	H	H	H	L	L	H
X	X	X	X	L	H	H	H	H	H	L	H	L
X	X	X	L	H	H	H	H	H	H	L	H	H
X	X	L	H	H	H	H	H	H	H	H	L	L
X	L	H	H	H	H	H	H	H	H	H	L	H
L	H	H	H	H	H	H	H	H	H	H	H	L

'148, 'LS148
FUNCTION TABLE

\multicolumn{9}{c}{INPUTS}									OUTPUTS				
EI	0	1	2	3	4	5	6	7	A2	A1	A0	GS	EO
H	X	X	X	X	X	X	X	X	H	H	H	H	H
L	H	H	H	H	H	H	H	H	H	H	H	H	L
L	X	X	X	X	X	X	X	L	L	L	L	L	H
L	X	X	X	X	X	X	L	H	L	L	H	L	H
L	X	X	X	X	X	L	H	H	L	H	L	L	H
L	X	X	X	X	L	H	H	H	L	H	H	L	H
L	X	X	X	L	H	H	H	H	H	L	L	L	H
L	X	X	L	H	H	H	H	H	H	L	H	L	H
L	X	L	H	H	H	H	H	H	H	H	L	L	H
L	L	H	H	H	H	H	H	H	H	H	H	L	H

H = high logic level, L = low logic level, X = irrelevant

FIGURE B-46

**SN54150, SN54151A, SN54LS151, SN54S151,
SN74150, SN74151A, SN74LS151, SN74S151
DATA SELECTORS/MULTIPLEXERS**

DECEMBER 1972—REVISED MARCH 1988

- '150 Selects One-of-Sixteen Data Sources
- Others Select One-of-Eight Data Sources
- All Perform Parallel-to-Serial Conversion
- All Permit Multiplexing from N Lines to One Line
- Also For Use as Boolean Function Generator
- Input-Clamping Diodes Simplify System Design
- Fully Compatible with Most TTL Circuits

TYPE	TYPICAL AVERAGE PROPAGATION DELAY TIME DATA INPUT TO W OUTPUT	TYPICAL POWER DISSIPATION
'150	13 ns	200 mW
'151A	8 ns	145 mW
'LS151	13 ns	30 mW
'S151	4.5 ns	225 mW

description

These monolithic data selectors/multiplexers contain full on-chip binary decoding to select the desired data source. The '150 selects one-of-sixteen data sources; the '151A, 'LS151, and 'S151 select one-of-eight data sources. The '150, '151A, 'LS151, and 'S151 have a strobe input which must be at a low logic level to enable these devices. A high level at the strobe forces the W output high, and the Y output (as applicable) low.

The '150 has only an inverted W output; the '151A, 'LS151, and 'S151 feature complementary W and Y outputs.

The '151A and '152A incorporate address buffers that have symmetrical propagation delay times through the complementary paths. This reduces the possibility of transients occurring at the output(s) due to changes made at the select inputs, even when the '151A outputs are enabled (i.e., strobe low).

SN54150 . . . J OR W PACKAGE
SN74150 . . . N PACKAGE
(TOP VIEW)

```
       ___ ___
E7  [ 1    24 ] VCC
E6  [ 2    23 ] E8
E5  [ 3    22 ] E9
E4  [ 4    21 ] E10
E3  [ 5    20 ] E11
E2  [ 6    19 ] E12
E1  [ 7    18 ] E13
E0  [ 8    17 ] E14
G̅   [ 9    16 ] E15
W   [ 10   15 ] A
D   [ 11   14 ] B
GND [ 12   13 ] C
```

SN54151A, SN54LS151, SN54S151 . . . J OR W PACKAGE
SN74151A . . . N PACKAGE
SN74LS151, SN74S151 . . . D OR N PACKAGE
(TOP VIEW)

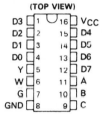

```
       ___ ___
D3  [ 1    16 ] VCC
D2  [ 2    15 ] D4
D1  [ 3    14 ] D5
D0  [ 4    13 ] D6
Y   [ 5    12 ] D7
W   [ 6    11 ] A
G   [ 7    10 ] B
GND [ 8     9 ] C
```

SN54LS151, SN54S151 . . . FK PACKAGE
(TOP VIEW)

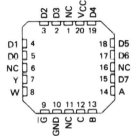

NC No internal connection

**SN54153, SN54LS153, SN54S153
SN74153, SN74LS153, SN74S153
DUAL 4-LINE TO 1-LINE DATA SELECTORS/MULTIPLEXERS**

DECEMBER 1972 — REVISED MARCH 1988

- Permits Multiplexing from N lines to 1 line
- Performs Parallel-to-Serial Conversion
- Strobe (Enable) Line Provided for Cascading (N lines to n lines)
- High-Fan-Out, Low-Impedance, Totem-Pole Outputs
- Fully Compatible with most TTL Circuits

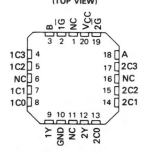

SN54153, SN54LS153, SN54S153 . . . J OR W PACKAGE
SN74153 . . . N PACKAGE
SN74LS153, SN74S153 . . . D OR N PACKAGE
(TOP VIEW)

1$\overline{G}$	1	16	V$_{CC}$
B	2	15	2$\overline{G}$
1C3	3	14	A
1C2	4	13	2C3
1C1	5	12	2C2
1C0	6	11	2C1
1Y	7	10	2C0
GND	8	9	2Y

TYPE	TYPICAL AVERAGE PROPAGATION DELAY TIMES			TYPICAL POWER DISSIPATION
	FROM DATA	FROM STROBE	FROM SELECT	
'153	14 ns	17 ns	22 ns	180 mW
'LS153	14 ns	19 ns	22 ns	31 mW
'S153	6 ns	9.5 ns	12 ns	225 mW

SN54LS153, SN54S153 . . . FK PACKAGE
(TOP VIEW)

NC – No internal connection

description

Each of these monolithic, data selectors/multiplexers contains inverters and drivers to supply fully complementary, on-chip, binary decoding data selection to the AND-OR gates. Separate strobe inputs are provided for each of the two four-line sections.

FUNCTION TABLE

SELECT INPUTS		DATA INPUTS				STROBE	OUTPUT
B	A	C0	C1	C2	C3	$\overline{G}$	Y
X	X	X	X	X	X	H	L
L	L	L	X	X	X	L	L
L	L	H	X	X	X	L	H
L	H	X	L	X	X	L	L
L	H	X	H	X	X	L	H
H	L	X	X	L	X	L	L
H	L	X	X	H	X	L	H
H	H	X	X	X	L	L	L
H	H	X	X	X	H	L	H

Select inputs A and B are common to both sections.
H = high level, L = low level, X = irrelevant

absolute maximum ratings over operating free-air temperature range (unless otherwise noted)

Supply voltage, V$_{CC}$ (See Note 1) . . . . 7 V
Input voltage: '153, 'S153 . . . . 5.5 V
'LS153 . . . . 7 V
Operating free-air temperature range: SN54' . . . . −55°C to 125°C
SN74' . . . . 0°C to 70°C
Storage temperature range . . . . −65°C to 150°C

NOTE 1: Voltage values are with respect to network ground terminal.

FIGURE B-48

**SN54157, SN54LS157, SN54LS158, SN54S157, SN54S158,
SN74157, SN74LS157, SN74LS158, SN74S157, SN74S158**
QUADRUPLE 2-LINE TO 1-LINE DATA SELECTORS/MULTIPLEXERS
MARCH 1974 — REVISED MARCH 1988

- Buffered Inputs and Outputs
- Three Speed/Power Ranges Available

TYPES	TYPICAL AVERAGE PROPAGATION TIME	TYPICAL POWER DISSIPATION
'157	9 ns	150 mW
'LS157	9 ns	49 mW
'S157	5 ns	250 mW
'LS158	7 ns	24 mW
'S158	4 ns	195 mW

applications

- Expand Any Data Input Point
- Multiplex Dual Data Buses
- Generate Four Functions of Two Variables (One Variable Is Common)
- Source Programmable Counters

description

These monolithic data selectors/multiplexers contain inverters and drivers to supply full on-chip data selection to the four output gates. A separate strobe input is provided. A 4 bit word is selected from one of two sources and is routed to the four outputs. The '157, 'LS157, and 'S157 present true data whereas the 'LS158 and 'S158 present inverted data to minimize propagation delay time.

SN54157, SN54LS157, SN54S157,
SN54LS158, SN54S158 . . . J OR W PACKAGE
SN74157 . . . N PACKAGE
SN74LS157, SN74S157,
SN74LS158, SN74S158 . . . D OR N PACKAGE
(TOP VIEW)

```
           ___
   A/B  [1    16]  VCC
   1A   [2    15]  G
   1B   [3    14]  4A
   1Y   [4    13]  4B
   2A   [5    12]  4Y
   2B   [6    11]  3A
   2Y   [7    10]  3B
   GND  [8     9]  3Y
```

SN54LS157, SN54S157, SN54LS158,
SN54S158 . . . FK PACKAGE
(TOP VIEW)

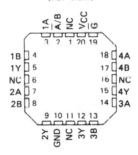

NC - No internal connection

FUNCTION TABLE

INPUTS				OUTPUT Y	
STROBE $\overline{G}$	SELECT $\overline{A}/B$	A	B	'157, 'LS157, 'S157	'LS158 'S158
H	X	X	X	L	H
L	L	L	X	L	H
L	L	H	X	H	L
L	H	X	L	L	H
L	H	X	H	H	L

H = high level, L = low level, X = irrelevant

absolute maximum ratings over operating free-air temperature range (unless otherwise noted)

Supply voltage, V_{CC} (See Note 1) . . . . 7 V
Input voltage: '157, 'S158 . . . . 5.5 V
            'LS157, 'LS158 . . . . 7 V
Operating free-air temperature range: SN54' . . . . −55°C to 125°C
                       SN74' . . . . 0°C to 70°C
Storage temperature range . . . . −65°C to 150°C

NOTE 1: Voltage values are with respect to network ground terminal.

FIGURE B-49

SN54160 THRU SN54163, SN54LS160A THRU SN54LS163A,
SN54S162, SN54S163, SN74160 THRU SN74163,
SN74LS160A THRU SN74LS163A, SN74S162, SN74S163
SYNCHRONOUS 4-BIT COUNTERS
OCTOBER 1976 — REVISED MARCH 1988

'160,'161,'LS160A,'LS161A . . . SYNCHRONOUS COUNTERS WITH DIRECT CLEAR
'162,'163,'LS162A,'LS163A,'S162,'S163 . . . FULLY SYNCHRONOUS COUNTERS

- **Internal Look-Ahead for Fast Counting**
- **Carry Output for n-Bit Cascading**
- **Synchronous Counting**
- **Synchronously Programmable**
- **Load Control Line**
- **Diode-Clamped Inputs**

SERIES 54', 54LS' 54S' . . . J OR W PACKAGE
SERIES 74' . . . N PACKAGE
SERIES 74LS', 74S' . . . D OR N PACKAGE
(TOP VIEW)

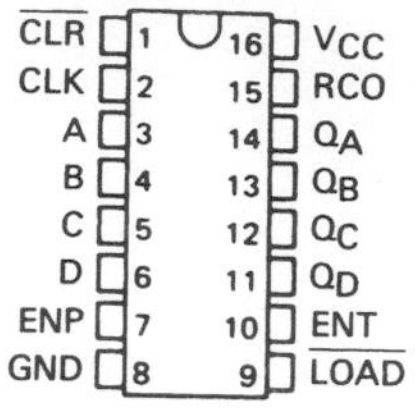

CLR	1	16	V$_{CC}$
CLK	2	15	RCO
A	3	14	Q$_A$
B	4	13	Q$_B$
C	5	12	Q$_C$
D	6	11	Q$_D$
ENP	7	10	ENT
GND	8	9	LOAD

NC—No internal connection

TYPE	TYPICAL PROPAGATION TIME, CLOCK TO Q OUTPUT	TYPICAL MAXIMUM CLOCK FREQUENCY	TYPICAL POWER DISSIPATION
'160 thru '163	14 ns	32 MHz	305 mW
'LS162A thru 'LS163A	14 ns	32 MHz	93 mW
'S162 and 'S163	9 ns	70 MHz	475 mW

description

These synchronous, presettable counters feature an internal carry look-ahead for application in high-speed counting designs. The '160,'162,'LS160A,'LS162A, and 'S162 are decade counters and the '161,'163,'LS161A,'LS163A, and 'S163 are 4-bit binary counters. Synchronous operation is provided by having all flip-flops clocked simultaneously so that the outputs change coincident with each other when so instructed by the count-enable inputs and internal gating. This mode of operation eliminates the output counting spikes that are normally associated with asynchronous (ripple clock) counters, however counting spikes may occur on the (RCO) ripple carry output. A buffered clock input triggers the four flip-flops on the rising edge of the clock input waveform.

SERIES 54LS', 54S' . . . FK PACKAGE
(TOP VIEW)

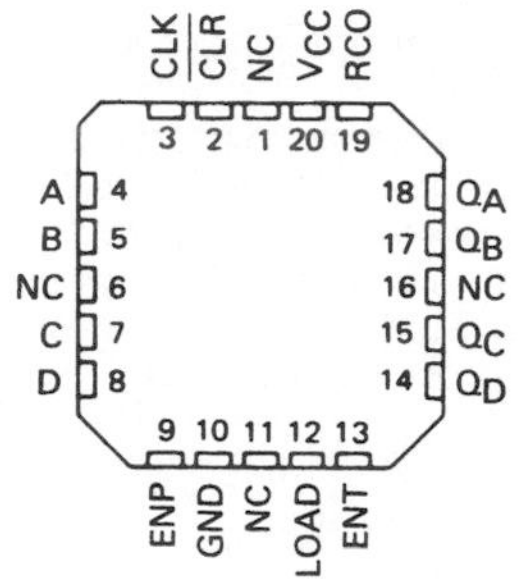

	CLK	CLR	NC	V$_{CC}$	RCO
	3	2	1	20	19
A	4				18	Q$_A$
B	5				17	Q$_B$
NC	6				16	NC
C	7				15	Q$_C$
D	8				14	Q$_D$
	9	10	11	12	13
	ENP	GND	NC	LOAD	ENT

NC—No internal connection

These counters are fully programmable; that is, the outputs may be preset to either level. As presetting is synchronous, setting up a low level at the load input disables the counter and causes the outputs to agree with the setup data after the next clock pulse regardless of the levels of the enable inputs. Low-to-high transitions at the load input of the '160 thru '163 should be avoided when the clock is low if the enable inputs are high at or before the transition. This restriction is not applicable to the 'LS160A thru 'LS163A or 'S162 or 'S163. The clear function for the '160, '161,'LS160A, and 'LS161A is asynchronous and a low level at the clear input sets all four of the flip-flop outputs low regardless of the levels of clock, load, or enable inputs. The clear function for the '162,'163,'LS162A,'LS163A, 'S162, and 'S163 is synchronous and a low level at the clear input sets all four of the flip-flop outputs low after the next clock pulse, regardless of the levels of the enable inputs. This synchronous clear allows the count length to be modified easily as decoding the maximum count desired can be accomplished with one external NAND gate. The gate output is connected to the clear input to synchronously clear the counter to 0000 (LLLL). Low-to-high transitions at the clear input of the '162 and '163 should be avoided when the clock is low if the enable and load inputs are high at or before the transition.

FIGURE B-50

SN54164, SN54LS164, SN74164, SN74LS164
8-BIT PARALLEL-OUT SERIAL SHIFT REGISTERS

MARCH 1974 — REVISED MARCH 1988

- **Gated Serial Inputs**
- **Fully Buffered Clock and Serial Inputs**
- **Asynchronous Clear**

TYPE	TYPICAL MAXIMUM CLOCK FREQUENCY	TYPICAL POWER DISSIPATION
'164	36 MHz	21 mW per bit
'LS164	36 MHz	10 mW per bit

description

These 8-bit shift registers feature gated serial inputs and an asynchronous clear. The gated serial inputs (A and B) permit complete control over incoming data as a low at either input inhibits entry of the new data and resets the first flip-flop to the low level at the next clock pulse. A high-level input enables the other input which will then determine the state of the first flip-flop. Data at the serial inputs may be changed while the clock is high or low, but only information meeting the setup-time requirements will be entered. Clocking occurs on the low-to-high-level transition of the clock input. All inputs are diode-clamped to minimize transmission-line effects.

The SN54164 and SN54LS164 are characterized for operation over the full military temperature range of −55°C to 125°C. The SN74164 and SN74LS164 are characterized for operation from 0°C to 70°C.

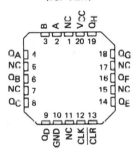

SN54164, SN54LS164 . . . J OR W PACKAGE
SN74164 . . . N PACKAGE
SN74LS164 . . . D OR N PACKAGE
(TOP VIEW)

SN54LS164 . . . FK PACKAGE
(TOP VIEW)

NC — No internal connection

FUNCTION TABLE

INPUTS				OUTPUTS			
CLEAR	CLOCK	A	B	Q_A	Q_B	$\cdots$	Q_H
L	X	X	X	L	L		L
H	L	X	X	Q_{A0}	Q_{B0}		Q_{H0}
H	↑	H	H	H	Q_{An}		Q_{Gn}
H	↑	L	X	L	Q_{An}		Q_{Gn}
H	↑	X	L	L	Q_{An}		Q_{Gn}

H = high level (steady state), L = low level (steady state)
X = irrelevant (any input, including transitions)
↑ = transition from low to high level.
Q_{A0}, Q_{B0}, Q_{H0} = the level of Q_A, Q_B, or Q_H, respectively, before the indicated steady-state input conditions were established.
Q_{An}, Q_{Gn} = the level of Q_A or Q_G before the most-recent ↑ transition of the clock; indicates a one bit shift.

schematics of inputs and outputs

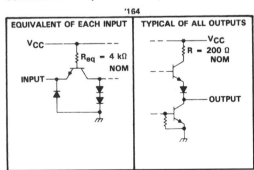

'164

EQUIVALENT OF EACH INPUT	TYPICAL OF ALL OUTPUTS
V_{CC} R_{eq} = 4 kΩ NOM INPUT	V_{CC} R = 200 Ω NOM OUTPUT

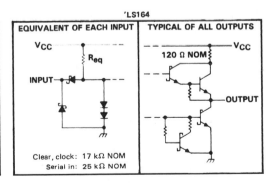

'LS164

EQUIVALENT OF EACH INPUT	TYPICAL OF ALL OUTPUTS
V_{CC} R_{eq} INPUT	V_{CC} 120 Ω NOM OUTPUT

Clear, clock: 17 kΩ NOM
Serial in: 25 kΩ NOM

FIGURE B-51

SN54165, SN54LS165A, SN74165, SN74LS165A
PARALLEL-LOAD 8-BIT SHIFT REGISTERS

OCTOBER 1976 – REVISED MARCH 1988

- **Complementary Outputs**
- **Direct Overriding Load (Data) Inputs**
- **Gated Clock Inputs**
- **Parallel-to-Serial Data Conversion**

TYPE	TYPICAL MAXIMUM CLOCK FREQUENCY	TYPICAL POWER DISSIPATION
'165	26 MHz	210 mW
'LS165A	35 MHz	90 mW

description

The '165 and 'LS165A are 8-bit serial shift registers that shift the data in the direction of Q_A toward Q_H when clocked. Parallel-in access to each stage is made available by eight individual direct data inputs that are enabled by a low level at the shift/load input. These registers also feature gated clock inputs and complementary outputs from the eighth bit. All inputs are diode-clamped to minimize transmission-line effects, thereby simplifying system design.

Clocking is accomplished through a 2-input positive-NOR gate, permitting one input to be used as a clock-inhibit function. Holding either of the clock inputs high inhibits clocking and holding either clock input low with the shift/load input high enables the other clock input. The clock-inhibit input should be changed to the high level only while the clock input is high. Parallel loading is inhibited as long as the shift/load input is high. Data at the parallel inputs are loaded directly into the register while the shift/load input is low independently of the levels of the clock, clock inhibit, or serial inputs.

SN54165, SN54LS165A . . . J OR W PACKAGE
SN74165 . . . N PACKAGE
SN74LS165A . . . D OR N PACKAGE
(TOP VIEW)

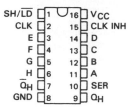

SN54LS165A . . . FK PACKAGE
(TOP VIEW)

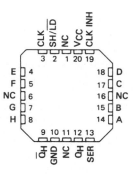

logic symbol[†]

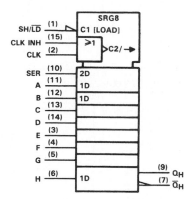

[†]This symbol is in accordance with ANSI/IEEE Std. 91-1984 and IEC Publication 617-12.

Pin numbers shown are for D, J, N, and W packages.

FUNCTION TABLE

INPUTS					INTERNAL OUTPUTS		OUTPUT
SHIFT/ LOAD	CLOCK INHIBIT	CLOCK	SERIAL	PARALLEL A . . . H	$\overline{Q}_A$	$\overline{Q}_B$	Q_H
L	X	X	X	a . . . h	a	b	h
H	L	L	X	X	Q_{A0}	Q_{B0}	Q_{H0}
H	L	↑	H	X	H	Q_{An}	Q_{Gn}
H	L	↑	L	X	L	Q_{An}	Q_{Gn}
H	H	X	X	X	Q_{A0}	Q_{B0}	Q_{H0}

FIGURE B-52

SN54178, SN74178
4-BIT PARALLEL-ACCESS SHIFT REGISTERS

DECEMBER 1972—REVISED MARCH 1988

- **Typical Maximum Clock Frequency . . . 39 MHz**
- **Three Operating Modes:**
  Synchronous Parallel Load
  Right Shift
  Hold (Do Nothing)
- **Negative-Edge-Triggered Clocking**
- **D-C Coupling Symplifies System Designs**

description

These shift registers utilize fully d-c coupled storage elements and feature synchronous parallel inputs and parallel outputs.

Parallel loading is accomplished by taking the shift input low, applying the four bits of data, and taking the load input high. The data is loaded into the associated flip-flop synchronously and appears at the outputs after a high-to-low transition of the clock. During loading, serial data flow is inhibited.

Shift right is also accomplished on the falling edge of the clock pulse when the shift input is high regardless of the level of the load input. Serial data for this mode is entered at the serial data input.

When both the shift and load inputs are low, clocking of the register can continue; however, data appearing at each output is fed back to the flip-flop input creating a mode in which the data is held unchanged. Thus, the system clock may be left free-running without changing the contents of the register.

SN54178 . . . J OR W PACKAGE
(TOP VIEW)

```
        ┌───┬─∪─┬───┐
    B  ─┤1      14├─ V_CC
    A  ─┤2      13├─ C
   SER ─┤3      12├─ D
   Q_A ─┤4      11├─ SHIFT
   CLK ─┤5      10├─ Q_D
   Q_B ─┤6       9├─ LOAD
   GND ─┤7       8├─ Q_C
        └──────────┘
```

logic symbol[†]

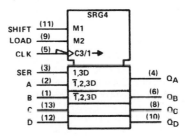

[†]This symbol is in accordance with ANSI/IEEE Std 91 1984 and IEC Publication 617.12.

FUNCTION TABLE

INPUTS								OUTPUTS			
SHIFT	LOAD	CLOCK	SERIAL	PARALLEL				Q_A	Q_B	Q_C	Q_D
				A	B	C	D				
X	X	H	X	X	X	X	X	Q_{A0}	Q_{B0}	Q_{C0}	Q_{D0}
L	L	↓	X	X	X	X	X	Q_{A0}	Q_{B0}	Q_{C0}	Q_{D0}
L	H	↓	X	a	b	c	d	a	b	c	d
H	X	↓	H	X	X	X	X	H	Q_{An}	Q_{Bn}	Q_{Cn}
H	X	↓	L	X	X	X	X	L	Q_{An}	Q_{Bn}	Q_{Cn}

H = high level (steady state), L = low level (steady state)
X = irrelevant (any input, including transitions)
↓ = transition from high to low level
a, b, c, d = the level of steady-state input at inputs A, B, C, or D, respectively.
Q_{A0}, Q_{B0}, Q_{C0}, Q_{D0} = the level of Q_A, Q_B, Q_C, or Q_D, respectively, before the indicated steady-state input conditions were established.
Q_{An}, Q_{Bn}, Q_{Cn} = the level of Q_A, Q_B, or Q_C, respectively, before the most-recent ↓ transition of the clock.

FIGURE B-53

SN54180, SN74180
9-BIT ODD/EVEN PARITY GENERATORS/CHECKERS

DECEMBER 1972 – REVISED MARCH 1988

FUNCTION TABLE

INPUTS			OUTPUTS	
Σ OF H's AT A THRU H	EVEN	ODD	Σ EVEN	Σ ODD
EVEN	H	L	H	L
ODD	H	L	L	H
EVEN	L	H	L	H
ODD	L	H	H	L
X	H	H	L	L
X	L	L	H	H

H = high level, L = low level, X = irrelevant

SN54180 . . . J OR W PACKAGE
SN74180 . . . N PACKAGE
(TOP VIEW)

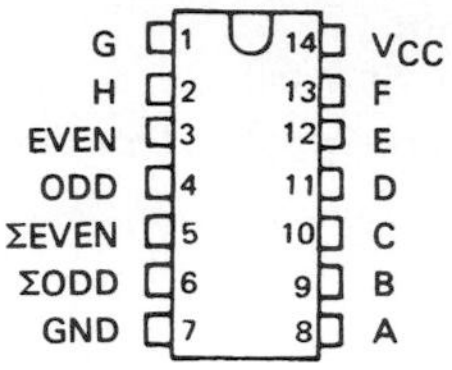

```
          ___ ___
    G  |1   U  14|  VCC
    H  |2      13|  F
 EVEN  |3      12|  E
  ODD  |4      11|  D
ΣEVEN  |5      10|  C
 ΣODD  |6       9|  B
  GND  |7       8|  A
```

description

These universal, monolithic, 9-bit (8 data bits plus 1 parity bit) parity generators/checkers, utilize familiar Series 54/74 TTL circuitry and feature odd/even outputs and control inputs to facilitate operation in either odd or even-parity applications. Depending on whether even or odd parity is being generated or checked, the even or odd inputs can be utilized as the parity or 9th-bit input. The word-length capability is easily expanded by cascading.

The SN54180/SN74180 are fully compatible with other TTL or DTL circuits. Input buffers are provided so that each data input represents only one normalized series 54/74 load. A full fan-out to 10 normalized series 54/74 loads is available from each of the outputs at a low logic level. A fan-out to 20 normalized loads is provided at a high logic level to facilitate the connection of unused inputs to used inputs. Typical power dissipation is 170 mW.

The SN54180 is characterized for operation over the full military temperature range of $-55°C$ to $125°C$; and the SN74180 is characterized for operation from $0°C$ to $70°C$.

absolute maximum ratings over operating free-air temperature range (unless otherwise noted)

Supply voltage, V_{CC} (see Note 1) . . . . 7 V
Input voltage . . . . 5.5 V
Operating free-air temperature range: SN54180 Circuits . . . . $-55°C$ to $125°C$
                            SN74180 Circuits . . . . $0°C$ to $70°C$
Storage temperature range . . . . $-65°C$ to $150°C$

NOTE 1: Voltage values are with respect to network ground terminal.

recommended operating conditions

	SN54180			SN74180			UNIT
	MIN	NOM	MAX	MIN	NOM	MAX	
Supply voltage, V_{CC}	4.5	5	5.5	4.75	5	5.25	V
High-level output current, I_{OH}			−800			−800	μA
Low-level output current, I_{OL}			16			16	mA
Operating free-air temperature, T_A	−55		125	0		70	°C

FIGURE B-54

- Full Look-Ahead for High-Speed Operations on Long Words

- Input Clamping Diodes Minimize Transmission-Line Effects

- Darlington Outputs Reduce Turn-Off Time

- Arithmetic Operating Modes:
    Addition
    Subtraction
    Shift Operand A One Position
    Magnitude Comparison
    Plus Twelve Other Arithmetic Operations

- Logic Function Modes:
    Exclusive-OR
    Comparator
    AND, NAND, OR, NOR
    Plus Ten Other Logic Operations

SN54LS181, SN54S181 . . . J OR W PACKAGE
SN74LS181, SN74S181 . . . DW OR N PACKAGE
(TOP VIEW)

$\overline{B0}$	1	24 V_{CC}
$\overline{A0}$	2	23 $\overline{A1}$
S3	3	22 $\overline{B1}$
S2	4	21 $\overline{A2}$
S1	5	20 $\overline{B2}$
S0	6	19 $\overline{A3}$
C_n	7	18 $\overline{B3}$
M	8	17 $\overline{G}$
$\overline{F0}$	9	16 C_{n+4}
$\overline{F1}$	10	15 P
$\overline{F2}$	11	14 A = B
GND	12	13 $\overline{F3}$

SN54LS181, SN54S181 . . . FK PACKAGE
(TOP VIEW)

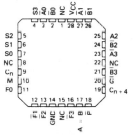

NC - No internal connection

TYPICAL ADDITION TIMES

NUMBER OF BITS	ADDITION TIMES		PACKAGE COUNT		CARRY METHOD BETWEEN ALUs
	USING 'LS181 AND 'S182	USING 'S181 AND 'S182	ARITHMETIC/ LOGIC UNITS	LOOK-AHEAD CARRY GENERATORS	
1 to 4	24 ns	11 ns	1		NONE
5 to 8	40 ns	18 ns	2		RIPPLE
9 to 16	44 ns	19 ns	3 or 4	1	FULL LOOK-AHEAD
17 to 64	68 ns	28 ns	5 to 16	2 to 5	FULL LOOK-AHEAD

description

The 'LS181 and 'S181 are arithmetic logic units (ALU)/function generators that have a complexity of 75 equivalent gates on a monolithic chip. These circuits perform 16 binary arithmetic operations on two 4-bit words as shown in Tables 1 and 2. These operations are selected by the four function-select lines (S0, S1, S2, S3) and include addition, subtraction, decrement, and straight transfer. When performing arithmetic manipulations, the internal carries must be enabled by applying a low-level voltage to the mode control input (M). A full carry look-ahead scheme is made available in these devices for fast, simultaneous carry generation by means of two cascade-outputs (pins 15 and 17) for the four bits in the package. When used in conjunction with the SN54S182 or SN74S182 full carry look-ahead circuits, high-speed arithmetic operations can be performed. The typical addition times shown above illustrate the little additional time required for addition of longer words when full carry look-ahead is employed. The method of cascading 'S182 circuits with these ALUs to provide multi-level full carry look-ahead is illustrated under typical applications data for the 'S182.

If high speed is not of importance, a ripple-carry input (C_n) and a ripple-carry output (C_{n+4}) are available. However, the ripple-carry delay has also been minimized so that arithmetic manipulations for small word lengths can be performed without external circuitry.

FIGURE B-55

**SN54190, SN54191, SN54LS190, SN54LS191,
SN74190, SN74191, SN74LS190, SN74LS191
SYNCHRONOUS UP/DOWN COUNTERS WITH DOWN/UP MODE CONTROL**

DECEMBER 1972—REVISED MARCH 1988

- **Counts 8-4-2-1 BCD or Binary**
- **Single Down/Up Count Control Line**
- **Count Enable Control Input**
- **Ripple Clock Output for Cascading**
- **Asynchronously Presettable with Load Control**
- **Parallel Outputs**
- **Cascadable for n-Bit Applications**

TYPE	AVERAGE PROPAGATION DELAY	TYPICAL MAXIMUM CLOCK FREQUENCY	TYPICAL POWER DISSIPATION
'190,'191	20ns	25MHz	325mW
'LS190,'LS191	20ns	25MHz	100mW

SN54190, SN54191, SN54LS190,
SN54LS191 . . . J PACKAGE
SN74190, SN74191 . . . N PACKAGE
SN74LS190, SN74LS191 . . . D OR N PACKAGE
(TOP VIEW)

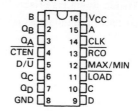

SN54LS190, SN54LS191 . . . FK PACKAGE
(TOP VIEW)

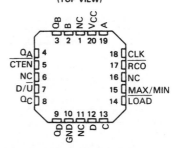

NC - No internal connection

description

The '190, 'LS190, '191, and 'LS191 are synchronous, reversible up/down counters having a complexity of 58 equivalent gates. The '191 and 'LS191 are 4-bit binary counters and the '190 and 'LS190 are BCD counters. Synchronous operation is provided by having all flip-flops clocked simultaneously so that the outputs change coincident with each other when so instructed by the steering logic. This mode of operation eliminates the output counting spikes normally associated with asynchronous (ripple clock) counters.

The outputs of the four master-slave flip-flops are triggered on a low-to-high transition of the clock input if the enable input is low. A high at the enable input inhibits counting. Level changes at the enable input should be made only when the clock input is high. The direction of the count is determined by the level of the down/up input. When low, the counter count up and when high, it counts down. A false clock may occur if the down/up input changes while the clock is low. A false ripple carry may occur if both the clock and enable are low and the down/up input is high during a load pulse.

These counters are fully programmable; that is, the outputs may be preset to either level by placing a low on the load input and entering the desired data at the data inputs. The output will change to agree with the data inputs independently of the level of the clock input. This feature allows the counters to be used as modulo-N dividers by simply modifying the count length with the preset inputs.

The clock, down/up, and load inputs are buffered to lower the drive requirement which significantly reduces the number of clock drivers, etc., required for long parallel words.

Two outputs have been made available to perform the cascading function: ripple clock and maximum/minimum count. The latter output produces a high-level output pulse with a duration approximately equal to one complete cycle of the clock when the counter overflows or underflows. The ripple clock output produces a low-level output pulse equal in width to the low-level portion of the clock input when an overflow or underflow condition exists. The counters can be easily cascaded by feeding the ripple clock output to the enable input of the succeeding counter if parallel clocking is used, or to the clock input if parallel enabling is used. The maximum/minimum count output can be used to accomplish look-ahead for high-speed operation.

Series 54' and 54LS' are characterized for operation over the full military temperature range of −55°C to 125°C; Series 74' and 74LS' are characterized for operation from 0°C to 70°C.

FIGURE B-56

SN54192, SN54193, SN54LS192 SN54LS193, SN74192, SN74193, SN74LS192, SN74LS193
SYNCHRONOUS 4-BIT UP/DOWN COUNTERS (DUAL CLOCK WITH CLEAR)

DECEMBER 1972 – REVISED MARCH 1988

- **Cascading Circuitry Provided Internally**
- **Synchronous Operation**
- **Individual Preset to Each Flip-Flop**
- **Fully Independent Clear Input**

TYPES	TYPICAL MAXIMUM COUNT FREQUENCY	TYPICAL POWER DISSIPATION
'192,'193	32 MHz	325 mW
'LS192,'LS193	32 MHz	95 mW

SN54192, SN54193, SN54LS192,
SN54LS193 . . . J OR W PACKAGE
SN74192, SN74193 . . . N PACKAGE
SN74LS192, SN74LS193 . . . D OR N PACKAGE
(TOP VIEW)

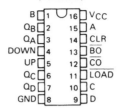

SN54LS192, SN54LS193 . . . FK PACKAGE
(TOP VIEW)

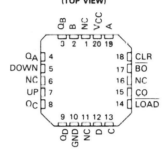

NC - No internal connection

description

These monolithic circuits are synchronous reversible (up/down) counters having a complexity of 55 equivalent gates. The '192 and 'LS192 circuits are BCD counters and the '193 and 'LS193 are 4-bit binary counters. Synchronous operation is provided by having all flip-flops clocked simultaneously so that the outputs change coincidently with each other when so instructed by the steering logic. This mode of operation eliminates the output counting spikes which are normally associated with asynchronous (ripple-clock) counters.

The outputs of the four master-slave flip-flops are triggered by a low-to-high-level transition of either count (clock) input. The direction of counting is determined by which count input is pulsed while the other count input is high.

All four counters are fully programmable; that is, each output may be preset to either level by entering the desired data at the data inputs while the load input is low. The output will change to agree with the data inputs independently of the count pulses. This feature allows the counters to be used as modulo N dividers by simply modifying the count length with the preset inputs.

A clear input has been provided which forces all outputs to the low level when a high level is applied. The clear function is independent of the count and load inputs. The clear, count, and load inputs are buffered to lower the drive requirements. This reduces the number of clock drivers, etc., required for long words.

These counters were designed to be cascaded without the need for external circuitry. Both borrow and carry outputs are available to cascade both the up- and down-counting functions. The borrow output produces a pulse equal in width to the count-down input when the counter underflows. Similarly, the carry output produces a pulse equal in width to the count-up input when an overflow condition exists. The counters can then be easily cascaded by feeding the borrow and carry outputs to the count-down and count-up inputs respectively of the succeeding counter.

absolute maximum ratings over operating free-air temperature range (unless otherwise noted)

	SN54'	SN54LS'	SN74'	SN74LS'	UNIT
Supply voltage, V_CC (see Note 1)	7	7	7	7	V
Input voltage	5.5	7	5.5	7	V
Operating free-air temperature range	−55 to 125		0 to 70		°C
Storage temperature range	−65 to 150		−65 to 150		°C

NOTE 1: Voltage values are with respect to network ground terminal.

FIGURE B-57

SN54194, SN54LS194A, SN54S194,
SN74194, SN74LS194A, SN74S194
4-BIT BIDIRECTIONAL UNIVERSAL SHIFT REGISTERS
MARCH 1974—REVISED MARCH 1988

- **Parallel Inputs and Outputs**
- **Four Operating Modes:**

  Synchronous Parallel Load
  Right Shift
  Left Shift
  Do Nothing

- **Positive Edge-Triggered Clocking**
- **Direct Overriding Clear**

TYPE	TYPICAL MAXIMUM CLOCK FREQUENCY	TYPICAL POWER DISSIPATION
'194	36 MHz	195 mW
'LS194A	36 MHz	75 mW
'S194	105 MHz	425 mW

SN54194, SN54LS194A, SN54S194 . . . J OR W PACKAGE
SN74194 . . . N PACKAGE
SN74LS194A, SN74S194 . . . D OR N PACKAGE
(TOP VIEW)

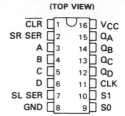

SN54LS194A, SN54S194 . . . FK PACKAGE
(TOP VIEW)

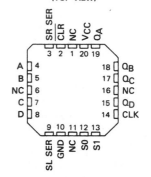

NC – No internal connection

description

These bidirectional shift registers are designed to incorporate virtually all of the features a system designer may want in a shift register. The circuit contains 46 equivalent gates and features parallel inputs, parallel outputs, right-shift and left-shift serial inputs, operating-mode-control inputs, and a direct overriding clear line. The register has four distinct modes of operation, namely:

Inhibit clock (do nothing)
Shift right (in the direction Q_A toward Q_D)
Shift left (in the direction Q_D toward Q_A)
Parallel (broadside) load

Synchronous parallel loading is accomplished by applying the four bits of data and taking both mode control inputs, S0 and S1, high. The data are loaded into the associated flip-flops and appear at the outputs after the positive transition of the clock input. During loading, serial data flow is inhibited.

Shift right is accomplished synchronously with the rising edge of the clock pulse when S0 is high and S1 is low. Serial data for this mode is entered at the shift-right data input. When S0 is low and S1 is high, data shifts left synchronously and new data is entered at the shift-left serial input.

Clocking of the shift register is inhibited when both mode control inputs are low. The mode controls of the SN54194/SN74194 should be changed only while the clock input is high.

logic symbol†

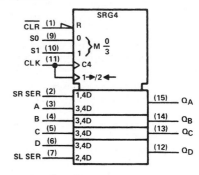

†This symbol is in accordance with ANSI/IEEE Std. 91-1984 and IEC Publication 617-12.
Pin numbers shown are for D, J, N, and W packages.

FIGURE B–58

SN54198, SN54199 SN74198, SN74199
8-BIT SHIFT REGISTERS

DECEMBER 1972—REVISED MARCH 1988

description

These 8-bit shift registers are compatible with most other TTL and MSI logic families. All inputs are buffered to lower the drive requirements to one normalized Series 54/74 load, and input clamping diodes minimize switching transients to simplify system design. Maximum input clock frequency is typically 35 megahertz and power dissipation is typically 360 mW.

Series 54 devices are characterized for operation over the full military temperature range of −55°C to 125°C; Series 74 devices are characterized for operation from 0°C to 70°C.

SN54198 and SN74198

These bidirectional registers are designed to incorporate virtually all of the features a system designer may want in a shift register. These circuits contain 87 equivalent gates and feature parallel inputs, parallel outputs, right-shift and left-shift serial inputs, operating-mode-control inputs, and a direct overriding clear line. The register has four distinct modes of operation, namely:

Inhibit Clock (Do nothing)
Shift Right (In the direction Q_A toward Q_H)
Shift Left (In the direction Q_H toward Q_A)
Parallel (Broadside) Load

Synchronous parallel loading is accomplished by applying the eight bits of data and taking both mode control inputs, S0 and S1, high. The data is loaded into the associated flip-flop and appears at the outputs after the positive transition of the clock input. During loading, serial data flow is inhibited.

Shift right is accomplished synchronously with the rising edge of the clock pulse when S0 is high and S1 is low. Serial data for this mode is entered at the shift-right data input. When S0 is low and S1 is high, data shifts left synchronously and new data is entered at the shift-left serial input.

Clocking of the flip-flop is inhibited when both mode control inputs are low. The mode controls should be changed only while the clock input is high.

```
SN54198 . . . J OR W PACKAGE
SN74198 . . . N PACKAGE
          (TOP VIEW)

    S0  [ 1    24 ]  V_CC
SR SER  [ 2    23 ]  S1
     A  [ 3    22 ]  SL SER
   Q_A  [ 4    21 ]  H
     B  [ 5    20 ]  Q_H
   Q_B  [ 6    19 ]  G
     C  [ 7    18 ]  Q_G
   Q_C  [ 8    17 ]  F
     D  [ 9    16 ]  Q_F
   Q_D  [ 10   15 ]  E
   CLK  [ 11   14 ]  Q_E
   GND  [ 12   13 ]  CLR
```

'198
FUNCTION TABLE

INPUTS							OUTPUTS			
CLEAR	MODE		CLOCK	SERIAL		PARALLEL	Q_A	Q_B ...	Q_G	Q_H
	S_1	S_0		LEFT	RIGHT	A ... H				
L	X	X	X	X	X	X	L	L	L	L
H	X	X	L	X	X	X	Q_{A0}	Q_{B0}	Q_{G0}	Q_{H0}
H	H	H	↑	X	X	a ... h	a	b	g	h
H	L	H	↑	X	H	X	H	Q_{An}	Q_{Fn}	Q_{Gn}
H	L	H	↑	X	L	X	L	Q_{An}	Q_{Fn}	Q_{Gn}
H	H	L	↑	H	X	X	Q_{Bn}	Q_{Cn}	Q_{Hn}	H
H	H	L	↑	L	X	X	Q_{Bn}	Q_{Cn}	Q_{Hn}	L
H	L	L	X	X	X	X	Q_{A0}	Q_{B0}	Q_{G0}	Q_{H0}

H = high level (steady state), L = low level (steady state)
X = irrelevant (any input, including transitions)
↑ = transition from low to high level
a . . . h = the level of steady-state input at inputs A thru H, respectively.
Q_{A0}, Q_{B0}, Q_{G0}, Q_{H0} = the level of Q_A, Q_B, Q_G, or Q_H, respectively, before the indicated steady-state input conditions were established.
Q_{An}, Q_{Bn}, etc. = the level of Q_A, Q_B, etc., respectively, before the most-recent ↑ transition of the clock.

FIGURE B-59

SN54LS242, SN54LS243, SN74LS242, SN74LS243
QUADRUPLE BUS TRANSCEIVERS

APRIL 1985—REVISED MARCH 1988

- **Two-Way Asynchronous Communication Between Data Buses**
- **PNP Inputs Reduce D-C Loading**
- **Hysteresis (Typically 400 mV) at Inputs Improves Noise Margin**

description

These four-data-line transceivers are designed for asynchronous two-way communications between data buses. The SN74LS' can be used to drive terminated lines down to 133 ohms.

The SN54' family is characterized for operation over the full military temperature range of −55°C to 125°C. The SN74' family is characterized for operation from 0°C to 70°C.

SN54LS242, SN54LS243 . . . J OR W PACKAGE
SN74LS242, SN74LS243 . . . D OR N PACKAGE
(TOP VIEW)

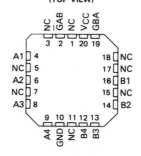

```
        ___
 GAB [1  U  14] VCC
  NC [2     13] GBA
  A1 [3     12] NC
  A2 [4     11] B1
  A3 [5     10] B2
  A4 [6      9] B3
 GND [7      8] B4
```

SN54LS242, SN54LS243 . . . FK PACKAGE
(TOP VIEW)

```
            N  G  N  V  G
            C  A  C  C  B
               B     C  A
            3  2  1  20 19
   A1 [ 4              18 ] NC
   NC [ 5              17 ] NC
   A2 [ 6              16 ] B1
   NC [ 7              15 ] NC
   A3 [ 8              14 ] B2
            9  10 11 12 13
            A  G  N  B  B
            4  N  C  4  3
               D
```

NC–No internal connection

FUNCTION TABLE (EACH TRANSCEIVER)

INPUTS		'LS242	'LS243
$\overline{\text{GAB}}$	GBA		
L	L	$\overline{\text{A}}$ to B	A to B
H	H	$\overline{\text{B}}$ to A	B to A
H	L	Isolation	Isolation
L	H	Latch A and B ($A = \overline{B}$)	Latch A and B ($A = B$)

schematics of inputs and outputs

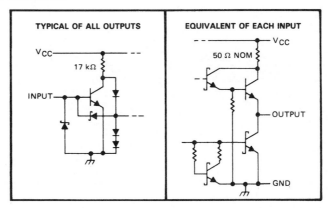

TYPICAL OF ALL OUTPUTS	EQUIVALENT OF EACH INPUT

FIGURE B-60

SN54LS245, SN74LS245
OCTAL BUS TRANSCEIVERS WITH 3-STATE OUTPUTS

OCTOBER 1976—REVISED MARCH 1988

- **Bi-directional Bus Transceiver in a High-Density 20-Pin Package**

- **3-State Outputs Drive Bus Lines Directly**

- **PNP Inputs Reduce D-C Loading on Bus Lines**

- **Hysteresis at Bus Inputs Improve Noise Margins**

- **Typical Propagation Delay Times, Port-to-Port . . . 8 ns**

TYPE	I_{OL} (SINK CURRENT)	I_{OH} (SOURCE CURRENT)
SN54LS245	12 mA	−12 mA
SN74LS245	24 mA	−15 mA

description

These octal bus transceivers are designed for asynchronous two-way communication between data buses. The control function implementation minimizes external timing requirements.

The devices allow data transmission from the A bus to the B bus or from the B bus to the A bus depending upon the logic level at the direction control (DIR) input. The enable input ($\overline{G}$) can be used to disable the device so that the buses are effectively isolated.

The SN54LS245 is characterized for operation over the full military temperature range of −55°C to 125°C. The SN74LS245 is characterized for operation from 0°C to 70°C.

SN54LS245 . . . J OR W PACKAGE
SN74LS245 . . . DW OR N PACKAGE
(TOP VIEW)

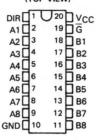

DIR	1	20 V_{CC}
A1	2	19 $\overline{G}$
A2	3	18 B1
A3	4	17 B2
A4	5	16 B3
A5	6	15 B4
A6	7	14 B5
A7	8	13 B6
A8	9	12 B7
GND	10	11 B8

SN54LS245 . . . FK PACKAGE
(TOP VIEW)

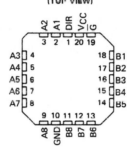

schematics of inputs and outputs

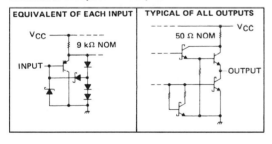

EQUIVALENT OF EACH INPUT	TYPICAL OF ALL OUTPUTS
V_{CC} 9 kΩ NOM INPUT	V_{CC} 50 Ω NOM OUTPUT

FUNCTION TABLE

ENABLE $\overline{G}$	DIRECTION CONTROL DIR	OPERATION
L	L	B data to A bus
L	H	A data to B bus
H	X	Isolation

H = high level, L = low level, X = irrelevant

FIGURE B-61

SN54LS295B, SN74LS295B
4-BIT RIGHT-SHIFT LEFT-SHIFT REGISTERS
WITH 3-STATE OUTPUTS
OCTOBER 1976 — REVISED MARCH 1988

- **'LS295B Offers Three Times the Sink-Current Capability of 'LS295A**

- **Schottky-Diode-Clamped Transistors**

- **Low Power Dissipation . . . 80 mW Typical (Enabled)**

- **Applications:**
  N-Bit Serial-To-Parallel Converter
  N-Bit Parallel-To-Serial Converter
  N-Bit Storage Register

description

These 4-bit registers feature parallel inputs, parallel outputs, and clock (CLK), serial (SER), mode (LD/$\overline{\text{SH}}$), and outputs control (OC) inputs. The registers have three modes of operation:

Parallel (broadside) load
Shift right (the direction Q_A toward Q_D)
Shift left (the direction Q_D toward Q_A)

Parallel loading is accomplished by applying the four bits of data and taking the mode control input high. The data is loaded into the associated flip-flops and appears at the outputs after the high-to-low transition of the clock input. During parallel loading, the entry of serial data is inhibited.

Shift right is accomplished when the mode control is low; shift left is accomplished when the mode control is high by connecting the output of each flip-flop to the parallel input of the previous flip-flop (Q_D to input C, etc.) and serial data is entered at input D.

When the output control is high, the normal logic levels of the four outputs are available for driving the loads or bus lines. The outputs are disabled independently from the level of the clock by a low logic level at the output control input. The outputs then present a high impedance and neither load nor drive the bus line; however, sequential operation of the registers is not affected.

The SN54LS295B is characterized for operation over the full military temperature range of −55°C to 125°C; the SN74LS295B is characterized for operation from 0°C to 70°C.

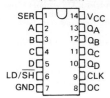

SN54LS295B . . . J OR W PACKAGE
SN74LS295B . . . D OR N PACKAGE
(TOP VIEW)

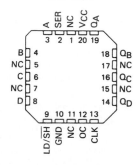

SN54LS295B . . . FK PACKAGE
(TOP VIEW)

NC - No internal connection

logic symbol[†]

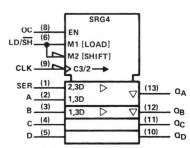

[†]This symbol is in accordance with ANSI/IEEE Std. 91-1984 and IEC Publication 617-12.
Pin numbers shown are for D, J, N, and W packages.

FIGURE B-62

SN54390, SN54LS390, SN54393, SN54LS393, SN74390, SN74LS390, SN74393, SN74LS393
DUAL 4-BIT DECADE AND BINARY COUNTERS
OCTOBER 1976 — REVISED MARCH 1988

- Dual Versions of the Popular '90A, 'LS90 and '93A, 'LS93

- '390, 'LS390 . . . Individual Clocks for A and B Flip-Flops Provide Dual ÷ 2 and ÷ 5 Counters

- '393, 'LS393 . . . Dual 4-Bit Binary Counter with Individual Clocks

- All Have Direct Clear for Each 4-Bit Counter

- Dual 4-Bit Versions Can Significantly Improve System Densities by Reducing Counter Package Count by 50%

- Typical Maximum Count Frequency . . . 35 MHz

- Buffered Outputs Reduce Possibility of Collector Commutation

description

Each of these monolithic circuits contains eight master-slave flip-flops and additional gating to implement two individual four-bit counters in a single package. The '390 and 'LS390 incorporate dual divide-by-two and divide-by-five counters, which can be used to implement cycle lengths equal to any whole and/or cumulative multiples of 2 and/or 5 up to divide-by-100. When connected as a bi-quinary counter, the separate divide-by-two circuit can be used to provide symmetry (a square wave) at the final output stage. The '393 and 'LS393 each comprise two independent four-bit binary counters each having a clear and a clock input. N-bit binary counters can be implemented with each package providing the capability of divide-by-256. The '390, 'LS390, '393, and 'LS393 have parallel outputs from each counter stage so that any submultiple of the input count frequency is available for system-timing signals.

Series 54 and Series 54LS circuits are characterized for operation over the full military temperature range of −55°C to 125°C; Series 74 and Series 74LS circuits are characterized for operation from 0°C to 70°C.

SN54390, SN54LS390 . . . J OR W PACKAGE
SN74390 . . . N PACKAGE
SN74LS390 . . . D OR N PACKAGE
(TOP VIEW)

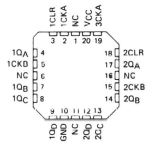

SN54LS390 . . . FK PACKAGE
(TOP VIEW)

SN54393, SN54LS393 . . . J OR W PACKAGE
SN74393 . . . N PACKAGE
SN74LS393 . . . D OR N PACKAGE
(TOP VIEW)

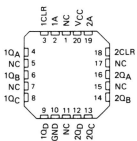

SN54LS393 . . . FK PACKAGE
(TOP VIEW)

NC - No internal connection

FIGURE B-63

Overview of
IEEE Standard 91-1984

Explanation of Logic Symbols

F.A. Mann
Semiconductor Group

TEXAS
INSTRUMENTS

Note: The information in this Appendix is reprinted
by permission of Texas Instruments.

If you have questions on this Explanation of Logic Symbols, please contact:

Texas Instruments Incorporated
F.A. Mann, MS 49
P.O. Box 225012
Dallas, Texas 75265

Telephone (214) 995-2867

IEEE Standards may be purchased from:

Institute of Electrical and Electronics Engineers, Inc.
IEEE Standards Office
345 East 47th Street
New York, N.Y. 10017

International Electrotechnical Commission (IEC) publications may be purchased from:

American National Standards Institute, Inc.
1430 Broadway
New York, N.Y. 10018

Contents

List of Tables

List of Illustrations

1.0 INTRODUCTION

The International Electrotechnical Commission (IEC) has been developing a very powerful symbolic language that can show the relationship of each input of a digital logic circuit to each output without showing explicitly the internal logic. At the heart of the system is dependency notation, which will be explained in Section 4.

The system was introduced in the USA in a rudimentary form in IEEE/ANSI Standard Y32.14-1973. Lacking at that time a complete development of dependency notation, it offered little more than a substitution of rectangular shapes for the familiar distinctive shapes for representing the basic functions of AND, OR, negation, etc. This is no longer the case.

Internationally, Working Group 2 of IEC Technical Committee TC-3 has prepared a new document (Publication 617-12) that consolidates the original work started in the mid 1960's and published in 1972 (Publication 117-15) and the amendments and supplements that have followed. Similarly for the USA, IEEE Committee SCC 11.9 has revised the publication IEEE Std 91/ANSI Y32.14. Now numbered simply IEEE Std 91-1984, the IEEE standard contains all of the IEC work that has been approved, and also a small amount of material still under international consideration. Texas Instruments is participating in the work of both organizations and this document introduces new logic symbols in accordance with the new standards. When changes are made as the standards develop, future editions will take those changes into account.

The following explanation of the new symbolic language is necessarily brief and greatly condensed from what the standards publications will contain. This is not intended to be sufficient for those people who will be developing symbols for new devices. It is primarily intended to make possible the understanding of the symbols used in various data books and the comparison of the symbols with logic diagrams, functional block diagrams, and/or function tables will further help that understanding.

2.0 SYMBOL COMPOSITION

A symbol comprises an outline or a combination of outlines together with one or more qualifying symbols. The shape of the symbols is not significant. As shown in Figure 1, general qualifying symbols are used to tell exactly what logical operation is performed by the elements. Table I shows general qualifying symbols defined in the new standards. Input lines are placed on the left and output lines are placed on the right. When an exception is made to that convention, the direction of signal flow is indicated by an arrow as shown in Figure 11.

All outputs of a single, unsubdivided element always have identical internal logic states determined by the function of the element except when otherwise indicated by an associated qualifying symbol or label inside the element.

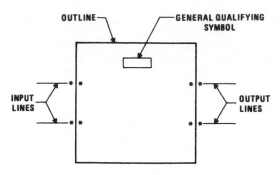

*Possible positions for qualifying symbols relating to inputs and outputs

Figure 1. Symbol Composition

The outlines of elements may be abutted or embedded in which case the following conventions apply. There is no logic connection between the elements when the line common to their outlines is in the direction of signal flow. There is at least one logic connection between the elements when the line common to their outlines is perpendicular to the direction of signal flow. The number of logic connections between elements will be clarified by the use of qualifying symbols and this is discussed further under that topic. If no indications are shown on either side of the common line, it is assumed there is only one connection.

When a circuit has one or more inputs that are common to more than one element of the circuit, the common-control block may be used. This is the only distinctively shaped outline used in the IEC system. Figure 2 shows that unless otherwise qualified by dependency notation, an input to the common-control block is an input to each of the elements below the common-control block.

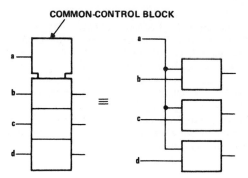

Figure 2. Common-Control Block

A common output depending on all elements of the array can be shown as the output of a common-output element. Its distinctive visual feature is the double line at its top. In addition the common-output element may have other inputs as shown in Figure 3. The function of the common-output element must be shown by use of a general qualifying symbol.

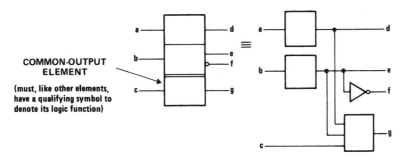

COMMON-OUTPUT
ELEMENT

(must, like other elements,
have a qualifying symbol to
denote its logic function)

Figure 3. Common-Output Element

3.0 QUALIFYING SYMBOLS

3.1 General Qualifying Symbols

Table I shows general qualifying symbols defined by IEEE Standard 91. These characters are placed near the top center or the geometric center of a symbol or symbol element to define the basic function of the device represented by the symbol or of the element.

3.2 Qualifying Symbols for Inputs and Outputs

Qualifying symbols for inputs and outputs are shown in Table II and will be familiar to most users with the possible exception of the logic polarity and analog signal indicators. The older logic negation indicator means that the external 0 state produces the internal 1 state. The internal 1 state means the active state. Logic negation may be used in pure logic diagrams; in order to tie the external 1 and 0 logic states to the levels H (high) and L (low), a statement of whether positive logic (1 = H, 0 = L) or negative logic (1 = L, 0 = H) is being used is required or must be assumed. Logic polarity indicators eliminate the need for calling out the logic convention and are used in various data books in the symbology for actual devices. The presence of the triangular polarity indicator indicates that the L logic level will produce the internal 1 state (the active state) or that, in the case of an output, the internal 1 state will produce the external L level. Note how the active direction of transition for a dynamic input is indicated in positive logic, negative logic, and with polarity indication.

The internal connections between logic elements abutted together in a symbol may be indicated by the symbols shown in Table II. Each logic connection may be shown by the presence of qualifying symbols at one or both sides of the common line and if confusion can arise about the numbers of connections, use can be made of one of the internal connection symbols.

Table I. General Qualifying Symbols

SYMBOL	DESCRIPTION	CMOS EXAMPLE	TTL EXAMPLE
&	AND gate or function.	'HC00	SN7400
≥ 1	OR gate or function. The symbol was chosen to indicate that at least one active input is needed to activate the output.	'HC02	SN7402
= 1	Exclusive OR. One and only one input must be active to activate the output.	'HC86	SN7486
=	Logic identity. All inputs must stand at the same state.	'HC86	SN74180
2k	An even number of inputs must be active.	'HC280	SN74180
2k + 1	An odd number of inputs must be active.	'HC86	SN74ALS86
1	The one input must be active.	'HC04	SN7404
▷ or ◁	A buffer or element with more than usual output capability (symbol is oriented in the direction of signal flow).	'HC240	SN74S436
⎍	Schmitt trigger; element with hysteresis.	'HC132	SN74LS18
X/Y	Coder, code converter (DEC/BCD, BIN/OUT, BIN/7-SEG, etc.).	'HC42	SN74LS347
MUX	Multiplexer/data selector.	'HC151	SN74150
DMUX or DX	Demultiplexer.	'HC138	SN74138
Σ	Adder.	'HC283	SN74LS385
P−Q	Subtracter.	*	SN74LS385
CPG	Look-ahead carry generator	'HC182	SN74182
π	Multiplier.	*	SN74LS384
COMP	Magnitude comparator.	'HC85	SN74LS682
ALU	Arithmetic logic unit.	'HC181	SN74LS381
⎍⎍	Retriggerable monostable.	'HC123	SN74LS422
1⎍⎍	Nonretriggerable monostable (one-shot)	'HC221	SN74121
$\underset{\text{⎍⎍⎍}}{G}$	Astable element. Showing waveform is optional.	*	SN74LS320
$\underset{\text{⎍⎍⎍}}{!G}$	Synchronously starting astable.	*	SN74LS624
$\underset{\text{⎍⎍⎍}}{G!}$	Astable element that stops with a completed pulse.	*	*
SRGm	Shift register. m = number of bits.	'HC164	SN74LS595
CTRm	Counter. m = number of bits; cycle length = 2^m.	'HC590	SN54LS590
CTR DIVm	Counter with cycle length = m.	'HC160	SN74LS668
RCTRm	Asynchronous (ripple-carry) counter; cycle length = 2^m.	'HC4020	*
ROM	Read-only memory.	*	SN74187
RAM	Random-access read/write memory.	'HC189	SN74170
FIFO	First-in, first-out memory.	*	SN74LS222
I = 0	Element powers up cleared to 0 state.	*	SN74AS877
I = 1	Element powers up set to 1 state.	'HC7022	SN74AS877
Φ	Highly complex function; "gray box" symbol with limited detail shown under special rules.	*	SN74LS608

*Not all of the general qualifying symbols have been used in TI's CMOS and TTL data books, but they are included here for the sake of completeness.

Table II. Qualifying Symbols for Inputs and Outputs

Logic negation at input. External 0 produces internal 1.

Logic negation at output. Internal 1 produces external 0.

Active-low input. Equivalent to ──◁ in positive logic.

Active-low output. Equivalent to ▷── in positive logic.

Active-low input in the case of right-to-left signal flow.

Active-low output in the case of right-to-left signal flow.

Signal flow from right to left. If not otherwise indicated, signal flow is from left to right.

Bidirectional signal flow.

	POSITIVE LOGIC	NEGATIVE LOGIC	POLARITY INDICATION
Dynamic inputs active on indicated transition	1 → 0	0 → 1 (not used)	not used
	not used		H → L
	0 → 1	0 → 1	L → H

Nonlogic connection. A label inside the symbol will usually define the nature of this pin.

Input for analog signals (on a digital symbol) (see Figure 14).

Input for digital signals (on an analog symbol) (see Figure 14).

Internal connection. 1 state on left produces 1 state on right.

Negated internal connection. 1 state on left produces 0 state on right.

Dynamic internal connection. Transition from 0 to 1 on left produces transitory 1 state on right.

Internal input (virtual input). It always stands at its internal 1 state unless affected by an overriding dependency relationship.

Internal output (virtual output). Its effect on an internal input to which it is connected is indicated by dependency notation.

The internal (virtual) input is an input originating somewhere else in the circuit and is not connected directly to a terminal. The internal (virtual) output is likewise not connected directly to a terminal. The application of internal inputs and outputs requires an understanding of dependency notation, which is explained in Section 4.

Table III. Symbols Inside the Outline

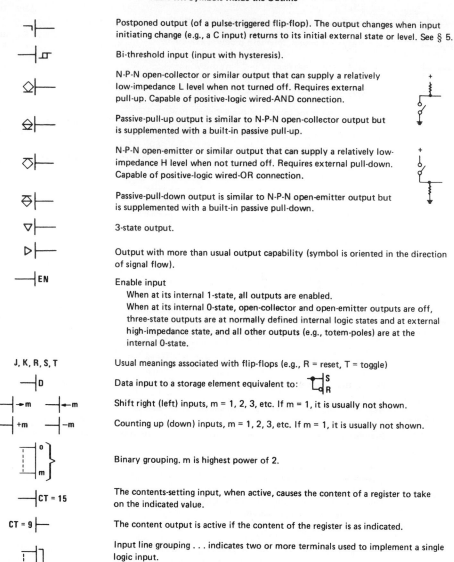

Postponed output (of a pulse-triggered flip-flop). The output changes when input initiating change (e.g., a C input) returns to its initial external state or level. See § 5.

Bi-threshold input (input with hysteresis).

N-P-N open-collector or similar output that can supply a relatively low-impedance L level when not turned off. Requires external pull-up. Capable of positive-logic wired-AND connection.

Passive-pull-up output is similar to N-P-N open-collector output but is supplemented with a built-in passive pull-up.

N-P-N open-emitter or similar output that can supply a relatively low-impedance H level when not turned off. Requires external pull-down. Capable of positive-logic wired-OR connection.

Passive-pull-down output is similar to N-P-N open-emitter output but is supplemented with a built-in passive pull-down.

3-state output.

Output with more than usual output capability (symbol is oriented in the direction of signal flow).

Enable input
  When at its internal 1-state, all outputs are enabled.
  When at its internal 0-state, open-collector and open-emitter outputs are off, three-state outputs are at normally defined internal logic states and at external high-impedance state, and all other outputs (e.g., totem-poles) are at the internal 0-state.

J, K, R, S, T Usual meanings associated with flip-flops (e.g., R = reset, T = toggle)

Data input to a storage element equivalent to:

Shift right (left) inputs, m = 1, 2, 3, etc. If m = 1, it is usually not shown.

Counting up (down) inputs, m = 1, 2, 3, etc. If m = 1, it is usually not shown.

Binary grouping. m is highest power of 2.

The contents-setting input, when active, causes the content of a register to take on the indicated value.

The content output is active if the content of the register is as indicated.

Input line grouping . . . indicates two or more terminals used to implement a single logic input.

e.g., The paired expander inputs of SN7450.

Fixed-state output always stands at its internal 1 state. For example, see SN74185.

In an array of elements, if the same general qualifying symbol and the same qualifying symbols associated with inputs and outputs would appear inside each of the elements of the array, these qualifying symbols are usually shown only in the first element. This is done to reduce clutter and to save time in recognition. Similarly, large identical elements that are subdivided into smaller elements may each be represented by an unsubdivided outline. The SN54HC242 or SN54LS440 symbol illustrates this principle.

3.3 Symbols Inside the Outline

Table III shows some symbols used inside the outline. Note particularly that open-collector (open-drain), open-emitter (open-source), and three-state outputs have distinctive symbols. Also note that an EN input affects all of the outputs of the circuit and has no effect on inputs. When an enable input affects only certain outputs and/or affects one or more inputs, a form of dependency notation will indicate this (see 4.9). The effects of the EN input on the various types of outputs are shown.

It is particularly important to note that a D input is always the data input of a storage element. At its internal 1 state, the D input sets the storage element to its 1 state, and at its internal 0 state it resets the storage element to its 0 state.

The binary grouping symbol will be explained more fully in Section 8. Binary-weighted inputs are arranged in order and the binary weights of the least-significant and the most-significant lines are indicated by numbers. In this document weights of input and output lines will be represented by powers of two usually only when the binary grouping symbol is used, otherwise decimal numbers will be used. The grouped inputs generate an internal number on which a mathematical function can be performed or that can be an identifying number for dependency notation (Figure 28). A frequent use is in addresses for memories.

Reversed in direction, the binary grouping symbol can be used with outputs. The concept is analogous to that for the inputs and the weighted outputs will indicate the internal number assumed to be developed within the circuit.

Other symbols are used inside the outlines in accordance with the IEC/IEEE standards but are not shown here. Generally these are associated with arithmetic operations and are self-explanatory.

When nonstandardized information is shown inside an outline, it is usually enclosed in square brackets [like these].

4.0 DEPENDENCY NOTATION

4.1 General Explanation

Dependency notation is the powerful tool that sets the IEC symbols apart from previous systems and makes compact, meaningful, symbols possible. It provides the means of denoting the relationship between inputs, outputs, or inputs and outputs without actually showing all the elements and interconnections involved. The information provided by dependency notation supplements that provided by the qualifying symbols for an element's function.

In the convention for the dependency notation, use will be made of the terms "affecting" and "affected." In cases where it is not evident which inputs must be considered as being the affecting or the affected ones (e.g., if they stand in an AND relationship), the choice may be made in any convenient way.

So far, eleven types of dependency have been defined and all of these are used in various TI data books. X dependency is used mainly with CMOS circuits. They are listed below in the order in which they are presented and are summarized in Table IV following 4.12.

Section	Dependency Type or Other Subject
4.2	G, AND
4.3	General Rules for Dependency Notation
4.4	V, OR
4.5	N, Negate (Exclusive-OR)
4.6	Z, Interconnection
4.7	X, Transmission
4.8	C, Control
4.9	S, Set and R, Reset
4.10	EN, Enable
4.11	M, Mode
4.12	A, Address

4.2   G (AND) Dependency

A common relationship between two signals is to have them ANDed together. This has traditionally been shown by explicitly drawing an AND gate with the signals connected to the inputs of the gate. The 1972 IEC publication and the 1973 IEEE/ANSI standard showed several ways to show this AND relationship using dependency notation. While ten other forms of dependency have since been defined, the ways to invoke AND dependency are now reduced to one.

In Figure 4 input **b** is ANDed with input **a** and the complement of **b** is ANDed with **c**. The letter G has been chosen to indicate AND relationships and is placed at input **b**, inside the symbol. A number considered appropriate by the symbol designer (1 has been used here) is placed after the letter G and also at each affected input. Note the bar over the 1 at input **c**.

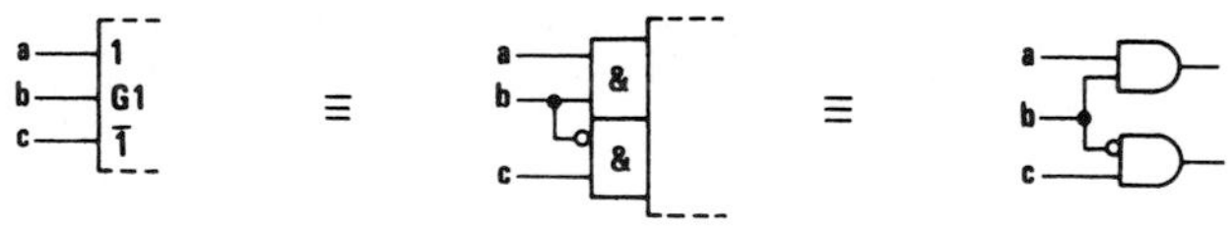

Figure 4. G Dependency Between Inputs

In Figure 5, output **b** affects input **a** with an AND relationship. The lower example shows that it is the internal logic state of **b**, unaffected by the negation sign, that is ANDed. Figure 6 shows input **a** to be ANDed with a dynamic input **b**.

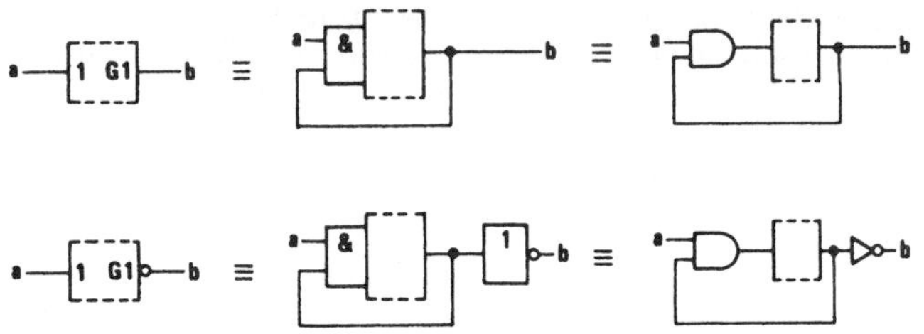

Figure 5. G Dependency Between Outputs and INputs

Figure 6. G Dependency with a Dynamic Input

The rules for G dependency can be summarized thus:

> When a Gm input or output (m is a number) stands at its internal 1 state, all inputs and outputs affected by Gm stand at their normally defined internal logic states. When the Gm input or output stands at its 0 state, all inputs and outputs affected by Gm stand at their internal 0 states.

4.3 Conventions for the Application of Dependency Notation in General

The rules for applying dependency relationships in general follow the same pattern as was illustrated for G dependency.

Application of dependency notation is accomplished by:

1) labeling the input (or output) *affecting* other inputs or outputs with the letter symbol indicating the relationship involved (e.g., G for AND) followed by an identifying number, appropriately chosen, and

2) labeling each input or output *affected* by that affecting input (or output) with that same number.

If it is the complement of the internal logic state of the affecting input or output that does the affecting, then a bar is placed over the identifying numbers at the affected inputs or outputs (Figure 4).

If two affecting inputs or outputs have the same letter and same identifying number, they stand in an OR relationship to each other (Figure 7).

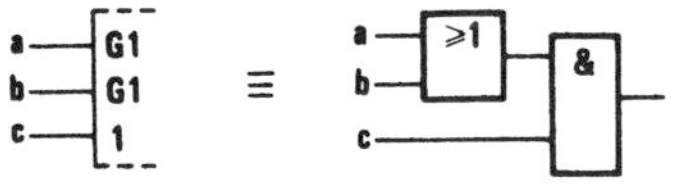

Figure 7. ORed Affecting Inputs

If the affected input or output requires a label to denote its function (e.g., ''D''), this label will be *prefixed* by the identifying number of the affecting input (Figure 15).

If an input or output is affected by more than one affecting input, the identifying numbers of each of the affecting inputs will appear in the label of the affected one, separated by commas. The normal reading order of these numbers is the same as the sequence of the affecting relationships (Figure 15).

If the labels denoting the functions of affected inputs or outputs must be numbers (e.g., outputs of a coder), the identifying numbers to be associated with both affecting inputs and affected inputs or outputs will be replaced by another character selected to avoid ambiguity, e.g., Greek letters (Figure 8).

Figure 8. Substitution for Numbers

4.4 V (OR) Dependency

The symbol denoting OR dependency is the letter V (Figure 9).

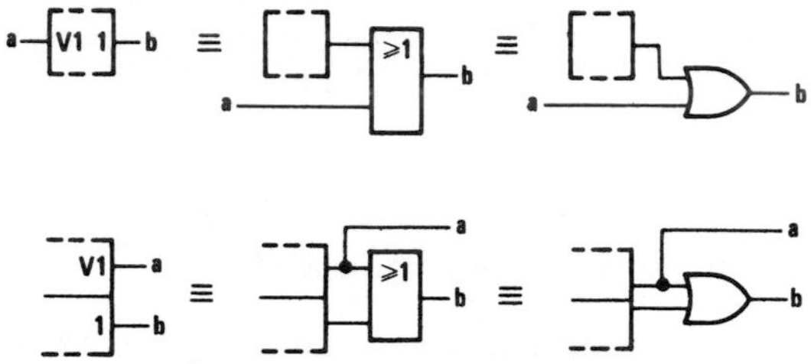

Figure 9. V (OR) Dependency

When a Vm input or output stands at its internal 1 state, all inputs and outputs affected by Vm stand at their internal 1 states. When the Vm input or output stands at its internal 0 state, all inputs and outputs affected by Vm stand at their normally defined internal logic states.

4.5 N (Negate) (Exclusive-OR) Dependency

The symbol denoting negate dependency is the letter N (Figure 10). Each input or output affected by an Nm input or output stands in an Exclusive-OR relationship with the Nm input or output.

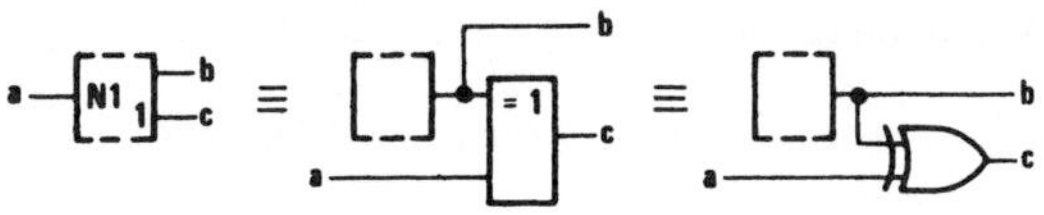

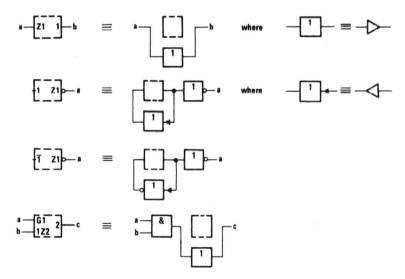

If a = 0,   then c = b
If a = 1,   then c = b

Figure 10. N (Negate) (Exclusive-OR) Dependency

When an Nm input or output stands at its internal 1 state, the internal logic state of each input and each output affected by Nm is the complement of what it would otherwise be. When an Nm input or output stands at its internal 0 state, all inputs and outputs affected by Nm stand at their normally defined internal logic states.

4.6   Z (Interconnection) Dependency

The symbol denoting interconnection dependency is the letter Z.

Interconnection dependency is used to indicate the existence of internal logic connections between inputs, outputs, internal inputs, and/or internal outputs.

The internal logic state of an input or output affected by a Zm input or output will be the same as the internal logic state of the Zm input or output, unless modified by additional dependency notation (Figure 11).

Figure 11. Z (Interconnection) Dependency

4.7 X (Transmission) Dependency

The symbol denoting transmission dependency is the letter X.

Transmission dependency is used to indicate controlled bidirectional connections between affected input/output ports (Figure 12).

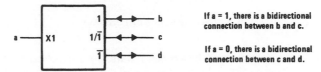

Figure 12. X (Transmission) Dependency

When an X*m* input or output stands at its internal 1 state, all input-output ports affected by this X*m* input or output are bidirectionally connected together and stand at the same internal logic state or analog signal level. When an X*m* input or output stands at its internal 0 state, the connection associated with this set of dependency notation does not exist.

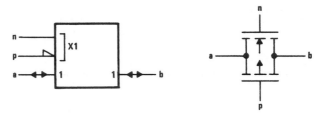

Figure 13. CMOS Transmission Gate Symbol and Schematic

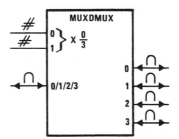

Figure 14. Analog Data Selector (Multiplexer/Demultiplexer)

Although the transmission paths represented by X dependency are inherently bidirectional, use is not always made of this property. This is analogous to a piece of wire, which may be constrained to carry current in only one direction. If this is the case in a particular application, then the directional arrows shown in Figures 12, 13, and 14 would be omitted.

4.8 C (Control) Dependency

The symbol denoting control dependency is the letter C.

Control inputs are usually used to enable or disable the data (D, J, K, R, or S) inputs of storage elements. They may take on their internal 1 states (be active) either statically or dynamically. In the latter case the dynamic input symbol is used as shown in the third example of Figure 15.

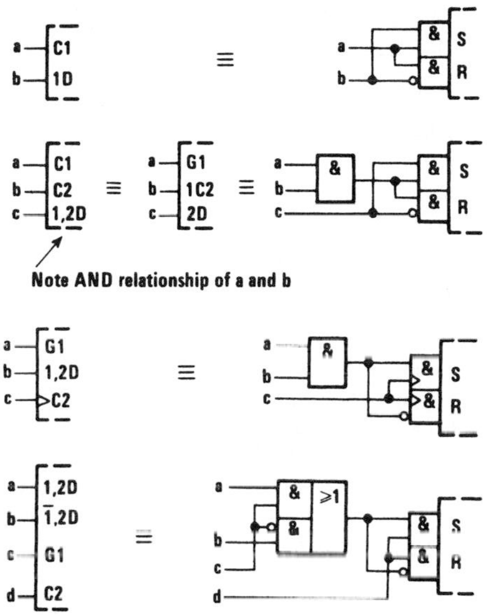

Input c selects which of a or b is stored when d goes low.

Figure 15. C (Control) Dependency

When a *Cm* input or output stands at its internal 1 state, the inputs affected by *Cm* have their normally defined effect on the function of the element, i.e., these inputs are enabled. When a *Cm* input or output stands at its internal 0 state, the inputs affected by *Cm* are disabled and have no effect on the function of the element.

4.9 S (Set) and R (Reset) Dependencies

The symbol denoting set dependency is the letter S. The symbol denoting reset dependency is the letter R.

Set and reset dependencies are used if it is necessary to specify the effect of the combination $R = S = 1$ on a bistable element. Case 1 in Figure 16 does not use S or R dependency.

When an Sm input is at its internal 1 state, outputs affected by the Sm input will react, regardless of the state of an R input, as they normally would react to the combination $S = 1$, $R = 0$. See cases 2, 4, and 5 in Figure 16.

When an Rm input is at its internal 1 state, outputs affected by the Rm input will react, regardless of the state of an S input, as they normally would react to the combination $S = 0$, $R = 1$. See cases 3, 4, and 5 in Figure 16.

When an Sm or Rm input is at its internal 0 state, it has no effect.

Note that the noncomplementary output patterns in cases 4 and 5 are only pseudo stable. The simultaneous return of the inputs to $S = R = 0$ produces an unforeseeable stable and complementary output pattern.

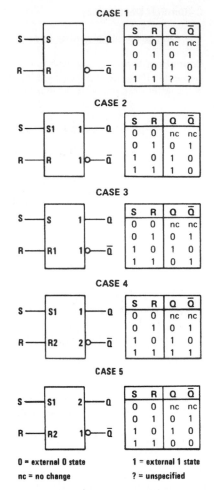

CASE 1

S	R	Q	$\overline{Q}$
0	0	nc	nc
0	1	0	1
1	0	1	0
1	1	?	?

CASE 2

S	R	Q	$\overline{Q}$
0	0	nc	nc
0	1	0	1
1	0	1	0
1	1	1	0

CASE 3

S	R	Q	$\overline{Q}$
0	0	nc	nc
0	1	0	1
1	0	1	0
1	1	0	1

CASE 4

S	R	Q	$\overline{Q}$
0	0	nc	nc
0	1	0	1
1	0	1	0
1	1	1	1

CASE 5

S	R	Q	$\overline{Q}$
0	0	nc	nc
0	1	0	1
1	0	1	0
1	1	0	0

0 = external 0 state   1 = external 1 state

nc = no change   ? = unspecified

Figure 16. S (Set) and R (Reset) Dependencies

4.10 EN (Enable) Dependency

The symbol denoting enable dependency is the combination of letters EN.

An ENm input has the same effect on outputs as an EN input, see 3.1, but it affects only those outputs labeled with the identifying number m. It also affects those inputs labeled with the identifying number m. By contrast, an EN input affects all outputs and no inputs. The effect of an ENm input on an affected input is identical to that of a Cm input (Figure 17).

When an ENm input stands at its internal 1 state, the inputs affected by ENm have their normally defined effect on the function of the element and the outputs affected by this input stand at their normally defined internal logic states, i.e., these inputs and outputs are enabled.

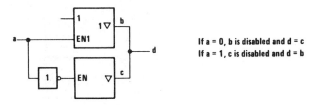

If a = 0, b is disabled and d = c
If a = 1, c is disabled and d = b

Figure 17. EN (Enable) Dependency

When an ENm input stands at its internal 0 state, the inputs affected by ENm are disabled and have no effect on the function of the element, and the outputs affected by ENm are also disabled. Open-collector outputs are turned off, three-state outputs stand at their normally defined internal logic states but externally exhibit high impedance, and all other outputs (e.g., totem-pole outputs) stand at their internal 0 states.

4.11 M (MODE) Dependency

The symbol denoting mode dependency is the letter M.

Mode dependency is used to indicate that the effects of particular inputs and outputs of an element depend on the mode in which the element is operating.

If an input or output has the same effect in different modes of operation, the identifying numbers of the relevant affecting Mm inputs will appear in the label of that affected input or output between parentheses and separated by solidi (Figure 22).

4.11.1 M Dependency Affecting Inputs

M dependency affects inputs the same as C dependency. When an Mm input or Mm output stands at its internal 1 state, the inputs affected by this Mm input or Mm output have their normally defined effect on the function of the element, i.e., the inputs are enabled.

When an Mm input or Mm output stands at its internal 0 state, the inputs affected by this Mm input or Mm output have no effect on the function of the element. When an affected input has several sets of labels separated by solidi (e.g., C4/2→/3+), any set in which the identifying number of the Mm input or Mm output appears has no effect and is to be ignored. This represents disabling of some of the functions of a multifunction input.

The circuit in Figure 18 has two inputs, **b** and **c**, that control which one of four modes (0, 1, 2, or 3) will exist at any time. Inputs **d**, **e**, and **f** are D inputs subject to dynamic control (clocking) by the **a** input. The numbers 1 and 2 are in the series chosen to indicate the modes so inputs **e** and **f** are only enabled in mode 1 (for parallel loading) and input **d** is only enabled in mode 2 (for serial loading). Note that input **a** has three functions. It is the clock for entering data. In mode 2, it causes right shifting of data, which means a shift away from the control block. In mode 3, it causes the contents of the register to be incremented by one count.

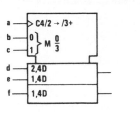

Note that all operations are synchronous.

In MODE 0 (b = 0, c = 0), the outputs remain at their existing states as none of the inputs has an effect.

In MODE 1 (b = 1, c = 0), parallel loading takes place thru inputs e and f.

In MODE 2 (b = 0, c = 1), shifting down and serial loading thru input d take place.

In MODE 3 (b = c = 1), counting up by increment of 1 per clock pulse takes place.

Figure 18. M (Mode) Dependency Affecting Inputs

4.11.2 M Dependency Affecting Outputs

When an Mm input or Mm output stands at its internal 1 state, the affected outputs stand at their normally defined internal logic states, i.e., the outputs are enabled.

When an Mm input or Mm output stands at its internal 0 state, at each affected output any set of labels containing the identifying number of that Mm input or Mm output has no effect and is to be ignored. When an output has several different sets of labels separated by solidi (e.g., 2,4/3,5), only those sets in which the identifying number of this Mm input or Mm output appears are to be ignored.

Figure 19 shows a symbol for a device whose output can behave like either a 3-state output or an open-collector output depending on the signal applied to input **a**. Mode 1 exists when input **a** stands at its internal 1 state and, in that case, the three-state symbol applies and the open-element symbol has no effect. When **a** = 0, mode 1 does not exist so the three-state symbol has no effect and the open-element symbol applies.

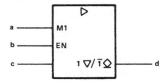

Figure 19. Type of Output Determined by Mode

In Figure 20, if input **a** stands at its internal 1 state establishing mode 1, output **b** will stand at its internal 1 state only when the content of the register equals 9. Since output **b** is located in the common-control block with no defined function outside of mode 1, the state of this output outside of mode 1 is not defined by the symbol.

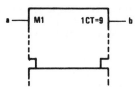

Figure 20. An Output of the Common-Control Block

In Figure 21, if input **a** stands at its internal 1 state establishing mode 1, output **b** will stand at its internal 1 state only when the content of the register equals 15. If input **a** stands at its internal 0 state, output **b** will stand at its internal 1 state only when the content of the register equals 0.

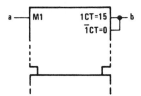

Figure 21. Determining and Output's Function

In Figure 22 inputs **a** and **b** are binary weighted to generate the numbers 0, 1, 2, or 3. This determines which one of the four modes exists.

At output **e** the label set causing negation (if **c** = 1) is effective only in modes 2 and 3. In modes 0 and 1 this output stands at its normally defined state as if it had no labels. At output **f** the label set has effect when the mode is not 0 so output **e** is negated (if **c** = 1) in modes 1, 2, and

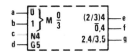

Figure 22. Dependent Relationships Affected by Mode

3. In mode 0 the label set has no effect so the output stands at its normally defined state. In this example $\overline{0}$,4 is equivalent to (1/2/3)4. At output **g** there are two label sets. The first set, causing negation (if **c** = 1), is effective only in mode 2. The second set, subjecting **g** to AND dependency on **d**, has effect only in mode 3.

Note that in mode 0 none of the dependency relationships has any effect on the outputs, so **e**, **f**, and **g** will all stand at the same state.

4.12 A (Address) Dependency

The symbol denoting address dependency is the letter A.

Address dependency provides a clear representation of those elements, particularly memories, that use address control inputs to select specified sections of a multildimensional arrays. Such a section of a memory array is usually called a word. The purpose of address dependency is to allow a symbolic presentation of the entire array. An input of the array shown at a particular element of this general section is common to the corresponding elements of all selected sections of the array. An output of the array shown at a particular element of this general section is the result of the OR function of the outputs of the corresponding elements of selected sections.

Inputs that are not affected by any affecting address input have their normally defined effect on all sections of the array, whereas inputs affected by an address input have their normally defined effect only on the section selected by that address input.

An affecting address input is labeled with the letter A followed by an identifying number that corresponds with the address of the particular section of the array selected by this input. Within the general section presented by the symbol, inputs and outputs affected by an A*m* input are labeled with the letter A, which stands for the identifying numbers, i.e., the addresses, of the particular sections.

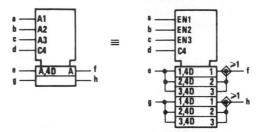

Figure 23. A (Address) Dependency

Figure 23 shows a 3-word by 2-bit memory having a separate address line for each word and uses EN dependency to explain the operation. To select word 1, input **a** is taken to its 1 state, which establishes mode 1. Data can now be clocked into the inputs marked "1,4D." Unless words 2 and 3 are also selected, data cannot be clocked in at the inputs marked "2,4D" and "3,4D." The outputs will be the OR functions of the selected outputs, i.e., only those enabled by the active EN functions.

The identifying numbers of affecting address inputs correspond with the addresses of the sections selected by these inputs. They need not necessarily differ from those of other affecting dependency-inputs (e.g., G, V, N, . . .), because in the general section presented by the symbol they are replaced by the letter A.

If there are several sets of affecting A*m* inputs for the purpose of independent and possibly simultaneous access to sections of the array, then the letter A is modified to 1A, 2A, ... Because they have access to the same sections of the array, these sets of A inputs may have the same identifying numbers. The symbols for 'HC170 or SN74LS170 make use of this.

Figure 24 is another illustration of the concept.

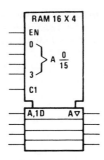

**Figure 24. Array of 16 Sections of Four Transparent Latches with 3-State Outputs
Comprising a 16-Word × 4-Bit Random-Access Memory**

Table IV. Summary of Dependency Notation

TYPE OF DEPENDENCY	LETTER SYMBOL*	AFFECTING INPUT AT ITS 1-STATE	AFFECTING INPUT AT ITS 0-STATE
Address	A	Permits action (address selected)	Prevents action (address not selected)
Control	C	Permits action	Prevents action
Enable	EN	Permits action	Prevents action of inputs ◇ outputs off ▽ outputs at external high impedance, no change in internal logic state Other outputs at internal 0 state
AND	G	Permits action	Imposes 0 state
Mode	M	Permits action (mode selected)	Prevents action (mode not selected)
Negate (Ex-OR)	N	Complements state	No effect
Reset	R	Affected output reacts as it would to S = 0, R = 1	No effect
Set	S	Affected output reacts as it would to S = 1, R = 0	No effect
OR	V	Imposes 1 state	Permits action
Transmission	X	Bidirectional connection exists	Bidirectional connection does not exist
Interconnection	Z	Imposes 1 state	Imposes 0 state

*These letter symbols appear at the AFFECTING input (or output) and are followed by a number. Each input (or output) AFFECTED by that input is labeled with that same number. When the labels EN, R, and S appear at inputs without the following numbers, the descriptions above do not apply. The action of these inputs is described under "Symbols Inside the Outline," sec 3.0.

5.0 BISTABLE ELEMENTS

The dynamic input symbol, the postponed output symbol, and dependency notation provide the tools to differentiate four main types of bistable elements and make synchronous and asynchronous inputs easily recognizable (Figure 25). The first column shows the essential distinguishing features; the other columns show examples.

Transparent latches have a level-operated control input. The D input is active as long as the C input is at its internal 1 state. The outputs respond immediately. Edge-triggered elements accept data from D, J, K, R, or S inputs on the active transition of C. Pulse-triggered elements require the setup of data before the start of the control pulse; the C input is considered static since the data must be maintained as long as C is at its 1 state. The output is postponed until C returns to its 0 state. The data-lock-out element is similar to the pulse-triggered version except that the C input is considered dynamic in that shortly after C goes through its active transition, the data inputs are disabled and data does not have to be held. However, the output is still postponed until the C input returns to its initial external level.

Notice that synchronous inputs can be readily recognized by their dependency labels (1D, 1J, 1K, 1S, 1R) compared to the asynchronous inputs (S, R), which are not dependent on the C inputs.

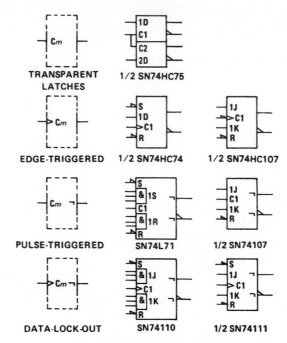

Figure 25. Four Types of Bistable Circuits

6.0 CODERS

The general symbol for a coder or code converter is shown in Figure 26. X and Y may be replaced by appropriate indications of the code used to represent the information at the inputs and at the outputs, respectively.

Figure 26. Coder General Symbol

Indication of code conversion is based on the following rule:

> Depending on the input code, the internal logic states of the inputs determine an internal value. This value is reproduced by the internal logic states of the outputs, depending on the output code.

The indication of the relationships between the internal logic states of the inputs and the internal value is accomplished by:

1) labeling the inputs with numbers. In this case the internal value equals the sum of the weights associated with those inputs that stand at their internal 1-state, or by
2) replacing X by an appropriate indication of the input code and labeling the inputs with characters that refer to this code.

The relationships between the internal value and the internal logic states of the outputs are indicated by:

1) labeling each output with a list of numbers representing those internal values that lead to the internal 1-state of that output. These numbers shall be separated by solidi as in Figure 27. This labeling may also be applied when Y is replaced by a letter denoting a type of dependency (see Section 7). If a continuous range of internal values produces the internal 1 state of an output, this can be indicated by two numbers that are inclusively the beginning and the end of the range, with these two numbers separated by three dots (e.g., 4 . . . 9 = 4/5/6/7/8/9) or by
2) replacing Y by an appropriate indiction of the output code and labeling the outputs with characters that refer to this code as in Figure 28.

Alternatively, the general symbol may be used together with an appropriate reference to a table in which the relationship between the inputs and outputs is indicated. This is a recommended way to symbolize a PROM after it has been programmed.

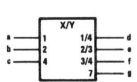

Figure 27. An X/Y Code Converter

FUNCTION TABLE

INPUTS			OUTPUTS			
c	b	a	g	f	e	d
0	0	0	0	0	0	0
0	0	1	0	0	0	1
0	1	0	0	0	1	0
0	1	1	0	1	1	0
1	0	0	0	1	0	1
1	0	1	0	0	0	0
1	1	0	0	0	0	0
1	1	1	1	0	0	0

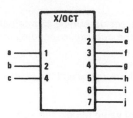

FUNCTION TABLE

INPUTS			OUTPUTS						
c	b	a	j	i	h	g	f	e	d
0	0	0	0	0	0	0	0	0	0
0	0	1	0	0	0	0	0	0	1
0	1	0	0	0	0	0	0	1	0
0	1	1	0	0	0	0	1	0	0
1	0	0	0	0	0	1	0	0	0
1	0	1	0	0	1	0	0	0	0
1	1	0	0	1	0	0	0	0	0
1	1	1	1	0	0	0	0	0	0

Figure 28. An X/Octal Code Converter

7.0   USE OF A CODER TO PRODUCE AFFECTING INPUTS

It often occurs that a set of affecting inputs for dependency notation is produced by decoding the signals on certain inputs to an element. In such a case use can be made of the symbol for a coder as an embedded symbol (Figure 29).

If all affecting inputs produced by a coder are of the same type and their identifying numbers shown at the outputs of the coder, Y (in the qualifying symbol X/Y) may be replaced by the letter denoting the type of dependency. The indications of the affecting inputs should then be omitted (Figure 30).

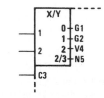

Figure 29. Producing Various Types of Dependencies

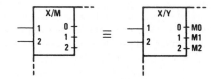

Figure 30. Producing One Type of Dependency

8.0   USE OF BINARY GROUPING TO PRODUCE AFFECTING INPUTS

If all affecting inputs produced by a coder are of the same type and have consecutive identifying numbers not necessarily corresponding with the numbers that would have been shown at the outputs of the coder, use can be made of the binary grouping symbol. k external lines effectively generate 2^k internal inputs. The bracket is followed by the letter denoting the type of dependency followed by m1/m2. The m1 is to be replaced by the smallest identifying number and the m2 by the largest one, as shown in Figure 31.

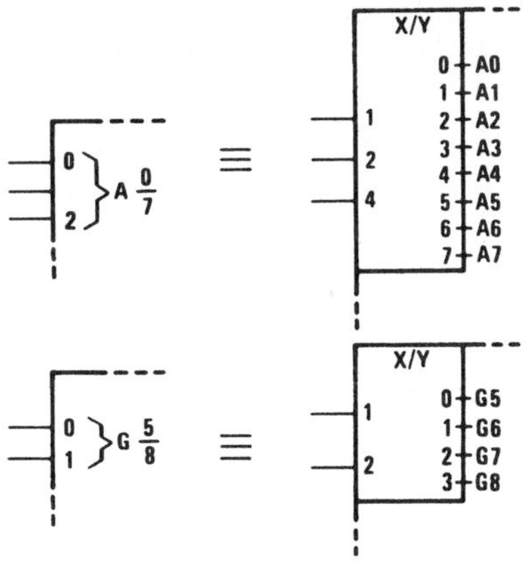

Figure 31. Use of the Binary Grouping Symbol

9.0 SEQUENCE OF INPUT LABELS

If an input having a single functional effect is affected by other inputs, the qualifying symbol (if there is any) for that functional effect is preceded by the labels corresponding to the affecting inputs. The left-to-right order of these preceding labels is the order in which the effects or modifications must be applied. The affected input has no functional effect on the element if the logic state of any one of the affecting inputs, considered separately, would cause the affected input to have no effect, regardless of the logic states of other affecting inputs.

If an input has several different functional effects or has several different sets of affecting inputs, depending on the mode of action, the input may be shown as often as required. However, there are cases in which this method of presentation is not advantageous. In those cases the input may be shown once with the different sets of labels separated by solidi (Figure 32). No meaning is attached to the order of these sets of labels. If one of the functional effects of an input is that of an unlabeled input to the element, a solidus will precede the first set of labels shown.

If all inputs of a combinational element are disabled (caused to have no effect on the function of the element), the internal logic states of the outputs of the element are not specified by the symbol. If all inputs of a sequential element are disabled, the content of this element is not changed and the outputs remain at their existing internal logic states.

Labels may be factored using algebraic techniques (Figure 33).

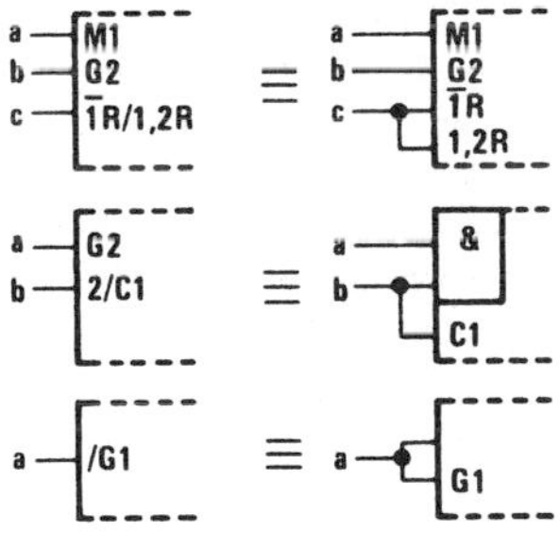

Figure 32. Input Labels

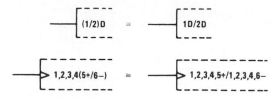

Figure 33. Factoring Input Labels

10.0 SEQUENCE OF OUTPUT LABELS

If an output has a number of different labels, regardless of whether they are identifying numbers of affecting inputs or outputs or not, these labels are shown in the following order:

1) If the postponed output symbol has to be shown, this comes first, if necessary preceded by the indications of the inputs to which it must be applied
2) Followed by the labels indicating modifications of the internal logic state of the output, such that the left-to-right order of these labels corresponds with the order in which their effects must be applied
3) Followed by the label indicating the effect of the output on inputs and other outputs of the element.

Symbols for open-circuit or three-state outputs, where applicable, are placed just inside the outside boundary of the symbol adjacent to the output line (Figure 34).

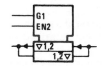

Figure 34. Placement of 3-State Symbols

If an output needs several different sets of labels that represent alternative functions (e.g., depending on the mode of action), these sets may be shown on different output lines that must be connected outside the outline. However, there are cases in which this method of presentation is not advantageous. In those cases the output may be shown once with the different sets of labels separated by solidi (Figure 35).

Two adjacent identifying numbers of affecting inputs in a set of labels that are not already separated by a nonnumeric character should be separated by a comma.

If a set of labels of an output not containing a solidus contains the identifying number of an affecting Mm input standing at its internal 0 state, this set of labels has no effect on that output.

Labels may be factored using algebraic techniques (Figure 36).

Figure 35. Output Labels

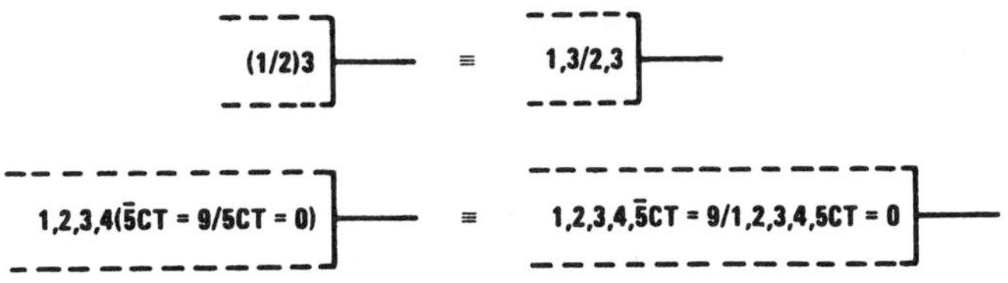

Figure 36. Factoring Output Labels

F.A. Mann received his Bachelor of Science degree from the United States Naval Academy, Annapolis, MD in 1953 and his Master's degree in Engineering Administration from Southern Methodist University, Dallas, TX in 1970. He joined Texas Instruments In 1957 and has worked as a Manufacturing Engineer, Product Marketing Engineer, and Semiconductor Data Sheet Manager. Currently, he serves as Technical Advisor for data sheets and other technical documentation for semiconductors. He has served on many JEDEC committees since 1963 and has been a USA delegate to the technical committee for semiconductor devices of the International Electrotechnical Commission since 1970. He is a member of the IEEE committees for Logic Symbols and for Logic Diagrams, and participates in meetings of an international working group on logic symbols.

Mr. Mann can be reached at (214) 995-2867 in Dallas.

GLOSSARY

Active Clock Transition: The high-to-low or low-to-high transition (change) of an incoming clock pulse.

Active-High Latch: A bistable storage device that is set or cleared when a high is applied to the SET or CLEAR input respectively and the other input is inactive.

Active-High Signal: A signal that must be at its Logic 1 level to accomplish a specified task.

Active-Low Latch: A bistable storage device that is set or cleared when a low is applied to the SET or CLEAR input respectively and the other input is inactive.

Active-Low Signal: A signal that must be at its Logic 0 level to accomplish a specified task.

Address Bus: Unidirectional bus used by the microprocessor to address a specific memory or I/O location.

Aliasing: Low-frequency distortion caused by a sampling rate that is not at least twice the sampled signal's highest frequency component.

Alternate Logic Gate Symbols: An alternative equivalent symbol that performs the same logical function. The alternate symbol replaces active-high symbols with active-low symbols and vice versa.

American Standard Code for Information Interchange (ASCII): An alphanumeric code that allows binary representation of letters, numbers, punctuation marks, and other special symbols.

Analog-to-Digital Conversion: The process of converting a signal in analog form to one in digital form.

Analog-to-Digital Converter (ADC): A circuit designed to convert analog signals to digital signals.

AND Gate: A logic circuit whose output is high only when all of its inputs are high.

ANSI: American National Standards Institute.

Antialiasing Filter: A filter designed to prevent aliasing distortion.

Antifuse: A one-time programmable, nonvolatile, two-terminal device used as the programmable interconnect switches in many FPGAs.

Application-Specific Integrated Circuit (ASIC): Custom-fabricated ICs designed for a specific application.

Arithmetic Logic Unit (ALU): A logic circuit implemented on an IC that can perform a multitude of arithmetic and logical operations.

Assert: To bring a digital signal to its active level.

Associative Property of Addition: The property of real numbers that indicates the order of ORing variables is the same no matter how they are grouped: for example, $(A + B) + C = A + (B + C) = A + B + C$.

Associative Property of Multiplication: The property of real numbers that indicates the order of ANDing variables is the same no matter how they are grouped: for example, $(AB)C = A(BC) = ABC$.

Asynchronous: Not happening at the same time; not clock dependent.

Asynchronous Counter: A binary counter that has the external clock input signal applied only to the LSB flip-flop—also called a ripple counter.

Asynchronous Preloading: The capability to put a number in a programmable counter without a clock pulse. This capability establishes the counter's initial count, which allows flexibility in setting up various MOD numbers.

BCD Adder: A binary adder set up to add binary-coded decimal numbers.

BCD Adjust: The correction made to a BCD number by adding 0110 when an invalid sum is generated by the adder or when a carry is generated out of the sum when adding two BCD numbers.

Binary: A numbering system based on two digits (0 and 1).

Binary-Coded Decimal (BCD): A decimal number that has had each digit coded in a 4-bit binary group (BCD code format).

Binary-Coded Octal (BCO): An octal number that has had each digit coded in a 3-bit binary group (BCO code format).

Binary-Coded Hexadecimal (BCH): A hexadecimal number that has had each digit/character coded in a 4-bit binary group (BCH code format).

Binary Counter: A digital circuit used to produce binary number counts and/or frequency division.

Bi-Quinary: Bi(2)-quinary(5); a bi-quinary counter is a MOD-10 (divide-by-10) symmetrical output counter.

Bistable: A circuit/device that has two stable states.

Bit: A binary digit (0 or 1).

Boolean Addition: The OR operation in Boolean algebra.

Boolean Algebra: A form of logic that describes logical processes with mathematical symbols.

Boolean Multiplication: The AND operation in Boolean algebra.

Boot-Up: The process of loading a program into a computer's memory to initialize the system.

Broadside: Loading a circuit such as a register in parallel.

Buffer: 1. A circuit with one data input and one data output that is used for isolation or to increase current levels to increase fan-out. 2. A block of memory used for temporary data storage.

Bus: A conductor (line) or set of conductors that connects two or more devices in a system.

Bus Contention: A condition caused when two or more digital devices are allowed on-line simultaneously; the condition may damage the circuits connected to the bus line, and it will usually produce an invalid output on the line.

Bus Cycle: Length of time it takes to execute an instruction.

Bus Transceiver: A circuit in a digital system that allows asynchronous two-way communications between two data busses.

Byte: Eight (8) bits of data.

Central Processing Unit: Microprocessor.

CLEAR (RESET) State: The output condition of a latch or flip-flop where $Q = 0$ and $\overline{Q} = 1$.

CMOS Technology: A logic family that uses unipolar transistors—n-type and p-type MOSFETs.

Code: A group of bits used to represent a digit, letter, or symbol.

Code Converter: A type of decoder that detects a binary code and converts it to another code.

Combinational Logic Circuit: A logic circuit whose output depends entirely upon the input levels applied.

Common Control Block: The block at the top of an ANSI/IEEE symbol that is separate from the remaining blocks. The inputs to this block are common to all of the circuits in the IC.

Commutative Property of Addition: The property of real numbers and Boolean algebra that indicates $A + B = B + A$.

Commutative Property of Multiplication: The property of real numbers and Boolean algebra that indicates $AB = BA$.

Complement: To reverse the binary state. For example, the complement of Logic 1 is Logic 0 and vice versa.

Complement Property: The property of Boolean algebra that implies $A \cdot \overline{A} = 0$ and $A + \overline{A} = 1$.

Complementary Metal-Oxide Semiconductor (CMOS): A logic family that uses p- and n-channel E-MOSFETs to implement logic functions.

Control Bus: Bidirectional bus used to indicate a specific operation such as a *read*, *write*, or *interrupt*. It is also used to monitor status of certain devices or to acknowledge an event.

Current Sinking: Current flowing into a logic gate whose output is low.

Current Sourcing: Current flowing out of a logic gate when its output is high.

Data: Digital system information (Logic 0s and 1s).

Data Bus: Bidirectional bus used to transfer data.

Data Lockout: A feature of some master–slave flip-flops that employs edge triggering instead of pulse triggering at the clock input.

Decade Counter: A counter that has a modulus of 10.

Decode: The process of identifying and/or converting a binary number or code.

Decoder: A logic circuit designed to decode.

Decrement: To decrease the value of a number by 1.

DeMorgan's Theorems: Theorems that indicate $\overline{A + B} = \overline{A}\,\overline{B}$ and $\overline{AB} = \overline{A} + \overline{B}$. Break the line and change the sign.

Demultiplex: Distribute data.

Demultiplexer (DEMUX): A circuit designed to distribute data.

Dependency Notation: Using letters on an ANSI/IEEE symbol to denote the relationship between inputs, outputs, or inputs and outputs without actually showing all of the interconnections involved.

Destination Register: The receiving register in data transfer.

Differential Nonlinearity: The worst case deviation from the ideal 1-LSB step in data conversion.

Digital-to-Analog Conversion: The process of changing the format of information by converting a binary signal to an analog value.

Digital-to-Analog Converter (DAC): A circuit designed to convert digital data to an analog value.

DIP: Dual-in-line package used as a carrier for an IC.

Direct Memory Access (DMA): A high-speed memory access where the CPU has relinquished control of the data bus and data are transferred directly from I/O to memory, memory to I/O, or memory to memory by a peripheral device.

Distributive Property of Multiplication Over Addition: The mathematical property used to clear parentheses by following the order of precedence: $A(B + C) = AB + AC$.

Divide-by-*n* Counter: Binary counter—the frequency divide-by capability at the MSB output is equal to its MOD number.

Don't Care Signals: Logic levels that can be low (0) or high (1).

Double Negation Property: The property of Boolean algebra that implies if a logic level is double inverted, or inverted any even number of times, it is equal to the original input logic level.

Drive Gate: A logic gate whose output is connected to the input of another gate.

D-Type Flip-Flop: A type of flip-flop that has only one data (*D*) input and three modes of operation–SET, CLEAR, and NO CHANGE.

Dynamic Input Indicator: An indicator used at the clock input of flip-flops to signify the circuit is edge triggered.

Dynamic Operation: Logic circuit/gate operation with fast-changing logic levels.

Dynamic RAM (DRAM): Random-access, volatile memory that retains stored data as long as power is applied and data is refreshed. DRAM uses MOSFET capacitance as the storage cell.

Edge Detector: A circuit that will produce a sharp, short-duration, positive output pulse on the active clock transition (PGT or NGT).

Edge Triggered: Circuits that are triggered on the positive- or negative-going transition of an incoming pulse.

8-4-2-1 Code: BCD code.

Electrically Erasable PROM (EEPROM/E^{2}PROM): A nonvolatile, static storage device whose contents may be selectively electronically erased.

Emitter-Coupled Logic: High-speed logic circuit family that uses bipolar transistor technology.

Enable: To activate a logic gate/circuit or allow its operation; the opposite of inhibit.

Enable Property: The property of Boolean algebra that implies $A \cdot 1 = A$ and $A + 0 = A$.

Enabler: The logic level that enables a logic gate/circuit.

Encode: The process of producing a binary number or code.

Encoder: A circuit designed to encode.

End-Around Carry: The final carry that is added to the sum in unsigned subtraction using addition of a 9s or 1s complemented subtrahend.

Erasable Programmable ROM (EPROM): An ultraviolet erasable, electrically reprogrammable ROM sometimes called a UVPROM.

Excess-Three (XS3) Code: A self-complementing binary code that is produced by adding 3 to each decimal digit and then coding the sum in a group of four bits.

Exclusive-NOR (XNOR) Gate: A 2-input logic circuit whose output is high only when its inputs are the same.

Exclusive-OR (XOR) Gate: A 2-input logic circuit whose output is high only when its inputs are complementary.

Fan-In: Unit load.

Fan-Out: The maximum number of inputs that the output of a logic gate can reliably drive.

Fetch Operation: Microprocessor read operation.

Field-Programmable Gate Array (FPGA): A user-programmable, high-density IC that is register rich when compared to its logic-rich PLD counterpart.

Firmware: A hardware component such as a ROM chip with software programmed into it.

Flag: Special-purpose bits (flip-flops) used to indicate microprocessor status and control conditions.

Flash ADC: A circuit designed to convert analog values to digital data—sometimes called a simultaneous ADC.

Flash Memory: A nonvolatile, high-density, fast-access-time, reprogrammable, low-cost memory device. The chip incorporates EPROM high density and EEPROM in-circuit programmability in one technology.

Flip-Flop: A bistable sequential logic circuit that can store a Logic 0 or Logic 1 bit.

Floating Inputs: Inputs to logic circuits that are disconnected or open.

Full Adder: A combinational logic circuit designed to add 3 bits—A (addend) + B (augend) + Carry In—and produce a sum and a carry output.

Gated Latch: A synchronous latch circuit that has a control (enable) input, which determines when the latch can change states.

Generic Array Logic (GAL): A type of PLD that can perform combinational and sequential logic operations. These devices were developed by Lattice Semiconductor Corp. and employ E^{2}CMOS technology combined with PLD technology.

Gray Code: A binary code designed so that successive counts result in only one bit changing states.

Half-Adder: A combinational logic circuit designed to add 2 bits and produce a sum and a carry output.

Hardware: Integrated circuits (ICs), drive motors, and other physical parts of a digital system.

Hexadecimal: A numbering system based on sixteen characters (0, 1, 2, 3, 4, 5, 6, 7, 8, 9, A, B, C, D, E, and F).

High-Impedance (Hi-Z) State: A state where the output of a circuit is electrically disconnected from a bus (line).

Hold Time: The interval of time from the active clock transition of a flip-flop's CLK input to when the input data is no longer required to ensure proper circuit operation.

Hybrid Counter: A binary counter that incorporates both synchronous and asynchronous clocking methods.

Hysteresis: The difference between the two threshold voltage levels of a Schmitt-trigger-input circuit.

IEEE: Institute of Electrical and Electronic Engineers.

Increment: To increase the value of a number by 1.

Inhibit: To deactivate a logic gate/circuit or prevent its operation.

Inhibit Property: The property of Boolean algebra that implies $A \cdot 0 = 0$ and $A + 1 = 1$.

Inhibitor: The logic level that inhibits a logic gate/circuit.

Input/Output (I/O) Ports: The channels of interface through which the microprocessor communicates with I/O devices.

Interface: To connect circuits together with the extra components/circuits necessary to make the connection compatible.

Interface Circuit: Gates or components used to condition a signal so that it is compatible to drive a load.

Interrupt: The CPU stops executing its current instruction, stores remaining data in the stack if necessary, and executes the interrupt service routine to fulfill the interrupt request.

Interrupt Request: A signal from a peripheral device indicating the device wants the CPU to execute an interrupt service routine.

Interrupt Service Routine: A special program designed to service an interrupting device in order to clear an interrupt.

Invalid/Ambiguous State: The undesired state of a latch or flip-flop when both Q and $\overline{Q}$ are at the same level.

Inverter: A single-input logic circuit whose output is the complement of its input—often called a NOT gate.

J-K Flip-Flop: A type of flip-flop that has two data inputs (*J* and *K*) and four modes of operation–SET, CLEAR, NO CHANGE, and TOGGLE.

J-K Master-Slave Flip-Flop: A type of pulse-triggered flip-flop that employs two J-K flip-flops connected together.

Johnson Counter: A shift register counter whose $\overline{Q}$ output is connected back to its own input; sometimes called a twisted ring counter.

Karnaugh Map: A graphical representation of a truth table that can be used to simplify Boolean expressions.

K-Map Looping: The looping together of horizontally and/or vertically adjacent ones into pairs, quads, or octets on a K-map.

Latch: A bistable circuit that can retain (store) a Logic 0 or Logic 1 output after the data input has been removed.

Leading Zero Suppression: A method used to blank out the leading (most significant) zero (zeros) in a number when using multiple 7-segment indicators.

Least Significant Bit (LSB)/Digit (LSD): The bit in a binary number or digit in a decimal number that carries the least weight (value).

Load: 1. Inputs being driven by the output of a gate/circuit. 2. To put data into a circuit.

Load Gate: A logic gate whose input is connected to the output of another gate.

Logic Gate: A gate that implements the logical AND, OR, NAND, NOR, XOR, X-NOR, or NOT invert functions.

Logic Level: A low (Logic 0) or high (Logic 1) level applied to or taken from the output of a logic gate/circuit.

Logic Probe: A tester used to indicate whether a line or input/output pin on a digital IC is low, high, pulsing, or dead.

Logic Pulser: A clocking device used to inject logic pulses into a digital circuit/IC.

Logical Inversion: The logical operation performed by a NOT gate that converts a low to a high and vice versa.

Logical Product: The logical operation performed by an AND gate.

Logical Sum: The logical operation performed by an OR gate.

Look-Ahead Carry: A method used in some adders and counters to increase the speed of the carry output.

Machine-Level Language: Binary bit patterns applied to digital circuits.

Magnitude Bits: All bits in a binary number except the sign/parity bit.

Magnitude Comparator: A combinational logic circuit designed to compare two binary quantities to determine which of the quantities, if either, is larger.

Mask ROM (MROM): Semiconductor memory that is programmed using a photo mask at the manufacturer's facility.

Memory Address: The binary address used to enable a memory chip and one specific memory location within the enabled chip.

Memory Capacity: The storage-handling capability of a memory chip.

Memory Cell: A device designed to store one bit.

Memory Map: A tablelike map that shows all memory addresses in a digital system and identifies address ranges assigned to specific memory areas such as RAM, ROM, video RAM, and so on.

Microcomputer: A digital system that uses ICs controlled by digital logic signals to retrieve, process, and store data.

Microprocessor: An IC that performs the math and logical operations and generates the timing and control signals required to retrieve, process, and store data in a computer.

Mnemonic: Letters used to name or designate a circuit line or input/output signal in a system.

Modulus (MOD): The maximum number of states of a counter.

Monotonic: Operation of a DAC when the output has a slope whose sign does not change for an increasing input signal.

Most Significant Bit (MSB)/Digit (MSD): The bit in a binary number or digit in a decimal number that carries the most weight (value) in the number.

Mutiplex: Select data.

Mutiplexer (MUX): A circuit designed to select data.

NAND Gate: A logic circuit whose output is low only when all of its inputs are high.

***n*-Bit Encoder:** An encoder that can take one active input line and encode it to *n* bits.

Negation: To perform the NOT operation; to complement a logic signal.

Negation Indicator: A bubble (standard symbol) or slanted line (ANSI/IEEE symbol) that indicates negation.

Negative-Going Transition (NGT): The high-to-low transition of a signal.

Nibble: Four (4) bits of data.

Nines Complement: The *unsigned* representation of a negative decimal number. The 9s complement allows subtraction by adding the complemented subtrahend and adding the carry generated to the initial sum.

Noise Immunity: A digital circuit's ability to tolerate noise signals on its inputs and still operate reliably.

Noise Margin: A measure of noise immunity.

NOR Gate: A logic circuit whose output is low when any of its inputs is high.

NOT Gate (Inverter): A single-input logic circuit whose output is the complement of its input.

Nyquist Sampling Theorem: A theorem that indicates reproduction of a periodic signal requires a sample rate greater than twice the sampled signal's highest frequency component.

Octal: A numbering system based on eight digits (0, 1, 2, 3, 4, 5, 6, and 7).

1-of-*n* Decoder: A type of decoder that activates only one of its output lines for a given input number or code.

Ones Complement: The *unsigned* representation of a negative binary number.

Open-Collector Output: A TTL digital circuit that does not have the top totem-pole output transistor—allows wire-ANDing the circuit's output provided a pull-up resistor is added to the circuit.

OR Gate: A logic circuit whose output is high when any (or all) of its inputs is high.

Overbar: A NOT sign over a digital variable that indicates that variable has been NOTed.

Parallel Data Transfer: The movement of data several bits at a time.

Parallel-In/Parallel-Out: A register whose data is loaded in parallel format and read in parallel format.

Parallel-In/Serial-Out: A shift register whose data is loaded in parallel format and read serially.

Parity: The state of being equal.

Parity Bit: A bit attached to the data stream by the parity generator that will establish the parity scheme.

Parity Checker: A circuit at the receiving end that will check data and parity bit received to ensure no error has occurred.

Parity Error: An error caused by one bit in a data stream being high when it should be low or vice versa.

Parity Generator: A circuit used to generate a parity bit that will be sent along with data to a receiving circuit/system.

Parity Scheme: An ODD or EVEN scheme is determined, and the parity generator/checker then checks for errors using the established scheme. The scheme established causes the total number of high bits transmitted to always be even or odd.

Positional-Weighted Numbering System: A system where each digit/character position has a specific weight (value).

Positive-Going Transition (PGT): The low-to-high transition of a signal.

Postponed Output Indicator: An inverted-backward L over the Q and $\overline{Q}$ outputs of a master–slave flip-flop that indicates the output is delayed.

Primary Memory: On-board computer semiconductor memory such as RAM and ROM.

Priority Encoder: An encoder that responds only to the highest-order number input when more than one input is asserted (activated) at the same time.

Product of Sums (POS): One form of a logical equation/circuit that shows the logical sum outputs of OR gates are ANDed together.

Program: A set of instructions used by a CPU to accomplish certain tasks.

Programmable Array Logic (PAL): A PLD that contains a programmable AND-gate array and a fixed OR-gate array. PAL is a registered trademark of Advanced Micro Devices, Inc.

Programmable Logic Array (PLA): A PLD that contains a programmable AND-gate array and a programmable OR-gate array.

Programmable Logic Device (PLD): A general term that refers to ICs with uncommitted AND/OR logic arrays that can be programmed by blowing fusible links.

Programmable Logic Element (PLE): A PLD that contains a fixed AND-gate array and a programmable OR-gate array.

Programmable ROM (PROM): A semiconductor memory designed with user-programmable fusible links.

Propagation Delay Time: The interval that elapses between when an input signal is applied and when the output signal responds to the input request.

Pseudo-Static RAM (PSRAM): A type of DRAM with total internal refresh operations.

Pull-Up Resistor: A resistor connected to V_{CC} to increase input/output voltages to greater than $V_{IH(MIN)}/V_{OH(MIN)}$. Required on all open-collector output ICs.

Pulse Triggered: Circuits that are level triggered; circuits enabled on the positive duration of an input pulse or on the negative duration.

Quantization: Limiting the possible values of a quantity to a discrete set of values by quantum mechanical rules.

Radix: The base of a numbering system. The base identifies how many numbers/characters are used in the numbering system.

Radix Division: The process used to convert a decimal number to another base number by repeated division by the radix of the numbering system being converted to.

Radix Multiplication: The process used to convert a binary, octal, or hexadecimal number to a decimal number by repeated multiplication using the radix of the numbering system being converted from.

Random-Access Memory (RAM): Read/write memory that is volatile.

Read Bus Cycle: Cycle when microprocessor is reading memory or an I/O port—fetch operation.

Read-Only Memory (ROM): Permanent memory not subject to loss or change if power is removed—nonvolatile.

Redundant Property: The property of Boolean algebra that implies $A \cdot A = A$ and $A + A = A$.

Refresh: Periodic rewriting of data in a DRAM chip to prevent memory loss.

Register: A group of latches or flip-flops that is used to store, transfer, or shift data.

RESET (CLEAR) State: The output condition of a latch or flip-flop where $Q = 0$ and $\overline{Q} = 1$.

Resolution: The smallest analog incremental or decremental change in output voltage of a *DAC* that results from a 1 LSB converter code change. It is further defined for *ADCs* as the change in input voltage required to increment or decrement the output of the ADC 1 LSB when the input is incremented or decremented by a value equal to 1 LSB.

RETAIN State: The state of a latch or flip-flop that holds the previous condition when its inputs are brought inactive. Sometimes called the NO CHANGE state.

Ring Counter: A shift register whose Q output is connected back to its own input so that a single Logic 1 will be continuously shifted through the register.

Ripple Blanking: The signals used by a decoder/driver to suppress leading zeros when several decoders are used to drive multiple seven-segment indicators.

Ripple Counter: Asynchronous counter.

R–2R Ladder: A resistive network designed to convert digital data to analog.

Schmitt-Trigger-Input Circuit: A special input circuit that uses positive feedback to speed up slow rise and fall time signals.

Secondary Memory: Mass computer memory such as hard disks, floppy disks, and magnetic tapes.

Security Bit: A bit used in PLDs that prevents copying or examining the data stored in the device.

Sequential Logic Circuit: A logic circuit whose output depends on its previous state (condition) in addition to its current inputs.

Serial Data Transfer: The movement of data one bit at a time.

Serial-In/Parallel-Out: A shift register whose data is loaded in serial format and read in parallel format.

Serial-In/Serial-Out: A shift register whose data is loaded in serial format and read serially.

SET State: The output condition of a latch or flip-flop where $Q = 1$ and $\overline{Q} = 0$.

Setup Time: The interval of time the data input to a flip-flop must be held constant prior to the active clock transition.

Shift Register: A register in which data is shifted internally one flip-flop per clock pulse—serial shift.

Short Logic: A set of precepts designating the relationship between a logic gate's inputs and output; basically, short logic is the "any rule" of a logic gate's operation.

Short Logic: The "all rule" of a logic gate's operation. It is derived by NOTing the short logic input and output logic levels and changing the word "any" to "all."

Sigma (Σ) : Sum.

Sign Bit: The binary digit used to represent a positive binary number (0) or a negative binary number (1).

Signs of Grouping: Parentheses, brackets, and braces used to indicate an exception to normal mathematical precedence. The parentheses, brackets, or braces in an expression must be removed first.

Simultaneous ADC: Flash ADC.

Sink Current: The current that flows into a logic circuit at an input or output pin.

Software: Programs consisting of Logic 0s and Logic 1s.

Source Current: The current that flows out of a logic circuit on an input or output pin.

Source Register: The transmitting register in data transfer.

State Indicator: A bubble/slanted line or the lack of a bubble/slanted line on the input/output of a device that indicates the input/output is active low or active high respectively.

State Table: A table that indicates that state of a latch or flip-flop circuit with various input combinations.

Static Operation: Logic circuit/gate operation with steady-state inputs.

Static RAM (SRAM): Random-access, volatile memory that retains stored data without refresh as long as power is applied.

Steering Gates: A pair of logic gates that controls when a latch or flip-flop can change states synchronously.

Store Operation: Microprocessor write operation.

Successive-Approximation ADC: A circuit designed to convert analog values to digital data.

Sum of Products (SOP): One form of a logical equation/circuit that shows the logical product outputs of AND gates are ORed together.

Sum of Weights: Adding together the weights of each position of a number.

Switch Debouncer: A circuit added to the output of a switch that prevents its output level from changing levels as the result of contact bounce.

Switched Current-Source DAC: A circuit designed to convert digital data to analog information.

Synchronous: Happening at the same time; in time with a clock signal.

Synchronous Counter: A binary counter that has the external clock input signal applied to all flip-flops in the counter.

Tens Complement: The nines complement of a decimal number plus 1; the 10s complement allows subtraction by adding the complemented subtrahend and ignoring the end-around carry.

Terminal Count: The maximum up count of a binary up-counter (1111...) or the count of 0000 of a binary down-counter.

Toggle: To change states from Logic 0 to Logic 1 or Logic 1 to Logic 0.

TOGGLE Mode of Operation: The output of a flip-flop changes states on each active clock transition; output frequency is half the input clock frequency.

Totem-Pole Output: Output circuit of a TTL device.

Transistor-to-Transistor Logic (TTL): A logic family that uses bipolar transistors to implement logic functions.

Transparent Latch: A D-type latch.

Tristate: A circuit that can produce a high logic level, low logic level, or Hi-Z condition at its output—sometimes referred to as 3-state logic. The circuit is used to prevent bus contention.

Truncate: To shorten by or as if by cutting off; short counting a counter.

Truth Table: A table that lists every binary input combination that can exist for a logic gate function and the resulting binary outputs for each of those input combinations.

TTL Technology: See *Transistor-to-Transistor Logic*.

Twos Complement: The 1s complement of a binary number plus 1; the 2s complement allows subtraction by adding the complemented subtrahend and ignoring the end-around carry.

Unit Load: A measure of the load capability of a logic gate; the amount of input current necessary to drive a load device.

Universal Register: A register whose input data can be loaded in serial or parallel format, and may be retrieved from the register in either format.

Unused Inputs: Inputs to logic gates/circuits that are not needed nor used. For example, a 3-input logic gate is available, but only 2 inputs are needed—the third is unused.

UVPROM: Erasable PROM.

Variable: The symbolic designation or mnemonic of a digital signal: for example, Read/$\overline{\text{Write}}$, A, B, and so on.

Vinculum: A bar (line) drawn over two or more variables in a Boolean expression.

Volatile: A memory device characteristic that means the data contents are lost when power is removed.

Wire-ANDing: The connecting together of two or more open-collector logic gate outputs to produce the AND function.

Word Size: Number of bits of data processed simultaneously in a digital system.

Write Bus Cycle: Cycle when microprocessor is writing data into a memory or I/O location—store operation.